CONSTRUCTION A
PAYMENTS

CONSTRUCTION ADJUDICATION AND PAYMENTS HANDBOOK

DOMINIQUE RAWLEY QC
KATE WILLIAMS
MERISSA MARTINEZ
PETER LAND

Great Clarendon Street, Oxford, OX2 6DP,
United Kingdom

Oxford University Press is a department of the University of Oxford.
It furthers the University's objective of excellence in research, scholarship,
and education by publishing worldwide. Oxford is a registered trade mark of
Oxford University Press in the UK and in certain other countries

First Edition published in 2013

Impression: 1

British Library Cataloguing in Publication Data
Data available

ISBN 978–0–19–955159–0

Printed and bound in Great Britain by
CPI Group (UK) Ltd, Croydon, CR0 4YY

FOREWORD

It is now fifteen years since the Housing Grants Construction and Regeneration Act 1996 came into force and introduced statutorily backed adjudication and minimum payment conditions in construction contracts. From the first court decision in the United Kingdom, the *Macob* case, the courts have had to deal on a regular and frequent basis with challenges to the enforceability of adjudicators' decisions on numerous grounds. Some 15 per cent of the business in the High Court in the Technology and Construction Court is concerned with adjudication related disputes.

It is high time that the reported cases were brought together in a book such as this. The particular feature of this book is that each relevant topic is discussed but is thereafter supported by extracts from the key cases and by chronological tables of all relevant cases on the particular topic (with a brief summary of the point of interest against each). This will be of particular use to both legal and other professional practitioners not only in the court process but equally or more importantly in the drafting of contracts and during the adjudication process itself. To have a brief description of the relevant cases will be most helpful in circumstances in which the reported cases are found in a wide range of publications and websites. This will enable users of the book to understand what the case is about and what propositions the particular judgment does or does not address before having to trawl through a large number of authorities seeking to find the relevant part of the relevant authority.

The book considers in a logical way the 1996 Act and the later amendments introduced by the Local Democracy, Economic Development and Construction Act 2009 and reviews the statutory basis for adjudication in construction contracts before moving on to consider the effect, enforcement, and enforceability of adjudicators' decisions in the courts followed by a review of the provisions and cases which address payment and withholding of payment under construction contracts. There is now, after fifteen years of judicial decisions, virtually no part of the 1996 Act which has not been considered by the courts.

This book should appeal to solicitors, barristers, claims consultants, construction professionals, and not least to adjudicators. I very much hope that it will also find a permanent place on even judges' shelves.

The Honourable Mr Justice Akenhead
Judge in Charge of the Technology and Construction Court
London 2013

CONTENTS

DETAILED CONTENTS

IV PAYMENT UNDER CONSTRUCTION CONTRACTS

TABLE OF CASES

Tables are referenced by page numbers in brackets

ABB Power Construction v Norwest Holst Engineering Ltd (2000) 77 Con LR 20 . . . . 1.19, 1.58, 1.62, T.1.2 (23)

ABB Zantingh Ltd v Zedal Building Services Ltd [2001] BLR 66. . . . 1.63, 7.59, T.1.2 (23)

A. C. Yule & Son Ltd v Speedwell Roofing & Cladding Ltd [2007] EWHC 1360 . . . . T 5.1 (156), T 9.5 (314)

A&D Maintenance and Construction Services Ltd v Pagehurst Construction Services Ltd, 23 Jun. 1999 (TCC) . . . . 2.24, T 2.1 (41), T 3.5 (93)

ART Consultancy v Naveda Trading Ltd [2007] All ER (D) 157 (Jul), [2007] EWHC 1375 (TCC) . . . . 7.105, T 2.1 (41), T 7.7 (260)

A & S Enterprises Ltd v Kema Holdings Ltd [2005] BLR 76 (TCC), [2004] EWHC 3365 (TCC) . . . . 11.08, 11.25, T 11.1 (389)

ALE Heavy Lift v MSD (Darlington) Ltd [2006] EWHC 2080 (TCC) . . . . 2.27, T 2.1 (43), T 8.1 (280)

AWG Construction Services Ltd v Rockingham Motor Speedway Ltd [2004] CILL 2154; [2004] EWHC 888 (TCC). . . . 3.32, 3.58, 3.61, 3.62, 7.81, 9.52, 8.11, 8.12, 10.17, 10.25, 10.30, T 3.2 (75), T 3.4 (86), T 7.5 (242), T 8.1 (279), T 9.6 (322), T 10.2 (338)

Able Construction (UK) Ltd v Forest Property Development Ltd [2009] All ER (D) 176 (Feb), [2009] EWHC 159 (TCC) . . . 6.18, 7.105, 7.109, T 7.1 (215), T 7.7 (261)

Absolute Rentals v Gencor [2000] CILL 1637 (TCC) . . . 1.81, 8.21, T 1.3 (30), T 8.1 (278), T 8.2 (285)

Adonis Construction v O'Keefe Soil Remediation [2009] CILL 2784, [2009] EWHC 2047 (TCC) . . . . 7.78, T 2.1 (44), T 7.5 (243)

Aedas Architects Ltd v Skanska Construction UK Ltd [2008] CSOH 64 (Outer House Court of Session). . . . 13.103, 13.107–13.110, T 13.2 (464)

Aedifice Partnership Ltd v Ashwin Shah (2010) 132 Con LR 100; [2010] EWHC 2106 (TCC) . . . . 2.05, 5.12, 5.13, 5.15, 5.21, 5.22, T 2.1 (45), T 5.1 (160)

Air Design (Kent) Ltd v Deerglen (Jersey) Ltd (2009) CILL 2657; [2008] EWHC 3047 (TCC) . . . . 3.17, 8.20, 9.12, T 2.1 (45), T 3.1 (65), T 5.1 (160), T 8.1 (280), T 9.1 (297)

Alan Auld Associates Ltd v Rick Pollard Associates and anor [2008] EWCA Civ 655 (CA) 12.26

All in One Building & Refurbishments Ltd v Makers (UK) Ltd (2005) CILL 2321; [2005] EWHC 2943 (TCC) . . . . 3.37, 10.95, 12.24, 15.23, T 3.2 (76), T 10.4 (354), T 10.7 (374), T 12.1 (405), T 15.1 (484)

All Metal Roofing v Kamm Properties Ltd [2010] EWHC 2670 (TCC). . . . T 2.1 (45), T 5.1 (160)

Allen Wilson Joinery v Privetgrange Construction Ltd 123 ConLR 1; [2008] EWHC 3160 (TCC) . . . . 2.15, 2.36, T 2.1 (45)

Allen Wilson Shopfitters v Anthony Buckingham (2005) 102 Cn LR 154; [2005] EWHC 1165 (TCC). . . . 1.92, 2.17, 12.38, T 1.3 (31), T 2.1 (42), T 7.1 (214)

Allied London & Scottish Properties plc v Riverbrae Construction Ltd [1999] BLR 346 . . . . 13.139, T 13.2 (463)

Allied P&L v Paradigm Housing Group Ltd [2010] BLR 59; [2009] EWHC 2890 (TCC) . . . . 5.13, 9.14, T 3.4 (88), T 5.1 (159)

Alstom Signalling Ltd (t/a Alstom Transport Information Solutions) v Jarvis Facilities Ltd (2004) 95 Con LR 55, [2004] EWHC 1232 . . . . 16.05, 16.30 T 7.1 (214), T 8.2 (285)

Alstom Signalling Ltd (t/a Alstom Transport Information Solutions) v Jarvis Facilities Ltd [2004] EWHC 1285 (TCC) . . . . 6.09, 7.04, 7.66, 7.71–7.72, 12.19, 12.34–12.36, 13.102, 13.104–13.106,16.05,16.30 T 6.2 (178), T 7.4 (234), T 12.1 (405), T 13.2 (463–4), T 16.1 (494)

AMEC Capital Projects Ltd v Whitefriars City Estates Ltd (2005) 1 All ER 723; [2004] EWCA Civ 1418 (CA). . . . 3.88, 4.30, 6.65, 7.06, 7.07, 9.11, 9.27, 9.47, 10.11, 10.57, 11.01, 11.05, 11.07, 11.10, 11.32–11.34, T 6.4 (201), T 8.3 (291), T 9.3 (303), T 10.4 (355), T 10.6 (367), T 11.1 (389)
Amec Civil Engineering Ltd v The Secretary of State for Transport [2005] 1 WLR 2339; [2005] EWCA Civ 291. . . . 3.30, 3.34–3.36, 3.39–3.45, T 3.2 (75)
Amec Group Ltd v Thames Water Utilities Ltd [2010] EWHC 419 (TCC) . . . 3.18, 4.51, 7.87, 7.94, 10.19, 10.20, T 3.1 (65), T 7.6 (252), T 10.1 (333), T 10.4 (355)
Amoco (UK) Exploration Co v British American Offshore Ltd [2002] BLR 135, [2001] EWHC 484 (Comm). . . . 7.104
Amsalem (t/a MRE Building Contractors) v Raivid [2008] EWHC 3226 (TCC) . . . 8.42
Andrew Wallace Ltd v Artisan Regeneration Ltd [2006] EWHC 15 (TCC) . . . T 7.1 (214)
Andrew Wallace Ltd v Jeff Noon [2009] BLR 158 . . . 4.35, 11.18, 11.19, T 11.1 (390)
Anglian Water Services Ltd v Laing O'Rourke Utilities Ltd (2010) 131 Con LR 94; [2010] EWHC 1529 (TCC). . . . 6.21, 17.05, 17.06, 17.13, 17.20–17.22, T 6.2 (180), T 17.2 (507)
Ardmore Construction v Taylor Woodrow Construction [2006] CSOH 3. . . . 3.60, 10.93, T 3.2 (76), T 3.4 (87), T 10.7 (374)
Ashley House Plc v Galliers Southern Ltd [2002] Adj LR 02/15 [2002] EWHC 274 (TCC) . . . 8.11
Ashville Investments v Elmer Contractors Ltd [1987] 37 BLR 55. . . . 3.12, T 3.1 (64)
Atlas Ceiling & Partition Co. Ltd v Crowngate Estates (Cheltenham) Ltd (2002) 18 Const LJ 49 . . . 1.07, T 1.1 (13)
Austin Hall Building Ltd v Buckland Securities Ltd [2001] BLR 272, [2001] EWHC 434 (TCC) . . . 10.04, 10.05, 10.14, 10.22, T 10.2 (338)
Aveat Heating Ltd v Jerram Falkus Construction Limited (2007) 113 Con LR 13; [2007] EWHC 131 (TCC). . . . 3.98, 4.04, 4.38, 9.39, 11.25, T 9.5 (314), T 11.1 (390)
Avoncroft Construction Ltd v Sharba Homes (CN) Ltd (2008) 119 Con LR 130; [2008] EWHC 933 (TCC) . . . 8.13, 8.19, 8.32–8.34, 8.37, T 6.3 (193), T 8.1 (280), T 8.2 (286)

BAL (1996) Ltd v Taylor Woodrow Construction Ltd [2004] All ER (D) 218 (Feb) . . . 10.80, T 10.6 (367)
Baldwins Industrial Services plc v Barr Ltd [2003] BLR 176, [2002] EWHC 2915 (TCC) . . . 8.11, 8.14, T 8.1 (279)
Balfour Beatty Building Ltd v Chestermount Properties Ltd [1993] 62 BLR 1; [1993] CILL 821 . . . 15.40
Balfour Beatty Construction Ltd v Serco Ltd [2004] EWHC 3336 (TCC) . . . 5.28, 6.27, 6.43, 6.52–655, T 6.3 (191–4)
Balfour Beatty Construction Ltd v London Borough of Lambeth [2002] BLR 288; [2002] EWHC 597 (TCC) . . . 9.49, 10.04, 10.18, 10.26, 10.80–10.82, 10.92, 10.95, T 3.4 (87), T 10.7 (373)
Balfour Beatty Construction Northern Ltd v Modus Corovest (Blackpool) Ltd [2009] CILL 2660, [2008] EWHC 3029 (TCC) . . . 6.42, 6.58–6.60, 8.04, 8.39, 10.29, 10.65, 10.66, 10.73, 13.44, 13.49, 13.50, 13.59, 13.112, T 8.2 (286), T 10.2 (340), T 10.4 (355), T 10.5 (360), T 13.2 (464)
Balfour Beatty Engineering Services (HY) Ltd v Shepherd Construction Ltd (2009) 127 Con LR 110, [2009] EWHC 2218 (TCC) . . . 7.101, 10.63, 10.72–10.74, 11.25, T 7.7 (261), T 10.5 (360), T 11.1 (391)
Ballast plc v Burrell Company (Construction Management) Ltd [2001] BLR 529; [2002] Scot CS 324 . . . 3.98, 9.06, 9.50, 9.52, T 3.1 (64), T 5.1 (151), T 9.6 (320), T 10.3 (348)
Banner Holdings Ltd v Colchester Borough Council [2010] EWHC 139 (TCC) . . . 3.14, 3.75, 3.98, 4.89, 7.11, 12.41, T 3.5 (94), T 7.1 (215), T 9.6 (324), T 12.2 (411)
Banque des Marchands de Moscou (Koupetschesky) v Kindersley [1950] 2 All ER 549, [1951] 1 Ch 112 . . . 7.84
Banque Financiere de la Cite v Parc (Battersea) Ltd [1999] 1 AC 221 . . . 4.136
Barnes & Elliott Ltd v Taylor Woodrow Holdings Ltd [2004] BLR 111, [2003] EWHC 3100 (TCC) . . . 9.39, T 9.5 (313)

Barr Ltd v Klin Investment UK Ltd (2009) CILL 2787; [2009] CSOH 104 (OH) 9.52, 11.25
T 6.4 (202) T 10.7 (375), T 11.1 (391)
Barr Ltd v Law Mining Ltd (2001) 80 Con LR 134, [2001]
Scot CS 152 . . . . 3.23, 3.51, 7.81, T 3.1 (64), T 3.3 (80), T 7.5 (240)
Baune v Zduc Ltd [2002] All ER (D) 55 . . . . T 10.6 (366)
Beck Interiors Ltd v Classic Decorative Finishing Ltd [2012] EWHC 1956 (TCC) . . . 6.48, T 6.3 (194)
Beck Interiors Ltd v UK Flooring Contractors Ltd [2012]
EWHC 1808 (TCC) . . . . 3.38, 7.80, 7.82, 7.83, T 3.2 (77), T 7.5 (244)
Beck Interiors Ltd v Russo (2009) 132 Con LR 56, [2009]
EWHC 3861 (TCC) . . . . 7.54, T 7.4 (235)
Beck Peppiat v Norwest Holst Construction Ltd [2003] BLR 316, (TCC) . . . . 3.33, T 3.2 (74)
Behzadi v Shaftesbury Hotels [1992] Ch 1 (CA) 12.26
Benfield Construction Ltd v Trudson (Hatton) Ltd [2008] CILL 2633, [2008]
EWHC 2333 (TCC) . . . . 6.69, T 6.4 (202)
Bennett Electrical Services v Inviron Ltd [2007] EWHC 49. . . . . 2.48, T 2.2 (51), T 2.3 (57)
Bennett v FMK Construction Ltd (2005) 101 Con LR 92, [2005] EWHC 1268 (TCC). . . . T 9.3 (303)
Berry Piling Systems Ltd v Sheer Projects Ltd [2012] EWHC 241 (TCC) 4.63, 8.12,
8.16, T 8.1 (282), T 10.7 (376)
Birmingham City Council v Paddison Construction Ltd [2008] BLR 622, [2008]
EWHC 2254 (TCC) . . . . 7.58, T 6.4 (202)
Bloor Construction (UK) Ltd v Bowmer & Kirkland (London) Ltd [2000]
BLR 314. . . . . 4.109, 4.118, 4.119, 9.49, T 4.2 (127), T 9.5 (313)
Boardwell v k3D Property Partnership Ltd [2006] Adj CS 04/21. . . . . 10.38, T 10.3 (349)
Bothma v Mayhaven Health Care Ltd [2007] EWCA Civ 527 (CA) . . . . 5.16, T 5.1 (157)
Bouygues (UK) Ltd v Dahl-Jensen (UK) Ltd [2000] BLR 49; [2000] BLR 522; [2001]
1 All ER (Comm) 1041 . . . . 4.111, 6.05, 7.05, 7.06, 7.121, 8.09, 8.12,
8.15, 9.48, 9.49, 9.54, 9.55, T 6.2 (180), T 7.1 (212),
T 7.1 (216), T 7.4 (235), T 8.1 (279), T 9.6 (320)
Bovis Lend Lease Ltd v Cofely Engineering Services [2009]
EWHC 1120 (TCC). . . . . 4.30, T 5.1 (159), T 9.1 (297), T 9.3 (303–4)
Bovis Lend Lease Ltd v The Trustees of the London Clinic (2009) 123 Con LR 15, [2009]
EWHC 64 . . . . 3.37, 10.26, 10.28, T 3.2 (77), T 7.5 (242), T 10.2 (340)
Bovis Lend Lease Ltd v Triangle Development Ltd [2003] BLR 31; [2002]
EWHC 3123 (TCC) . . . . 6.14, 6.36, 6.39, 7.03, 7.14, 7.43, T 6.3 (191)
Bracken v Billinghurst [2003] CILL 2039, [2003] EWHC 1333 (TCC) . . . . 6.18, T 6.2 (178)
Brenton (t/a Manton Electrical Components) v Palmer, unreported, 19 January 2001
(TCC) . . . . 8.41, T 7.1 (212), T 8.2 (285), T 9.2 (299)
Bridgeway Construction Ltd v Tolent Construction Ltd [2000]
CILL 1662 . . . . 4.143, 4.147, T 4.4 (138)
British Steel Corporation v Cleveland Bridge and Engineering Company Ltd [1984]
1 All ER 504. . . . . 2.48
Brown & Sons v Crosby North West homes [2009] EWHC 3503 (TCC) . . . . 3.15, T 3.1 (64)
Bryen & Langley Ltd v Martin Boston [2005] BLR 508; [2005] EWCA Civ 973 . . . . 1.92, 1.99,
1.100, 2.49, 2.50, 2.54, T 1.3 (31), T 2.3 (56)
Butler Contractors Ltd v Merewood Homes Ltd [2002] 18 Cont LJ 74 . . . . 3.22
Buxton Building Contractors v Governors of Durand Primary School [2004] BLR 374; [2004]
EWHC 733 (TCC) . . . . 10.38, 10.55, 13.99, T 3.4 (86), T 10.3 (348), T 13.2 (463)

C&B Scene Concept Design Ltd v Isobars Ltd [2002] BLR 93; [2002]
EWCA Civ 46 . . . . 2.17, 3.22, 7.03, 7.05, 7.06, 7.15, 7.16, 9.49,
9.50, 12.32, 12.38, 12.42, 12.50–12.52, T 3.1 (64),
T 3.4 (86), T 7.1 (213), T 9.6 (321), T 12.2 (410)
CIB Properties Ltd v Birse Construction Ltd [(2005) BLR 173; 2005] 1 WLR 2252; [2004]
EWHC 2365 (TCC) . . . . 4.109, 4.111, 10.18, 10.23, 10.26, T 4.2 (127),
T 10.1 (333), T 10.2 (339)
CJP Builders Ltd v William Verry Ltd [2008] BLR 545 . . . . 4.70, 4.79, 7.6110.38,
10.41, 10.47–10.49, T 5.1 (157), T 10.3 (350)

CN Associates v Holbeton [2011] BLR 261, [2011] EWHC 43 (TCC) . . . 2.05, T 2.1 (45), T 5.1 (161)
CRJ Services Ltd v Lanstar Ltd (t/a CSG Lanstar) [2011] EWHC 972 (TCC) . . . . T 2.1 (46), T 10.6 (368)
CSC Braehead Leisure Ltd v Laing O'Rourke Scotland Ltd [2009] BLR 49 . . . . 4.104, T 3.3 (81), T 10.7 (375)
Cable & Wireless plc v IBM United Kingdom Ltd [2003] BLR 89, [2002] EWHC 2059 (Comm) 8.04
Camillin Denny Architects Limited v Adelaide Jones & Co. Limited [2009] BLR 606; [2009] EWHC 2110 (TCC) . . . . 3.13, 7.106, 9.12, 11.25, T 5.1 (158–9), T 7.7 (260), T 9.1 (297), T 9.2 (300), T 11.1 (391)
Canterbury Pipe Lines Ltd v Christchurch Drainage Board [1979] 16 BLR 76 (CA) 15.02
Cantillon Ltd v Urvasco Ltd [2008] BLR 250; [2008] EWHC 282 (TCC) . . . . 3.59, 3.65, 3.66, 7.76, 9.06, 9.52, 10.08, 10.91, 10.95–10.97, T 3.2 (76), T 3.4 (87), T 7.5 (242), T 9.1 (297), T 9.6 (322), T 10.7 (375)
Capital Structures v Time & Tide Construction Ltd [2006] BLR 226, [2006] EWHC 591 (TCC) . . . . 3.24, T 3.1 (64)
Captiva Estates Ltd v Rybarn Ltd (In Administration) [2005] EWHC 2744 . . . . 1.93, T 1.3 (31)
Carillion Construction Ltd v Devonport Royal Dockyard [2005] BLR 310; [2005] EWCA Civ 1358; [2006] BLR 15 . . . . 2.14, 2.19, 2.33, 2.38, 2.43, 2.44, 4.82, 4.91, 4.94–4.96, 4.107, 5.08, 7.04, 7.06, 7.07, 7.17, 7.18, 9.47, 10.10, 10.41, 10.55, 10.56, 10.58, 10.62, 10.65, 10.92, 11.31, T 2.2 (50), T 3.2 (74), T 3.4 (86), T 5.1 (156), T 7.1 (214), T 7.5 (242), T 9.6 (322), T 10.4 (354), T 10.5 (359)
Carillion Construction Ltd v Smith (2011) 141 Con LR 117; [2011] EWHC 2910 . . . . T 6.4 (203)
Carillion Utility Services Ltd v SP Power Systems Ltd [2011] CSOH 139 . . . . 10.95, T 7.5 (243), T 10.7 (376)
Castle Inns (Stirling) Ltd v Clark Contracts Ltd [2005] Scots CS CSOH 178 . . . . T 4.3 (134), T 6.2 (178), T 6.3 (192)
Chamberlain Carpentry & Joinery Ltd v Alfred MacAlpine Construction [2002] EWHC 514 . . . . 3.50, T 3.3 (80), T 4.1 (103)
Channel Tunnel Group Ltd & France Manche SA v Balfour Beatty Construction Ltd [1993] AC 334; [1993] 2 WLR 262 . . . . 8.04, 15.02
Checkpoint Ltd v Strathclyde Pension Fund [2003] All ER (D) 56 (Feb),[2003] EWCA (Civ) 84 . . . . 10.66
Christiani & Nielsen Ltd v Lowry Centre Development Company Ltd, unreported, 26 June 2000 (TCC) . . . . 1.07, 5.09, 5.14, T 1.1 (12), T 5.1 (151)
Citex Professional Services v Kenmore Developments [2004] Scot CS 20 . . . . 6.13
City Inn Ltd v Shepherd Construction Ltd [2002] SLT 781; (2001) SCLR 961 . . . . 6.13, 6.30, 6.31, T 6.2 (177)
Clark Contracts v The Burrell Co [2002] SLT 103 . . . . 13.48, T 13.1 (447)
Clark Electrical Ltd v JMD Developments (UK) Ltd [2012] EWHC 2627 (TCC) . . . . T 5.1 (161)
Clark v Associated Newspapers [1998] WLR 1558 . . . . 7.104
Clarke Quarries Ltd v PT McWilliams Ltd [2009] IEHC 403, [2010] BLR 520 . . . . 7.13
Cleveland Bridge (UK) Ltd v Whessoe-Volker Stevin Joint Venture [2010] BLR 415, [2010] EWHC 1076 (TCC) . . . . 1.19, 7.79, T 1.1 (15), T.1.2 (24), T 7.5 (240, 242–3)
Collier v Williams and ors [2006] 1 WLR 1945; [2006] EWCA Civ 20 . . . . T 17.2 (506)
Collins (Contractors) Ltd v Baltic Quay Management (1994) Ltd [2004] EWCA Civ 1757 . . . . 3.30, 3.35, 3.36, T 3.2 (75)
Collins (Contractors) Ltd v Baltic Quay Management (1994) Ltd [2005] BLR 63; [2004] EWCA Civ 1757 . . . . 8.48
Company (No. 1299 of 2001), Re [2001] CILL 1745 . . . . 7.30
Comsite Project Ltd v AAG [2001] Scot CS 150 . . . . 1.64
Comsite Projects Ltd v Andritz AG [2004] 20 Const LJ 24; [2003] EWHC 958 . . . . T.1.2 (23)
Connex South Eastern Ltd v M J Building Services Group plc [2005] 1 WLR 3323, [2005] BLR 201 (CA) . . . . 3.78, 3.85, 3.86, T 3.5 (94)

Connex v MJ Builders [2004] BLR 333, [2004] EWHC 1518 (TCC); [2005] EWCA Civ 193; [2005] 1WLR 3323 . . . . 2.33, T 2.2 (50), T 3.5 (94)
Conor Engineering Ltd v Les Constructions Industrielles de la Mediterranee SA [2004] BLR 212, [2004] EWHC 899 . . . . 6.48
Construction Centre Group Ltd v Highland Council 2003 SLT 623, [2003] Scot CS 114 . . . . 6.40, T 6.3 (191)
Construction Partnership UK Ltd v Leek Developments Ltd (2006) CILL 2357 (TCC) . . . . 17.05, T 17.2 (506)
Costain Ltd v Strathclyde Builders Ltd [2003] Scot CS 316; 2004 SLT 102 . . . . 5.27, 10.07, 10.80, T 10.6 (367)
Costain Ltd v Wescol Steel Ltd [2003] EWHC 312 . . . . T 9.4 (308)
Coventry Scaffolding Co. (London) Ltd v Lancsville Construction Ltd [2009] All ER (D) 93 (Dec), [2009] EWHC 2995 . . . . 7.51, T 7.4 (235)
Cowlin Construction Ltd v CFW Architects (a firm) [2003] BLR 241; [2002] EWHC 2914 . . . . 1.17, 3.33, T 1.1 (13), T 3.2 (74), T 5.1 (154), T 9.6 (321)
Cruden Construction v. Commission for the New Towns [1995] 2 Lloyd's Rep 37. . . . . 3.45
Cubitt Building & Interiors Ltd v Fleetglade Ltd (2007) 110 Con LR 36; [2006] EWHC 3413 . . . . 3.77, 4.05, 4.42, 4.109, 6.14, 9.32, 9.39, 17.01, 17.16, T 4.2 (127), T 4.3 (134), T 6.2 (178), T 9.4 (308), T 9.5 (314)
Cubitt Building & Interiors Ltd v Richardson Roofing (Industrial) Ltd [2008] BLR 354 . . . . 3.82, T 3.5 (94)
Cygnet Healthcare v Higgins City Ltd (2000) 16 Const LJ 394 . . . . 8.47, T 3.5 (93), T 8.3 (291)

DGT Steel and Cladding Ltd v Cubitt Building and Interiors Ltd [2007] EWHC 1584 . . . . 3.80, T 3.5 (94)
D. R. Bradley (Cable Jointing) Ltd v Jefco Mechanical Services Ltd (1988) 6-CLD-07–19 . . . . . 15.05
Dairy Containers Ltd v Tasman Orient Line CV 1 WLR 215 . . . . .16.29
Dalkia Energy & Technical Services Ltd v Bell Group Ltd (2009) 122 Con LR 66; [2009] EWHC 73 (TCC) . . . . 3.22, 4.28, 7.58, 9.28, 9.39, 9.41, T 3.1 (64), T 5.1 (159), T 9.3 (304), T 9.5 (315)
David & Theresa Bothma (in partnership) t/a DAB Builders v Mayhaven Healthcare (2007) 114 Con LR 131; [2007] EWCA 527. . . . . 3.52, 3.55, 3.56, T 3.3 (80)
David McLean Contractors Ltd v The Albany Building Ltd,[2005] Adj LR 11/10. . . 6.40, T 6.3 (192)
David McClean Housing Ltd v Swansea Housing Association [2001] EWHC 830; [2002] BLR 125 (TCC) . . . . 3.50, 3.53, 3.54, 4.04, 4.30, 6.04, 6.14, 6.45, 8.48, T 3.3 (80), T 6.2 (178), T 6.4 (201), T 9.3 (303)
Dawnays Ltd v FG Minter Ltd [1971] 1 WLR 1205 . . . . 7.112
Dean and Dyball Construction Ltd v Kenneth Grubb Associates Ltd (2003) 100 Con LR 92; [2003] EWHC 2465 . . . . 2.08, 2.15, 2.16, 5.25, 10.03, 10.77, 10.78, 10.86, 10.87, T 2.1 (42), T 3.2 (75), T 10.6 (367)
Debeck Ductwork Installation Ltd v T&E Engineering Ltd, unreported, 14 October 2002 (TCC). . . . . 2.39, 7.53, T 2.1 (41), T 7.4 (234)
Decro-Wall SA v Practitioners in Marketing Ltd [1971] 1 WLR 361 . . . . 12.26
Deko Scotland v Edinburgh Royal Joint Venture and anor [2003] SLT 727. . . . . . . 4.145, T 4.4 (139)
Director General of Fair Trading v First National Bank plc [2002] 1 AC 481 . . . . 1.88
Discain Project Services Ltd v Opecprime Development Ltd [2000] BLR 402. . . . . 10.04, 10.05, 10.07, 11.08, 11.23
Discain Project Services Ltd v Opecprime Development Ltd (No. 2) [2001] BLR 285 . . . . 9.49, 10.07, 10.76, 11.09, T 10.6 (366), T 11.1 (388)
Dorchester Hotel Ltd v Vivid Interiors Ltd [2009] BLR 135, [2009] EWHC 70 (TCC) . . . . 7.59, 10.22, 10.26, 10.33–10.35, T 10.2 (340)
Dr Peter Rankilor v Perco Engineering Services Ltd, 27 January 2006 . . . . T 4.3 (134)
Drake & Scull Engineering Ltd v McLaughlin & Harvey Plc [1992] CILL 768 . . . . 7.35
Dunlop & Ranken Limited v Hendall Steel Structures Ltd [1957] 1 WLR 1102 . . . . 7.126
Durham County Council v Jeremy Kendall (t/a HLB Architects) [2011] EWHC 780; [2011] BLR 425 (TCC) . . . . 2.16, 2.23, 5.16, 7.91, 9.14, T 2.1 (45), T 5.1 (161), T 9.2 (300)

EQ Projects Ltd v Alavi (t/a Merc London) [2006] BLR 130, [2006] EWHC 29 (TCC) 7.104
Ealing London Borough Council v Jan [2002] EWCA Civ 329 . . . . .11.06
Earl's Terrace Properties Ltd v Waterloo Investments Ltd [2002] CILL 1889
(TCC) . . . . 1.09, 1.29, T 1.1 (13)
Edenbooth Ltd v CRE8 Developments Ltd [2008] EWHC 570
(TCC) . . . . 1.81, 10.24, T.1.2 (23), T 1.3 (31), T 10.2 (340)
Edmund Nuttall Ltd v R. G. Carter Ltd [2002] BLR 312
(TCC) . . . . .3.32, 9.49, 9.52, T 3.2 (74), T 9.6 (321)
Edmund Nuttall Ltd v Sevenoaks District Council [2000] Adj LR 04/14
(TCC) . . . . T 6.3 (190)
Elanay Contracts Ltd v The Vestry [2001] BLR 33 (TCC) 10.14
Ellerine Brothers (Pty) Ltd v Klinger [1982] 1 WLR 1375 . . . . 3.45
Ellis Building Contractors Ltd v Vincent Goldstein (2011) CILL 3049; [2011]
EWHC 269 (TCC) . . . . 11.27, T 10.7 (376), T 11.1 (392)
Emcor Drake & Scull Ltd v Costain Construction Ltd (2004) 97 Con LR 142, [2004]
EWHC 2439 (TCC) . . . . 6.26, 6.68, 10.24, T 6.4 (201), T 10.2 (339)
Emcor Drake & Scull Ltd v Sir Robert McAlpine Ltd [2004] EWHC 1017
(TCC) . . . . 13.06, 13.11, 13.60, T 13.1 (447)
Emson Easton Ltd v EME Developments Ltd [1991] 55 BLR 114. . . . . .12.71
Enterprise Managed Services Ltd v Tony McFadden Utilities Ltd [2010] BLR 89; [2009]
EWHC 3222 (TCC) . . . . 9.23, T 5.1 (158)
Epping Electrical Company Ltd v Briggs & Forrester (Plumbing Services) Ltd [2007]
BLR 126, [2007] EWHC 4 (TCC). . . . . 9.41, T 9.5 (314)
Estor Ltd v Multifit (UK) Ltd (2009) 126 Con LR 40; [2009] EWHC 2565
(TCC) . . . . .2.05, 9.20, T 2.1 (44), T 9.2 (300)
Euro Construction Scaffolding Ltd v SLLB Construction Ltd [2008] EWHC 3160
(TCC) . . . . .T 2.2 (51), T 5.1 (157)
Excelsior Commercial & Industrial Holdings Ltd v Salisbury Hamer Aspden & Johnson [2002]
EWCA Civ 879 . . . . 7.104

FG Minter v Dawnays 13 BLR 1. . . . . 4.94
F. W. Berk & Co. Ltd v Knowles and Foster [1962] 1 Lloyd's Rep 430 . . . . 5.02
FW Cook Ltd v Shimizu (UK) Ltd [2000] BLR 199 (TCC) . . . . T 4.1 (102)
Farebrother building Services Ltd v Frogmore Investments Ltd (2001) CILL 1762 . . . . .4.85, 9.12,
T 7.1 (211), T 7.5 (241), T 9.1 (297), T 10.3 (347)
Fastrack Contractors Ltd v Morrison Construction (1999) 75 Con LR 33, [2000] BLR 168;
(TCC) . . . . 3.32, 3.57, 7.75, T 3.2 (73), T 3.3 (80), T 5.1 (150)
Felton v Wharrie (1906) 2 Hudson's BC (4th edn) 398 . . . . .15.02
Fence Gate Ltd v J. R. Knowles Ltd (2001) 84 Con LR 206 [2001]
(TCC) . . . . 1.16, 1.18, 1.31–1.33, T 1.1 (13), T 5.1 (153)
Fenice Investments Inc. v Jerram Falkus Construction Ltd (2009) 128 Con LR 124, [2009]
EWHC 3272 (TCC) . . . . .7.105, 7.109, T 7.7 (261)
Fenice Investments Inc. v Jerram Falkus Construction (2011) 141 Con LR 206, [2011]
EWHC 1678 (TCC) . . . . .4.130, 4.133, T 4.3 (135)
Ferson Contractors Ltd v Levolux AT Ltd (2003) 86 Con LR 98, [2003] BLR 118; [2003]
EWCA Civ 11 (CA) . . . . 5.28, 6.39, 6.47, 6.49–6.51, 8.15, T 6.3 (190–1)
Fileturn Ltd v Royal Garden Hotel Ltd [2010] EWHC 605 (TCC); [2010]
BLR 512. . . . . .4.34, 11.08, 11.16, T 11.1 (391)
Fillite (Runcorn) Ltd v Aqua-Lift [1989] 45 BLR 27 . . . . .3.12, T 3.1 (64)
Fiona Trust & Holding Company and ors v Yuri Privalov (2007) 4 All ER 951(HL) 3.13
Fitzpatrick Contractors Ltd v Tyco Fire and Integrated Solutions (UK) Ltd (2008)
119 Con LR 155, [2008] EWHC 1301 (TCC). 7.104
Fleming Buildings Ltd v Forrest or Hives [2008] CSOH 103 . . . . 5.27, 5.29, 13.44, 13.56,
13.57, T 13.1 (448)
Forest Heath District Council v ISG Jackson Ltd [2010] All ER (D) 16 (Nov), [2010]
EWHC 322 (TCC) . . . . 7.65, T 7.4 (236)

GPS Marine Contractors Ltd v Ringway Infrastructure Ltd [2010] BLR 377; [2010] EWHC 283 (TCC) . . . . 3.25, 5.16, 7.09, 9.14, T 3.1 (66), T 3.3 (81), T 5.1 (158), T 7.1 (215), T 10.4 (355)
Gabrielle & Christopher Shaw v Massey Foundation & Pilings Ltd [2009] EWHC 493 (TCC) . . . . 1.80, T 1.3 (32)
Galliford Try Construction Ltd v Michael Heal Associates Ltd (2003) 99 Con LR 19; [2003] EWHC 2886 . . . . 2.35, 7.89, T 2.1 (42), T 5.1 (154), T 7.6 (250)
General Horticultural Co. ex parte Whitehouse, Re [1886] 32 Ch D 512 . . . . 7.126
Geoffrey Osborne Ltd v Atkins Rail Ltd [2010] BLR 363; [2009] EWHC 2425 (TCC) . . . . 4.111, 6.12, 7.65, 7.67, T 6.2 (180), T 7.1 (216), T 7.4 (235)
George Parke v Fenton Gretton Partnership [2001] CILL 1712 . . . . 7.31, T 7.2 (223)
Geris Handelsgesellschaft GmbH v Les Constructions Industrielles de la Mediterrannee SA [2005] EWHC 499 (TCC) . . . . T 6.3 (192)
Gibson Lea Retail Interiors Ltd v Makro Self Service Wholesalers Ltd [2001] BLR 407 . . . . 1.19, 1.41, 1.48, 1.49, 1.71, 1.72, T.1.2 (23)
Gibson v Imperial Homes [2002] All ER (D) 367 (Feb) . . . . T 9.2 (299)
Gillies Ramsay Diamond v PJW Enterprises Ltd [2003] BLR 48 . . . . 1.13, 1.16, 1.17, 10.64, 10.73, T 1.1 (13), T 10.4 (354), T 10.5 (359, 361)
Gillies v Secretary of State for Work & Pensions [2006] 1 WLR 781, [2006] UKHL 2 . . . . 11.05
Gipping Construction Ltd v Eaves Ltd [2008] EWHC 3134 (TCC) . . . . 7.106, 8.42, T 7.7 (260), T 8.2 (287)
Glencot Development and Design Co Ltd v Ben Barrett & Son (Contractors) Ltd [2001] BLR 207 (TCC) . . . . 6.07, 6.14, 7.48, 7.50, 10.05, 10.15, 11.04, 11.34, T 6.2 (177), T 11.1 (388)
Glencot Development and Design Co Ltd v Ben Barrett & Son (Contractors) Ltd [2001] All ER (D) 384 (TCC) . . . . T 7.4 (234)
Gray & Sons Builders (Bedford) Ltd v Essential Box Company Ltd (2006) 108 Con LR 49, [2006] EWHC 2520 (TCC) . . . . 7.103, 7.105, 7.110–7.113, T 7.4 (234), T 7.7 (260), T 8.2 (286)
Greenmast Shipping Co SA v Jean Lion & Cie SA, The Saronikos [1986] 2 Lloyd's Rep 277 . . . 12.65
Grovedeck Ltd v Capital Demolition Ltd [2000] BLR 181 . . . . 2.24, T 2.1 (41), T 3.3 (80), T 5.1 (151)
Guardi Shoes Ltd v Datum Contracts [2002] CILL 1934 . . . . 7.30, T 7.2 (224)
Gulf International Bank v Al Ittefaq Steel Products Co [2010] EWHC 2601 (QB) 8.42

HG Construction Ltd v Ashwell Homes (East Anglia) Ltd [2007] BLR 175; EWHC 144 (TCC) . . . . 4.48, 6.63, 6.66, T 6.4 (202)
HS Works Ltd v Enterprise Managed Services Ltd [2009] BLR 378; [2009] EWHC 729 (TCC) . . . . 6.08, 6.27, 6.28, 7.11, 9.52, 10.39, T 6.2 (179), T9.6 (323), T 10.1 (333), T 10.3 (350)
Halki Shipping Corporation v Sopex Oils Ltd [1998] 1 WLR 726; [1998] 1 Lloyd's Rep 49 (CA) . . . . 3.33, 3.45, 8.48, T 3.2 (73)
Hanak v Green [1958] 2 QB 9 . . . . 13.32
Harlow & Milner Ltd v Teasdale [2006] All ER (D) 382, [2006] EWHC 535 . . . . 7.43, 7.103, 7.105, 8.46, T7.4 (234), T7.8 (266), T 8.2 (285)
Harlow & Milner Ltd v Teasdale [2006] BLR 359, [2006] EWHC 1708 (TCC) . . . . 7.121–7.122, T 7.8 (266)
Harlow & Milner Ltd v Teasdale [2006] BLR 359, [2006] EWHC 54 . . . . 7.22, 7.103, T 7.7 (259)
Harris Calnan Construction Company Ltd v Ridgewood (Kensington) Ltd [2008] BLR 132; [2007] EWHC 2738 (TCC) . . . . 2.52, 7.105, T 2.3 (57), T 5.1 (152, 157), T 7.7 (260)
Hart (t/a DW Hart & Son) v Smith [2009] All ER (D) 29 (Sep), [2009] EWHC 2223 (TCC) . . . . 6.29, T 6.2 (179)
Hart Investments Ltd v Fidler and anor [2006] BLR 30; [2006] EWHC 2857 (TCC) . . . . 2.47, 2.53, 2.54, 4.42, 9.17, 9.32, 9.35, 9.36, 9.38, T 2.3 (57), T 9.4 (308)
Harvey Shopfitters Ltd v ADI Ltd [2003] EWCA Civ 1757; [2004] 2 All ER 982 . . . . 2.49
Harwood Construction Ltd v Lantrode Ltd, unreported, 24 November 2000 (TCC) . . . . . 8.14, 8.23, 8.24, 13.124, T 8.1 (278), T 8.2 (284)

Hayter v Nelson and Home Insurance (No. 2) [1990] 2 Lloyd's Rep 265 .... 3.45, T 3.2 (73)
Healy-Upright v Bradley [2007] All ER (D) 29 (Nov), [2007] EWHC 3161(Ch)............ 7.104
Heifer International Inc v Christiansen [2007] EWHC 3015 ....1.88
Herbosh-Kiere Marine Contractors Ltd v Dover Harbour Board [2012] BLR 177; [2012]
EWHC 84 (TCC) ....3.60, 4.63, 10.95, 10.98, 10.99, T 3.4 (88), T 9.6 (325), T 10.7 (376)
Herschel Engineering Ltd v Breen Property Ltd [2000] BLR 272 ....3.79, 3.83, 3.84,
8.39 T 3.5 (93), T 8.2 (285)
Highlands and Islands Airports Ltd v Shetland Islands Council [2012]
CSOH 12...10.07, 10.80, T 10.6 (368)
Hills Electrical & Mechanical Plc v Dawn Construction [2004]
SLT 477...................................12.44, 12.53, 12.54, 12.65, T 12.2 (410)
Hillview Industrial Developments (UK) Ltd v Botes Building Ltd [2006] All ER (D) 280 (Jun);
[2006] EWHC 1365 (TCC)............6.36, 6.40, 8.39, 8.51, 8.52, T 6.3 (193), T 8.2 (286)
Hitec Power Protection BV v MCI Worldcom Ltd [2002] EWHC 1953
(TCC) ....3.32, T 3.2 (74), T 7.5 (241)
Holt Insulation Ltd v Colt International Ltd [2001] EWHC 451 (TCC) ....T 6.4 (200)
Homer Burgess v Chirex (Annan) Ltd (2000) CILL 1580; 71 Con LR 245; [2000]
BLR 124............................. 1.61, 1.66, 5.10, 7.75, T.1.2 (23–4), T 7.5 (240)
Hortimax Ltd v Hedon Salads Ltd, unreported, 15 October 2004 (TCC) .... T 5.1 (155)
Humes Building Contracts Ltd v Charlotte Homes (Surrey) Ltd, unreported, 4 January 2007
(TCC) .... 10.39, T 10.3 (349), T 10.7 (374)
Hyder Consulting (UK) Ltd v Carillion Construction Ltd (2011) 138 Con LR 212; [2011]
EWHC 1810.....................3.60, 6.25, 6.32, 6.33, 10.95, T 6.2 (181), T 10.7 (376)

IDE Contracting Ltd v R. G. Carter Cambridge Ltd [2004] BLR 172; [2004]
EWHC 36 (TCC) .... 3.88, 4.28, 9.10, 9.27, 9.28, T 5.1 (155), T 9.3 (303)
Integrated Building Services Engineering Consultants Ltd (t/a Operon) v Pihl UK Ltd [2010]
BLR 622; [2010] CSOH 80 .... 8.07, 8.14, 8.15, T 8.1 (281)
Interserve Industrial Services Ltd v Cleveland Bridge UK Ltd [2006] All ER (D) 49 (Feb); [2006]
EWHC 741 (TCC) ....6.36, 6.41, 7.81, 8.37, T 6.3 (193), T 7.5 (242), T 8.2 (285)
Investors Compensation Scheme v West Bromwich Building Society [1998] 1WLR 886 ....16.29

J. Spurling Ltd v Bradshaw [1956] 1 WLR 461 .... 1.86, 1.89
JPA Design & Build Ltd v Sentosa (UK) Ltd [2009] All ER (D) 06 (Oct), [2009]
EWHC 2312 (TCC) ....8.11, 8.14, T 8.1 (280)
JW Hughes Building Contractors Ltd v GB Metalwork Ltd [2003] Adj LR 10/03, [2003]
EWHC 2421 .... 8.03, 8.13, T 5.1 (149), T 8.1 (279), T 10.2 (338)
Jacques (t/a C&E Jacques Partnership) v Ensign Contractors Ltd [2009] EWHC 3383
(TCC) ....6.65, 7.106, 8.11, 8.42, 10.61, 10.62, T 6.4 (203), T 7.7 (262),
T 8.1 (281), T 8.2 (287), T 10.4 (355)
Jerome Engineering Ltd v Lloyd Morris Electrical Ltd [2002] CILL 1827 ....9.05, T 4.1 (103)
Jerram Falkus Construction Ltd v Fenice Investments (No. 4) [2011] BLR 644, [2011]
EWHC 1935 (TCC).....................6.18, 6.21, 6.34, 6.35, T 6.2 (181), T 6.4 (204)
Jim Ennis Construction Ltd v Premier Asphalt Ltd (2009) 125 Con LR 141, [2009]
EWHC 1906 (TCC) ....6.15, T 6.2 (179)
John Mowlem v Hydra-Tight Ltd [2002] 17 Const LJ 358............3.73, 5.24, 12.41, T 3.5 (93)
John Roberts Architects Ltd v Parkcare Homes (No. 2) Ltd [2006] BLR 106, [2006]
EWCA Civ 64 ....4.144, T 4.4 (139), T 9.6 (322)
John Stirling v Westminster Properties Scotland Ltd [2007] Scot CS CSOH 117 ....T 3.2 (76)
Johnstone v Bloomsbury Health Authority [1992] QB 333 .... 2.34
Joinery Plus Ltd v Laing Ltd [2003] BLR 184; [2003] EWHC 3513 (TCC)............ 4.111, 9.17,
9.49, 9.52, T 7.1 (213), T 9.6 (321)

KNS Industrial Services (Birmingham) Ltd v Sindall Ltd (2000) 75 Con LR 71, [2000]
All ER (D) 1153 (TCC) .... 3.57, 3.59, 9.05, 9.06, 9.51, 13.31, 13.36–13.37, T 6.3 (190),
T 7.5 (240), T 9.1 (296), T 9.6 (320), T 10.3 (347), T 13.1 (446),

KNS Industrial Services (Birmingham) Ltd v Sindall Ltd (2001) 75 Con LR 4.72, 6.51, 15.14, 15.32, T 15.1 (483), 71, [2001] Const LJ 170 (TCC) . . . . T 3.2 (74), T 3.4 (86), T 4.1 (102)
Karl Construction (Scotland) Ltd v Sweeney Civil Engineering (Scotland) Ltd (2002) 85 Con LR 59, [2002] SCLR 766 . . . . 10.90, 12.21, 12.37, 12.46–12.49, T 7.1 (212), T 10.7 (373), T 12.2 (410)
Ken Griffin and anor (t/a K&D Contractors) v Midas Homes Ltd (2000) 78 Con LR; [2000] EWHC 182 (TCC) . . . . 4.16, 4.17, 4.131, 7.81, 9.05, 9.52, T 3.2 (73), T 4.1 (102), T 4.3 (133), T 7.5 (241), T 9.6 (319)
Kiam v MGN Ltd No 2 [2002] 1 WLR 2810, [2002] EWCA Civ 66. 7.104
Kier Regional Ltd (t/a Wallis) v City & General (Holborn) Ltd [2006] EWHC 848; [2008] EWHC 2454 (TCC) . . . . .7.123, 7.125–7.128, 8.39, 10.55, 10.56, 10.59, 10.60, T 7.1 (211), T 7.8 (266), T 8.2 (286), T 10.4 (354)
Kier Regional Ltd (t/a Wallis) v City & General (Holborn) Ltd [2006] EWHC 848 (TCC), [2006] BLR 315 . . . . . . 10.55, 10.56, 10.59–10.60, T7.1 (214), T10.4 (355)
Kier Regional Ltd (t/a Wallis) v City & General (Holborn) Ltd [2008] EWHC 2454 (TCC) 7.125–7.128, 8.39, T8.2 (284)

L. Brown & Sons Ltd v Crosby Homes (North West) Ltd [2005] EWHC 3503 (TCC) . . . . 1.24, 1.36, 1.37, T 1.1 (14)
LPL Electrical Services Ltd v Kershaw Mechanical Services Ltd, unreported, 2 February 2001 (TCC) . . . . T 9.6 (320)
Lanes Group plc v Galliford Try Infrastructure Ltd (t/a Galliford Try Rail) [2011] EWCA Civ 1617 (CA) . . . . 4.39, 5.31, 6.65, 7.63, 9.09, 9.33, 9.59, 9.61, 9.63, 11.02, 11.05, 11.07, 11.12, 11.17, 11.29–11.31, T 9.4 (309), T 11.1 (392)
Lathom Construction Ltd v Brian Cross and Ann Cross [2000] CILL 1568. . . . . 1.24, T 1.1 (12)
Lead Technical Services Ltd v CMS Medical Ltd [2007] BLR 251; [2007] EWCA Civ 316 . . . . 2.39, T 2.2 (51), T 9.3 (304)
Leading Rule v Phoenix Interiors Ltd [2007] EWHC 2293 (TCC) . . . . 15.17, 15.21, T 7.4 (234), T 15.1 (484)
Leander Construction Ltd v Mulalley & Company Ltd [2012] BLR 152; [2011] EWHC 3449 . . . . 4.147, 13.90, 13.96, T 13.2 (465)
Ledwood Mechanical Engineering Ltd v Whessoe Oil and Gas Ltd [2008] BLR 198, [2007] EWHC 2743 (TCC) . . . . 6.29, 6.61, 6.62, T 6.3 (193)
Lee v Chartered Properties (Building) Ltd [2010] BLR 500; [2010] EWHC 1540 (TCC) . . . . 9.39, T 9.5 (315), T 10.4 (356)
Letchworth Roofing Company v Sterling [2009] EWHC 1119; (2009) CILL 2717 (TCC) . . . . 3.59, 4.15, 4.73, 4.74, 10.40, 13.90, 13.124, T 3.4 (87), T 9.5 (315), T 9.6 (323), T 10.3 (350), T 13.1 (449), T 13.2 (464)
Levolux AT Ltd v Ferson Contractors [2003] BLR 118; [2003] EWCA Civ 11 . . . . 7.06, 8.07
Lidl UK GmbH v RG Carter Colchester Ltd [2013] CILL 3276, [2012] EWHC 3138 (TCC) . . . . T 7.5 (244)
Linnett v Halliwells [2009] BLR 312; [2009] EWHC 319 (TCC) . . . . 2.08, 4.78, 4.126, 4.131, 4.132, 4.135, 4.136, 7.94, T 2.1 (44), T 4.3 (133–5), T 7.6 (251), T 9.4 (308)
Lloyd Projects v John Malnick, unreported, 5 September 2005 . . . . 2.15, T 2.1 (42)
Lloyd v McMahon [1987] 2 WLR 821, [1987] AC 625. . . . . 10.03
Locabail (UK) Ltd v Bayfield Properties Ltd [2000] 2 WLR 870, [2000] QB 451 . . . . 11.06
London & Amsterdam Properties Ltd v Waterman Partnership Ltd [2004] BLR 179, [2003] EWHC 3059 (TCC) . . . . 10.05, 10.17, 10.23, 10.25, 10.27, 10.31, 10.32, 11.21, T 3.2 (75), T 3.4 (86), T 7.1 (213), T 9.3 (303), T 10.2 (338), T 11.1 (389)
London & Scottish Properties plc v Riverbrae Construction Ltd [1999] BLR 346 . . . . 13.124
Lovell Projects Ltd v Legg and Carver [2003] 1BLR 452 . . . . 1.90, 1.92, 5.23, T 1.3 (31)
Lubenham Fidelities and Investments Co. Ltd v South Pembrokeshire District Council (1986) 33 BLR 39, 6 Con LR 85 (CA) 15.02

MBE Electrical Contractors Ltd v Honeywell Control Systems Ltd [2010] BLR 561, [2010] EWHC 2244 (TCC) . . . . 7.94, 8.46, T 7.2 (224), T 8.3 (291)
MJ Gleeson Group plc v Devonshire Green Holdings Ltd [2004] Adj LR 03/19 (TCC) . . . . 6.40, T 6.3 (192)
McAlpine PPS Pipeline Systems Joint Venture v Transco plc [2004] BLR 352; [2004] EWHC 2030 (TCC) . . . . 3.58, 3.95, 4.58, 9.52, 9.56–9.58, 10.25, T 3.4 (86), T 9.6 (322), T 10.2 (338)
McConnell Dowell Constructors (Aust) Pty Ltd v National Grid Gas plc [2007] BLR 92, [2006] EWHC 2551 (TCC). . . . 7.107, 8.41, T 1.1 (14), T 7.1 (215), T 7.7 (260), T 8.2 (286)
Macob Civil Engineering Ltd v Morrison Construction Ltd [1999] BLR 93 (TCC). . . . 7.03, 7.12, 7.13, 7.86, 7.94, 7.96, 7.97, 7.121, 8.46, 8.48, 8.49, 8.50, 10.07, 10.90, T 7.1 (211), T 7.2 (223), T 7.6 (249), T 8.3 (291)
Magill v Weeks [2002] 2 AC 357, [2001] UKHL 67 . . . . 11.04
Mair v Arshad, [2007] ScotSC 60 . . . . 12.15, 12.16, T 12.1 (405)
Makers (UK) Ltd v London Borough of Camden [2008] EWHC 1836; [2008] BLR 470 . . . . 4.34, 11.03, 11.08, 11.14, 11.23, T 9.3 (304), T 11.1 (390)
Management Solutions & Professional Consultants Ltd v Bennett (Electrical) Services Ltd [2006] EWHC 1720 (TCC) . . . . 2.38, 2.40, 2.45, 2.46, T 2.1 (43), T 2.2 (50)
Mast Electrical Services v Kendall Cross Holdings [2007] NPC 70; [2007] EWHC 1296 (TCC) . . . . T 2.1 (43)
Maxi Construction Management Ltd v Mortons Rolls Ltd(2001) CILL 1784, [2001] Scot HC 78 . . . . 12.19, 12.30–12.33, 12.43, 12.69, T 12.1 (404)
Mayhaven Healthcare Ltd v Bothma and Bothma [2010] BLR 154, [2009] EWHC 2634 (TCC) . . . . 15.32, 15.42, 15.43, T 15.1 (484)
Maymac Environmental Services Ltd v Farraday Building Services Ltd [2001] CILL 1685 (TCC) . . . . 5.08, T 5.1 (152)
Mayor & Burgesses of Camden LBC v Makers UK Ltd [2009] EWHC 605 (TCC) . . . . 3.79, T 3.5 (94)
Mead General Building Ltd v Dartmoor Properties Ltd [2009] BLR 225, [2009] EWHC 200 (TCC) . . . . 8.13, T 8.1 (280)
Mecright v TA Morris Developments Ltd [2001] Adj LR 06/22. . . . 9.05, T 4.1 (103), T 10.3 (348)
Medicaments and Related Classes of Goods (No. 2), Re, *sub nom* Director General of Fair Trading v Proprietary Association for Great Britain [2001] 1 WLR 700 . . . . 11.01, 11.03, 11.04, 11.05, 11.07
Melville Dundas Ltd v George Wimpey UK Ltd [2007] 1 WLR 1136; [2007] 3 All ER 889; [2007] BLR 257; [2007] UKHL18. . . . 1.26, 8.15, 13.66, 13.70–13.75, 14.01, 14.13, 16.21, T 13.1 (447)
Melville Dundas Ltd v Hotel Corporation of Edinburgh Ltd [2006] BLR 474, [2006] CSOH B136. . . . T 1.1 (14)
Mentmore Towers Ltd v Packman Lucas Ltd [2010] BLR 393, [2010] EWHC 457 (TCC) 7.62, 8.40
Mersey Docks Property Holdings v Kilgour [2004] BLR 412; [2004] EWHC 1638 (TCC) . . . . 17.01, 17.12, T 17.2 (505)
Metropolitan Properties v Lannon [1969] 1 QB 577. . . . 4.35
Michael John Construction v Golledge and ors [2006] TCLR 3; [2006] EWHC 71 (TCC) . . . . 3.20, 8.20, 8.21, 11.15, T 3.1 (65), T 3.3 (80), T 6.2 (178), T 8.1 (280), T 8.1 (280), T 11.1 (390)
Midland Expressway v Carillion Construction Ltd and ors (No. 2) (2005) 106 Con LR 154; [2005] EWHC 2963 (TCC) . . . . 3.73, 16.01, 16.06, 16.11, 16.14–16.16, 16.22–16.24, 16.33, T 16.1 (494)
Midland Expressway Ltd and ors v Carillion Construction Ltd and ors (No. 3) [2006] BLR 325; [2006] EWHC 1505 (TCC). . . . 1.96, 3.46–3.48, 6.23, 9.59, T 3.2 (67), T 6.2 (178)
Millers Specialist Joinery Co. Ltd v Nobles Construction [2001] CILL 1770 (TCC) . . . . T 13.1 (446)
Mitsui Babcock Energy Services Ltd v Foster Wheeler Energia OY [2003] EWHC 958 (TCC) 1.64, T.1.2 (23)

Mivan Ltd v Lighting Technology Projects Ltd [2001] Adj CS 04/09 . . . . T 6.4 (200)
Moschi v Lep Air Services Ltd (1972) 2 WLR 1175 (HL) 15.05
Mott MacDonald Ltd v London & Regional Properties Ltd (2007) 113 Con LR 33; [2007]
EWHC 1055 (TCC). . . . . 2.24, 2.51, 4.102, 9.45, 9.46, 11.07, T 2.1 (44), T 2.3 (57), T 9.5 (314), T 11.1 (390)
Multiplex Construction (UK) Ltd v West India Quay Development Company Eastern Ltd [2006]
EWHC 1569 (TCC) . . . . 10.63, 10.95, T 3.4 (87), T 10.5 (360), T 10.7 (374)
Multiplex Constructions (UK) Ltd v Mott MacDonald Ltd [2007] 110 Con LR 63;
EWHC 20 (TCC) . . . . 3.57, 3.60, 4.63, 7.12, 7.29, 7.367.40, 7.53, T 3.2 (76), T 3.3 (81), T 3.4 (87), T 7.2 (224), T 7.4 (234)
Murray Building Services v Spree Developments, unreported, 30 July 2004. . . . . T 2.1 (42)
Mylcrist Builders Ltd v G. Buck [2008] BLR 611 . . . . 1.88, T 1.3 (32)

NAP Anglia Ltd v Sun-Land Development Co Ltd [2011] EWHC 2846 (TCC); [2012]
BLR 195. . . . . 4.66, T 8.1 (282), T 10.2 (340), T 10.5 (360)
Nageh v Giddings [2007] CILL 2420, [2006] EWHC 3240 (TCC) . . . . . . 17.06, 17.11, T 17.2 (506)
Naylor (t/a Powerfloated Concrete Floors) v Greenacres Curling Ltd [2001]
Scot CS 163 . . . . T 6.4 (200)
Nickleby FM Ltd v Somerfield Stores Ltd (2010) 131 Con LR 203; [2010]
EWHC 1976 (TCC) . . . . 7.91, T 2.1 (45), T 7.6 (252)
Nikko Hotels (UK) Ltd v MEPC plc [1991] 2 EGLR 103 (Ch D) 9.49
Nolan Davis Ltd v Catton [2001] All ER (D) 232 (Mar) . . . 8.16, T 5.1 (152), T 7.1 (212), T 8.1 (279)
Nordot Engineering Services Ltd v Siemens [2000] CILL September 2001 . . . . 1.84, 5.07, T.1.2 (22), T 5.1 (151)
North Midland Construction plc v AE&E Lentjes UK Ltd [2009] BLR 574; [2009]
EWHC 1371 (TCC). . . . . 1.19, 1.57, 1.65, 1.73–1.76, T 1.1 (15), T.1.2 (24)
Northern Developments (Cumbria) Ltd v J&J Nichol [2000]
BLR 158 (TCC) . . . . 4.140, 4.148, 4.149, 5.08, 13.36, T 4.4 (138), T 5.1 (150), T 7.1 (212), T 9.6 (319), T 13.1 (445)
Nottingham Community Housing Association Ltd v Powerminster Ltd [2000]
BLR 309 . . . . 1.46, 1.69, 1.70, T.1.2 (22)

OSC Building Services Ltd v Interior Dimensions Contracts Ltd [2009]
EWHC 248(TCC). . . . . 4.14, 9.52, T 3.2 (77), T 3.4 (88), T 4.1 (104), T 5.1 (159), T 9.6 (323)
O'Donnell Developments Ltd v Build Ability Ltd (2009) 28 Con LR 141; [2009]
EWHC 3388 (TCC) . . . . . 4.112, 4.122, 4.123, 7.105, 8.16, T 4.2 (127), T 7.7 (262), T 8.1 (281)
Orange EBS Ltd v ABB Ltd [2003] BLR 323. . . . . 3.33, T 3.2 (74)
Outwing Construction Ltd v H. Randell & Son Ltd [1999] BLR 156, [1999]
EWHC 100 (TCC) . . . . 7.26, 7.43, 13.136, T 7.2 (223), T 7.4 (233)

PC Harrington Contractors Ltd v Multiplex Constructions (UK) Limited [2008]
BLR 16; [2007] EWHC 2833 (TCC). . . . . 13.32, 13.63, 13.64, T 13.1 (448)
PC Harrington Contractors Ltd v Tyroddy Construction Ltd [2011] All ER (D) 162 (Apr),
[2011] EWHC 813 (TCC) . . . . 10.38, 10.93, T 3.4 (88), T 10.3 (351), T 10.6 (368)
PC Harrington Ltd v Systech International Ltd [2012]
EWCA Civ 1371. . . . . 4.127, 4.132, 4.137, 4.138, T 4.3 (133–5)
PT Building Services Ltd v ROK Build Ltd [2008]
EWHC 3434 (TCC) . . . . 4.40, 7.86, 7.87, 7.94, 9.34, T 7.6 (251), T 9.4 (308)
Palmac Contracting Ltd v Park Lane Estate Ltd [2005] EWHC 919; [2005]
BLR 301 (TCC) . . . . 4.28, 4.78, 9.28, T 9.3 (304), T 10.7 (373)
Palmers Ltd v ABB Power Construction Ltd (1999)BLR 426; 68 Con LR 52
(TCC) . . . . 1.18, 1.19, 1.44, 1.55, 1.58, 1.59, 15.24, T.1.2 (22, 24), T 15.1 (483)
Parsons Plastic (Research & Development) Ltd v Purac Ltd [2002]
BLR 334 (CA) . . . . 5.28, 6.47, T 5.1 (153), T 6.3 (191)
Partner Projects Ltd v Corinthian Nominees Ltd [2012] BLR 97, [2011]
EWHC 2989 (TCC) . . . . T 8.1 (282)

Pegram Shopfitters Ltd v Tall y Weijl (UK) Ltd [2003] EWCA Civ 1750; [2004] 1 All ER 818; [2004] 1 WLR 2082 . . . . 5.10, 7.06, 7.07, 7.98, 7.99, 9.18, 9.47, T 5.1 (155), T 7.1 (213), T 7.6 (250), T 9.1 (297), T 9.3 (303)
Peter Mair v Mohammed Arshad [2007] Sheriffdom of Tayside, Central & Fife at Cupar, A538/04, October 2007 . . . . 12.18, 12.28, 12.29
Picardi v Cuniberti and Cuniberti [2003] 1 BLR 487 . . . . .1.86, 1.89, 1.90, T 1.3 (31)
Pierce Design International Ltd v Mark Johnson and Deborah Johnson [2007] BLR 381, [2007] EWHC 1691 (TCC) . . . . .13.68, 13.76–13.79, T 13.1 (448)
Pihl UK Ltd v Ramboll UK Ltd [2012] CSOH 139 . . . . T 10.5 (360)
Pilon Ltd v Breyer Group plc [2010] BLR 452; [2010] EWHC 837 (TCC) . . . . 3.59, 4.111, 5.10, 7.78, 7.94, 8.13, 8.20, 8.22, 9.06, 10.38, 10.52, 10.53, T 5.1 (160), T 7.5 (243), T 7.6 (252), T 8.1 (281), T 10.3 (351)
President of India v La Pintada C. Nav. S.A [1985] AC 104 . . . . 4.94
Primus Build Ltd v Pompey Centre Ltd [2009] BLR 437; [2009] EWHC 1487 (TCC) . . . . 9.52, 10.89, 10.91, 10.93, 17.05, 17.06, T 9.6 (324), T 10.7 (375), T 17.2 (506)
Pring & St Hill Limited v C. J. Hafner (t/a Southern Erectors) [2002] EWHC 1775 (TCC) . . . . .4.21, 10.76, 11.20, T 10.6 (366), T 11.1 (389)
Pring Island Ltd v Castle Contracting Ltd, unreported, 15 December 2003 . . . . .4.129, T 4.3 (134)
Pro-Design Ltd v New Millennium Experience Company Ltd, unreported, 26 September 2001, TCC . . . . T 7.1 (211)
Profile Projects Ltd v Elmwood (Glasgow) Ltd [2011] CSOH 64 .T 9.3 (305)
Project Consultancy Group v The Trustees of the Gray's Trust [1999] BLR 37 . . . . 1.06, 5.02, 5.11, 5.17, 5.18, 9.17, T 1.1 (12), T 5.1 (150), T 7.1 (211)

Quality Street Properties (Tradings) Ltd v Elmwood (Glasgow) Ltd unreported, 8 February 2002 (Scotland) 1.34
Quarmby Construction Co. Ltd v Larraby Land Ltd unreported, 14 April 2003 (TCC) . . . . 1.27, T 1.1 (14)
Quartzelec Ltd v Honeywell Control Systems Ltd [2009] BLR 328, [2008] EWHC 3315 (TCC) . . . . 7.77, 7.80, 9.06, 10.38, 10.50, 10.51, T 7.5 (242), T 9.1 (297), T 10.3 (350), T 13.1 (448)
Quietfield Ltd v Vascroft Contractors Ltd [2006] EWHC 174 (TCC); [2007] BLR 67 (CA); [2006] EWCA Civ 1737 . . . . 4.47, 10.38, 6.63, 6.64, 6.66, 6.69–6.72, 10.44–10.46, T 6.4 (201), T 10.3 (349)

R&C Electrical Engineers Ltd v Shaylor Construction Ltd [2012] BLR 373, [2012] EWHC 1254 (TCC) . . . . 6.48, T 6.3 (194), T 16.1 (495)
R. C. Pillar & Son v The Camber (Portsmouth) Ltd (2007) 115 Con LR 102 [2007] (TCC) . . . . T 5.1 (156)
RBG Ltd v SGL Carbon Fibers Ltd [2010] BLR 631, [2010] CSOH 77 . . . . 10.38, T 10.3 (351)
R. Durtnell & Sons Ltd v Kaduna Ltd [2003] BLR 225; [2003] EWHC 517 (TCC) . . . . 5.15, 7.87, 7.93, T 3.2 (75), T 5.1 (154), T 7.6 (250)
RG Carter Ltd v Clarke [1990] 1 WLR 578 . . . . 7.70
RG Carter v Edmund Nuttall Ltd, unreported, 21 June 2000 (TCC). . . . . 2.17, 3.22, 3.73, T 2.1 (42), T 3.1 (64), T 3.5 (94), T 11.1 (388)
RJ Knapman Ltd v Richards (2006) 108 Con LR 64; [2006] EWHC 2518 (TCC) . . . . . . 6.27, 6.40, 7.86, T 6.3 (193), T 7.6 (250)
RJT Consulting Engineers Ltd v DM Engineering (Northern Ireland) Ltd [2002] 1 WLR 2344; [2002] BLR 217; [2002] EWCA Civ 270 (CA) . . . . .2.12, 2.14, 2.17, 2.22, 2.28, 2.31, 2.32, 2.35, 2.47, 3.22, T 2.1 (41)
RSL (South West) Limited v Stansell Limited [2003] EWHC 1390 (TCC) . . . . 6.04, 6.25, 10.12, 10.75, 10.83–10.85, T 6.2 (178), T 7.5 (242), T 10.6 (367)
RWE NPower v Alstom Power Ltd (2010) 133 Con LR 155, [2010] EWHC 3061 (TCC) . . . T 7.6 (251)
R v Cripps ex p. Muldoon [1984] QB 686 . . . . .4.110
R v Sussex Justices ex p. McCarthy [1924] 1 KB 256 (KBD) 11.02

Rainford House Ltd v Cadogan Ltd [2001] All ER (D) 144, [2001] EWHC 18 (TCC) . . . . 8.11, 8.14, 8.17, 8.25, 8.26, T 8.1 (279)
Redwing Construction Ltd v Charles Wishart [2010] EWHC 3366 (TCC); [2011] BLR 186; [2011] EWHC 19 . . . . 4.109, 6.24, 7.102, 7.114–7.116, T 4.2 (128), T 6.2 (181), T 6.4 (203), T 7.7 (262)
Redworth Construction Ltd v Brookdale Healthcare Ltd [2006] BLR 366, [2006] EWHC 1994 (TCC) . . . . 7.84, 7.90, 7.91, T 2.1 (43), T 7.6 (250, 253), T 9.2 (300)
Reid Minty v Taylor [2002] 1 WLR 2800, [2001] EWCA Civ 1723 . . . 7.103, 7.104, 7.111, T 7.7 (259)
Reinwood Ltd v L. Brown & Sons Ltd [2008] 1 WLR 696; [2008] 2 All ER 885; [2008] BLR 219 . . . . 12.14, 12.23, 12.26, 13.127, 13.130–13.135, T 12.1 (405), T 13.2 (464)
Rice (t/a The Garden Guardian) v Great Yarmouth Borough Council [2000] All ER (D) 902 (CA) 12.26
Ringway Infrastructure Services Ltd v Vauxhall Motors Ltd (2007) 115 Con LR 149, [2007] EWHC 2507 (TCC) . . . . 7.108, 7.117–7.118, T 7.7 (260)

Ringway Infrastructure Service Ltd v Vauxhall Motors Ltd [2007] All ER (D) 333, [2007] EWHC 2421 (TCC) . . . . 12.17, 12.74, 13.04, 13.19, 13.21–13.23, 13.58, T 12.1 (405), 16.05, 16.30, 16.32, T 13.1 (448), T 16.1 (495)
Ritchie Brothers (PWC) Ltd v David Philip (Commercials) Ltd [2004] BLR 379 . . . 4.38, T 9.5 (313)
Roberts Petroleum Ltd v Bernard Kenny Ltd [1983] 1 WLR 301 . . . . 7.127
Rohde (t/a M Rohde Construction) v Markham-David [2006] BLR 291 (TCC), [2006] EWHC 814 . . . . 17.01, 17.11, 17.18, 17.19, 10.25, T 10.2 (339), T 17.2 (506)
ROK Build Ltd v Harris Wharf Development Company Ltd [2006] EWHC 3573 (TCC) . . . . T 9.2 (300)
ROK Building Ltd v Bestwood Carpentry Ltd [2011] EWHC 1409 (TCC) . . . . 2.36, T 2.1 (45), T 2.2 (51)
ROK Building Ltd v Celtic Composting Systems Ltd (No. 2) (2010) 130 Con LR 74; [2010] EWHC 66 (TCC) . . . . 4.113, 7.07, T 4.2 (128), T 6.3 (194), T 7.1 (215)
Rossco Civil Engineering Ltd v Dwr Cymru Cyfyngedic, 5 August 2004 . . . . T 5.1 (155)
Rossiter v Miller (1878) 3 App Cas 1124 . . . . 2.49
Rover International Ltd v Cannon Film Sales (No. 3) [1989] 3 All ER 423, [1989] 1 WLR 912 (CA) 12.65
Rupert Morgan Building Services (LLC) Ltd v David Jervis and Harriet (2004) 91 Con LR 81; [2004] 1 WLR 1867, [2003] EWCA Civ 1563 (CA) . . . . 13.40, 13.42, 13.43, 13.45–13.48, 13.59, 13.65, 13.92, 14.06, T 13.1 (447, 449), T 13.2 (463)
Ruttle Plant Hire Ltd v Secretary of State for Environment Food & Rural Affairs (No. 3) [2009] 1 All ER 448, [2008] EWHC 238 (TCC) 7.109

S. G. South Ltd v (1) King's Head Cirencenster LLP and (2) Corn Hall Arcade Ltd [2009] EWHC 2645 (TCC) . . . . 3.26, 7.09, 8.20, 8.21, T 3.1 (65), T 8.1 (281), T 8.2 (287)
St Andrews Bay Development Ltd v HBG Management [2003] Scots CS 103 . . . . 4.101, T 9.5 (313)
Samuel Thomas Construction v Anon, unreported, 28 January 2000 . . . . 1.79, 1.83, T 1.3 (30)
SG South Ltd v Swan Yard (Cirencester) Ltd [2010] EWHC 376 (TCC) . . . . 2.28, T 2.1 (44), T 7.1 (216), T 8.2 (287)
Shaw v Massey Foundation & Pilings Ltd [2009] EWHC 493 (TCC) . . . . 8.20, 8.39, T 1.3 (32), T 8.1 (280), T 8.3 (291)
Shaw v MFP Foundations & Piling Ltd [2010] CILL 2831, [2010] EWHC 9 (Ch) . . . . 7.32, T 7.2 (224)
Shepherd Construction Ltd v Mecright Ltd [2000] BLR 489 . . . . 1.24, 1.34, 1.35, 1.39, 3.25, 3.27, 3.28, T 1.1 (12), T 3.1 (64)
Sherwood & Casson v MacKenzie (2000) CILL 1577; (2000) 2 TCLR 418 (TCC) . . . . 4.46, 7.06, T 6.4 (199), T 7.1 (211)
Shimizu Europe Ltd v Automajor Ltd [2002] EWHC 1571; [2002] BLR 113 (TCC) . . . . 4.85, 4.114, 6.48, 7.06, 7.94, 9.49, 9.52, 10.93, T 4.2 (127), T 5.1 (153), T 6.3 (191), T 7.1 (213), T 7.5 (241), T 7.6 (249), T 9.6 (322), T 10.7 (373)

Simons Construction Ltd v Aardvark Developments Ltd [2004] BLR 117, [2003] EWHC 2474 . . . . . T 9.5 (313)
Sindall Ltd v (1) Abner Solland (2) Grazyna Solland (3) Solland Interiors (a firm) (4) Solland Interiors Ltd (2001) 80 Con LR 152; [2001] 3 TCLR 30 (TCC) . . . . . 3.32, 3.33, 3.50, 7.11, T 3.2 (74), T 3.3 (80), T 3.4 (86), T 7.2 (223), T 9.6 (320)
Skanska Construction UK Ltd v The ERDC Group Ltd [2002] Scot CS 307 . . . . . T 6.4 (201)
SL Timber Systems Ltd v Carillion Construction Ltd [2001] BLR 516, 2002 SLT 997. . . . . 9.52, 12.64, 13.29, 13.38, 13.39, 13.64, T 7.1 (212), T 9.6 (321), T 13.1 (446)
Smith v Kvaerner Cementation Foundations Ltd [2006] 3 All ER 593. . . . . 4.35, 11.18
South West Contractors Ltd v Birakos Enterprises Ltd [2006] All ER (D) 63 (Nov), [2006] EWHC 2794 (TCC) 10.41
Specialist Ceiling Services Northern Ltd v ZVI Construction (UK) Ltd [2004] BLR 403 . . . . . T 11.1 (389)
Speymill Contracts Ltd v Baskind [2010] BLR 257, [2010] EWCA Civ 120 . . . . . 7.09, T 3.1 (65), T 7.1 (216)
Sprunt Ltd v London Borough of Camden [2012] BLR 83; [2011] EWHC 3191 (TCC). . . . . 2.28, 3.98, 4.33, T 2.1 (46), T 9.3 (305), T 11.1 (392)
Squibb Group Ltd v Vertase FLI Ltd [2012] EWHC 1958 (TCC) . . . . . 5.28, 6.48, T 6.3 (194)
Staveley Industries plc v Odebrecht Oil & Gas Services Ltd [2001] 98(10) LSG 46 . . . 1.10, 1.42, 1.45, 1.48, 1.67, 1.68, T 1.1 (13)
Stent Foundations Ltd v Carillion Construction (Contracts) Ltd (formerly Tarmac Construction (Contracts) Ltd) (2000) 78 Con LR 188 . . . . . 2.49
Steve Domsalla (t/a Domsalla Building Services) v Kenneth Dyason [2008] BLR 348; [2007] EWHC 1174 (TCC). . . . . 1.91, 5.23, 5.26, 7.05, 9.47, 13.85, 13.86, T 1.3 (31), T 7.1 (215), T 10.3 (349), T 13.1 (447)
Strathmore Building Services Ltd v Colin Scott Greig (t/a Hestia Fireside Design) (2001) 17 Const LJ 72, [2000] Scot CS 133 . . . . . 13.97, 13.117, 13.120–13.122, 14.12, 17.14, T 13.2 (463), T 17.2 (505)
Straw Realisations (No. 1) Ltd (formerly known as Haymills (contractors) Ltd (in administration)) v Shaftsbury House (Developments) Ltd [2011] BLR 47, [2010] EWHC 2597 (TCC) . . . . . 6.22, 8.14, 8.15, T 6.2 (181), T 6.3 (194), T 8.1 (282), T 10.7 (375)
Stubbs Rich Architects v W. H. Tolley & Son Ltd, unreported, 8 August 2011. . . . . 4.130, 4.133, T 4.3 (134)
Sughra Sulaman v AXA Insurance plc [2009] All ER (D) 116 (Dec), [2009] EWCA Civ 1331. . . . . 7.106
Supablast (Nationwide) Ltd v Story Rail Ltd [2010] EWHC 56; [2010] BLR 211. . . . . 3.19, 7.106, 9.12, T 2.1 (45), T 3.1 (66), T 5.1 (159), T 7.7 (262), T 9.1 (297)
Swain v Hillman [2001] 1 All ER 91 . . . . . 7.47, 7.53
Systech International Ltd v P. C. Harrington Ltd [2011] EWHC 2722 (TCC) . . . . . T 4.3 (133, 135)

Taylor v Lawrence [2003] QB 528 . . . . . 4.35, 11.18
Thames Iron Works & Shipbuilding Co. Ltd v R [1869] 10B & S 33 . . . . . 5.02
Thermal Energy Construction Ltd v AE & E Lentjes UK Ltd [2009] All ER (D) 271 (Jan), [2009] EWHC 408 (TCC) . . . . . 10.38, 10.41, 10.65, 10.67–10.71, 10.73, T 10.3 (350), T 10.5 (360)
Thomas Frederic (Construction) Ltd v Keith Wilson (CA) [2004] BLR 23 . . . . . 5.19, 5.20, 9.20, T 5.1 (155), T 7.1 (213)
Thomas Vale Construction plc v Brookside Syston Ltd (2009) 25 Const LJ 675; [2006] EWHC 3637 . . . . . 13.101, 13.108, T 13.2 (464)
Tim Butler Contractors Ltd v Merewood Homes Ltd [2002] 18 Const LJ 74. . . . . 7.07, 12.10, 12.38, T 7.1 (211), T 12.1 (405)
Total M&E Services Ltd v ABB Building Technologies Ltd (2002) 87 Con LR 154; [2002] EWHC 248 (TCC) . . . . . 2.20, 2.42, 4.142, 4.150, 4.151, 8.13, 8.16, 8.27–8.29, T 2.2 (50), T 4.4 (138), T 8.1 (279), T 9.2 (299)
Treasure & Son Ltd v Martin Dawes [2008] BLR 24; [2007] EWHC 2420 (TCC) . . . . 2.08, 4.104, 8.18, 8.21, 8.30, 8.31, T 2.1 (43), T 8.1 (280), T 9.5 (315)
Trollope & Colls Ltd v North West Metropolitan Regional Hospital Board [1973] 2 All ER 260. . . . . 2.35

Trustees of Stratfield Saye Estate v AHL Construction Ltd [2004] All ER (D) 77 (Dec); [2004] EWHC 3286 . . . . . 2.14, 2.16, 2.28, T 2.1 (42)
Try Construction Ltd v Eton Town House Group Ltd [2003] BLR 286, [2003] EWHC 60 (TCC) . . . . . 10.05, 10.78, 10.80, T 10.6 (366)
Turner & Goudy v McConnell [1985] 1 WLR 898 . . . . . 8.48

Urang Commercial Ltd v (1) Century Investments Ltd and (2) Eclipse Hotels (Luton) Ltd [2011] EWHC 1561 . . . . . 4.74, 10.40, 13.44, 13.51–13.55, 13.114, 13.124, 13.126, T 7.1 (216), T 9.6 (324), T 10.3 (351), T 13.1 (449), T 13.2 (465)

VGC Construction Ltd v Jackson Civil Engineering Ltd [2008] EWHC 2082 (TCC) . . . . . T 3.2 (76), T 5.1 (157)
VHE Construction plc v RBSTB Trust Co. Ltd [2000] BLR 187 . . . . . 6.04, 6.07, 6.38, 7.53, 7.56, 7.68–7.70, 13.98, T 6.3 (190), T 7.4 (233), T 13.1 (445), T 13.2 (463)
Vakauta v Kelly (1989) 167 CLR 568, (1988) 13 NSWLR 502 . . . . . 11.06
Vaultrise Ltd v Paul Cook [2004] ADJCS 04/06 . . . . . 4.87
Vertase FLI Ltd v Squibb Group Ltd [2012] EWHC 3194 (TCC) . . . . . T 6.4 (204)
Vision Homes Ltd v Lancsville Construction Ltd [2009] BLR 525; [2009] EWHC 2042 (TCC) . . . . . 4.28, 4.50, 6.65, 9.10, 9.28, 9.44, 10.93, 10.95, T 6.4 (203), T 9.3 (304), T 9.6 (324), T 10.5 (360), T 10.7 (375)
Vitpol Building Service v Samen [2008] EWHC 2283 (TCC) . . . . . 7.43, 7.59, 7.65, T 7.4 (235)
Volker Stevin Ltd v Holystone Contracts Ltd [2010] EWHC 2344 (TCC) . . . . . 4.59, 9.52, 11.26, T 4.1 (104), T 9.6 (324), T 11.1 (391)
Von Essen Hotels 5 Ltd v Vaughan and anor [2007] EWCA Civ 1349 . . . . . 17.06, 17.13, T 17.2 (506)
Von Hatzfeldt-Widenburg v Alexander [1912] 1 Ch 284 . . . . . 2.49

WW Gear Construction Ltd v McGee Group Ltd [2012] BLR 355, [2012] EWHC 1509 . . . . . 7.64
Walter Lilly & Co Ltd v DMW Developments Ltd [2008] All ER (D) 214 (Dec), [2008] EWHC 3139 (TCC) . . . . . 7.65, T 7.4 (235)
Walter Llewellyn & Sons Ltd v Excel Brickwork Ltd [2011] CILL 2978, [2010] EWHC 3415 (TCC) . . . . . 8.44, 8.46, T 8.3 (292)
Wates Construction Ltd. v HGP Greentree Allchurch Evans Ltd [2006] BLR 45, [2005] EWHC 2174 . . . . . 7.104, 7.111, T 7.7 (259)
Watkin Jones & Son Ltd v Lidl UK GMBH, unreported, 27 December 2001 (TCC) . . . . . 6.65, T 6.4 (200)
Watson Building Services Ltd v Harrison [2001] SLT 846 . . . . . 3.88, T 5.1 (149), T 9.3 (302)
Webb v Stenton (1883) 11 QBD 518 . . . . . 7.126
West Country Renovations Ltd v McDowell [2013] WLR 416, [2012] EWHC 307 (TCC) 7.23
Westdawn Refurbishment Ltd v Roselodge Ltd [2006] Adj LR 04/25 . . . . . 2.15, T 2.1 (43)
Westminster Building Company v Beckingham [2004] BLR 163; [2004] EWHC 138 (TCC) . . . . . 1.38, 1.39, 1.90, 1.91, T 1.1 (14), T 1.3 (31)
Westminster Chemicals & Produce Ltd v Eicholz & Loeser [1954] 1 LLR 99 . . . . . 5.02, 5.18, 5.20, T 5.1 (149–50)
Whiteways Contractors (Sussex) Ltd v Impresa Castelli Construction UK Ltd (2000) 75 Con LR 92 EWHC Technology 61 (TCC) . . . . . T 5.1 (152), T 13.1 (446)
William Hare Ltd v Shepherd Construction Ltd [2009] BLR 447; (2009) EWHC 1603; [2010] BLR 358, [2010] All ER (D) 168; EWCA Civ 283 (CA) . . . . . 16.08, 16.10, 16.19, 16.25–16.29, T 16.1 (495)
William Oakley & David Oakley v Airclear Environmental Ltd and Airclear TS Ltd [2002] CILL 1824 (Ch D) . . . . . T 5.1 (153)
William Verry (Glazing Systems) Ltd v Furlong Homes Ltd [2005] EWHC 138 . . . . . 3.32, 3.51, 3.59, 3.63, 3.64, 4.81, 6.42, 6.56, 6.57, 6.59, 6.60, 10.18, 10.24, 10.26, 10.42, 10.43, 13.83, 13.84, T 3.2 (75), T 3.4 (87), T 4.1 (103), T 10.1 (333), T 10.2 (339), T 10.3 (348)
William Verry Ltd v North West London Communal Mikvah [2004] BLR 3008; [2004] EWHC 1300 (TCC) . . . . . 4.42, 9.32, T 7.1 (213), T 8.2 (285), T 9.4 (308), T 10.3 (348)

William Verry Ltd v London Borough of Camden [2006] All ER (D) 292 (Mar), [2006] EWHC 761 (TCC) . . . . 6.42, 6.56, 6.57, 6.59, 6.60, 8.07, 8.37, T 6.3 (193), T 8.2 (286)
Williams (t/a Sanclair Construction) v Noor (t/a India Kitchen) [2007] All ER (D) 51 (Dec), [2007] EWHC 3467 (TCC). . . . . T 9.2 (299)
Wimbledon Construction Company 2000 Ltd v Vago [2005] BLR 374, [2005] EWHC 1086 (TCC) . . . . 8.10, 8.20, 8.21, 8.29, T 8.1 (279)
Windglass Windows Ltd v Capital Skyline Construction Ltd (2009) 126 Con LR 118; [2009] EWHC 2022 (TCC) . . . . 9.52, 13.125, T 9.6 (324), T 13.1 (449), T 13.2 (464)
Witney Town Council v Beam Construction [2011] EWHC 2332; [2011] BLR 707 (TCC). . . . . 3.52, 3.67, 3.68, 4.13, 4.14, T 3.3 (81), T 3.4 (88), T 4.1 (104)
Woodar v Wimpey Ltd [1980] 1 WLR 277 (HL) 12.26, 15.43
Woods Hardwick Ltd v Chiltern Air Conditioning Ltd [2001] BLR 23 (TCC) . . . . 10.76, 11.08, 13.31, 13.34, 13.35, T 10.6 (366), T 11.1 (388), T 13.1 (446)
Working Environments Ltd v Greencoat Construction Ltd [2012] EWHC 1039 (TCC) . . . . 3.69, 3.70, 7.80, 9.07, T 3.2 (77), T 3.4 (88), T 7.5 (244)
Workplace Technologies v E Squared Ltd Unreported (TCC) 7.59
Workspace Management Ltd v YJL London Ltd [2009] BLR 497; [2009] EWHC 2017 (TCC) . . . . 6.28, 9.52, T 4.1 (104), T 6.2 (179), T 9.6 (323)
World Trade Corporation Ltd v C Czarnikow Sugar Ltd [2004] 2 All ER (Comm) 813, [2004] EWHC 2332 (Comm). 10.66

YCMS Ltd v Stephen & Miriam Grabiner [2009] EWHC 127 (TCC); [2009] BLR 211 . . . . 4.109, 4.120, 4.121, T 4.2 (127–8), T 6.3 (194), T 9.6 (323)
Yarm Road Ltd v Costain Ltd, unreported, 30 July 2001 . . . . 1.09, T 1.1 (13)
Yates Building Company v RJ Pulleyn & Sons (York) Ltd [1976] 1 EGLR 157 . . . . 17.05
Yuanda (UK) Co. Ltd v WW Gear Construction Ltd [2010] EWHC 720; [2010] BLR 435 (TCC). . . . . 3.98, 4.143, 4.147, 12.40, 12.45, T 4.4 (139), T 12.2 (411)

Zealander & Zealander v Laing Homes [2000] TCLR 724 . . . . 1.88

TABLE OF LEGISLATION

European Convention on Human Rights is tabled under Sch 1 to Human Rights Act 1998

STATUTES

STATUTORY INSTRUMENTS

Part I

STATUTORY AND CONTRACTUAL ADJUDICATION

1

CONSTRUCTION CONTRACTS AND CONSTRUCTION OPERATIONS

(1) The 1996 and 2009 Acts

1.01 The provisions of the Housing Grants Construction and Regeneration Act 1996 ('the 1996 Act') Part II, apply only to construction contracts within the meaning contained in ss. 104, 105 and 106. This chapter considers what construction contracts are caught by the legislation. Section 107 provided that the 1996 Act applied to construction contracts that were in writing and gave rise to considerable difficulties and much case law. That requirement has since been removed by the 2009 Act, but still applies to contracts covered by the old law, namely those contracts entered into before October 2011. The requirement for writing is discussed in detail in Chapter 2.

1.02 The Local Democracy, Economic Development and Construction Act 2009 ('the 2009 Act') makes certain amendments to the 1996 Act but makes no changes to ss. 104 and 105 and retains the exemption for contracts made with residential occupiers at s. 106. The only change introduced to these provisions by the 2009 Act concerns the ability of the Secretary of State to make an order excluding any description of construction contract from the operation of the Act. The original provision at s. 106(2) has been deleted and replaced with a new s. 106A whereby the power is now divided between the Secretary of State, the Welsh Minister, and the Scottish Ministers in relation to their own territories.[1]

(2) What Is a Construction Contract?

1.03 Section 104(1) provides a wide definition of 'construction contract' as being an agreement with a person for (a) the carrying out of construction operations; (b) arranging for the carrying out of construction operations by others, whether under subcontract to him or otherwise; or (c) providing his own labour, or labour of others, for the carrying out of construction operations. As a matter of statutory interpretation, a person includes a body of persons corporate or unincorporated.[2]

1.04 Section 104(2) broadens the categories of construction contract to include professional service contracts. Construction contracts thus also include an agreement to do architectural, design, or surveying work[3] or to provide advice on building, engineering, interior or exterior decoration, or on the laying-out of landscape, subject to the overriding caveat that activities described in s. 104(2) must be 'in relation to construction operations'.[4]

1.05 The meaning of 'construction operations' is thereafter defined in s. 105(1) by detailed list of operations covered by the Act. Operations which are exempt from being treated as construction operations for the purposes of the legislation are identified in s. 105(2). Contracts with residential occupiers are said to be wholly outside the operation of Part II of the Act and the Secretary of State may, by order, exclude any other description of construction contract from the operation of the Act (s. 106).[5]

Contracts Affected by the 1996 Act

1.06 In England Wales and Scotland Part II of the 1996 Act came into effect on 1 May 1998[6] and applies only to construction contracts entered into after that date.[7] The Construction Contracts (Northern Ireland) Order 1997 No. 274 (NI 1) brought Part II of the Act into operation in Northern Ireland. The commencement date for Northern Ireland was 1 June 1999.[8] An adjudicator has no jurisdiction to determine a dispute arising out of a construction contract made before the 1996 Act came into effect[9] or where the construction operations are not carried out in a qualifying territory (s. 104(6)).

1.07 It was decided in *Christiani & Nielsen Ltd v Lowry Centre Development Company Ltd* (2000)[10] that a contract entered into after the commencement of the Act but taking effect from an earlier date before the commencement of the Act was nevertheless governed by the provisions of the Act. The date a contract is executed is not necessarily the date it is entered into. In *Atlas Ceiling & Partition Co. Ltd v Crowngate Estates (Cheltenham) Ltd* (2002)[11] a contract was signed in April 1998 but was not entered into until certain outstanding matters were resolved at a later date, thus bringing the agreement within the scope of the Act.

---

[1] Power of Secretary of State to amend ss. 104–6.
[2] Interpretation Act 1978, Ch. 30, Sch. 1.
[3] s. 104 (2)(a).
[4] s. 104 (2)(b).
[5] See, for example, SI 1998 648.
[6] See SI 1998 650 (England and Wales) and SI 1998 894 (Scotland).
[7] s. 104(6).
[8] See SI 1999 No. 34 (NI).
[9] *Project Consultancy Group v Grey Trust Trustees* [1999] BLR 377.
[10] 29 Jun. 2000 (TCC) Lawtel. See also *Atlas Ceiling & Partition Co. Ltd v Crowngate Estates (Cheltenham) Ltd* (2002) 18 Const LJ 49.
[11] (2002) 18 Const LJ 49.

Contracts Affected by the 2009 Act

1.08 The amendments will apply to construction contracts entered into after the 2009 Act came into force in the respective territories.[12] The new provisions apply from 1 October 2011 in England[13] and Wales[14] and, from 1 November 2011, in Scotland.[15] The location of the construction operations determines which commencement order applies. The 1996 Act continues to apply to contracts made after 1 May 1998 and before the commencement of the new Act.

Changes to Existing Contracts after Commencement of the Act

1.09 After the 1996 Act was passed there were a number of cases about whether the Act applied to contracts made before the Act came into effect but varied afterwards. The same principles will apply to the application of the 2009 Act and so are considered here. In *Earl's Terrace Properties Ltd v Waterloo Investments Ltd* (2002) (Key Case)[16] it was decided that where parties entered into a construction contract prior to 1 May 1998 but varied the contract after that date, the variation would usually not bring the contract within the scope of the Act. However, where the variation is capable of being construed as a construction contract in its own right, then the variation itself could attract the provisions of the Act. Judge Seymour in *Earls Terrace* left open the question as to whether a variation made to a contract outside the provisions of the Act can ever have the effect of bringing the whole contract within the statutory scheme. In contrast, a novated contract is a new agreement so that a contract made before 1 May 1998 could fall within the scheme of the Act by virtue of a novation agreement made after 1 May 1998; *Yarm Road Ltd v Costain Ltd* (2001).[17]

Construction Operations in Qualifying Territory

1.10 The Act applies to construction contracts for the carrying out of construction operations in England, Wales, or Scotland. The Construction Contracts (Northern Ireland) Order 1997 No. 274 (NI 1) brought Part II of the Act into operation in Northern Ireland. It is the location of the construction operations which will determine which commencement order applies. Operations carried out on offshore installations are not within the scope of the Act: *Staveley Industries plc v Odebrecht Oil & Gas Services Ltd* (2001).[18]

Crown Contracts

1.11 The 1996 Act applies to construction contracts entered into by the Crown, save for where the contract is made by Her Majesty in a private capacity, and to contracts made by the Duchy of Cornwall. Special provisions are made regarding how the Crown or Duchy is to be represented in any adjudication proceedings.[19]

Contracts for the Carrying Out of Construction Operations

1.12 Construction contracts are dealt with in two broad categories by s. 104: those concerned directly with the carrying out of the construction operations (s. 104(1)) and those connected

---

[12] s. 149(3) and (4).

[13] S.I. 2011 No. 1582 (C. 59).

[14] S.I. 2011 No. 1597 (W.185) (C.61) Construction, Wales.

[15] S.I. 2011 No. 291 (C27). In Northern Ireland the commencement date was 14 Nov. 2012 pursuant to the Construction Contracts (2011 Act) (Commencement) Order (Northern Ireland) 2012.

[16] [2002] CILL 1889 (TCC).

[17] 30 Jul. 2001 (unreported).

[18] [2001] 98(10) LSG 46.

[19] s. 117 of the 1996 Act.

agreements made 'in relation to the construction operations' such as agreements for architectural, design, surveying or advice work (s. 104(2)).

1.13 'Construction contracts' in the first category are agreements with a person for the actual performance of construction operations and include agreements solely to provide labour to perform construction operations.[20] Also in this first category are agreements which constitute 'arranging for the carrying out of construction operations by others, whether under subcontract to him or otherwise'.[21] There is no explanation in the statute as to what activities fall within this latter type of contract. It appears to be directed towards main contracts and management contracts where the contractor does none of the work itself but arranges for others to perform the construction operations. It has been found to be wide enough to cover a surveyor's agreement to administer a building contract.[22]

1.14 The common ingredient in the three subsections of s. 104(1) is the 'carrying out of construction operations' which suggests that only those individuals who have a direct connection with the work itself are intended to be caught. It is submitted that contracts which have a more indirect role in arranging for the performance of construction operations, such as certain funding agreements, are not intended to be covered by this provision. This is consistent with the fact that finance agreements and development agreements are expressly excluded from the ambit of the Act pursuant to SI 1998 648.

Professional Agreements in Relation to Construction Operations

1.15 Professional services contracts are brought within the ambit of the Act by s. 104(2). There are two limbs to s. 104(2). First the contract must be one to perform work of the character described in sub-ss (a) or (b) and secondly that work must be work 'in relation to construction operations'. Consequently a construction contract includes an agreement (a) to do architectural, design, or surveying work, or (b) to provide advice on building, engineering, interior or exterior decoration or on the laying-out of landscape, where such agreements are 'in relation to construction operations'.

1.16 Work is performed in relation to construction operations when it is incidental to the actual performance of the construction operation itself, or is directly connected to it. Accordingly giving factual evidence in an arbitration is an activity which is incidental to the arbitration and not to the actual performance of the construction operation: *Fence Gate Ltd v J. R. Knowles Ltd* (2001) (Key Case).[23] An agreement to perform contract administration services qualifies as surveying work in the terms of s. 104(2)(a): *Gillies Ramsay Diamond v PJW Enterprises Ltd* (2002).[24]

1.17 Pursuant to s. 104(2) a professional may seek adjudication for outstanding fees,[25] and professional negligence allegations can be submitted to adjudication, as was the case in *Gillies Ramsay Diamond v PJW Enterprises Ltd* (2002).[26]

Agreement Relating to Construction Operations and Other Matters

1.18 If a contract relates to construction operations as well as other matters the Act applies to such an agreement 'only so far as it relates to construction operations' (s. 104(5)). The qualifying

---

20 s. 104(1)(a) and (c).

21 s. 104(1)(b).

22 *Gillies Ramsay Diamond v PJW Enterprises Ltd* [2003] BLR 48 at para. 45.

23 84 Con LR 206 [2001] (TCC).

24 [2003] BLR 48.

25 As was the case in *Cowlin Construction Ltd v CFW Architects (a firm)* [2003] BLR 241,[2002] EWHC 2914 (TCC).

26 [2003] BLR 48.

construction operations are to be treated as severable from the other matters so that the adjudication and payment provisions of Part II will apply to the construction operations alone.[27]

1.19 This means an adjudicator may have jurisdiction in a statutory adjudication to determine disputes over part of a contract but not over other issues. It is not difficult to see how this could give rise to difficulties if, for instance, an adjudicator was asked to decide a dispute about payment which concerned both construction operations and other matters and the issues could not be severed. In *North Midland Construction plc v AE&E Lentjes UK Ltd* (2009)[28] Ramsey J suggested that in such a situation it might be impossible to apply the adjudication provisions of the Act to only part of a dispute so that ultimately the Act could not be applied at all.[29] In *Cleveland Bridge (UK) Ltd v Whessoe-Volker Stevin Joint Venture* (2010)[30] the adjudicator's decision concerned matters covered by the Act and matters exempt under s. 105(2). On the facts the decision could not be severed with the result that the entire decision was unenforceable.

1.20 In many cases however contracts will contain a written agreement to adjudicate disputes under that contract, such as is to be found in many standard forms. These agreements rarely differentiate between the type of dispute that may or may not be adjudicated, and the problem identified above will not arise. In particular cases this may mean the contractual adjudication provisions are wider than required by statute but this will not undermine their efficacy as the parties are entitled to agree to adjudicate whatever matters they choose. As a consequence an adjudicator acting in an adjudication brought under a contractual adjudication provision is unlikely to have to consider the difficulties raised by s. 104(5) of the Act.

Contract of Employment

1.21 A contract of employment (within the meaning of the Employment Rights Act 1996) is not a construction contract for the purposes of the Act.[31]

Agreements to Vary, Supplement, or Settle a Construction Contract

1.22 The question has arisen as to whether an agreement made after a construction contract and related to it will itself be governed by the Act's provisions. Most commonly this issue will occur in relation to variation agreements, supplemental contracts, side agreements, or settlements of issues arising out of the construction contract itself. It is a matter of construction in each case whether the secondary agreement is itself a construction contract.

1.23 In general terms, if an agreement is a variation to the original construction contract a dispute arising thereunder will fall within the scope of the dispute resolution provisions of the construction contract. Consequently it will attract the adjudication provisions of the Act or any express adjudication clause that is contained in the contract.

1.24 A stand-alone settlement agreement made in full and final settlement of all claims will not automatically be a construction contract simply because it compromises a construction

[27] *Fence Gate Ltd v J. R. Knowles Ltd* [2002] Con LR 206 at para. 7 of the judgment; *Palmers Ltd v ABB Power Construction Ltd* [1999] BLR 426 at 435.

[28] [2009] BLR 574, [2009] EWHC 1371 (TCC) at para. 51 of the judgment.

[29] See also *Palmers v ABB Power Construction Ltd* [1999] BLR 426 at para. 43; *ABB Power Construction v Norwest Holst Engineering Ltd* [2000] 77 Con LR 20 at para. 16; *Gibson Lea Retail Interiors Ltd v Makro Self Service Wholesalers Ltd* [2001] BLR 407 at paras. 21 and 24.

[30] [2010] BLR 415, [2010] EWHC 1076 (TCC).

[31] s. 104(3).

contract, nor will it be covered by the adjudication provisions in the primary contract[32] (unless those are drafted broadly enough).[33] It will be a matter of interpretation in such a case as to whether a stand-alone settlement agreement is itself a construction contract within Part II of the Act.

1.25 Side agreements that can be construed as variations to the construction contracts will usually be subject to the right to adjudicate in that agreement. A side agreement that is independent of the construction contract will not be subject to the adjudication provisions in the Act unless the side agreement itself is a qualifying construction contract. However if the side agreement is a qualifying construction contract it follows that disputes arising under that agreement will have to be adjudicated separately from those under the original construction contract, unless the parties widen the jurisdiction of the adjudicator or there is an express term in the applicable adjudication rules which enables the adjudicator to hear related disputes under different contracts (as is found in paragraph 8(2) of the Scheme for Construction Contracts—but this requires the consent of the parties).

1.26 The reported cases on this subject deal primarily with the issue of the true nature of the supplemental agreement, whether a variation or stand-alone agreement, rather than whether it was a qualifying construction contract itself. In *Melville Dundas Ltd v Hotel Corporation of Edinburgh Ltd* (2006),[34] Lord Drummond Young made the distinction between settlements that were independent of the construction contract and other forms of settlement which purported only to agree a sum due under the contract terms, such as an agreement that contractual retention should be fixed at a specified sum. In that case an agreement of the latter variety was not independent of the construction contract and would be given effect through the mechanisms available under the construction contract. As a consequence it was subject to the provisions of the Act. Lord Drummond Young recognized that some settlements could be a combination of both forms.[35]

1.27 An adjudicator is entitled to consider whether the dispute referred has been compromised by a stand-alone settlement agreement but as that issue goes to his jurisdiction he cannot make a binding decision on this point unless given power to do so by the parties (this topic is discussed in more detail in Chapters 5 and 9). The court can consider the question of jurisdiction afresh: *Quarmby Construction Co. Ltd v Larraby Land Ltd* (2003).[36] In the circumstances of that case the claim referred to adjudication had not been compromised by the settlement, the adjudicator had jurisdiction and his decision was valid.

1.28 It is a question of construction whether a collateral warranty is a construction contract within the meaning of the Act. It is thought that where such an agreement simply warrants compliance with a building contract, as is commonly the case in standard-form warranties between employer and subcontractor, that is not an agreement for the carrying out of construction operations. It would be odd if the Act operated to imply the payment provisions of ss. 109–11 into such a warranty, where the original agreement has no obligations as to payment.

[32] *Shepherd Construction Ltd v Mecright Ltd* [2000] BLR 489 (Key Case); *Lathom Construction Ltd v Brian Cross and Ann Cross* [2000] CILL 1568.

[33] Such as was the case in *L. Brown & Sons Ltd v Crosby Homes (North West) Ltd* [2005] EWHC 3503 (TCC) (Key Case).

[34] [2006] BLR 474, [2006] CSOH 136.

[35] *Ibid.* at para. 33.

[36] 14 Apr. 2003 (TCC) (unreported).

Key Cases: s. 104

Earl's Terrace Properties Ltd v Waterloo Investments Ltd [2002] CILL 1889 (TCC)

1.29 **Facts:** The parties entered into an agreement dated 4 December 1996 under which Waterloo Investments undertook to manage a building development. It was agreed that this was a construction contract within the meaning of the Act. The contract was varied by an agreement that affected the provisions as to fees only. A dispute arose about the payment of fees and Waterloo Investments served a notice of adjudication. The claimant sought declarations that (inter alia) the Deed of Variation was not a construction contract within the meaning of the Act.

1.30 **Held:** The Deed of Variation concerned payment issues alone and was not itself a construction contract within the meaning of the Act. In those circumstances Parliament would not have intended that its effect would have been to bring the entire contract within the scope of the Act when it was not covered by the Act when entered into by the parties (at paragraphs 26 and 28).

Fence Gate Ltd v J. R. Knowles Ltd [2002] Con LR 206 (TCC)

1.31 **Facts:** Fence Gate engaged J. R. Knowles to carry out architectural and surveying works in relation to a dispute between Fence Gate and its building contractor. The contract contemplated that litigation support services might be required. J. R. Knowles prepared a report on a defective kitchen floor and subsequently were called to provide witness-of-fact evidence in the arbitration between Fence Gate and its building contractor. An adjudicator awarded J. R. Knowles fees for both services. Fence Gate then applied to the court to challenge the decision on the ground that the adjudicator lacked jurisdiction on the basis that the giving of factual evidence in an arbitration was not a service 'in relation to construction operations' and so was not covered by the 1996 Act.

1.32 **Held:** The judge accepted that reporting back to the client of what was discovered during a survey was an essential part of the service he was asked to provide to the client, that it should be regarded as part of the survey work, and thus included in the 'doing' of the work. However the judge found that giving of factual evidence was a significantly different activity from actually surveying the property or reporting back to the client and was not incidental thereto. Likewise assisting at an arbitration was not advice in relation to construction operations but rather was advice in relation to the legal proceedings. The judge rejected the argument that giving of factual evidence fell within the words of s. 104(2)(a) of the Act because it was the 'doing' of architectural or surveying work itself that was encompassed by that provision.

1.33 In reaching this decision the judge considered that whether an activity was to be regarded as one performed 'in relation to construction operations' was a question of the degree of connection between the activity and the construction operations:

> although the preparation of the report on the defective kitchen floor is not itself a construction operation it is sufficiently connected with construction operations so that

it could properly be said to relate to construction operations, but if the giving of factual evidence or the provision of advice at an arbitration is not sufficiently connected with the construction operations so that it does not relate to construction operations, the contract can be severed and adjudication will be available in relation to any disputes in connection with the preliminary report.[37]

Shepherd Construction Ltd v Mecright Ltd [2000] BLR 489[38]

1.34 **Facts:** Shepherd sought a declaration that an adjudicator, who had begun to adjudicate on a dispute referred by Mecright, had no jurisdiction. The referral notice alleged a failure on the part of Shepherd properly to value and pay Mecright for works performed. That dispute had been the subject of a settlement agreement under which Mecright had agreed to accept £366,000 in full and final settlement of its claims against Shepherd. Shepherd relied on the settlement agreement to argue no jurisdiction, and Mecright alleged in response that the settlement had been obtained under duress.

1.35 **Held:** On the question of the settlement agreement Judge LLoyd found that a dispute about whether a settlement agreement is binding was not a dispute which arose out of a construction contract and was not therefore caught by s. 108 of the 1996 Act. In the course of his judgment he said:

> 12. ... the settlement agreement is an agreement which, but for the plea of economic duress would have the effect of extinguishing all disputes that then existed on 15 March so that there could be no dispute capable of being referred to adjudication thereafter in relation to valuation.
>
> 13. The sub-contract incorporated the terms of DOM/1. Clause 38A.1 of DOM/1 applies where a party exercises its right under Article 3 to refer 'any dispute or difference arising under this Subcontract to adjudication'. In my judgment where parties have reached an agreement which settles their disputes there can thereafter be no dispute about what had been the subject matter of the settlement capable of being referred to adjudication under a provision such as Clause 38A.1 or otherwise for the purpose of Section 108 of the Housing Grants Construction and Regeneration Act 1996. The prior disputes have gone and no longer exist. Therefore on 3 July there was no dispute about any of the matters which were the subject of the notice of adjudication (just as there was no dispute about whether the agreement had been entered into under duress). Thus Mecright had no right to apply for adjudication and the adjudicator had no authority or jurisdiction to deal with the notice of 3 July.
>
> ...
>
> 14. ... I should make it clear that in my judgment a dispute about a settlement agreement of this kind could not be a dispute under the sub-contract since the effect of the settlement agreement is one which replaces the original agreement to the extent to which it applies. Here the agreement has the effect of replacing Shepherd's obligations to value and to pay Mecright under the sub-contract the value of the work. The only subsisting obligation to pay that apparently was not extinguished was the obligation to

[37] Para. 7 of the judgment.

[38] Also followed in *Quality Street Properties (Tradings) Ltd v Elmwood (Glasgow) Ltd* 8 Feb. 2002 (Scotland) (unreported).

> release retention as and when the time arose. So there could be no dispute under the sub-contract. Indeed it was also part of Shepherd's case, and in my judgment correctly accepted by Mr Bartle, that the effect of a settlement agreement is that a dispute about it is outside Section 108, since a settlement agreement is not a construction contract within the meaning of Section 108 … A dispute about an agreement which settles a dispute or disputes is not a *dispute* under that contract. The word 'under' in the Act was plainly chosen deliberately. It was being followed in this sub-contract. It is not, nor is it, accompanied by words such as 'in connection with' or 'arising out of' which have a well established wider reach.

L. Brown & Sons Ltd v Crosby Homes (North West) Ltd [2005] EWHC 3503 (TCC)

1.36 **Facts:** The claimant and defendant had contracted under an amended version of the JCT Standard Form of Building Contract (with Contractor's Design) 1988. Clause 39A.1 had been amended so as to permit the reference of 'any dispute or difference arising under out of or in connection with' the contract. The disputes referred to adjudication concerned the right to retain liquidated damages and the obligation to pay a completion bonus which arose under side agreements. The issue in this case therefore turned on the particular wording of the adjudication agreement and whether disputes under the side agreements were 'in connection with' the original contract.

1.37 **Held:** The side agreements were variations to the construction contract and were therefore caught by the provisions of the existing adjudication agreement. In any event, even if that were not the case, the dispute under the side agreement was caught by the wide wording of the amended adjudication agreement because it was a dispute 'in connection with' the original construction contract agreement. Therefore in this case it was not necessary to consider whether the side agreements themselves were construction contracts under the 1996 Act and subject to the statutory right to adjudicate (at paragraphs 49–54).

Westminster Building Company v Beckingham [2004] BLR 163, [2004] EWHC 138 (TCC)

1.38 **Facts:** At an enforcement hearing it was argued that the effect of a capping agreement was to preclude the adjudicator from having jurisdiction over the dispute.

1.39 **Held:** The judge rejected that argument, deciding that the capping agreement was not a settlement agreement which stood as a stand-alone agreement, and consequently distinguished the decision of *Shepherd Construction Ltd v Mecright Ltd* on the facts in the following terms:

> That case is, however, not relevant to this one. First and foremost, the agreement of 20 February 2003 was not a settlement agreement settling all disputes or a stand alone agreement. It was clearly intended to be a variation agreement varying the terms of

the underlying contract. It is to be read with and as part of that underlying contract. Furthermore, it does not settle all disputes, it merely provides a new contract sum or cap, albeit that cap is subject to unspecified deductions. Thus, a dispute as to whether it is enforceable is one arising under the contract since its terms form part of and are to be read with the underlying contract.

Table 1.1 Table of Cases: s. 104

Title	Citation	Issue
Project Consultancy Group v Trustees of Grey Trust	[1999] BLR 377, 65 Con LR 146, Dyson J	**Jurisdiction issue** Whether a contract falls within the provisions of the 1996 Act is a question that goes to the jurisdiction of an adjudicator.
Shepherd Construction Ltd v Mecright Ltd	[2000] BLR 489, Judge LLoyd	**Settlement not construction contract** Section 104: a stand-alone settlement agreement was itself not a construction contract and therefore there was no right to adjudicate pursuant to s. 108, nor was it subject to the adjudication clause in the original contract.
Christiani & Nielsen Ltd v Lowry Centre Development Company Ltd	26 June 2000 (TCC) (unreported), Judge Thornton	**Contract made after commencement of Act, relating to work performed earlier** A contract was entered into after 1 May 1998 but was stated as having retrospective effect to cover works carried out before the commencement of the Act. The contract was governed by the provisions of the Act and was not affected by the fact that it covered works carried out pre 1 May 1998.
Lathom Construction Ltd v Brian Cross and Ann Cross	[2000] CILL 1568 (TCC), Judge Mackay	**Settlement not construction contract** In an adjudication it was alleged there had been a binding compromise which deprived the adjudicator of jurisdiction. The adjudicator found that there was a binding compromise and it was not a construction contract as defined by s. 104.

Title	Citation	Issue
		However, he made an award based on his interpretation of the compromise. Summary judgment of the decision was refused, because whether the adjudicator had jurisdiction to interpret the compromise or whether it was outside the 1996 Act, and his jurisdiction, was a triable issue.
Fence Gate Ltd v J. R. Knowles Ltd	[2001] 84 Con LR 206 (TCC), Judge Gilliland	**Section 104(2): witness statement** The giving of factual evidence in arbitration was not an activity 'in relation to construction contracts' but was rather in relation to arbitration.
Staveley Industries plc v Odebrecht Oil & Gas Services plc	[2001] 98(10) LSG 46 (TCC), Judge Havery	**Offshore installations** Offshore installations were found to be outside ss. 104(6) and 105(1)(c) and so not covered by the 1996 Act.
Atlas Ceiling & Partition Co. Ltd v Crowngate Estates (Cheltenham) Ltd	(2002) 18 Const LJ 49 (TCC), Judge Thornton	**Date contract entered into** A contract was signed in April 1998 but was not entered into until certain outstanding matters were resolved at a later date, thus bringing the agreement within the scope of the Act.
Earl's Terrace Properties Ltd v Waterloo Investments Ltd	[2002] CILL 1889 (TCC), Judge Seymour	**Section 104(6): contracts entered into after the commencement of the Act** A contract made before 1 May 1998 was not brought within the 1996 Act by a variation to the fee provisions made after that date.
Gillies Ramsay Diamond v PJW Enterprises Ltd	[2003] BLR 48, Scottish OH, Lady Paton	**Section 104(2): architect's appointment** Architect's appointment is a construction contract caught by the Act.
Cowlin Construction Ltd v CFW Architects (a firm)	[2003] BLR 241,[2002] EWHC 2914 (TCC), Judge Kirkham	**Section 104(2): architect's appointment** Architect's appointment is a construction contract caught by the Act.
Yarm Road Ltd v Costain Ltd	30 Jul. 2001 (unreported), Judge Havery	**Section 104(6): contracts entered into after the commencement of the Act** Contract made before 1 May 1998 was novated after 1 May 1998. The novation was a new agreement and was covered by the Act.

(*Continued*)

Table 1.1 *Continued*

Title	Citation	Issue
Quarmby Construction Co. Ltd v Larraby Land Ltd	14 Apr. 2003 (TCC) (unreported), Judge Grenfell	**Settlement not construction contract** A settlement agreement was a stand-alone agreement, but the claim referred had not been compromised and the adjudication decision was valid.
Westminster Building Company v Beckingham	[2004] BLR 163, [2004] EWHC 138 (TCC), Judge Thornton	**Variation was construction contract** A settlement agreement was a variation to the construction contract and the dispute as to whether it was enforceable arose under the terms of the construction contract. Accordingly the dispute fell within the provisions of the 1996 Act.
L. Brown & Sons Ltd v Crosby Homes (North West) Ltd	[2005] EWHC 3503 (TCC), Ramsey J	**Side agreement was construction contract** Side agreements were variations to the original construction contract and caught by its adjudication provision. Further and in any event, the terms of the adjudication agreement which provided for 'any dispute or difference arising under out of or in connection with [the contract]' was wide enough to cover side agreements which were not variations.
Melville Dundas Ltd v Hotel Corporation of Edinburgh Ltd	[2006] BLR 474, [2006] CSOH 136, Lord Drummond Young	**When settlement is/not construction contract** A settlement agreement contained, in part, agreements which merely gave effect to the contract terms (such as agreeing the final account and retention amount) and were to be operated through the contract. Those parts of the settlement agreement which were stand-alone settlements of disputed items were not referable to the contract but stood apart, and were not caught by the provisions of the Act.
McConnell Dowell Constructors (Aust) Pty v National Grid Gas plc	[2007] BLR 92, [2006] EWHC 2551 (TCC), Jackson J	**Supplemental agreement was construction contract** A supplemental agreement was a variation to the primary construction contract and the adjudicator had jurisdiction under the adjudication clause to decide whether sums claimed were comprised within the supplemental agreement.

Title	Citation	Issue
North Midland Construction plc v AE&E Lentjes UK Ltd	[2009] BLR 574, [2009] EWHC 1371 (TCC), Ramsey J	**Section 104(5): contracts for both construction operations and work not covered by 1996 Act** Where the dispute concerned valuation for work where part is covered by the Act and part is outside the Act, it might render impossible the application of the adjudication provisions to the part covered by the Act. The result would be the Act did not apply at all.
Cleveland Bridge (UK) Ltd v Whessoe-Volker Stevin Joint Venture	[2010] BLR 415, [2010] EWHC 1076 (TCC), Ramsey J	**Section 104(5): contracts for both construction operations and work not covered by 1996 Act** An adjudicator's decision concerned work, part of which was within the ambit of the 1996 Act and part of which was outside the Act. The adjudicator did not have jurisdiction over the work outside the Act and as that part of the decision could not be severed, the entire decision was unenforceable.

(3) What Are Construction Operations?

1.40 Construction operations are defined by a detailed list of qualifying activities in s. 105(1). That is followed in s. 105(2) by a list of more specific operations that are exempt from the operation of the Act. The descriptions in s. 105(1) are widely drafted and encompass all activities which may fall within those descriptions.

Buildings and Structures Forming Part of the Land

1.41 Under s. 105(1)(a) construction operations include:

> construction, alteration, repair, maintenance, extension, demolition or dismantling of buildings, or structures forming, or to form, part of the land (whether permanent or not).

The condition that the structures or buildings must be ones that form part of the land is a requirement repeated in s. 105(1)(b) and (c). Thus a structure that sits on the land and does not become part of the land, such as a portacabin, will not be covered by this provision.

The proviso ('whether permanent or not') refers to the buildings or structures in question rather than the attachment to the land: *Gibson Lea Retail Interiors v Makro Self Service Wholesalers Ltd* (2001) (Key Case).[39]

1.42 The use of the definite article suggests that the land referred to is not any land but is the land on which the relevant construction operation is to take place.[40] Thus where a structure is being assembled in a yard with the intention of moving it and fixing it to land at another location, the relevant question is whether it will form part of the land at the site of its installation.

1.43 The manufacture of building or engineering components, equipment, plant, or machinery is not a construction operation at all unless included in a contract which also provides for their installation: s. 105(2)(d). This is consistent with s. 105(1)(a) in that a contract for manufacture of such equipment does not itself result in a structure forming part of the land.

1.44 However, a relevant structure that is constructed on site before being lifted into its final resting place may qualify as a structure to form part of the land: *Palmers Ltd v ABB Power Construction Ltd* (1999) (Key Case).[41]

1.45 The Act applies only to construction operations carried out in England, Wales or Scotland (s. 104(6)) and no mention is made of offshore activities. It is submitted that the construction activities must be carried out on land or on the shore within the tidal reach. Thus construction activities carried out on an oil rig are not covered by the 1996 Act as an oil rig is not land and the construction of a structure on a rig will not form part of the land: *Staveley Industries v Odebrecht Oil & Gas Services Ltd* (2001) (Key Case).[42]

1.46 It seems that 'buildings or structures' in s. 105 should be construed as including fixtures such as the heating systems. This was the conclusion of Dyson J in *Nottingham Community Housing Association Ltd v Power Minster Ltd* (2000)[43] where he construed the maintenance and repair of buildings referred to in s. 105(1)(a) as including maintenance and repair of heating systems within that building.

Any Works Forming Part of the Land

1.47 Section 105(1)(b) widens the net beyond buildings and structures to 'any works forming or to form part of land' so that a construction operation also includes:

> construction, alteration, repair, maintenance, extension, demolition or dismantling of any works forming, or to form, part of the land, including (without prejudice to the foregoing) walls, roadworks, power-lines, telecommunication apparatus, aircraft runways, docks and harbours, railways, inland waterways, pipe-lines, reservoirs, water-mains, wells, sewers, industrial plant and installations for purposes of land drainage, coast protection or defence.

Installation of Fittings

1.48 Section 105(1)(c) focuses its attention on fittings forming part of the land. Unlike sub-ss (a) and (b), this provision does not expressly say it is sufficient that the fittings will subsequently form part of the land. The different wording was presumably chosen deliberately but efforts to determine the purpose behind them from parliamentary debates were not fruitful in the

---

[39] [2001] BLR 407.

[40] *Staveley Industries v Odebrecht Oil & Gas Services Ltd*, 28 Feb. 2001 (TCC) (2001) 98(10) LSG 46, at paras. 9–12

[41] [1999] BLR 426 at 432.

[42] See n. 41.

[43] [2000] BLR 309 at paras. 13 and 14).

case of *Staveley Industries v plc v Odebrecht Oil & Gas Services Ltd* (2001) (Key Case).[44] Construction operations thus include work of:

> installation in any building or structure of fittings forming part of the land, including (without prejudice to the foregoing) systems of heating, lighting, air-conditioning, ventilation, power supply, drainage, sanitation, water supply or fire protection, or security or communications systems.

1.49 In *Gibson Lea Retail Interiors Ltd v Makro Self Service Wholesalers Ltd* (2001)[45] (Key Case) the claimant sought a declaration that work to install shop fittings (such as business counter islands and moveable gondolas) was a construction operation. Judge Seymour found that the phrase 'forming or to form part of land' imported concepts from the law relating to fixtures and fittings, and consequently he found that the determinative issue was the degree of permanence of the attachment. He decided that the shop installations in question were only intended to be temporarily attached and were not fittings within the meaning of s. 105(1)(c).

Cleaning

1.50 Construction operations also include 'external or internal cleaning of buildings and structures, so far as carried out in the course of their construction, alteration, repair, extension or restoration' (s. 105(1)(d)). Hence the cleaning of stonework on the façade of a building during its restoration will undoubtedly be a construction operation. It may also be possible to say this is maintenance of a building and so a construction operation within s. 105(1)(a). The final tidy-up of a site before handover will also be covered as cleaning carried out in the course of construction.

Integral, Preparatory, or Completion Activities

1.51 Section 105(1)(e) introduces a wide catch-all provision aimed at bringing a number of ancillary works within the definition of construction operations. The provision appears to be directed predominantly at enabling works, temporary works, and completion/landscaping works, and provides that construction operations will include:

> operations which form an integral part of, or are preparatory to, or are for rendering complete, such operations as are previously described in this subsection, including site clearance, earth-moving, excavation, tunnelling and boring, laying of foundations, erection, maintenance or dismantling of scaffolding, site restoration, landscaping and the provision of roadways and other access works.

1.52 There are several limbs to this provision. First, it requires a construction operation within sub-ss (a)–(d) be established as the primary activity. Secondly, for the ancillary activity to qualify as a construction operation, it must form an integral part of, be preparatory to, or be for the purpose of rendering complete the primary activity. What is a preparatory or completion activity is probably a fairly straightforward question of fact.

1.53 Less clear is what is meant by 'forms an integral part of' and no explanation is contained within the Act. This provision would seem to introduce a very broad and general category which may cover a whole range of activities which take place on construction or engineering projects but which are not specifically mentioned in sub-ss. (a)–(d).

Painting and Decorating

1.54 By s. 105(1)(f) construction operations includes 'painting or decorating the internal or external surfaces of any building or structure'.

---

[44] 28 Feb. 2001 (TCC) Lawtel at paras 7–8. See also *Gibson Lea Retail Interiors Ltd v Makro Self Service Wholesalers Ltd* [2001] BLR 407.

[45] *Ibid.*

The Exemptions in s. 105(2)

1.55 It is generally known that a limited number of contracting organizations representing specific sections of the construction and engineering industry persuaded the government to exclude their contracts from the ambit of the 1996 Act. This was because these sections of the industry were said to be already operating satisfactory contractual arrangements concerned with payment and did not need the protection provided by the legislation.[46]

1.56 Whilst the history of s. 105(2) is well known, its scope and meaning has proved more obscure. Of particular difficulty has been the potential overlap between activities apparently covered both by s. 105(1) and (2) and the scope of the matters intended to be encompassed by the exemption.

1.57 Most of the cases on this issue have concerned s. 105(2)(c) and whether specific activities are 'assembly, installation or demolition of plant or machinery'. Whether an individual operation is 'assembly, installation ... of plant and machinery' is a question of fact and degree which turns of the facts of each case, as illustrated by the decisions discussed below.[47]

1.58 The cases have not always yielded consistent decisions and it has been said that there are effectively two lines of authority represented by a narrow construction of s. 105(2) as in *Palmers Ltd v ABB Power Construction Ltd* (1999) (Key Case)[48] and a broader construction of s. 105(2) adopted in *ABB Power Construction v Norwest Holst Engineering Ltd* (2000) (Key Case).[49]

1.59 In *Palmers Ltd v ABB Power Construction Ltd* (1999) (Key Case)[50] ABB were engaged to supply and install a boiler at a power plant, an activity clearly exempt under s. 105(2)(c). ABB entered into a subcontract with Palmers for the supply of scaffolding to provide temporary access and support to the structural frame within which the boiler was supported during erection. The scaffolding was clearly within the definition of preparatory works in s. 105(1)(e). The judge had to decide whether scaffolding to the exempt operation would itself be exempt under s. 105(2)(c) or whether it would remain a construction operation by virtue of s. 105(1)(e). The judge decided that s. 105(2)(c) operated only to exempt the assembly of the boiler from the operation of the Act. It did not operate to also exclude scaffolding.

1.60 The court in *Palmers* had been referred to passages from Hansard where the government minister had explained that s. 105(2)(c) was intended only to apply to the assembly of plant and machinery and that ancillary operations on the same site were intended to remain construction operations under the Act.[51]

1.61 Lord Macfadyen appeared to give s. 105(2)(c) a wider construction in *Homer Burgess Ltd v Chirex (Annan) Ltd* (1999) (Key Case)[52] when he found that 'assembly and installation of plant or machinery' in s. 105(2)(c) was wide enough to cover a contract for the assembly and installation of pipework which formed the link between various pieces of equipment. However the reasoning behind this conclusion appears to be that, unlike the scaffolding in *Palmers*, the pipework actually formed part of the plant.

[46] See *Palmers Ltd v ABB Power Construction Ltd* [1999] BLR 426 at para. 29, *ABB Power Construction Ltd v Norwest Holst Engineering Ltd* [2000] 77 ConLR 20 at para. 12.
[47] *North Midland Construction Ltd v AE&E Lentjes UK Ltd* [2009] EWHC 1371 (TCC).
[48] [1999] BLR 426 (TCC).
[49] 77 Con LR 20 [2000] (TCC).
[50] [1999] BLR 426.
[51] At paras. 31 and 32 of the judgment.
[52] [2000] BLR 124.

1.62 Judge LLoyd went further still in *ABB Power Construction Ltd v Norwest Holst Engineering Ltd* (2000) (Key Case).[53] That case concerned a subcontract to install insulation to clad pipework and various parts of a boiler. ABB's work to install the boiler was exempt under s. 105(2)(c). The judge decided the subcontract to install the insulation also fell within the exemption. The judge's reasons included the fact that the provision of insulation is an integral part of the construction of pipework and boilers to achieve power generation. He said the effect of s. 105(2)(c) was that any work that would be a construction operation within s. 105(1) which is necessary for the full and proper assembly or installation of plant so that it will fulfil the purpose or purposes for which it is intended would be exempt by reason of s. 105(2)(c).

1.63 In *ABB Zantingh Ltd v Zedal Building Services Ltd* (2000)[54] ABB were employed for the construction of two diesel-powered generation stations to supply power to printing plants. ABB subcontracted with Zedal for the supply and installation of related field wiring, metal containment systems, and secondary steel support at the two sites. Judge Bowsher decided that the primary activity at the site was printing not power generation and so s. 105(2)(c)(i) did not apply. However, as to whether the relevant activities were the 'assembly or installation of plant or machinery' he expressed the view that one should look at the nature of the work broadly and not subject the contract to a minute analysis of the activities involved:

> The drum of cable on the floor is just a piece of material. When the cable is worked into the plant or even joins two pieces of plant it becomes part of the plant and is properly referred to as plant. Similarly, a screw is just a screw when it is in an engineer's pocket, but it becomes part of the plant when he screws it into the generator. In keeping with the sense of the earlier authorities to which I have referred, it seems to me that one cannot make sense of the Act by a minute analysis of the work to see what was plant and what was not. One must look at the nature of the work broadly. Adjudication cannot be divided in its jurisdiction between minute parts of a subcontractor's work. Looking at the work overall, and regardless of any disputes about the ambit or nature of that work, I have no doubt that Zedal were employed to install plant.

1.64 In *Mitsui Babcock Energy Services Ltd v Foster Wheeler Energia OY* (2001)[55] and *Comsite Project Ltd v AAG* (2003)[56] the court appeared to follow the broad interpretation approach expressed in *Norwest Holst*.

1.65 In *North Midland Construction Ltd v AE&E Lentjes UK Ltd* (2009) (Key Case)[57] Ramsey J was asked to decide whether various enabling and civil works preparatory to the installation of a flue-gas desulphurization plant at a power station were within the exemption in respect of 'assembly, installation or demolition of plant or machinery' in s. 105(2)(c)(i). The judge decided that the works in question were not caught by the exemption. In reaching this conclusion the judge was asked to resolve the difference of approach between *Palmers* and *Norwest Holst* and he decided that the narrow construction adopted in *Palmers* was to be preferred.

1.66 The judge also found that whether an individual operation is 'assembly, installation … of plant and machinery' is a question of fact and degree which turns on the facts of each case. Consequently the judge found that on the facts of *Homer Burgess*, *Norwest Holst*, and *Zedal* the works would have fallen within the exemption in s. 105(2)(c), even construing the provision narrowly.

---

[53] 77 Con LR 20 [2000].
[54] [2001] BLR 66.
[55] [2003] EWHC 958 (TCC).
[56] [2001] Scot CS 150.
[57] [2009] EWHC 1371 (TCC).

Key Cases: s. 105

Staveley Industries plc v Odebrecht Oil & Gas Services plc [2001] 98(10) LSG 46

1.67 **Facts:** The case concerned a subcontract for the design and installation of instrumentation, fire and gas, and electrical and telecommunications equipment. The equipment was to be installed in living-quarter modules for operatives at an oil and gas rig. The modules were to be welded onto platforms supported by legs founded on the sea bed. It was common ground that the work was of a kind described in s. 105 (1)(c) being 'installation in any ... structure of fittings forming part of the land including ... systems of heating, lighting, air conditioning, ventilation, power supply', save for the issue of whether the modules were to form part of the land. The claimant contended that the modules formed part of land (i) when in the yard at Teeside or (ii) when welded to the rigs because the rigs were founded on the sea bed and that the Interpretation Act 1978 defined land as including land covered by water. The defendant contended, inter alia, that (i) land was not any land but was land on which the relevant construction operation was carried out and (ii) the rigs were not 'land'.

1.68 **Held:** (i) The modules did not form part of the land whilst standing in the yard at Teeside. The use of the definite article in s. 105(1)(c) suggested the land referred to was the land to which the building or structure was fixed when it became part of the land. (ii) By virtue of s. 104(6) of the Act, the operations did not fall within the 1996 Act unless they were carried out in England, Scotland, or Wales. The wording of s. 105(1) was derived from the Income Corporation Taxes Act 1988 which provided specifically for structures 'including offshore installations'. The absence of a corresponding provision indicated no intention to include offshore installations in s. 105(1). The structures were founded in the sea bed below the water mark and were not structures forming, or to form, part of land for the purposes of s. 105(1). The 1996 Act did not apply.

Nottingham Community Housing Association Ltd v Powerminster Ltd [2000] BLR 309, 75 Con LR 65

1.69 **Facts:** Powerminster entered into a contract for the annual service and repair of gas appliances (gas central-heating systems, gas fires, and gas cookers) in Nottingham's properties. A dispute arose whereby Nottingham claimed a substantial counterclaim against sums invoiced by Powerminster. Powerminster served a notice of adjudication. The issue for the court was whether the contract between the parties was for construction operations within s. 105(1) of the 1996 Act. Nottingham argued that heating systems were neither 'buildings' nor 'structures' and did not fall within s. 105(1)(a). Particular reliance as placed on s. 105(1)(c) which covers installation of fittings forming part of land including heating systems but makes no reference to the repair or maintenance thereof.

1.70 **Held:** The installation, alteration, repair, and maintenance of heating systems in buildings fell within s. 105(1)(a) of the Act and was a construction operation.

Gibson Lea Retail Interiors Ltd v Makro Self Service Wholesalers Ltd [2001] BLR 407

1.71 **Facts:** This was a claim for a declaration that four contracts fell within s.105(1). The claimant was engaged under four contracts to install shop fittings at four cash-and-carry stores. The shop fittings included moveable furniture and gondolas, business counter islands, corner boxing, and column cladding fixed to the walls by screw fittings, primarily for stability.

Held: Installation of shop fittings does not amount to 'construction operations' unless they are 'construction ... of ... structures forming or to form part of land (whether permanent or not)' (s. 105(1)(a)) or 'installation in any building or structure of fittings forming part of land' (s. 105(1)(c)). Judge Seymour held that the reference in s. 105(1)(a) to 'whether permanent or not' related to the buildings or structures, and that the reference to 'forming part of the land' imported the concepts and tests of the law of real property relating to fixtures. A factor relevant to the determination of whether an item formed part of the land is whether the attachment is to be permanent or not. On the evidence the judge decided that none of the items were fixtures because they were not intended to be permanently attached and as a consequence none of the works were 'construction operations' within the meaning of s. 105(1). **1.72**

North Midland Construction Ltd v AE&E Lentjes UK Ltd [2009] BLR 574, [2009] EWHC 1371 (TCC)

Facts: NMC sought declarations from the court that works did not fall within the exemptions in s. 105(2)(c) but were instead construction operations. AEE was appointed turnkey contractor to provide flue-gas desulphurization units (FGDs) to two coal-fired power stations. AEE entered into four contracts with NMC for enabling and civil works at each of the power stations. The enabling works included demolition of existing structures and tarmac surfaces and construction of temporary fencing and gates, temporary roads, temporary drainage, temporary services to site offices, foundations for site offices, and a car park. The civil works were the construction of foundations for components of plant items necessary to the FGD units (involving piling, earthworks, and concreting) and for the steelwork which supported components of plant and provided access to plant. **1.73**

It was common ground that the enabling and civil works fell within the definition of construction operations in s. 105(1)(a) and (e). The issue was whether the works were excluded by s. 105(2)(c)(i). It was also common ground that the works were carried out on a site where the primary activity was power generation. Therefore the issue was whether the enabling and civil works were 'assembly, installation or demolition of plant or machinery, or erection or demolition of steelwork for the purposes of providing access to plant or machinery'. **1.74**

The judge was asked to resolve the difference of approach in construing s. 105(2)(c) represented in *Palmers v ABB* (the narrow construction) and *ABB v Norwest Holst* (the broader construction). In summary, in *Palmers v ABB* the scope of s. 105(2) was construed narrowly so that construction of buildings and concrete foundations for use with the plant would not come within the exclusion nor would any painting of the internal or external surfaces to the plant. In *ABB v Norwest Holt* the scope of s.105(2) was construed broadly so that all the construction operations necessary to achieve the aims and purposes of the owner or of the principal contractors would be exempt so that a subcontractor providing paint systems or cathodic protection systems necessary to protect plant against corrosion or erosion would be exempt. **1.75**

Held: The judge decided that the narrow construction adopted by Judge Thornton in *Palmers v ABB* was the correct approach. The judge gave the following reasons for this decision: **1.76**

(1) The operations described in s. 105(2) can generally be brought within the description of operations in s. 105(1) so that the intention was to exclude a specific limited operation

from a more general description of operations. In these circumstances the narrow approach to the construction of s. 105(2) would generally be appropriate (judgment paragraphs 23 and 49).

(2) It is to be expected that the exclusions to the provisions of the Act were introduced in respect of specific operations for particular reasons applying to those operations (judgment paragraph 25).

(3) The exclusions in s. 105(2)(a)–(c) are aimed at specific industries. The exclusions are limited to particular operations and do not include all operations at a site where the primary activity is one of those industries (judgment paragraph 26).

(4) Section 105(2) contains no general exclusion for works 'which form an integral part of, or are preparatory to, or are for rendering complete, such operations' as is found in s. 105(1)(e) (judgment paragraph 27).

(5) The focus of s. 105(2)(c) is plant and machinery and there are indications that this refers to components or items of plant rather than the whole industrial plant (judgment paragraph 28).

(6) The practical effect of the narrow construction means that civil works will not be exempt and this is envisaged by s. 104(5) of the Act (judgment paragraphs 50–1).

(7) The intent of the Act was that it should generally apply to construction operations and the broad construction would deprive the Act of effect in many cases (judgment paragraph 53).

(8) Material from parliamentary debates clearly indicated that the intention of the Act was that s. 105(2)(c) was to be construed narrowly (judgment paragraphs 57–65). The particular operations were to be excluded because the specialist process engineering industry was perceived to be well organized and did not need protection.

Table 1.2 Table of Cases: Construction Operations within s. 105

Title	Citation	Issue
Palmers Ltd v ABB Power Construction Ltd	[1999] BLR 426, Judge Thornton	**Section 105(2)(c): scaffolding to plant** Scaffolding to an exempted operation (a boiler on a power generation site) was not caught by the exemption with the result that it was still a construction operation under s. 105(1)(e).
Nordot Engineering Services Ltd v Siemens	[2000] CILL (TCC Salford), Sep. 2001, Judge Gilliland	**Agreement to adjudicate non-construction operation** Agreement to adjudicate on matter excluded from statute will be enforced by court.
Nottingham Community Housing Association Ltd v Powerminster Ltd	[2000] BLR 309, Dyson J	**Section 105(1)(a): maintenance of heating in building** The installation, maintenance, and repair of heating systems was construction operation within meaning of s. 105(1)(a) which refers to 'construction ... repair, maintenance ... of buildings'. Thus the reference to maintenance of buildings in this section included the fixtures within the buildings.

Title	Citation	Issue
ABB Power Construction Ltd v Norwest Holst Engineering Ltd	[2000] TCLR 831, Judge LLoyd	**Section 105(2)(c): insulation to plant** The case concerned whether insulation or cladding to boilers was caught within the exemption which covered the installation of the boiler itself. The relevant exemption was 'assembly, installation ... of plant' in s. 105(2)(c). Held: the insulation and cladding was caught by the exemption as it was essential to permit the boiler to function.
Homer Burgess Ltd v Chirex (Annan) Ltd	[2000] BLR 124, 71 Con LR 245, Lord Macfadyen	**Section 105(2)(c): pipework to plant** The scope of the exemption in s. 105(2)(c) was considered by the court. It was decided that pipework was plant within the meaning of s. 105(2)(c) and so was caught by the exemption.
Gibson Lea Retail Interiors Ltd v Makro Self Service Wholesalers Ltd	[2001] BLR 407, Judge Seymour	**Section 105(1): shop fittings** Section 105(1) not to be construed as including everything that is not exempted by s. 105(2). Installation of shop fittings is not a construction operation within s. 105(1).
ABB Zantingh v Zedal Building Services Ltd	[2001] BLR 66, Judge Bowsher	**Section 105(2)(c): primary activity on site** Construction of diesel-powered generation stations to supply power to printing plants. Issue whether contract to install field wiring, metal containment, and secondary steel was within exemption of s. 105(2)(c). Held primary activity on site was printing and not power generation. Further, when considering whether activity was within description of 'assembly or installation of plant or machinery', one should look at nature of work broadly and not subject it to minute analysis of activity involved.
Mitsui Babcock Energy Services Ltd v Foster Wheeler Energia OY	[2001] Scot CS 150	**Section 105(2)(c): primary activity** Pipework and equipment to boilers providing steam to petrochemical plant on adjacent site. Whether primary activity on site within s. 105(2)(c) referred to whole petrochemical complex, or site on which boiler was located only. Held: broad interpretation appropriate because exemption was directed at primary activity on site.
Comsite Projects Ltd v Andritz AG	[2004] 20 Const LJ 24, [2003] EWHC 958 (TCC), Judge Kirkham	**Section 105(2)(c): electric systems for building housing plant** Works to install electrics, fire and gas alarms, heating, and ventilation systems to the dryer-plant building was integral to the operation of the building and not to the operation of the plant itself. Consequently it was not caught by the exemption in s. 105(2)(c).
Edenbooth Ltd v CRE8 Developments Ltd	[2008] EWHC 570 (TCC), Coulson J	**Groundworks and drainage** Clearly construction operations within s. 105(1)(a) or (e).

(*Continued*)

Table 1.2 *Continued*

Title	Citation	Issue
North Midland Construction Ltd v AE&E Lentjes UK Ltd	[2009] BLR 574, [2009] EWHC 1371 (TCC), Ramsay J	**Section 105(2)(c): enabling works and foundations for plant** The judge construed the scope of the exemption in s. 105(2)(c) narrowly, as not intended to cover works beyond the installation of the plant and machinery itself. The result was that ancillary work was not caught by the exemption and so remained governed by s. 105(1) and was a construction operation within the meaning of the 1996 Act (*Palmers v ABB* followed, *ABB v Norwest Holst* not followed). Whether an individual operation is 'assembly, installation … of plant and machinery' is a question of fact and degree which turns of the facts of each case. On the facts of *Homer Burgess, ABB v Norwest Holst* and *ABB v Zedal* the pipework, insulation, and electrical wiring would fall within s. 105(2)(c), even construing them narrowly.
Cleveland Bridge (UK) Ltd v Whessoe-Volker Stevin Joint Venture	[2010] BLR 415, [2010] EWHC 1076 (TCC), Ramsey J	**Section 105(2)(c): erection of steelwork** Where a subcontract covered design, manufacture and erection of steelwork, only the erection of steelwork was covered by the s. 105(2)(c) exemption. The adjudicator had jurisdiction for part of the dispute as contemplated by s. 104(5) of the Act. However it was not possible to sever the decision so as to enforce only the valid part, and the entire decision was rendered unenforceable.

(4) Residential Occupiers and Other Excluded Agreements

Residential Occupiers

1.77 A construction contract with a residential occupier is excluded from the scope of the 1996 Act (s. 106(1)). This means an agreement which 'principally relates to operations on a dwelling which one of the parties to the contract occupies, or intends to occupy, as his residence'.[58] A dwelling is either a dwelling-house or a flat[59] but does not include a building containing a flat.[60]

---

[58] s. 106(2).

[59] The meaning of flat is also defined as 'separate self contained premises constructed or adapted for use for residential purposes and forming part of a building from some other part of which the premises are divided horizontally'.

[60] s. 106(2).

Principally Relates to Operations on a Dwelling

1.78 It is thought that the phrase 'principally relates to' qualifies the operations and not the dwelling. In other words, it is the operations which must principally be operations on a dwelling. They may also be operations on something other than a dwelling, as long as they principally relate to the dwelling. What the phrase does not connote is that the operations must be on a residence which is principally a dwelling.

1.79 The test for determining whether an agreement 'principally relates to operations on a dwelling' is the subject to little authority. In *Samuel Thomas Construction v Anon* (2000)[61] a married couple purchased some agricultural buildings with the purpose of converting a barn into their residence. They had been unable to buy the barn as a separate unit and having purchased several agricultural buildings on the same site they entered into a construction contract to convert barn A to a dwelling for selling on, barn B into a dwelling for their occupation, and to convert further buildings into a garage block for both barns. There was uncontested evidence that the value of the barn B works was about 65 per cent of the total. The judge rejected an argument based on value that the principal objective of the contract was converting barn B into the employer's residence. The judge decided that it was impossible to conclude that the contract principally related to works on a dwelling when it comprised work to barn A, barn B, the shared garage block, and the external courtyard and drainage.[62]

1.80 A similar issue arose in *Gabrielle & Christopher Shaw v Massey Foundation & Pilings Ltd* (2009)[63] where the employer was the owner of a large residence and entered a contract for works to be carried out at a separate building on their land known as East Lodge. The employer did not occupy or intend to occupy the lodge at the date the contract was made. In distinction to the facts in *Samuel Thomas*, there was no intention to sell the lodge. Nevertheless the TCC judge rejected the argument that the lodge was part of the employer's dwelling. The key facts were that it was a separate dwelling as a point of fact, it was not occupied by the employer and they had no intention of moving in at the relevant time. Consequently s. 106 of the 1996 Act did not apply.

Company as Residential Occupier?

1.81 In *Edenbooth Ltd v CRE8 Developments Ltd* (2008)[64] the contract concerned ground works, foundations, and drainage works at two adjacent residential properties. One was owned and both were occupied by directors of the defendant, a father and son. The defendant sought to resist enforcement at a hearing in the TCC on the ground that the works were not construction operations and/or the contract was made with a residential occupier. The TCC judge rejected the argument stating that he found it difficult to imagine how a company could ever be a residential occupier. The phrase conveys a requirement, for the exemption to bite, that a real person must be living in the house or flat in question. In those circumstances the judge did not see how the defendant company, a property development company, could ever be a residential occupier.[65]

1.82 Whether a residents' association that engages a contractor to renovate the common parts of a block of flats is a residential occupier for the purposes of the Act has yet to be tested in the courts.

---

[61] *Samuel Thomas Construction v Anon*, 28 Jan. 2000, Lawtel report LTL 4/9/2000, Judge Overend.
[62] The TCC judge granted permission to appeal but none appears to have been pursued.
[63] [2009] EWHC 493 (TCC).
[64] [2008] EWHC 570 (TCC).
[65] See also *Absolute Rentals v Gencor* [2000] CILL 1637 (TCC) at para. 4 of the judgment.

Conversions

1.83 Section 106 is not limited to buildings which are dwellings at the start of the construction work in question. It is sufficient if the building starts off as a derelict barn and the construction operations are intended to convert the property into a dwelling for the occupation of one of the parties to the contract. The residential-occupier exemption will apply even though occupation is not possible until completion.[66]

Express Adjudication Clause in Contract with Residential Occupier

1.84 Express adjudication agreements within construction contracts with residential occupiers will be effective despite the terms of s. 106 of the Act, as parties are free to contract on whatever terms they choose. If parties contract into adjudication the courts will enforce the outcome even though it is otherwise outside the Act[67] (unless the terms of a bespoke adjudication agreement provide that the decision is not binding and enforceable through the courts such that it cannot be enforced in a summary judgment application in the same manner as could a statutory adjudication).

1.85 As many construction contracts are based on standard forms such as the JCT Agreement for Minor Building Works, the adjudication agreement will have to be deleted from the terms and conditions if a residential occupier wishes to avoid adjudication altogether. Another option that has been argued, with limited success, to escape from the effect of an adjudication agreement in a contract with a residential occupier, is reliance on the Unfair Terms in Consumer Contract Regulations to argue that the adjudication agreement is unfair and should be struck down.

Unfair Terms in Consumer Contract Regulations

1.86 If, despite the statutory exclusion, the parties included an adjudication clause in a contract relating to a dwelling, it has been successfully argued that this would be an usual provision which would need to be brought to the specific attention of the lay party if it was to be validly invoked.[68] However, although Unfair Terms in Consumer Contracts Regulations 1999 ('the 1999 Regulations') have been cited in a number of reported cases, the arguments in favour of striking out the adjudication clauses have rarely been successful.

1.87 A party acting as a consumer and seeking to rely on the 1999 Regulations must first establish that the adjudication clause in the contract was a term which was not individually negotiated, and this will include terms drafted in advance where the consumer has not been able to influence the substance of the term.[69] It is for any seller or supplier who claims that a term was individually negotiated to show that it was.[70]

1.88 A term which has not been individually negotiated shall be regarded as unfair if, contrary to the requirement of good faith, it causes a significant imbalance in the parties' rights and obligations arising under the contract, to the detriment of the consumer.[71] Such an unfair term in a contract between consumer and seller or supplier will not be binding on the

---

[66] *Samuel Thomas Construction v Anon*, 28 Jan. 2000 (unreported).

[67] *Nordot Engineering Services Ltd v Siemens* [2000] CILL Sep. 2001 (TCC Salford).

[68] See the test for incorporation of unreasonable clauses in *J. Spurling Ltd v Bradshaw* [1956] 1 WLR 461 at 466, and see *Picardi v Cuniberti and Cuniberti* [2003] 1 BLR 487 obiter at paras. 124–7. However in this latter case the guidance notes to the standard RIBA form of contract identified the adjudication clause as being an unusual clause which must be brought properly and fairly to the attention of the other party, and this did not happen in this case.

[69] Unfair Terms in Consumer Contracts Regulations 1999, Reg. 5(1) and 5(2).

[70] Unfair Terms in Consumer Contracts Regulations 1999, Reg. 5(4).

[71] Unfair Terms in Consumer Contracts Regulations 1999, Reg. 5(1).

consumer.[72] Thus, to effectively strike down the adjudication clause the consumer must satisfy twin requirements, showing both the significant imbalance and that the clause is contrary to the requirement of good faith. In *Director General of Fair Trading v First National Bank plc* (2001)[73] the House of Lords considered the meaning of 'significant imbalance' under the 1994 Regulation.[74]

1.89 In *Picardi v Cuniberti and Cuniberti* (2002)[75] the TCC judge considered that an adjudication clause in a standard RIBA form of contract had not been incorporated into the contract as a matter of common law because it had not been adequately brought to the attention of the consumer, as contemplated by the test for incorporation of unreasonable clauses in *J. Spurling Ltd v Bradshaw*.[76] The failure of the architect (seller) to draw the adjudication clause to the consumers' attention, contrary to the advice to that effect in the RIBA guidance notes, coupled with the failure to go through the terms with the consumers as promised, was central to the judge's finding that the adjudication agreement was an unfair term. The consumers were unfamiliar with standard forms and adjudication clauses and did not have the benefit of professional advice on the contract terms.

1.90 As to whether an adjudication clause could be said to cause a significant imbalance between the parties' rights and obligations under the contract, the judge in *Picardi* considered, *obiter*, that because the adjudication procedure gave rise to irrecoverable expenditure, it might hinder the consumer's ability to take legal action. However in *Lovell Projects Ltd v Legg and Carver* (2003)[77] the court took a different view finding that the adjudication clause in a JCT Minor Works Agreement did not hinder the consumer's right to take legal action or exercise any other legal remedy and that in the circumstances it was not unfair in the manner contemplated by Schedule 2 (q) of the 1999 Regulations.[78]

1.91 Similarly in both *Westminster Building Company v Beckingham* (2004) and *Steve Domsalla v Kenneth Dyason* (2007)[79] it was decided that an adjudication clause did not cause a significant imbalance in the rights of the parties under the contract. The fact that the adjudication involves a speedy timetable by an adjudicator who may have no legal training and makes no provisions for payment of costs to the successful party was not enough to render the clauses unfair.

1.92 The 'significant imbalance' point was not considered by the Court of Appeal in *Bryen & Langley Ltd v Martin Boston* (2005) (Key Case).[80] In that case, as in *Lovell*, the adjudication agreement was contained in a standard-form agreement proposed by the consumer's own agent and the Court of Appeal was able to dispose of the claim of unfairness on the ground that it was repugnant to common sense to suggest there was any lack of good faith or fair dealing on the part of the supplier in these circumstances.[81]

---

[72] Unfair Terms in Consumer Contracts Regulations 1999, Reg. 8(1).

[73] [2002] 1 AC 481.

[74] Whether an arbitration clause in a consumer contracts is unfair has been considered in a number of TCC decisions: *Zealander & Zealander v Laing Homes* [2000] TCLR 724 (clause struck out as unfair), *Heifer International Inc. v Christiansen* [2007] EWHC 3015 (clause not unfair), and *Mylcrist Builders Ltd v G. Buck* [2008] BLR 611 (clause struck out as unfair).

[75] [2003] BLR 487.

[76] [1956] 1 WLR 461 at 466.

[77] [2003] 1BLR 452.

[78] The same view was taken by the TCC judge in *Westminster Building Company Ltd v Andrew Beckingham* [2004] BLR 163 at para. 31.

[79] [2007] BLR 348.

[80] [2005] BLR 508.

[81] Those were also the factual circumstances in *Westminster*, and in *Allen Wilson Shopfitters v Antony Buckingham* (2005) 102 Con LR 154.

Development Agreements

1.93 Development agreements are excluded from the ambit of the Act by Statutory Instrument SI 1998 648[82] if they include provision for the grant or disposal of a relevant interest in the land on which takes place the principal construction operations to which the contract relates. A relevant interest is a freehold or leasehold which expires not less than 12 months after completion of the construction operations. This includes options for the grant of leases: *Captiva Estates Ltd v Rybarn Ltd (In Administration)* (2005).[83] It may be possible therefore to contract out of the statute if the parties to a development agreement are willing to include in the agreement an option in relation to a relevant interest in part of the development.

PFI and Finance Agreements

1.94 Private Finance Initiatives and Finance Agreements are also excluded from the ambit of the Act by the Construction Contracts (England and Wales) Exclusion Order Statutory Instrument (SI 1998 648). To some extent this clarifies ambiguities inherent in s. 104(1) discussed above at paragraph 1.14, at least to the extent that it is clear from this Exclusion Order that a finance agreement will not qualify as 'arranging for the carrying out of construction operations by others'.

1.95 In relation to PFI projects therefore the project agreement and concession agreements with the government are excluded from the ambit of the 1996 Act but the construction contract down the line (between project company and contractor) is not caught by this exclusion. Many such contracts include provisions for 'equivalent project relief' and a parallel claims procedure whereby the contractor's entitlement to additional payment will be determined by reference to what the project company has been paid or is entitled to.

1.96 In *Midland Expressway Ltd v CAMBA* (2006)[84] the Secretary of State for Transport granted Midland Expressway Ltd ('MEL') the right to design, construct, and operate the Birmingham northern relief road. The Secretary of State issued a significant variation relating to the detailed road layouts and the contractor claimed over £13 million as a result of the change. MEL accepted that they were entitled to extra payment for the change, but the quantum of the change was disputed and the contractor commenced adjudication proceedings. The construction contract contained provisions for equivalent project relief, making the contractor's entitlement to payment dependent upon sums paid to MEL by the employer and confirming that the contractor could take no steps to enforce any right pending the determination of any payment under the project agreement.

1.97 MEL sought declarations to the effect that the contractor was not entitled to proceed to adjudication because the express provisions of the contract debarred them from pursuing any claim in advance of the determination of the matter under the project agreement. Jackson J found that the provisions of the concession agreement were contrary to s. 108 of the Act in purporting to prevent the contractor from adjudicating at any time and accordingly the Scheme applied. Further, it was held that the provisions in question offended the 'pay when paid' provisions of the Act at s. 113.

Other Excluded Agreements

1.98 Certain agreements specified by statutes are excluded from the ambit of the Act by paragraph 3 of Statutory Instrument SI 1998 648. This list includes agreements made

[82] A copy of which is contained in the Appendices.
[83] [2005] EWHC 2744, [2006] BLR 66.
[84] [2006] EWHC 1505, [2006] BLR 325.

pursuant to the Highways Act 1980 (ss. 38 and 278 agreements), the Town and Country Planning Act 1990 (ss. 106, 106A, 299A agreements), the Water Industry Act 1991 (s. 104 agreements), and externally financed development agreements within the meaning of s. 1 of the National Health Service (Private Finance) Act 1997 (powers of NHS Trusts to enter into agreements).

Key Case: s. 106

Bryen & Langley Ltd v Martin Boston [2005] BLR 508, [2005] EWCA Civ 973

1.99 **Facts:** The Court of Appeal considered whether, pursuant to the Unfair Terms in Consumer Contract Regulations 1999, an adjudication clause in a consumer contract was unfair.

1.100 **Held:** The adjudication agreement was not an unfair term because it had been proferred by the consumer, through his own agent, and therefore there could be no suggestion that the seller's reliance upon it was contrary to the requirement of good faith:

> 44. I do not propose to engage in any consideration of whether, on some sort of attempted objective assessment, the particular provisions to which Mr Bowsher refers do or do not cause a 'significant imbalance' in the respective rights of B & L and Mr Boston to the detriment of Mr Boston. That is because, in my judgment, the performance of such an exercise will not, by itself, provide an answer to the question raised by Mr Bowsher's submission. As Regulation 5(1) makes clear, a term which has not been individually negotiated will only be relevantly 'unfair' if it causes the relevant imbalance 'contrary to the requirements of good faith.' Regulation 6(1) requires the assessment of the unfairness of a contractual term to take account (inter alia) of 'all the circumstances attending the conclusion of the contract'. Lord Bingham of Cornhill explained in *Director General of Fair Trading v. First National Bank plc* [2002] 1 AC 481, at 491, that the 'object of the [1994] Regulations and the Directive is to protect consumers against the inclusion of unfair and prejudicial terms in standard-form contracts into which they enter,' and at page 494 he further explained the requirement of 'good faith' in the predecessor of Regulation 5(1) in the 1994 Regulations. He said:
>
> 'The requirement of good faith in this context is one of fair and open dealing. Openness requires that the terms should be expressed fully, clearly and legibly, containing no concealed pitfalls or traps. Appropriate prominence should be given to terms which might operate disadvantageously to the customer. Fair dealing requires that a supplier should not, whether deliberately or unconsciously, take advantage of the consumer's necessity, indigence, lack of experience, unfamiliarity with the subject matter of the contract, weak bargaining position or any factor listed in or analogous to those listed in Schedule 2 to the Regulations. Good faith in this context is not an artificial or technical concept; nor, since Lord Mansfield was its champion, is it a concept wholly unfamiliar to British lawyers. It looks to good standards of commercial morality and practice. Regulation 4(1) [whose terms were essentially the same as those of Regulation 5(1) of the 1999 Regulations] lays down a composite test, covering both the making and the substance of the contract, and must be applied bearing in mind the objective which the Regulations are designed to promote.'
>
> 45. It follows, in my view, that in assessing whether a term that has not been individually negotiated is 'unfair' for the purposes of Regulation 5(1) it is necessary to consider not merely the commercial effects of the term on the relative rights of the parties but, in particular, whether the term has been imposed on the consumer in circumstances which justify a conclusion that the supplier has fallen short of the requirements of fair dealing. The situation at which Regulation 5(1) is directed is one in which the supplier, who will normally be presumed to be in the stronger bargaining position, has imposed a standard-form contract

on the consumer containing terms which are, or might be said to be, loaded unfairly in favour of the supplier. The *Picardi* case was one in which the terms had been imposed by the claimant architect (in that case, the supplier). In the *Lovell* case the terms had been imposed on the supplier by the employers' (i.e. the consumers') architect, the judge finding not only that they caused no significant imbalance to the employers, but that nor in the circumstances in which the contract came to be made was there any question of any lack of good faith or fair dealing by the supplier contractor. HH Judge Thornton QC arrived at a similar result, in like circumstances, in *Westminster Building Company Limited v. Beckingham* [2004] 1 BLR 265.

46. In my judgment, Mr Boston faces exactly the same difficulties in relation to his Regulation 5(1) argument as did the consumers in the *Lovell* and *Beckingham* cases. His problem is that the relevant provisions were not imposed upon him by B & L, the supplier. It was Mr Boston (the consumer), acting through his agent Mr Welling, who imposed them on the supplier, since they were specified in Mr Welling's original invitation to tender. I am prepared to assume that, in practice, Mr Boston played no part in the preparation of that invitation and that he did not receive any advice from Mr Welling on the provisions now in question; and it is clear that there was no individual negotiation over them with B & L. In principle, however, Mr Boston had the opportunity to influence the terms on which the contractors were being invited to tender, even though he may not have taken it up; and there is therefore at least an argument available to B & L under Regulation 5(2) to the effect that the terms of which he now complains are not terms which fall within the first nine words of Regulation 5(1) at all. I specifically express no view on that last point, on which we had no argument, and assume that the terms do so fall. Even so, in light of the fact that it was Mr Boston, by his agent, who imposed these terms on B & L, I regard the suggestion that there was any lack of good faith or fair dealing by B & L with regard to the ultimate incorporation of these terms into the contract as repugnant to common sense. If they were to tender at all, B & L were being asked by Mr Boston to tender on (inter alia) the very terms of which Mr Boston now complains. It was not for B & L to take the matter up with him and ensure that he knew what he was doing: they knew that he had the benefit of the services of a professional, Mr Welling, to advise him of the effects of the terms on which he was inviting tenders. In my judgment, there was no lack of openness, fair dealing or good faith in the manner in which the June 2001 contract came to be made and in those circumstances I, like the judge, regard Mr Boston's case under the 1999 Regulations as not made out.

47. It follows that I would allow B & L's appeal against the judge's dismissal of its summary judgment application and claim.

Table 1.3 Tables of Cases: s. 106

Title	Citation	Issue
Samuel Thomas Construction v Anon	28 Jan. 2000 (unreported) (TCC), Judge Overend	**Section 106(1): contract to renovate two properties, one for resale** The work could not be said to be work that 'principally relates to operations on a dwelling' when only one property was to be occupied by the defendant and the other (farm building conversions) were to be sold.
Absolute Rentals v Gencor	[2000] CILL 1637 (TCC), Judge Wilcox	**Company as residential occupier** A company can never be a residential occupier (at para. 4 of judgment).

Title	Citation	Issue
Picardi v Cuniberti and Cuniberti	[2003] BLR 487, [2002] EWHC 2923 (TCC), Judge Toulmin	**Unfair Terms in Consumer Contracts Regulations** Adjudication agreement with residential occupier was unfair.
Lovell Projects Ltd v Legg and Carver	[2003] 1BLR 452, Judge Moseley	**Unfair Terms in Consumer Contracts Regulations** Adjudication agreement with residential occupier was not unfair.
Westminster Building Company v Beckingham	[2004] BLR 163, [2004] EWHC 138 (TCC), Judge Thornton	**Unfair Terms in Consumer Contracts Regulations** Adjudication agreement with residential occupier was not unfair.
Bryen & Langley Ltd v Martin Boston	[2005] BLR 508, [2005] EWCA Civ 973 (CA)	**Unfair Terms in Consumer Contracts Regulations** Not contrary to requirement of good faith if consumer (or his agent) was the party proffering the adjudication clause.
Captiva Estates Ltd v Rybarn Ltd (In Administration)	[2006] BLR 66, [2005] EWHC 2744 (TCC), Judge Wilcox	**Development agreement** A relevant interest is a freehold or leasehold which expires not less than 12 months after completion of the construction operations and this provision includes options for the grant of leases.
Allen Wilson Shopfitters v Anthony Buckingham	102 Con LR 154, (2005) CILL 2249, [2005] EWHC 1165 (TCC), Judge Coulson	**Unfair Terms in Consumer Contracts Regulations** Adjudication clause in JCT 1998 Private Without Quantities was not unfair as it was proffered by the consumer's agent and accepted by the seller.
Steve Domsalla v Kenneth Dyason	[2007] BLR 348, [2007] EWHC 1174 (TCC), Judge Thornton	**Unfair Terms in Consumer Contracts Regulations** Adjudication agreement with residential occupier was not unfair as it did not cause significant imbalance in parties' rights.
Edenbooth Ltd v CRE8 Developments Ltd	[2008] CILL 2592, [2008] EWHC 570 (TCC), Coulson J	**Company as residential occupier** The defendant development company was not a residential occupier, even though its directors owned one and occupied both properties. It was difficult to imagine how a company could ever be a residential occupier. **Construction operations** The contract for ground works, foundations and drainage was for construction operations as per s. 105(1) (a) and (e).

(*Continued*)

Table 1.3 *Continued*

Title	Citation	Issue
Mylcrist Builders Ltd v Buck	[2008] BLR 611, [2008] EWHC 2172 (TCC), Ramsey J	**Arbitration clause struck out as unfair** Ramsey J found that an arbitration clause was unfair because contrary to the requirement of good faith it caused a significant imbalance between the rights and obligations of the parties. Although dealing with arbitration clauses, the judge set down useful guidance for when a dispute resolution provision might be judged unfair.
Gabrielle & Christopher Shaw v Massey Foundations & Pilings Ltd	[2009] EWHC 493 (TCC), Coulson J	**Where separate lodge on estate, owner not residential occupier** Contract to renovate separate 'lodge' on land of defendant's residence. No intention to sell lodge but no intention to occupy it at time of contract. The Act only exempts 'a dwelling house' whereas this was two dwellings. The defendants were not residential occupiers of the lodge and the exemption did not apply.

2

SECTION 107 AND THE REQUIREMENT FOR WRITING

(1) The 1996 and the 2009 Acts

2.01 Section107(1) of the Housing Grants Construction and Regeneration Act 1996 (the '1996 Act') states that 'The provisions of this Part apply only where the construction contract is in writing, and any other agreement between the parties as to any matter is effective for the purposes of this Part only if in writing.'[1] Accordingly this requirement is a precondition for the application of the other provisions of Part II of the Act, and as such it is one of the gateways into Part II along with ss. 104 and 105.

2.02 In its origin s. 107 of the 1996 Act was an attempt to enforce the construction industry to submit to a standard form of contract. That did not succeed. However the requirement for writing was maintained because writing provides certainty. Certainty was thought to be important in the context of the adjudication process which is a summary procedure to be undertaken under a demanding short timetable. The requirement that the contract be in writing means the adjudicator who is obliged to decide a dispute arising under a contract within a rapid timeframe starts with some certainty as to what the terms of the contract actually are.

2.03 However s. 107 gave rise to jurisdictional arguments thought by some to undermine expeditious and effective adjudication of disputes. Where a contract is not recorded in writing, unless there is an express agreement to adjudicate a dispute, there is no right to adjudicate as s. 108 of the 1996 Act will not apply. Accordingly, in the past respondents have been able to rely

[1] Section 107 is identical to s. 5 of the Arbitration Act 1996.

on the absence of writing to defeat the adjudication process. Jurisdiction arguments based on s. 107 have been deployed during the adjudication, during enforcement proceedings, and sometimes have been subject of pre-emptive declaratory relief actions.

2.04 Against this background it may not be surprising that s. 107 has been repealed by s. 139 of the Local Democracy, Economic Development and Construction Act 2009 (the '2009 Act'). Accordingly contracts entered into after the 2009 Act came into force will no longer need to be evidenced in writing in order for the other provisions of the 1996 and 2009 Acts to apply.[2]

2.05 This chapter therefore considers the law as it applies to contracts entered into before the 2009 Act came into effect. There is no shortage of judicial decisions on the meaning and effect of s. 107 as can be seen from the Tables of Cases set out in this chapter. Most of the cases concern enforcement applications at which the defendant challenged the validity of an adjudicator's decision on the ground that it was made without jurisdiction because the contract did not satisfy s. 107. In that context therefore the judicial consideration of s. 107 has frequently been from the perspective of an application for summary judgment. It is important to remember when reading these authorities that at a summary judgment application the defendant does not have to prove that it is right but needs only to show that its arguments have a reasonable prospect of success and therefore need to be tried fully. This means that at a summary judgment application to enforce an adjudicator's decision, where the defendant has been able to show that there is a triable issue concerning whether a contract was concluded, summary enforcement of the adjudicator's decisions have failed.[3] When this happens the court's judgment about the enforcement of the adjudicator's decision is deferred until after the full trial of those issues regarding the formation of contract. Effectively this can mean the intended summary enforcement of the adjudicator's decisions is frustrated.

2.06 However the repeal of s. 107 will not necessarily remedy that problem. Without the requirement for writing more agreements will be brought within the sphere of operation of the 1996 Act (as amended by the 2009 Act) and potentially this means more adjudications. However, because the existence of an agreement within the meaning of ss. 104 and 105 remains a gateway into the statute, a defendant may still argue that no such contract was in fact concluded. If the contract relied on by the referring party was made orally, then the court at the enforcement hearing will likely be asked to decide if any oral contract was ever concluded. A dispute based on oral testimony is usually unsuitable for summary judgment and may be deferred to full trial where oral evidence may be given. In other words, whilst the removal of the requirement for writing may bring more contracts within the Act, it is foreseeable that summary enforcement of decisions based on those non-written agreements may prove difficult.

2.07 Having said that, the consequence of the matter proceeding to a full trial need not lead to significant delay to enforcement. The issues regarding oral agreements may be limited and can be tried relatively quickly, or as a preliminary issue. The Technology and Construction division of the High Court will usually aim to give directions to ensure the jurisdiction issue which cannot be determined summarily will nevertheless be tried quickly.

Relevance of s. 107 to Contracts with Adjudication Agreements

2.08 There is an important distinction between contracts that incorporate express adjudication provisions and those construction contracts that do not. In a contract with an express adjudication

[2] 1 Oct. 2011 in England and Wales, 1 Nov. 2011 for Scotland and 14 Nov. 2012 in Northern Ireland, see Ch. 1 at 1.08.

[3] See e.g. *CN Associates v Holbeton* [2011] BLR 261, [2011] EWHC 43 (TCC); *Aedifice Partnership Ltd v Ashwin Shah* (2010) 132 Con LR 100, [2010] EWHC 2106 (TCC); *Estor Ltd v Multifit (UK) Ltd* (2009) 126 Con LR 40, [2009] EWHC 2565 (TCC).

clause the availability of adjudication as a mechanism for interim dispute resolution does not depend on s. 107 of the Act but on the terms of the adjudication agreement (subject to the proviso that in a contract governed by the Act where the adjudication clause fails to comply with s. 108, the Scheme for Construction Contracts ('the Scheme') will replace the agreed provision[4]). In principle therefore a construction contract with an express adjudication agreement may be varied orally and the oral variation need not satisfy s. 107 and a dispute thereunder may be the subject of a contractual adjudication as long as it falls within the scope of the adjudication agreement.[5]

(2) Section 107 and the Requirement for Writing

2.09 Prior to its amendment in 2009 the provisions of Part II of the 1996 Act applied only where the construction contract was in writing, and only where 'any other agreement between the parties as to any matter' was in writing. As discussed below, the interpretation of s. 107 gave rise to a fairly large body of case law concerning what constitutes an agreement in writing for the purpose of s. 107, what constitutes writing and how much of the agreement needs to be evidenced in writing.

Section 107(2): What Constitutes an Agreement in Writing?

2.10 There are three categories described in s. 107(2) where the agreement is to be treated as being in writing. The first is where the agreement itself is made in writing (whether signed by the parties or not).[6] This contemplates a written document containing the agreement itself. The second is where an agreement is made by communications in writing.[7] In other words the offer, acceptance and terms are set out in correspondence between the parties. The third category is where the agreement is evidenced in writing.[8] This latter provision has given rise to some difficulties regarding what consists of evidence in writing and whether the whole or part of the agreement must be evidenced in writing.

What Is Writing?

2.11 References in Part II of the 1996 Act to anything being written or in writing include its being recorded by any means.[9] There are no reported cases on this provision but it is suggested this will include drawings and may also include tape recordings of agreements.

What Must Be Evidenced in Writing?

2.12 It is the terms and not merely the existence of a construction contract which must be evidenced in writing. *Ex post facto* evidence of the existence of an agreement will be insufficient for the purposes of s. 107 of the Act (unless the parties can rely on s.107(5), as discussed below). In *RJT Consulting Engineers Ltd v DM Engineering (Northern Ireland) Ltd* (2002) (Key Case)[10] the Court of Appeal were divided over whether the whole of the agreement needs to be in writing or whether it is sufficient for those material terms relevant to the issues in the adjudication to be in writing. The majority of the Court of the Appeal gave a literal interpretation

[4] This is discussed in Chs. 3 and 4.

[5] *Dean and Dyball Construction Ltd v Kenneth Grubb Associates Ltd* (2003) 100 Con LR 92, [2003] EWHC 2465 (TCC) at paras. 16–18; *Treasure & Sons v Dawes* [2008] BLR 24, [2007] EWHC 2420 (TCC); *Linnett v Halliwells* [2009] BLR 312, [2009] EWHC 319 (TCC).

[6] s. 107(2)(a).

[7] s. 107 (2)(b).

[8] s. 107(2)(c).

[9] s. 107(6).

[10] [2002] 1 WLR 2344, [2002] BLR 217, [2002] EWCA Civ 270 (CA).

to the Act holding that 'what has to be evidenced in writing is, literally, the agreement, which means all of it, not a part of it' (per Ward LJ at paragraph 19). The reasoning was based not just on the wording of the statutory provisions themselves, which require 'the agreement' to be in writing, but also on policy grounds. It was said that disputes about oral agreements are not readily susceptible to resolution by a summary procedure such as adjudication, because the written agreement is the foundation from which a dispute may spring and the least the adjudicator has to be certain about is the terms of the agreement which are giving rise to the dispute.

2.13 Auld LJ, expressing the minority view, considered that it was the *material terms* which needed to be in writing. He feared that jurisdictional wrangling on this issue would clog the adjudicative process and defeat the purpose of the Act.

2.14 For a short time after the decision in *RJT Consulting Engineers* there was some doubt as to the weight to be given to the minority reasoning of Auld LJ.[11] Since then however, that the High Court is bound by the majority view was unequivocally stated by Jackson J in *Trustees of Stratfield Saye Estate v AHL Construction Ltd* (2004):[12] 'In my view, it is not possible to regard the reasoning of Auld LJ as some kind of gloss upon or amplification of the reasoning of the majority. The reasoning of Auld LJ, attractive though it is, does not form part of the ratio of RJT.' The overwhelming majority of cases have followed this approach, applying the test that all the terms must be in writing or evidenced in writing.

2.15 Regarding minor or trivial matters, Ward LJ left the door very slightly ajar when he said: 'No doubt adjudicators will be robust in excluding the trivial from the ambit of the agreement and the matter must be entrusted to their common sense.'[13] In this context it has been suggested that an item may be considered trivial or minor if, considered objectively, it is not a material term of the contract and therefore falls within the category of items with which an adjudicator should deal robustly.[14]

Acceptance by Conduct

2.16 Despite the requirement in s. 107(1) that 'any agreement between the parties as to any matter' is effective only if in writing, the courts do not seem to have had a problem allowing acceptance by conduct, as was the case in *Trustees of Stratfield Saye Estate v AHL Construction Ltd* (2004).[15] In that case the judge found that all the terms that had been negotiated between the parties had been recorded in writing thus satisfying the test laid down by the Court of Appeal in *RJT Consulting*[16] for compliance with s. 107(2)(c). Similarly in *Dean and Dyball Construction Ltd v Kenneth Grubb Associates Ltd* (2003)[17] and confirmed in *Durham County Council v Jeremy Kendall* (2011)[18] the courts have considered acceptance by conduct sufficient for a s. 107 compliant contract to exist. Acceptance by conduct is also probably valid pursuant to s. 107(3) as an agreement other than in writing by reference to terms that are in writing.

---

[11] See Judge Bowsher in *Carillion Construction Ltd v Devonport Royal Dockyard* [2003] BLR 79 at para. 25.

[12] [2004] EWHC 3286 (TCC).

[13] At para. 19 of the judgment.

[14] *Westdawn Refurbishment Ltd v Roselodge Ltd* [2006] Adj LR 04/25, citing *Dean and Dyball Construction Ltd v Kenneth Grubb Associates Ltd*, (2003) 100 Con LR 92, [2003] EWHC 2465 (TCC). See also *Lloyd Projects v John Malnick*, 5 Sep. 2005 (unreported), Judge Kirkham in the Birmingham TCC; and *Allen Wilson Joinery v Privetgrange Construction Ltd* 123 ConLR 1, [2008] EWHC 3160 (TCC).

[15] [2004] All ER (D) 77 (Dec), [2004] EWHC 286 (TCC).

[16] [2002] BLR 217.

[17] 100 Con LR 92, [2003] EWHC 2465 (TCC) at para. 12.

[18] [2011] BLR 425, [2011] EWHC 780 (TCC) at para. 35.

Incorporation of Terms

2.17 Once it is established that the adjudicator has jurisdiction, a question about whether a particular term has or has not been incorporated into an agreement is a matter which the adjudicator has power to decide:

> Thus so long as it is established or agreed that there is a contract in existence between the parties, [and] that it is a construction contract...any other dispute as to the terms of the construction contract is as much a dispute arising under the contract as would be a dispute as to the working through of any terms to the valuation machinery.[19]

2.18 If however the incorporation argument is really a question about an oral agreement of a term of the contract which has not been evidenced by writing, that is a s. 107 matter which goes to jurisdiction.

Section 107(3): The Effect of Referring to Terms in Writing

2.19 Where parties agree otherwise than in writing by reference to terms which are in writing, they make an agreement in writing (s. 107(3)). In *Carillion Construction Ltd v Devonport Royal Dockyard* (2002)[20] Carillion argued that the oral agreement made reference to the existing written contract which was sufficient for the purpose of s. 107(3). Judge Bowsher rejected that argument. In that case the variation in question was a change to the payment terms, from contractually defined 'Actual Cost' plus a fee and a gainshare arrangement, to a cost-reimbursable basis. The judge relied on the fact that s. 107(3) uses the words 'by reference to terms that are in writing' and not 'by reference to a previous agreement that is in writing'. The judge said that if oral variations were brought within the 1996 Act purely because they referred to the original written contract, the intent of the Act, that adjudicators would not have to decide on the existence and terms of oral agreements, would be defeated. Accordingly he found that a fundamental variation to a construction contract made orally and without writing was outside the Act and takes the entire contract outside the ambit of the Act.

2.20 Judge Bowsher found that it is necessary for any oral agreement itself to be evidenced by the written terms referred to.[21] The judge postulated that this would include a verbal agreement in relation to a draft written contract which had already been prepared. It could also cover an oral agreement that the terms of a standard form contract will apply.

Section 107(4): Oral Agreement Recorded with Authority

2.21 An oral agreement recorded by one party, or by a third party, with the authority of the parties will constitute an agreement evidenced in writing for the purpose of the s. 107.[22] This provision does not stipulate the only method by which an agreement may be evidenced in writing[23] and s. 107(6) provides that reference to anything being written includes it being recorded by any means.

Section 107(5): Exchange of Written Submissions

2.22 An exchange of written submissions in adjudication proceedings or in arbitral or legal proceedings in which the existence of an agreement otherwise than in writing is alleged by one

---

[19] *R. G. Carter Ltd v E. Nuttall Ltd*, 21 Jun. 2000 (unreported), approved in *RJT Consulting Engineers v DM Engineering* [2002] 1 WLR 2344, [2002] 5 BLR 217, [2002] EWCA Civ 270 (CA). See also *C&B Scene v Isobars* [2002] BLR 93, [2002] EWCA Civ 46 and *Allen Wilson Shopfitters v Anthony Buckingham* (2005) 102 Con LR 154, [2005] EWHC 1165 (TCC).

[20] [2003] BLR 79 at para. 32.

[21] However, see the differing approach in *Total M E Services Ltd v ABB Building Technologies Ltd* (2002) [2002] EHWC 248 (TCC) discussed in more detail at 2.42 below.

[22] s. 107(4).

[23] *Durham County Council v Jeremy Kendal* [2011] BLR 425, EWHC 780 (TCC) at para. 34.

party against another party and not denied by the other party in his response, constitutes, as between those parties, an agreement in writing to the alleged effect. Section 107(5) provides the only exception to the general requirement of s. 107 that all the agreement must be evidenced in writing.[24]

2.23 The Departmental Advisory Committee (DAC) report on the equivalent provision in the Arbitration Act provided some useful guidance about the effect of s. 107(5):

(a) The provision operates to create a form of statutory estoppel where one party admits the existence of an arbitration clause in its response document. This is equivalent to an ad hoc conferral of jurisdiction onto the arbitrator.
(b) Interestingly, if there is no response served at all there is no statutory estoppel created by silence.
(c) The DAC report on the Arbitration Act advised that not every written response would comprise a submission for the purpose of the section, only formal submissions would suffice.

2.24 The last four words of the provision are important. The exchange constitutes an agreement in writing which does more than evidence the existence of the agreement. It also evidences *the effect of the agreement alleged* and that must mean it evidences the terms of the contract material to the purposes of that particular adjudication. This was the conclusion of the TCC judge in *Grovedeck Ltd v Capital Demolition Ltd* (2000).[25] It was also the conclusion of Judge Thornton in *Mott MacDonald Ltd v London & Regional Properties Ltd* (2007).[26] In that case the respondent admitted an agreement existed, but contended it was a different agreement from the one alleged by the referring party and was not fully evidenced in writing. On the facts the judge decided there had been no effective admission because the respondent continued to allege the contract failed to comply with s. 107. These cases suggest that s. 107(5) only applies where the terms of the agreement have also been admitted.[27]

2.25 Section 107(5) would prevent a defendant who had accepted that there was a written construction contract in the adjudication from subsequently arguing at the enforcement hearing that no written agreement existed. This is a relatively straightforward proposition consistent with the common law rules of estoppel and those authorities that deal with the conferral of jurisdiction.[28]

2.26 It is not clear whether s. 107(5) is intended to remove from a party the right to challenge jurisdiction on the ground that the contract, whilst admitted, was not one that was in writing.[29] The judge in *Grovedeck* decided that s. 107(5) cannot have been intended to have that effect and concluded that s. 107(5) applied only to oral agreements that had been admitted in *other preceding* adjudications.

2.27 The opposite view was expressed in *ALE Heavy Lift v MSD (Darlington) Ltd* (2006).[30] The claimant's referral notice cited a contract evidenced in writing by a letter and its enclosures. The response alleged the contract incorporated oral agreements that were referred to in the letter. The respondent therefore did not deny the contract in the referral notice and did not

[24] *RJT Consulting Engineers Ltd v DM Engineering (Northern Ireland) Ltd* [2002] 1 WLR 2344, [2002] BLR 217, [2002] EWCA Civ 270 at para. 19 (Key Case).

[25] [2000] BLR 181.

[26] 113 ConLR 33, [2007] EWHC 1055 (TCC).

[27] However see *A&D Maintenance and Construction Services Ltd v Pagehurst Construction Services Ltd*, 23 Jun. 1999 (TCC) at para. 15. Judge Wilcox stated that s. 107(5) applied where the contract in writing was admitted but the terms of the contract were in dispute. The point was not considered in detail and was *obiter dicta*.

[28] Discussed in more detail in Ch. 5.

[29] Jurisdiction challenges are discussed Ch. 9 and reserving the right to challenge jurisdiction is discussed in Chs. 5 and 9.

[30] [2006] EWHC 2080 (TCC).

take a jurisdiction point that the contract was part oral. There had been no denial of jurisdiction and so on the facts of the case the judge decided that any challenge to jurisdiction had been waived. However, in dealing with the claimant's additional argument that as a matter of statute there was also jurisdiction pursuant to s. 107(5) Judge Toulmin said:

> 88. If I had to disagree with His Honour Judge Bowsher Q.C's interpretation of section 107 (5) of the Act in Grovedeck I would do so, but I simply need to follow Ward LJ's proposition in paragraph 19 of his judgment that where the material relevant parts of a contract alleged in the written submissions in the adjudication are not denied, that is sufficient. That applies both to submissions relating to an alleged written agreement as well as submissions relating to an alleged oral agreement. If I have to take this route I conclude that applying the Statute the Adjudicator had jurisdiction. The simple answer is that as a matter of contract where the jurisdiction of the Adjudicator is not challenged on a particular ground, a challenge of jurisdiction on that ground has been waived, and the parties have agreed to proceed with the adjudication despite the possibility of that challenge.

2.28 The Court of Appeal in *RJT Consulting Engineers* expressly chose not to rule on the judge's finding in Grovedeck (at paragraph 16 of the judgment). In *Trustees of Stratfield Saye Estate v AHL Construction Ltd* (2004)[31] Jackson J decided that the existence of a s. 107 compliant contract had been admitted by the Estate in two out of three adjudications and so the contractor was entitled to rely on this pursuant to s. 107(5) of the Act. Similarly in *SG South Ltd v Swan Yard (Cirencester) Ltd* (2010)[32] the TCC judge also found that s. 107(5) applied to admissions made in the submissions in the adjudication itself.[33]

2.29 The judge in *ALE Heavy Lift* did not consider whether his interpretation of the statute would apply even where there had been a denial of jurisdiction and reservation of rights by the defendant. This point and this provision remain to be considered by an appellate court.

2.30 Although there have been a great many cases on whether the requirement for writing has been satisfied, almost all turned on the facts of the individual contractual negotiations between the parties. As far as a statement of principle goes, there is really only one authority for the proposition that all the terms of the agreement need to be evidenced in writing.

Key Case: The Requirement for Writing

RJT Consulting Engineers Ltd v DM Engineering (Northern Ireland) Ltd [2002] 1 WLR 2344, [2002] BLR 217, EWCA Civ 270

2.31 **Facts:** The claimant 'RJT' was engaged by the defendant 'DM Engineering' to complete the design of mechanical and electrical systems at a Holiday Inn. DM Engineering referred a dispute to adjudication claiming damages for professional negligence. The agreement between the parties was not itself in writing. The adjudicator decided that there was sufficient evidence in the form of fee notes, letters, drawing schedules, and meeting minutes that referred to the existence of an agreement and that this was sufficient for the purpose of s. 107 of the Act. RJT sought a declaration from the High Court that there was no agreement in writing which satisfied s. 107 of the Act. Judge Mackay agreed with the adjudicator and dismissed the application but gave permission to appeal.

[31] [2004] All ER (D) 77 (Dec).
[32] [2010] EWHC 376 (TCC).
[33] In *Sprunt Ltd v London Borough of Camden* [2012] BLR 83, [2011] EWHC 3191 (TCC) at para. 41 of the judgment the TCC judge rejected an assertion that s. 107(5) did not apply to adjudication proceedings that were subject of court proceedings.

2.32 **Held:** Evidence in writing of the existence of an agreement is insufficient for the purpose of s. 107. The whole of the agreement needs to be evidenced in writing, not just part of it (per Ward and Walker LJJ). On the facts the terms of the agreement were not evidenced in writing:

> 12. I turn to the construction of section 107. Section 107(1) limits the application of the Act to construction contracts which are in writing or to other agreements which are effective for the purposes of that part of the Act only if in writing. This must be seen against the background which led to the introduction of this change. In its origin it was an attempt to force the industry to submit to a standard form of contract. That did not succeed but writing is still important and writing is important because it provides certainty. Certainty is all the more important when adjudication is envisaged to have to take place under a demanding timetable. The adjudicator has to start with some certainty as to what the terms of the contract are.
>
> 13. Section 107(2) gives three categories where the agreement is to be treated in writing. The first is where the agreement, whether or not it is signed by the parties, is made in writing. That must mean where the agreement is contained in a written document which stands as a record of the agreement and all that was contained in the agreement. The second category, an exchange of communications in writing, likewise is capable of containing all that needs to be known about the agreement. One is therefore led to believe by what used to be known as the *eiusdem generis* rule that the third category will be to the same effect namely that the evidence in writing is evidence of the whole agreement.
>
> 14. Sub-section (3) is consistent with that view. Where the parties agree by reference to terms which are in writing, the legislature is envisaging that all of the material terms are in writing and that the oral agreement refers to that written record.
>
> 15. Sub-section (4) allows an agreement to be evidenced in writing if it (the agreement) is recorded by one of the parties or by a third party with the authority of the parties to the agreement. What is there contemplated is, thus, a record (which by sub-section (6) can be in writing or a record by any means) of everything which has been said. Again it is a record of the whole agreement.
>
> 16. Sub-section (5) is a specific provision. Where there has been an exchange of written submissions in the adjudication proceedings in which the existence of an agreement otherwise than in writing is alleged by one party and not denied by the other, then that exchange constitutes 'an agreement in writing to the effect alleged'. The last few words are important. The exchange constitutes an agreement in writing which does more than evidence the existence of the agreement. It also evidences the effect of the agreement alleged, and that must mean such terms which it may be material to allege for the purpose of that particular adjudication. It is not necessary for me to form a view about *Grovedeck Ltd v Capital Demolition Ltd* 2000 Building Law Reports 181. Dealing with section 107(5) His Hon. Judge Bowsher Q.C. said:-
>
> > 'Disputes as to the terms, express and implied, of *oral* construction agreements are surprisingly common and are not readily susceptible of resolution by a summary procedure such as adjudication. It is not surprising that Parliament should have intended that such disputes should not be determined by adjudicators under the Act, ...'
>
> (Emphasis added by me.)
>
> I agree. That is why a record in writing is so essential. The written record of the agreement is the foundation from which a dispute may spring but the least the adjudicator has to be certain about is the terms of the agreement which is giving rise to the dispute.
>
> ...
>
> 19. On the point of construction of section 107, what has to be evidenced in writing is, literally, the agreement, which means all of it, not part of it. A record of the agreement also suggests a complete agreement, not a partial one. The only exception to the generality of that construction is the instance falling within sub-section 5 where the material or relevant

parts alleged and not denied in the written submissions in the adjudication proceedings are sufficient. Unfortunately, I do not think sub-section 5 can so dominate the interpretation of the section as a whole so as to limit what needs to be evidenced in writing simply to the material terms raised in the arbitration. It must be remembered that by virtue of section 107(1) the need for an agreement in writing is the precondition for the application of the other provisions of Part II of the Act, not just the jurisdictional threshold for a reference to adjudication. I say 'unfortunately' because, like Auld L.J. whose judgment I have now read in draft, I would regard it as a pity if too much 'jurisdictional wrangling' were to limit the opportunities for expeditious adjudication having an interim effect only. No doubt adjudicators will be robust in excluding the trivial from the ambit of the agreement and the matter must be entrusted to their common sense. Here we have a comparatively simple oral agreement about the terms of which there may be very little, if any, dispute. For the consulting engineers to take a point objecting to adjudication in those circumstances may be open to the criticism that they were taking a technical point but as it was one open to them and it is good, they cannot be faulted. In my judgment they were entitled to the declaration which they sought and I would accordingly allow the appeal and grant them that relief.

Table 2.1 Table of Cases: The Requirement for Writing

Title	Citation	Issue
A&D Maintenance and Construction Ltd v Pagehurst Construction Services Ltd	[1999] CILL 1518 (TCC), Judge Wilcox	**Section 107(5)** The exchange of submissions in an adjudication in which the existence of an agreement in writing was asserted and not denied, although the terms were disputed, was an agreement evidenced in writing within the meaning of s. 107(5).
Grovedeck Ltd v Capital Demolition Ltd	[2000] BLR 181, (2000) 2 TCLR 689 (TCC), Judge Bowsher	**Section 107(5)** The section only applies when the oral agreement and its terms are admitted by the defendant. It does not apply to exchange of submissions in the existing adjudication, only to preceding adjudications (but see discussion at 2.22–2.29 above).
Debeck Ductwork Installation Ltd v T&E Engineering Ltd	14 Oct. 2002 (TCC) (unreported), Judge Kirkham	**All agreement must be evidenced in writing** The whole of the contract was not evidenced in writing because the letter relied on by the claimant as evidencing the contract failed to state, even in summary terms, the scope of work, matters as to quality and issues as to timing which were agreed orally. It would be quite wrong for a claimant to be able to rely on a document containing some of the terms agreed and ask the court to ignore the additional terms which were agreed orally.
RJT Consulting Engineers Ltd v DM Engineering (Northern Ireland) Ltd	[2002] 1 WLR 2344, [2002] BLR 217, EWCA Civ 270 (CA)	**Section 107(2) and (4): evidenced in writing** The adjudicator found an agreement was evidenced by drawing schedules and a letter dated 31 January 2001. 1. Evidence in writing of the existence of an agreement is insufficient for the purpose of s. 107. 2. The whole of the agreement needs to be evidenced in writing, not just part of it. 3. On the facts the terms of the agreement were not evidenced in writing.

(*Continued*)

Table 2.1 *Continued*

Title	Citation	Issue
R. G. Carter Ltd v Edmund Nuttall Ltd	[2002] BLR 359 (TCC), Judge Thornton	**Disagreements as to terms** Once the adjudicator has jurisdiction he is entitled to decide incorporation of terms.
Dean and Dyball Construction Ltd v Kenneth Grubb Associates Ltd	100 Con LR 92, [2003] EWHC 2465 (TCC), Judge Seymour	**Acceptance by conduct** A written proposal, followed by a counter-proposal was accepted by the conduct of getting on with the work. This created a contract which was evidenced in writing as required by s. 107.
Galliford Try Construction Ltd v Michael Heal Associates Ltd	99 Con LR 19, [2003] EWHC 2886 (TCC), Judge Seymour	**No contract** Adjudicator's decision could not be enforced as it was based on an erroneous conclusion that there was a contract between the parties.
Murray Building Services v Spree Developments	30 Jul. 2004 (unreported), Judge Raynor	**Price may not need to be recorded in writing if can be construed** Where price can be determined by the construction of the contract terms, that will not offend against the need for writing. However a term that needs to be implied will offend against s. 107. (This last point doubted in *ROK v Bestwood* [2010] at para. 29 of the judgment.)
Trustees of Stratfield Saye Estate v AHL Construction Ltd	[2004] All ER (D) 77 (Dec), [2004] EWHC 286 (TCC), Jackson J	**Acceptance by conduct** A construction contract was formed when the Estate's offer was accepted by the Contractor's conduct in starting work. This was evidenced in writing within the meaning of s. 107 in three drawings, a letter of 2 September, a minute of meeting of 8 September. Every term negotiated was recorded in writing. **Section 107(5)** Alternatively the existence of a s. 107 compliant contract had been admitted by the Estate in two out of three adjudications. The contractor was entitled to rely on this pursuant to s. 107(5) of the Act.
Allen Wilson Shopfitters v Anthony Buckingham	102 Con LR 154, (2005) CILL 2249, [2005] EWHC 1165 (TCC), Judge Coulson	**Incorporation of standard terms and adjudication clause** Letter of intent incorporated JCT 1998, Private without Quantities 1998. Work was performed in excess of letter intent value limit pursuant to JCT variation clauses. The JCT Adjudication provision clause 41A was incorporated into letter of intent. The JCT clause 41A was not unfair. The adjudicator's employment of the payment terms of the Scheme rather than the JCT was not a matter which went to jurisdiction but was an error of law (*C&B Scene* considered).
Lloyd Projects v John Malnick	5 Sep. 2005 (TCC) (unreported), Judge Kirkham	**All terms must be in writing** An oral agreement evidenced in writing failed to include agreed matters such as the standard of work, who was to bear the risk of unforeseen events and scope of work. These were not trivial matters and were not items that the adjudicator should deal with robustly.

Title	Citation	Issue
Management Solutions & Professional Consultants Ltd v Bennett (Electrical) Services Ltd (2006)	[2006] EWHC 1720 (TCC), Judge Thornton	**Oral variations pursuant to written variations clause** An adjudicator had jurisdiction to deal with oral variation instructions which were given under the authority of a written variations clause in a contract which was entirely in writing.
ALE Heavy Lift v MSD (Darlington) Ltd	[2006] EWHC 2080 (TCC), Judge Toulmin	**Section 107(5): contract admitted in submissions** Where the material parts of the contract alleged in the written submissions are not denied, that is sufficient for s. 107(5).
Redworth Construction Ltd v Brookdale Healthcare Ltd	[2006] BLR 366, [2006] EWHC 1994 (TCC), Judge Havery	**All terms not in writing** An adjudicator had no jurisdiction to determine a construction dispute over deductions, as the parties had not evidenced all the terms in writing, and the agreement did not satisfy s. 107 and further the contract had not incorporated JCT terms so the express adjudication agreement was not incorporated. **Election principle** On the basis of the principle of election, the claimant could not go beyond the matters it relied on in the adjudication in order to support the adjudicator's decision that he had jurisdiction. (In *Nickleby* [2010] the TCC judge doubted this was an invariable rule).
Westdawn Refurbishment Ltd v Roselodge Ltd	[2006] Adj LR 04/25, Judge McCahill	**All agreement must be evidenced in writing** The whole of the contract was not evidenced in writing because oral agreements concerning the trigger events for completion and time for payment of invoices were not recorded in writing.
A.R.T. Consultancy Ltd v Nevada Trading Ltd	[2007] EWHC 1375 (TCC), Judge Coulson	**All terms in writing** An oral agreement to perform design was independent of a written contract for the construction operations (which incorporated the JCT Minor Works Agreement). An adjudication decision regarding construction works was valid.
Treasure & Sons Ltd v Dawes	[2008] BLR 24, [2007] EWHC 2420 (TCC), Akenhead J	**Effect of contractual agreement to adjudicate on s. 107** Where the contract contains an express agreement to adjudicate, s. 107 is overtaken by that fact. Accordingly an oral variation to such a contract will not be undermined by s. 107 and a dispute about the variation will usually fall within the scope of the dispute resolution agreement (unless there was a contract term that oral variations were invalid).
Mast Electrical Services v Kendall Cross Holdings	[2007]NPC 70, [2007] EWHC 1296 (TCC), Jackson J	**All terms not in writing** The documents relied upon by the claimant did not satisfy the requirements of s. 107 in that they had failed to set out, evidence or record all the material terms of its subcontract, particularly in respect of any agreed payment rates. It was probable that there was no contract at all. The claimant was not entitled to the declarations it sought and was unable to refer its payment disputes to adjudication.

(*Continued*)

Table 2.1 *Continued*

Title	Citation	Issue
Mott Macdonald Ltd v London & Regional Properties Ltd	113 Con LR 33, [2007] EWHC 1055 (TCC), Judge Thornton	**All terms not in writing** The letter of intent had expired no later than October 2000. Mott's claims did not arise solely from the agreement in writing set out in the letter of intent, but from a separate and different contract that had been the subject of substantial amendments that were partly evidenced in writing, partly formed by conduct and partly to be inferred from the conduct of the parties. Many of the amendments must have been in writing or evidenced in writing, but Mott had not relied on them or adduced them in evidence. Thus, the construction contract that Mott was entitled to rely on was enforceable in law but was not sufficiently in writing to fall within s. 107(2). **Section 107(5) did not apply** This was because the respondent continued to allege the contract failed to comply with s. 107.
Allen Wilson Joinery Ltd v Privetgrange Construction Ltd	123 Con LR 1, [2008] EWHC 2802 (TCC), Akenhead J	**Oral agreements re trivial matters** Adjudicators should be robust in determining whether trivial matters said to have been agreed only orally between the parties can prevent what would otherwise be a written contract for the purpose of s. 107. The exercise of determining what is trivial must be an objective one in relation to the particular contract and parties concerned. What may be 'trivial' in one contract may not be in another. Thus, for example, an oral agreement on a £1m project as to which of two mildly differing shades of light blue paint might be used may be trivial on one development but not on another (also see Ward LJ in *RJT*).
Linnett v Halliwells	[2009] BLR 312, [2009] EWHC 319 (TCC), Ramsey J	**Section 107 and adjudication clauses** Where a contract contains a written agreement to adjudicate (such as clause 41A of the JCT form) s. 107 of the 1996 Act does not need to be complied with (at para. 108 of the judgment).
Adonis Construction v O'Keefe Soil Remediation	[2009] EWHC 2047 (TCC), Clarke J	**Enforcement refused—doubt over contract formation** Summary enforcement was refused where there was an arguable case that there had been no contract in writing.
Estor Ltd v Multifit (UK) Ltd	126 Con LR 40, [2009] EWHC 2565 (TCC), Akenhead J	**Challenge re correct party—triable issue** The adjudicator decided the contract was made with Estor. On the evidence before the court the judge decided there was a realistic prospect of Estor establishing that it was not the company which entered into the contract with Multifit. The adjudicator had no jurisdiction to decide these issues in a binding way.
SG South Ltd v Swan Yard (Cirencester) Ltd	[2010] BLR 47, [2010] EWHC 376 (TCC), Coulson J	**Section 107(5): contract admitted in exchange of submissions** (*Obiter*) Where there was no written agreement but the referring party had alleged an oral contract and the responding party had not denied it, the adjudicator had jurisdiction.

Title	Citation	Issue
ROK Building Ltd v Bestwood Carpentry Ltd	[2010] EWHC 1409 (TCC), Akenhead J	**Implied terms as to reasonable sum may not need to be recorded in writing** Whilst all the material terms needed to be in writing, it was not necessary that the price should be in writing if it could be determined by construing the contract. Accordingly an implied term for a reasonable sum could suffice. But an oral agreement of a defined price that was not recorded in writing would not satisfy s. 107 rendering the whole contract outside s. 107.
Nickleby FM Ltd v Somerfield Stores Ltd	131 Con LR 203, [2010] EWHC 1976 (TCC), Akenhead J	**Jurisdiction argument in court different from before adjudicator** The judge doubted it was an invariable rule that a party could not alter the basis of its challenge to jurisdiction at the enforcement (doubting *Redworth v Brookdale* [2006] EWHC 1994, above). Guidance was given at para. 28 of judgment. In this case the judge concluded that the adjudicator did in any event have jurisdiction and so did not need to decide the *Redworth* point.
All Metal Roofing v Kamm Properties Ltd	[2010] EWHC 2670 (TCC), Akenhead J	**Allegation that time for completion orally agreed** The judge rejected a challenge based on the assertion that a material term of the contract (the time obligation) was not in writing. On the facts the judge decided the written purchase order set out the obligation as to time. Decision enforced.
Aedifice Partnership Ltd v Ashwin Shah	132 Con LR 100, [2010] EWHC 2106 (TCC), Akenhead J	**Section 107: contract/no contract debate a triable issue** *Obiter*: there undoubtedly were issues between the parties as to whether there was even a contract between Aedifice and Mr Shah as opposed to the company, if so on what terms, and whether all terms were evidenced in writing. But for the fact that it was conceded, the judge would have found on the evidence that these raised triable issues which could not be resolved on a summary application.
CN Associates (a firm) v Holbeton Ltd	[2011] BLR 261, [2011] EWHC 43 (TCC), Akenhead J	**Section 107: contract/no contract debate a triable issue** The contract was challenged on the ground that there had been no acceptance by conduct. The defendant passed the threshold for permission to defend, but only just, because it was inappropriate to determine the issues on a summary application.
Durham County Council v Jeremy Kendall	[2011] BLR 425, [2011] EWHC 780 (TCC), Akenhead J	**Section 107(4): status of agreed minute of meeting** An agreed minute of meeting can be a written record for the purpose of s. 107(4) if it records agreement in writing of a material term. But this is not the only method by which an agreement may be evidenced in writing. Section 107(4) does not purport to be so restrictive.
Supablast (Nationwide) Ltd v Story Rail Ltd	[2010] BLR 211, [2010] EWHC 56 (TCC), Akenhead J	**Claim for extra work (overlap with s. 107 issue)** Most contracts permit extra work to be instructed. Work instructed under the contract mechanism does not need to be in writing. However it is always open to one party to say that work was not a variation envisaged or permitted by the contract. That may give rise to both a substantive defence and a jurisdiction challenge as discussed in *Air Design v Deerglen* [2008] EWHC 3047 at 22–4 and *Camillin Denny v Adelaide Jones* [2009] EWHC 2110 at 30).

(*Continued*)

Table 2.1 *Continued*

Title	Citation	Issue
CRJ Services Ltd v Lanstar Ltd	[2011] EWHC 972 (TCC), Akenhead J	**Section 107: contract/no contract debate a triable issue** The court decided at the Part 24 enforcement that there was not enough evidence to decide the question of whether the contract had or had not been concluded (having been signed by an employee without actual or ostensible authority). Triable issue referred for full hearing. Decision not enforced summarily at the Part 24 hearing.
Sprunt Ltd v LB Camden	[2012] BLR 83, [2011] EWHC 3191 (TCC), Akenhead J	**Section 107(5): contract admitted in submissions** The Response submission in the adjudicaton admitted an agreement had been made. At the enforcement the court rejected the suggestion that s. 107(5) did not refer to adjudication proceedings that were subject of court proceedings but found it related to any adjudication proceeding (judgment para. 41).

(3) Implied Terms, Variations, and Other Non-written Agreements

Implied Terms

2.33 There is no Court of Appeal authority regarding the status of implied terms, which by their nature are not recorded in writing. As to first-instance consideration of the point: in *Carillion Construction Ltd v Devonport Royal Dockyard* (2002) (Key Case),[34] in a statement that was *obiter dicta*, Judge Bowsher expressed the view that an adjudicator could decide what implied terms were incorporated into a contract. Judge Havery, in *Connex v MJ Builders* (2004),[35] accepted (at paragraph 24 of the judgment) an agreed proposition that a construction contract would not be outside the operation of the Act just because it contained implied terms. The point was not in issue and his statement was also *obiter*.

2.34 There are different types of implied terms. Some are implied by statute such as the Supply of Goods and Services Act. Whether those terms are implied is a matter of law. For the purpose of s. 107 one might say the terms are recorded in writing in the statute. Similarly, implied terms can arise by operation of a principle of common law to a particular type of contract as a matter of course. For example, the contract of employment of a doctor will include an implied term that the doctor will take care for the patient's safety: *Johnstone v Bloomsbury Health Authority* (1992).[36] Whether such a term is to be implied is also a question of established law.

2.35 However, implied terms alleged to arise by necessity to achieve 'business efficacy' or to give effect to the obvious but unexpressed intent of the parties, are, it is submitted, in another class. Such terms are to be implied if it is found that the parties 'must have intended that term to form part of their contract... and are necessary to give business efficacy to the contract': *Trollope & Colls Ltd v North West Metropolitan Regional Hospital Board* (1973).[37] Consideration

[34] [2003] BLR 79 at para. 34 of the judgment.

[35] [2004] BLR 333, [2004] EWHC 1518 (TCC). The case was appealed but not on this point, see [2005] EWCA Civ 193, [2005] 1WLR 3323.

[36] [1992] QB 333.

[37] [1973] 2All ER 260.

of whether an implied term of this type would arise involves an analysis of factual matters occurring at the time and the conduct of the parties. This is precisely the uncertainty that s. 107 of the Act was designed to prevent, at least according to the Court of Appeal in *RJT Consulting Engineers*.[38] It is submitted that this latter category of implied terms could fail to meet the requirements of s. 107. This is supported by the analysis of Judge Seymour in *Galliford Try Construction Ltd v Michael Heal Associates Ltd*[39] in which he said:

> The Court of Appeal did not expressly consider what the position would be if a contract included terms which were to be implied. The focus of the concerns of the majority (Ward LJ at page 222 and Robert Walker LJ at page 223) was that Parliament had decided that it was inappropriate for an adjudicator to have to deal with finding the terms of an oral contract. It may be that the mischief which Parliament was anxious to avoid does not arise in a case in which terms fall to be implied into a contract as a matter of law, regardless of the actual intention of the parties. However, it could arise in an acute form if it were suggested that a contract, not otherwise complete, could be completed after it had been executed by the implication of terms which were said to represent the actual, but unexpressed, intention of the parties.

2.36 However in *Allen Wilson Joinery v Privetgrange Construction Ltd* (2008)[40] Akenhead J came to the opposite conclusion. The judge said that terms are implied by law, even when they arise out of factual relationships or the history between the parties. Accordingly the judge found that implied terms will not fall foul of s. 107 and will not take a written contract outside the operation of the Act. The point was subsequently repeated in *ROK Building Ltd v Bestwood Carpentry Ltd* (2011).[41] Further, the judge stated in that case that where the price of a contract had not been agreed in writing it could be determined by construing the agreement to include an entitlement to a 'reasonable sum', thus avoiding the constraints imposed by s. 107.

Oral Variations to Written Agreements

2.37 It is not only the original construction contract which must be in writing but s. 107(1) also says that any other agreement between the parties as to any matter is effective for the purpose of Part II of the Act only if in writing. This has given rise to difficulties where a written contract is subsequently varied by an oral agreement.

2.38 An oral agreement to vary a material term of a written construction contract will usually take the contract outside the operation of the Act unless it is evidenced in writing within the meaning of s. 107: *Carillion Construction Ltd v Devonport Royal Dockyard* (2002) (Key Case).[42] In that case an alleged oral agreement to alter the remuneration terms of a written agreement to a cost-reimbursable basis were not evidenced in writing and so were outside the scope of the Act. Consequently the adjudicator had no jurisdiction to consider a claim for payment founded on that alleged oral variation even though the original construction contract complied with s. 107. The judge drew a distinction between variations to the written agreement and variations made pursuant to the terms of the contract. The former had to be evidenced in writing whereas the latter have been found not to require evidence in writing: *Management Solutions & Professional Consultants Ltd v Bennett (Electrical) Services Ltd* (2006) (Key Case).[43]

---

[38] As explained by the Court of Appeal in *RJT Consulting Engineers Ltd v DM Engineering (Northern Ireland) Ltd* [2002] 1 WLR 2344, [2002] BLR 217, EWCA Civ 270 (Key Case).

[39] 99 Con LR 19, [2003] EWHC 2886.

[40] [2008] EWHC 2802 (TCC).

[41] [2011] EWHC 1409 (TCC) at para. 29.

[42] [2003] BLR 79.

[43] [2006] EWHC 1720 (TCC), discussed below at 2.40 onwards.

2.39 In the reverse situation a referring party will not be entitled to rely on a written construction contract where there was a subsequent oral agreement which materially changed its terms. In *Lead Technical Services Ltd v CMS Medical Ltd* (2007)[44] a claim for fees based on a written construction contract was challenged by the respondent to the adjudication on the grounds that there had been an oral agreement to cap fees at £20,000. The adjudicator decided that he had jurisdiction under the written contract and since the oral variation to cap fees was not evidenced in writing he did not accept it existed. The Court of Appeal held that this was a dispute which went to the jurisdiction of the adjudicator. There were three important items of evidence which supported the contention that there had been such an oral agreement. In the circumstances it was found there was a real prospect of succeeding on this issue at trial and the adjudicator's decision was not summarily enforced.[45]

Variations Pursuant to Terms of Construction Contracts

2.40 A different approach has been taken by the courts in relation to oral variations to the scope of work, instructed pursuant to the express terms of a construction contract.

2.41 Many construction contracts contain machinery which govern changes in the scope of work: how it is to be instructed, measured, valued, and ultimately paid for. In *Management Solutions & Professional Consultants Ltd v Bennett (Electrical) Services Ltd* (2006) (Key Case)[46] oral variations had been instructed pursuant to a written variations clause in a sub-subcontract that was entirely in writing. The TCC judge decided that that the oral variations did not vary the contract but were merely instructions issued under the contract with the authority of the written contractual provisions that related to the carrying out of the contract.[47] As a consequence the court found the adjudicator had jurisdiction to consider the claim based on those oral variations.

2.42 In the earlier case of *Total ME Services Ltd v ABB Building Technologies Ltd* (2002)[48] the TCC judge went further, deciding that oral variations to the scope of work under a written contract did not take the contract outside the scope of s. 107, despite there being no express variations provision in the contract. The justification for this decision was said to be s. 107(3) of the Act which provides that oral agreements made by reference to terms which are in writing constitute an agreement in writing.[49] The judgment provides little detail regarding the oral variations or how they were agreements 'otherwise than in writing by reference to terms which are in writing' (s. 107(3)). This decision is perhaps to be contrasted with the decision of Judge Bowsher later the same year in *Carillion Construction Ltd v Devonport Royal Dockyard* (2002) (Key Case) regarding the interpretation of s. 107(3).[50] It is submitted that the reliance on s. 107(3) in *Total M&E Services* was probably misplaced which must cast doubt on the correctness of the judgment.

Key Cases: Oral Variations

Carillion Construction Ltd v Devonport Royal Dockyard [2003] BLR 79

2.43 **Facts:** The parties entered a written construction contract for the refurbishment of a dockyard. The Scheme applied. The payment terms provided for payment of Actual Cost

[44] [2007] BLR 251, [2007] EWCA Civ 316.
[45] See also *Debeck Ductwork Installation Ltd v T&E Engineering Ltd*, 14 Oct. 2002 (TCC) (unreported).
[46] [2006] EWHC 1720 (TCC).
[47] At paras. 13–17 of the judgment, and see Key Cases below.
[48] [2002] EHWC 248 (TCC).
[49] At para. 34 of the judgment.
[50] [2003] BLR 79 at para. 32, discussed below. It is to be noted that this case post-dated the Court of Appeal's decision in *RJT Consulting*, whereas *Total M E Services Ltd v ABB Building Technologies Ltd* predated it.

as defined in that agreement, plus accruals and a fee. There were further terms whereby overspend above Target Cost was shared between the parties (the gainshare agreement). Carillion alleged the payment terms were orally varied to a cost-reimbursable agreement, and referred their payment dispute to adjudication. Devonport contended the oral variation was not agreed, was not an agreement in writing, nor evidenced in writing within the meaning of s. 107. Carillion contended the oral variation was referred to in correspondence which satisfied the requirements of s. 107(3) as being an agreement made 'otherwise than in writing by reference to terms which are in writing'.

2.44 **Held:** The oral variation to the written construction contract was not evidenced in writing within the meaning of s. 107(2)(c) or (3). The adjudicator did not have jurisdiction to consider the existence or effect of that alleged oral agreement simply because it follows and amends a written agreement:

> 34. It is a simple proposition, and easy to accept, that once a construction agreement in writing is before an adjudicator he has the jurisdiction to construe its express terms and to decide what, if any, terms are to be implied or incorporated by reference. But it is quite a different thing to suggest that once a construction agreement is before an adjudicator he has jurisdiction to decide on the existence of an oral agreement not evidenced in writing just because it follows and amends the written agreement. I am not considering what in the construction industry would come under the normal heading of 'Variations made pursuant to a term of the contract'. What is in issue is an alleged oral agreement that radically changed the written agreement (if it was made). I do not believe that the citations from the judgments of Judge Thornton and the Court of Appeal were intended to or did apply to the sort of agreement now under consideration.

Management Solutions & Professional Consultants Ltd v Bennett (Electrical) Services Ltd [2006] EWHC 1720 (TCC)

2.45 **Facts:** The case concerned two separate applications for summary judgment to enforce two separate decisions of two different adjudicators. Management Solutions, the claimant in the first and respondent in the second, specialized in the installation of IT and electrical systems and Bennett specialized in electrical installations. In the first application Management Solutions sought to enforce an adjudicator's decision in relation to their unpaid fee claim. The subcontract had definitely been a written contract. However the claim included sums for extra work instructed verbally. Bennett contended that the effect of the oral instructions to extend the scope of work was to take the whole contract outside s. 107 of the Act.

2.46 **Held:** Where there is a written variations clause and the extra work is ordered under that written provision, that does not vary the contract but they are instructions issued under the contract with the authority of the contractual provisions:

> 12. This case raised squarely the question as to whether the words 'made in writing' or 'evidenced in writing' extend to variations ordered orally under a contract by virtue of an express contractual term allowing the ordering of such variations. In deciding this issue, I take account of the fact that virtually all construction contracts are subject to variations. A variation includes additional or omitted work, the provision of additional details to allow for ill-defined work to be executed or the ordering of work whose scope or quantity is only provisional in the contract documents. Very often these variations are ordered orally and not subsequently fully evidenced in writing.
>
> 14. I also take account of the fact that, if Bennett's contention is correct, this sub-sub-contract was originally, once accepted, a contract in writing subject to the adjudication provisions

of the HGCRA. However, once the first variation was instructed orally, subject only to a de minimis argument, the sub-sub-contract changed its nature to become one which was neither in writing nor subject to the adjudication provisions of the Act. Although such a result is possible, it is not one which makes business sense nor is one which gives full effect to the nature and purpose of the compulsory statutory adjudication scheme provided for by the HGCRA.

15. In my judgment, Management Solutions' contentions are to be preferred. The entirety of this sub-sub-contract was in writing and took effect once Bennett's written acceptance was sent off to Management Solutions. That written sub-sub-contract allowed the scope of the work to be changed within the limits provided for by the written contractual provisions. Within those limits, the sub-sub-contract works could be varied. Such variations were not varying the contract, they were merely instructions issued under the contract and with the authority of the contractual provisions that related to the carrying out of the contract.

16. It follows that the disputed variations were undertaken under the terms of the original sub-sub-contract and were within its scope. Even if these variations were oral and not evidenced in writing, the work required by them was carried out and if it can be established that the instructions were issued by or on behalf of Bennett, the resulting work was carried out by agreement with the result that the contractual requirement that the variations should be evidenced in writing was waived by both parties.

17. The conclusion is that the adjudicator had jurisdiction to embark upon and decide the dispute referred to him that arose out of the sub-sub-contract. As it happens, he decided that Management Solutions was entitled to be remunerated for the disputed variations as well as for outstanding sums due for the original sub-sub-contract works. These decisions are not capable of being challenged in these enforcement proceedings and Management Solutions are entitled to the full sum that the adjudicator directed Bennett to pay.

Table 2.2 Table of Cases: Oral Variations and Implied Terms

Title	Citation	Issue
Total M and E Services Ltd v ABB Building Technologies Ltd	87 ConLR 154, [2002] EWHC 248 (TCC), Judge Wilcox	**Oral variation** The oral instruction of variations to the scope of work did not take the written contract outside s. 107, despite there being no variations clause, because s. 107(3) allowed oral variations to be reference to terms that are in writing (but see the discussion of this decision at 2.40–2.42 above).
Carillion Construction Ltd v Devonport Royal Dockyard	[2003] BLR 79 (TCC), Judge Bowsher	**Oral variation** Oral variation to payment terms of a written construction contract was not evidenced in writing as required by s. 107(2)(c) of the Act. Consequently adjudicator had no jurisdiction and the decision was unenforceable.
Connex v MJ Building Services Group Plc	[2004] BLR 333, [2004] EWHC 1518 (TCC), Judge Havery	**Implied terms** The judge accepted an agreed proposition that a construction contract would not be outside the operation of the Act just because it contained implied terms.
Management Solutions & Professional Consultants Ltd v Bennett (Electrical) Services Ltd	[2006] EWHC 1720 (TCC), Judge Thornton	**Oral variations pursuant to written variations clause** An adjudicator had jurisdiction to deal with oral variation instructions which were given under the authority of a written variations clause in a contract which was entirely in writing.

Title	Citation	Issue
Bennett Electrical Services v Inviron Ltd	[2007] EWHC 49 (TCC), Judge Wilcox	**Letter of intent** A letter of intent headed 'subject to contract' was not intended to create a contract and the parties did not intend the letter of intent to have contractual effect. The adjudicator had no jurisdiction. **Oral variations to agreement not recorded in writing** In any event, the letter of intent did not contain all the terms of the agreement. The letter of intent referred to a fixed price 'for the duration of the contract'. However a great deal of extra work was instructed and it was evident that the original agreement was subject to additional oral terms. There were no written terms for rates and prices or the method of assessing or the timing of payments for these extra works. These matters could not be regarded as trivial or immaterial.
Lead Technical Services Ltd v CMS Medical Ltd	[2007] BLR 251, [2007] EWCA Civ 316 (CA)	**Oral variation to written agreement** A claim for fees by a consulting engineer. The client respondent alleged (1) an agreement made December 2002 was supplanted by a valid written deed of appointment dated June 2003, and the adjudicator's appointed pursuant to the earlier agreement by the ICE was invalid; (2) the written deed of appointment was varied by an oral agreement to cap fees at £20,000, taking it outside s. 107 of the Act. The Court of Appeal decided the judge was wrong to enforce the adjudicator's decision on a summary basis: 1. On the available evidence there was a real prospect of establishing that the deed of appointment was valid and enforceable and the adjudicator should have been appointed under the TeCSA rules by TeCSA not the ICE (paras. 14–15). 2. On the available evidence there was a real prospect of success on establishing there had been an oral variation to cap fees. 3. Consequently there was an arguable jurisdiction defence and summary judgment enforcing the adjudicator's decision should not have been granted by the judge at first instance.
Euro Construction Scaffolding Ltd v SLLB Construction Ltd	[2008] EWHC 3160 (TCC), Akenhead J	**Section 107: oral variation argument rejected at Part 24 hearing** At a Part 24 enforcement hearing the court rejected the evidence of the defendant and its argument that there had been an oral variation that took the contract outside the ambit of s. 107.
ROK Building Ltd v Bestwood Carpentry Ltd	[2010] EWHC 1409 (TCC), Akenhead J	**Implied terms as to reasonable sum may not need to be recorded in writing** Whilst all the material terms needed to be in writing, it was not necessary that the price should be in writing if it could be determined by construing the contract. Accordingly an implied term for a reasonable sum could suffice. But an oral agreement of a defined price that was not recorded in writing would not satisfy s. 107 rendering the whole contract outside s. 107.

(4) Letters of Intent

Incomplete Agreements

2.47 Whether a letter of intent creates a contract within s. 107 of the Act is a question of construction of the letter itself. Consideration of this issue was undertaken in *Hart Investments Ltd v Fidler and anor* (2006) (Key Case).[51] The judge found that the letter in that case did not create a binding and enforceable contract and more importantly did not provide clarity of terms envisaged by s. 107(2) and the Court of Appeal in *RJT Consulting.*[52] In that context the judge considered that the biggest difficulty arose out of the description of the work scope. According to the letter of intent the work scope comprised work which might be ordered in the future, whether orally or in writing. Such a work scope was, according to the judge, a recipe for confusion and dispute of the very sort s. 107(2)(c) is designed to avoid.[53]

'Subject to Contract'

2.48 A letter of intent that is headed 'subject to contract' may be outside s. 107 if the words are used in their usual sense so as to indicate an intention not to create a binding contract. In *Bennett Electrical Services v Inviron Ltd* (2007)[54] it was decided that a letter of intent headed 'subject to contract' was not intended to have contractual effect. This was evidenced by the fact that the approval of a third party was required before a contract could be created and the fact that the parties had agreed remuneration on a restitutionary basis should no formal contract be agreed.[55]

Pending Execution of Contract Documents

2.49 This is to be distinguished from cases in which the parties have reached written agreement on all matters and the letter of intent requests the contractor to proceed pending the execution of the contract documentation. The mere fact that two parties propose that their agreement should be contained in a formal contract to be drawn and signed in the future does not preclude the conclusion that they have already informally contractually committed themselves on exactly the same terms: *Bryen & Langley Ltd v Martin Boston* (2005) (Key Case).[56] As the Court of Appeal observed in *Harvey Shopfitters Ltd v ADI Ltd* (2003):[57]

> The recorder was entitled to conclude, as Dyson J had done in [*Stent Foundations Ltd v. Carillion Construction (Contracts) Ltd (formerly Tarmac Construction (Contracts) Ltd)* (2000) 78 Con LR 188], that the mere fact that the letter giving instructions to proceed envisages the execution of further documentation, does not preclude the court from concluding that a binding contract was none the less entered into, provided that all the necessary ingredients of a valid contract are present. Having concluded that the parties had agreed to a fixed-sum contract under IFC 84 conditions, it is not surprising that the recorder held that the words in question, construed conjunctively, mean what they say. In other words, the only circumstance in which the appellants were to be entitled to a quantum meruit was if the contract did not proceed and was not finalised. The contract did proceed.

---

[51] [2007] BLR 30, [2006] EWHC 2857 (TCC).
[52] [2002] 1 WLR 2344, [2002] BLR 217, EWCA Civ 27 (CA).
[53] See Key Cases for a fuller citation from the judgment.
[54] [2007] EWHC 49.
[55] *British Steel Corporation v Cleveland Bridge and Engineering Company Ltd* [1984] 1 All ER 504 at 510–11.
[56] [2005] BLR 508, [2005] EWCA Civ 973 at paras. 36–7, citing *Rossiter v Miller* (1878) 3 App Cas 1124 at 115 and *Von Hatzfeldt-Widenburg v Alexander* [1912] 1 Ch. 284 at 288, 289.
[57] [2003] EWCA Civ 1757, [2004] 2 All ER 982.

2.50 In *Bryen & Langley Ltd v Martin Boston* (2005) (Key Case)[58] the parties had reached agreement on all terms and a letter of intent was issued which stated:

> 1. Further to our recent meeting, I can now confirm on behalf of our Client, Mr Martin Boston, that it is his intention to proceed with the works with your Company in accordance with your Tender and subsequent amendments as appended in the sum of £436,923 for a Contract Period of 16 weeks, possession 18th June 2001 ... 2. The Contract will be executed under the Standard Form of Building Contract 1998 Edition, Private With Quantities and, should the project not proceed, your reasonable and ascertainable costs will be recoverable from the Client but will not include any loss of profit or overhead recovery... The Contract Documents will be drawn up shortly.

It was common ground between the parties that the letter comprised an offer which was accepted by conduct by the prompt commencement of the works. The issue between the parties concerned whether the contract incorporated the JCT terms. The Court of Appeal found that the contract incorporated the JCT terms, the adjudication clause was incorporated, and hence the adjudicator's decision was valid.[59]

2.51 In *Mott MacDonald Ltd v London & Regional Properties Ltd* (2007)[60] the dispute concerned fees under a letter of intent that had expired. The adjudicator held that the letter of intent had not lapsed and that it had continued to regulate the parties' relationship. At the enforcement hearing the TCC judge found that the letter of intent had expired and that the claims did not arise solely from the agreement in writing set out in the letter of intent, but from a separate and different contract that had been the subject of substantial amendments partly evidenced in writing, partly formed by conduct, and partly to be inferred from the conduct of the parties. Thus, the construction contract that the claimant was entitled to rely on was enforceable in law but was not sufficiently in writing to fall within s. 107(2).[61]

2.52 However where a letter of intent contains all material terms for a binding contract to exist, it can be a construction contract for the purposes of s. 107: *Harris Calnan Construction Company Ltd v Ridgewood (Kensington) Ltd* (2007).[62]

Key Cases: Letters of Intent

Hart Investments Ltd v Fidler & Larchpark [2007] BLR 30, [2006] EWHC 2857 (TCC)

2.53 **Facts:** The claimant ('Hart') engaged the second defendant ('Larchpark') to carry out extensive building works to a property in Muswell Hill. Larchpark brought adjudication proceedings regarding a claim for unpaid sums, and the adjudicator's decision awarded sums to Larchpark. At the enforcement hearing Hart argued the adjudicator had no jurisdiction to decide the dispute because (1) the referral notice was served out of time, and (2) there was no contract in writing within the meaning of s. 107(2) of the Act. Larchpark alleged the contract in writing was a letter of intent in the following terms:

> We write as agent on behalf of the client, Hart Investments Limited, to advise you that it is their intention to enter into a contract with you for the structural works required to be

[58] See Footnote 56.

[59] If the JCT form had not been incorporated the contract would have been exempt from the effect of the Act by virtue of s. 106, it being with a residential occupier. The attempt to have the adjudication clause struck out as an unfair term in a consumer contract also failed; see discussion of this in Ch. 2.

[60] 113 Con LR 33, [2007] EWHC 1055 (TCC).

[61] At para. 49 of the judgment.

[62] [2008] BLR 132, [2007] EWHC 2738 (TCC).

> carried out at Queen's Lodge, 53–55 Queen's Avenue, Muswell Hill, London N.10. The contract to be entered into will be the JCT intermediate form of building contract 1998 edition, incorporating amendments 1–4 inclusive . . .
>
> Upon receipt of your acceptance of the terms set out in this letter, Larchpark Ltd are authorised to proceed with all activities to comply with the requirements of the overall programme together with any necessary placement of orders for materials, goods and services subject to the client's liability for costs arising from such activities being limited to a maximum of £20,000 or such other increased sum as is subsequently confirmed in writing by ourselves pending issue of the contract documentation. Also let us have, as a matter of expedition, your detailed method statement for the works and any statutory pre-commencement submissions that you are required to make.
>
> The terms and conditions of the proposed contract shall govern retrospectively the work carried out by you and any monies paid to you in respect of the work performed pursuant to this authorisation shall form part of the amounts due under the contract.
>
> This letter of intent will automatically terminate on 2nd December 2002 unless it is renewed by or on behalf of the client or when the building contract is duly executed by both parties. The client reserves the right to terminate the letter of intent by written notice at any time before it expires. If, for any reason, the building contract is not entered into or this letter of intent is terminated or terminates and is not renewed then the following terms will apply to the whole of the works carried out by Larchpark Ltd.
>
> 1. The client will reimburse the reasonable costs together with VAT properly and reasonably incurred in connection with the work done and orders placed under the authority of this letter subject to all such costs being verified by and recommended for payment by ourselves and subject to liability being limited to the amounts stated above or subsequently increased.
> 2. No compensation will be due in respect of the termination of this instruction. In particular, you will have no claim for breach or loss of contract, loss of profit or loss of expectation.
> 3. Larchpark Limited will promptly vacate the site with as little disruption as possible removing all plant and waste materials and leaving the site clean and tidy . . .

There was a second letter of intent which purported to extend the validity of the first letter by two weeks and remove the limit on the client's liability for expenditure. No further formal contract was finalized or executed.

2.54 **Held:** Even if the letter of intent created a binding and enforceable contract, which the judge did not think they did, the sort of clarity of terms envisaged by s. 107(2) and the Court of Appeal in *RJT* was wholly absent. There were numerous problems including uncertainty over the identity of the parties, an absence of an agreement as to time and an inadequately described scope of work. The fact that the three numbered paragraphs of the letter of intent were designed to be a fall-back position, only relevant at all if no formal/full contract was ever concluded, also militates against the submission that this was a contract in writing containing all the terms that had been agreed by the parties. The three paragraphs in the letter of intent were not designed to be a complete record of the parties agreement. The letter of intent did not comply with s. 107(2) of the Act:

> 61. However, the biggest difficulty comes with a consideration of the contract workscope. The workscope, according to the letter, is work which will, or might be, the subject of orders in the future, whether written or oral. That might be sufficient for a binding contract, although I do not think it is and, as I have indicated, enforcement of it would be next to impossible. More importantly, such a definition of workscope is a recipe for confusion and dispute of the very sort which s. 107(2)(c) is designed to avoid. This point can be

emphasised by reference to Hart's own pleading in this case. In para.3 of the particulars of claim Hart defined the contractual workscope as including:

> 'The retention and preservation of the front and side facades of the property, the removal of the main part of the building and the construction of the basement and the reconstruction of the building above the new constructed basement area.'

This workscope is plainly not discernible from the letter of 1st November. It is based on subsequent orders, instructions and the like which may, or may not, have been reduced to writing. If the contract document does not even begin to define the contract workscope it seems to me impossible to say that all the terms, or even all the material terms, are set out in writing.

Bryen & Langley Ltd v Martin Boston [2005] BLR 508, [2005] EWCA Civ 973

Facts: The claimant had successfully tendered for a contract to carry out building works for the respondent. The respondent's quantity surveyor had written to the claimant stating that the contract would be executed under the JCT Standard Form of Contract 1998, but the respondent never signed his copy of the contract. The claimant referred its claim for payment due under the last certificate to an adjudicator who found that the contract had incorporated the JCT form and so he had jurisdiction. At the enforcement hearing the judge found that the contract evidenced by the letter did not constitute a contract in the JCT form and accordingly that the adjudicator had not had jurisdiction and dismissed the application to enforce the decision by summary judgment. The claimant appealed.

Held: The letter of intent was a contract that incorporated the JCT terms and conditions:

> 29. Mr Sampson's submission was, therefore, that by the time of the letter of 12 June all the terms of the building contract had been agreed. His initial submission was that the letter amounted to an acceptance of an offer made by B & L's tender, as subsequently varied by agreement, and so created the building contract upon which B & L sued. He accepted, however, that paragraph 2 of the letter introduced a new provision which had not been the subject of prior agreement, and he revised his submission to one to the effect that the letter amounted to a contractual counter-offer which B & L accepted by its conduct in embarking upon the project. He of course recognised that the letter expressly envisaged the future signing of a formal contract in JCT Form but pointed out that there was nothing either in that, or in the parties' prior negotiations, to suggest that they were operating on a 'subject to contract' basis. If they had been, that would of course have been fatal to B & L's case. But Mr Sampson submitted that the mere fact that the parties envisage the formalisation of their agreement by way of a formal contract does not preclude the conclusion, if the facts justify it, that they are intending an immediate contractual commitment to each other on the terms later to be incorporated into that formal contract.
>
> 30. For that last proposition, Mr Sampson referred us to the decision of this court in *Harvey Shopfitters Ltd v. ADI Ltd* [2003] EWCA Civ 1757; [2004] 2 All ER 982. The facts of that case bore a striking similarity to those of the present one. The appellant builders had tendered to carry out certain works in accordance with a 1984 JCT form of contract. The employers' architects told the appellants that the tender was acceptable, the appellants started work on 6 July 1998 and on the next day the architects wrote them a letter confirming the employers' intention to enter into a contract with them at the tender price. The letter said the contract documents were being prepared and it also specified the main terms. The letter continued: 'I have been instructed by our client to request that you accept this letter as authority to proceed. If, for any unforeseen reason, the contract should

fail to proceed and be formalised, then any reasonable expenditure by you in connection with the above will be reimbursed on a quantum meruit basis. Any such payment would strictly form the limit of our client's commitment and our client would not be subject to any further payment of compensation for damages for breach of contract.'

31. The work continued but no formal IFC 84 contract was ever signed. It was common ground that there was nothing left for the parties to agree. The work proceeded to completion and the appellants obtained an adjudicator's award in for their final account. The litigation arose because on the appellants' bid to enforce the award they amended to assert for the first time a claim to a quantum meruit. The Recorder rejected the claim, holding that, by the letter of 7 July 1998, the parties had entered into a contract in IFC 84 form. This court dismissed the appeal, Latham LJ saying in paragraph 9:

> 'The recorder was entitled to conclude, as Dyson J had done in [*Stent Foundations Ltd v. Carillion Construction (Contracts) Ltd (formerly Tarmac Construction (Contracts) Ltd)* (2000) 78 Con LR 188], that the mere fact that the letter giving instructions to proceed envisages the execution of further documentation, does not preclude the court from concluding that a binding contract was none the less entered into, provided that all the necessary ingredients of a valid contract are present. Having concluded that the parties had agreed to a fixed-sum contract under IFC 84 conditions, it is not surprising that the recorder held that the words in question, construed conjunctively, mean what they say. In other words, the only circumstance in which the appellants were to be entitled to a quantum meruit was if the contract did not proceed and was not finalised. The contract did proceed.'

32. Mr Sampson said the same principle applies here. He might perhaps have added that the present case is arguably a stronger one than the *Harvey* case. In *Harvey*, the quantum meruit entitlement was to exist if 'the contract should fail to proceed and be formalised'. On one view, that could be said to mean that the quantum meruit entitlement was to exist if formal contracts were not signed, which they were not. This court appears, however, to have agreed with the recorder that the two conditions in the relevant phrase had to be construed conjunctively and that it was sufficient that the contract had proceeded, even if it had not been formalised (or finalised, as Latham LJ put it). In the present case, paragraph 2 of the letter of 12 June was in simpler form. It referred to B & L's right to recover certain costs 'should the project not proceed' but neither counsel suggested that that was a reference to a failure to sign formal contracts. Mr Bowsher, for Mr Boston, accepted that the 'project' *did* proceed, and that it did so as soon as B & L started work on it in June, with which Mr Sampson agreed. Mr Bowsher's further submission was that the costs referred to in paragraph 2 were merely costs in respect of any preparatory work carried out by B & L prior to commencement of such work. Again, I understood Mr Sampson to agree.

Table 2.3 Table of Cases: Letters of Intent

Title	Citation	Issue
Bryen & Langley Ltd v Martin Boston	[2005] BLR 508, [2005] EWCA Civ 973 (CA)	**Letter of intent** The parties had reached agreement on all terms and a letter of intent recorded that the contract documents would be executed under the Standard Form of Contract 1998, Private with Quantities, to be drawn up shortly. The statement that the formal contract documents were pending did not deprive the letter of having contractual effect, and the terms of the JCT form were incorporated into that contract.

Title	Citation	Issue
Bennett Electrical Services v Inviron Ltd	[2007] EWHC 49 (TCC), Judge Wilcox	**Letter of intent** A letter of intent headed 'subject to contract' was not intended to create a contract and the parties did not intend the letter of intent to have contractual effect. The adjudicator had no jurisdiction.
Hart Investments Ltd v Fidler and ors	[2007] BLR 30, [2007] EWHC 2857 (TCC), Coulson J	**Letter of intent** Decision not enforced because referral notice was served outside the seven days stipulated by the Scheme. Alternatively the letter of intent was not a binding and enforceable contract, alternatively, if it was, the clarity of terms envisaged by s. 107(2) was not present.
Mott Macdonald Ltd v London & Regional Properties Ltd	113 Con LR 33, [2007] EWHC 1055 (TCC), Judge Thornton	**All terms not in writing** The letter of intent had expired no later than October 2000. Mott's claims did not arise solely from the agreement in writing set out in the letter of intent, but from a separate and different contract that had been the subject of substantial amendments that were partly evidenced in writing, partly formed by conduct, and partly to be inferred from the conduct of the parties. Many of the amendments must have been in writing or evidenced in writing, but Mott had not relied on them or adduced them in evidence. Thus, the construction contract that Mott was entitled to rely on was enforceable in law but was not sufficiently in writing to fall within s. 107(2).
Harris Calnan Construction Company Ltd v Ridgewood (Kensington) Ltd	[2008] BLR 132, [2007] EWHC 2738 (TCC), Coulson J	**Letter of intent** Where all that needed to be agreed had been agreed and committed to writing it did not matter that there had not yet been a formal execution of documents. The adjudicator was right and there was a contract in writing (at paras. 11–12 of the judgment).

3

SECTION 108 AND THE RIGHT TO ADJUDICATE

(1) Introduction

3.01 Section 108 of the Housing Grants Construction and Regeneration Act 1996 ('the 1996 Act') contains the right to adjudicate. It sets out the minimum requirements that a contractual adjudication scheme needs to contain in a construction contract governed by the 1996 Act.

3.02 This chapter considers the provisions of s. 108 concerning what disputes may be adjudicated and when. Certain aspects of s. 108 are dealt with in more detail elsewhere in this book and so are touched on only briefly below, in particular; the duty to act impartially (s. 108(2)(e)) is dealt with in Chapters 10 and 11, the adjudicator's power to take the initiative in ascertaining the facts and the law (s. 108(2)(f)) is also considered in Chapters 4 and 10. The binding nature of the adjudicator's decision (s. 108(3)) is the subject of Chapter 6.

3.03 The focus of this chapter is the statutory right to adjudicate. It does not discuss adjudication schemes in contracts not governed by the 1996 Act. In practice many construction contracts contain more detailed adjudication procedures than the Act requires. A party wishing to adjudicate a dispute under a construction contract must first consider whether any contractual adjudication scheme complies with the Act. If so, it is a valid adjudication procedure and any adjudication will be governed by the rules contained in that contractual adjudication scheme. Any supplementary provisions in that adjudication scheme will still be valid in principle, as long as they do not conflict with the statutory requirements.

3.04 Naturally there are contracts which are outside the ambit of the Act but that nevertheless contain an adjudication scheme. A contract which is not a construction contract within the meaning of Part II of the Act is not required to have an adjudication procedure. Nor does a contract with a residential occupier need an adjudication procedure. If such contracts contain an adjudication scheme, it does not have to comply with the statutory requirements. In such a situation, the absence of a matter that the 1996 Act considers to be fundamental will not usually render that contractual adjudication scheme invalid. Nevertheless many of the legal doctrines developed in relation to statutory adjudications are applicable to these pure contractual adjudications. Chapter 5 contains a discussion of whether the principles applicable to adjudication under the 1996 Act apply also to adjudication agreements in other contracts.

Changes Made by the 2009 Act

3.05 One of the main changes made to the 1996 Act by the Local Democracy, Economic Development and Construction Act 2009 ('the 2009 Act') is the deletion of s. 107 and the requirement that construction contracts be in writing (as discussed in Chapter 2). At the same time s.108 was amended so that the adjudication provisions required by s. 108(2)–(4) must now be in writing,[1] and this means that in contracts entered into after October 2011 it is only the adjudication procedure that must be in writing.[2] Failure to properly express any of the adjudication procedures in writing will result in the Scheme for Construction Contracts ('the Scheme') becoming applicable.[3]

3.06 The only other change to s. 108 is the introduction at new s. 108(3A) of the requirement that the construction contract must include, in writing, a provision permitting the adjudicator to correct his decision 'so as to remove a clerical or typographical error arising by accident or omission'. The power to correct a decision, both before and after the 2009 Act, is discussed in Chapter 4.

(2) Issues Arising Out of s. 108(1)

3.07 Section 108(1) provides that a party to a construction contract (as defined by ss. 104 and 105) has the right to refer a dispute arising under the contract for adjudication under a procedure complying with s. 108.[4]

3.08 The meaning of s. 108(1) has been tested in numerous cases where the enforcement of the adjudicator's decision has been challenged. The most common issues arising out of s. 108(1) are dealt with in the sections that follow and in summary are:

(1) whether there is a dispute 'arising under the contract' as required by s. 108(1);
(2) whether the matter referred to the adjudicator had crystallized into 'a dispute';
(3) whether s. 108(1) requires a single dispute only to be referred;
(4) whether the scope of the dispute referred was different from that which had crystallized or changed during the adjudication and the effect of that.

3.09 These four questions are considered below. These questions are matters that may go to the jurisdiction of an appointed adjudicator, because, in a statutory adjudication, an adjudicator

---

[1] By s. 139 of the 2009 Act—this rule therefore applies to construction contracts made after the 2009 Act came into effect.

[2] See Ch. 1 at 1.08.

[3] Pursuant to s. 108(5).

[4] s. 108(1). The term 'construction contract' is used to mean a contract within the ambit of Part II of the Act.

is appointed to decide 'a dispute' that 'arises under the contract'. If the issue referred to him is not 'a dispute' or does not arise under the contract, then it is not a matter that the statute empowers him to decide.[5] Accordingly, s. 108(1) must be satisfied before a party may make use of the right to adjudicate contained within it.

Disputes Arising under the Contract

3.10 It goes without saying that the statutory right to adjudicate only arises in respect of construction contracts within the meaning of Part II of the Act. Therefore there must be a qualifying construction contract in existence which satisfies ss. 104–7[6] in order for a valid adjudication to be started. These provisions are the subject matter of Chapters 1 and 2.

3.11 Section 108(1) provides that the right to adjudicate concerns disputes 'arising under' the contract. Much of the attention given to this section has focused on what constitutes a 'dispute' but it is equally important for a referring party to ensure that the claim it wishes to adjudicate does 'arise under' the relevant construction contract.

3.12 *Ashville Investments v Elmer Contractors Ltd* (1987) [7] is authority for the proposition that an arbitration clause which covers disputes arising under the contract but also includes the words 'in connection with' should be given a wide interpretation and will cover related claims for rectification, negligent misstatement, and the like. When the Court of Appeal applied that decision in the later case of *Fillite (Runcorn) Ltd v Aqua-Lift* (1989),[8] they concluded that an arbitration clause which encompassed all disputes 'under' the contract (but did not contain the additional words 'in connection with'), embraced claims for breach of contact but, in the words of Slade LJ, was 'not wide enough to include disputes that do not concern obligations created by or incorporated in that contract'.

3.13 In relation to arbitration clauses, in *Fiona Trust & Holding Company and ors v Yuri Privalov* (2007)[9] the House of Lords said it was to be assumed that the parties, as rational businessmen, were likely to have intended any dispute arising out of the contract, including disputes over the validity of the agreement itself, were to be decided by the arbitrator, unless express words excluded certain disputes from the arbitrator's jurisdiction. However, the 2009 Act has not amended s. 108(1) to include disputes arising 'in connection' with the contract and it is submitted the ambit of s. 108(1) remains limited to disputes under the contract. In *Camillin Denny Architects Ltd v Adelaide Jones & Company Ltd* (2009)[10] Akenhead J expressed uncertainty as to whether *Fiona Trust* was authority for any proposition about an adjudicator's jurisdiction.

3.14 Whether a particular type of dispute 'arises under' a contract is discussed below. Other than this and the requirement that a single dispute is referred (see 3.31 below) s. 108 contains no limit on the nature, scope or extent of the disputes that can be referred to adjudication under a construction contract; *Banner Holdings Ltd v Colchester Borough Council* (2010).[11] In that case the TCC judge said that the contract could not validly prevent a dispute about the validity of a contractual determination from being a matter that could be referred to adjudication.[12]

[5] It may, however, be a matter that a particular contractual adjudication agreement empowers an adjudicator to decide where the contract has a valid adjudication agreement that is wider than that required by s. 108. For a more detailed discussion about the adjudicator's jurisdiction and what the adjudicator is empowered to decide see Chapters 4 and 9.

[6] Save for in respect of contracts governed by the 2009 Act to which s. 107 will not apply.

[7] [1987] 37 BLR 55.

[8] [1989] 45 BLR 27.

[9] (2007) 4 All ER 951(HL).

[10] [2009] EWHC 2110 (TCC) and [2009] BLR 606 at para. 30

[11] [2010] EWHC 139 (TCC), 131 Con LR 77 at para. 39.

[12] As to the validity of clauses that try to prevent issues or disputes from being referable to adjudication, see 3.72 below. As to whether a dispute can be too complex for adjudication see the discussion in Ch. 10 at 10.17.

Side Agreements

3.15 Whether an adjudication agreement will be wide enough to cover disputes about side agreements will depend on its wording: *Brown & Sons v Crosby North West Homes* (2005).[13] The agreement in this case had been amended to refer to disputes arising under, out of, or in connection with the contract and was found to be wide enough to cover terms in side agreements.

Dispute under More than One Contract

3.16 If the dispute arises under more than one contract, then it may be that part of the dispute does not arise under the contract and cannot be referred to adjudication. If it is two disputes they cannot be referred to the same single adjudication without the consent of the parties (see 3.49 below) or unless this is permitted by the adjudication procedure. However in the cases in which this question has arisen the courts have usually avoided that result.

3.17 In *Air Design (Kent) Ltd v Deerglen (Jersey) Ltd* (2008)[14] there was an original contract and a number of supplementary agreements. The adjudicator decided the subsequent agreements were variations of the original contract and that he had jurisdiction under the original contract adjudication agreement to consider disputes arising under that contract or variations to it. At the enforcement hearing Akenhead J decided that the adjudicator had been empowered to decide that issue (because it was a question of fact and/or law) and so the decision was enforceable whether right or wrong.

3.18 In *Amec Group v Thames Water Utilities Ltd* (2010)[15] the written adjudication agreement was contained in a framework agreement and applied to disputes 'arising under or in connection with' that framework agreement. Under the framework agreement work was instructed as separate works packages forming separate contracts. The TCC judge decided that the dispute in question arose under the framework agreement but that in any event the wording of the adjudication clause was wide enough to cover a dispute arising under the individual works orders.

3.19 In *Supablast (Nationwide) Ltd v Story Rail Ltd* (2010)[16] the TCC judge rejected the challenge to enforcement on the ground that the dispute arose under two separate contracts, finding that the adjudicator had had jurisdiction to decide that the steelworks were instructed as a variation to the subcontract.

Identity of Contracting Party

3.20 In *Michael John Construction v Richard Henry Golledge and ors* (2006)[17] the defendant argued that a dispute concerning the correct identity of the employer, which required consideration of documents such as the club constitution, was a dispute 'in connection with' but not 'under' the construction contract. The judge decided that the issue, as to the individuals who were liable (on behalf of the employer) to make proper payment to the claimant *under* the contract, was part and parcel of the single dispute referred to the adjudicator. In this case the employer under the contract was named as the Club. The claimant was simply seeking to be paid by the Club under the contract by reference to the Trustees at the time that the contract was made. The judge found that this was plainly a dispute concerned with the obligations owed under the contract by the employer to the contractor and that accordingly the adjudicator was empowered to decide this question.[18]

---

[13] [2009] EWHC 3503 (TCC).

[14] [2008] EWHC 3047 (TCC).

[15] [2010] EWHC 419 (TCC).

[16] [2010] EWHC 56 (TCC), [2010] BLR 211.

[17] [2006] EWHC 71 (TCC).

[18] See Ch. 9 at 9.19 (and following) for further discussion about the jurisdictions implications when the adjudication is brought by or against the wrong party.

Contract Formation

3.21 A dispute about the formation of the contract itself is not a dispute arising under the contract. It is a dispute about whether a relevant contract came into existence for the purpose of the 1996 Act or s. 107. This is discussed in more detail in Chapter 2.

Incorporation of Terms

3.22 A question about whether a particular term has or has not been incorporated into an agreement is a matter which the adjudicator[19] has power to decide:

> Thus so long as it is established or agreed that there is a contract in existence between the parties, that it is a construction contract... any other dispute as to the terms of the construction contract is as much a dispute arising under the contract as would be a dispute as to the working through of any terms to the valuation machinery.[20]

The same decision was reached in *Dalkia Energy & Technical Services Ltd v Bell Group Ltd* (2009).[21] This is consistent with the rule that an adjudicator does have the power to interpret or construe the contract. In *C&B Scene Concept Design Ltd v Isobars Ltd* (2001)[22] the Court of Appeal decided that, even if the adjudicator had been wrong in his decision as to which of two competing sets of terms were incorporated into the contract, it was a decision he was empowered to make and the decision was binding. If however the incorporation argument is really a question of an oral agreement of a term of the contract which has not been evidenced by writing, that is a s. 107 matter which goes to jurisdiction.

Rescission

3.23 In *Barr Ltd v Law Mining Ltd* (2001)[23] the responding party argued that the contract had been rescinded and the claim for payment for work done after that rescission could not be determined by the adjudicator as it did not arise under the contract. The adjudicator declined to answer the rescission issue but awarded sums for the relevant work. At the enforcement hearing Lord Macfadyen held that the adjudicator would only have had jurisdiction if he had first decided that there had been no rescission. The court was of the view that liabilities incurred after a contract has been rescinded are not within the jurisdiction of the adjudicator.

Duress

3.24 If the contract itself, such as a settlement agreement, was obtained as the result of economic duress by one party, it is a voidable agreement. In *Capital Structures v Time & Tide Construction Ltd* (2006)[24] Judge Wilcox decided that it was arguable that a settlement agreement had been avoided before the adjudicator assumed jurisdiction. If correct this would have meant there was no contract at all. As that was a dispute that went to the formation of the contract it was not one the adjudicator could make a binding decision about. Summary enforcement of the decision was refused.

Settlement Agreement

3.25 A dispute about the validity of a settlement agreement is not a dispute under the construction contract itself: *Shepherd Construction v Mecright Ltd* (2000).[25] However a claim that the

[19] Assuming he otherwise has jurisdiction.

[20] *R. G. Carter Ltd v E. Nuttall Ltd*, 21 June 2000 (unreported), approved in *RJT Consulting Engineers v DM Engineering* [2002] 5 BLR 217. See also *Butler Contractors Ltd v Merewood Homes Ltd* [2002] 18 Cont LJ 74 where the court upheld the adjudicator's decision that the contract was not one of less than 45 days (thus bringing it within s. 109 of the Act), as being a decision made within jurisdiction.

[21] 122 Con LR 66, [2009] EWHC 73 (TCC).

[22] [2002] BLR 93, [2002] EWCA Civ 46.

[23] [2001] Scot CS 152, 80 Con LR 134.

[24] [2006] BLR 226, [2006] EWHC 591 (TCC).

[25] [2000] BLR 489.

dispute has been compromised is an issue that goes to the the jurisdiction of the adjudicator and may be a ground for challenging enforcement.[26]

Fraud

3.26 In *S. G. South Ltd v Kings Head Cirencester LLP and anor* (2009)[27] a developer challenged the summary enforcement of adjudicator's decisions on the ground, inter alia, that there had been fraud on the part of the contractor. The allegations had been raised before the adjudicator who had decided that fraud was not an issue within his jurisdiction because it was a matter for the police and the courts. The TCC judge stated *obiter* that there is nothing in the 1996 Act to suggest that an adjudicator does not have jurisdiction to decide issues of fraud which arise under a contract. However the judge suggested that a claim in the tort of deceit would probably not arise under contract, but thought that a claim for fraudulent misrepresentation might do. This was confirmed by the Court of Appeal in *Speymill Contracts Ltd v Baskind* (2010),[28] which confirmed the approach taken in the two previous TCC decisions.[29] For a discussion of when a party may rely on fraud as a defence to enforcement of an adjudicator's decision, see Chapter 7 at 7.09.

Key Case: 'Arising Under'

Shepherd Construction v Mecright Ltd [2000] BLR 489 (TCC)

3.27 **Facts:** Mecright referred a dispute about the failure on the part of Shepherd properly to value and pay for works carried out. In fact, the value of those works had been the subject of a settlement agreement between the parties. Mecright had agreed to accept the sum of £366,600 in full and final settlement of all its claims. Shepherd took the point before the adjudicator and Mecright then alleged for the first time that the settlement had been procured as a result of economic duress, so as to be liable to be set aside.

3.28 **Held:** The settlement may be voidable if Mecright had entered into the agreement under economic duress. However that dispute as to whether the settlement agreement was binding was not 'a dispute or difference arising under this subcontract' and therefore could not be referred to the adjudicator pursuant to clause 38A.1 of DOM/1. Nor was the settlement agreement itself a construction contract under the Act covered by s. 108. Thus the settlement agreement was one which, but for the plea of economic duress, had the effect of extinguishing the payment disputes which existed. As the settlement was effective before the adjudicator, because he could not decide the duress point, there was no payment dispute for him to decide:

> in my judgment a dispute about a settlement agreement of this kind could not be a dispute under the subcontract since the effect of the settlement agreement is one which replaces the original agreement to the extent to which it applies. Here the agreement has the effect of replacing Shepherd's obligations to value and to pay Mecright under the subcontract the value of the work ... the effect of the settlement agreement is that a dispute about it is outside Section 108, since a settlement agreement is not a construction contract within the meaning of Section 108 ... A dispute about an agreement which settles a dispute or disputes under a contract is not a dispute under that contract. The word 'under' in the Act was plainly chosen deliberately. It has been followed in this subcontract. It is not, nor is it accompanied by words such as 'in connection with' or 'arising out of' which have a well-established wider reach.

[26] *GPS Marine Contractors Ltd v Ringway Infrastructure Ltd* [2010] BLR 377, [2010] EWHC 283 (TCC).
[27] [2010] BLR 47, [2009] EWHC 2645 (TCC) at paras. 19–21 of the judgment.
[28] [2010] BLR 257, [2010] EWCA Civ 120.
[29] *S. G. South Ltd v King's Head Cirencester LLP* [2010] BLR 47, [2009] EWHC 2645 (TCC); *GPS Marine Contractors Ltd v Ringway Infrastructure Services Ltd* [2010] BLR 377, [2010] EWHC 283 (TCC).

Table 3.1 Table of Cases: 'Arising Under'

Title	Citation	Issue
Ashville Investments v Elmer Contractors Ltd	[1987] 37 BLR 55 (CA)	**Non-adjudication case** An arbitration clause which covers disputes 'arising under or in connection with the contract' should be given a wide interpretation to include claims for misrepresentation or rectification due to mistake.
Fillite (Runcorn) Ltd v Aqua Lift	[1989] 45 BLR 27 (CA)	**Non-adjudication case** An arbitration clause which encompassed disputes arising under but not in connection with a contract was wide enough to cover claims for breach of contract but did not embrace claims that did not arise out of the obligations created by or incorporated into the contract.
R. G. Carter v Edmund Nuttall Ltd	21 June 2000 (TCC) (unreported)	**Dispute about incorporation of terms arises under contract** As long as the existence of the contract is proven, a dispute as to the term of the contract (such as whether the term is incorporated) is a dispute arising under the contract.
Shepherd Construction Ltd v Mecright	[2000] BLR 489 (TCC), Judge LLoyd	**Dispute about settlement agreement does not arise under contract itself** A dispute about a stand-alone settlement agreement, which settled payment issues under a contract, does not arise under the contract itself.
Barr Ltd v Law Mining Ltd	[2001] Scot CS 152, 80 Con LR 134 (CSOH), Lord Macfadyen	**Dispute about liabilities arising after contract ended do not arise under contract** Liabilities incurred after a contract has been rescinded do not arise under the contract.
Ballast plc v the Burrell Company	[2001] BLR 529 (CSOH), Lord Reid	**Under the contract does not mean variations must be ignored** The adjudicator appeared to have been of the view that he was only empowered to order payment under 'the contract' and that that expression had to mean the standard JCT contract was unaffected by any departure from it. That approach was wrong in law and the adjudicator's error had led him to decline jurisdiction where he should not have done so.
C&B Scene Concept Design v Isobars	[2002] BLR 93, [2002] EWCA Civ 46 (CA)	**Dispute about incorporation of terms arises under the contract** Dispute about what terms are incorporated in a contract arises under the contract.The adjudicator had to decide a question of law as to whether the contract payment terms or the statutory Scheme applied. That was a question of law within his jurisdiction. These points did not affect his jurisdiction.
Brown & Sons v Crosby North West Homes	[2005] EWHC 3503 (TCC), Ramsey J	**Agreement to refer disputes arising under or in connection with contract** Agreement to refer disputes 'arising out of, under or in connection with the contract' is wide enough to cover disputes under side agreements.
Capital Structures v Time & Tide Construction Ltd	[2006] BLR 226 [2006] EWHC 591 (TCC), Judge Wilcox	**Dispute about whether contract void for duress** If it is arguable that the contract was reached by economic duress, and so was voidable, and has been avoided, that is not a dispute arising under the contract which the adjudicator can make a binding decision about. It is a dispute about the formation/validity of the contract itself.

Title	Citation	Issue
Michael John Construction v Richard Henry Golledge and ors	[2006] EWHC 71 (TCC), Coulson J	**Identity of contracting party** The identity of the contracting party was part of a larger dispute about what payment was due, which of itself could be referred to adjudication. The identity of the adjudicator could therefore be decided by the adjudicator (but see Ch. 9 at 9.19 and following for a fuller discussion about whether the adjudicator has jurisdiction to decide such a dispute).
Air Design (Kent) Ltd v Deerglen (Jersey) Ltd	[2008] EWHC 3047 (TCC), Akenhead J	**Dispute under main agreement about series of separate agreements** Where a dispute arose under a series of agreements and the adjudicator decided there was one main contract and a series of variations to it, the judge found the adjudicator had jurisdiction to decide this issue (even though it went to jurisdiction) because it was also a matter he needed to decide as part of the substantive dispute. Accordingly, having been given jurisdiction to decide the point as part of the substantive dispute, the corollary to that was he had jurisdiction to decide it for the purpose of the jurisdiction argument.
Dalkia Energy & Technical Services Ltd v Bell Group Ltd	[2009] EWHC 73 (TCC), Coulson J	**Dispute about incorporation of terms did arise under the contract** Dispute about whether standard terms were incorporated into a contract or not was a matter which the adjudicator would have power to decide normally (i.e. it was a matter arising under the contract).
S. G. South Ltd v Kings Head Cirencester LLP and anor	[2009] EWHC 2645 (TCC), [2010] BLR 47 Akenhead J	**Fraud** The judge said that allegations of fraud relating to contract claims under the contract could be decided by an adjudicator, and (*obiter*) that fraudulent misrepresentation claims might also be within the jurisdiction but claims in the tort of deceit probably could not be. See also *Speymill Contract Ltd v Baskind*. The adjudicator's decision was binding and only fraud arising after the decision could therefore be raised as a challenge to enforcement.
Speymill Contract Ltd v Baskind	[2010] EWCA Civ 120, [2010] BLR 257 (CA)	**Theft** The dispute referred to the adjudicator included an issue about whether withholding notices had been served. The responding party said he had served them but the employees of the referring party had stolen the files with them inside. The challenge based on this alleged theft was unsuccessful. The Court of Appeal affirmed the decision in *South* and in so doing effectively said the adjudicator was entitled to rule on the theft allegation in a binding manner.
Amec Group Ltd v Thames Water Ltd	[2010] EWHC 419 (TCC), Coulson J	**Agreement to refer disputes under or in connection with framework agreement** A framework agreement with a clause permitting adjudication of disputes 'arising under or in connection with the agreement' was wide enough to cover a dispute about work under an individual works order. The TCC judge also decided the dispute under the work order was a dispute under the framework agreement.

(*Continued*)

Table 3.1 *Continued*

Title	Citation	Issue
Supablast (Nationwide) Ltd v Storyrail Ltd	[2010] EWHC 56, (TCC), [2010] BLR 211, Akenhead J	**Allegation that work arose out of separate agreements on same project** Challenge at enforcement on the basis that dispute arose under separate contracts for different aspect of works on the same project. The judge rejected the argument as artificial and decided that as a matter of fact the works arose under one subcontract.
GPS Marine Contractors Ltd v Ringway Infrastructure Ltd	[2010] BLR 377, [2010] EWHC 283 (TCC), Ramsey J	**Allegation that dispute compromised was a triable issue** An allegation that a dispute had been compromised was an issue that went to the jurisdiction of the adjudicator and accordingly he could not make a binding decision about it (unless empowered to do so by the parties). In this case the allegation that there had been a compromise turned on oral evidence and raised a triable issue and therefore summary enforcement was refused.

What Constitutes a 'Dispute' for s. 108(1)?

3.29 Section 108(1) provides that a party to a construction contract has the right to refer a dispute arising under the contract for adjudication under a procedure complying with s. 108. For this purpose s. 108(1) tells us that a 'dispute' includes 'any difference'. It is thus a statutory precondition to the right to adjudicate that there is in existence a dispute capable of being referred. It is significant that the Act does not say that the contracting parties may refer any claim or issue to adjudication. It must be a claim which has become a dispute.

3.30 There have been a large number of cases on the subject of when a claim or issue becomes a dispute, as can be seen from Table 3.2 at the end of this section. However, since the Court of Appeal has given guidance on the subject in *Amec Civil Engineering Ltd v The Secretary of State for Transport* (2005) (Key Case)[30] and *Collins (Contractors) Ltd v Baltic Quay Management (1994) Ltd* (2004),[31] the early cases are relevant as historical background to what is now the settled position. The following paragraphs deal briefly with that background.

The Early Cases: The High-threshold Test

3.31 After the 1996 Act was passed, there were not infrequent challenges to the enforcement of adjudication decisions on the ground that the matters referred to the adjudicator had not yet crystallized into a 'dispute' that could be referred, and so (it was argued) the decision was invalid, the condition precedent of s. 108(1) not having been fulfilled. The 'no dispute' arguments followed several typical lines which included (1) respondents claiming they had not had sufficient time to consider the issue before the notice of intention to refer was served; (2) respondents arguing they had not been given sufficient information about the issue to understand it before the notice of intention to refer was served; (3) respondents arguing that the issue referred to adjudication differed in material ways to that which had been previously discussed by the parties.

3.32 The view was expressed in the early cases that a dispute arose only after the claim had been notified to the opposing party who had been given the opportunity of 'considering and

[30] [2005] EWCA Civ 291.

[31] [2004] EWCA Civ 1757 CA.

admitting, modifying or rejecting the claim or assertion': *Fastrack Contractors Ltd v Morrison Construction Ltd* (1999).[32] In *Sindall Ltd v Abner Solland and ors* (2001)[33] Judge LLoyd considered that for there to be 'a dispute' for the purpose of exercising the statutory right to adjudication it must be clear that 'a point has emerged from the process of discussion or negotiation has ended and that there is something that needs to be decided'.[34] The high-water mark of that trend is found in *Edmund Nuttall Ltd v R. G. Carter Ltd* (2002)[35] and a later decision of the same judge in *Hitec Power Protection BV v MCI Worldcom Ltd* (2002).[36] In those cases Judge Seymour expressed his view that, for the purposes of adjudication, a dispute crystallized when both sides had been given sufficient opportunity to consider the arguments of the other, and also that the dispute which crystallized included 'the whole package of arguments advanced and facts relied on by each side'. The rationale behind this early line of authority was to prevent respondents from being ambushed in adjudication by large claims of which they had little notice such that they were disadvantaged in preparing their defence. Unfortunately what also happened was that respondents sought to take advantage of the prevailing trend by cynically trying to prevent disputes from crystallizing. This delayed adjudications and frustrated the purpose of the Act.

The Adoption of the Halki *Low-threshold Test*

3.33 In the Birmingham TCC however, Judge Kirkham decided in November 2002[37] that there was no special meaning to the word 'dispute' in the context of adjudication, and that the Court of Appeal decision in *Halki Shipping Corporation v Sopex Oils Ltd* (1998)[38] was binding in the context of adjudication. The Court of Appeal in *Halki*, dealing with an arbitration agreement, had set the threshold standard required to establish there was a dispute at a relatively low level: 'There is a dispute once money is claimed unless and until the defendants admit that the sum is due and payable.'[39] Putting it another way, if letters written by the claimant making some request or demand are not responded to by the defendant, then there is a dispute. The majority of the Court of Appeal also decided that a dispute exists whether or not the issue is disputable as a matter of fact or law. Thus a denial of liability by the respondent, no matter that the denial is based on a demonstrably incorrect interpretation of the law, is still a dispute. Subsequently in *Beck Peppiat v Norwest Holst Construction Ltd* (2003)[40] Forbes J also held that the decision in *Halki* was binding in the context of adjudication.[41] Following *Cowlin* and *Beck Peppiat* the High Court cases tended to adopt the lower and more conventional threshold test for identifying a dispute capable of being referred to adjudication.[42]

The Court of Appeal Guidance: The Middle Ground

3.34 In *Amec Civil Engineering Ltd v The Secretary of State for Transport* (2004) (Key Case) Jackson J attempted to distil the effect of the rapidly growing jungle of decisions on the meaning of 'a

[32] [2000] BLR 168 at para. 27.

[33] [2001] TCLR 712.

[34] At para. 15.

[35] [2002] BLR 312 (TCC) at para. 36 of the judgment. This decision has subsequently not been followed or has been distinguished, see *AWG Construction Services Ltd v Rockingham Motor Speedway Ltd* [2004] EWHC 888, [2004] CILL 2154 at para. 146 of the judgment, and *William Verry (Glazing Systems) Ltd v Furlong Homes Ltd* [2005] EWHC 138 at para. 42 of the judgment.

[36] [2002] EWHC 1953 (TCC).

[37] *Cowlin Construction v CFW Architects* (2003) CILL 1961.

[38] [1998] 1 Lloyd's Rep 49 (CA).

[39] At p. 761F.

[40] [2003] BLR 316, (TCC).

[41] This decision represented something of a compromise in that it also approved *Sindall v Solland* above, subsequently doubted by the Court of Appeal in *Collins.*

[42] See for example *Orange EBS Ltd v ABB Ltd* [2003] BLR 323.

dispute'. That was a case concerning whether a dispute existed within the meaning of clause 66 of the ICE conditions that could be referred to the engineer for decision, as required by that clause, and then subsequently referred to arbitration. The judge conducted a review of the principal authorities, including those arising in the context of arbitration and adjudication, from which he derived seven propositions. Those are set out in full in the Key Cases at the end of this section. In essence the judge decided that the mere fact that one party notifies the other party of a claim does not automatically and immediately give rise to a dispute. A dispute does not arise unless and until it emerges that the claim is not admitted. The circumstances from which it may emerge that a claim is not admitted are changeable, depending on the circumstances of the case. The circumstances may involve express rejection, implied rejection after discussions, or implied rejection after a period of silence or prevarication.

3.35 Jackson J's seven propositions were broadly accepted by the Court of Appeal in *Collins (Contractors) Ltd v Baltic Quay Management (1994) Ltd* (2004).[43] Subsequently the Court of Appeal were asked to consider an appeal from Jackson J's judgment in *Amec* and once again his seven propositions were endorsed by the members of the court.[44] May and Rix LJJ each added their own additional observations on the issue which are set out verbatim in the Key Case extracts below. In so far as the existence of a dispute involves affording a party a reasonable time to respond to a claim, what may constitute a reasonable time depends on the facts of the case and the relevant contractual structure.[45] On the facts in *Amec*, May LJ agreed with Jackson J that the short deadline for response to the claim imposed by the Secretary of State was reasonable in the circumstances, not least because limitation was in danger of expiring imminently.[46] In his additional observations, Rix LJ resurrected the idea that, in the context of adjudication, the reasonable time to respond may be influenced by the need to avoid prematurely plunging parties into expensive adjudications before they are ready:

> 68. Thirdly, and significantly, the problem over 'dispute' has only really arisen in recent years in the context of adjudication for the purposes of Part II of the Housing Grants Construction and Regeneration Act 1996. Jackson J referred below to some of the burgeoning jurisprudence to which the need for a 'dispute' in order to trigger adjudication has given rise. In this new context, where adjudication is an *additional* provisional layer of dispute resolution, pending final litigation or arbitration, there is, as it seems to me, a legitimate concern to ensure that the point at which this additional complexity has been properly reached should not be too readily anticipated. Unlike the arbitration context, adjudication is likely to occur at an early stage, when in any event there is no limitation problem, but there is the different concern that parties may be plunged into an expensive contest, the timing provisions of which are tightly drawn, before they, and particularly the respondent, are ready for it. In this context there has been an understandable concern that the respondent should have a reasonable time in which to respond to any claim.

3.36 The cases decided after *Amec* and *Collins*, as set out in Table 3.2 at the end of this section, have followed the guidance given by the Court of Appeal in those decisions and the meaning of 'dispute' in s. 108 of the Act now seems to be reasonably well settled.

3.37 An example of the robust approach that the courts take in relation to this type of challenge is illustrated by *All In One Building & Refurbishments v Makers UK Ltd* (2006).[47] Prior to the adjudication, the claimant did no more than assert its claim for loss of overheads and

---

[43] [2004] EWCA Civ 1757.

[44] *Amec Civil Engineering Ltd v The Secretary of State for Transport* [2005] 1 WLR 2339 (CA) (Key Case).

[45] *Ibid.*, per May LJ at para. 31(5).

[46] The facts of the case are set out in the Key Case section below.

[47] [2005] EWHC 2943 (TCC), Apr. 2006 CILL 2321.

profits against the defendant and then particularized the claim in the adjudication itself. The defendant argued that that there was no dispute in relation to this element of the claim when the adjudication was commenced. Judge Wilcox rejected this argument, stating that a common-sense approach needed to be adopted and 'there is no warrant for being legalistic and overly technical'. Similarly in *Bovis Lend Lease v The Trustees of the London Clinic* (2009)[48] Akenhead J said it is important to distinguish between the dispute that has crystallized and the evidence provided to support or contest it, which may legitimately change as the dispute develops.

3.38 However, it remains a matter of fact as to whether a dispute has crystallized and this also applies to whether an additional claim can be added onto an existing dispute or whether it needs to crystallize. In *Beck Interiors Ltd v UK Flooring Contractors Ltd* (2012)[49] Beck claimed damages for repudiation against a subcontractor. In a letter served after close of business on the day before Good Friday they added a claim for delay and liquidated damages. They served an adjudication notice on the first working day after Easter. UK Flooring asserted there was no jurisdiction in relation to the liquidated damages claim as that dispute had not crystallized. The TCC judge agreed and accordingly the liquidated damages claim was severed from the rest of the adjudication decision, and not enforced.[50]

Key Cases: Meaning of 'Dispute'

Amec Civil Engineering Ltd v The Sec of State For Transport [2004] EWHC 2339 (TCC)

3.39 **Facts:** Amec were the main contractor employed by the Secretary of State to renovate a viaduct. The contract incorporated the ICE conditions (5th edn) subject to certain amendments. Clause 66 of the conditions required there to be a decision of the engineer before a dispute was referred to arbitration.

3.40 The works were substantially completed on 23 December 1996. In June 2002 defects in the roller bearings came to light. The matter needed to be decided by the engineer and referred to the arbitrator before limitation expired. Between 29 July and December 2002 the Secretary of State notified the claim in outline detail, the defects discovered, and the work undertaken. The engineer issued a decision on 18 December 2002 that the defects were caused by the use of materials or workmanship not in accordance with the contract.

3.41 On 19 December 2002 the Treasury Solicitor wrote to Amec asking for confirmation that the engineer's decision was accepted. The letter imposed a deadline of 5.00 pm that same day for a response. The letter stated that in the absence of that confirmation the employer would deem Amec to be dissatisfied with the decision and would take the necessary steps to protect the employer. No response was forthcoming and at the end of that day the employer referred the matter to arbitration. Amec claimed no dispute existed which was capable of being referred to arbitration.

3.42 **Held:** In Part 4 of the judgment Jackson J considered what constituted a 'dispute' for the purposes of clause 66 of the ICE conditions. In so doing he referred to the rapidly growing jungle of decisions on this subject and set out to distil the effect of the principal authorities.

[48] [2009] EWHC 64 (TCC).

[49] [2012] BLR 417, [2012] EWHC 1808 (TCC).

[50] The question of when the courts will entertain a complaint of 'ambush' is considered in Ch. 10 at 10.21 and following regarding challenges based on grounds of natural justice.

The judge considered authorities arising both in the context of arbitration and adjudication. From those authorities he derived the following seven propositions:[51]

1. The word 'dispute' which occurs in many arbitration clauses and also in section 108 of the Housing Grants Act should be given its normal meaning. It does not have some special or unusual meaning conferred upon it by lawyers.
2. Despite the simple meaning of the word 'dispute', there has been much litigation over the years as to whether or not disputes existed in particular situations. This litigation has not generated any hard-edged legal rules as to what is or is not a dispute. However, the accumulating judicial decisions have produced helpful guidance.
3. The mere fact that one party (whom I shall call 'the claimant') notifies the other party (whom I shall call 'the respondent') of a claim does not automatically and immediately give rise to a dispute. It is clear, both as a matter of language and from judicial decisions, that a dispute does not arise unless and until it emerges that the claim is not admitted.
4. The circumstances from which it may emerge that a claim is not admitted are Protean. For example, there may be an express rejection of the claim. There may be discussions between the parties from which objectively it is to be inferred that the claim is not admitted. The respondent may prevaricate, thus giving rise to the inference that he does not admit the claim. The respondent may simply remain silent for a period of time, thus giving rise to the same inference.
5. The period of time for which a respondent may remain silent before a dispute is to be inferred depends heavily upon the facts of the case and the contractual structure. Where the gist of the claim is well known and it is obviously controversial, a very short period of silence may suffice to give rise to this inference. Where the claim is notified to some agent of the respondent who has a legal duty to consider the claim independently and then give a considered response, a longer period of time may be required before it can be inferred that mere silence gives rise to a dispute.
6. If the claimant imposes upon the respondent a deadline for responding to the claim, that deadline does not have the automatic effect of curtailing what would otherwise be a reasonable time for responding. On the other hand, a stated deadline and the reasons for its imposition may be relevant factors when the court comes to consider what is a reasonable time for responding.
7. If the claim as presented by the claimant is so nebulous and ill-defined that the respondent cannot sensibly respond to it, neither silence by the respondent nor even an express non-admission is likely to give rise to a dispute for the purposes of arbitration or adjudication.

In applying those propositions to the facts of the case the judge found at paragraphs 74–80 that a dispute had existed at the relevant time.

Amec Civil Engineering Ltd v The Secretary of State For Transport [2005] 1 WLR 2339 (CA), [2005] EWCA Civ 291

3.43 **Facts:** The facts are set out in the case summary above.

3.44 **Held:** The judge at first instance was correct to find that a dispute existed at the relevant date and the appeal was dismissed. All three members of the Court of Appeal accepted the seven propositions cited by Jackson J in his judgment at first instance but both May LJ and

[51] At para. 68 of the judgment.

Rix LJ added their own observations. May LJ entirely agreed with the judge's analysis of the facts and made the following additional observations on the points of principal:

> 31. Each of the parties has accepted in this court that the judge's propositions correctly state the law. I am broadly content to do so also, but with certain further observations, as follows:
>
> 1. Clause 66 refers, not only to a 'dispute', but also to a 'difference'. 'Dispute or difference' seems to me to be less hard-edged than 'dispute' alone. This accords with the view of *Danckwerts LJ in F & G Sykes v. Fine Fare* [1967] 1 LLR 53 at 60 where he contrasted a difference, being a failure to agree, with a dispute.
> 2. In many instances, it will be quite clear that there is a dispute. In many of these, it may be sensible to suppose that the parties may not expect to challenge the Engineer's decision in subsequent arbitration proceedings. But major claims by either party are likely to be contested and arbitration may well be probable and necessary. Commercial good sense does not suggest that the clause should be construed with legalistic rigidity so as to impede the parties from starting timely arbitration proceedings. The whole clause should be read in this light. This leads me to lean in favour of an inclusive interpretation of what amounts to a dispute or difference.
> 3. The main circumstances in which it may matter whether there was a dispute or difference which has been referred to and settled by the Engineer include (a) where one party contends that this has occurred without due reference to arbitration, so that the Engineer's decision has become final and binding; and (b) where, as in the present case, one party wishes to contend that arbitration proceedings have not been started within a statutory period of limitation.
> 4. If the due operation of the mechanism of clause 66 really is to be seen as a condition precedent to the ability to start arbitration proceedings within a period of limitation, the parties cannot have intended to afford one another opportunistic technical obstacles to achieving this beyond those which the clause necessarily requires.
> 5. I agree with the judge that, insofar as the existence of a dispute may involve affording a party a reasonable time to respond to a claim, what may constitute a reasonable time depends on the facts of the case and the relevant contractual structure. The facts of the case here included that: (a) Major defects in very substantial works emerged relatively shortly before the perceived end of the limitation period. These required detailed investigation. In consequence, the formulation of a precisely detailed claim was impossible within a short period. (b) Liability for the defects was bound to be highly contentious, but Amec were bound to be a first candidate for responsibility. (c) Amec (and others) were inevitably going to resist liability well beyond the perceived end of the limitation period.

3.45 The additional observations of Rix LJ were as follows:

> 63. Like Lord Justice Clarke in *Collins (Contractors) v. Baltic Quay Management (1994) Limited* and Lord Justice May in the present case, I am broadly content to accept the propositions set out by Jackson J in para 68 of his judgment below. I would also agree with Lord Justice Clarke's and Lord Justice May's further observations, and would hazard these remarks of my own.
>
> 64. First, I would wish to be somewhat cautious about the concept of 'a reasonable time to respond' to a claim. The facts of the present case demonstrate to my mind the difficulty of that test. In many ways Amec were not left with a reasonable time to respond. They were effectively given a period of one day in circumstances where they had been previously told (in the Highways Agency's letter to them dated 2 October 2002) that a 'formal response' would be appreciated 'once you have received the factual report referred to earlier'. That factual report was never sent to them. It was true that the limitation period was expiring, but that was not Amec's fault.
>
> 65. The words 'dispute' and 'difference' are ordinary words of the English language. They are not terms of art. It may be useful in many circumstances to determine the existence of

a dispute by reference to a claim which has not been admitted within a reasonable time to respond; but it would be a mistake in my judgment to gloss the word 'dispute' in such a way. I would be very cautious about accepting that either a 'claim' or a 'reasonable time to respond' was in either case a condition precedent to the establishment of a dispute.

66. Secondly, however, like most words, 'dispute' takes its flavour from its context. Where arbitration clauses are concerned, the word has on the whole caused little trouble. If arbitration has been claimed and it emerges that there is after all no dispute because the claim is admitted, there is unlikely to be any dispute about the question of whether there had been any dispute to take to arbitration. And if the claim is disputed, any argument that the arbitration had not been justified because at the time it was invoked there had not been any dispute is, it seems to me, unlikely to find a receptive audience (although it appears that it did in *Cruden Construction v Commission for the New Towns* [1995] 2 Lloyd's Rep 37). So it is that in this arbitration context the real challenge to the existence of a 'dispute' has arisen where a party seeking summary judgment in the courts has been met by a request for a stay to arbitration and the claimant has wanted to argue that an unanswerable claim cannot be a real dispute. In that context it was held in *Hayter v. Nelson and Home Insurance* [1990] 2 Lloyd's Rep 265 that for the purposes of section 1 of the Arbitration Act 1975 'there is not in fact any dispute' where a claim is unanswerable, even if disputed. However, for the purposes of section 9 of the Arbitration Act 1996, from which that particular language had been dropped, this court held, applying *Ellerine Brothers (Pty) Ltd v. Klinger* [1982] 1 WLR 1375, that an un admitted claim gave rise to a dispute, however unanswerable such a claim might be: *Halki Shipping Corporation v. Sopex Oils Ltd* [1998] 1 WLR 726.

67. It follows that in the arbitration context it is possible and sensible to give to the word 'dispute' a broad meaning in the sense that a dispute may readily be found or inferred in the absence of an acceptance of liability, a fortiori because the arbitration process itself is the best place to determine whether or not the claim is admitted or not.

68. Thirdly, and significantly, the problem over 'dispute' has only really arisen in recent years in the context of adjudication for the purposes of Part II of the Housing Grants Construction and Regeneration Act 1996. Jackson J referred below to some of the burgeoning jurisprudence to which the need for a 'dispute' in order to trigger adjudication has given rise. In this new context, where adjudication is an *additional* provisional layer of dispute resolution, pending final litigation or arbitration, there is, as it seems to me, a legitimate concern to ensure that the point at which this additional complexity has been properly reached should not be too readily anticipated. Unlike the arbitration context, adjudication is likely to occur at an early stage, when in any event there is no limitation problem, but there is the different concern that parties may be plunged into an expensive contest, the timing provisions of which are tightly drawn, before they, and particularly the respondent, are ready for it. In this context there has been an understandable concern that the respondent should have a reasonable time in which to respond to any claim.

69. Fourthly, the question might arise as to whether the prior existence of a dispute is a condition precedent to a reference. The parties are agreed in this case that it is. Since they are agreed, I am content to assume that that is so. Ultimately, it would be a question of construction of any particular clause.

Midland Expressway Ltd and anor v Carillion Construction Ltd and ors [2006] BLR 325, [2006] EWHC 1505 (TCC)

3.46 **Facts:** Midland had been granted by the Secretary of State the right to design, construct, and operate the M6 toll road. Midland contracted the design and construction to the defendants. Midland asked the defendants for an estimate of the time and money impacts of proposed changes. The response was expressly limited to direct costs as the defendants were in the process of assessing the indirect costs and the time consequences. Time impact details were submitted subsequently. Later the defendants submitted a claim for indirect costs. The defendants served notice of intention to refer to adjudication their disputed claim for direct costs for the changes. Midland then sought to include the entitlement to indirect costs. The defendants withdrew the claim for indirect costs. The adjudicator ruled that there was no dispute as to the defendants' indirect costs because the defendants had accepted that, as presented, the indirect cost claim would not succeed. The adjudicator issued his decision on the direct costs arising from the changed plans only.

3.47 Midland sought declarations (1) that the adjudicator's decision that there was no dispute regarding indirect costs was wrong, (2) that the adjudicator had jurisdiction to deal with the indirect costs, and (3) that the claim for indirect costs should have failed.

3.48 **Held:** The adjudicator had been right to conclude that there was no dispute capable of being adjudicated arising out of the claim for indirect costs. At no stage during the adjudication reference did there exist a quantified claim for indirect costs that was capable of being disputed. The adjudication was correctly limited to direct costs. In any event the defendants were entitled to withdraw their claim for indirect costs.[52]

Table 3.2 Table of Cases: Meaning of 'Dispute'

Title	Citation	Issue
Hayter v Nelson (No. 2)	[1990] 2 Lloyd's Rep 265, Saville J	**Non-adjudication case** If there were a disagreement as to the result of the Boat Race: '*The fact that it can be easily and immediately demonstrated beyond any doubt that one is right and the other is wrong does not and cannot mean that the dispute did not in fact exist. Because one man can be said to be indisputably right and another indisputably wrong does not, in my view, entail that there was therefore never any dispute between them.*'
Halki Shipping Corporation v Sopex Oils Ltd	[1998] 1 Lloyd's Rep 49 (CA)	**Low-threshold test for 'dispute' (non-adjudication case)** A dispute exists for the purpose of an arbitration clause whether or not the issue is disputable as a matter of fact or law. 'There is a dispute once money is claimed unless and until the defendants admit that the sum is due and payable' (at page 761F).
Fastrack Contractors Ltd v Morrison Construction Ltd	[2000] BLR 168, Judge Thornton	**High-threshold test for 'dispute'** 'A "dispute" can only arise once the subject matter of a claim, issue or other matter has been brought to the attention of the opposing party and that party has had an opportunity of considering and admitting, modifying or rejecting the claim or assertion.' Cf. *Amec* and *Collins* (below).
Ken Griffin & John Tomlinson (t/a K&D Contractors) v Midas Homes Ltd	21 Jul. 2000, 78 Con LR 152, Judge LLoyd	**High-threshold test for 'dispute'** Whether a dispute has arisen is a question of fact. Adjudication is intended to resolve what has not been settled by the normal process of discussion and agreement. A dispute is not lightly to be inferred. There must come a time when a dispute will arise, usually where a claim or assertion is rejected in clear language without the possibility of further discussion, and such rejection might conceivably be by way of an obvious and outright refusal to consider a particular claim at all. Cf. *Amec* and *Collins* (below).

(*Continued*)

[52] For a more detailed consideration of when a party can withdraw a claim from an adjudicator and start again, see Ch. 9 at 9.59.

Table 3.2 *Continued*

Title	Citation	Issue
Sindall Ltd v Abner Solland and ors	[2001] TCLR 712 (TCC), Judge LLoyd	**High-threshold test for 'dispute'** For there to be 'a dispute' it must be clear that 'a point has emerged from the process of discussion or that negotiation has ended and that there is something that needs to be decided'. In this case the judge found there was no dispute because the claimant had provided substantial new information and the respondent did not have time to consider it and reject before the adjudication notice was served. Cf. *Amec* and *Collins* (below).
KNS Industrial Services v Sindall	[2001] 17 Const LJ 170, Judge LLoyd	**High-threshold test for 'dispute'** Judge LLoyd adopted the test set out by Judge Thornton in *Fastrack* that 'the dispute is whatever claims, heads of claims, issues or contentions or causes of action that are then in dispute which the referring party has chosen to crystallize into an adjudication reference'. A party to a dispute who identifies the dispute in simple or general terms has to accept that any ground that exists which might justify the action complained of is comprehended within the dispute for which adjudication is sought. The adjudicator had decided the dispute referred and the decision was enforced.
Edmund Nuttall Ltd v R. G. Carter Ltd	[2002] BLR 312 (TCC) at 322, Judge Seymour	**High-threshold test for 'dispute'** Dispute arises only when a party has had an opportunity to consider the position of his opponent and to formulate responses to it. The scope of the dispute will include the whole package advanced and facts relied on by each side. This was necessary because the whole concept of adjudication requires the parties to first try and resolve their disputes, and this approach would avoid unnecessary premature adjudications. Cf. *Amec* and *Collins* (below).
Hitec Power Protection BV v MCI Worldcom Ltd	[2002] EWHC 1953 (TCC), Judge Seymour	**High-threshold test for 'dispute'** A dispute in the context of adjudication arises when a claim is made, there has been an opportunity for the parties to consider the arguments of the other and formulate arguments of a reasoned kind. Cf. *Amec* and *Collins* (below).
Carillion Construction Ltd v Devonport Royal Dockyard	[2003] BLR 79 (TCC), Judge Bowsher	**No dispute had crystallized** Devonport had asked why payment was demanded and on what grounds. They had neither denied the claim nor ignored it but were still seeking clarification. The response they received was so close in time to the notice of adjudication that they had no opportunity to respond to it.
Cowlin Construction Ltd v CFW Architects (a firm)	[2003] BLR 240 (TCC), Judge Kirkham	**Low threshold for 'dispute'** There is no need to construe the word 'dispute' more narrowly for the purposes of adjudication than its ordinary English meaning. The test in *Halki* applied.
Beck Peppiat v Norwest Holst Construction Ltd	[2003] BLR 316, Forbes J	**Low threshold for 'dispute'** The word dispute was an ordinary English word which should be given ordinary English meaning and each case is determined on its facts. The test in *Halki* applied.
Orange EBS Ltd v ABB Ltd [2003] BLR 323	[2003] BLR 323 (TCC), Judge Kirkham	**Low threshold for 'dispute'** The test in *Halki* applied.

Title	Citation	Issue
London & Amsterdam Properties v Waterman Partnership Ltd	[2004] BLR 179 (TCC), Judge Wilcox	**Low threshold for 'dispute'** The word 'dispute' should be given its ordinary meaning and included any claim which the other party refused to admit or did not pay, whether or not there was an answer to the claim in fact or law. *Halki* applied.
Dean and Dyball Construction Ltd v Kenneth Grubb Associates Ltd	100 Con LR 92, [2003] EWHC 2465 (TCC), Judge Seymour	**Change to quantum of claim** Dispute about liability had crystallized. That dispute did not cease to be crystallized simply because the quantum of the claim was revised. Neither was the dispute referred a new dispute.
R. Durtnell & Sons v Kaduna Ltd	[2003] BLR 225 (TCC), Judge Seymour	**Time for consideration of claim: contract term** It could not be said that there was a 'dispute' as to entitlement to extensions of time when the issue had been referred to the architect, the time allowed by the standard form for him to make his determination had not expired, and no determination had been made. Until the architect had made his assessment there was nothing to argue about and no dispute.
AWG Construction Services v Rockingham Motor Speedway Ltd	[2004] EWHC 888 (TCC), Judge Toulmin	**Low threshold for 'dispute'** Applying the test in *Halki* a wide interpretation should be given to the word 'dispute' so that the adjudicator's jurisdiction was preserved wherever possible. **Scope of dispute/new material?** The adjudicator should not be confined to considering rigidly only the package of issues, facts, and arguments which had been referred to him.
Amec Civil Engineering Ltd v The Secretary of State for Transport	[2004] EWHC 2339 (TCC), Jackson J	**Guidance on when a dispute exists—Key Case** Seven propositions distilled from the principal authorities (see Key Case for verbatim citation). The mere notification of a claim does not give rise to a dispute. A dispute does not arise until it emerges that the claim is not admitted.
Amec Civil Engineering Ltd v The Secretary of State for Transport	[2005] 1WLR 2339, [2005] BLR 227 (CA)	**CA guidance on when a dispute exists—Key Case** The CA upheld Jackson J's judgment at first instance and added observations on what may be a reasonable time for responding and how it may be adjudged.
Collins (Contractors) Ltd v Baltic Quay Management (1994) Ltd	[2004] EWCA Civ 1757 (CA), [2005] BLR 63 (CA)	**CA guidance on when a dispute exists** Approved Jackson J in *Amec* and added that the seven propositions did not include the Sindall v Solland suggestion that the dispute exists when negotiations or discussions have been concluded. The CA in *Amec* also so stated.
William Verry (Glazing Systems) Ltd v Furlong Homes Ltd	[2005] EWHC 138, Judge Coulson	**Scope of dispute/new material?** A response, submitted by one of the parties, containing further detail in relation to the claim and which requested an extension of time, did not constitute a new claim; it was a fuller explanation of the claim originally made.
Midland Expressway Ltd and anor v Carillion Construction Ltd and ors	[2006] BLR 325, [2006] EWHC 1505 (TCC), Jackson J	**No dispute existed concerning indirect cost claim** At no time during the adjudication was there a quantified claim for indirect costs capable of being disputed. There did exist a dispute of principle concerning entitlement to recover indirect costs, but neither party had referred that issue of principle to the adjudicator. Consequently no dispute regarding indirect costs had been referred and, even if it had, the referring party was entitled to withdraw the claim from the adjudicator.

(*Continued*)

Table 3.2 *Continued*

Title	Citation	Issue
All In One Building & Refurbishments v Makers UK Ltd	Apr. 2006 CILL 2321, [2005] EWHC 2943 (TCC), Judge Wilcox	**Low-threshold test for 'dispute'** Claim for loss of profits and overheads asserted before adjudication, but particularized in adjudication. Argument that no dispute existed rejected. Common sense and not legalistic approach needed.
Multiplex Constructions (UK) Ltd v Mott MacDonald Ltd	110 Con LR 63, [2007] EWHC 20 (TCC), Jackson J	**Scope of dispute wide enough to allow adjudicator own formulation** Where parties to a construction agreement differed as to what constituted 'pertinent records' liable to inspection, an adjudicator had been entitled to formulate his own interpretation of the term because of the wide ranging nature of the question referred. Whether he was right or wrong he was determining a pre-existing dispute between the parties and was, thus, within his jurisdiction. Cf. *Ardmore Construction v Taylor Woodrow Construction* [2006] CSOH 3.
John Stirling v Westminster Properties Scotland Ltd	[2007] Scot CS CSOH 117, Lord Drummond	**Silence of respondent** Claim had been asserted, no sum paid, the defenders had failed to give a reason to explain non-payments. That was sufficient for the inference that the defenders disputed the pursuers claim. Guidance in Amec followed.
Ringway Infrastructure Services Ltd v Vauxhall Motors Ltd	[2007] All ER (D) 333, [2007] EWHC 2421 (TCC), Akenhead J	**Question of fact whether dispute crystallized** Dispute regarding extension of time submitted as interim application after practical completion had crystallized. Guidance as to when dispute crystallizes given in para. 55 of judgment.
Cantillon Ltd v Urvasco Ltd	[2008] BLR 250, [2008] EWHC 282 (TCC), Akenhead J	**Dispute not limited to arguments/evidence already cited** The adjudication had concerned Cantillon's claim for an extension of time and loss and expense. Urvasco contended the adjudicator had exceeded his jurisdiction or breached the rules of natural justice because he had awarded prolongation expenses in relation to a different period of time than that claimed. The TCC judge gave guidance about when a dispute crystallizes (para. 55 of judgment). In particular the judge said the courts should not take an overly legalistic approach but should construe the crystallized dispute broadly. The dispute is not necessarily defined by evidence or arguments previously submitted by the parties.
VGC Construction Ltd v Jackson Civil Engineering Ltd	[2008] EWHC 2082 (TCC), Akenhead J	**Nebulous claim?** The claim for payment, which included an item for delay and disruption, was not so nebulous or ill-defined that it warranted being struck out.
		Withdrawal of claim and estoppel If a party represents that it will withdraw a claim which is disputed as to jurisdiction, and the respondent relies on that representation to its detriment, then it may be arguable that the potential dispute ceases to be a dispute and that the claiming party will be estopped from asserting it is a dispute within the jurisdiction of the adjudicator.

Title	Citation	Issue
OSC Building Services Ltd v Interior Dimensions Contracts Ltd	[2009] EWHC 248 (TCC), Ramsey J	**Whether decision broadened dispute beyond its narrow description in the notice of adjudication** Whether the scope had been broadened beyond notice of adjudication. The responding party construed the notice of adjudication narrowly as referring only to the narrow issue of non-payment of a certificate and not the related question of the value of the final account. The judge rejected this interpretation and construed the notice of adjudication and referral against the background of prior communications between parties. The dispute referred was what further sum was due.
Bovis Lend Lease v The Trustees of the London Clinic	123 ConLR 15, [2009] EWHC 64 (TCC), Akenhead J	**Dispute crystallized not same as evidence in support** Once it was clear that there was a crystallized dispute, it was necessary to differentiate between the substance of the dispute which was then referred to adjudication (or arbitration) and the evidence needed to support or contest that disputed claim. It was clear that in the instant case there was an expanding dispute between the parties as to the responsibility for delays. The detailed claim served was not an 'ambush'.
Working Environments Ltd v Greencoat Construction Ltd	[2012] BLR 309, [2012] EWHC 1039 (TCC), Akenhead J	**Dispute about valuation crystallized before witholding notice served** The defendant sought to resist enforcement of an adjudicator's decision relating to an interim valuation on the ground that the dispute had not crystallized because it was referred before the valid withholding notice was served. The TCC judge decided that, on the facts, the dispute about the interim valuation crystallized when the defendant valued the claimant's application in a lower amount, and that 10 of the 12 issues in the withholding notice were already part of the dispute that had already crystallized. Two items in the withholding notice were new and had not been referred to the adjudicator. His decision about them was severed and the valid parts of his decision about the value of the interim application were enforced.
Beck Interiors Ltd v UK Flooring Contractors Ltd	[2012] BLR 417, [2012] EWHC 1808 (TCC), Akenhead J	**Dispute not crystallized re late addition to claim** Beck claimed damages for repudiation against a subcontractor. In a letter served after close of business on the day before Good Friday they added a delay claim for LADs. They served an adjudication notice on the first working day after Easter. UK Flooring asserted there was no jurisdiction in relation to the LAD claim as the dispute had not crystallized. The TCC judge agreed for two reasons: (1) Beck must have known that UK Flooring would have no opportunity in working hours to consider the LAD claim or produce a response to it, and (2) in any event the LAD claim in the adjudication notice was very different from that in the earlier letter (judgment paras. 27–8). The LAD claim was severed from the rest of the adjudication decision, and not enforced: '*It is a matter of fact and degree as to whether amendments to a disputed claim fall within the umbrella of the original dispute*' (para. 29).

Must Only a Single Dispute Be Referred?

3.49 It is to be noted that the 1996 Act refers to 'a dispute' and not 'disputes'. Read literally this means that the referring party ought to limit the notice of adjudication to a single dispute, although the Scheme says that the parties are free to agree to extend the reference to cover more than one dispute under the same contract and related disputes under different contracts. It is not difficult to imagine the problems this could cause if interpreted narrowly. Construction contracts frequently give rise to numerous interrelated claims, for example an unexpected event could give rise to a claim for additional payment, for an extension of time, and for loss and expense. In such circumstances identifying where one dispute stops and another starts can potentially be difficult and could make s. 108 unworkable in some cases.

3.50 In practice this has not proved to be problematic due to the willingness on the part of the courts to adopt a robust approach to utilize a 'benevolent interpretation' of the adjudication notice, invariably to conclude that whilst there may have been numerous issues or claims in the adjudication, there was but one dispute: *David McLean Housing Contractors Ltd v Swansea Housing Association Ltd* (2001) (Key Case)[53] and *Chamberlain Carpentry & Joinery Ltd v Alfred MacAlpine Construction* (2002).[54] In *Sindall Ltd v Solland* (2001),[55] Judge LLoyd said:

> Where a dispute is referred there is comprehended within it all its constituent elements, including sub-disputes, contentions, issues (some of which might have been referred separately)—in other words all the ingredients which go into the dispute referred.

3.51 This has been the approach adopted by the courts from the outset and the cases on this aspect of s. 108 are consistent on this point. Thus it is not uncommon for one party to refer a final account dispute to adjudication which includes just about every claim that can arise under a building contract. This 'kitchen sink' type of claim will be considered as a single dispute characterized as 'what sums are due on the final account' as long as the notice of adjudication is drafted carefully in wide terms. Despite the fact that many legal commentators have stated that the adjudication process is fundamentally unsuited to this type of complex wide-ranging dispute,[56] such a dispute can nevertheless be referred to adjudication. This led Lord Macfadyen in *Barr Ltd v Law Mining Ltd* (2001)[57] to comment that:

> There was a danger that, if a multiplicity of unrelated issues could be focused in one adjudication, the supposedly speedy process of adjudication would be overburdened. A party had a legitimate interest to withhold consent to several disputes being determined in a single adjudication, since multiplication of issues might be incompatible with the speedy and summary nature of the process.

3.52 If however the notice of adjudication refers to different issues which are unrelated, a court may decide that more than one dispute was referred in contravention of s. 108(1) and the reference was invalid. In *David & Theresa Bothma (in partnership) t/a DAB Builders v Mayhaven Healthcare* (2007)(Key Case)[58] the court decided that the notice to refer included unrelated issues about the entitlement to an extension of time and the correct figure for an interim valuation. In the circumstances the court refused to summarily enforce the adjudicator's

---

[53] [2002] BLR 125.

[54] [2002] EWHC 514.

[55] [2001] 3 TCLR 30.

[56] See e.g. the comments of the TCC judge in *William Verry (Glazing Systems) Ltd v Furlong Homes Ltd* [2005] EWHC 138 at paras. 11 and 17 of the judgment.

[57] [2001] Scot CS 152, 80 Con LR 134, paras. 13–15 of the judgment.

[58] 114 Con LR 131 (CA), [2007] EWCA 527 Civ.

decision. However in *Witney T own Council v Beam Construction (Cheltenham) Ltd* (2011)[59] Akenhead J rejected such an argument and found that despite the inclusion of different issues in the adjudication notice, it was all part of one single dispute which concerned how much was due to Beam Construction under their final account.

Key Cases: Single Dispute

David McLean Housing Contractors Ltd v Swansea Housing Association Ltd
[2002] BLR 125 (TCC),

3.53 **Facts:** In 1998 the parties entered into a contract in the JCT Standard Form of Building Contract (with Contractor's Design) 1981 edn. McLean contended that it was entitled to an extension of time and Swansea gave notice of intention to deduct liquidated damages. McLean gave notice of adjudication for monies due under a payment application which had included claims for direct loss and expense, valuation of variations, money due in respect of measured work, and adjustments in relation to the expenditure of provisional sums. The notice also asked for determination of its right to an extension of time. In making his award, the adjudicator determined both the extension of time claim and the amount of the liquidated damages to which Swansea was entitled, but did not deduct the latter from the amount of the award. Swansea paid the amount of the decision after deducting the liquidated damages. McLean brought an action to enforce the award and Swansea counterclaimed for the amount of the liquidated damages. Several issues arose on the enforcement, one of which was whether the notice of adjudication was invalid as referring to more than one dispute.

3.54 **Held:** In determining whether a notice of adjudication referred to one or more than one dispute, it was appropriate to apply a 'sensible' or 'benevolent' construction, bearing in mind that a single dispute might well consist of several discrete elements. In the present case the dispute was as to the amount due to McLean under the payment application. In determining that amount, it was necessary to consider a discrete element as to the extension of time, if any, to which McLean was entitled. However, the existence of such an element or elements did not mean that there was more than one dispute.

David & Theresa Bothma (in partnership) t/a DAB Builders v Mayhaven Healthcare Ltd
[2007] 114 Con LR 131, [2007] EWCA 527 Civ (CA)

3.55 **Facts:** The Bothmas applied for leave to appeal against a refusal to enforce an adjudicator's decision. The adjudication notice identified four issues which had arisen between the parties namely the date for completion of the contract, the scope and validity of the architect's instructions, the issue and non-withdrawal of notice of non-completion, and the sum of one of the interim valuations.

3.56 **Held:** In reality two separate disputes had been referred to the adjudicator. Mayhaven had not consented to more than one dispute being argued nor did it waive jurisdiction issues. The argument that the claim for the hire of the Portakabin was a time-related item which was sufficient to provide a bridge between the claim in respect of the interim valuation and the time issues raised by the claim for extension of time was rejected. Permission to appeal was refused.

[59] [2011] EWHC 2332 (TCC).

Table 3.3 Table of Cases: Single Dispute

Title	Citation	Issue
Fastrack Contractors Ltd v Morrison Construction Ltd	[2000] BLR 168 (TCC), Judge Thornton	**Multiple issues but one dispute** If the dispute referred is characterized as 'what sum is due?' then that is a single dispute which can encompass a multitude of individual and complex sub-issues.
Grovedeck Ltd v Capital Demolition Ltd	[2000] BLR 181 (TCC), Judge Bowsher	**Two sites/two disputes** *Obiter*: claimant's reference to adjudication of claims arising out of two different sites was two disputes and not one and a ground for refusal to enforce.
Sindall Ltd v Abner Solland and ors	[2001] 3 TCLR 30 (TCC), Judge LLoyd	**Scope of dispute includes sub-disputes** '16. Where a dispute is referred, there is comprehended within it all its constituent elements, including sub disputes, contentions, issues (some of which might have been referred separately) in other words all the ingredients which go into the dispute referred.'
Barr Ltd v Law Mining Ltd	[2001] 80 Con LR 134 (CSOH), Lord Macfadyen	**Single dispute** Whether what is in issue is a dispute or several disputes is a matter of circumstance which the adjudicator must decide for himself (if the point is raised). It is easy to subdivide and analyse what was in substance one dispute into its component parts and to label each a separate dispute. That was not the correct approach because a realistic view had to be taken. In neither case was the adjudicator wrong to treat all of the matters referred to him as one dispute.
Chamberlain Carpentry and Joinery Ltd v Alfred McAlpine Construction Ltd	[2002] EWHC 514 (TCC), Judge Seymour	Interpretation of notice of adjudication A letter was free-standing as a notice of adjudication. The request for payment of a sum, calculated by reference to eight particular matters listed, was a single dispute as to what sum should be paid.
David McLean Housing Contractors Ltd v Swansea Housing Association Ltd	[2002] BLR 125 (TCC), Judge LLoyd	**Single dispute** In determining whether a notice of adjudication referred to one or more than one dispute it was appropriate to apply a 'sensible' or 'benevolent' construction, bearing in mind that a single dispute might well consist of several discrete elements.
Michael John Construction v Golledge	[2006] EWHC 71 (TCC), Judge Coulson	**Single dispute** 'It would be contrary to the whole purpose of adjudication if such a simple dispute could then be broken down into its component parts, to enable the Defendants to be able to say that, because the dispute incorporates more than one issue, there must somehow be more than one dispute. That would be contrary to all of the authorities to which I have referred at paragraphs 26–28 above. It would be untenable as a matter of commercial common-sense.'
David & Theresa Bothma (in partnership) t/a DAB Builders v Mayhaven Healthcare Ltd	114 Con LR 131, [2007] EWCA 527 Civ (CA)	**More than one dispute** Multiple disputes were included in the notice of intention to refer to adjudication. The judge decided they fell into two unconnected categories: valuation matters and unconnected issues about the time for completion. Thus there were two independent disputes, which meant the notice to refer failed to comply with s. 108, and the decision was not enforced.

Title	Citation	Issue
Multiplex Constructions (UK) Ltd v Mott MacDonald Ltd	110 Con LR 63, [2007] EWHC 20 (TCC), Jackson J	**Dispute included wider question** Although an adjudicator's jurisdiction is circumscribed by the scope of the pre-existing dispute, and the referring party cannot widen what is in dispute by drafting an unduly broad or optimistic notice of adjudication, nevertheless on a fair reading of the correspondence in the case, one issue in contention was the true meaning of the phrase 'all records pertinent to the services'.
CSC Braehead Leisure Ltd v Laing O'Rourke Scotland Ltd	[2009] BLR 49, [2008] CSOH 119, Lord Menzies	**Single dispute referred** The enforcement of a decision of an adjudicator was challenged on the ground that he had only issued an interim decision on some of the questions referred and so had failed to exhaust his jurisdiction. One of the issues for the court to determine was whether this was one dispute or several disputes. The judge found there was only a single dispute referred about how much was due for payment. 'I am in no doubt that a single dispute may contain sub-disputes or heads of claim which may themselves be the subject of dispute, but this does not necessarily result in more than one dispute being referred to the arbiter.'
GPS Marine Contractors Ltd v Ringway Infrastructure Services	[2010] BLR 377, [2010] EWHC 283 (TCC) Ramsey J	**Allegation that more than one dispute referred** The court rejected a jurisdiction challenge based on the assertion that two disputes had been referred when only one was permitted. The court found that only a single dispute had been referred.
Witney Town Council v Beam Construction (Cheltenham) Ltd	[2011] BLR 707, [2011] EWHC 2332 (TCC), Akenhead J	**Allegation that more than one dispute referred** Enforcement of an adjudicator's decision was challenged on the ground that more than one dispute had been referred. The TCC judge rejected the challenge and found that on the facts a single dispute had been referred which asked how much money was due and owing to the contractor. Although the adjudication notice was worded as if there were four disputes referred, in reality they were steps along the way to considering the overall question of how much was due (which was the first question asked in the notice).

Determining the Scope of the Dispute Referred

3.57 The adjudicator is empowered to decide the dispute properly referred by the notice of adjudication. Where a dispute has crystallized between the parties, that is the dispute that must be referred to the adjudicator and not another different dispute. Thus the referring party should not materially change the existing dispute by its adjudication notice: *Fastrack Contractors Ltd v Morrison Construction Ltd and anor* (1999),[60] *KNS Industrial Services (Birmingham) Ltd v Sindall Ltd* (2000),[61] *Multiplex Constructions (UK) Ltd v Mott MacDonald Ltd* (2007) (Key Case).[62]

3.58 Similarly the dispute properly referred in the adjudication notice should not be materially changed by the referring party during the course of the adjudication: *AWG Construction*

[60] [2000] BLR 168 (TCC).

[61] 75 Con LR 1.

[62] 110 Con LR 63, [2007] EWHC 20 (TCC).

Services v Rockingham Motor Speedway Ltd (2004) (Key Case),[63] *McAlpine PPS Pipeline Systems Joint Venture v Transco plc* (2004).[64] Whether the dispute has been materially changed in the adjudication notice or during the reference is a matter of fact to be determined by reference to all the circumstances.

3.59 A respondent to an adjudication is entitled to rely on any point legally available to it for its defence whether or not it has notified the point previously: *KNS Industrial Services (Birmingham) Ltd v Sindall Ltd* (2000),[65] *William Verry (Glazing Systems) Ltd v Furlong Homes Ltd* (2005)(Key Case),[66] *Cantillon Ltd v Urvasco Ltd* (2008),[67] *Pilon v Breyer* (2010).[68] Accordingly a respondent may widen the dispute by serving a legitimate defence even if that defence is not referred to in the notice of adjudication and even if it has not been previously disclosed.[69] This does not mean however that the respondent may rely on a defence or counterclaim when it should have but failed to serve a withholding notice. Hence the respondent is only entitled to rely on a defence not prohibited by the absence of a withholding notice or other equivalent restriction.[70]

3.60 Where the adjudicator is asked to decide whether the referring or responding party's case is correct, it may be appropriate for the adjudicator to reach a decision that differs from the cases put. In *Multiplex Constructions (UK) Ltd v Mott MacDonald Ltd* (2007)[71] both parties put forward their own interpretations of a contract term referring to 'all pertinent records to the services'. The adjudicator was unable to accept either parties' formulation and reached his own interpretation of the clause. The judge at enforcement decided that one of the issues in contention was the true meaning of the phrase and so it was open to the adjudicator to reach his own answer to that question.[72]

Key Cases: Determining the Scope of the Dispute

AWG Construction Services v Rockingham Motor Speedway Ltd [2004] EWHC 888 (TCC)

3.61 **Facts:** A dispute arose between the parties regarding three disputes which were agreed to be heard together. One concerned the retention of surface water on the racetrack. Rockingham alleged in its referral notice that the racetrack as constructed was unfit for the purpose of racing cars and that AWG was negligent in choosing stabilized subsoil (the Oval dispute). During the adjudication Rockingham widened the Oval dispute to allege AWG had failed to implement a proper track drainage scheme. AWG objected to the attempt to introduce additional material. The adjudicator found for Rockingham on the Oval dispute on the new material. At enforcement AWG argued the adjudicator did not have jurisdiction to make the decision that he had on the Oval dispute and had breached the rules of natural justice.

[63] [2004] EWHC 888 (TCC).
[64] [2004] BLR 352, [2004] EWHC 2030 (TCC).
[65] 75 Con LR 1 [2000].
[66] [2005] EWHC 138 (TCC).
[67] [2008] BLR 250, [2008] EWHC 282 (TCC).
[68] [2010] BLR 452, [2010] EWHC 837 (TCC).
[69] This topic is also discussed in Ch. 9 at 9.06 and in Ch. 10 at 10.36 and following.
[70] *Letchworth Roofing Company v Sterling* [2009] EWHC 1119 (TCC).
[71] 110 Con LR 63, [2007] EWHC 20 (TCC).
[72] As to whether an adjudicator is entitled to reach a formulation different from those proposed by either party, see *Hyder Consulting (UK) Ltd v Carillion Construction Ltd* (2011) 138 Con LR 212, [2011] EWHC 1810 (TCC) which discusses the law on this topic at paras. 60–70, and also in the Table of Cases below see *Ardmore Construction v Taylor Woodrow Construction* [2006] CSOH 3, *Multiplex Construction Ltd v West India Quay Developments* [2006] EWHC 1569 (TCC), and *Herbosh-Kiere Marine Contractors Ltd v Dover Harbour Board* [2012] BLR 177. This topic is discussed in more detail in Ch. 10 at 10.88 and following.

Held: The adjudicator did not have the jurisdiction to make the decision that he had on the Oval dispute. A fundamental point was what constituted a dispute referred. The court had to approach that question with robust common sense (judgment paragraph 144). The additional drainage point upon which the adjudicator principally reached his decision in relation to the Oval dispute was not part of the claim or notice. The adjudicator did not have jurisdiction to reach the decision that he had on those grounds. The adjudicator in making his decision on the Oval dispute had breached natural justice because AWG was not given sufficient time to consider the new material presented by Rockingham before the adjudication: **3.62**

> 141. In my view each case must depend on the circumstances and the context in which the Referral is made. In some cases the issues referred are very specific. In other cases it is clear that the issues are more general and have been so treated by the parties and that there is significantly more room for the case to be developed. The test in each case is first what dispute did the parties agree to refer to the Adjudicator? And, secondly, on what basis? If the basis which is argued in the Adjudication is wholly different to that which a Defendant has had an opportunity to respond in advance of the Adjudication, this may constitute a different dispute not referred to the Adjudicator or, put another way, in so far as the Adjudicator reaches a decision on the new issues it is not responsive to the issues referred to him.

William Verry (Glazing Systems) Ltd v Furlong Homes Ltd [2005] EWHC 138 (TCC)

Facts: Verry submitted a draft final account and a claim for an extension of time to 24 June 2004. Furlong refused to grant additional time. Furlong commenced adjudication against Verry to deal with the entirety of the dispute concerning the final account. Verry submitted a response which contained a newly prepared extension of time claim which requested an extension to 31 July 2004. The headings under which Verry sought an extension of time remained largely unchanged, but the details describing the background for each event were new and certain dates had been changed. Furlong objected to the response as being outside the jurisdiction of the adjudicator. **3.63**

Held: It was excessively legalistic to classify Verry's response as a new claim. It was a fuller explanation of the claim originally made. Even if the response was a new extension of time claim, it fell within the wide dispute referred by Furlong which asked 'what was an appropriate extension of time?' Verry was entitled to take whatever points it needed to defend itself and the adjudicator had been obliged to consider them (judgment paragraphs 47–9). **3.64**

Cantillon Ltd v Urvasco Ltd [2008] BLR 250, [2008] EWHC 2008 (TCC)

Facts: Urvasco sought to resist enforcement of an adjudicator's decision on the ground that he had awarded prolongation costs for a period later than that contained in the dispute referred. **3.65**

Held: The TCC judge gave guidance as to how the scope of the dispute was to be determined. At paragraph 54 of the judgment he confirmed that whatever dispute is referred to the adjudicator, it includes and allows for any ground open to the responding party which would amount in law or in fact to a defence of the claim with which it is dealing: **3.66**

> 55. There has been substantial authority, both in arbitration and adjudication, about what the meaning of the expression 'dispute' is and what disputes or differences may arise on

the facts of any given case. Cases such as *Amec Civil Engineering Ltd -v- Secretary of State for Transport* [2005] BLR 227 and *Collins (Contractors) Ltd -v- Baltic Quay Management (1994) Ltd* [2004] EWCA (Civ) 1757 address how and when a dispute can arise. I draw from such cases as those the following propositions:

(a) Courts (and indeed adjudicators and arbitrators) should not adopt an over legalistic analysis of what the dispute between the parties is.
(b) One does need to determine in broad terms what the disputed claim or assertion (being referred to adjudication or arbitration as the case may be) is.
(c) One cannot say that the disputed claim or assertion is necessarily defined or limited by the evidence or arguments submitted by either party to each other before the referral to adjudication or arbitration.
(d) The ambit of the reference to arbitration or adjudication may unavoidably be widened by the nature of the defence or defences put forward by the defending party in adjudication or arbitration. It will follow from the above that I do not follow the judgment of HHJ Seymour, QC, in *Edmund Nuttall Ltd -v- R. G. Carter Ltd* [2002] BLR 312 where the learned judge said at paragraph 36:

'However, where a party has an opportunity to consider the position of the opposite party and to formulate arguments in relation to that position, what constitutes a "dispute" between the parties is not only a "claim" which has been rejected, if that is what the dispute is about, but the whole package of arguments advanced and facts relied upon by each side'.

In my view, one should look at the essential claim which has been made and the fact that it has been challenged as opposed to the precise grounds upon which that it has been rejected or not accepted. Thus, it is open to any defendant to raise any defence to the claim when it is referred to adjudication or arbitration. Similarly, the claiming party is not limited to the arguments, contentions and evidence put forward by it before the dispute crystallized. The adjudicator or arbitrator must then resolve the referred dispute, which is essentially the challenged claim or assertion but can consider any argument, evidence or other material for or against the disputed claim or assertion in resolving that dispute.

Witney Town Council v Beam Construction Cheltenham Ltd [2011] BLR 707, [2011] EWHC 2332 (TCC)

3.67 **Facts:** The defendant resisted enforcement of an adjudicator's decision on the ground that more than one dispute had been referred.

3.68 **Held:** In considering the issues raised by the challenge, the judge had to consider the nature and scope of the dispute that had arisen and that had been referred to adjudication. Having considered the relevant authorities on the subject the judge set out this guidance:

(i) A dispute arises generally when and in circumstances in which a claim or assertion is made by one party and expressly or implicitly challenged or not accepted.
(ii) A dispute in existence at one time can in time metamorphose in to something different to that which it was originally.
(iii) A dispute can comprise a single issue or any number of issues within it. However, a dispute between parties does not necessarily comprise everything which is in issue between them at the time that one party initiates adjudication; put another way, everything in issue at that time does not necessarily comprise one dispute, although it may do so.

(iv) What a dispute in any given case is will be a question of fact albeit that the facts may require to be interpreted. Courts should not adopt an over legalistic analysis of what the dispute between the parties is, bearing in mind that almost every construction contract is a commercial transaction and parties cannot broadly have contemplated that every issue between the parties would necessarily have to attract a separate reference to adjudication.

(v) The Notice of Adjudication and the Referral Notice are not necessarily determinative of what the true dispute is or as to whether there is more than one dispute. One looks at them but also at the background facts.

(vi) Where on a proper analysis, there are two separate and distinct disputes, only one can be referred to one adjudicator unless the parties agree otherwise. An adjudicator who has two disputes referred to him or her does not have jurisdiction to deal with the two disputes.

(vii) Whether there are one or more disputes again involves a consideration of the facts. It may well be that, if there is a clear link between two or more arguably separate claims or assertions, that may well point to there being one dispute. A useful if not invariable rule of thumb is that, if disputed claim No 1 cannot be decided without deciding all or parts of disputed claim No 2, that establishes such a clear link and points to there being only one dispute.

Working Environments Ltd v Greencoat Construction Ltd [2012] BLR 309, [2012] EWHC 1039 (TCC)

3.69 **Facts:** The case concerned a dispute about the valuation of an interim payment. At the enforcement hearing the defendant said that there was no jurisdiction to make the decision in the claimant's favour because disputes arising out of the defendants' withholding notice had not crystallized and were not part of the adjudication. In particular the notice of adjudication had been served after the defendant's valuation but before the final date for payment and before any withholding notice was due. A withholding notice was served during the adjudication but the defendant expressly refused to give the adjudicator jurisdiction to consider the 12 items set out in it.

3.70 **Held:** The judge cited *Cantillon v Urvasco* at paragraphs 54 and 55 but added the corollary that, if a defending party has not put forward a defence prior to the adjudication and does not put forward a particular defence in the adjudication, the adjudicator does not have jurisdiction to address such a defence even if it seems a sensible thing to do to save time and cost later. However, if the crystallized dispute referred to adjudication encompasses a particular defence, the defending party cannot withdraw that defence during the adjudication to fight another day, so to speak, on that particular defence (paragraph 24 of the judgment). In this case the dispute about the interim valuation crystallized when the defendant provided its valuation and it was illogical to say that there can be no dispute about an interim valuation until after it falls due (paragraph 26 of the judgment). The dispute about the value of the interim valuation (that had been referred to adjudication) included the issues that were also later contained in the withholding notice as reasons for reducing the valuation. However two items in the withholding notice that were entirely new and not part of the dispute that had crystallized earlier had been validly withheld from the adjudicator and he had no jurisdiction to make decisions about them (paragraph 32 of the judgment). The decision dealing with the two items was severed from the rest and the balance of the decision on valuation was enforced.

Table 3.4 Table of Cases: Determining the Scope of the Dispute

Title	Citation	Issue
KNS Industrial Services (Birmingham) Ltd v Sindall Ltd	[2001] 17 Const LJ (TCC), Judge LLoyd	**Scope of dispute included any ground for non payment** The reference to adjudication of a dispute as to non-payment of monies allegedly due to a subcontractor conferred jurisdiction to consider any ground that justified non-payment.
Sindall Ltd v Abner Solland and ors	[2001] 3 TCLR 30 (TCC), Judge LLoyd	**Scope of dispute includes sub-disputes** 'Where a dispute is referred, there is comprehended within it all its constituent elements, including sub-disputes, contentions, issues (some of which might have been referred separately) in other words all the ingredients which go into the dispute referred.'
C&B Scene Concept Design v Isobars	[2002] BLR 93, [2002] EWCA Civ 46 (CA)	**Question of contract interpretation not one of jurisdiction** Where the scope of the dispute was agreed between the parties as the employer's obligation to make payment and contractor's entitlement to receive payment, the adjudicator had to decide a question of law as to whether the contract payment terms or the statutory Scheme applied. That was a question of law within his jurisdiction. These points did not affect his jurisdiction.
London & Amsterdam Property Ltd v Waterman Partnership Ltd	[2004] BLR 179, [2003] EWHC 3059 (TCC), Judge Wilcox	**Late service new material: ambush** Adjudicator had breached the rules of natural justice by failing to exclude additional material put forward by claimant at a late stage, or to give sufficient opportunity to deal with it.
AWG Construction Services Ltd v Rockingham Motor Speedway Ltd	[2004] EWHC 888, [2004] CILL 2154 (TCC), Judge Toulmin	**Late introduction: new issues** The referring party introduced a new claim during the adjudication about negligent design and inadequate drainage construction. The adjudicator's decision found in favour on the new issue. This was outside his jurisdiction. This is cited in Key Cases above.
McAlpine PPS Pipeline Systems Ltd v Transco PLC	[2004] BLR 352, [2004] EWHC 2030 (TCC), Judge Toulmin	**Nature of dispute altered in reply: ambush** The referring party served substantial evidence in its reply to make good a lack of substantiation. This changed the nature of the dispute to such an extent that there was a reasonable prospect of proving the dispute was not the same. The decision was not summarily enforced.
Buxton Building Contractors Ltd v The Governors of Durand Primary School	[2004] BLR 374, [2004] EWHC 733 (TCC), JudgeThornton	**Scope of dispute included all questions referred** Adjudicator decided sum certified was due as no withholding notice had been served. On enforcement judge decided that the adjudicator's decision showed that he had not considered all issues including whether D's correspondence amounted to a valid withholding notice. The cross-claim could and should have been set off. The adjudicator had not fulfilled his statutory duty imposed by s. 108(2)(c) of the Act to decide the dispute referred to him in its entirety. The failures constituted serious irregularities in the adjudication procedure and were unfair. The decision was unenforceable on a summary judgment. Doubted by CA, Chadwick LJ, in *Carillion Construction Ltd v Devonport Royal Dockyard Ltd* [2005] EWHC Civ 1358.

Title	Citation	Issue
William Verry (Glazing Systems) Ltd v Furlong Homes Ltd	[2005] EWHC 138 (TCC), Judge Coulson	**New defences** The respondent's alterations to its extension of time claim was not a new claim but a fuller explanation of that claim. In any event the respondent was entitled to rely on any point available to it to defend itself, irrespective of whether notified previously. This is cited in Key Cases above.
Ardmore Construction Ltd v Taylor Woodrow Construction Ltd	[2006] CSOH 3, Lord Clarke	**Findings not part of case argued** An adjudicator decided that a party was entitled to payment for overtime work based on an alleged acquiescence or verbal instructions. Neither contention had been contained in the material placed before the adjudicator. The court found there had been a breach of natural justice.
Multiplex Constructions Ltd v West India Quay Developments	111 Con LR 33, [2006] EWHC 1569 (TCC), Ramsey J	**Adjudicator's conclusions on EOT claim departed from case presented** WIQ challenged an adjudicator's decision on the grounds (inter alia) that the findings on individual claims for extensions of time differed from the way the case was presented. This was rejected on the facts. Further, looking at the matter more broadly, the adjudicator had to decide the question of an extension of time. WIQ contended that the date should remain at 29 March 2004. Multiplex contended that it should be extended to 11 June 2004, and the adjudicator held, based on claims raised by Multiplex, that the date should be 9 May 2004. The adjudicator is obliged to make his decision and in doing so he has to assess the case put forward by each party. He did not, as in *Balfour Beatty v Lambeth*, create his own as-built programme and then derive his own critical path, thereby adopting his own methodology. Rather, he came to conclusions on the basis of the evidence, analysis, and submissions put before him.
Multiplex Constructions (UK) Ltd v Mott MacDonald Ltd	[2007] 110 Con LR 63, [2007] EWHC 20 (TCC), Jackson J	**Dispute included wider question** Although an adjudicator's jurisdiction is circumscribed by the scope of the pre-existing dispute, and the referring party cannot widen what is in dispute by drafting an unduly broad or optimistic notice of adjudication, nevertheless on a fair reading of the correspondence in the case, one issue in contention was the true meaning of the phrase 'all records pertinent to the services'. That was the question answered by the adjudicator and the challenge to enforcement failed.
Cantillon Ltd v Urvasco Ltd	[2008] BLR 250, [2008] EWHC 282 (TCC), Akenhead J	**Dispute includes all defences available at law** The TCC judge confirmed that the scope of the dispute referred was wide enough to include any defence available as a matter of law to the responding party (confirmed *KNS Industrial*). At paras. 54–5 of the judgment the TCC judge set out four propositions regarding how to define the scope of the dispute (see Key Cases above).
Letchworth Roofing Company v Sterling Building Company	[2009] EWHC 1119 (TCC), Coulson J	**But dispute cannot include cross-claim if necessary witholding notice had not been served** Although a responding party may rely on defences available as a matter of law, it did not mean that a cross-claim could be relied on in the absence of a withholding notice where one would otherwise have been required. That was a defence not available at law. When a point is taken in adjudication the adjudicator is entitled to decide the cross-claim is excluded because no withholding notice was served.

(*Continued*)

Table 3.4 *Continued*

Title	Citation	Issue
OSC Building Services Ltd v Interior Dimensions Contracts Ltd	[2009] EWHC 248 (TCC), Ramsey J	**Scope of dispute not widened by referral, background communications relevant to construe dispute** The judge rejected the allegation that the Adjudication Notice had been limited to a claim about the absence of a withholding notice and that this had been widened to a valuation dispute in the referral. The judge found the Adjudication Notice was referring the broader question of the sum due to the contractor. The scope of the dispute referred was determined by reading the Notice in the context of the background communications between the parties.
Allied P&L Ltd v Paradigm Housing Group Ltd	[2010] BLR 59, [2009] EWHC 2890 (TCC), Akenhead J	**Scope of crystallized dispute** Although a dispute had crystallized about whether termination was justified, there was no dispute crystallized as to the financial consequences. However no valid reservation of rights was made and jurisdiction was effectively conferred on the adjudicator to decide that question.
PC Harrington Ltd v Tyroddy Construction Ltd	[2011] EWHC 813 (TCC), Akenhead J	**Adjudicator mistaken as to scope of dispute** The adjudicator decided that the scope of the dispute referred (a claim for repayment of retention) was not wide enough to permit him to consider the defence of the responding party as to the true value of the final account (which PCH said was less then the amount paid to date). The TCC judge found that the adjudicator was mistaken about the scope of the dispute, which had validly included the defences raised, and through his honest mistake had committed a breach of natural justice. The decision was not enforced.
Witney Town Council v Beam Construction (Cheltenham) Ltd	[2011] BLR 707, [2011] EWHC 2332 (TCC), Akenhead J	**Guidance on how to determine scope of dispute** The TCC judge gave further guidance about how to determine the scope of the dispute (see para. 38 of the judgment cited in Key Cases above).
Herbosh-Kiere Marine Contractors Ltd v Dover Harbour Board	[2012] BLR 177, [2012] EWHC 84 (TCC), Akenhead J	**Adjudicator's formula different from that in referral or defence** The adjudicator decided a claim using a formula that was not proposed by either party. The court decided the decision was unenforceable both because the decision was outside the scope of the dispute referred and because there had been a breach of natural justice. The adjudicator went on a frolic of his own.
Working Environments Ltd v Greencoat Construction Ltd	[2012] BLR 309, [2012] EWHC 1039 (TCC), Akenhead J	**Dispute includes all defences unless validly withheld from adjudicator** The scope of the dispute will include all defences available as a matter of law unless they are new and have neither been referred to the adjudicator in the adjudication notice nor raised by the responding party as a defence in the adjudication (see Key Cases above).

(3) The Procedure and Time Limits Required by s. 108(2)

3.71 Section 108 makes it obligatory for all construction contracts to contain an adjudication procedure, which conforms at least to the Act's minimum requirements, which are:

- The contract shall enable a party to give notice at any time of his intention to refer a dispute to adjudication (s. 108(2)(a)).

- The contract shall provide a timetable with the object of securing the appointment of the adjudicator and referral of the dispute to him within seven days of such notice (s. 108(2)(b)).
- The contract shall require the adjudicator to reach a decision within 28 days of referral or such longer period as is agreed by the parties after the dispute has been referred (s. 108(2)(c)).
- The contract shall allow the adjudicator to extend the period of 28 days by up to 14 days, with the consent of the party by whom the dispute was referred (s. 108(2)(d)).
- The contract shall impose a duty on the adjudicator to act impartially (s. 108(2)(e)).
- The contract shall enable the adjudicator to take the initiative in ascertaining the facts and the law (s. 108(2)(f)).
- The contract shall provide that the decision of the adjudicator is binding until the dispute is finally determined by legal proceedings, arbitration, or agreement (s. 108(3)).
- The contract shall provide that the adjudicator is not liable for anything done or omitted in the discharge of his functions unless done in bad faith (s. 108(4)).

3.72 Challenges to the enforcement of the adjudicators' decisions have been made on the grounds that the procedural time limits of s. 108(2) have not been complied with. The consequence of a procedural contravention of the time limits in s. 108(2) may be to deprive the adjudicator of jurisdiction and invalidate a decision reached. These challenges to enforcement are discussed more fully in Chapter 9. The sections that follow consider briefly the meaning of the time limits contained within s. 108(2), what the construction contract must provide to comply with s. 108(2), and the consequences of non-compliance.

'At Any Time': s. 108(2)(a)

3.73 The construction contract must contain an adjudication procedure that enables a party to give notice of intention to refer a dispute to adjudication at any time.[73] If the contract contains a stipulation that a notice of intention to refer may only be served after the happening of some event, that provision will offend s. 108(2)(a) of the Act and will be invalid. The effect will be to invalidate the entire contractual adjudication scheme. In such a situation the statutory Scheme will apply in place of the contractual adjudication scheme.[74] In *John Mowlem & Co. plc v Hydra-Tight & Co. plc* (2001)[75] the court declared invalid a precondition that required the parties to serve a notice of dissatisfaction before adjudicating. The court said that this would delay the start of adjudication by four weeks and that contravened the requirement that the parties be able to adjudicate at any time.[76] Equally, a provision that the parties must mediate before a dispute could be referred to adjudication contravenes the right to refer a dispute at any time rendering the whole contractual adjudication scheme invalid; *R. G. Carter v Edmund Nuttall Ltd* (2000).[77]

3.74 Following the decision in *Hydra-Tight* the NEC 3 adjudication provisions were changed to remove this precondition. The first version of the ICE Conditions of Contract (7th edn) included a similar provision seeking to prevent a 'dispute' arising until after a 'matter of dissatisfaction' had been referred to and decided by the Engineer. The form has now been revised.

3.75 In *Banner Holdings Ltd v Colchester Borough Council* (2010)[78] the TCC judge considered a contract term that said the adjudicator could not vary or overrule the employer's decision to

[73] s. 108(2)(a).
[74] s. 108(5).
[75] 17 Const LJ 358.
[76] See also *Midland Expressway Ltd v Carillion Construction Ltd and ors* (2005) 106 Con LR 154, [2005] EWHC 2963 (TCC).
[77] 21 Jun. 2000 (TCC) (unreported), Judge Thornton.
[78] [2010] EWHC 139 (TCC).

operate a contractual provision to determine the contract. The judge said the clause was not objectionable because once the decision to determine had been taken it was too late to vary or reverse it. The issue that remained was whether the decision taken was valid and the clause did not restrict the right of the parties to refer that dispute to adjudication. However the judge said that if the clause had restricted the right of the parties to refer the disputes about the validity or conseqences of the determination, it would have been invalid as contrary to s. 108 of the Act.

3.76 If however there is no dispute because, for example, the issue has been compromised, then there will be no right to adjudicate. Hence, the right to adjudicate a dispute at any time does not operate to trump a legal defence the result of which is there is no dispute at all. Arguably the same result may be achieved by time bar provisions in a contract. However it is submitted the question of whether the settlement or the time bar provision operates to extinguish the dispute is a question that a party may choose to refer to an adjudicator for a decision. However as the issue goes to the jurisdiction of the adjudicator it is submitted that it is a matter that the adjudicator may not make a binding decision about unless given that authority by the parties.[79]

3.77 Some forms of contract provide that a Final Certificate shall be conclusive evidence that effect has been given to contract terms that add or subtract amounts from the Contract Sum. This may extinguish any dispute about matters which the contract says are decided conclusively by the certificate such that one party may object to an adjudication on the ground that there is no longer a dispute. Such clauses usually contain a period during which the certificate can be challenged. The JCT Standard Form of Contract (1998) at Clause 30.9.3 provides that a Final Certificate will not be binding on matters challenged by adjudication (and other legal proceedings) if commenced within 28 days after the Final Certificate is issued. In *Cubitt Building & Interiors Ltd v Fleetglade Ltd*[80] the responding party, having taken part in the adjudication, pointed out that the referral notice was provided on the eighth day instead of within seven days of the notice of adjudication. If correct, the referring party would have needed to start the adjudication again but that would have been more than 28 days after a Final Certificate had been issued. The judge rescued that adjudication decision by holding that the referral notice was provided in time in the circumstances of the case.

3.78 'At any time' means what it says and there is no time restriction on the right to refer a dispute to adjudication. It is, therefore, permissible to refer a dispute arising under the contract to adjudication after the work has been completed or the after contract has been repudiated or terminated. It is even possible to refer a dispute to adjudication after the expiry of the limitation period, although the referring party runs the risk that the limitation defence will be relied on and the adjudicator will make an award in the respondent's favour: *Connex South Eastern Ltd v M J Building Services Group plc* (2005).[81]

3.79 A dispute may be referred to adjudication even after court proceedings have been started in relation to the same matter, and no stay of enforcement of the adjudicator's decision will be warranted pending determination of the court action: *Herschel Engineering Ltd v Breen Property Ltd* (2000).[82] Save in exceptional circumstances the court will not impose

[79] However it is noted that in some cases the courts have taken the view that such a question is an issue of law and that the adjudicator may make a binding enforceable decision about such an issue of law. There is a tension between such cases and the line of authority that says an adjudicator cannot decide his own jurisdiction in a binding manner.

[80] [2007] 110 Con LR 36; [2006] EWHC 3413.

[81] [2005] 1 WLR 3323, [2005] BLR 201 (CA).

[82] [2000] BLR 272.

conditions that prevent a party from pursuing an adjudication, whether or not there were concurrent court proceedings on the same issue: *Mayor & Burgesses of Camden LBC v Makers UK Ltd* (2009).[83]

3.80 In the case of *DGT Steel and Cladding Ltd v Cubitt Building and Interiors Ltd* (2007)[84] Judge Coulson considered the related question of whether court proceedings once issued should be stayed to allow for an adjudication to proceed. Having lost an adjudication in which they claimed £193,815, DGT issued a High Court action for a larger claim. There was a dispute between the parties as to the degree of overlap between the original adjudication and the court proceedings. Cubitt relied on a clause in the contract that any dispute 'shall' be referred to adjudication first and argued that the court proceedings should be stayed to allow for an adjudication to be conducted. The court had to consider whether the disputes were the same for the purposes of a further adjudication and then to determine whether a temporary stay should be granted to restrain court proceedings to allow the adjudication to take place. Judge Coulson determined that there are three principles relevant to this issue:

1. The court will not grant an injunction to prevent one party from commencing and pursuing adjudication proceedings even if there is already court or arbitration proceedings in respect of the same dispute.
2. The court has an inherent jurisdiction to stay court proceedings issued in breach of an agreement to adjudicate.
3. The court's discretion as to whether or not to grant a stay should be exercised in accordance with the principles noted above. If a binding adjudication agreement has been identified then the persuasive burden is on the party seeking to resist the stay to justify their stance.

3.81 Where there was a mandatory adjudication agreement, if it was not the same dispute as in the previous adjudication, the matter should be referred to adjudication. The court also commented *obiter* that even if there was no such mandatory agreement, s. 108 of the Act provided for adjudication and there should still be a stay unless there was a very good reason why not.

3.82 However, absent a contract adjudication provision expressed in mandatory terms, s. 108 (1) contains a right and not an obligation to adjudicate. Where the right to adjudicate proceeds under the Act, or a clause framed in the similar language as s. 108(1), a stay of court or arbitration proceedings will not usually be granted for the sole purpose of allowing adjudication to proceed: *Cubitt Building & Interiors Ltd v Richardson Roofing (Industrial) Ltd* (2008).[85]

Key Cases: 'At Any Time'

Herschel Engineering Ltd v Breen Property Ltd [2000] BLR 272 (TCC)

3.83 **Facts:** Herschel issued county court proceedings in relation to unpaid invoiced sums. The county court judge stayed the matter for 28 days for adjudication to be considered. Herschel started an adjudication and Breen took no part. A decision was given ordering Breen to pay £17,355 plus VAT. At the enforcement Breen argued that the court should not countenance two concurrent proceedings and/or that Herschel had waived its right to adjudicate or had repudiated the adjudication agreement by starting the county court proceedings.

[83] [2009] EWHC 605 (TCC), applying *Herschel*.
[84] [2007] EWHC 1584.
[85] [2008] BLR 354.

3.84 **Held:** (1) A reference to adjudication could be made at any time including after the issue of court proceedings relating to the same dispute. (2) Considerations of waiver and repudiation did not apply; a party was not put to its election and did not waive its right to adjudicate by starting court proceedings. (3) No stay of enforcement was warranted pending determination of the county court proceedings.

Connex South Eastern Ltd v MJ Building Services Group plc [2005] 1 WLR 3323, [2005] BLR 201 [2005] EWCA Civ 193 (CA)

3.85 **Facts:** A contract between MJ Building and two sister companies, Connex South Eastern Ltd and Connex South Central Ltd, concerned a rolling programme of works at railway stations. South Central was bought by another company and in February 2002 agreed with MJ to restructure the project. South Eastern denied liability to pay for materials purchased by MJ for the original works. In November 2002 MJ alleged a repudiatory breach of contract which they accepted. An adjudication was started in February 2004. The court was asked by South Eastern to grant declarations: (1) that the February 2002 restructuring agreement released both South Eastern and South Central from their obligations, and (2) that it was an abuse of process to start adjudication proceedings so long after the purported acceptance of repudiation.

3.86 **Held:** On the abuse of process argument: the Act at s. 108(2) provided that a party was entitled to give notice 'at any time' of an intention to refer a dispute to adjudication. Those words were to be given their literal and ordinary meaning and there was nothing to prevent a party from referring a dispute to adjudication at any time, even after the expiry of the relevant limitation period. Adjudicators had no powers, either under the 1996 Act, or under the Scheme to strike out or stay an adjudication as an abuse of process. Dyson LJ said:

> 37. Mr Ashton submits that 'Parliament was content for adjudication to take place after the cessation of work because this was seen in the context of a procedure which was (a) quick, (b) cheap and (c) a temporary decision. Once this quick, cheap and temporary decision had been taken, it could then be followed by a permanent decision via arbitration or the courts'. But he argues that if, as a result of the passage of time, it is no longer possible to have a quick, cheap and temporary adjudication, then it is an abuse of process to permit an adjudication to take place.
>
> 38. I cannot accept these submissions. The phrase 'at any time' means exactly what it says. It would have been possible to restrict the time within which an adjudication could be commenced, say, to a period by reference to the date when work was completed or the contract terminated. But this was not done. It is clear from Hansard that the question of the time for referring a dispute to adjudication was carefully considered, and that it was decided not to provide any time limit for the reasons given by Lord Lucas. Those reasons were entirely rational.
>
> 39. There is, therefore, no time limit. There may be circumstances as a result of which a party loses the right to refer a dispute to adjudication: the right may have been waived or the subject of an estoppel. But subject to considerations of this kind, there is nothing to prevent a party from referring a dispute to adjudication at any time, even after the expiry of the relevant limitation period. Similarly, there is nothing to stop a party from issuing court proceedings after the expiry of the relevant limitation period. Just as a party who takes that course in court proceedings runs the risk that, if the limitation defence is pleaded, the claim will fail (and indeed may be struck out), so a party who takes that course in an adjudication runs the risk that, if the limitation defence is taken, the adjudicator will make an award in favour of the respondent.

> 40. In the civil courts, the concept of 'abuse of process' is well understood. It applies in a number of different contexts: see Civil Procedure Volume 1, para 3.4.3. But neither the Act nor the Scheme for Construction Contracts (England and Wales) Regulations (S1 1998/649) gives an adjudicator the power to strike out or stay an adjudication for abuse of process. Indeed, they contain no reference to 'abuse of process'. In my judgment, the only question is whether there is any limit on the time within which a party may refer a dispute to adjudication. The answer to that question depends on a proper interpretation of section 108(2) of the Act, and not on an application of the principles developed by the courts to control their own process so as to prevent abuse. In my judgment, there is nothing in the Act which indicates that the words 'at any time' should be construed as bearing other than their literal and ordinary meaning.
>
> 41. I confess that I had some difficulty in understanding the logic of Mr Ashton's submission. I can accept that Parliament intended adjudication to be quick and (relatively) cheap, although it may not have been entirely successful in bringing this about. But that says nothing about when the quick and (relatively) cheap adjudication may be commenced. There is no link between the speed and expense of an adjudication and the time when it starts. An adjudication started before practical completion may be complex, slow and expensive. Conversely, an adjudication started long after practical completion may be simple, quick and cheap. Nor do I understand why an adjudication conducted long after practical completion cannot on that account result in a decision which has provisional or temporary effort only.

Table 3.5 Table of Cases: 'At Any time'

Title	Citation	Issue
A&D Maintenance and Construction Ltd v Pagehurst Construction Services Ltd	July 1999 CILL 1518 (TCC), Judge Wilcox	**Dispute may be adjudicated after contract determined** The fact that the subcontract was determined in November 1999 did not prevent the dispute being a 'dispute arising under the contract' and did not affect the right of the claimant to issue notices of adjudication subsequently. The adjudication provisions of a contract survive determination of a contract.
Cygnet Healthcare plc v Higgins City Ltd	[2000] 16 Const LJ 394 (TCC), Judge Thornton	**Adjudication decision delayed to await arbitrator ruling on jurisdiction question** Existence of contract referred to arbitration. Before the arbitration award was made a dispute under the 'contract'was referred to adjudication. In the circumstances (this was an issue which went to jurisdiction) the adjudicator's decision held in abeyance in anticipation of arbitrator's award.
Herschel Engineering Ltd v Breen Property Ltd	[2000] BLR 272 (TCC), Dyson J	**Adjudication permitted even when court proceedings already being pursued** A reference to adjudication may be made at any time including after the issue of court proceedings in respect of the same dispute.
John Mowlem & Co. plc v Hydra-Tight Ltd	[2002] 17 Const LJ 358 Judge Toulmin	**Preconditions to right to adjudicate invalid** The contract required a notice of dissatisfaction and meeting before the parties could adjudicate a dispute. That would have delayed the start of an adjudication by four weeks. That conflicted with the statutory right to adjudicate at any time. It rendered the whole adjudication procedure invalid and it was replaced with the Scheme.

(Continued)

Table 3.5 *Continued*

Title	Citation	Issue
R. G. Carter Ltd v Edmund Nuttall Ltd	21 Jun. 2000 (TCC) (unreported), Judge Thornton	**Precondition to right to adjudicate invalid** A term that made it mandatory to mediate before starting an adjudication was invalid because it was contrary to the 1996 Act and the right to adjudicate at any time.
Connex South Eastern Ltd v MJ Building Services Group plc	[2005] 1 WLR 3323, [2005] BLR 201, [2005] EWCA Civ 193 (CA), Dyson LJ	**'At any time' includes after repudiation** Section108(2) of the 1996 Act enabled a party to refer a dispute to adjudication at any time and a notice of adjudication given by a party long after it had accepted a repudiatory breach of contract was not an abuse of process.
DGT Steel and Cladding Ltd v Cubitt Building and Interiors Ltd	[2007] EWHC 1584 (TCC), Judge Coulson	**Stay of High Court action to allow adjudication** A stay of High Court proceedings was granted to allow adjudication where the adjudication clause was framed in mandatory terms.
Cubitt Building & Interiors Ltd v Richardson Roofing (Industrial) Ltd	[2008] BLR 354 (TCC), Akenhead J	**Adjudication is right but not obligation** Clear words in a contract are required to make adjudication a prerequisite for arbitration or litigation. There was neither a binding agreement under the particular contract nor under the s. 108 of the 1996 Act to adjudicate prior to other forms of proceedings. Adjudication is a right under the Act, but not an obligation.
Mayor & Burgesses of Camden LBC v Makers UK Ltd	124 Con LR 32, [2009] EWHC 605 (TCC), Akenhead J	**Inappropriate to impose conditions that would restrict right to adjudicate** At an application to set aside judgment in default, the claimant asked for conditions to be imposed on the defendant that they be prevented from adjudicating the dispute. The claimant argued that Makers UK Ltd were in a very poor financial condition and enforcement of any adjudication decision would in any event be stayed on the grounds of their impecuniosity, and so adjudication would be a waste of time and money. The court refused the condition because it would be inappropriate to fetter the right to adjudicate.
Banner Holdings Ltd v Colchester Borough Council	[2010] EWHC 139 (TCC), Coulson J	**Contract may not prevent adjudicator from deciding certain questions** The contract said the adjudicator could not vary or overrule a decision by the Council about contract determination. The judge found this did not offend against the 'at any time' rule because (i) the decision had been taken by the Council and all the clause was saying was that the adjudicator did not have power to reverse that decision but (ii) the adjudicator could still be asked whether the Council's decision was a valid one—which was what the dispute was about. If the clause had prevented the adjudicator from deciding a dispute he would otherwise have been allowed to decide it would have been invalid as contrary to the Act.

Appointment and Referral: s. 108(2)(b)

3.87 A construction contract must provide a timetable which has the object of securing the appointment of the adjudicator and the referral of the dispute to him within seven days of the notice of intention to refer.[86] The seven-day period begins immediately after the date the notice of intention to refer is given.[87] If the construction contract fails to comply with this

[86] s. 108(2)(b).

[87] s. 116(2) deals with the reckoning periods of time.

provision then its entire adjudication scheme will be invalidated and replaced by the statutory Scheme.[88]

3.88 Equally, if the contract fails to include a valid appointment mechanism the Scheme applies. If written adjudication provisions have not been incorporated into the contract then the Scheme applies and an adjudicator appointed under the Scheme in such a situation will be validly appointed: *Watson Building Services Ltd v Harrison* (2001).[89] If the contract provisions are unworkable then the Scheme applies.[90] However if the referring party fails to comply with the agreed mechanism for appointing the adjudicator, such that an adjudicator is appointed by the wrong nominating body, the adjudicator will not have jurisdiction: *IDE Contracting v R. G. Carter Cambridge Ltd* (2004).[91]

3.89 The Scheme (at paragraphs 2–6) contains the various options as to how and when the adjudicator is to be appointed. In any particular case the time it takes to secure an appointment will differ and therefore the Scheme provides that where an adjudicator has been selected, the referring party shall refer the dispute to that adjudicator not later than seven days from the notice of adjudication. In other words the time allowed for the referral document to be served is measured from the date of the notice of intention to refer to adjudication. In this way the seven-day limit of s. 108(2)(b) is maintained and is not affected by how long it takes to appoint the adjudicator. An adjudicator appointed on day seven must therefore have the referral notice sent to him on the same day.

3.90 The seven-day time limit is mandatory. Chapter 9 contains a detailed discussion about the consequences of there being either a late appointment or late referral of the dispute to the adjudicator.[92]

Decision within 28 Days: s. 108(2)(c)

3.91 A construction contract must include a procedure that requires the adjudicator to reach a decision within 28 days of the referral or such longer period as is agreed by the parties after the dispute has been referred.[93] The 28-day period starts on the day after the date on which the referral notice is served.[94] The effect of this provision is that the adjudicator's jurisdiction lasts only 28 days unless validly extended.

3.92 The contract must also provide that the referring party can extend the period of 28 days to 42 days.[95] Any extension after that must be consented to by both parties. Any construction contract may include individual provisions as long as they do not conflict with s. 108(2)(c). A contract term that the decision must be reached within 45 days would be invalid as s. 108(2)(c) says the parties may only reach such agreement 'after the dispute has been referred'.

3.93 A decision reached outside the 28-day period is a nullity unless there is a valid extension of that period. Chapter 9 below contains a detailed discussion about the consequences of the decision being provided after the 28 or 42 days.

---

[88] s. 108(5).
[89] [2001] SLT 846.
[90] *Amec Projects Ltd v Whitefriars City Estates Ltd* [2005] 1 All ER 723, [2004] EWCA Civ 1418.
[91] [2004] BLR 172, [2004] EWHC 36 (TCC).
[92] Chap 9 at 9.27–9.28 and 9.29 to 9.34.
[93] s. 108(2)(c).
[94] s. 116(2).
[95] s. 108(2)(d).

(4) Section 108(2)(e) and (f)

3.94 The construction contract must impose a duty to act impartially on the adjudicator.[96] The scope of the duty to act impartially is considered in detail in Chapters 10 and 11.

3.95 The construction contract must also enable the adjudicator to take the initiative in ascertaining the facts and the law.[97] This is subject to the overriding duty to act fairly and within the ambit of the dispute which the adjudicator has been asked to decide: *McAlpine v Transco* (2004).[98] The scope of the power to ascertain the facts and the law is discussed in detail in Chapters 4 and 10 .

(5) The Binding Nature of the Decision: s. 108(3)

3.96 The construction contract must provide that the decision of the adjudicator is binding until the dispute is finally determined by legal proceedings, by arbitration (if the contract provides for arbitration or the parties agree to arbitration), or by agreement (s. 108(3)). This means that the adjudicator's decision is to be complied with by the parties until the dispute is subsequently resolved by final proceedings, or until the parties agree to set it aside. This is often described as 'temporary finality'. If the adjudicator's decision is not complied with either party may apply to the High Court for an enforcement order. The Technology and Construction Courts have a special speedy procedure for such enforcement applications and details can be found in the TCC Guide. This is discussed in more detail in Chapter 7 .

(6) The Adjudicator's Immunity: s. 108(4)

3.97 The contract shall also provide that the adjudicator is not liable for anything done or omitted in the discharge or purported discharge of his functions as adjudicator unless the act or omission is in bad faith (s. 108(4)). The employee or agents of the adjudicator shall also be similarly protected.[99]

(7) Non-compliance with Statutory Requirements: s. 108(5)

3.98 If the construction contract conflicts with or fails to comply with the minimum requirements contained in s. 108(1)–(4) then, according to s. 108(5), the entire contract adjudication scheme is replaced with the Scheme: *Aveat Heating Ltd v Jerram Falkus Construction Ltd.*[100] This means that any single non compliance will result in the entirety of the contractual adjudication procedure being replaced by the Scheme. This was confirmed in *Banner*

[96] s. 108(2)(e).

[97] s. 108(2)(f).

[98] [2004] EWHC 2030 (TCC), at para. 124.

[99] The cases about an adjudicator's entitlement to his fees touch on the question of the adjudicator's immunity, see Ch. 4 at 4.125–4.138.

[100] 113 Con LR 13 [2007]; however, cf. *Ballast plc v The Burrell Company (Construction Management) Ltd* [2001] BLR 529, in which Lord Reid considered that an adjudication might be governed partly by the express terms of the contract and partly by s. 108(1)–(4). Whilst this may be the case in respect of the payment provisions of a contract, it is thought not to be the effect of s. 108(5).

Holdings Ltd v Colchester Borough Council (2010),[101] and *Yuanda (UK) Co. Ltd v WW Gear Construction Ltd* (2010).[102]

3.99 This policy of complete replacement contrasts with the approach adopted in ss. 109–111 of the Act. Those provisions stipulate that if the contract does not contain the Act's obligatory payment terms then it is the *relevant* provisions from the Scheme for Construction Contracts that will apply.[103] In other words in the case of payment provisions the contract may be supplemented by some aspects of the Scheme, whereas the adjudication scheme will only be applied wholesale.

[101] [2010] EWHC 139 (TCC), [2010] 131 Con LR 77.
[102] [2010] EWHC 720 (TCC), [2010] BLR 435 (TCC). See also *Sprunt Ltd v LB Camden* [2011] EWHC 31919 (TCC) at paras. 28–31.
[103] See ss. 109(3), 110(3), 111(3).

4

THE STATUTORY SCHEME

(1) Introduction

4.01 This chapter examines the statutory adjudication scheme contained in Part 1 of the Scheme for Construction Contracts ('the Scheme').[1] Originally created to work together with the Housing Grants Construction and Regeneration Act 1996 ('the 1996 Act') the Scheme has been amended to accommodate the changes introduced by the Local Democracy, Economic Reform and Construction Act 2009 ('the 2009 Act'). The amended Scheme came into force

[1] The original Scheme referred to in this chapter is contained in the Statutory Instrument 1998/649 'The Scheme for Construction Contracts (England and Wales) Regulations 1998'. Throughout this chapter the paragraph numbers of the Scheme referred to are those in Part 1. A separate statutory instrument contains the Scheme for Scotland and for Northern Ireland.

in England on 1st October 2011.[2] For the purpose of this chapter they will be referred to as the Scheme or the amended Scheme.

4.02 The Scheme contains provisions that give effect to the mandatory requirements in s. 108 of the 1996 and 2009 Acts, plus additional provisions which will govern any adjudication conducted under the Scheme.

When Does the Scheme Apply?

4.03 The Scheme will apply if it is so stated in the contract, if the parties have chosen it to be the applicable adjudication procedure, or if there is no agreed adjudication procedure in a construction contract.[3] In the latter situation the terms of the Scheme have effect as implied terms of the contract.[4]

4.04 If an adjudication scheme included in a construction contract (or otherwise agreed) conflicts with or fails to comply with the minimum requirements contained in s. 108(1)–(4) then according to s. 108(5) that entire contract adjudication scheme is replaced with the statutory Scheme for Construction Contracts: *David McLean Housing Contractors Ltd v Swansea Housing Association Ltd* (2001),[5] *Aveat Heating Ltd v Jerram Falkus Construction Limited* (2007).[6] This is discussed in more detail in Chapter 3 at 3.98–3.99.

4.05 Where a contract has an adjudication procedure that complies with the Act the juridical nature of adjudication is contractual rather than statutory. This means that, assuming the contractual provisions are in accordance with the Act, it is those provisions which have to be construed and operated by the parties and must be at the forefront of the consideration of the parties' respective rights and liabilities in an adjudication: *Cubitt Building Interiors Ltd v Fleetglade Ltd* (2006).[7] This is important as parties sometimes focus on the provisions of the Act instead of their own procedure which may contain more onerous terms than the Act itself.

Changes Introduced by the 2009 Act

4.06 The main changes made to the Scheme are as follows:

- To give effect to the new s. 108(3A) the amended Scheme now includes at paragraph 22A an express power to correct the decision to remove clerical or typographical errors. This is discussed below at 4.109–4.117.
- The new s. 108A renders ineffective any provision about the allocation of 'costs relating to the adjudication' unless it is made in writing, contained in the construction contract, and confers power on the adjudicator to allocate his fees and expenses between the parties or, alternatively, it is an agreement made in writing after the notice of adjudication has been given. The Scheme paragraphs 9(4), 11(1) and 25, which concern the adjudicator's fees, have all been amended and the following words introduced: 'Subject to any contractual provision pursuant to section 108A(2) of the Act, the adjudicator may determine how the payment is to be apportioned and the parties are jointly and severally liable for any sum which remains outstanding following the making of any such determination.' These changes are discussed at 4.125–4.147 below.

---

[2] Statutory Instrument 2011/2333 – 'The Scheme for Construction Contracts (England and Wales) Regulations 1998 (Amendment) (England) Regulations 2011'. Equivalent statutory instruments were made for Wales (Regulation 2011 No. 1715) and for Scotland (Regulation 2011 No. 371) and Northern Ireland (Regulation 2012 No. 365).

[3] s. 108(5).

[4] s. 114(4).

[5] [2002] BLR 125.

[6] [2007] EWHC 131 (TCC), 113 Con LR 13 [2007].

[7] [2006] EWHC 3413 (TCC), 110 Con LR 36.

- The power to order peremptory relief at paragraph 23(1) has been deleted, as has the express application to the Scheme of s. 42 of the Arbitration Act 1996 in paragraph 24.
- There are relatively minor drafting changes made to paragraphs 1(1), 7(3), 15(b), 19(1), 20(b), and 21.

4.07 It would appear from the language of the new s. 108A that it is intended to render ineffective only an offending clause as opposed to the whole adjudication procedure. This is consistent with s. 108(5) whereby the Scheme is imported when the construction contract does not comply with s. 108(1)–(4) of the Act – compliance with the new s. 108A is not included within that mechanism. If this is right then standard-form procedures will not have to be amended to give effect to the new s. 108A, but any offending term concerning costs will simply be struck out.

4.08 The same may not be true for the new s. 108(3A). Many contract adjudication procedures do contain such a power to correct slips but those that do not will need to be amended to include this new power or face the risk that this omission will cause the entire adjudication procedure being struck down altogether pursuant to s. 108(5) and replaced by the Scheme.

(2) Starting the Adjudication: Scheme Paragraph 1

The Notice of Adjudication

4.09 The adjudication process is commenced by a notice of intention to refer a dispute to adjudication.[8] This document is most commonly referred to as the notice of adjudication, as it is in the Scheme, although terminology in individual contracts differs.

4.10 The Act does not contain any requirements for the form or content of the notice of adjudication. However the Scheme and most standard-form adjudication procedures require the notice to be in writing and some specify what information ought to be contained in it.

4.11 The Scheme requires the notice shall give a brief description of the dispute and of the parties involved, the details of when and where the dispute has arisen, the nature of the redress which is sought, and the names and addresses of the parties.[9] The various standard-form procedures differ in approach and care must be taken to ensure that the notice of adjudication conforms to the stipulations of the relevant adjudication procedure.

4.12 The notice needs to be drafted so as to avoid some common problems that can arise. In particular:

(1) If the contract adjudication scheme mirrors s. 108 and permits the reference of a single dispute then the notice of adjudication should be drafted in such a way as to define the issues between the parties as a single dispute. It may be possible to raise a jurisdictional challenge against a notice of adjudication that refers to several disputes if only a single dispute is permitted. It is usually possible to avoid this outcome by drafting the dispute in wide terms.[10] However there seems to be no reason why a contract could not go further than s. 108 and permit more than one dispute to be referred at a time. The Scheme 8(1) permits the adjudicator to consider more than one dispute as long as the parties consent.

(2) If the notice of adjudication describes the dispute in narrow specific terms there is a risk that changes in the dispute, which commonly occur as arguments develop, will be subject to the jurisdictional challenge that they fall outside the definition in the notice. The

[8] See s. 108(2)(a).

[9] Scheme para. 1(3).

[10] See also 4.18–4.20 for a fuller discussion of the single dispute issue.

notice of adjudication ought to describe the dispute in wide terms to protect against such arguments.[11]

(3) The notice of adjudication ought to comply with the requirements of the relevant adjudication scheme. Failure to comply with the formalities required may run the risk of an argument that the notice of adjudication is invalid. To avoid jurisdictional challenges based on inadequacies of the notice it ought to be drafted to contain (i) a brief description of the contract, (ii) a brief description of the issues, framed widely as one dispute, and (iii) the nature and extent of the remedy requested from the adjudicator.

(4) Where a claim is made for a sum of money the notice of adjudication should ask in the alternative for 'such other sum as the adjudicator thinks fit'. This gives the adjudicator jurisdiction to make a decision for a lesser sum if necessary. Without such a request the adjudicator's jurisdiction will most likely be restricted to finding wholly in favour of the referring party's case or not at all.

4.13 The notice of adjudication plays an important role in defining the scope of the dispute over which the adjudicator has jurisdiction. The adjudicator is appointed to decide the dispute set out in the notice of adjudication. It is then only that the dispute should be referred to him. Unless there is a later agreement to widen the dispute contained in the notice of adjudication, or this is permitted by the terms of the adjudication procedure, the scope of the dispute for adjudication is determined by that set out in the notice of adjudication.[12]

4.14 Hence a decision by the adjudicator on matters outside those in the notice of adjudication will exceed his jurisdiction unless the parties confer additional jurisdiction on him, one party waives the lack of jurisdiction and allows him to deal with the widened dispute, or there is an express term in the adjudication procedure which permits the dispute to be altered in the relevant way. The cases in Table 4.1 concerned arguments that the decision reached by the adjudicator went outside what was stated in the notice of adjudication. In some cases the court was able to construe the notice as being wide enough to cover the decision reached. The court will sometimes be willing to interpret the scope of the dispute contained in the notice of adjudication by reference to the prior communications between the parties on the subject, as in *OSC Building Services Ltd v Interior Dimensions Contracts Ltd* (2009)[13] where Ramsey J said it was the substance of the dispute that mattered and not the label given to it in the notice of adjudication.[14] However where the decision clearly goes beyond the scope of the notice it will be unenforceable for want of jurisdiction.

4.15 As discussed in more detail at 3.59 and 9.06, the scope of the dispute referred will include any legitimate defence available to the responding party, even though not expressly cited in the notice of adjudication. However, this only applies to defences that are legally available. So when it is argued before the adjudicator that a withholding notice would have been required but had not been served, the adjudicator is entitled to decide that the cross-claim is excluded as a consequence: *Letchworth Roofing Company v Sterling* (2009).[15]

---

[11] See also 3.57–3.60 for a discussion about determining the scope of the dispute referred.

[12] The interpretation of the character and scope of the dispute contained in the notice of adjudication may need to be conducted against the background facts: *Witney Town Council v Beam Construction* [2011] EWHC 2332 (TCC), [2011] BLR 707, at para. 38(v) where the TCC judge said it may be necessary to look at the background facts to understand what the true dispute was and whether there was a single dispute or more than one dispute.

[13] [2009] EWHC 248(TCC) at paras. 14(1) and 32.

[14] See also proposition (v) in para. 38 of the judgment in *Witney Town Council v Beam Construction Cheltenham Ltd* [2011] EWHC 2332 (TCC), [2011] BLR 707.

[15] [2009] EWHC 1119 (TCC). When an adjudicator may and may not exclude a defence is considered in Ch. 10 at 10.36–10.41.

Key Case: Notice of Adjudication

K. Griffin and anor (t/a K&D Contractors) v Midas Homes Ltd [2000] EWHC 182 (TCC), 78 Con LR

4.16 **Facts:** In an adjudication under the Scheme, the referring party served notice of adjudication in a letter dated 3 May 2000. The respondent alleged the notice was inadequate because it failed to identify which of the numerous items they were intending to refer to the adjudicator or the grounds on which they did so.

4.17 **Held:** The real dispute could not be seen from the notice. The 3 May letter took the short cut of referring to other correspondence as containing the details of the dispute. Whilst not objectionable it meant that, to be effective, the other correspondence needed to be sufficiently clear and record the dispute with clarity:

> Paragraph 1(3) of the Scheme provides that the notice of adjudication has to set out briefly –
>
> '(a) the nature and a brief description of the dispute and the parties involved, (b) details of where and when the dispute has arisen, (c) the nature of redress which is sought, and (d) the names and addresses of the parties to the contract ...'
>
> The purposes of such a notice are first, to inform the other party of what the dispute is; secondly, to inform those who may be responsible for making the appointment of an adjudicator, so that the correct adjudicator can be selected; and finally, of course, to define the dispute of which the party is informed, to specify precisely the redress sought, and the party exercising the statutory right and the party against whom a decision may be made so that the adjudicator knows the ambit of his jurisdiction.

The TCC judge decided that, in this particular case, the letter failed to comply with paragraph 1(3) of the Scheme in respect of all but one issue.

Table 4.1 Table of Cases: Notice of Adjudication

Title	Citation	Issue
K. Griffin and anor v Midas Homes Ltd	[2000] EWHC 182 (TCC), [2000] 78 Con LR (TCC), Judge LLoyd	**Purpose of notice** The case concerned the terms of the Scheme. The purpose of the adjudication notice is to (1) enable the other side to know what dispute is referred, (2) enable the parties to select the right adjudicator, and (3) let the adjudicator know the scope of his jurisdiction.
F. W. Cook Ltd v Shimizu (UK) Ltd	[2000] BLR 199 (TCC), Judge LLoyd	**When relief given not asked for in notice** The notice of adjudication did not ask for a decision as to how much the responding party was obliged to pay and the responding party said the decision to that effect was made without jurisdiction. The TCC judge interpreted the decision as having provided opinion on specific issues (as requested in the notice) and not a decision that £X was due and owing which would have been unenforceable.

Title	Citation	Issue
KNS Industrial Services Ltd v Sindall Ltd	[2001] 17 Const LJ 170, Judge LLoyd	**Adjudication notice defines dispute referred** The judge expressed the view that the referral notice and subsequent submissions do not either cut down or expand the dispute as set out in the notice of adjudication (unless there is an agreement to do so).
Mecright Ltd v T&A Morris Developments Ltd	26 Jun. 2001 (TCC) (unreported), Judge Seymour	**Decided issue not in notice** Case concerned the Scheme paras. 1 and 20. The notice of adjudication did not include dispute about how much M was entitled to be paid. The decision to that effect was therefore reached without jurisdiction and was not enforced.
Jerome Engineering Ltd v Lloyd Morris	[2002] CILL 1827, Judge Cockcroft	**Contract terms permit matter left for referral** The case concerned DOM/2 clause 38A. The notice of adjudication did not specifically ask for a decision that the responding party pay a certain sum. The judge found that there was no need to set out the amount claimed because the DOM/2 conditions allow for the specifics to be set out in the referral notice. The absence of a statement of relief was found not to be fatal in this case.
Chamberlain Carpentry and Joinery Ltd v Alfred McAlpine Construction Ltd	[2002] EWHC 514 (TCC), Judge Seymour	**Letter was notice of adjudication** The case concerned McAlpine's own adjudication rules and the interpretation of the notice of adjudication. It was held that a letter was free-standing as a notice of adjudication. The letter request for payment of a sum, calculated by reference to eight particular matters listed, was a single dispute as to what sum was payable.
William Verry (Glazing Systems) Ltd v Furlong Homes Ltd	[2005] EWHC 138 (TCC), Judge Coulson	**Issue part of wide-ranging dispute in notice** Dispute referred was a final account claim including an extension of time claim. The dispute notified was wide ranging. The challenge at enforcement was that the extension of time claim advanced in the adjudication differed from those made before. The challenge was unsuccessful. The TCC judge considered that the questions framed in the notice of adjudication required the adjudicator to look afresh at the entitlement to extensions of time and accordingly the adjudicator was allowed to consider all aspects of that question. Further the defendant was allowed in any event to run the new case re extensions of time as part of its defence.

(*Continued*)

Table 4.1 *Continued*

Title	Citation	Issue
Workspace Management Ltd v YJL London Ltd	[2009] EWHC 2017 (TCC), [2009] BLR 497, Coulson J	**Scope of dispute about 'true valuation'** The claiming party asked for a decision about how much it was due under an interim valuation. The adjudicator decided the claiming party owed money to the responding party. The judge rejected the argument that the decision went beyond the dispute referred. The question of the proper valuation was wide enough to include a decision that the claiming party had been overpaid.
OSC Building Services Ltd v Interior Dimensions Contracts Ltd	[2009] EWHC 248 (TCC), Ramsey J	**May be necessary to interpret adjudication notice against background** Challenge to enforcement on the ground that the dispute in the adjudication notice was expanded in the referral. The challenge was rejected. It was the substance of the dispute referred that mattered and not the label it was given in the adjudication notice which had to be read in the context and against the background of the prior communications between the parties, which included the process of submissions, comments, and assessments which had taken place by that time (para. 14(1)).
Volker Stevin Ltd v Holystone Contracts Ltd	[2010] EWHC 2344 (TCC), Coulson J	**Sums in decision different from adjudication notice** The claim was reduced, and the adjudicator had the jurisdiction to deal with the reduced figure, not least because the original notice of intention to refer did not confine the dispute to a particular figure. Parties given notice of the new figures and commented on them. Challenge rejected.
Witney Town Council v Beam Construction (Cheltenham) Ltd	[2011] EWHC 2332 (TCC), [2011] BLR 707, Akenhead J	**May be necessary to interpret adjudication notice against background** The judge set out a number of propositions about how to determine the scope of a dispute (which are set out in Ch. 3 at 3.68) and said the notice of adjudication and referral notice are not necessarily determinative of the true dispute between the parties and it may be necessary to consider the background facts as well as the notices served.

A Single Dispute

4.18 The first paragraph of the Scheme permits a party to refer a dispute, expressed in the singular, to adjudication. The meaning of 'dispute' has been considered in numerous cases which are discussed fully in Chapter 3.[16] That line of authority applies equally to both

[16] See in particular 3.29–3.38.

statutory and contractual adjudications. The concept of a single dispute is also discussed at 3.49–3.52.

4.19 Like s. 108(1), the Scheme also refers to a dispute 'arising under the contract'. The scope of this phrase has been considered at 3.10–3.26. A wider contractual adjudication clause that permits disputes arising under and 'in connection' with the contract will still be valid. Such a provision will permit a wider class of disputes to be referred, as discussed at 3.12.

4.20 Whilst paragraph 1 of the Scheme permits a single dispute to be referred there seems to be no reason why a contract could not go further than this and expressly permit more than one dispute to be referred at a time.

Additional Disputes and Additional Parties

4.21 Paragraph 8(1) of the Scheme permits the adjudicator to consider more than one dispute as long as the parties consent. It also permits related disputes under different contracts between other parties to be adjudicated upon at the same time. Consent is essential for there to be consolidation of these other claims and it cannot be achieved by the back door without consent: *Pring & St Hill Ltd v C. J. Hafner t/a Southern Erectors* (2002).[17] In *Pring* the party found liable in adjudication 1 requested appointment of the same adjudicator in adjudications 2 and 3 in which it sought to pass on its liability to its own subcontractors. The TCC judge held that where the adjudicator was to decide related disputes under the same contract in parallel adjudications, consent had to be obtained. The adjudicator was found to have erred in going ahead without the consent of one party who objected to his appointment.

4.22 Some standard procedures also permit joinder of other parties, subject always to consent, for instance:

- The ICE 1997 Adjudication Procedure (paragraph 5.7) says that other parties can be joined to the adjudication subject to the agreement of the adjudicator and all the parties.
- The NEC 3 adjudication procedure Option W2 provides (at paragraph W2.3(3)) where the matter in dispute under the main contract is also disputed between the contractor and a subcontractor, then, subject to the consent of the subcontractor, the contractor may refer the connected dispute to the same adjudicator to be decided at the same time as the main contract referral.
- The CIC Model Adjudication Procedure (4th edn) at paragraph 23 allows additional parties to be joined into the adjudication, subject to the consent of the parties and the adjudicator.

Notice at Any Time

4.23 The drafting of the original Scheme has been amended to introduce the words 'at any time' into paragraph 1. Accordingly paragraph 1 of the amended Scheme now reads that 'Any party to a construction contract may give written notice (the 'notice of adjudication') at any time of his intention to refer any dispute arising under the contract to adjudication'. This change was to tidy up the drafting of the existing Scheme to make absolutely clear that notice could be given at any time as required by s. 108(2)(a) of the 1996 Act. For a fuller

[17] [2002] EWHC 1775 (TCC), [2004] Const LJ.

discussion of the meaning of the phrase 'at any time' see 3.73–3.82 above. In summary 'at any time' means what it says and there is no time restriction on the right to refer a dispute to adjudication. It is, therefore, permissible to refer a dispute arising under the contract to adjudication after the work has been completed or after the contract has been repudiated or terminated.

(3) Appointment and Referral within Seven Days: Scheme Paragraphs 2–11

The Adjudicator's Appointment: Scheme Paragraphs 2–6

4.24 Once notice of an intention to adjudicate has been given the referring party needs to secure the appointment of an adjudicator and refer the dispute to him or her. The contract procedure should provide a timetable with the object of securing both the appointment and the referral within seven days of the notice of adjudication.[18] The seven-day period begins immediately after the date the notice of intention to refer is given.[19]

4.25 The Act contains no requirements as to how the adjudicator shall be appointed, hence this is left up to the parties. The appointments procedures contained at paragraphs 2–6 of the Scheme are subject to the overriding right of the parties to agree who shall act as adjudicator in relation to a dispute; this is expressed in the introductory words of paragraph 2(1): 'Following the giving of notice of adjudication and subject to any agreement between the parties to the dispute as to who shall act as adjudicator.' If no agreement has been reached the Scheme provides that:

(1) The referring party must request an individual named in the contract to act as adjudicator (Scheme paragraph 2(1)(a)).
(2) If there is no named individual or he/she is unavailable then the referring party shall make an application to any nominating body named in the contract (Scheme paragraph 2(1)(b)).
(3) If neither option applies then the referring party shall ask an adjudicator nominating body to select an adjudicator (Scheme paragraph 2(1)(c)).

4.26 The Scheme contains various measures designed to ensure that the adjudicator's appointment is made in time to enable the dispute to be referred within seven days. Where a person is requested to act pursuant to paragraph 2(1) he or she must respond within two days of receiving the request saying whether he or she is willing or able to act (Scheme paragraphs 2(2) and 5(3)). If the answer is no, or if no answer is received in the time stipulated, then the referring party may start again and the alternative adjudicator must again respond within two days (Scheme paragraph 6).

4.27 The requisite response period is five days where it is an adjudication nominating body which is asked to select an adjudicator (Scheme paragraph 5(1)). If the first-choice nominating body fails to appoint an adjudicator then the referring party may agree with the other party to the dispute to request a specified person to act or may request any other adjudicating body to appoint an adjudicator (Scheme paragraph 5(2)). However, as five days will have already passed, this time there are only two days allowed for the person selected to indicate whether he or she is willing to act (Scheme paragraph 5(3)).

[18] s. 108(2)(b).
[19] s. 116(2) of the 1996 Act deals with the reckoning of periods of time.

4.28 It is a strict requirement of the Scheme that the request for appointment of an adjudicator is not made until after the notice of adjudication is served. A failure to comply with this requirement may render the adjudicator's appointment invalid and the decision as being unenforceable: *IDE Contracting Ltd v R. G. Carter Cambridge Ltd* (2004),[20] *Vision Homes Ltd v Lancsville Construction Ltd* (2009).[21] However, bespoke adjudication schemes need not adopt this approach and it is not necessary for such a procedure to require the first step be the service of the notice of adjudication. Nor is it necessary for the options to be sequential and a series of options that can be selected by choice is valid; *Palmac Contracting Ltd v Park Lane Estates* (2005),[22] *Dalkia Energy & Technical Services Ltd v The Bell Group Ltd* (2009).[23]

4.29 In principle an adjudicator appointed in contravention of the agreed appointment procedures will not have jurisdiction to decide the dispute. This is discussed in more detail in Chapter 9 at 9.27–9.28 and Table 9.3.

4.30 If the appointment procedure in the contract is unworkable the Scheme will apply: *David McClean Housing Ltd v Swansea Housing Association* (2001),[24] *AMEC Capital Projects Ltd v Whitefriars City Estates Ltd* (2004).[25] If the construction contract contains no provisions at all for the appointment of an adjudicator, the Scheme will apply: *Bovis Lend Lease v Cofely Engineering Services* (2009).[26]

4.31 Objecting to the appointment of a certain adjudicator will not invalidate his appointment, as long as the relevant appointment procedures are followed. This point is made expressly in the Scheme at paragraph 10.

4.32 Whilst the Scheme does not provide a pro forma agreement to be entered into between the parties and the adjudicator, many standard-form adjudication procedures do provide an adjudicator's agreement form. It is also common for adjudicators themselves to require the parties to agree to their own terms and conditions.

4.33 The Scheme does not say that the adjudicator shall be independent but paragraph 4 requires that a person requested or selected to act as an adjudicator shall not be an employee of any party to the dispute and shall declare 'any interest, financial or otherwise, in any matter relating to the dispute'. This means that the appointment of an employee of either party will be an invalid appointment. Such an adjudicator will have no jurisdiction to decide the matter and his decision will not be enforceable.[27] It is less clear what remedy is available to a party if the adjudicator validly selected and appointed by a nominating body declares an interest in the dispute that is objectionable to one party but he will not resign. It is submitted that it will depend on whether the facts give rise to an argument of apparent bias on the part of the adjudicator (as discussed in Chapter 11).

4.34 In *Makers (UK) Ltd v London Borough of Camden* (2008)[28] one party suggested to RIBA to appoint a named adjudicator (because he was qualified both as an architect and as a lawyer) and RIBA acceded to that requested. An enforcement challenge based on bias was rejected

[20] [2004] EWHC 36 (TCC), [2004] BLR 172.
[21] [2009] EWHC 2042 (TCC), [2009] BLR 525.
[22] [2005] EWHC 919 (TCC), [2005] BLR 301.
[23] [2009] EWHC 73 (TCC),122 Con LR 66.
[24] [2001] EWHC 830 (TCC), [2002] BLR 125.
[25] [2005] 1 All ER 723, [2005] EWCA Civ 1418 (CA).
[26] [2009] EWHC 1120 (TCC).
[27] In *Sprunt Ltd v LB Camden* [2012] BLR 83 the court found that a clause stating that a party to the contract, in this case Camden, was the specified nominating body offended against the 1996 Act and the statutory policy of having impartial adjudicators (para. 51).
[28] [2008] EWHC 1836, [2008] BLR 470.

and the court found there was no duty to consult the other party where suggestions about nomination were made. In *Fileturn Ltd v Royal Garden Hotel Ltd* (2010)[29] the TCC judge rejected a similar allegation of bias and in so doing said there was no objection in principle to the fact that the adjudicator was well known to one or both of the parties.

4.35 The fact that an adjudicator has previously acted as dispute resolver in proceedings involving one of the parties is not of itself sufficient grounds to allege apparent bias or a lack of impartiality: *Andrew Wallace Ltd v Jeff Noon* (2008).[30] In that case the adjudicator had been a mediator of an unrelated dispute with AWL only two days before being appointed as adjudicator by RIBA. The problem partly arose because when RIBA asked if the adjudicator had an existing relationship with either of the parties, the adjudicator said no. Whilst it could be said that the bias accusation might have been avoided had he made the disclosure before being appointed, the editors of the Building Law Reports have noted the Court of Appeal guidance in *Taylor v Lawrence*[31] that 'judges should be circumspect about declaring the existence of a relationship where there is no real possibility of it being regarded by a fair-minded and informed observer as raising the possibility of bias'.[32] Ultimately, the TCC judge in *Jeff Noon* found the allegation of bias unsubstantiated and was led to this conclusion by the fact that the adjudicator (a) had no personal knowledge of the parties, (b) was a professionally qualified arbitrator, (c) was appointed by RIBA as opposed to being a party appointee, and (d) had no current relationship with either party.

4.36 Allegations of bias on the part of the adjudicator are considered in more detail in Chapter 11.

The Referral Notice: Scheme Paragraph 7

4.37 Once the adjudicator is appointed the dispute must be referred to him. This is done by the service of a statement of case usually called the referral notice. Paragraph 7(1) of the Scheme makes it clear that it is by sending the referral notice to the adjudicator that the dispute is actually referred to him or her (as required by s. 108(2)(b) of the Act).

4.38 The date of 'the referral' itself is significant because it starts the clock ticking for the adjudicator to reach a decision within 28 days of that date: s. 108(2)(c).[33] For this purpose the 'date of the referral' is usually when he receives the referral notice itself: *Aveat Heating Ltd v Jerram Falkus Construction Ltd* (2007).[34] Indeed many of the standard-form procedures define the date of 'the referral' as being the date it is received. The amended Scheme now contains two sets of amendments which clarify this:

- The following words are inserted into paragraph 7(3): 'Upon receipt of the referral notice the adjudicator must inform every party to the dispute of the date it was received.'
- Paragraph 19 has been amended as follows:

  19. — (1) The adjudicator shall reach his decision not later than —

  (a) twenty eight days after the date receipt of the referral notice mentioned in paragraph 7(1), or

---

[29] [2010] EWHC 605 (TCC), [2010] BLR 512.

[30] [2009] BLR 158.

[31] [2003] QB 528 at para. 64.

[32] However, see Lord Phillips in *Smith v Kvaerner Cementation Foundations Ltd* [2006] 3 All ER 593 citing Lord Denning's proposition in *Metropolitan Properties v Lannon* [1969] 1 QB 577, that no man could act as advocate or adviser for or against a party in one dispute and at the same time sit as judge of that party in another proceeding without people inevitably thinking he was biased.

[33] s. 116(2) of the Act deals with reckoning periods of time.

[34] [2007] EWHC 131 (TCC), 113 Con LR 13, cf. *Ritchie Brothers (PWC) Ltd v David Philip (Commercials) Ltd* [2004] BLR 379.

(b) forty two days after the date receipt of the referral notice if the referring party so consents, or

(c) such period exceeding twenty eight days after receipt of the referral notice as the parties to the dispute may, after the giving of that notice, agree.

4.39 Unless otherwise stated in his appointment, an adjudicator has no power to proceed or give directions until he has received the referral notice: *Lanes Group Plc v Galliford Try Infrastructure Ltd* (2011).[35]

4.40 The Act does not stipulate any particular form or content for the referral notice and this is left to the applicable adjudication procedure. The Scheme simply says the referral notice shall be accompanied by copies of the contract and documents relied on. In contrast many of the standard adjudication procedures are more prescriptive, expressly requiring the referring party to provide a full explanation of the claim and the supporting information. The referring party ought to ensure the referral does not expand or materially change the dispute set out in the notice of adjudication (as discussed above at 4.14) to avoid jurisdictional challenges being raised by the responding party. Similarly the referral notice should comply with the formal requirements of the Scheme or other applicable adjudication rules although not every breach of the rules will result in there being a valid jurisdiction challenge: *PT Building Services Ltd v ROK Euro Build Ltd* (2008).[36] In that case the contract was provided a day after the referral notice contrary to the Scheme requirement that it be served with the referral. The TCC judge rejected this as a jurisdiction challenge.

Within Seven Days

4.41 After the valid appointment of the adjudicator the referring party shall, not later than seven days from the date of the notice of adjudication, refer the dispute in writing (the 'referral notice') to the adjudicator (Scheme paragraph 7(1)). This gives effect to s. 108(2)(b) of the Act.

4.42 It has been decided in several cases that the seven-day time limit in paragraph 7(1) of the Scheme is mandatory: *Hart Investments Ltd v Fidler and anor* (2006).[37] In *Hart* the referral notice served more than seven days after the notice of adjudication was found to be invalid and the adjudicator's decision was not enforced for want of jurisdiction.[38] For consideration of the cases in which enforcement of adjudication decisions have been challenged on the ground that the referral was out of time, see Chapter 9 at 9.29–9.34.

4.43 An adjudicator appointed pursuant to the Scheme does not have power to extend time for service of the referral notice before it arrives for two reasons: first, the adjudicator is not seized of the adjudication until the referral notice is provided and has no power until then; secondly, as decided in *Hart*, the only express power to extend time is in paragraph 13 of the Scheme and that provision does not permit the adjudicator to ignore the express time limits unless the extension is consented to.

Resignation or Revocation of Appointment: Scheme Paragraphs 9 and 11

4.44 The Scheme at paragraph 9(1) stipulates that an adjudicator may resign at any time on giving written notice. It is common practice for a responding party making a jurisdiction objection

[35] [2011] EWCA Civ 1617, [2012] BLR 121 at para. 40 of the judgment.

[36] [2008] EWHC 3434 (TCC).

[37] [2007] BLR 30.

[38] See also *Cubitt Building Interiors Ltd v Fleetglade Ltd* [2006] EWHC 3413, 110 Con LR 36 [2006], but cf. *William Verry v North West London Community Mikvah* [2004] EWHC 1300 (TCC), (2004) BLR 308 in which Judge Thornton found that s. 108(2)(b) was directory only.

to invite the adjudicator to resign. If the adjudicator comes to the conclusion that the jurisdiction objection is a good one then he ought to resign. However 9(1) is not limited to those situations and the adjudicator may resign for any reason at all.

4.45 Equally the parties may at any time agree to revoke the appointment of an adjudicator: Scheme paragraph 11.

4.46 Pursuant to paragraph 9(2), the adjudicator *must* resign if the dispute he is asked to decide is the same (or substantially the same) as one already referred to an adjudication in which a decision has been taken. This is because if the dispute has already been decided, and is binding until finally determined, there is no dispute that is capable of being referred to the second adjudicator, with the result that the second adjudicator will have no jurisdiction to decide the second dispute: *Sherwood Casson Ltd v MacKenzie* (2000).[39]

4.47 Whether one dispute is substantially the same as another dispute is a question of fact and degree: *Quietfield Ltd v Vascroft Construction Ltd* (2006).[40]

4.48 Standard-form contract adjudication schemes do not always mention an express obligation of an adjudicator to resign in these circumstances. However, whether or not the contract has a term equivalent to paragraph 9(2) of the Scheme, the same jurisdictional question will arise: *HG Construction Ltd v Ashwell Homes (East Anglia)* (2007).[41]

4.49 Cases in which the court has been asked not to enforce an adjudicator's decision because it overlapped with a previous adjudicator's decision are considered in Chapter 6 at 6.64–6.69.

4.50 Paragraph 9(2) of the Scheme does not operate until after a decision has been reached by the first adjudicator. Accordingly if there are two competing adjudications occurring at the same time concerning the same dispute, paragraph 9(2) does not mean that one of the adjudicators must resign: *Vision Homes Ltd v Lancsville Construction Ltd*.[42]

4.51 If the adjudicator considers the dispute is too complex or substantial for him to deal with fairly within the time available, he may ask for further time and if it is refused he ought to resign: *Amec Group Ltd v Thames Water Utilities Ltd* (2010).[43] There have been various attempts to challenge enforcement of an adjudicator's decision on the grounds that the dispute was too complex for adjudication. Those cases are considered in Chapter 10 at 10.17–10.18.

4.52 Whether the adjudicator is entitled to be paid fees following resignation or revocation of the appointment is considered in 4.125–4.134 below.

4.53 Where an adjudicator has resigned pursuant to Scheme paragraph 9(1) the referring party may serve a fresh notice and start again (paragraph 9(3)). The parties shall supply the replacement adjudicator with documents provided to the first adjudicator but only 'if requested by the new adjudicator and insofar as it is reasonably practicable'. As a matter of principle however, paragraph 9(3) does suggest that a party should not be prevented from starting a fresh adjudication where the first adjudication has not resulted in a binding decision. A similar point is made in paragraph 19(2) of the Scheme which says that where an adjudicator fails, for any reason, to reach his decision in accordance with paragraph 19(1) (i.e. within the

[39] (2000) 2 TCLR 418 (TCC).

[40] [2006] EWCA Civ 1737 CA, [2007] BLR 67, at para. 46 of the judgment of Dyson LJ.

[41] [2007] BLR 175 (TCC) at para. 39.

[42] [2009] EWHC 2042, [2009] BLR 525, 126 Con LR 95 [2009] at para. 70 of the judgment.

[43] [2010] EWHC 419 (TCC).

time limits stated) any of the parties to the dispute may serve a fresh notice and request an adjudicator to act. The question of whether a party may withdraw a dispute or part of a dispute from an adjudicator and start again has been debated in a number of cases that are discussed in Chapter 9 at 9.49 onwards.

(4) The Powers of the Adjudicator: Scheme Paragraphs 12–18

4.54 As far as the powers of the adjudicator are concerned, the only obligatory provisions for a contract adjudication scheme are those at s. 108(2)(e) and (f) of the Act which stipulate that the contract must: (e) impose on a the adjudicator a duty to act impartially, and (f) enable the adjudicator to take the initiative in ascertaining the facts and the law. The Scheme goes beyond the minimum requirements of the Act and sets out in detail what the adjudicator may do.

Duty to Act Impartially: Scheme Paragraph 12

4.55 To act impartially is to act without bias towards any party. Arguably the adjudicator's duty to act impartially is implicit in all adjudication procedures as an adjudicator is obliged to conduct the adjudication fairly. The failure to act impartially is a ground for challenging the validity of an adjudicator's decision. This topic is dealt with in detail in Chapter 11 on the subject of bias. Neither the 1996 Act nor the Scheme expressly state that the adjudicator must be independent although some contractual procedures do say this. Ultimately, any perceived lack of independence of the adjudicator will be a factor in the consideration of whether there is a real danger that the adjudicator may appear to be biased.[44]

Power to Take Initiative to Ascertain Facts and Law: Scheme Paragraph 13

4.56 In contrast to the imposition of the positive duty in s. 108(2)(e), the 1996 Act only requires that the adjudicator is enabled to ascertain the facts and the law. In other words he must be given the power to do so, but he is not obliged to use it. Consequently the Scheme at paragraph 13 says that the adjudicator *may* take the initiative. An adjudicator is empowered therefore to conduct the adjudication in an inquisitorial fashion. Alternatively he may prefer to allow the parties or their representatives to take responsibility for this. Equivalent powers are bestowed on adjudicators by most standard contractual procedures.

4.57 Paragraph 13 of the Scheme sets out the various ways in which a Scheme adjudicator may take the initiative:

> 13. The adjudicator may take the initiative in ascertaining the facts and the law necessary to determine the dispute, and shall decide on the procedure to be followed in the adjudication. In particular he may –
>
> (a) request any party to the contract to supply him with such documents as he may reasonably require including, if he so directs, any written statement from any party to the contract supporting or supplementing the referral notice and any other documents given under paragraph 7(2),
> (b) decide the language or languages to be used in the adjudication and whether a translation of any document is to be provided and if so by whom,

[44] See also the brief discussion at 4.33 onwards above concerning the appointment of the adjudicator.

(c) meet and question any of the parties to the contract and their representatives,
(d) subject to obtaining any necessary consent from a third party or parties, make such site visits and inspections as he considers appropriate, whether accompanied by the parties or not,
(e) subject to obtaining any necessary consent from a third party or parties, carry out any tests or experiments,
(f) obtain and consider such representations and submissions as he requires, and, provided he has notified the parties of his intention, appoint experts, assessors or legal advisers,
(g) give directions as to the timetable for the adjudication, any deadlines, or limits as to the length of written documents or oral representations to be complied with, and
(h) issue other directions relating to the conduct of the adjudication.

4.58 In performing these functions the adjudicator must always be careful to act fairly and impartially and must not introduce matters which are outside the scope of the adjudication and thus outside his jurisdiction. If he invites a party to alter or increase the scope of the dispute referred he may ultimately be judged to have reached a decision outside his jurisdiction: *McAlpine PPS Pipeline Systems Joint Venture v Transco plc* (2004).[45]

4.59 In *Volker Stevin Ltd v Holystone Contracts Ltd* (2010)[46] the enforcement of a decision was challenged on a number of grounds, one of which was that the adjudicator had exceeded his jurisdiction by asking for more material after the deadline he had set. The TCC judge found that the adjudicator had been entitled to make enquiries to obtain further information and to take account of responses. If the adjudicator needed further information in order to allow him to answer questions properly, he was entitled, indeed obliged, to ask for it. An adjudicator should not stand mutely by, hoping that one side or the other gives him the information that he wants; if he considers that he lacks vital information, he must take the initiative and ask for it directly. Further, in Volker the adjudicator had ensured that the process was fair by showing the other side the new material and giving it the opportunity to comment. The challenge was rejected.

4.60 Jurisdictional issues that may arise from the approach taken by the adjudicator are discussed more fully in Chapter 9.

Using Own Knowledge or Experience

4.61 The Scheme does not expressly say that the adjudicator may use his own experience or expertise, although paragraph 13(f) does enable him to appoint experts, assessors or legal advisers to help him as long as he has notified the parties of his intention to do so.

4.62 Other standard-form procedures expressly provide that an adjudicator may act as an expert himself and/or may use his own knowledge and experience: the GC Works contract scheme says the adjudicator shall act as an expert adjudicator (clause 59(6)); the CIC Model Adjudication Procedure (4th edn) states he may use his own knowledge and experience (clause 3); JCT clause 41A.5.5.1 says he may use his own experience or knowledge; the ICE 1997 Adjudication Procedure says that the adjudication is 'neither expert determination nor arbitration but the adjudicator may rely on his own expert knowledge and experience' (General Principles).

4.63 It is submitted that, in principle, the Scheme would allow an adjudicator to use his own knowledge as long as he complies with paragraph 17 which requires him to make available to the parties any information to be taken into account when making his decision. He should

[45] [2004] BLR 352.
[46] [2010] EWHC 2344 (TCC).

also provide the parties with a reasonable opportunity to make submissions on that material, as discussed in more detail in Chapter 10. If the adjudicator fails to give the parties an opportunity to consider and respond to material obtained by him and relied upon in the decision it may result in a breach of natural justice that renders the decision unenforceable.[47] Equally, the adjudicator must disclose advice provided by third parties and invite the parties to make submissions on it.[48] The adjudicator is not permitted to rely on his own knowledge or experience to reformulate the case and decide it on a basis different from the dispute referred. In *Herbosh-Kiere Marine Contractors Ltd v Dover Harbour Board* (2012)[49] the adjudicator's decision relied on a formula that had neither been part of the dispute in the referral nor the response. The court found that this was both a breach of natural justice, as the parties were not given a chance to consider it, and was a decision made without jurisdiction given that it was outside the dispute referred. Effectively the adjudicator had answered the wrong question. In different circumstances, it may be valid to decide on a basis that neither party argued where the adjudicator is answering a broad question posed by the dispute as referred, as was the case in *Multiplex Constructions (UK) Ltd v Mott MacDonald Ltd* (2007).[50] Alternatively, a court may dismiss a complaint made about an argument raised for the first time in the decision if it was not a central point in the decision nor was the basis on which the decision was made.[51]

Power to Decide the Procedure

4.64 Save for the time limit for the adjudicator to reach his decision, nothing is said in the 1996 Act about how the adjudication shall be conducted after the referral notice is provided. This is left for the specific adjudication procedure in the contract or the Scheme, whichever is applicable. Thus individual contract schemes are free to predetermine what procedures and timetables will be applicable.

4.65 Pursuant to the Scheme there is no predetermined procedure and the adjudicator *must* decide the procedure to be followed himself (paragraph 13). This means he may make directions as to what submissions shall be provided or may direct that he will hear the arguments and obtain evidence by any of the methods set out in paragraph 13(a)–(g). He is not limited by the specific options of paragraph 13 for he has a broad power to issue other directions relating to the conduct of the adjudication as he sees fit (paragraph 13(h)). He has a complete discretion as to the timetable (subject to meeting any time limit for reaching his decision) and may impose deadlines for the provision of submissions and may even require their length to be restricted (paragraph 13(g)).

4.66 However, the adjudicator's powers are always subject to his overarching duty to act fairly and impartially and accordingly any timetable imposed must be fair to both parties. In *NAP Anglia Ltd v Sun-Land Development Co Ltd* (2011)[52] the court rejected an allegation that the submissions timetable ordered by the adjudicator was unfair on the respondent or was a breach of natural justice. The procedure may also be subject to any compulsory deadlines contained in the adjudication agreement or any applicable adjudication rules. As outlined at 4.41–4.43 and 4.97–4.104 of this chapter and more fully in Chapter 9 at 9.29 and following, the deadlines in the Scheme for the provision of the referral and making the decision are

[47] Discussed in more detail in Ch. 10 at 10.75–10.78

[48] See 10.79–10.80 below.

[49] [2012] BLR 177, [2012] EWHC 84 (TCC).

[50] [2007] EWHC 20 (TCC), 110 Con LR 63, discussed in Chapter 3 at 3.64 and Key Case at 3.69

[51] This was what happened in *Berry Piling Systems Ltd v Sheer Projects Ltd* [2012] EWHC 241 (TCC).

[52] [2011] EWHC 2846 (TCC), [2012] BLR 195.

obligatory. The time for referral may not be extended and the adjudicator may not extend time for reaching his decision without consent of one or both parties.

4.67 The adjudication procedure may, as discussed below, include a deadline for the service of a defence. The question has arisen whether the adjudicator may extend the time for the defence and the effect of a failure to do so. This is discussed below and in Chapter 10 at 10.36–10.41.

Defence and Counterclaims

Form and Timing

4.68 The 1996 Act says nothing about the provision of a defence and so this is left to the individual adjudication procedures. The Scheme does not specifically provide that the respondent shall provide a written defence. It is for the adjudicator to decide the procedure but he is specifically empowered to obtain and consider such representations and submissions as he shall require (paragraph 13(f)), make directions about service of written documents (paragraph 13(g)), or make any other direction relating to the conduct of the adjudication (paragraph 13(h)). Although the respondent usually is required to put its defence in written form, this is a matter for the adjudicator who may prefer to elicit information by other means such as meeting and questioning the parties himself (paragraph 13(c)). This must all be done within the overarching duty to act fairly and impartially.

4.69 Other standard adjudication procedures take a different approach and predetermine that the responding party may submit a written response to the claim within a specified period. It is not uncommon for the responding party to have just seven days from the date of the referral to provide its response.

4.70 These predetermined timescales are not, however, necessarily a final date for the submission of the defence that cannot be extended: *CJP Builders Ltd v William Verry Ltd* (2008).[53] In that case the contract was in the DOM/2 form, of which clause 38A.5.1.2 provides that the respondent *may* send the adjudicator a response within seven days of the referral. Akenhead J said this was not the latest date a defence could be considered. Furthermore he went on to say that, because the contract provided that the adjudicator *shall* set his own procedure, the adjudicator had power to extend time and failure to do so was a breach of natural justice in this case.

4.71 The question of what happens if the respondent fails completely to put in a written submission requested by the adjudicator is discussed at 4.75–4.79 below.

Content

4.72 As discussed at 4.13 above every dispute referred to an adjudicator is defined by reference to the notice of adjudication. The dispute cannot be cut down or enlarged by subsequent documents without agreement between the parties. However, whatever dispute is referred, it includes and allows any ground open to the responding party which would amount in law or in fact to a defence of the claim: *KNS Industrial Services Limited v. Sindall* (2001).[54] The responding party is not restricted to (a) arguments that have been aired previously and/or (b) matters mentioned in the notice of adjudication.

4.73 Whilst the general principle is that a defendant can raise defences in adjudication, that does not permit the defendant to run defences or counterclaims that should have been the subject

[53] [2008] EWHC 2025 (TCC), [2008] BLR 545.
[54] [2001] 17 Const LJ 170.

of a withholding notice: *Letchworth Roofing Company v Sterling* (2009).[55] However, as long as the defence in question is not one that needs a withholding notice to have been served, then the general principle is that a respondent is entitled to run any available defence to defend itself against the assertions made in the claim. The relevant cases are discussed in more detail in Chapter 10 at 10.36–10.53.

4.74 An adjudicator has to decide whether or not a withholding notice is required to permit a cross-claim to be raised as a defence and, if so, whether or not there has been a valid notice. If he concludes that no notice was required, or that a notice was required and that there was a valid notice, then he must take the cross-claim into account in arriving at his decision. If he concludes that a notice was required but was not served then he is not entitled to take the cross-claim into account when reaching his conclusion.[56] However where an adjudicator wrongly decided that a withholding notice was required and so wrongly excluded a defence of set off, the court decided the error made was one of law. As errors of law are errors made within jurisdiction the decision was nevertheless enforceable; *Urang Commercial Ltd v Century Investments Ltd* (2011).[57]

Procedural Breach Does Not Invalidate Decision: Scheme Paragraph 15

4.75 What is the consequence of a party failing to comply with a procedural direction made by the adjudicator, or failing to take part in the adjudication at all? The Scheme at paragraph 15 expressly empowers the adjudicator to continue with the adjudication in any event. He may:

(a) continue the adjudication in the absence of that party or of the document or written statement requested,
(b) draw such inferences from that failure to comply as circumstances may, in the adjudicator's opinion, be justified, and
(c) make a decision on the basis of the information before him attaching such weight as he thinks fit to any evidence submitted to him outside any period he may have requested or directed.

4.76 This falls short of saying that an adjudicator may ignore information submitted after a directed deadline and subparagraph (c) chimes with paragraph 17 of the Scheme which provides that the adjudicator *shall* consider *any* relevant information submitted to him by any of the parties. This mandatory obligation apparently requires the adjudicator to consider late information, as long as it is relevant to the dispute. However it is not always a breach of natural justice for an adjudicator to refuse to consider an argument or evidence particularly when it is submitted very late; see Chapter 10.[58]

4.77 The effect of paragraph 15 of the Scheme is that a failure of the type identified (namely, where a party fails to comply with any request, direction, or timetable of the adjudicator made in accordance with his powers; fails to produce any document or written statement requested by the adjudicator, or in any other way fails to comply with a requirement under the Scheme provisions relating to the adjudication) will not invalidate the decision of the adjudicator.

---

[55] [2009] EWHC 1119 (TCC).

[56] Para. 33 of the judgment in *Letchworth Roofing Co. v Sterling Building Co.* (2009) CILL 2717, [2009] EWHC 1119 (TCC).

[57] [2011] EWHC 1561 (TCC).

[58] In particular see the discussion about when late submission of material/evidence may be rejected by an adjudicator at 10.21–10.30 (ambush) and the discussion at 10.54–10.62 about when the adjudicator is entitled to reject arguments or evidence that are not material or relevant. But if the adjudicator relies on material submitted at the last minute which the other party has not had an opportunity to consider, that may result in a breach of natural justice; see 10.75–10.78.

4.78 Some contract clauses, such as clause 41A.5.6 of the JCT contract, expressly state that any failure by either party to comply with a requirement of the adjudicator or a term of the adjudication procedure 'shall not invalidate the decision of the adjudicator'. Such a term however cannot be used to cure failures which rob the adjudicator of jurisdiction—such as the late provision of the referral notice, as discussed in more detail at 9.28 onwards: *Palmac Contracting Ltd v Park Lane Estates Ltd* (2005).[59]

4.79 As the decision in *CJP Builders Ltd v William Verry Ltd* (2008)[60] illustrates, a failure to grant an extension of time to serve a defence may be judged to be a breach of natural justice. The same conclusions might not follow if the late information was less crucial. However an adjudicator conducting an adjudication under the Scheme would be unwise to completely ignore relevant information on the sole ground that it was late. Information may however validly be dismissed on the grounds of irrelevance, as discussed at 4.82 below. Chapter 11 considers in more detail the circumstances in which procedural failures may give rise to a breach of natural justice that may affect the enforceability of a decision.

Obligation to Consider any Relevant Information: Scheme Paragraph 17

4.80 The adjudicator shall consider any relevant information submitted to him by any of the parties to the dispute and shall make available to them any information to be taken into account in reaching his decision (Scheme paragraph 17).

4.81 As stated above at 4.72–4.74 a responding party in an adjudication is not limited to points rehearsed before the start of the adjudication: *William Verry (Glazing Systems) Ltd v Furlong Homes Ltd* (2005).[61]

4.82 However, if an adjudicator declines to consider evidence which, on his analysis of the facts or the law, is irrelevant, that is neither (a) a breach of the rules of natural justice nor (b) a failure to consider relevant material which undermines his decision on Wednesbury grounds or for breach of paragraph 17 of the Scheme: *Carillion Construction Ltd v Devonport Royal Dockyard* (2005).[62] The extent to which a failure to consider particular evidence may lead to a breach of natural justice is considered in more detail in Chapter 10 at 10.54– 10.62.

Confidentiality: Scheme Paragraph 18

4.83 There is a prohibition imposed on the adjudicator and any party to the dispute by paragraph 18 of the Scheme from disclosing to any other person any information or document provided in connection with the adjudication which the supplying party has indicated is to be treated as confidential, except to the extent that it is necessary for the purposes of, or in connection with, the adjudication.[63]

(5) The Adjudicator's Decision: Scheme Paragraphs 19–22

Matters Necessarily Connected with the Dispute

4.84 Paragraph 20 of the Scheme describes the matters which may be dealt with by the adjudicator's decision. The adjudicator is obliged to decide the matters in dispute and in so doing he is empowered to take into account any other matters which the parties to the dispute agree

59 [2005] EWHC 919 (TCC), [2005] BLR 301; *Linnett v Halliwells* [2009] BLR 312.
60 [2008] BLR 545.
61 [2005] EWHC 138 (TCC).
62 [2005] BLR 310, approved in the Court of Appeal [2005] EWCA Civ 1358, [2006] BLR 15.
63 See Ch. 6 at 6.03–6.04.

should be within the scope of the adjudication or which are matters under the contract which he considers are necessarily connected with the dispute. In particular he may:

(a) open up, revise and review any decision taken or any certificate given by any person referred to in the contract unless the contract states that the decision or certificate is final and conclusive,
(b) decide that any of the parties to the dispute is liable to make a payment under the contract (whether in sterling or some other currency) and, subject to s. 111(4) of the Act, when that payment is due and the final date for payment,
(c) having regard to any term of the contract relating to the payment of interest decide the circumstances in which, and the rates at which, and the periods for which simple or compound rates of interest shall be paid.

4.85 In a term similar to paragraph 20 of the Scheme, the TeCSA Adjudication Rules (Version 2.0) empower the adjudicator to decide the matters in dispute and any other matter that the Adjudicator determines should be included 'in order that the Adjudication may be effective or meaningful'. This is backed by a term that the adjudicator may decide upon his own substantive jurisdiction.[64] In *Farebrother Building Services Ltd v Frogmore Investments Ltd* (2001)[65] it was said that this provision was unambiguous and meant that a decision on the scope of an adjudication by the adjudicator could not be challenged in court. However, in *Shimizu Europe Ltd V LBJ Fabrications Ltd* (2003)[66] the TCC judge decided that the TeCSA power to take into consideration other matters was not unlimited. In that case the adjudicator had decided how much ought to have been the amount of an interim payment and said it should be paid 'without set-off'. Because the contract required a VAT invoice be issued before payment was due the adjudicator said that the invoice should be issued after the adjudication. The judge found that the adjudicator could not by his decision prevent the paying party from serving a withholding notice against that invoice in accordance with the contract terms, and the statement that the sum was payable 'without set-off' could only refer to those matters which had been before the adjudicator which he had rejected as the adjudicator could not decide a future right of set-off.

Power to Open Up, Review, and Revise Certificates: Scheme Paragraph 20(a)

4.86 The Scheme gives an adjudicator an express power to open up, revise, and review any decision taken or any certificate given by any person referred to in the contract unless the contract states that the decision or certificate is final and conclusive.

4.87 It has been argued (unsuccessfully) that where the contract reserves the power to open up, review, and revise certificates to the contract administrator then paragraph 20 of the Scheme does not empower an adjudicator to issue a certificate. In *Vaultrise Ltd v Paul Cook* (2004)[67] a dispute arose under a contract in the JCT standard form of contract IFC 98. The court held that an adjudicator can consider whether or not a certificate should have been issued and if a missing certificate was due he could determine the sum. There was no reason why a dispute as whether or not a certificate should be issued, and if so what it should contain, should not be referred to adjudication.

4.88 There was a proposal to include in the 2009 Act a provision making ineffective any contract term that purported to make an interim certificate binding so as to prevent a subsequent review by an adjudicator. This was widely criticized and was dropped.

[64] Para. 14 of the TeCSA Adjudication Rules (Version 2.0).
[65] (2001) CILL 1762. This case was decided on Version 1.3 of the TeCSA Rules in which the same provision appeared.
[66] [2003] EWHC 1229 (TCC), [2003] BLR 381.
[67] [2004] ADJCS 04/06.

4.89 A contract term may not prevent the adjudicator from considering a dispute about a certificate or decision made by a certifier. In *Banner Holdings Ltd v Colchester Borough Council* (2010)[68] the TCC judge considered a contract term that said the adjudicator could not vary or overrule the employer's decision to operate a contractual provision to determine the contract. The judge said the clause was not objectionable because once the decision to determine had been taken it was too late to vary or reverse it. The issue that remained was whether the decision taken was valid and the clause did not restrict the right of the parties to refer that dispute to adjudication. However the judge said that if the clause had restricted the right of the parties to refer the disputes about the validity or conseqences of the determination, it would have been invalid as contrary to s.108 of the Act.[69]

Power to Award Interest: Scheme Paragraph 20(c)

4.90 There is no requirement pursuant to s. 108 of the Act for the construction contract to confer a power on the adjudicator to award interest. Some contracts include an express right to interest on late payments and some do not. The Scheme says:

> 20 The adjudicator shall decide the matters in dispute. He may take into account any other matters which the parties to the dispute agree should be within the scope of the adjudication or which are matters under the contract which he considers are necessarily connected with the dispute. In particular he may ... (c) having regard to any term of the contract relating to the payment of interest decide the circumstances in which, and the rates at which, and the periods for which simple or compound rates of interest shall be paid.

4.91 This provision contains some ambiguity which has given rise to difficulties. In particular it has been debated whether it creates a free-standing power for the adjudicator to award interest on late payments when the contract contains no express right. In *Carillion Construction Ltd v Devonport Royal Dockyard Ltd* (2005)[70] Jackson J at first instance found that, in adjudications conducted under the Scheme, pursuant to paragraph 20(c) there was a free-standing right to award interest. He found that 'having regard to any term of the contract relating to the payment of interest' meant that if there was any such term the adjudicator must have regard to it. In other words, the free-standing right conferred by paragraph 20(c) did not override any express term of the contract. The Court of Appeal disagreed (see Key Case below).

4.92 The award of interest was upheld by the Court of Appeal on different grounds, namely that the respondent had acquiesced in the adjudicator having jurisdiction to consider it. This aspect of the Court of Appeal's decision is discussed in further detail in Chapter 5 concerning ad hoc submissions.

4.93 It follows from the Court of Appeal's decision that the adjudicator is empowered to award interest if it is one of the matters in dispute referred to him and (a) there is a contractual right to claim interest or (b) the defendant nevertheless has agreed that the adjudicator shall have the power. It is not however clear what is meant by the words 'or which are matters under the contract which he considers are necessarily connected with the dispute' and whether this empowers the adjudicator to consider interest where there is a contract right to it but it has not been included in the referring parties' claim.

---

[68] [2010] EWHC 139 (TCC).

[69] For further consideration of restrictions of the right to adjudicate see Chapter 3 at 3.73–3.82, and also the effect of a final certificate on the right to adjudicate at 3.77. Where a certificate is said by the contract to be final and binding it may be the adjudicator has no jurisdiction to consider it on the basis that there is no dispute.

[70] [2005] EWHC Civ 1358, [2006] BLR 1.

4.94 The Court of Appeal decision in *Carillion* was limited to whether paragraph 20(c) of the Scheme conferred a free-standing power to award interest. It does not consider the wider question of whether a power to award interest may be derived from other sources. In the absence of a contractual term allowing interest claims to be made a referring party may be able to rely on the Late Payment of Commercial Debts (Interest) Act 1998 or may be able to claim interest as damages for late payment in the manner contemplated in *FG Minter v Dawnays*.[71] It has been suggested that an adjudicator may have an implied power to award interest by analogy with the arbitration precedents.[72] This argument was not considered in the judgments in *Carillion* and remains to be considered by the courts.

Key Case: Interest

Carillion Construction Ltd v Devonport Royal Dockyard Ltd [2005] EWHC Civ 1358, [2006] BLR 1

4.95 **Facts:** Application for permission to appeal against the decision of Jackson J (TCC) enforcing an adjudicator's decision by giving summary judgment for Carillion and appeal on the issue of whether it had been within the jurisdiction of the adjudicator to award interest on the sum that he had found to be due.

4.96 **Held:** Permission to appeal was refused on all grounds save that of interest. On the issue of interest, the Court of Appeal found that the judge was wrong to hold that an adjudicator had a 'free-standing' power to award interest under paragraph 20(c) of the Scheme for Construction Contracts. In the instant case, the parties had conferred jurisdiction to decide whether interest should be paid on moneys outstanding and so the decision on interest was within the adjudicator's jurisdiction. Appeal dismissed:

> 91. The real question, as it seems to us, is what effect is to be given to the words 'In particular' which precede the three subparagraphs (a) to (c). It is necessary to have regard to the structure of paragraph 20 as a whole. There are three sentences: (1) The adjudicator shall decide the matters in dispute; (2) [In deciding those matters] he may take into account other matters (which are specified); (3) In particular [in deciding those matters] he may (a) open up, revise and review decisions already taken or certificates already given (unless the contract otherwise provides), (b) decide that any of the parties is liable to make payment and if so when and in what currency and (c) decide the circumstances in which (and the rates at which and the periods for which) interest is to be paid. Within that structure effect has to be given to the words 'In particular' at the beginning of the third sentence. We can see no reason why those words should not bear their usual and natural meaning. What comes after them is intended to be a particularisation of what has gone before. What comes after elaborates and explains what has gone before; it does not add to what has gone before.
>
> So the adjudicator may decide questions as to interest if, but only if, (i) those questions are 'matters in dispute' which have been properly referred to him or (ii) those are questions which the parties to the dispute have agreed should be within the scope of the adjudication or (iii) those are questions which the adjudicator considers to be 'necessarily connected with the dispute'. Questions which do not fall within one or other of those categories are not within the scope of paragraph 20(c) of the Scheme. There is no freestanding power to award interest.

---

[71] 13 BLR 1.

[72] The arbitration precedents including *President of India v La Pintada C. Nav. S.A* [1985] AC 104, per Lord Brandon at pp. 119A–B. This is the suggestion of Andrew Bartlett QC in his chapter on adjudication in *Emden's Construction Law*, Vol. 1, Pt II (773).

Reaching and Communicating of the Decision: Scheme Paragraph 19

4.97 Section 108(2)(c) of the 1996 Act stipulates that the contract shall require the adjudicator to reach a decision within 28 days of referral or such longer period as is agreed by the parties after the dispute has been referred. By s. 108(2)(d) the adjudicator may be given a power to extend the period by another 14 days with the consent of the referring party. This is given effect in paragraph 19 of the Scheme.

4.98 Similar provisions are found at in clause 41A.5.3 of the JCT 1998 standard form, in the NEC 3 procedure Option W2 clause W2.3(8), the TeCSA Adjudication Rules Version 2.0 at clause 24, and in the CIC Model Adjudication Procedure (4th edn) at clause 16.

4.99 The time limit of 28 or 42 days is mandatory, unless extended by the further agreement of the parties. This means that the adjudicator's power to reach a decision is limited to his reaching a decision within that mandatory time limit. A decision reached out of time is made outside his jurisdiction and is a nullity.

4.100 It is clear from paragraph 19(3) of the Scheme (and confirmed by the decided cases) that there is a distinction between reaching the decision and delivering the decision. The Scheme requires that the decision must be delivered as soon as possible after it is reached. Many contractual procedures include a provision that the decision shall be communicated forthwith.

4.101 Where there is an obligation to communicate the decision forthwith, there would appear to be little doubt that this means it is to be sent immediately or as quickly as possible by what is currently regarded as conventional and universally available methods of business communication. In particular therefore there would appear to be no reason why any such decision cannot be immediately transmitted to interested parties by fax transmission or electronically: *St Andrews Bay Development Ltd v HBG Management* (2003).[73]

4.102 An adjudicator fails to comply with the requirement in paragraph 19(3) to deliver his decision to each of the parties as soon as it was reached if he refuses to release his decision until a party had paid his fees: *Mott Macdonald v Land & Regional* (2007).[74] The same facts were also held to be a failure to act impartially and the failure to communicate the decision as soon as possible was an excess of jurisdiction.

4.103 The time limits for reaching and communicating the decision are discussed in more detail in Chapter 9 at 9.37 onwards.

4.104 The adjudicator's decision need not be signed unless the wording of the adjudication agreement requires it: *Treasure & Son v Dawes* (2007),[75] *CSC Braehead Leisure Ltd v Laing O'Rourke Scotland Ltd* (2008).[76]

Obligation to Give Reasons: Scheme Paragraph 22

4.105 The Act does not require that an adjudication scheme should require the adjudicator to provide reasons.

4.106 Paragraph 22 of the Scheme provides that, if requested by one of the parties to the dispute, the adjudicator shall provide reasons for his decision. Unsuccessful parties have tried to argue that the failure to provide reasons renders the decision a nullity because the reader could have no confidence that he had asked himself the right questions.

[73] [2003] Scots CS 103 at para. 17 of the judgment.
[74] [2007] EWHC 1055 (TCC).
[75] [2007] EWHC 2420 (TCC). [2008] BLR 24 at paras. 45–8 of the judgment.
[76] [2009] BLR 49; see also *Treasure & Sons v Dawes*.

4.107 If an adjudicator is requested to give reasons, a brief statement of those reasons will suffice. The reasons should be sufficient to show that the adjudicator has dealt with the issues remitted to him and what his conclusions are on those issues. It will only be in extreme circumstances, such as those described by Clerk LJ in *Gillies Ramsay*, that the court will decline to enforce an otherwise valid adjudicator's decision because of the inadequacy of the reasons given. The complainant would need to show that the reasons were absent or unintelligible and that, as a result, he had suffered substantial prejudice: *Carillion Construction Ltd v Devonport Royal Dockyard* (2005).[77]

4.108 However not all contract procedures include a requirement to provide reasons and where no such term exists the adjudicator is not obliged to provide reasons. For a fuller discussion of this topic see Chapter 10 at 10.63 onwards.

Power to Correct Slips: Scheme Paragraph 22A

The 1996 Act

4.109 Neither the 1996 Act as originally enacted nor the Scheme stipulate that the adjudicator has the power to make corrections to his decision after it has been communicated to the parties. Many contract adjudication procedures do contain such a power. Absent an express term in the contract adjudication procedure conferring such a power on the adjudicator, the position at law was summarized by Akenhead J in *YCMS Ltd v Stephen & Miriam Grabiner* (2009)[78] as follows:

(a) An adjudicator can only revise a decision if it is an implied term of the contract by which adjudication is permitted to take place that permits it. It does not follow that, if it is purely a statutory [adjudication] under the HGCRA (if there is no contractual adjudication clause), such implication can be said to arise statutorily.[79]
(b) If there is such an implied term, it can and will only relate to 'patent errors'. A patent error can certainly include the wrong transposition of names or the failing to give credit for sums found to have been paid or simple arithmetical errors.
(c) The slip rule cannot be used to enable an adjudicator who has had second thoughts and intentions to correct an award. Thus for example, if an adjudicator decides that the law is that there is no equitable right of set off but then changes his mind having read some cases feeling that he has got that wrong, such a change would not be permitted because that would be having second thoughts.
(d) The time for revising a decision by way of the slip rule will be what is reasonable in all the circumstances.[80] In the *Bloor* case, the Adjudicator revised his decision within several hours and before the time for issuing a decision had been given. It will be an exceptional and rare case in which the revision can be made more than a few days after the decision. The reason for this is that, unlike a court judgment or an arbitration award, a principal purpose of the 1996 Act is to facilitate cash flow. If an adjudicator was able to revise his decision, say, 21 or 28 days later, that would necessarily slow down and interfere with the speedy enforcement of adjudicators' decisions. That would in broad terms be contrary to the policy of the Act.

---

[77] [2005] EWHC 778 (TCC), [2005] BLR 310.

[78] [2009] EWHC 127 (TCC), [2009] BLR 211 at paragraph 50, this approach was also applied in *Redwing Construction Ltd v Wishart* [2010] EWHC 3366 (TCC), 135 Con LR 119.

[79] Cf. *Bloor Construction v Bowmer & Kirkland* [2000] BLR 314 (TCC) where the judge said that such a term will usually be implied into agreements for adjudication under the Act; and see *CIB Properties Ltd v Birse Construction* [2005] BLR 173 paras. 33–5.

[80] On this point see: *Cubitt Building & Interiors Ltd v Fleetglade Ltd* [2006] EWHC 3413 (TCC), 110 Con LR 36: The decision will not be considered to be issued in breach of the required time limits when issued within the allotted time but a corrected version is issued outside the time limit, as long as the corrected version is communicated within a reasonable time.

4.110 As to what errors will constitute a 'slip' which may be corrected, Akenhead J cited Sir John Donaldson's analysis in *R v Cripps ex p. Muldoon*[81] where it was said:

> It is a distinction between having second thoughts and intentions and correcting an award to give effect to first thoughts or intentions which creates the problem. Neither an arbitrator nor a judge can make any claim to infallibility. If he assesses the evidence wrongly or fails to understand the law the resulting award or judgment will be erroneous but it cannot be corrected under section 17 (of the Arbitration Act 1950) or under the old Order 20 Rule 11. It cannot normally be corrected under section 22 (where the arbitrator has made a mistake). The remedy is to appeal if the right of appeal exists. The skilled arbitrator or judge may be tempted to describe this as an accidental slip but this is a natural form of self-exculpation.

4.111 It is clear that fundamental errors which go to the heart of the decision cannot be corrected under the slip rule.[82] In general the courts are loath to embark upon a lengthy review of the adjudicator's decision. In *CIB Properties Ltd v Birse Ltd* (2004)[83] JudgeToulmin concluded that the courts would give effect to an amendment if two conditions were satisfied:

> (1) is the Adjudicator prepared to acknowledge that he has made a mistake and correct it?[84]
> (2) is the mistake a genuine slip which failed to give effect to his first thoughts? If the answer to both questions is 'Yes' then, subject to the important question of the time within which the correction is made and questions of prejudice, the court can if the justice of the case so requires give effect to the amendment to rectify the slip.

4.112 It appears that the courts will not entertain an argument that the adjudicator exceeded his jurisdiction where he seeks to correct an error under the slip rule and in doing so makes an error of fact or law. This point was addressed by Ramsey J in *O'Donnell Developments Ltd v Build Ability Ltd* (2009)[85] in which he accepted submissions that an erroneous exercise of a power does not constitute an act in excess of powers so as to fall outside of the jurisdiction of the adjudicator.

4.113 In *ROK Building Ltd v Celtic Composting Systems Ltd (No. 2)* (2010)[86] Akenhead J dealt with an argument that the adjudicator failed properly to exercise his jurisdiction in declining to make certain 'corrections'. Although the adjudicator in that case accepted that he had made certain clerical errors and corrected these, he considered that the further changes suggested went beyond mere correction and went to the substance of his decision. Akenhead J considered that errors, if there were any, were caused by the complicated way in which the defendant had put its case in the adjudication, and that it was simply unarguable that the adjudicator acted unfairly or in breach of natural justice. In the light of this decision, it is submitted that

---

[81] [1984] QB 686, first cited by Judge Toulmin in *Bloor*.

[82] See *Joinery Plus Ltd v Laing Ltd* [2003] BLR 184, [2003] EWHC 3513 (TCC). The decision in *Geoffrey Osborne Ltd v Atkins Rail Ltd* [2010] BLR 363, [2009] EWHC 2425 (TCC), in which substantial errors made by the adjudicator were opened up and corrected by the court, has been considered to be a decision on its own particular facts which does not lay down any wider principle; see *Pilon Ltd v Breyer Group plc* [2010] BLR 452, [2010] EWHC 837 (TCC).

[83] [2005] 1 WLR 2252, [2004] EWHC 66 (TCC).

[84] In *Bouygues (UK) Ltd v Dahl-Jensen (UK) Ltd*, reported at first instance at [2000] BLR 49 and in the Court of Appeal at [2000] BLR 522 the adjudicator was not prepared to recognize that he had made a slip or clerical error and the court declined to interfere as it considered he had made an error within his jurisdiction.

[85] 28 Con LR 141, [2009] EWHC 3388 (TCC) at paragraphs 33–35.

[86] 130 Con LR 74, [2010] EWHC 66 (TCC).

a court is unlikely to entertain an argument that the adjudicator wrongly refused to exercise his power to correct errors, unless the errors are patent and uncontroversial.

4.114 In addition to the above, any party who participates in an adjudication but reserves its right to later challenge the adjudicator's jurisdiction should be aware that a request to correct an award may be taken as an acceptance that the award is valid.[87]

The 2009 Act

4.115 By s. 140 of the Local Democracy, Economic Development and Construction Act 2009,[88] a new s. 108(3A) is inserted into the 1996 Act. This gives statutory effect to the slip rule in the following terms:

> The contract shall include provision in writing permitting the adjudicator to correct his decision so as to remove a clerical or typographical error arising by accident or omission.

This provision is mandatory and if the contract fails to include such a power then it appears that by virtue of s. 108(5) the Scheme will apply.

4.116 It is notable that s. 108(3A) contains no time limit for making such corrections. It is also unclear whether a contract term that goes beyond this, permitting correction of errors in wider circumstances, would be considered to contravene s. 108(3A).

4.117 The power to correct slips is now contained in new paragraph 22A of the Scheme which provides:

> 22A.—(1) The adjudicator may on his own initiative or on the application of a party correct his decision so as to remove a clerical or typographical error arising by accident or omission.
> (2) Any correction of a decision must be made within five days of the delivery of the decision to the parties.
> (3) As soon as possible after correcting a decision in accordance with this paragraph, the adjudicator must deliver a copy of the corrected decision to each of the parties to the contract.
> (4) Any correction of a decision forms part of the decision.

Key Cases: Power to Correct Slips

Bloor Construction (UK) Ltd v Bowmer & Kirkland (London) Ltd [2000] BLR 314 (TCC)

4.118 **Facts:** An adjudicator communicated his decision which contained a miscalculation in that he had failed to allow for payments on account made by Bowmer. This was pointed out to him, and he sent a revised version by fax within three hours on the same day, which after allowing for the advance payments, concluded that no additional money was due to Bloor. Both versions of the decision were wrongly dated 9 February 2000. Bloor sought to enforce the earlier decision, on the grounds that the adjudicator had no powers to amend or vary his decision as the adjudication provisions contained no provision analogous to the 'slip rule'.

4.119 **Held:** There was to be implied into the adjudication agreement a power for an adjudicator to correct any accidental error or omission or clarify or remove any ambiguity provided

[87] See *Shimizu Europe Ltd v Automajor Ltd* [2002] EWHC 1571 (TCC).

[88] This affects construction contracts entered into from 1 Oct. 2011.

that it was done within a reasonable time and did not prejudice any party. The corrected decision was valid and the application to enforce the uncorrected version was dismissed:

> It is clear that the error in this case falls into the category of a slip. Mr Bastone was giving effect to his first thoughts and intentions in his amended ruling. In my view, in the absence of any specific agreement to the contrary, a term can and should be implied into the contract referring the dispute to adjudication, that the Adjudicator may, on his own initiative or on the application of a party, correct an error arising from an accidental error or omission. The purpose of the adjudication is to enable broad justice to be done between the parties. Parties acting in good faith would be bound to agree at the start of the adjudication that the Adjudicator could correct an obvious mistake of the sort which he made in this case. Clearly, there must be a time limit within which such an amendment can be made, but in this case the amendment was made within three hours of the communication of the original decision. This must in the circumstances of this case be within any acceptable time limit. I bear in mind that both parties agree that the revised decision corrected a manifest error and that there is no suggestion that Bloor was prejudiced by the amendment.
>
> I note that the time limits under s.57(3) of the Arbitration Act 1996 stipulate a period of 28 days within which any application for the correction of an arbitrator's award must be made. I am not prepared to say that such a long time limit is necessarily appropriate for an adjudication. An additional reason for holding that the slip rule applies is the lack of ability of the High Court to correct obvious errors in adjudications except in very restricted circumstances, even where such errors cause manifest injustice The primary reason for my decision is that, in the absence of a specific agreement by the parties to the contrary, there is to be implied into the agreement for adjudication the power of the Adjudicator to correct an error arising from an accidental error or omission or to clarify or remove any ambiguity in the decision which he has reached, provided this is done within a reasonable time and without prejudicing the other party.

YCMS Ltd v Stephen & Miriam Grabiner [2009] BLR 211, [2009] EWHC 127 (TCC)

4.120 **Facts:** The case concerned three successive adjudication decisions. The issue concerning the slip rule arose out of the first decision (and this summary does not therefore consider the other points in issue in the case). The dispute in the first adjudication concerned the amount due from the defendants under draft interim certificate 13. The dispute was referred to adjudication. The defendants argued that certificate 13 had been replaced with interim certificate 14 that had since been paid and so no further monies were due. The adjudicator found in favour of the claimant and decided the defendants should pay £25,942.74. The claimant informed the adjudicator that he had made an arithmetical error and that in fact a greater sum was due. The adjudicator decided that both his and the claimant's calculations were wrong. He amended his decision by inserting an even greater sum due, which he had reached by using an alternative method of calculation.

4.121 **Held:** The adjudicator's correction was more than an arithmetical correction. The new calculation method introduced new elements and constituted 'second thoughts'. The defendants were prejudiced by the amendment using the new method of calculation:

> 57. The next issue relates to whether the Adjudicator was entitled to revise the First Decision. I have formed the view that the Adjudicator was not entitled in all the circumstances to revise the First Decision as he did:
>
> (a) It is clear that in the First Decision (Unrevised) the Adjudicator made an inexplicable arithmetical error. Having decided at Paragraph 5.1(d) that the sum due was

£132,578.70 and at Paragraph 5.1(e) that £91,020.51 had been paid, he then deducted £106,635.96 (for some unaccountable reason) from the sum due to leave a balance of £25,942.74. Applying his logic, the sum which he should have found due was £132,578.70 less £91,020.51, namely a balance of £41,558.19 (which is exactly the figure which YCMS said in its letter of 29 October 2007 to the Adjudicator was the correct figure).

(b) If the Adjudicator had revised his Decision to produce this figure £41,558.19, the correction (subject to time) would have fallen within the ambit of the adjudication slip rule and its legitimate application.

(c) However, what the Adjudicator seems to have done in his revision was to decide that the net sum certified in Certificate No. 14 should be taken into account both in terms of a gross sum due as well as a sum paid. This was in effect another and a serious error for the Adjudicator to make because he overlooked the fact that Certificates 13 and 14 effectively and necessarily duplicated themselves because Certificate No. 14 was for a lesser sum than that certified in Certificate No. 13. Put another way, by allowing £18,348.15 in the sum invoiced and due column for Certificate No. 14, he was necessarily duplicating what was in the same column for Certificate No.13, because Certificate No. 13 was overall for a greater sum than Certificate No. 14. Thus, arithmetically if one takes the sum certified in Certificate No. 14 both in the sum invoiced and sum paid column whilst also allowing the gross sum certified as due in Certificate No. 13, the sum due on those figures is bound to be overstated by the amount certified as due in Certificate No. 14.

(d) I must conclude that the Adjudicator thought carefully about the changes which he introduced arithmetically. He had been pointed clearly by YCMS to the arithmetical error. He decided (and it was clearly in my judgment second thoughts) that he needed to bring in the sum due and sum paid on Certificate No. 14. He thus rejected any correction of a simple arithmetical error (which would have produced a total due to YCMS of some £41,000) in favour of a further calculation, the logic of which must be known only to the Adjudicator.

(e) In the ordinary course of events the operation of the slip rule does not result in any prejudice to either party because the Tribunal is simply putting right a mistake which it has made which it would not otherwise have made. Here, the Defendants are materially prejudiced by the amendment because the Adjudicator simply got it wrong the second time round.

O'Donnell Development Ltd v Build Ability Ltd, (2009) 128 Con LR 141, [2009] EWHC 3388 (TCC)

4.122 **Facts:** The case concerned enforcement of two adjudicator's awards. The day after the adjudicator issued his decision the claimant wrote to the adjudicator to highlight two errors: (i) in calculating the sum due for the value of the works in valuation 25 the adjudicator had deducted a sum paid for loss and expense awarded in a previous adjudication dated 28 August 2009, which was after the date of valuation; (ii) the adjudicator had incorrectly calculated the retention figure as 3 per cent of the gross value of the works, including loss and expense rather than excluding loss and expense. On the following day the defendant objected to the corrections on the basis that they did not fall within the definition of slips that it was permissible for the adjudicator to correct. The adjudicator wrote to the parties informing them that he considered he did have the power to correct the errors and he issued a corrected version of his decision. The defendant subsequently accepted that the adjudicator had the power to correct the mistake in respect of retention. However it maintained its objection to the correction of the other error on the basis that it had been induced by misinformation which had been mistakenly provided by the claimant. In the circumstances the defendant argued that the correction was not within the slip rule and that the adjudicator therefore had no power to amend the decision.

4.123 **Held:** The adjudicator was asked to correct a slip and accepted that he had made an error within the slip rule. In the circumstances it was not open to the court to interfere with the exercise of the adjudicator's powers within his jurisdiction.

> 33. Ms. Finola O'Farrell QC, on behalf of ODD, refers to the decision of the House of Lords in *Lesotho Highlands Development Authority v Impregilo SpA* [2006] 1 AC 221 which was cited by the court in argument. She submits that it is authority for the proposition that an erroneous exercise of a power does not constitute an act in excess of powers so as to fall outside of the jurisdiction of the adjudicator. She refers to the following passage from Mustill & Boyd on Commercial Arbitration, cited with approval by Lord Steyn at [25]: '*if ... [the arbitrator] applies the correct remedy, but does so in an incorrect way – for example by miscalculating the damages which the submission empowers him to award – then there is no excess of jurisdiction. An error, however gross, in the exercise of his powers does not take an arbitrator outside his jurisdiction and this is so whether his decision is on a matter of substance or procedure.*'
>
> 34. She submits that, as set out by Lord Steyn at [31] and [32], the concept of a tribunal exceeding its powers necessarily assumes that the tribunal was acting within its substantial jurisdiction.
>
> 35. I accept her submission that an erroneous exercise of a power does not fall outside the jurisdiction of an arbitrator or adjudicator. However, the distinction between disputes as to the jurisdiction of an adjudicator and disputes as to ways in which that jurisdiction should be exercised is not an easy one to draw as the decision in Lesotho Highlands shows. This can be illustrated in the case of the slip rule as follows. First if the adjudicator were to exercise a slip rule when there was no express or implied slip rule, that would clearly be a decision which was outside his jurisdiction. Secondly, if the adjudicator is asked by one party to correct a slip and he accepts that an error has been made within the slip rule then if the adjudicator makes an error of fact or law in so doing, I consider that such an error does not take the exercise of the slip rule outside his jurisdiction. Finally, if the adjudicator is asked by one party to correct a slip which the other party agrees is a slip within the slip rule but in operating the slip rule he makes an error of fact or law, then I do not consider that the court can interfere in that decision.
>
> 36. The dividing line between exercising a wrong jurisdiction which does not exist and exercising a jurisdiction which does exist, wrongly is difficult. Each case obviously has to be considered on its facts to decide whether it is a decision within or outside the adjudicator's jurisdiction.
>
> 37. As Dyson J said in Bouygues at [36]: ' ... *in deciding whether the adjudicator has decided the wrong question rather than given a wrong answer to the right question, the court should bear in mind that the speedy nature of the adjudication process means that mistakes will inevitably occur, and, in my view, it should guard against characterising a mistaken answer to an issue that lies within the scope of the reference as excess of jurisdiction.*'
>
> 38. In considering whether the adjudicator was acting within his jurisdiction in operating the slip rule the court should similarly guard against characterising a mistaken application of the slip rule as a decision in excess of, and therefore, outside his jurisdiction.

Table 4.2 Table of Cases: Slip Rule

Case Name	Citation	Issue
Bloor Construction v Bowmer & Kirkland	[2000] BLR 314 (TCC), Judge Toulmin	**Power to correct slips usually implied** The judge said that there will usually be a term implied into adjudication agreements that the adjudicator has the power to correct slips.
Shimizu Europe Ltd v Automajor Ltd	[2002] EWHC 1571 (TCC), [2002] BLR 113, Judge Seymour	**Asking for corrections may prevent later challenge** Where a party participates in an adjudication but reserves the right to challenge the decision later on grounds of a lack of jurisdiction, that party may lose the right to object if it asks for the decision to be corrected because that action may be interpreted as accepting the decision as valid.
CIB Properties Ltd v Birse Construction	[2005] 1 WLR 2252, [2005] BLR 173, [2005] EWHC 2365 (TCC), Judge Toulmin	**Conditions required for court to give effect to corrections** The adjudicator declined to make a correction asked of him. The parties asked for the Bloor principle to be extended so that the court could make the correction or require the adjudicator to do so. The judge said the court would only entertain corrections if (1) the adjudicator acknowledged he had made a mistake and was willing to correct it, (2) if the mistake was a genuine slip which failed to give effect to his first thoughts, (3) if the corrections were made within a reasonable time and without prejudice to the other party. If the adjudicator did not make the corrections then the error would usually fall to be treated as an error made within jurisdiction which did not affected the validity of the decision.
Cubitt Building Interiors Ltd v Fleetglade Ltd	110 Con LR 36, [2006] EWHC 3413 (TCC), Coulson J	**Corrected version issued out of time not fatal** The decision will not be invalid because it is in breach of strict time limits if it is initially issued within the time but then the corrected version is issued outside the time limits. As long as the corrected version is issued within a reasonable time it will be valid.
YCMS Ltd v Stephen & Miriam Grabiner	[2009] EWHC 127 (TCC), [2009] BLR 211, Akenhead J	**Slip rule cannot be used for second thoughts** The power to correct slips will only arise when there is an express or implied term of the contract that the adjudicator may correct errors. The power will be limited to clerical or arithmetic errors which are patent. The slip rule cannot be used to enable an adjudicator to give effect to second thoughts. The rule must be exercised within a reasonable time of the decision having been made.
O'Donnell Developments Ltd v Build Ability Ltd	28 Con LR 141, [2009] EWHC 3388 (TCC), Ramsey J	**When correction in itself includes mistake** The judge said that the erroneous exercise of a power does not constitute an excess of power so as to fall outside the jurisdiction of the adjudicator. Therefore a mistake of fact or law inherent in a correction otherwise validly made is not a ground for challenge.

Table 4.2 *Continued*

Case Name	Citation	Issue
ROK Building Ltd v Celtic Composting Systems Ltd (No. 2)	130 Con LR 74, [2010] EWHC 66 (TCC), Akenhead J	**Adjudicator declines to make corrections** The adjudicator was willing to make some corrections requested but refused to consider others. The TCC judge said the errors were the result of the complicated way the case had been put in the adjudication and it was simply unarguable that the adjudicator acted unfairly or in breach of natural justice in refusing to correct matters that were unclear and complicated.
Redwing Construction Ltd v Wishart	135 Con LR 119, [2010] EWHC 3366 (TCC)	**Slip rule only when clerical errors** The TCC judge adopted the same approach as in *YCMS* that the slip rule can only be used to correct minor clerical errors. The judge found that the adjudicator had made an obvious arithmetical error and was not correcting his reasoning but was simply correcting a very obvious error.

Power to Make Peremptory Decision: Scheme Paragraph 23(1)

4.124 Paragraph 23(1) which empowers the adjudicator to require the parties to comply peremptorily with his decision has been deleted in the amended Scheme. Paragraph 24 of the Scheme is also deleted. For a discussion about these two provisions see Chapter 7 at 7.125.

(6) Costs and Fees

The Adjudicator's Fees: Scheme Paragraph 25

The 1996 Act

4.125 There is no requirement under the 1996 Act for the construction contract to deal with the matter of the adjudicator's fees and expenses, and there is nothing in the 1996 Act which gives the adjudicator a right to payment.

4.126 The basis of adjudication is contractual and the basis of an adjudicator's fee claim is also contractual. The adjudication process consists of two agreements. First there is the adjudication agreement made between the parties to the construction contract. It is made expressly or impliedly where s.108 of the Act implies terms to this effect. Secondly, there is the agreement made between the adjudicator and one or both parties that he will adjudicate upon the dispute.[89] The right of an adjudicator to be paid his fees and expenses depends on the terms of the contractual relationship between the adjudicator and the parties.[90]

4.127 The Scheme contains a number of provisions relevant to an adjudicator's entitlement to fees. The question has arisen as to whether an adjudicator in a Scheme adjudication is entitled to fees if the decision is unenforceable.

4.128 The starting point in paragraph 25 is that the adjudicator shall be entitled to reasonable fees and expenses incurred by him and following the making of a determination as to how those shall be apportioned the parties shall remain jointly and severally liable for the payment of the fees and expenses. However, given that this is implied into the contract between the

[89] *Linnett v Halliwells* [2009] BLR 312 at para. 32.
[90] *Ibid.*, para. 34.

parties and rather than directly into the contract with the adjudicator, it must be right that the parties may make a different agreement with the adjudicator about fees. Paragraph 25 is therefore the default position. The purpose of paragraph 25 is to make clear that an adjudicator cannot charge a fee that is unreasonably high. For example the adjudicator may not charge for an unreasonable number of hours.[91]

4.129 The Scheme provides at paragraph 9(4) that where an adjudicator is obliged to resign because the dispute is the same or substantially the same as one already decided, he is entitled to payment of such reasonable amount as he may determine by way of fees and expenses reasonably incurred by him. This also applies when he resigns because the dispute he is asked to decide varies from that referred. The adjudicator's fee entitlement is the same if the parties agree to revoke the appointment of the adjudicator (Scheme paragraph 11(1)) unless this is due to his default or misconduct (Scheme paragraph 11(2)). These are express exceptions to the general principle that under the Scheme the adjudicator's entitlement to his fees depends on the production of an enforceable decision (see 4.131 below).[92]

4.130 The adjudicator's right to payment does not arise under the adjudication decision. The decision may determine which party shall pay his fees and expenses and in what proportion, but his entitlement to be paid by either party turns on his contractual relationship with the parties.[93]

4.131 The Court of Appeal has said that the adjudicator's obligation under the Scheme is an entire agreement the object of which is to produce an enforceable decision. Accordingly if the adjudicator commits a breach of natural justice that renders the decision unenforceable, the adjudicator is not entitled to be paid; *P. C. Harrington Ltd v Systech International Ltd* (2012) (Key Case).[94] Whether the adjudicator is entitled to be paid fees and expenses when the decision is unenforceable for want of jurisdiction depends on the facts. In *K. Griffin v Midas Homes*[95] part of a decision was severed for want of jurisdiction and not enforced, but the balance was enforced. As to the adjudicator's decision that the responding party should pay his fees, the court said that the responding party could not be liable for work performed without jurisdiction and that only the claiming party could be liable for those fees. A different conclusion was reached by the TCC judge in *Linnett v Halliwells* (2009)[96] (Key Case). Notwithstanding that the responding party had expressly not accepted the adjudicator's terms of appointment and had challenged his jurisdiction, by asking in the alternative for the adjudicator to deal with the dispute the responding party was found liable for his fees either on the basis of a contract made by conduct or as a *quantum meruit*. The judge said that this result would be the same whether the adjudicator had had jurisdiction (as the judge found) or not. The judge distinguished *Griffin v Midas* on the ground that in that case the judge was not considering the relationship between the adjudicator and the losing party but was considering whether the claiming party could enforce that part of the adjudicator's decision relating to fees against the losing party.

---

[91] *PC Harrington Ltd v Systech Int. Ltd* [2012] EWCA Civ 1371.

[92] In *Pring Island Ltd v Castle Contracting Ltd* (2003) an unreported decision of R. A. Dunlop Sheriff of Tayside and Fife,the court decided that under para. 9(4) where the adjudicator acted in good faith and should have resigned but mistakenly decided to continue, he was still entitled to be paid his fees. However this case was decided before the Court of Appeal decision in *P. C. Harrington.*

[93] *Linnett v Halliwells*, para. 35 and *Fenice Investments Inc. v Jerram Falkus Construction* [2011] EWHC 1678 (TCC), 141 Con LR 206; but cf. *Stubbs Rich Architects v W. H. Tolley & Son Ltd*, 8 Aug. 2011, Gloucester County Court (unreported) where the court decided the decision about which party shall pay the adjudicator's fees was also temporarily binding and enforceable.

[94] [2012] EWCA Civ 1371.

[95] 78 Con LR 152 [2000] (TCC) at 159, para. 3.

[96] [2009] BLR 312 (TCC).

4.132 The issue that arose in *P. C. Harrington Ltd v Systech International Ltd* (2012)[97] (Key Case) concerned a situation where the adjudicator was found to have honestly and unwittingly made a mistake about the extent of his jurisdiction and had excluded a defence he should have considered. As a result the decision was unenforceable by reason of a breach of natural justice. The adjudicator pursued the responding party PCH for payment of his fees. The TCC judge, following *Linnett v Halliwells*, found the responding party liable for the adjudicator's fees. The Court of Appeal reversed the first instance decision holding that the adjudicator was not entitled to be paid in these circumstances. However the finding in *Linnett v Halliwells* was preserved as an exception to the rule that the adjudicator's entitlement to be paid under the Scheme payment depended on the production of an enforceable decision. The exception applies where one party raises a jurisdiction issue and the other party decides nevertheless to proceed with the adjudication. In that situation the parties are taken to have accepted the risk that the outcome may be a decision that is unenforceable.

4.133 In *Stubbs Rich Architects v W. H. Tolley & Son Ltd* (2011)[98] the court interpreted the adjudicator's immunity from suit contained in paragraph 26 of the Scheme as meaning that an adjudicator's fees could only be challenged on the grounds of bad faith. However that same argument was rejected in *Fenice Investments Inc v Jerram Falkus Construction Ltd* (2011)[99] where the TCC judge said that under the Scheme the touchstone for the adjudicator's fee entitlement was reasonableness and that this was not consistent with the suggestion that his fee could only be challenged on the grounds of bad faith.

The 2009 Act

4.134 Section 108A of the 2009 Act says that any provision about costs of an adjudication is ineffective unless made in writing, contained in the construction contract, and confers powers on the adjudicator to allocate his fees and expenses between the parties (or is agreed after the start of the adjudication). The new Scheme has been amended to give effect to this so that now paragraphs 9(4), 11(1), and 25 all contain the following words:

> Subject to any contractual provision pursuant to Section 108A(2) of the Act, the adjudicator may determine how the payment is to be apportioned and the parties are jointly and severally liable for any sum that remains outstanding following the making of any such determination.

Key Cases: Adjudicators' Fees

Linnett v Halliwells LLP [2009] BLR 312 (TCC)

4.135 **Facts:** The responding party had objected to the jurisdiction of the adjudicator, but having reserved its position it took part in the adjudication without prejudice to its contention that there was no jurisdiction. The adjudicator's decision found against the responding party and decided that it should pay his fees and expenses. He issued a claim in the High Court for payment of his fees by the responding party.

4.136 **Held:** Halliwells had invited the adjudicator to adjudicate the merits, albeit reserving the position on jurisdiction. In doing so, Halliwells had asked the adjudicator to proceed and carry out work. This was an acceptance that if the jurisdictional argument was rejected the adjudicator would carry out the work. This gave rise to a contract formed by conduct with

[97] [2012] EWCA Civ 1371.
[98] Unreported decision of Recorder Lane QC sitting in Gloucester County Court.
[99] [2011] EWHC 1678 (TCC), 141 Con LR 206.

an obligation on Halliwells to pay the adjudicator's reasonable fees and expenses. If that was wrong then Halliwells was liable to pay on the grounds of unjust enrichment:

> 76. In this case it is evident that Halliwells took the second route which I have identified above. They asked the adjudicator to withdraw but, in the alternative asked him to adjudicate the merits, albeit reserving the position on jurisdiction. In doing so, I consider that they asked the adjudicator to proceed and carry out work and, in my judgment, whatever the correct position on jurisdiction, there was an acceptance by Halliwells that, if the Adjudicator rejected the jurisdictional argument, he would carry out work in dealing with the merits which would involve considering the arguments of both sides. The Adjudicator then rejected the jurisdictional argument and proceeded to consider the merits, including Halliwells' arguments.
>
> 77. In such circumstances, the Adjudicator proceeded both in compliance with the request of Halliwells and pursuant to the agreement with ISG. In relation to Halliwells the adjudicator proceeded at their request and did so without any express agreement as to fees. I consider that, as submitted by Ms Monastiriotis, the request from Halliwells and the fact that the Adjudicator proceeded with the adjudication gave rise to a contract formed by conduct with an obligation by Halliwells to pay the Adjudicator's reasonable fees and expenses.
>
> 78. If that be wrong or if, in a particular case, the matter could not be characterised in terms of a contract, I consider that Ms Monastiriotis would be correct in her submissions that the principles identified by Lord Steyn in *Banque Financiere de la Cite v Parc (Battersea) Ltd* [1999] 1 AC 221 at 227 apply. First, a responding party such as Halliwells has benefited or been enriched by having a decision on the merits which it can seek to rely on if it wishes. Secondly, that enrichment was at the expense of the adjudicator who spent time and incurred cost in dealing with Halliwells' submissions and the arguments raised. Thirdly, that enrichment was unjust where a party accepts the benefit of the adjudicator's services without payment. Fourthly, there are no specific defences to the payment of the fees in this case.
>
> 79. In my judgment, therefore, whether of not the Adjudicator had jurisdiction Halliwells are liable to pay the Adjudicator his reasonable fees and expenses of conducting the adjudication and are jointly and severally liable to the Adjudicator for the payment of such reasonable fees and expenses with ISG, whose liability is limited to the fees and expenses agreed with the Adjudicator, as set out above.

P. C. Harrington Ltd v Systech International Ltd [2012] EWCA Civ 1371

4.137 **Facts:** The adjudication was between Harrington and its subcontractor Tyroddy. It concerned Tyroddy's claim for the return of retention monies withheld from interim payments. The claim was brought five years after the contract works had been completed. Harrington defended on the basis of a revaluation of the final account which showed there had been an overpayment made, so that no sums were due. The adjudicator honestly and unwittingly made a mistake about the scope of his jurisdiction. He decided that Harrington's revaluation defence, which Harrington accepted had only surfaced during the adjudication for the first time, was outside the scope of the adjudication. Ultimately the adjudicator's decision in Tyroddy's favour was found to be unenforceable. The adjudicator sued Harrington for his fees (Tyroddy were in liquidation). The judge at first instance found that the adjudicator was entitled to his fees. Harrington appealed.

4.138 **Held:** The adjudicator was not entitled to be paid his fees. The adjudicator's agreement with the parties was to perform the services of adjudicator in accordance with the Scheme.

Although the adjudicator's own terms and conditions said he was entitled to be paid on the basis of hours incurred, those terms had to be read together with the terms of the Scheme:

> 26. But the terms of engagement must be read together with the terms of the Scheme. The significance of the Scheme is that it contains important provisions which deal with the question of remuneration in the event that the adjudicator does not reach a decision in various circumstances. Para 8(4) provides that, where an adjudicator ceases to act because a dispute is to be adjudicated by another person, he is entitled to payment of his fees and expenses in accordance with para 25. Para 9(1) provides that an adjudicator may resign at any time on notice. Para 9(2) provides that an adjudicator must resign where the dispute is the same or substantially the same as one which has previously been referred to adjudication, and a decision has been taken in that adjudication. Para 9(4) provides that where an adjudicator resigns in the circumstances referred to in para 9(2), or where a dispute varies significantly from the dispute referred to him in the referral notice and for that reason he is not competent to decide it, he is entitled to payment of reasonable fees and expenses. It is significant that, if the adjudicator resigns by giving notice under para 9(1), he is not entitled to any remuneration. This shows that the adjudicator is entitled to fees and expenses where he does not complete his engagement by making a decision, but only in carefully defined circumstances. The contrast between the treatment of a resignation under para 9(1) and 9(2) is striking.
>
> 27. A similar contrast is made at para 11 in relation to the adjudicator's remuneration in the event of a revocation of his appointment. Para 11(1) provides that the parties may at any time agree to revoke the appointment for any reason. In that event, the adjudicator is entitled to payment of reasonable fees and expenses. But if the revocation is due to 'the default or misconduct of the adjudicator', para 11(2) provides that there is no entitlement to fees or expenses.
>
> 28. It can, therefore, be seen that the Scheme carefully defines the circumstances in which the adjudicator is entitled to remuneration where his appointment comes to an end before he has made a decision.
>
> ...
>
> 31. None of the circumstances mentioned in para 8(4), 9(2) or 11(1) existed in this case. It follows that the adjudicator had no discrete entitlement to his fees and expenses for the ancillary and anterior functions that he performed. I should add that I accept the submission of Mr Bowling that these functions, which included making directions, considering the papers and so on, had no discrete value to the parties. Even the adjudicator's decision on the jurisdiction issue to which I referred at para 3 above was of no value in itself. It did not produce a decision which was binding in any future adjudication: it is well established that an adjudicator does not have inherent power to decide his own jurisdiction: see *Coulson, Construction Adjudication* (2nd edition) at para 7.09.
>
> 32. I return to the question: what was the bargained-for performance? In my view, it was an enforceable decision. There is nothing in the contract to indicate that the parties agreed that they would pay for an unenforceable decision or that they would pay for the services performed by the adjudicator which were preparatory to the making of an unenforceable decision. The purpose of the appointment was to produce an enforceable decision which, for the time being, would resolve the dispute. A decision which was unenforceable was of no value to the parties. They would have to start again on a fresh adjudication in order to achieve the enforceable decision which Mr Doherty had contracted to produce.
>
> 33. Para 11(2) of the Scheme provides powerful support for PCH's case. If the adjudicator's appointment is revoked due to his default or misconduct, he is not entitled to any fees. It can hardly be disputed that the making of a decision which is unenforceable by reason of a breach of the rules of natural justice is a 'default' or 'misconduct' on the part of the adjudicator. It is a serious failure to conduct the adjudication in a lawful manner. If during the course of an adjudication, the adjudicator indicates that he intends to

act in breach of natural justice (for example, by making it clear that he intends to make a decision without considering an important defence), the parties can agree to revoke his appointment. In that event, the adjudicator is not entitled to any remuneration. It makes no sense for the parties to agree that the adjudicator is not entitled to be paid if his appointment is revoked for default or misconduct before he makes his purported decision, but to agree that he is entitled to full remuneration if the same default or misconduct first becomes manifest in the decision itself. I would not construe the agreement as having that nonsensical effect unless compelled to do so by express words or by necessary implication. I can find no words which yield such a meaning either expressly or by necessary implication.

34. The fact that the adjudicator was not liable for anything done or omitted to be done unless it was in bad faith (para. 26) lends further support to the view that the parties did not intend that the adjudicator should be paid for producing an unenforceable decision. If Miss Rawley is right, the adjudicator was entitled to be paid the same fee for producing an unenforceable decision as for producing one that was enforceable and yet, absent bad faith, the parties are not able to claim damages for the adjudicator's failure to produce an enforceable decision, regardless of the seriousness of the failure and the loss it has caused. That is a most surprising bargain for the parties to have made. I would be reluctant to impute to them an intention to make such a bargain unless compelled to do so. I can find nothing in the terms of engagement or the Scheme which compels the conclusion that this was their intention.

...

44. As to the special situation arising in an adjudication where one of the parties raises a challenge on jurisdiction before a decision is reached and then, having received the adjudicator's ruling on jurisdiction, elects that the adjudicator should proceed to a decision, that situation is in my view correctly addressed by Ramsey J at paragraphs 76 to 79 of his judgment in *Linnett v Halliwells LLP* [2009] EWHC 319 (TCC), [2009] BLR 312. The adjudicator's fees are then—subject of course to any express terms agreed—payable even if the Court subsequently were to declare the initial challenge to the jurisdiction to have been well-founded.

Table 4.3 Table of Cases: Adjudicator's Fees

Title	Citation	Issue
Griffin v Midas Homes	(2000) 78 Con LR 152 (TCC), Judge LLoyd	**Responding party fee liability where there was no jurisdiction** The case concerned a challenge to enforce an adjudicator's decision on the grounds that the dispute had not crystallized. The TCC judge found that part of the decision concerned a dispute that had not crystallized. He severed that from the valid part of the decision which was enforced. In an addendum to the judgment and without hearing full argument, the judge decided that the responding party could not be liable for the work which had been performed without jurisdiction and that only the claiming party could be liable for those fees (p. 159, para. 3). The court made the responding party for the fees that related only to the enforceable part of the decision. This was distinguished in *Linnett v Halliwells*, but now see *P.C. Harrington v Systech International Ltd.*

Title	Citation	Issue
Pring Island Ltd v Castle Contracting Ltd	15 Dec. 2003, R. A. Dunlop QC Sheriff of Tayside Central and Fife (unreported)	**Scheme adjudicator entitled to reasonable fee even when he should have but did not resign** The adjudicator had acted in good faith but had mistakenly continued when he should have resigned (because the dispute was the same as one previously adjudicated). The judge said the adjudicator was validly appointed and, as he remained in his post, was entitled to be paid his fees, pursuant to para. 9(4) of the Scheme, until he resigned. Even though 9(2) says the adjudicator must resign if the dispute is the same, 9(4) says he is entitled to 'payment of such reasonable amount as he may determine by way of fees and expenses reasonably incurred by him', which the judge found assumes the adjudicator must decide if the dispute is the same and may get the answer wrong and continue—but is still entitled to his fees.
Castle Inns (Stirling) Ltd v Clark Contracts Ltd	[2005] Scots CS CSOH 178	**Adjudicator fee apportionment cannot be challenged** An adjudicator's decision as to liability to pay his fees is final and not subject to challenge in subsequent arbitration/ litigation (but see *Fenice v Jerram Falkus* below where a different conclusion was reached).
Dr Peter Rankilor v Perco Engineering Services Ltd	27 Jan. 2006 (TCC), Judge Gilliland	**Fees when breach of natural justice?** *Obiter* comment by the judge that it was 'a surprising submission that if an adjudicator's decision had been reached in serious breach of the rules of natural justice and thus would not be enforced by the court, the adjudicator should nevertheless be entitled to claim payment for producing what was in effect a worthless decision without even any temporary binding legal effect'.
Cubitt Building & Interiors Ltd v Fleetglade Ltd	[2006] EWHC 3413 (TCC), [2006] 110 Con LR 36 Coulson J	**No fees if decision out of time?** *Obiter* comment by the judge that if an adjudicator failed to produce the decision within 28 days so that the decision was unenforceable, the adjudicator's immunity would not have protected him because he would have been a complete failure to discharge his functions at all (para. 91 of judgment)
Linnett v Halliwells	[2009] BLR 312 (TCC), Ramsey J	**Responding party liable for adjudicator fees when expressly or impliedly asks for dispute to be decided** The responding party had challenged jurisdiction, but had asked in the alternative for the adjudicator to consider its defence. Having lost the adjudication they refused to pay the adjudicator's fees. The judge found that even though they had not accepted the adjudicator's terms, by asking him to adjudicate on the dispute, they were liable to pay his fees under an implied contract or as a *quantum meruit*, whether or not his decision was ultimately one made with jurisdiction.
Stubbs Rich Architects v W. H. Tolley & Son Ltd	Adj LR 08/8, 8 August 2011, Gloucester County Court, Recorder Lane QC	**Bad faith only ground to challenge fees** The court rejected a challenge to adjudicator's fees on the grounds that only bad faith is a ground for challenging adjudicator's fees. This decision was not followed in *Fenice v Jerram Falkus*.

(*Continued*)

Table 4.3 *Continued*

Title	Citation	Issue
Fenice Investments Inc v Jerram Falkus Construction Ltd (2011)	[2011] EWHC 1678 (TCC), 141 Con LR 206	**Touchstone for fee entitlement under scheme is reasonableness** In an adjudication to which the Scheme applied the adjudicator decided that the losing party should pay his fees. The losing party said the fees were excessive and paid only a small portion. The balance was paid to the adjudicator by the winner who then issued a claim in the TCC for reimbursement. **Held:** pursuant to para. 25 of the Scheme, the touchstone for the question of entitlement is reasonableness. In this case the fees were reasonable and the defendant found liable for them. Disagreeing with *Stubbs* the court rejected the argument that the adjudicator's decision about his fees was part of his decision that was temporarily binding and summarily enforceable (like the decision on the substantive dispute—see judgment para. 22). Also rejected was the notion that the only way to challenge the adjudicator's fees was on the grounds of bad faith (para. 24). The court said that immunity from prosecution in para. 26 of the Scheme was to have nothing to do with the adjudicator's entitlement to fees (para. 28).
Systech International Ltd v P. C. Harrington Ltd	[2011] EWHC 2722 (TCC), Akenhead J	**Fees where decision unenforceable for breach of natural justice** The adjudication decision was found unenforceable because the adjudicator had made an innocent mistake and disregarded part of PCH's defence thinking it outside his jurisdiction. The adjudicator sued the losing party for his fees and they alleged in response he was not entitled because the failure to produce an enforceable decision was a total failure of consideration. Argument rejected, the losing party was liable for the adjudicator's fees as under the terms of the adjudicator's appointment the fees were payable for time incurred and an enforceable decision was not a condition precedent to the fee entitlement.
P. C. Harrington Ltd v Systech International Ltd	[2012] EWCA Civ 1371 (CA)	**Adjudicator under Scheme not entitled to fees where decision unenforceable due to breach of natural justice** The Court of Appeal reversed the first instance decision and found that on a true construction of the Scheme the adjudicator's entitlement to fees depended on the production of an enforceable decision. However the exception was the *Linnett v Halliwells* situation where a jurisdiction objection was taken and the parties decided to continue, they had by that action accepted the risk of an unenforceable decision.

Adjudicator's Power to Award Costs

The 1996 Act

4.139 Section 108 of the 1996 Act does not require that the construction contract empower an adjudicator to award legal costs of an adjudication, nor does it prohibit the construction contract from containing such a provision. In contracts governed by the 1996 Act[100] the parties are free to include terms concerning how legal costs shall be treated in the adjudication provisions of their contract. That has now been changed in the 2009 Act as is discussed

[100] See Ch. 2.

below. Absent such an agreement there is no free-standing power vested in the adjudicator to award legal costs.

4.140 The Scheme includes no term regarding the legal costs of the parties[101] and an adjudicator will have no jurisdiction to order legal costs to be paid by either party unless both parties have agreed, expressly or impliedly, that his jurisdiction is widened to include such a power: *Northern Developments (Cumbria) Ltd v J&J Nichol* (2000) (Key Case).[102] In that case there was no contractual term that the adjudicator could deal with legal costs, but Judge Bowsher found that because both parties had asked the adjudicator to award them their costs, and none had taken the point that there was no jurisdiction to deal with them, there had been an ad hoc conferral of jurisdiction on the adjudicator to decide that one party should pay the legal costs of the other.

4.141 Until the passing of the 2009 Act, standard-form adjudication procedures have frequently contained an express term dealing with the parties' costs of the adjudication, for example:

- GC/Works/1 (1998) includes a power to award legal and other costs and expenses (clause 59(6)).
- The ICE Adjudication Procedure 1997 states the parties shall bear their own costs and expenses incurred in the adjudication (clause 6.5).
- The TeCSA (Version 2.0) Adjudication Rules provide that if the parties so agree, the adjudicator shall have power to award costs to the successful party, and shall have no jurisdiction to require the referring party to pay the costs of the other party solely by reason of having referred the dispute (notwithstanding a term to the contrary in the contract) (clauses 28 and 29).
- The JCT Standard Form of Building Contract (1998) provides that the parties shall meet their own costs of the adjudication except that the adjudicator may direct who should pay the cost of any test or opening up if required (clause 41A.5.7).
- The CIC Model Adjudication Procedure (4th edn) states that the parties shall bear their own costs and expenses incurred in the adjudication.

4.142 In *Total M&E Services Ltd v ABB Building Technologies Ltd* (2002)[103] an argument that legal costs could be claimed as damages for breach of contract was rejected by Judge Wilcox.

4.143 In *Bridgeway Construction Ltd v Tolent Construction Ltd* (2000)[104] Judge Mackay decided there was no public policy reason for declaring void a provision in a construction contract which provided that the party serving a notice of adjudication should bear all the adjudicator's costs and fees, and all the costs and expenses incurred by both parties, in relation to any adjudication (these clauses are now often referred to as 'Tolent' clauses). However in *Yuanda (UK) Co. Ltd v WW Gear Construction Ltd* (2010)[105] the TCC judge decided that a clause similar to that in *Tolent* would discourage parties from referring disputes to adjudication and was therefore an inhibition on the parties' right to refer a dispute at any time and was contrary to the 1996 Act. Accordingly the judge found the contract procedure did not comply with the 1996 Act and was replaced entirely by the Scheme.

4.144 In *John Roberts Architects Ltd v Parkcare Homes (No. 2) Ltd* (2006)[106] the contract contained an express term that 'The Adjudicator may in his discretion direct the payment of

[101] Both the original Scheme and as amended.
[102] [2000] BLR 158.
[103] 87 Con LR 154, [2002] EWHC 248 (TCC).
[104] [2000] CILL 1662.
[105] [2010] EWHC 720 (TCC), [2010] BLR 435.
[106] [2006] BLR 106, [2006] EWCA Civ 64.

legal costs and expenses of one party by another as part of his decision. The Adjudicator may determine the amount of costs to be paid . . . '. The Court of Appeal decided that even though the dispute had been withdrawn from the adjudicator, the adjudicator nevertheless retained jurisdiction to give a decision about liability for the legal costs of the aborted adjudication.

4.145 In *Deko Scotland v Edinburgh Royal Joint Venture and anor* (2003)[107] the Official Referees Solicitors Association Adjudication Rules 1998 allowed for an adjudicator to require a party 'to pay the legal costs of another party arising in the adjudication' (r. 21A). Lord Drummond decided that the power under r. 21A was qualified in two respects: costs had to be legal and they had to arise in the adjudication. In other words, any express or implied power to award costs was restricted to expenses analogous to legal expenses unless there was a clear statement to the contrary. He thus excluded internal costs and sums paid to claims consultants and surveyors. He also decided that there was an implied power to have any award of costs taxed.

The 2009 Act

4.146 Important amendments have been made to the 1996 Act by the introduction of an entirely new provision at s. 108A which provides:[108]

> (1) This section applies in relation to any contractual provision made between the parties to a construction contract which concerns the allocation between those parties of costs relating to the adjudication of a dispute arising under the contract.
> (2) The contractual provision referred to in subsection (1) is ineffective unless:
> (a) it is made in writing, is contained in the construction contract and confers powers on the adjudicator to allocate his fees and expenses between the parties, or
> (b) it is made in writing after the giving of notice to refer the dispute to adjudication.

4.147 Section 108A was introduced to prevent clauses of the types seen in *Bridgeway v Tolent* and *Yuanda v WW Gear*. On one view it may appear to leave open the possibility that a clause can be included in the contract in writing that says the referring party must pay all legal costs of both parties save that the adjudicator may allocate his fees and expenses between them. However in *Leander Construction Ltd v Mulalley & Co. Ltd* (2011)[109] Coulson J said, *obiter*, that such a clause would be automatically invalid pursuant to s. 108A and it appears that giving s. 108A a purposive interpretation achieves that result.

Key Cases: Costs

Northern Developments (Cumbria) Ltd v J&J Nichol [2000] BLR 158 (TCC)

4.148 **Facts:** The claimant applied for a declaration that an adjudicator's decision was invalid and the defendant counterclaimed for the enforcement of decision. Each side in the adjudication claimed costs against the other, and neither side made any submissions to the adjudicator that he had no jurisdiction to deal with *inter partes* costs.

4.149 **Held:** Having rejected the challenges made to the substance of the award, the judge decided on the legal costs issue that an adjudicator had no jurisdiction under the 1996 Act to award *inter partes* costs, but the parties could confer that jurisdiction on him by consent, and their respective actions in claiming costs against each other and failing to address any argument that there was no such jurisdiction amounted to conferring that jurisdiction by consent.

---

107 [2003] SLT 727.
108 Introduced by s. 141 of the Local Democracy Economic Development and Construction Act 2009.
109 [2011] EWHC 3449 (TCC) at para. 12 of the judgment.

Total M&E Services Ltd v ABB Building Technologies Ltd (2002) 87 Con LR 154, [2002] EWHC 248 (TCC)

4.150 **Facts:** Application for summary judgment on a claim to enforce an adjudication award. The claim consisted not only of the amount of the award, but also the costs of the adjudication, which the claimant ('Total') sought to recover as a separate head of damage. Total's contract with the defendant ('ABB') was a fixed-price labour-only subcontract. The statutory scheme applied. ABB resisted the application for summary judgment on several gounds and in relation to the legal costs decision argued that the costs of the adjudication were not recoverable as damages.

4.151 **Held:** The costs of the adjudication were not recoverable as damages:

> 24. Mr Harding contends that the costs of the Adjudication are recoverable as a damages claim. He submits that if a defendant fails to pay under a construction sub-contract to which the Act applies it is foreseeable that the Claimant would seek adjudication and properly incur costs, and thereafter seek to recover them. Mr Coulson submits that since the Statutory Scheme does not make any provision for the Adjudicator to award costs unless the parties agree otherwise the Adjudicator has no jurisdiction to order that one party's adjudication costs should be paid by the other. There is no such agreement in the present case; indeed, the entire basis of the Claimant's claim is that there was no such agreement. I agree with Mr Coulson that since the Act does not provide for the recovery of costs the claim is misconceived. Furthermore, this claim is put as a claim for damages for breach of contract arising out of ABB's failure to pay. Because the Statutory Scheme envisages both parties may go to Adjudication and incur costs which they cannot, under the Act recover from the other side, it follows that such costs cannot therefore arise as damages for breach.
>
> 25. To permit such a claim would be to subvert the statutory scheme under the Act.

Table 4.4 Table of Cases: Power to Award Costs

Case Name	Reference	Issue
Northern Developments (Cumbria) Ltd v J&J Nicol	[2000] BLR 158, Judge Bowsher	**Asking adjudicator to deal with costs** There was no express power for the adjudicator to deal with costs but because both sides had asked him to award them costs they had conferred ad hoc jurisdiction on him to decide costs.
Bridgeway Construction Ltd v Tolent Construction Ltd	[2000] CILL 1662, Judge Mackay	**Clause making one party bear all costs** The judge said there was no public policy reason why a clause that provided that one party was liable for all costs and for the adjudicator's fees, should be invalid (but see *Yuanda* below).
Total M&E Services Ltd v ABB Building Technologies Ltd	87 ConLR 154, [2002] EWHC 248 (TCC), Judge Wilcox	**Costs as damages not legitimate** An argument that legal costs of an adjudication could be claimed by one party as damages for breach of contract was rejected by the judge.

(*Continued*)

Table 4.4 *Continued*

Case Name	Reference	Issue
Deko Scotland v Edinburgh Royal Joint Venture and anor	[2003] SLT 727, Lord Drummond	**Legal costs analogous to those in court proceedings** Interpretation of express term that adjudicator had power to award legal costs arising in adjudication. Lord Drummond found this to mean legal costs analogous to court proceedings and so disallowed internal surveyor's costs.
John Roberts Architects Ltd v Parkcare Homes (No. 2) Ltd	[2006] BLR 106, [2006] EWCA Civ 64 (CA)	**Power to award costs when claim withdrawn** The adjudication agreement gave the adjudicator power to order that legal costs be paid by one party to another. After a claim was withdrawn from the adjudicator, which the CA said was a valid withdrawal, the adjudicator was still entitled to make a costs order in respect of the aborted adjudication.
Yuanda (UK) Co. Ltd v WW Gear Construction Ltd	[2010] EWHC 720 (TCC), [2010] BLR 435	**Tolent clause invalid** A clause that provided that the legal costs would always be paid by the referring party would discourage parties from using the adjudication procedure. That was contrary to the requirement of the 1996 Act that the parties be able to refer a dispute at any time. The judge found the adjudication procedure was non-compliant with the Act and so had to be replaced by the Scheme.

Part II

AD HOC ADJUDICATION

5

AD HOC REFERENCES AND ADJUDICATIONS OUTSIDE THE ACT

(1) Ad Hoc References to Adjudications

5.01 The statutory right to adjudicate arises[1] only in respect of construction contracts within the meaning of ss. 104 and 105 of the Act.[2] Contracts not caught by the legislation may nevertheless contain an adjudication clause thus providing either contracting party with a right to adjudicate. Where neither situation applies and there is no pre-existing right to adjudicate, an adjudicator may be appointed to decide a dispute or an issue on the basis of a one-off agreement to adjudicate. Such agreements may be referred to as 'ad hoc' adjudications as the expression 'ad hoc' generally signifies a solution that has been custom designed for a specific problem. The situations in which the parties may expressly or impliedly confer ad hoc jurisdiction on an adjudicator are discussed in this chapter.

5.02 In the context of arbitration it is well established that the parties may agree to confer jurisdiction upon an arbitrator on an ad hoc basis.[3] In principle, if two people agree to submit a dispute to a third person for determination then the parties are conferring jurisdiction on that third party to decide the matter and are agreeing to be bound by his decision: *Westminster Chemicals & Produce Ltd v Eicholz & Loeser* (1954),[4] *Project Consultancy Group v The Trustees*

[1] Under the Housing Grants Construction Regeneration Act 1996 ('the 1996 Act') and/or the Local Democracy Economic Development and Construction Act 2009 ('the 2009 Act').

[2] See Ch. 2 for a discussion of what is a construction contract within the meaning of ss. 104 and 105. Further, in respect of contracts entered into before the 2009 Act came into force, they must satisfy the requirement for writing laid down by s. 107 of the 1996 Act, as discussed in Ch. 3.

[3] *Thames Iron Works & Shipbuilding Co. Ltd v R* [1869] 10B & S 33; *Westminster Chemicals & Produce Ltd v Eicholz & Loeser* [1954] 1 LLR 99, 105–6; *F. W. Berk & Co. Ltd v Knowles and Foster* [1962] 1 Lloyd's Rep 430.

[4] [1954] 1 LLR 99 at 105–6.

of the Gray's Trust (1999) (Key Case).[5] The agreement to confer special jurisdiction on an adjudicator may be express but it may also be implied from the conduct of the parties.

5.03 If one party considers that an issue or dispute is outside the agreement to adjudicate, or the scope of the reference, he may protest there is no jurisdiction to determine that matter. If he protests in a form that indicates he will not abide by the decision of the adjudicator on that matter, he may continue to take part in the adjudication without losing his right to argue the decision is not binding because it was made without jurisdiction.

5.04 Where the responding party does not wish to confer ad hoc jurisdiction on the adjudicator, the objection to jurisdiction ought to be taken at the earliest opportunity or there will be a real prospect that the responding party will be found to have waived the lack of jurisdiction and acquiesced in conferring ad hoc jurisdiction on the adjudicator.[6]

5.05 Typical situations in which the issue of ad hoc jurisdiction arises are where the referring party asks an adjudicator to determine a claim and:

1. An adjudication agreement exists but is not wide enough to cover the claim referred.
2. An adjudication agreement exists wide enough to cover the claim in question but one party asks for an issue to be decided which was not part of the original reference.
3. No adjudication agreement exists and there is no statutory right to adjudicate.

5.06 Whether the parties have conferred ad hoc jurisdiction on an adjudicator is a question of fact. An ad hoc submission may occur in a number of situations.

Express Agreements to Confer Jurisdiction

5.07 Conferring ad hoc jurisdiction on an adjudicator may be done expressly by agreement as was the case in *Nordot Engineering Services Ltd v Siemens plc* (2001).[7] The issue concerned whether the works were 'construction operations' under the Act. The respondent argued they were not but also said 'we will, however, abide by your decision in this matter and will comply with what ever direction you deem appropriate'. This was a clear and unequivocal agreement to be bound by the decision of the adjudicator that conferred jurisdiction to decide the jurisdiction question.

Implied Agreements to Confer Jurisdiction

5.08 For there to be an implied agreement to give the adjudicator jurisdiction over a dispute or a particular issue, one needs to look at everything material that was done and said to determine whether one can conclude that the parties must be taken to have agreed that the adjudicator had such jurisdiction. Conferral of jurisdiction has been implied from the circumstances in the following cases:

(1) In *Northern Developments (Cumbria) Ltd v J&J Nichol* (2000)[8] the respondent raised several grounds for challenging the validity of the decision, including that the adjudicator had no power to decide costs. On the issue of costs it was shown that both parties' submissions had asked the adjudicator for costs and neither had denied there was a power to deal with costs. Hence neither party had contended there was no jurisdiction to deal with costs. The judge found that it was implied from the facts that the parties had

[5] [1999] BLR 37.
[6] See 5.09–5.15 below.
[7] 2001 CILL 1778 (TCC).
[8] [2000] BLR 158 (TCC).

agreed the adjudicator had the authority to determine costs and hence they had conferred jurisdiction on the adjudicator to decide the issue.

(2) In *Carillion Construction Ltd v Devonport Royal Dockyard* (2005)[9] the referring party had asked for interest to be decided and the respondent had answered by saying the question of interest did not arise as no sum was due. The respondent did not argue that there was no jurisdiction to deal with interest. The Court of Appeal found that this amounted to acquiescence by the respondent in the conferral of ad hoc jurisdiction to make a decision about the claim to interest.

(3) *Maymac Environmental Services Ltd v Faraday Building Services Ltd* (2001)[10] concerned a dispute as to whether a relevant construction contract existed. The court found there was a construction contract and that there had been an agreement to adjudicate on the same terms. It was held that the respondent would be estopped by representation from asserting that there was no jurisdiction.

Objecting and Reserving the Right to Object

5.09 There will be no implied agreement if at any material stage before or during the adjudication a clear objection is raised by the party contesting jurisdiction of the adjudicator, coupled with a reservation of the right to maintain that objection. A party that does not wish to give the adjudicator power to decide a matter needs to raise the jurisdiction objection at the earliest opportunity. The objecting party may ask the adjudicator to enquire into the jurisdiction issue and to decide that there is none and that he or she should resign.

5.10 An adjudicator does not have power to give a binding decision upon his or her own jurisdiction unless the parties confer that power.[11] A decision of an adjudicator about the existence or scope of his or her own jurisdiction is therefore, in principle, a non-binding decision.[12] However, when raising a jurisdiction objection and asking the adjudicator to consider it, it is important that the objecting party does not give the adjudicator power to decide the jurisdiction issue in a binding way. Accordingly a party must reserve the right to maintain its jurisdiction objection at a later date. In the absence of an express reservation the court may find by implication that the party has permitted the adjudicator to make a binding decision on the issue: *Christiani & Neilsen Ltd v Lowry Centre Development Co. Ltd* (2000).[13] It is essential therefore that the objecting party makes it clear that, despite asking for the jurisdiction to be investigated, they do not agree to give the adjudicator power to make a binding decision either on the jurisdiction point or on the substantive dispute.[14]

5.11 Where an effective reservation of rights has been made the objecting party may continue to participate in the adjudication and yet still argue that any decision is unenforceable because it was made without jurisdiction: *Project Consultancy Group v The Trustees of the Gray's Trust* (1999) (Key Case).[15]

5.12 A clear reservation can be made by words expressed by or on behalf of the objecting party. Words such as 'I do not intend to confer on the adjudicator jurisdiction to make a binding

---

9 [2006] BLR 15 (CA).

10 [2001] CILL 1685 (TCC).

11 See *Homer Burgess v Chirex (Annan) Ltd* [2000] BLR 124; *Pegram Shopfitters Ltd v Tally Weijl (UK) Ltd* (2004) 1 WLR 2082; *Pilon v Breyer* [2010] EWHC 837 (TCC).

12 Where the jurisdiction issue overlaps with a substantive issue the adjudicator is to decide whether he or she may be empowered to decide the jurisdiction issue. This is an exceptional situation. See Ch. 9 at 9.12.

13 29 Jun. 2000 (unreported).

14 This topic is also discussed in Ch. 9 at 9.14–9.15 in the context of jurisdictional challenges to enforcement.

15 [1999] BLR 37 (TCC).

decision, I am only participating in the adjudication under protest and reserve the right to argue that any decision was made without jurisdiction' will usually suffice to make an effective reservation. One can, however, look at every relevant thing said and done during the course of the adjudication to see whether by words and conduct what was clearly intended was a reservation as to the jurisdiction of the adjudicator: *Aedifice Partnership Ltd v Ashwin Shah* (2010) (Key Case).[16]

5.13 If a party does not reserve its position effectively then generally it cannot avoid enforcement on the ground that it was made without jurisdiction. There may be unusual circumstances if the party making the challenge did not know or could not reasonably have ascertained the grounds of challenge before the decision was issued.[17] However where a party knows or should have known of grounds for a jurisdictional objection and participates in the adjudication without making an effective reservation then it may be said that it has submitted to the jurisdiction or waived the right to object.[18]

No Conferral of Jurisdiction where Continued Jurisdiction Objection

5.14 Whether the parties have agreed that the adjudicator could determine his own jurisdiction is a matter of fact. In *Christiani & Nielsen Ltd v The Lowry Centre Development Co. Ltd* (2000)[19] the respondent agreed to make submissions on the existence of a construction contract. That was found to be simply an agreement as to procedure. The respondent's submissions were made against the background of their continuing protest as to jurisdiction. Thus the judge decided that the respondent was not agreeing that the adjudicator could make a binding decision on jurisdiction.

5.15 A party will not be taken to have submitted to the jurisdiction of an adjudicator unless he knew of the matters which meant the adjudicator lacked jurisdiction: *R. Durtnell & Sons Ltd v Kaduna Ltd* (2003).[20] However, the court will consider whether a party actually knew or ought to have known that there was a jurisdiction issue to be made: *Aedifice Partnership Ltd v Ashwin Shah* (2010) (Key Case).[21]

General Reservations

5.16 Generally a party who wishes to do so may object to the jurisdiction of the adjudicator either in general terms or by objecting to jurisdiction in relation to a specific matter.[22] However where the general reservation was used a party had to be careful to ensure it was clearly worded to ensure it achieved its purpose. Where both a general and specific reservation is made at the outset but later only the specific reservation is argued, a party may be taken to have abandoned the general reservation.[23]

---

[16] [2010] EWHC 2106 (TCC).

[17] *Allied P&L Ltd v Paradigm Housing Group* [2009] EWHC 2890 (TCC) at para. 32 [2010] BLR 59.

[18] *Aedifice Partnership Ltd v Ashwin Shah* [2010] EWHC 2106 (TCC) at para. 15 and following (Key Case) 132 Con LR 100.

[19] 29 Jun. 2000 (TCC) (unreported).

[20] [2003] BLR 225.

[21] [2010] EWHC 2106 (TCC) at para. 21(e).

[22] *Bothma v Mayhaven Health Care Ltd* [2007] EWCA Civ 527 (CA); *GPS Marine Contractors Ltd v Ringway Infrastructure Services Ltd* [2010] BLR 377.

[23] *Durham County Council v Jeremy Kendall* [2011] BLR 425.

Key Cases: Ad Hoc Jurisdiction

Project Consultancy Group v The Trustees of the Gray Trust [1999] BLR 37 (TCC)

5.17 **Facts:** At an enforcement hearing the responding party argued that the contract was made prior to 1 May 1998 and as the Act did not apply the adjudication decision was invalid. The referring party argued that the contract was made after 1 May 1998, alternatively there had been an ad hoc conferral of jurisdiction on the adjudicator on the basis that the respondent had put forward their case to the adjudicator that the contract was made before 1 May 1998, and in so doing the defendants were submitting the question of jurisdiction to the adjudicator for his decision, and agreeing to be bound by it.

5.18 **Held:** In principle, a party could confer jurisdiction on the adjudicator on an ad hoc basis by asking the adjudicator to deal with the issue. However such a party was entitled to ask for an issue to be decided without prejudice to the overarching primary contention that the adjudicator had no jurisdiction to determine the dispute at all. The judge found that in this case the respondent's solicitors had reserved their right to maintain there was no jurisdiction. As a result the judge found that although the respondent had asked for the formation of the contract issue to be decided it had not agreed to be bound by the decision:

> 14. Ms Rawley submits that, by putting forward their case to the adjudicator that the contract was made before 1 May 1998, and that for that reason he had no jurisdiction, the defendants were submitting the question of jurisdiction to the adjudicator for his decision, and agreeing to be bound by it. She relies on the principles enunciated by Devlin J in *Westminster Chemicals & Produce Ltd v Eicholz & Loeser* [1954] 1 LLR 99, 105–6. Although that case concerned an arbitration, I agree that what Devlin J said was equally applicable to an adjudication. He said that if two people agree to submit a dispute to a third person, then the parties agree to accept the award of that person, or, putting it another way, they confer jurisdiction on that person to determine the dispute. If one of the parties thinks that the dispute is outside the agreement that they have made, then he can protest against the jurisdiction of the arbitrator.
>
> > 'If he protests against the jurisdiction of the arbitrator, which is merely an elaborate way of saying: "I have not agreed to abide by your award," if he protests in that form it is held that he can take part in the arbitration without losing his rights, and what he is doing, in effect, is that he is merely saying: "I will come before you, but I am not by my conduct in coming before you and arguing the case, to be taken as agreeing to accept your award, because I am not going to do so". In those circumstances he may or may not be allowed to take part in the arbitration. Customarily I think he is, but whether that be so or not, if he protests it is well settled that he enters into no agreement to abide by the award.'
>
> 15. In my view, the defendants' solicitors' letter of 9 March 1999 stated in the clearest terms that the defendants protested the adjudicator's jurisdiction, and that they would not recognise and comply with any decision to award money to the claimant. The letter also made it clear that, if the adjudication proceeded, they reserved their right to participate, but without prejudice to their contention that there was no jurisdiction. ... It is a question of fact whether a party submits to the jurisdiction of a third person ...

Thomas Frederic (Construction) Ltd v Keith Wilson (CA) [2004] BLR 23

5.19 **Facts:** Application to enforce an adjudicator's decision at which Keith Wilson argued the adjudicator had no jurisdiction over the dispute because it arose out of a contract not made with him personally but with his company. Keith Wilson had argued the same point before the adjudicator. By a letter of 2 January 2003 to the adjudicator Keith Wilson had alleged that he was not the developer in the contract and that there could be no referral to adjudication on the ground that the wrong party was named in the notice. The adjudicator ruled against Keith Wilson and found he was the contracting party and the adjudication could continue because he had the requisite jurisdiction. The TCC judge decided the adjudicator had jurisdiction to make a binding decision that Keith Wilson was a contracting party and all the evidence pointed that way in any event. Keith Wilson appealed.

5.20 **Held:** (1) Keith Wilson had a good arguable case (sufficient to resist summary enforcement) that the adjudicator's decision on jurisdiction was wrong. (2) Nevertheless the adjudicator's decision would be binding if the respondent had agreed to accept the ruling. On the facts Keith Wilson had not conferred ad hoc jurisdiction on the adjudicator to make a binding decision on the issue. The letter of 2 January 2003 squarely took the jurisdictional objection, and nothing that was said then or thereafter can reasonably have been understood as an agreement to be bound by the adjudicator's decision on the point, whatever it might be. The Court of Appeal said the case fell full-square within Devlin J's words in the *Westminster Chemicals* case, cited in paragraph 14 of Dyson J's judgment in *Project Consultancy*:

> 33. The position can I think be summarised in the following two propositions. (1) If a defendant to a Part 24(2) application has submitted to the adjudicator's jurisdiction in the full sense of having agreed not only that the adjudicator should rule on the issue of jurisdiction but also that he would then be bound by that ruling, then he is liable to enforcement in the short term, even if the adjudicator was plainly wrong on the issue. (2) Even if the defendant has not submitted to the adjudicator's jurisdiction in that sense, then he is still liable to a Part 24(2) summary judgment upon the award if the adjudicator's ruling on the jurisdictional issue was plainly right.
>
> 34. Applying those propositions in the instant case, I would hold that this appellant did not submit to the adjudicator's jurisdiction in the full sense and that the adjudicator's ruling was, on any view, not plainly right. Indeed, as already indicated, it seems to me that the adjudicator's ruling was, if anything, plainly wrong. I would accordingly allow the appeal, set aside the order made below and substitute for it an order dismissing the respondent's application for summary judgment.

Aedifice Partnership Ltd v Ashwin Shah [2010] EWHC 2106 (TCC)

5.21 **Facts:** In the course of an adjudication concerning Aedifice's fee claim, the defendant alleged the adjudicator had no jurisdiction to determine the dispute because there was no contract between him and Aedifice. The defendant served a response to the referral stating the adjudicator did not have jurisdiction. The adjudicator found that it was more likely than not that he had jurisdiction and determined that Aedifice was entitled to payment of its fees. At the enforcement the defendant contended that he had effectively challenged and reserved his position in relation to the jurisdiction issue and that, in any event, the adjudicator had not determined his jurisdiction but had merely enquired into it.

5.22 **Held:** On the facts the defendant had made a clear reservation of rights and had maintained that throughout the adjudication. His participation in the adjudication did not amount to a waiver of the jurisdiction point. Further the adjudicator indicated that he had enquired into his jurisdiction but had not undertaken a full analysis and had not purported to decide jurisdiction. In the course of the judgment Akenhead J reviewed the authorities regarding how an effective reservation of rights should be made (at paragraphs 15–20 of the judgment) and went on to summarize the following propositions from those cases:

21. I can draw these various strands together:

(a) An express agreement to give an adjudicator jurisdiction to decide in a binding way whether he has jurisdiction will fall into the normal category of any agreement; it simply has to be shown that there was an express agreement.
(b) For there to be an implied agreement giving the adjudicator such jurisdiction, one needs to look at everything material that was done and said to determine whether one can say with conviction that the parties must be taken to have agreed that the adjudicator had such jurisdiction. It will have to be clear that some objection is being taken in relation to the adjudicator's jurisdiction because otherwise one could not imply that the adjudicator was being asked to decide a non-existent jurisdictional issue which neither party had mentioned.
(c) One principal way of determining that there was no such implied agreement is if at any material stage shortly before or, mainly, during the adjudication a clear reservation was made by the party objecting to the jurisdiction of the adjudicator.
(d) A clear reservation can, and usually will, be made by words expressed by or on behalf of the objecting party. Words such as 'I fully reserve my position about your jurisdiction' or 'I am only participating in the adjudication under protest' will usually suffice to make an effective reservation; these forms of words whilst desirable are not absolutely essential. One can however look at every relevant thing said and done during the course of the adjudication to see whether by words and conduct what was clearly intended was a reservation as to the jurisdiction of the adjudicator. It will be a matter of interpretation of what was said and done to determine whether an effective reservation was made. A legitimate question to ask is: was it or should it have been clear to all concerned that a reservation on jurisdiction was being made?
(e) A waiver can be said to arise where a party, who knows or should have known of grounds for a jurisdictional objection, participates in the adjudication without any reservation of any sort; its conduct will be such as to demonstrate that its non-objection on jurisdictional grounds and its active participation was intended to be and was relied upon by the other party (and indeed the adjudicator) in proceeding with the adjudication. It would be difficult to say that there was a waiver if the grounds for objection on a jurisdictional basis were not known of or capable of being discovered by that party.

Table 5.1 Table of Cases: Ad Hoc Referrals and Reservations of Rights

Title	Citation	Issue
Westminster Chemicals & Produce v Eicholz & Loeser	[1954] 1 LLR 99, 105–6 (CA)	**General principles** If two people agree to submit a dispute to a third person, then the parties agree to accept the award of that person, or, putting it another way, they confer jurisdiction on that person to decide the dispute.

(Continued)

Table 5.1 *Continued*

Title	Citation	Issue
Project Consultancy Group v The Trustees of the Gray's Trust	[1999] BLR 377 (TCC), Dyson J	**Effective reservation of rights** Principles in the *Westminster Chemicals* case apply to adjudication. The parties had asked the adjudicator to decide whether a contract falling within the Act had been created, but the defendant's request was made without prejudice to its right to argue at enforcement that no contract had been created. In the circumstances the defendant was not agreeing to be bound by the adjudicator's decision on whether a relevant contract had been created. The defendant was entitled to take part in the adjudication having reserved its right to argue at enforcement that any decision was invalid because a relevant contract did not exist.
Fastrack Contractors v Morrison	[2000] BLR 168 (TCC), Judge Thornton	**How to proceed when jurisdiction lacking:** 1. widen adjudicator power to decide own jurisdiction, or 2. refer the jurisdiction dispute to a second adjudicator, or 3. ask the court in parallel to decide the jurisdiction issue,[24] or 4. reserve position on lack of jurisdiction, participate in the adjudication, and then challenge an attempt to enforce on jurisdictional grounds.
Northern Developments (Cumbria) Ltd v J&J Nichol	[2000] BLR 158 (TCC), Judge Bowsher	**When both parties ask for costs award** At an application to resist enforcement of an adjudicator's decision several grounds were raised challenging the validity of the decision. On the issue of costs, both parties had asked for their own costs and neither had denied there was a power to do so and hence neither party had contended there was no jurisdiction to deal with costs. **Held:** It was implied

[24] As to when a court might make a finding in parallel, see the discussion in Ch. 7 at 7.57–7.64.

Title	Citation	Issue
		from the facts that the parties had agreed the adjudicator had the authority to determine costs and hence they had conferred jurisdiction on the adjudicator to decide the issue.
Grovedeck Ltd v Capital Demolition Ltd	[2000] 2 TCLR 689, Judge Bowsher	**Effect of s. 107(5)** Respondent denied existence of relevant construction contract in writing. Submission then set out the defence that 'in the alternative ... if it is found that the construction contracts do comply with s. 107 of the Act'. The claimant argued that this amounted to admission of contract as per s. 107(5) of the Act. **Held:** It does not apply to exchange of submissions in the existing adjudication, only to preceding adjudications.[25]
Ballast v Burrell Company (Construction Management) Ltd	[2011] BLR 529, Lord Reid	**Cannot decide own jurisdiction** Adjudicator cannot make a binding determination of his own jurisdiction and his jurisdiction could not be widened by his own misconstruing of the scope of his jurisdiction. Adjudicator's decision was unenforceable as he had misconstrued jurisdiction and failed to determine the dispute actually referred.
Nordot Engineering Services Ltd v Siemens plc	2001 CILL 1778 (TCC), Judge Gilliland	**Express agreement to abide by adjudicator's decision** The issue of whether works were 'construction operations' under the Act. The respondent said 'we will, however, abide by your decision in this matter and will comply with what ever direction you deem appropriate'. The judge found this was a clear and unequivocal agreement to be bound by the decision of the adjudicator.
Christiani & Nielsen Ltd v The Lowry Centre Development Co. Ltd	29 Jun. 2000 (TCC), Judge Thornton (unreported)	**Participation against background of continuing protest** Whether the parties have agreed the adjudicator could

(Continued)

[25] The finding in this case has been doubted. See comments in Ch. 2 at 2.26–2.29 above.

Table 5.1 *Continued*

Title	Citation	Issue
		determine his own jurisdiction is a matter of fact. The respondent agreed to make submissions on the existence of a construction contract. That was just an agreement as to procedure. The respondent's submissions were made against the background of their continuing protest as to jurisdiction. Thus the defendant was not agreeing that the adjudicator could make a binding decision on jurisdiction (judgment at paras. 14–19).
Nolan Davies Ltd v Steven P. Catton	[2000] EWHC 590 (TCC), Judge Wilcox	The parties gave the adjudicator the power to decide his own jurisdiction.
Whiteways Contractors (Sussex) Ltd v Impresa Castelli Construction UK Ltd	[2000] 75 Con LR 92 (TCC), Judge Bowsher	**Respondent widened adjudicator's power** Impresa disputed the adjudicator's jurisdiction to deal with some matters in the referral. Impresa served written submissions in relation to the issue of jurisdiction and invited the adjudicator 'to decide on this issue as a matter of urgency as our Response to the … notice of referral will depend on your decision'. Impresa had clearly requested that the adjudicator determine his own jurisdiction, and hence had given him power to do so. (See *Harris Calnan* below, doubting the judge's conclusion that decisions on jurisdiction were binding.)
Maymac Environmental Services Ltd v Faraday Building Services Ltd	[2001] CILL 1685 (TCC), Judge Toulmin	**Respondent estopped from denying jurisdiction** Dispute as to whether relevant construction contract existed. Court found there was a construction contract. In any event the referral was made on the basis that an adjudication would take place by agreement on the same terms as the Act and the Scheme for Construction Contracts. Such an agreement was enforceable as if the Act had applied. The respondent would be estopped

Title	Citation	Issue
		(by representation) from asserting that the Act did not apply or that there was no jurisdiction.
Watson Builders Services v Harrison	[2001] Scots CSOH, Lady Paton	After initial reservations both parties gave adjudicator power to determine the validity of the contract which led to his appointment.
Fence Gate Ltd v J. R. Knowles Ltd	[2002] Con LR 206 (TCC), Judge Gilliand	**Conferral of ad hoc jurisdiction** Where part of a dispute was not covered by the Act (because it was outside the terms of s. 104(2)), there had been no ad hoc referral because the facts did not lead to the conclusion that Fence Gate had agreed to confer jurisdiction on the adjudicator to decide that part of the dispute.
William Oakley & David Oakley v Airclear Environmental Ltd and Airclear TS Ltd	[2002] CILL 1824 (Ch D), Etherton J	**Respondent estopped from denying jurisdiction** Both parties proceed on the common assumption that there was a contract governed by NAM/T and NAM/SC forms, including the express adjudication provisions. The first Judge found the respondent was estopped from arguing no contract and no jurisdiction at enforcement. However on appeal the finding of estoppel was reversed because there had not been evidence showing it was unconscionable to resile from the common assumption.
Shimizu Europe Ltd v Automajor Ltd	[2002] BLR 113 (TCC), Judge Seymour	**Effect of paying Adjudicator's decision** *Obiter*: paying part of decision and making application to correct it under the slip rule was electing to treat the decision as made with jurisdiction (judgment at paras. 29–30). See alternative view in *Joinery Plus (in Administration) Ltd v Laing Ltd [2003] BLR 184.*
Parsons Plastic (Research and Development) Ltd v Purac Ltd	[2002] BLR 334 (CA)	**Ad hoc referral binding** Dispute as to whether construction operations within

(Continued)

Table 5.1 *Continued*

Title	Citation	Issue
		Act, but agreement to submit to ad hoc adjudication. Although decision binding respondent resisted enforcement on the ground that under contract terms it was entitled to set off against adjudicator's decision.[26]
R. Durtnell Ltd v Kaduna Ltd	[2003] BLR 225 (TCC), Judge Seymour	**No objection raised during adjudication** By initially accepting the referring party's assertion as to the nature of the contract, the responding party waived any right to object to jurisdiction on the ground that there was no contract.
Cowlin Construction Ltd v CFW Architects	[2003] BLR 241 (TCC), Judge Kirkham	**Waiver of right to object to lack of jurisdiction** CFW had by their counter-notice accepted that the adjudicator had jurisdiction to determine the terms of the contract between the parties.
Galliford Try Construction Ltd v Michael Heal Associates Ltd	[2003] EWHC 2886 (TCC), Judge Seymour	**Reservation of rights overtaken by events** Court held (inter alia) that there had been an agreement (without qualification) to submit disputes to adjudication if mediation failed. The expression 'without prejudice to our contentions on jurisdiction' was insufficient to prevent a binding decision being made on the point in question. The reservation had been overtaken by the subsequent events. However enforcement refused on other grounds (Galliford no longer contended there had been a contract and so the decision was not binding).
J. W. Hughes v GB Metal Work Ltd	[2003] EWHC 2421 (TCC), Forbes J	**Ad hoc agreement** The conduct of the parties indicated they had agreed to give jurisdiction to adjudicator by way of ad hoc agreement (at para. 16).

[26] For a discussion of the topic of when a party may set off against an adjudicator's decision see Ch. 6 at 6.36–6.48.

Title	Citation	Issue
Pegram Shopfitters Ltd v Tally Wiejl (UK) Ltd	[2004] 1 WLR 2082 (CA)	**Effective reservation of rights** Claimant said simple contract with no adjudication terms and Scheme applied. Respondent said contract on different terms, with own adjudication clause, alternatively no contract. Respondent said adjudicator appointed under Scheme had no jurisdiction and made effective reservation of rights. It was held that because the respondent had made an effective reservation of rights it had not conceded construction contract existed sufficient for jurisdiction to exist. Two different contracts being contended for by parties. **No power to decide own jurisdiction** The CA said (para. 10, May LJ) that the adjudicator's decision on jurisdiction was not binding.
Rossco Civil Engineering Ltd v Dwr Cymru Cyfyngedic	5 Aug. 2004 Lawtel (TCC), Recorder Dermod O'Brien QC	**Estoppel** Responding party was estopped from challenging the jurisdiction of the adjudicator where it had accepted in the adjudication that the limited company, and not the partnership that had previously carried on the same business, was the correct party.
IDE Contracting Ltd v R. G. Carter Cambridge Ltd	[2004] BLR 172 (TCC), Judge Havery	**No ad hoc conferral** The TCC judge decided that the responding party had not agreed to be bound by the decision on jurisdiction made by the adjudicator, and hence that decision was capable of being reviewed by the court in the usual way.
Hortimax Ltd v Hedon Salads Ltd	15 Oct. 2004 (TCC), Judge Gilliland (unreported)	**Respondent expressly gave adjudicator power to decide own jurisdiction** Judgment para. 31.
Thomas Frederic (Construction) Ltd v Keith Wilson	[2004] BLR 23, [2003] EWCA Civ 1494 (CA)	**No implied conferral of jurisdiction** The respondent objected to jurisdiction on the grounds

(Continued)

Table 5.1 *Continued*

Title	Citation	Issue
		that he was the wrong party. After losing that argument before the adjudicator he took part in the adjudication. He made no specific reservation of rights. However, the Court of Appeal held it was impossible to conclude on the facts that he was submitting to the jurisdiction of the adjudicator, because nothing was said thereafter which could reasonably be understood to be an agreement to be bound by the adjudicator's decision (para. 30).
Carillion Construction Ltd v Devonport Royal Dockyard Ltd	[2006] BLR 15 (CA)	**Acquiescence where failure to dispute jurisdiction** There was no free-standing power to award interest under the Scheme. However Carillion had asked for interest to be awarded and Devonport had responded that the issue of interest did not arise, but Devonport did not specifically deny jurisdiction. The Court of Appeal found this was an acquiescence by Devonport in the conferral of jurisdiction, and there had been an ad hoc submission.[27]
R. C. Pillar & Son v The Camber (Portsmouth) Ltd	[2007] 115 Con LR 102 (TCC), Judge Thornton	**Cross-claim an offer to waive jurisdiction objection** During the adjudication the respondent had challenged jurisdiction, but also defended the substantive claims without prejudice to its jurisdiction argument. The respondent's defence included substantial cross-claims. The judge found these were more than defences of set-off or abatement and amounted to counter-claims. The cross-claim document was an offer to waive the jurisdiction objection in return for allowing the cross-claim to be heard. The offer was

[27] Some commentators have said that this stretches the doctrine of acquiescence almost to breaking point (see Andrew Bartlett QC in the Techbar Review (Spring 2008)). In relation to acquiescence, see also *A. C. Yule & Son Ltd v Speedwell Roofing & Cladding Ltd* [2007] EWHC 1360, where one party acquiesced in a delay to the decision.

Title	Citation	Issue
		accepted and ad hoc jurisdiction existed (paras. 10–14 of the judgment).
Bothma (t/a DAB Builders) v Mayhaven Healthcare Ltd	[2007] EWCA Civ 527 (CA)	**General reservation** Court of Appeal accepted that general reservation of rights was sufficient (at paras. 6 and 14).
Harris Calnan Construction company Ltd v Ridgewood (Kensington) Ltd	[2008] BLR 132 (TCC), Coulson J	**No reservation of rights** The defendant unsuccessfully objected to adjudicator's jurisdiction on the ground there was no agreement in writing. At enforcement the same argument was run. The judge found the defendant never reserved its position before the adjudicator and had therefore agreed to be bound by the adjudicator's decision on jurisdiction (para. 8 of the judgment).
CJP Builders Ltd v William Verry Ltd	[2008] BLR 545 (TCC), Akenhead J	**No reservation of rights** Failure to reserve position when challenging jurisdiction meant a party had agreed to be bound by the adjudicator's decision on the jurisdiction challenge.
VGC Construction Ltd v Jackson Civil Engineering Ltd	[2008] 120 Con LR 178 (TCC), Akenhead J	**Withdrawal of claim and estoppel** If a party represents that it will withdraw a claim which is disputed as to jurisdiction, and the respondent relies on that representation to its detriment, then it may be arguable that the potential dispute ceases to be a dispute and that the claiming party will be estopped from asserting it is a dispute within the jurisdiction of the adjudicator.
Euro Construction Scaffolding Ltd v SLLB Construction Ltd	[2008] EWHC 3160 (TCC), Akenhead J	**Effective reservation of rights on jurisdiction** The respondent argued no construction contract in writing within the Act meaning. Court found the respondent did not confer jurisdiction on adjudicator to decide the point because it had

(Continued)

Table 5.1 *Continued*

Title	Citation	Issue
		made an effective reservation of rights on the issue (at para. 29 of the judgment).
Air Design (Kent) Ltd v Deerglen (Jersey) Ltd	[2008] EWHC 3047 (TCC), Akenhead J	**Effective reservation of rights on jurisdiction** The respondent made a clear reservation of rights and only asked the adjudicator to enquire into jurisdiction and find he had none. No ad hoc submission (at para. 20 of the judgment). **Rare case: overlap with issue of substance** Where it was said that there were multiple subcontracts it was argued there was no jurisdiction to decide issues in one adjudication. Judge found three later subcontracts were all variations to first subcontract, and there was no doubt the adjudicator had been properly appointed under the first subcontract and under that subcontract had jurisdiction to decide whether other claims were variations to that subcontract. Therefore the answer to the question of substance, which the adjudicator had jurisdiction to decide, resulted in the adjudicator reaching a binding decision which also went to the jurisdiction challenge that the claims 2, 3, and 4 were separate contracts. This point was clarified by the same judge later in *Camillin Denny Architects v Adelaide Jones & Co. Ltd* [2009] EWHC 2110 at para. 30.
Enterprise Managed Services Ltd v Tony McFadden	[2009] EWHC 3222 (TCC), [2010] BLR 89, Coulson J	**Enquiry into jurisdiction** The TCC judge criticized the adjudicator for having failed to investigate the jurisdiction challenge at all which he thought was part of the adjudicator's failure to manage the adjudication adequately.

Title	Citation	Issue
Bovis Lend Lease v Cofely Engineering Services	[2009] EWHC 1120 (TCC), Coulson J	**No ad hoc conferral** An agreement to be bound by adjudicator's decision in adjudication number two was not wide enough to mean the parties had agreed to confer ad hoc jurisdiction in adjudication six.
OSC Building Services Ltd v Interior Dimensions Contracts Ltd	[2009] EWHC 248 (TCC), Ramsey J	**Failure to protest** No objection taken to adjudicator dealing with question of what sum due. Subsequent argument made at enforcement that there was a broadening of the dispute in the notice of adjudication. Failure to take the point at all in the adjudication meant the respondent had submitted to the jurisdiction (*obiter*).
Dalkia Energy & Technical Services Ltd v Bell Group UK Ltd	[2009] EWHC 73 (TCC), Coulson J	**Effective reservation at outset** The respondent made an effective reservation at the outset of the adjudication and repeatedly said the adjudicator did not have jurisdiction. At no point after that did the respondent agree to be bound by the decision as to jurisdiction.
Camillin Denny Architects v Adelaide Jones & Co. Ltd	[2009] BLR 606, [2009] EWHC 2110 (TCC), Akenhead J	**Power to decide own jurisdiction** If there is an overlap between the issues of substance and the jurisdiction issue, the adjudicator may be able to decide his or her own jurisdiction.
Allied P&L Ltd v Paradigm Housing Group Ltd	[2009] EWHC 2890 (TCC), [2010] BLR 59 Akenhead J	**Ineffective reservation of rights on jurisdiction** Dispute had not crystallized and jurisdiction challenged but no reservation of rights made.
Supablast (Nationwide) Ltd v Story Rail Ltd	[2010] EWHC 56, [2010] BLR 211 (TCC), Akenhead J	**Power to decide own jurisdiction** Where the dispute referred concerned variations under a construction contract there may be an overlap between issues as to his jurisdiction (under s. 107) and the substantive dispute. The adjudicator may have jurisdiction to decide

(Continued)

Table 5.1 *Continued*

Title	Citation	Issue
		jurisdictional issues when they overlap with substantive matters referred (citing *Air Design* and *Camillin* above).
Pilon Ltd v Breyer Group PLC	[2010] EWHC 837 (TCC), [2010] BLR 452, Coulson J,	**No agreement to be bound by jurisdiction decision** In order to be bound by an adjudicator's jurisdictional decision there needs to be either an express agreement between the parties that the adjudicator's decision on jurisdiction is to be binding or an implied agreement which may arise where the objecting party fails to reserve its position, or there has been a unilateral waiver of any jurisdictional objection (at para. 13).
GPS Marine Contractors Ltd v Ringway Infrastructure Services Ltd	[2010] BLR 377 (TCC), [2010] EWHC 283, Ramsey J	**General reservation** Generally a party who wishes to do so can object to the jurisdiction of an adjudicator either in general terms or by making a reservation on a specific matter. The general reservation in this case was sufficiently clear to prevent the subsequent participation in the adjudication from amounting to waiver of the lack of jurisdiction.
Aedifice Partnership Ltd v Ashwin Shah	[2010] EWHC 2106 (TCC), 132 Con LR 100, Akenhead J	**How reservations can be made** The judge reviewed the jurisprudence on how reservations as to jurisdiction should be made. On the facts a clear reservation had been made.
All Metal Roofing v Kamm Properties Ltd	[2010] EWHC 2670 (TCC), Akenhead J	**No waiver when short delay in making reservation** Where the reservation of rights as to jurisdiction was made two days after the response submission (which had not taken the adjudication point expressly) the judge stated (*obiter*) that it was doubtful

Title	Citation	Issue
		there had been a waiver because there was no suggestion that anyone had acted on the initial defence to their detriment.
CN Associates v Holbeton Ltd	[2011] BLR 261 (TCC), Akenhead J	**Reservation not express** The judge found that a denial of jurisdiction on the ground that there was no contract amounted to a reservation of rights. While there was no express reservation, judged objectively, the words used should have been understood as unequivocally conveying that Holbeton believed the adjudicator did not have jurisdiction. That was an effective reservation of jurisdiction. It was relevant that the adjudicator himself said he did not consider he had been given power to decide the jurisdiction issue in a binding way.
Durham County Council v Jeremy Kendall	[2011] BLR 425 (TCC), Akenhead J	**Failure to maintain general reservation** *Obiter*: where a party made both specific and general reservations at the outset but later made only specific reservations it was taken to have abandoned its general reservation.
Clark Electrical Ltd v JMD Developments (UK) Ltd	[2012] EWHC 2627 (TCC), Judge Behrens	**Paying fee and asking for guidance was not submission to jurisdiction** The responding party wrote an email to the adjudicator stating it was not yet represented and asking for guidance on the procedure and on extending the timetable and paid its half of the appointment fee. After obtaining representation it said the adjudicator had no jurisdiction and withdrew from the adjudication The court found there had been no submission to the adjudicator's jurisdiction and the decision was not enforced.

(2) Contractual Adjudications when the Act Does Not Apply

5.23 An express adjudication procedure will in principle be valid even if contained in a contract which is not within the ambit of the Act: *Lovell Projects v Legg & Carver* (2003),[28] *Steve Domsalla v Kenneth Dyason* (2007).[29] There has been some controversy as to whether legal principles developed in relation to statutory adjudications are applicable to this form of adjudication.

Preconditions to the Right to Adjudicate

5.24 As discussed at 3.73 above, a contractual term that imposes a precondition to the right to adjudicate will contravene the requirement of s. 108(2)(a) that the contract must enable the parties to give notice of adjudication at any time.[30] However, because contracts that fall outside the Act are not subject to s. 108(2)(a), the adjudication clause does not need to say that the notice of adjudication may be given 'at any time'. It follows that such contracts may contain valid preconditions to the right to adjudicate.

The Requirement for Writing

5.25 There is an important distinction between contracts that incorporate express adjudication provisions and those construction contracts that do not. In a contract with an express adjudication clause the availability of adjudication as a mechanism for interim dispute resolution does not depend on s. 107 of the Act but on the terms of the adjudication agreement (subject to the proviso that if the adjudication clause fails to comply with s. 108 of the Act, the Scheme for Construction Contracts ('the Scheme') will replace the agreed provision). In principle therefore a construction contract with an express adjudication agreement may be varied orally and the oral variation need not satisfy s. 107 of the Act but may be the subject of a contractual adjudication as long as it falls within the scope of the adjudication agreement: *Dean and Dyball Construction Ltd v Kenneth Grubb Associates Ltd* (2003).[31] Equally, a contract with an adjudication agreement but which is not governed by the Act does not need to comply with the requirement for writing in s. 107.

The Principle that Errors of Law Are Not Reviewable

5.26 An adjudicator who has been appointed under the Scheme derives his authority from the Act. It is established law that an adjudicator's error of fact or law made within jurisdiction is not a ground for challenging enforcement of the decision.[32] It has been argued that this rule does not apply to contractual adjudication schemes in contracts outside the ambit of the Act. In particular in *Steve Domsalla v Kenneth Dyason* (2007)[33] Judge Thornton said that the doctrine of unreviewable error made within jurisdiction was only applicable to statutory adjudications. As a result the judge found that the adjudicator had wrongly decided that he should not consider the defence of abatement and set-off because he had wrongly decided a withholding notice was needed. That judge found that this was both procedurally unfair and

[28] [2003] BLR 452.
[29] [2007] BLR 458.
[30] *John Mowlem & Co. plc v Hydrya-Tight & Co. plc* (2001) 17 Const LJ 358.
[31] 100 Con LR 92, [2003] EWHC 2465 at paras. 16–18.
[32] This is discussed more fully at 7.05–7.07 and 9.47–9.53. The central question is whether the adjudicator has answered the right question. If he or she has done so but has got the answer wrong, that is an error within jurisdiction and is not reviewable. If the adjudicator has asked the wrong question, that may result in an error made without jurisdiction, or may be a breach of natural justice, and can be reviewed.
[33] [2007] BLR 348 at paras. 99–100.

resulted in the ajudicator not deciding all matters referred to him for decision. The judge refused to enforce the decision summarily and gave permission to defend.

5.27 A different view was taken in *Fleming Builders Ltd v Forrest or Hives* (2008).[34] That was a case which concerned a domestic house building contract in the standard form of the SBCC (Scottish Building Contract Contractors) designed portion with quantities. At the enforcement hearing the respondent relied on *Domsalla* and on *Costain Ltd v Strathclyde Builders Ltd* (2004)[35] to argue that an adjudicator under a contract to which the Act does not apply must be regarded as a type of arbiter, and that the well-established rules that govern the judicial control of arbiters apply to adjudicators. The respondent argued that if it is properly viewed as a form of arbitration, the adjudicator's decision may be reviewed on grounds of misconduct, and if the adjudicator made an error of law so serious as to undermine the whole process this would constitute misconduct. The court found that there was no justification for a distinction between the decision of an adjudicator who has dealt with a dispute under the Act and one who has dealt with a dispute under a contractual procedure. The matter has yet to be considered by an appellate court.

Set-off against an Adjudicator's Award?

5.28 The circumstances in which a party may set off sums against those awarded by the decision of an adjudicator are dicussed at 6.36–6.48 in Chapter 6. The general principle is that because s. 108(3) of the Act requires the decision to be binding until resolved finally, this means there is no right of set-off against an adjudicator's decision.[36] Broadly there are three exceptions to this general rule: (1) when the consequence of the decision is that the employer is undeniably entitled to deduct liquidated damages they can be set off;[37] (2) when the contract contains a clause that says the parties may set off claims against an adjudicator's decision;[38] and (3) when the adjudicator's decision is advisory as to the proper operation of the contract payment machinery and a sum that should be included, and where a withholding notice may legitimately be served in accordance with the payment rules after the adjudicator's decision.[39]

5.29 As the general rule against setting off against an adjudicator's decision rests on the provision of s. 108(3) of the Act it has been argued that there is no equal restriction of the right of set-off regarding adjudications outside the Act. However that argument was rejected in *Fleming Builders Ltd v Forrest or Hives* (2008).[40] Where the contract does not have a term saying the adjudicator's decision is binding until finally determined the legal basis for restricting the right of usual contractual or equitable set-off is unclear. It remains to be considered by an appellate court.

Contracts with Residential Occupiers

5.30 Contracts with residential occupiers are excluded from the ambit of the Act. Where a contract contains an express adjudication agreement there have been a number of attempts to strike out the adjudication clause as unfair. Those efforts have not been successful, as discussed in Chapter 1. Accordingly the usual rule will be that an adjudication agreement in a contract with a residential occupier will be valid whether the contract is governed by the Act or not.

---

34 [2008] Scot CS CSOH 103.
35 2004 SLT 102 in which the view was expressed (at para. [7]).
36 *Ferson Contractors Ltd v Levolux AT Ltd* [2003] BLR 118 (CA).
37 *Balfour Beatty Construction Ltd v Serco* [2004] EWHC 3336 (TCC).
38 *Parsons Plastic (Research & Development) Ltd v Purac Ltd* [2002] BLR 334 (CA).
39 *Squibb Group Ltd v Vertase FLI Ltd* [2012] EWHC 1958 (TCC), [2012] BLR 408.
40 [2008] Scot CS CSOH 103.

Doctrine of Repudiation

5.31 *In Lanes Group Plc v Galliford Try Infrastructure Ltd* (2011)[41] Akenhead J stated (*obiter*) that, by analogy with the law relating to arbitration agreements, a party could repudiate a purely contractual adjudication agreement where it evinces an intention to no longer be bound by it (see paragraphs 23–24). However he decided the same did not apply to statutory adjudication agreements. The Court of Appeal upheld the judge's decision that repudiation does not apply to statutory adjudications but did not deal with the position of a purely contractual adjudication.[42]

[41] [2011] EWHC 1035 (TCC).
[42] [2012] BLR 121, [2011] EWCA Civ 1617.

Part III

EFFECT, ENFORCEMENT, AND ENFORCEABILITY

6

EFFECT OF AN ADJUDICATOR'S DECISION

(1) Overview

6.01 Once an adjudicator's decision has been delivered, the successful party will generally be expecting the unsuccessful party to comply with that decision, for example by payment of monies or compliance with a declaration. Whether the unsuccessful party complies with the adjudicator's decision, or whether the decision is summarily enforced by the courts, the decision of the adjudicator has many important effects. This chapter explores those general effects; identifies how an adjudicator's decision can become permanently binding on the parties; explores the circumstances in which there can be a set-off against a payment due pursuant to a decision; and discusses the prohibition on the readjudication of a dispute which is substantially the same as a dispute which has previously been determined.

(2) General Effects of an Adjudicator's Decision

6.02 An adjudicator's decision:

1. Is private between the parties and cannot be used as a precedent in any dispute between other parties unless there are express contract terms to the contrary.
2. Is temporarily binding, and this will extend to include matters which may properly be inferred from the decision.
3. Must be complied with by the parties until the dispute is finally determined (which may be by litigation, arbitration, or agreement), subject to any successful challenge to enforceability.
4. Does not alter a contractual right or liability but temporarily determines rights and liabilities until final determination.

5. Creates a new contractual cause of action for non-compliance.
6. Will ordinarily give rise to a separate contractual entitlement to immediate payment without set-off.
7. Prohibits any re-adjudication of a dispute which is substantially the same as the dispute determined.

Private between the Parties

6.03 Adjudication is a dispute resolution procedure that arises as a result of a contractual relationship and the outcome cannot therefore affect anyone who is not a party to the adjudication, either directly or by use as a precedent in other proceedings.[1] However, the temporarily binding nature of an adjudicator's decision does mean that it acts as a precedent in any subsequent disputes between the same parties. Unlike decisions of the courts, which are available for public review, decisions by adjudicators in the United Kingdom are not published.[2] Some adjudication agreements or construction contracts specifically make the adjudication proceedings and outcome confidential. In the absence of express contractual terms, any information or document provided in connection with the adjudication (including the decision itself) is not confidential unless the party supplying it has indicated it is to be treated as confidential.[3]

6.04 Adjudicators' decisions are therefore more akin to arbitral awards than court judgments, as arbitral awards are generally not available to the public.[4] However, adjudication decisions are not wholly analogous to decisions made under the Arbitration Act 1996. For example, there is no statutory equivalent in the Housing Grants Construction and Regeneration Act 1996 ('the 1996 Act'), or in its 2009 amendments, of s. 66 of the Arbitration Act 1996, which provides that an arbitral award may, with the leave of the court, be enforced in the same way as a judgment or order of the court. There is no procedure for the entry of an adjudicator's decision as a judgment.[5] Instead, the right to enforce an adjudicator's decision is purely contractual:

> 37. . . . The decision of an adjudicator is not like an award of an arbitrator or the judgment of a court and directly enforceable. It is enforceable at all simply because by their contract the parties have agreed to comply with it or to give effect to it. This is so whether the parties have expressly agreed in their contract to a procedure for adjudication, as in the present case, or whether the relevant provisions of *The Scheme for Construction Contracts* have been implied into their contract by virtue of the provisions of *Housing Grants, Construction and Regeneration Act 1996 s. 114(4)*.[6]

Temporarily Binding until Final Determination

6.05 Section 108(3) of the Act requires a construction contract to provide that the decision of the adjudicator is binding until the dispute is finally determined by legal proceedings, arbitration, or agreement of the parties. The Court of Appeal described the effect of this provision in the following way:

---

[1] In some large projects where there is a single employer and several contractors, the construction contracts may specifically anticipate that an adjudicator's decision is confidential outside the project but available to all parties in relation to that project, and so may form some kind of 'project precedent'.

[2] In some jurisdictions, this is not the case and a public register is kept of all adjudicators' decisions.

[3] Para. 18 of the Scheme for Construction Contracts ('the Scheme'), which also provides an exception to the extent that it is necessary to make disclosure to a third party for the purposes of, or in connection with, the adjudication.

[4] There is a considerable body of case law concerning confidentiality of arbitration and arbitrators' awards which is outside the scope of this book. See Merkin, *Arbitration Law*, (2004), 17.26–17.34; Mustill and Boyd, *Commercial Arbitration*, 2nd edn (2001), Companion pp. 112–13.

[5] *VHE Construction plc v RBSTB Trust Co. Ltd* [2000] BLR 187 (TCC) at 195; *David McLean Housing Contractors Limited v Swansea Housing Association Limited* [2002] BLR 125 (TCC).

[6] *RSL (South West) Limited v Stansell Limited* [2003] EWHC 1390 (TCC).

> 26. ... The adjudicator's decision, although not finally determinative, may give rise to an immediate payment obligation. That obligation can be enforced by the courts. But the adjudicator's determination is capable of being reopened in subsequent proceedings. It may be looked upon as a method of providing a summary procedure for the enforcement of payment provisionally due under a construction contract.[7]

6.06 If the adjudication regime in a construction contract does not contain clauses to satisfy the mandatory requirements of the Act, the rules of the Scheme apply.[8] Paragraph 23(2) of the Scheme encompasses the requirements of s.108(3) of the Act, requiring the parties to comply with a decision of an adjudicator until the dispute is finally determined.

6.07 The adjudicator's decision gives rise to a separate contractual obligation to comply with the decision.[9] Proceedings to enforce an adjudicator's decision are proceedings to enforce that contractual obligation.[10] Failure to comply with a valid adjudicator's decision is a new breach of contract and a cause of action arises at the date when the relevant party should have complied with the adjudicator's decision.[11]

Time for Compliance

6.08 As an adjudicator's decision is temporarily binding, the parties must comply with it, whether it relates to an order for payment or a declaratory decision by the adjudicator.[12] If conducted under the Scheme, in the absence of directions from the adjudicator as to time for performance, the decision is to be complied with immediately upon delivery.[13] This means that the parties should comply with the decision without debate or question. If a peremptory order is made under paragraph 23(1) of the Scheme that does not affect the time for compliance.[14]

6.09 However, if the losing party intends to challenge the enforceability of the decision, then there is no requirement that it should comply with the decision before being able to advance its case that the decision is unenforceable:

> 19. ... Naturally the Act assumes that such a final determination is likely to follow the decision. That is consistent with the concept of adjudication whereby a dispute would be resolved during the course of a contract and only resurrected for final determination, if required, at a later stage. 'Pay now; argue later', as some are wont to say. In my judgment there is nothing in the Act (or the Scheme, if applicable) which requires a party who wishes to challenge a decision of an adjudicator to comply with it before being able to advance its case, any more than a party is precluded from subsequently challenging a decision, having complied with it ...[15]

6.10 On the contrary, a party intending to resist enforcement on the basis that the adjudicator's decision is a nullity should not make any payment in respect of the decision, as to do so

---

[7] *Bouygues (UK) Ltd v Dahl-Jensen (UK) Ltd* [2001] All ER (Comm) 1041 (CA). In other cases it has been said the effect of s. 108(3) is to prevent set-off against an adjudicator's decision, see below at 6.39 and 6.42.

[8] ss. 108(5) and 114(4) of the Act.

[9] *Glencot Development and Design Co. Ltd v Ben Barrett & Son (Contractors) Ltd* [2001] BLR 207 (TCC).

[10] *VHE Construction v RBSTB Trust Co.* [2000] BLR 187 (TCC) at paras. 51–4.

[11] *Ringway Infrastructure Services Ltd v Vauxhall Motors Ltd (No. 2)* 115 Con LR 149, [2007] EWHC 2507 (TCC) at para. 14.

[12] *HS Works Ltd v Enterprise Managed Services Ltd* [2009] BLR 378, [2009] EWHC 729 (TCC).

[13] Scheme, para. 21. As to the mechanics of complying with the decision and issuing a relevant payment certificate, see further 13.136–13.137 below.

[14] See 7.25–7.27 for enforcement by peremptory order, but note that if the decision is one relating to a construction contract entered into after 1 Oct. 2011 to which the amended Scheme applies, there is no longer a provision for the making of a peremptory order.

[15] *Alstom Signalling Ltd v Jarvis Facilities Ltd* [2004] EWHC 1285 (TCC).

may result in the paying party being considered to have made an election that the decision is valid.[16]

Final Determination

6.11 Subject to any time bars (whether contractual or statutory), or an agreement to accept an adjudicator's decision as final, a party may commence legal proceedings or, if the contract contains an arbitration clause, an arbitration, for final determination of the dispute. In the meantime, unless the losing party has reason to resist enforcement of the adjudicator's decision,[17] it must comply with it, including making payment of any amounts awarded by the adjudicator.

6.12 Disputes referred to adjudication commonly involve the consideration of substantial factual material and any proceedings for final determination will be equally fact-heavy. A losing party seeking finality will therefore typically have to start a court claim (or arbitration if there is an arbitration agreement).[18] However, if there is no arbitration agreement and there is a point decided by the adjudicator that can be finally determined as a matter of law without a factual enquiry (for example, a point of contractual construction) then it may be possible, by the issuing of a Part 8 claim, to have that point finally determined, potentially at the same time as any summary judgment proceedings for enforcement.[19]

6.13 Where a party issues proceedings for an adjudicator's decision to be overturned, then the cause of action is heard anew and the adjudicator's decision has no bearing on the case unless the parties agree otherwise, nor does it reverse the burden of proof in any subsequent litigation. In other words, it is not incumbent on the losing party to show that the adjudicator's decision was not justified: *City Inn Ltd v Shepherd Construction Ltd* (2001) (Key Case).[20]

6.14 An adjudicator's decision does not create or modify a right or liability under the contract in question but is equivalent to a determination or upholding of rights or liabilities under the contract.[21] An adjudicator's decision is an expression as to liability (and usually quantum); it does not alter or replace the original cause of action that was the subject of the dispute, since the underlying cause of action survives and neither merges into, nor is superseded by, the disputed adjudicator's decision.[22] Therefore, if the adjudicator's decision is found to be a nullity, the parties do not lose their original cause of action, subject to any express finality provisions in the contract.[23]

[16] Discussed further at 7.84–7.95.

[17] See 7.08 below.

[18] See 7.65.

[19] *Geoffrey Osborne Ltd v Atkins Rail Ltd* [2010] BLR 363, [2009] EWHC 2425 (TCC), discussed in more detail in Ch. 7 at 7.65 onwards.

[20] [2002] SLT 781, (2001) SCLR 961 (Key Case). Lord Macfadyen's decision in relation to the onus of proof was later endorsed in another Scottish case: *Citex Professional Services v Kenmore Developments* [2004] Scot CS 20. However, in that case, the court declined to answer the 'academic questions about onus of proof' as the matter had not been fully focused in the pleadings and was held over until another time.

[21] *David McLean Housing Ltd v Swansea Housing Association Ltd* [2002] BLR 125 (TCC).

[22] *Glencot Development and Design Co. Ltd v Ben Barrett & Son (Contractors) Ltd* [2001] BLR 207 (TCC); *Bovis Lend Lease Ltd v Triangle Development Ltd* [2003] BLR 31, [2002] EWHC 3123 (TCC).

[23] *Cubitt Building & Interiors Ltd v Fleetglade Ltd* [2007] 110 Con LR 36, [2006] EWHC 3413 (TCC). See further 6.19–6.23 below.

6.15 In the event that it is found on final determination that there has been an overpayment as a result of a payment made pursuant to an adjudicator's decision, that overpayment can be recovered either by way of claim pursuant to an implied term that any overpayment is to be repaid in such circumstances or in restitution[24] (NB: not followed in [2013] EWHC 1322 (TCC)).

Finality by Agreement

6.16 Section 108(3) provides that the parties may agree to accept the decision of the adjudicator as finally determining the dispute.

6.17 There is no reason why parties to construction contracts cannot agree that they will be finally bound by the decision of the adjudicator either at any time before the adjudication is commenced (for example by express drafting in the construction contract itself or the adoption of particular adjudication rules which state that the decision would have that effect) or at any time during or after the adjudication.

6.18 Like any other contractual agreement, the courts would be unlikely to interfere with the contractual freedom of parties to agree the decision of an adjudicator was to be final.[25] Similarly, a court will not enforce an adjudicator's decision through summary judgment if the dispute that was the subject of the decision has been compromised after (or indeed before) an adjudicator has delivered his decision.[26] In such circumstances, the defendant will have a complete defence to the claimant's application, and the claimant will, in all likelihood, be liable for the costs of the summary judgment application and any hearing to determine the defence of compromise.[27] However, a settlement agreement will not be a defence to enforcement proceedings where a party has defaulted on that settlement agreement.[28]

Contractual Finality Provisions

6.19 One example of a finality provision in the construction contract is where the parties agree that a decision shall become final unless certain steps, such as commencing arbitration or litigation, are taken within an identified period of time. However, in the absence of any such agreement, the only possible time limit to a party seeking final determination of a matter previously referred to adjudication is the relevant statutory time bar to the right to bring a claim in respect of a cause of action imposed by the Limitation Act 1980.

6.20 Some of the standard-form construction contracts provide a time limit for the commencement of litigation or arbitration following an adjudicator's decision, as can be seen in Table 6.1.

---

[24] *Jim Ennis Construction Ltd v Premier Asphalt Ltd* (2009) 125 Con LR 141, [2009] EWHC 1906 (TCC), see paras. 28–9 for the restitution point. The paying party will usually have six years from the date of payment in which to bring legal proceedings to recover that overpayment (para. 26).

[25] *Jerram Falkus Construction Ltd v Fenice Investments* (No. 4) [2011] BLR 644, [2011] EWHC 1935 (TCC) (Key Case).

[26] *Bracken v Billinghurst* [2003] CILL 2039, [2003] EWHC 1333 (TCC).

[27] As they were in *Bracken and anor v Billinghurst* (costs awarded on the standard basis).

[28] *Able Construction (UK) Ltd v Forest Property Development Ltd* [2009] All ER (D) 176 (Feb), [2009] EWHC 159 (TCC).

Table 6.1 Standard-form Contract Time Limits

Contract type	Relevant clause
JCT 1998/2005 Design and Build	No limitation for adjudicators' decisions generally. However, where an adjudicator's decision is delivered after the submission of the final account, if either party wishes to have that dispute determined by arbitration or legal proceedings, they may (i.e must—see 6.34–6.35) be commenced within 28 days (**clause 1.9.4**).
NEC3	**Option W1.4(2) and (3)**[29] provides that a party who is dissatisfied with the decision of the adjudicator may notify the other party of its intention to refer the matter to the tribunal (which is often arbitration). However, 'it is not referable' unless that intention is notified within four weeks of the earlier of (a) notification of the adjudicator's decision[30] or (b) the time provided for notification if the adjudicator fails to notify a decision within that time.[31]
ICE 1997 (Measurement Version)	Where an adjudicator has given a decision under **clause 66(6)**, the notice to refer a dispute to arbitration must be given within three months of the giving of the adjudicator's decision, otherwise the adjudicator's decision shall be final as well as binding: **clause 66(9)(a)**.
PPC2000	No limitation.

6.21 The courts' approach to interpreting these clauses appears to favour certainty and finality. In *Jerram Falkus Construction Ltd v Fenice Investments* (No. 4) (2011) (Key Case),[32] Coulson J upheld the interpretation of clause 1.9.4 that said any adjudication decision given after the Final Account or Final Statement must be challenged within 28 days of the decision.[33] The failure to give notice within that 28 days meant the adjudication decision became final and could not be challenged. In *Anglian Water Services Ltd v Laing O'Rourke Utilities Ltd* (2010),[34] the finality provisions in clause 93.1 of NEC2 were found to be compatible with the 1996 Act as the requirement to adjudicate disputes as a mandatory first step before seeking final determination did not fetter the right to adjudicate at any time and could be operated consistently with the other provisions of the Scheme.

6.22 However, such finality provisions may be construed as not requiring strict compliance with the letter of the provision where one of the parties to the contract has become the subject of an administration order or is in liquidation. In *Straw Realisations (No. 1) Ltd (formerly known as Haymills (contractors) Ltd (in administration) v Shaftsbury House (Developments) Ltd*[35] there was a provision that a decision would become final unless a notice was served within three

[29] The equivalent provision in NEC2 is at Clause 93.1.

[30] This provision (a) was not part of NEC1 Clause 93.1.

[31] p. 98 of the Guidance Notes to the NEC 3 confirms the intended effect of these provisions: 'The effect of this clause is time-barring. . . . If neither Party does so within that period, the Adjudicator's decision becomes final as well as binding and it can no longer be referred to the tribunal. The stated period is only for the notification of dissatisfaction. The dispute can be, and normally is, referred to the tribunal at a later stage.'

[32] [2011] BLR 644, [2011] EWHC 1935 (TCC).

[33] See the distinction drawn by the Scottish Courts in *Castle Inns (Stirling) Ltd v Clark Contracts Ltd* [2007] CSOH 21.

[34] [2010] 131 Con LR 94; [2010] EWHC 1529 (TCC). This case also confirms the importance of complying with notice provisions, for example by ensuring it is sent to the correct address, and is discussed further in Ch. 14.

[35] [2011] BLR 47, [2011] EWHC 2597 (TCC) at para. 48 (Key Case).

months of the decision stating that the dispute was to be referred for final determination. A letter within three months of the decision that simply indicated that the claim for payment was considered unenforceable and rejected (but did not expressly give notice of intention to refer to legal proceedings) was held to satisfy the provision.

6.23 The calculation of the relevant period of time must be interpreted in accordance with the particular facts and contractual provision. Where an adjudicator provided an initial decision, for example on matters such as jurisdiction, before providing his decision on the substantive dispute, and the contract required notice of final determination within a period of time from the 'determination' of the adjudicator, that time period ran from the final decision by which he discharged his functions: *Midland Expressway Ltd and ors v Carillion Construction Ltd and ors (No. 3)* (2006).[36] Jackson J found that the natural interpretation of 'determination' in the contract did not refer to 'any earlier ruling or decision which the adjudicator may make along the way':

> 87. . . . The adjudication process, as provided for by the 1996 Act . . . is intended to be a speedy and efficient procedure. It would be contrary to the purpose of this procedure and it would not make commercial sense if time starts to run . . . at any point in the course of the adjudication. This might lead to the need for successive actions. It might also lead to the need for litigation while the adjudication is still in progress. Furthermore, such litigation might turn out to be fruitless or unnecessary after the adjudicator has given his final decision.

Effect of the Decision in Subsequent Adjudications

6.24 An adjudicator appointed in a subsequent adjudication has no jurisdiction to set aside, revise, or vary his previous decision(s) or those of any other adjudicator. Accordingly, findings of fact or law on issues in previous adjudication decisions will be binding on the parties and a subsequent adjudicator alike, as long as that issue was part of the dispute referred in the previous adjudication so that the adjudicator was given jurisdiction to decide the issue[37] and the decision is not otherwise unenforceable, or the parties agree otherwise.

6.25 It is possible for parties to agree to be bound by a constituent part of a decision or any process of reasoning adopted in the course of reaching a conclusion on the overall dispute.[38] However, where the reasoning within that decision is not an essential component of, or basis for that decision, it will not be binding.[39]

6.26 An adjudicator may consider facts and matters considered in previous adjudications in reaching a conclusion, without trespassing on the previous decision.[40] However, if he reaches a different conclusion based on these same facts and the same cause of action, then that decision will be unenforceable.[41]

Compliance with Matters Properly Inferred

6.27 The parties may also temporarily be bound by consequences that follow naturally from an adjudicator's decision. The most common consequences to flow from adjudicators' decisions

---

[36] [2006] BLR 325, [2006] EWHC 1505 (TCC).

[37] *Redwing Construction Ltd v Charles Wishart* [2010] EWHC 3366 (TCC) at para. 27(e).

[38] *RSL (South West) Limited v Stansell Limited* [2003] EWHC 1390 (TCC). However, in that case, the court decided that there was no opportunity to pay a sum other than the sum awarded.

[39] *Hyder Consulting (UK) Limited v Carillion Construction Limited* (2011) 138 Con LR 212, [2011] EWHC 1810 (TCC) (Key Case).

[40] *Emcor Drake & Scull Ltd v Costain Construction Ltd* (2004) 97 Con LR 142, [2004] EWHC 2439 (TCC).

[41] See further. 6.63–6.72 below in relation to 'double jeopardy'.

concern extensions of time and their effect on the employer's entitlement to levy liquidated and ascertained damages, or the contractor's entitlement to claim prolongation costs and/ or other loss and expense.[42] Thus, where an adjudicator has decided a ten-week extension of time was due, the parties would have to abide by that decision. This would mean that the employer could not deduct liquidated damages for that period; however, there could still be an argument between the parties as to what, if any, compensation was payable to the contractor.[43] Each case will turn on its facts and precisely what the adjudicator decided.[44] See also the discussion about setting off liquidated damages against an adjudicator's decision at 6.43 below (and following).

6.28 An obligation to comply with matters which 'follow logically' from an adjudicator's decision was enforced where an adjudicator's decision contained 'inexorable logic' which meant that the claimant had been overpaid by the defendant even though no sums were actually awarded;[45] and where the adjudicator made a declaration as to the net value of the final account without giving a direction to pay the amounts due.[46]

6.29 However, in another case the court declined to find that payment of liquidated damages flowed naturally from an adjudicator's decision that certificates of non-completion were to be issued. The adjudicator had declined to make any financial award or determination of the financial consequences in respect of liquidated damages in the absence of such certificates.[47] Similarly, whilst the natural corollary of an adjudicator's decision may be to increase the number of expended hours in a 'pain/gain' calculation, where there is disputed evidence as to how that 'pain' is calculated, a deduction for that sum will not necessarily follow logically from the adjudicator's decision.[48]

Key Cases: The General Effects of an Adjudicator's Decision

City Inn Ltd v Shepherd Construction Ltd [2002] SLT 781; (2001) SCLR 961 (Outer House, Court of Session)

6.30 **Facts:** An adjudicator had determined that the contractor, Shepherd, was entitled to an extension of time (EOT) five weeks longer than that granted by the architect. City Inn commenced proceedings to have the adjudicator's decision overturned, arguing that Shepherd needed to establish its entitlement to an EOT beyond that certified by the architect. City Inn relied on the JCT Scottish Building Contract with Quantities 1998 which described the substantive determination as 'a consideration of the dispute ... as if no decision has been made by the Adjudicator'. Shepherd argued that:

[42] See 6.43–6.62 and the the principles set out by Jackson J in *Balfour Beatty Construction v Serco Ltd* [2004] EWHC 3336 (TCC) (Key Case).

[43] This example is given by Akenhead J in *HS Works Ltd v Enterprise Managed Services* [2009] BLR 378, [2009] EWHC 729 (TCC) at para. 61.

[44] E.g. the adjudicator was found not to have reached any definitive conclusion as to the total extension of time due to the contractor in *Balfour Beatty Construction v Serco Ltd* [2004] EWHC 3336 (TCC) (Key Case); and *RJ Knapman Ltd v Richards* (2006) 108 Con LR 64, [2006] EWHC 2518 (TCC) in which it was concluded that it was not a logical consequence of the adjudicator's decision that liquidated damages were due. These cases are discussed in detail below in relation to set off.

[45] *Workspace Management Ltd v YJL London Ltd* [2009] BLR 497, [2009] EWHC 2017 (TCC).

[46] *HS Works Ltd v Enterprise Managed Services Ltd* [2009] BLR 378, [2009] EWHC 729 (TCC).

[47] *Hart (t/a DW Hart & Son) v Smith* [2009] All ER (D) 29 (Sep), [2009] EWHC 2223 (TCC).

[48] *Ledwood Mechanical Engineering Ltd v Whessoe Oil and Gas Ltd* [2008] BLR 198, [2007] EWHC 2743 (TCC) (Key Case). This case is discussed below at 6.61.

1. The effect of the adjudicator's decision was to 'throw onto the [employer] the burden' of showing the EOT awarded by the adjudicator was not justified.
2. Section 108(3) supported its argument that the binding quality of the adjudicator's decision continued, not merely until the dispute was made the subject of litigation, but until the court proceedings were finally determined.
3. During the court proceedings, the adjudicator's decision remained binding and had to be rebutted by the party arguing against it.

6.31 **Held:** Shepherd's arguments were rejected. The onus remained on the contractor to discharge its onus of proof to entitlement of a longer EOT than certified by the architect, as if there had been no adjudication.

Hyder Consulting (UK) Limited v Carillion Construction Limited (2011) 138 Con LR 212, [2011] EWHC 1810 (TCC)

6.32 **Facts:** Carillion was engaged as a main contractor to Network Rail on the East London Line Extension project. Carillion engaged Hyder to undertake design works under a contract that included provisions for disallowed cost and a pain/gain share mechanism by which Hyder shared 50 per cent of the profits if the actual cost was below the target cost, and 50 per cent of the losses if above. The adjudicator adopted his own methodology in relation to his assessment of the target cost. Whilst Carillion challenged the enforceability of the decision on grounds of breach of natural justice (as discussed in Chapter 10), it also sought a declaration that the adjudicator's assessment of target cost was not in any event binding on the parties.

6.33 **Held:** Edwards-Stuart J found that it was the decision of the adjudicator that was binding on the parties, not the reasoning. The adjudicator decided what was due to Hyder in respect of a particular application for payment, the interest that attached to that application for payment, and the amount and allocation of his fees and expenses. The adjudicator's conclusion as to the calculation of the target cost was not a decision that was binding on the parties:

> 36. Since it is the decision of the adjudicator that is binding on the parties, not his reasoning, one must consider what is meant by '*the decision of the adjudicator*'. In most cases the adjudicator will determine that a sum of money is due from one party to the other and the decision will therefore consist of a declaration that the particular sum is due, together with related declarations in relation to the amount of interest and questions of costs. In that type of decision, it is clear beyond doubt that the adjudicator's conclusion that A owes (and must pay) £X to B is binding until finally determined by litigation or arbitration.
>
> 37. However, suppose that the adjudicator's reason for deciding that the sum owed to B is £X is that he has decided that B was entitled to an extension of time of Y weeks with a weekly prolongation cost of £Z. In this situation, I find it difficult to see how it could be said that the amount of the extension of time to which B was found to be entitled was not also part of the decision and therefore not binding as between A and B (subject, of course, to B having the right to argue in a subsequent adjudication that he is entitled to a further extension of time on the grounds not put before the adjudicator in the first adjudication). In my judgment, in that situation an adjudicator's conclusion on the amount of the extension of time attributable to the stated events would also be binding on the parties (until finally determined otherwise).

38. Accordingly, I consider that an adjudicator's decision consists of (a) the actual award (i.e. that A is to pay £X to B) and (b) any other finding in relation to the rights of the parties that forms an essential component of or basis for that award (for example, in a decision awarding prolongation costs arising out of particular events, the amount of the extension of time to which the referring party was entitled in respect of those events)...

40. In relation to the Target Cost, the adjudicator concluded that, since the net fee claimed was less than the applicable Target Cost, '*no further consideration needs to be given to the Target Cost*'. The adjudicator determined that the net fee claimed was £15 million odd, nearly £2 million less than his derived figure for the Target Cost (of £17 million odd). Thus his conclusion that the Target Cost was irrelevant would have been the same if he had assessed its value at any figure above the figure that he found to be the net fee due. So, subject only to a finding that the Target Cost was greater than the net fee due, the actual value of the Target Cost played no part in the process by which the adjudicator arrived at his decision. It would have been the same if the adjudicator had found that the Target Cost was only £1 more than the net fee due.

41. In my judgment, therefore, the adjudicator's conclusion in relation to the value of the Target Cost is not a decision that is binding on the parties. Whilst there is probably nothing to prevent Hyder from starting a further adjudication in order to claim the gain based on the adjudicator's calculation of the Target Cost, there is nothing to prevent Carillion from disputing the correctness of that calculation since its actual value did not form part of the adjudicator's decision in this referral.

Jerram Falkus Construction Ltd v Fenice Investments (No. 4) [2011] BLR 644, [2011] EWHC 1935 (TCC)

6.34 **Facts:** An adjudication between the parties to an amended JCT Design & Build 2005 contract resulted in the decision that the contractor (Jerram Falkus) was responsible for delays, there had been no act of prevention by the employer (Fenice), and time was not at large. The contract contained clause 1.9.4, which provided:

> In the case of a dispute or difference on which an Adjudicator gives his decision on a date which is after the date of submission of the Final Account and Final Statement... if either Party wishes to have that dispute or difference determined by arbitration or legal proceedings, that Party may commence arbitration or legal proceedings within 28 days of the date on which the adjudicator gives his decision.

The referral to adjudication followed the issue of the final account and the contractor commenced proceedings more than 28 days after the decision, requesting declarations from the court that would reverse the adjudicator's findings.

6.35 **Held:** The court refused to grant the declarations sought. Coulson J reviewed the clause in question and found that, even though the literal wording was only permissive ('may commence... within 28 days') it had the intended effect of providing finality and certainty following the final account:

> 22. Superficially, JFC's best point arises out of the precise words of clause 1.9.4. Nowhere in that provision does it say in terms that the adjudicator's decision, if not challenged within 28 day, is conclusive on the matters with which it deals. In a form of contract which appears liberally to endorse the concept of conclusive decisions and statements, that might be thought to be something of an obstacle to the argument that, [the adjudicator's] decision was indeed conclusive. Furthermore, I note that the requirement to challenge that decision within 28 days was not said to be mandatory but merely permissive: 'that party *may* commence arbitration or legal proceedings within 28 days...'
>
> 23. However, on a proper analysis, it seems to me that these narrow points on the wording of the clause ignore two fundamental issues. The first is the purpose of clause 1.9 itself.

The clause is designed to provide for various circumstances in which, following the provision of the Final Account, the position between the parties can become conclusive, thereby precluding any further dispute. Clause 1.9.4 must therefore be read in that context; it is providing a deadline beyond which something—in this case the decision in an adjudication started after the provision of the Final Account—becomes conclusive.... [I]t was providing a 'last chance'. If it were not a provision relating to conclusivity, it would not be part of clause 1.9 at all.

24. Secondly, if clause 1.9.4 was not providing some form of deadline, beyond which the result in a post-Final Account adjudication could not be challenged, then the provision was entirely redundant. If the clause was simply recording that the losing party could challenge the adjudicator's decision within 28 days, but that there was no consequence if they did not do so, then the provision would be meaningless: it would simply be recording something which the losing party could do in a 28 day period, but which it could also do just as well after the 28 day period had expired. It is contrary to one of the canons of contractual interpretation to read a clause of this type as mere verbiage, without any consequence or effect [Lewison, *The Interpretation of Contracts*, 4th edn, section 7.03].

25. I acknowledge that there is a gap between the language of clause 1.9.4 and its intended purpose. But, for the reasons I have given, I conclude that clause 1.9.4 was plainly intended to ensure that, if there was an adjudication after the Final Account had been provided, the losing party had 28 days in which to challenge the result, or the result became conclusive. JFC failed to do that in this case, and as a result they cannot now raise any argument to the effect that time is at large, that contention having been expressly considered and rejected by the adjudicator.

26. Standing back from the clause for a moment, it seems to me that this result makes commercial common sense. It would be absurd for the parties to enter into a detailed adjudication on the issues between them, after the Final Account has been provided and thus a long time after the works were completed; for the losing party to allow the decision to rest unchallenged for months or even years; and for that losing party to endeavour, months or years later, to go over exactly the same course all over again. That would provide for neither finality nor certainty, both of which clause 1.9 was designed to provide.

Table 6.2 Table of Cases: The General Effects of an Adjudicator's Decision

Title	Citation	Issue
Bouygues (UK) Ltd v Dahl-Jensen (UK) Ltd	[2001] All ER (Comm) 1041 (CA) (TCC), Peter Gibson LJ; Chadwick LJ; Buxton LJ	**Payment provisionally due** An adjudicator's decision that gives rise to an immediate payment obligation, enforceable by the courts, can be reopened for final determination in subsequent proceedings. Adjudication provides a summary procedure for enforcement of a payment provisionally due. A decision may require an immediate payment that was not anticipated by the contract payment mechanism.
Glencot Development and Design Co. Ltd v Ben Barrett & Son (Contractors) Ltd	[2001] BLR 207 (TCC), Judge LLoyd	**Underlying cause of action remains** The underlying cause of action in contract is not superseded by an adjudicator's decision.
City Inn Ltd v Shepherd Construction Ltd	[2002] SLT 781; (2001) SCLR 961 (Outer House, Court of Session), Lord Macfadyen	**Onus of proof in legal proceedings or arbitration** An adjudicator's decision had no effect on the burden of proof in litigation, therefore the contractor had to justify the EOT it sought. Section 108(3) did not affect the burden of proof in arbitration or court proceedings. This approach was endorsed in principle in *Citex Professional Services v Kenmore Developments* [2004] Scot CS 20.

(Continued)

Table 6.2 *Continued*

Title	Citation	Issue
David McLean Housing Ltd v Swansea Housing Association Ltd	[2002] BLR 125 (TCC), Judge LLoyd	**Decision determines contractual right** The adjudicator determined that the contractor was entitled to an EOT that fell short of the actual delay in achieving practical completion. The employer made only partial payment, deducting liquidated and ascertained damages for the period of delay for which there was no EOT. An adjudicator's decision does not create or modify a right or liability under the contract, it simply determines or upholds such rights or liabilities.
Bracken v Billinghurst	[2003] CILL 2039, [2003] EWHC 1333, Judge Wilcox	**Binding until agreement** Where, following an adjudicator's decision, the parties agreed to compromise the dispute, there was no decision to enforce at summary judgment.
RSL (South West) Limited v Stansell Limited	[2003] EWHC 1390 (TCC), Judge Seymour	**Binding until agreement** The parties may agree to be bound by, and give effect to, any decision by an adjudicator or a constituent element in the eventual overall total, or any process or reasoning adopted in the course of reaching a conclusion on the overall dispute.
Alstom Signalling Ltd v Jarvis Facilities Ltd	[2004] EWHC 1285 (TCC), Judge LLoyd	**No need to comply with decision before advancing a challenge** An unsuccessful party in an adjudication does not need to comply with the adjudicator's decision before contesting its enforceability. Unless a party has waived its right to do so or is estopped, the party may obtain a final determination at any time.
Michael John Construction v Golledge and ors	[2006] EWHC 71 (TCC), Coulson J	**Consideration of matters determined by adjudicator** During enforcement proceedings, the court refused to consider a possible argument that had been rejected by the adjudicator, when considering whether to grant a stay.
Midland Expressway Ltd and ors v Carillion Construction Ltd and ors (No. 3)	[2006] BLR 325, [2006] EWHC 1505 (TCC), Jackson J	**Time runs from the adjudicator's final decision** The adjudicator's first decision was on 30 December 2005 and the last decision was on 10 February 2006—as revised on 21 February 2006. If the first decision was the relevant 'determination', the contractual 60-day time limit expired before the notice was given. The 'determination' was the final decision of the adjudicator on 10 February 2006. Accordingly, the time-bar defence was rejected and the claimant was entitled to pursue the action.
Cubitt Building & Interiors Ltd v Fleetglade Ltd	110 Con LR 36, [2006] EWHC 3413 (TCC), Coulson J	**Conclusivity of final account** It is conceivable that an adjudicator may, through the adjudicator's acts or omissions, remove a party's right to challenge a final account.
Castle Inns (Stirling) Ltd v Clark Contracts Ltd	[2007] CSOH 21, Lord Drummond Young	**Contractual finality provision** The form of contract (Scottish Building Contract, Contractor's Designed Portion, Without Quantities (January 2000 Revision)) provided that

Title	Citation	Issue
		proceedings for final determination of disputes on matters decided by an adjudicator after the date of issue of the final certificate may be brought within 28 days of the adjudicator's decision. The Scottish court decided that the adjudicator's decision issued after the final certificate had been on the loss and expense aspect of a delay claim which was not the same as the claim for loss of profit due to delay in completing the project in respect of which final determination was sought. The court therefore found that the dispute was not caught by the time bar in the contract.
Ringway Infrastructure Services Limited v Vauxhall Motors Limited (No. 2)	115 Con LR 149, [2007] EWHC 2507 (TCC), Akenhead J	**Failure to comply with decision** The nature of enforcement of adjudicators' decisions is contractual. The contract required the parties to comply with the adjudicator's decision. Failure to comply with an adjudicator's decision (whether right or wrong) is a new breach of contract and a cause of action arises at the date when the unsuccessful party should have complied with the adjudicator's decision.
HS Works Ltd v Enterprise Managed Services Ltd	[2009] BLR 378, [2009] EWHC 729 (TCC), Akenhead J	**Declaratory award by adjudicator** An adjudicator gave a declaration as to the net value of the final account but issued no direction to pay any balance back. Where an adjudicator's decision is declaratory, the parties must comply with the consequences of the decision.
Jim Ennis Construction Ltd v Premier Asphalt Ltd	125 Con LR 141, [2009] EWHC 1906 (TCC), Judge Davies	**Time limit for recovery of monies paid** Proceedings for final determination in respect of sums overpaid following compliance with an adjudicator's decision could be brought on the basis of an implied term in respect of repayment or alternatively in restitution. The unsuccessful party had six years from the date of the payment in respect of the adjudicator's decision to bring proceedings.
Workspace Management Ltd v YJL London Ltd	[2009] BLR 497, [2009] EWHC 2017 (TCC), Coulson J	**Inexorable logic that claimant had been overpaid** An adjudicator's decision contained the 'inexorable logic' that the claimant had been overpaid by the defendant by £56,143.34. However, the adjudicator did not consider he had jurisdiction to direct that the amount was due. Subsequent arbitral awards were made. Either the adjudicator's decision that the sum was due to the defendant was express or it was to be reasonably inferred from the inevitable and logical consequences of his valuation.
Hart (t/a DW Hart & Son) v Smith	[2009] All ER (D) 29 (Sep), [2009] EWHC 2223 (TCC), Judge Toulmin	**Liquidated damages did not logically flow** An adjudicator made a declaration that the Smiths were entitled to certificates of non-completion under the JCT Standard Building Contract with Quantities 2005. The Smiths sought payment of liquidated and ascertained damages (LADs) in reliance on the adjudicator's decision. This was refused for three main reasons:

(Continued)

Table 6.2 *Continued*

Title	Citation	Issue
		(i) the adjudicator had been asked to order payment to the Smiths under a notice of withholding which included alleged LADs. The adjudicator found that, until the certificates of non completion were issued, the Smiths could not request payment of LADs; (ii) there was a discrepancy between the sums claimed in the adjudication and court proceedings; and (iii) there was a significant difference between this case (where the result was that the contract administrator should issue the certificates of non-completion and nothing more) and circumstances where it is possible to calculate the sums which were a direct consequence of the adjudicator's award. In order to award LADs, the court would have to consider the effect of contractual provisions on which the adjudicator made no affirmative finding.
Geoffrey Osborne Ltd v Atkins Rail Ltd	[2010] BLR 363, [2009] EWHC 2425 (TCC), Edwards-Stuart J	**Temporarily binding until finally determined by legal proceedings: use of Part 8 proceedings** An adjudicator was invited to correct a significant arithmetical error in his decision but refused to do so. The party the error benefited sought to summarily enforce the decision and the court permitted the challenge to the arithmetic as long as the matter was one which could be finally determined on the material before the court. *Bouygues UK Ltd v Dahl-Jensen UK Ltd* distinguished on the basis that there was no agreement to arbitrate between the parties in the current action. If there is a part of the decision that can be isolated and determined by the court, if considered just and expedient to do so, that would give effect to the overriding objective of the CPR.
Anglian Water Services Ltd v Laing O'Rourke Utilities Ltd	131 Con LR 94, [2010] EWHC 1529 (TCC), Edwards-Stuart J	**Adjudicator's decision not final and binding as notice of dissatisfaction properly served** The parties agreed that the adjudication provisions of the NEC2 standard form (with Y(UK)2 amendment) offended the Act and therefore conducted an adjudication under the Scheme. Clause 13.2 requires exact compliance with the notice provisions, including sending to the correct recipient. However, the NEC2 service provision at Clause 13.2 is not within the adjudication provisions of the contract. Therefore it was not removed and/or replaced by the provisions of the Scheme as it could be operated consistently with the scheme provisions. In this case, the Clause 93.1 notice of intention to refer the dispute to final determination had been properly served on the party's solicitor and therefore the adjudicator's decision was not final and binding.

Title	Citation	Issue
Straw Realisations (No. 1) Ltd (formerly known as Haymills (Contractors) Ltd) (in administration) v Shaftsbury House (Developments) Ltd	[2011] BLR 47, [2010] EWHC 2597 (TCC), Edwards-Stuart J	**Notice requirements relaxed where one of the parties was insolvent** The contractor went into administration between two adjudication decisions. The relevant notice provision was found to contemplate a situation in which both parties were solvent. The court did not require the notice provision to be strictly followed where one party was insolvent. An indication as to whether the adjudicator's decision was going to be accepted or challenged was sufficient.
Redwing Construction Ltd v Charles Wishart	[2010] EWHC 3366 (TCC), Akenhead J	**When an issue in adjudication 1 impinges on adjudication 2** Where the dispute is materially different in both adjudications but an issue in the first adjudication impinges on the issues to be decided in the second adjudication, the parties may still be bound by the finding made in the first decision. It was held that it depended on whether the issue was part of the dispute actually referred for decision, but if not then it would not be binding in subsequent adjudications.
Hyder Consulting (UK) Limited v Carillion Construction Limited	138 Con LR 212, [2011] EWHC 1810 (TCC), Edwards-Stuart J	**The decision, not the reasoning, is temporarily binding** A dispute arose between a main contractor (Carillion) and subcontractor (Hyder) under a contract with target-cost provisions. The adjudicator's decision related to an application for payment. The adjudicator assessed the target cost but it did not directly affect his calculation of the application for payment. His reasoning in relation to the target cost therefore did not bind the parties.
Jerram Falkus Construction Ltd v Fenice Investments (No. 4)	[2011] BLR 644, [2011] EWHC 1935 (TCC), Coulson J	**Adjudicator's decision concerning a JCT final account was final and binding** The effect of Clause 1.9.4 of the JCT 2005 Design and Build form of contract is that an adjudicator's decision delivered following the issue of the final account becomes conclusive if proceedings for final determination are not commenced within 28 days of the decision.

(3) Set-off against an Adjudicator's Decision

6.36 As explained above, a decision of an adjudicator gives rise to a separate contractual entitlement to immediate payment. Ordinarily that payment should be made without deduction, set-off, withholding, reliance on a cross-claim or abatement.[49] Where there are subsequent adjudications, at the end of each adjudication (absent special circumstances), the losing party

[49] *Bovis Lend Lease Ltd v Triangle Development Ltd* [2003] BLR 31, [2002] EWHC 3123 (TCC).

must comply with the adjudicator's decision.[50] However, there are circumstances where it may be possible to raise a set-off in defence of a claim for payment pursuant to an adjudicator's decision.

6.37 Any proposed set-off can only be in respect of issues not considered in the adjudication. If the issue has been considered by the adjudicator, then it will be temporarily binding between the parties. Unless the cross-claim that has been decided by the adjudicator can be finally determined summarily (i.e. without any consideration of disputed factual matters) as a part of enforcement proceedings, it cannot be taken into account in defence of enforcement.

Set-off Not Generally Available

6.38 The question of whether set-off against an adjudicator's decision might in principle be allowed was first considered in *VHE Construction plc v RBSTB Trust Co. Ltd* (2000).[51] Judge Hicks rejected the employer's submissions that there remained a residual right to set off liquidated damages not claimed in the adjudication against sums that the adjudicator had decided were due. In his view, there was no such residual right where adjudication decisions were concerned. This was because the sums in question were not due 'under the contract', but by way of compliance (albeit contractually required) with the adjudicator's decision. In this context the judge said that to 'comply' with the decision meant to 'comply, without recourse to defences or cross-claims not raised in the adjudication'.

6.39 That general position was confirmed by the Court of Appeal in *Ferson Contractors Ltd v Levolux AT Ltd* (2003)[52] (Key Case). The court upheld the first-instance decision of Judge Wilcox[53] that provisions within the contractual scheme which assumed that no further monies would be paid following determination of the contract, did not apply to money payable pursuant to an adjudicator's award. The Court of Appeal found that *Bovis Lend Lease Ltd v Triangle Ltd* (2003)[54] was wrongly decided. In that case, it was held that the contractual determination provisions in that case could override the obligation to make payment pursuant to an adjudicator's decision. The Court of Appeal had no doubts that the general position as stated by Judge Wilcox was correct. It stated that the intended purpose of s. 108 was plain and any construction of such provisions that allowed the payment pursuant to the decision to be avoided would defeat that purpose. If that could not be achieved by way of construction, the Court of Appeal said the offending clause must be struck down.

6.40 The Court of Appeal's decision in *Ferson v Levolux* has been applied in relation to:

1. Events which occurred after an adjudication notice had been served.[55]
2. Cross-claims in relation to liquidated damages for delay which had not been raised in the adjudication.[56]
3. Claims for set-off in relation to defects which had been considered in the adjudication but which the adjudicator decided raised no immediate right to damages.[57]

---

50 *Interserve Industrial Services Ltd v Cleveland Bridge (UK) Ltd* [2006] All ER (D) 49 (Feb.), [2006] EWHC 741 (TCC); as endorsed in *Hillview Industrial Developments (UK) Ltd v Botes Building Ltd* [2006] All ER (D) 280 (Jun); [2006] EWHC 1365 (TCC).

51 [2000] BLR 187 (TCC) at para. 65.

52 [2003] 1 All ER (Comm) 385, [2003] EWCA Civ 11 (CA).

53 [2002] Adj LR 06/26 (TCC), <http:www.nadr.co.uk>.

54 [2003] BLR 31, [2002] EWHC 3123 (TCC) at para. 26.

55 *MJ Gleeson Group plc v Devonshire Green Holding Ltd* (2004) Adj LR 03/19 (TCC), <http:www.nadr.co.uk>.

56 *Construction Centre Group Ltd v Highland Council* 2003 SLT 623, [2003] Scot CS 114, (Inner House, Court of Session), <http:www.nadr.co.uk>; *David McLean Contractors Ltd v The Albany Building Ltd* [2005] Adj LR 11/10 (TCC), <http:www.nadr.co.uk>.

57 *RJ Knapman Ltd v Richards* (2006) 108 Con LR 64, [2006] EWHC 2518 (TCC).

4. Cases in which the employer's claims for set-off were the subject of current or impending adjudications.[58]

6.41 In *Interserve Industrial Services Ltd v Cleveland Bridge UK Ltd* (2006)[59] Jackson J said that, where the parties to a construction contract engage in successive adjudications, each adjudication should focus upon the parties' current rights and remedies. At the end of each adjudication, absent special circumstances, the losing party must comply with the adjudicator's decision and cannot withhold payment on the ground of anticipated recovery in a future adjudication based upon different issues.

6.42 In *William Verry Ltd v London Borough of Camden* (2006)[60] (Key Case), Ramsey J went a stage further and considered the case in which an interim certificate which had been the subject of the adjudicator's decision had, at the time the decision was made, been superseded by a final certificate. The later valuation took into account defects that might have undermined the very basis of the adjudicator's earlier assessment meaning that the sums awarded were never due. Nevertheless, the judge concluded that such arguments could not be permitted to defeat the intention of Parliament that adjudicators' awards should be enforced. Ramsey J stated that the effect of s. 108 of the Act was to operate like a restriction of the right of set-off. This approach has been widely accepted and subsequently applied in the decision of Coulson J in *Balfour Beatty Construction Northern Ltd v Modus Corovest (Blackpool) Ltd* (2008)[61] (Key Case).

Special Circumstances that May Give Rise to a Set-off

6.43 In *Balfour Beatty Construction Ltd v Serco Ltd* (2004)[62] (Key Case) Jackson J considered a claim to set off liquidated damages and considered two possible situations from his review of the authorities:

> 53. . . . a. Where it follows logically from an adjudicator's decision that the employer is entitled to recover a specific sum by way of liquidated and ascertained damages, then the employer may set off that sum against monies payable to the contractor pursuant to the adjudicator's decision, provided that the employer has given proper notice (insofar as required).
>
> b. Where the entitlement to liquidated and ascertained damages has not been determined either expressly or impliedly by the adjudicator's decision, then the question whether the employer is entitled to set off liquidated and ascertained damages against sums awarded by the adjudicator will depend upon the terms of the contract and the circumstances of the case.

6.44 The facts of *Balfour Beatty v Serco* were not sufficient to give rise to such a situation. There, the adjudicator had only awarded an extension of time for a proportion of the period of delay claimed and there were further issues which had not been decided which might have given rise to a further extension. Accordingly, it was not possible to conclude from the adjudicator's award that the employer was entitled to withhold an ascertainable sum in respect of the continuing delay.

6.45 These facts may be contrasted with those in the earlier case of *David McLean Housing Ltd v Swansea Housing Association Ltd* (2002)[63] in which the adjudicator had awarded an extension of time that fell short of the actual delay in achieving practical completion. Since the residual

[58] *Interserve Industrial Services Ltd v Cleveland Bridge UK Ltd* [2006] EWHC 741 (TCC); *Hillview Industrial Developments (UK) Ltd v Botes Building Ltd* [2006] All ER (D) 280, [2006] EWHC 1365 (TCC).

[59] [2006] All RE (D) 49 (Feb.). [2006] EWHC 741 at para. 43.

[60] [2006] All ER (D) 292 (Mar), [2006] EWHC 761 (TCC).

[61] [2009] CILL 2660, [2008] EWHC 3029 (TCC).

[62] [2004] EWHC 3336 (TCC).

[63] [2002] BLR 125 (TCC).

delay was not the subject of any further proceedings, and liquidated damages could be calculated to the actual date of practical completion, Judge LLoyd upheld the employer's right to serve a notice of withholding in respect of these amounts.

6.46 Jackson J's first exception concerning the set-off of liquidated damages will only apply when the amount of liquidated damages to be set off is indisputable. In addition, it is suggested that s. 111 would still require a withholding notice to be served within the very limited time available to comply with the adjudicator's decision. This might require construing any time limit for compliance set out in the adjudicator's decision as a 'final date for payment', and serving notice within the specified period prior to this date.[64]

6.47 Jackson J's second exception contemplates a contract term permitting set-off against any sum due, as was the case in *Parsons Plastics Ltd v Purac Ltd* (2002).[65] In that case the Court of Appeal decided that a contract term permitting set-off against any sums due was drafted broadly enough to mean that sums could be set off against the adjudicator's decision and so was found to trump the adjudicator's decision. However, the contract in *Parsons* was not governed by the 1996 Act and so this was not a decision that the contractual right of set-off overrode the effect of s. 108 of the Act. In *Ferson Contractors Ltd v Levolux AT Ltd* (2003) (Key Case)[66] the Court of Appeal said contracts governed by the Act had to be construed to give effect to the intention of Parliament and s. 108 and a clause inconsistent with the statute should be struck down. It distinguished *Parsons Plastics* on the ground that it concerned a contract outside the Act.

6.48 There is a third exception to the rule that there can be no set-off against an adjudicator's decision. This concerns the nature of the adjudicator's decision. If the adjudicator decides that one party owes another a certain sum and must pay that sum, then the usual rule is the sum must be paid without set-off. However if the decision is instead a declaration as to the proper operation of the contract payment machinery, and the adjudicator identifies the sum which should have been subject of that machinery, then it may leave open the possibility that following the decision a withholding notice can still be served. So, for instance if the adjudicator decides that the next interim certificate should include certain sums and subsequently, in compliance with the decision, that certificate is then issued, then the paying party may be able to serve a valid withholding notice under the contract terms before the final date for payment of that certificate. There are a number of cases in which this approach has been endorsed: *Shimizu Europe Ltd v LBJ Fabrications Ltd* (2003),[67] *Conor Engineering Ltd v Les Constructions Industrielles de la Mediterranee SA* (2004),[68] *R&C Electrical Engineers Ltd v Shaylor Construction Ltd* (2012),[69] *Squibb Group Ltd v Vertase FLI Ltd* (2012),[70] and *Beck Interiors Ltd v Classic Decorative Finishing Ltd* (2012).[71]

Key Cases: Set-off against an Adjudicator's Decision

Ferson Contractors Limited v Levolux AT Limited [2003] 1 All ER (Comm) 385, [2003] EWCA Civ 11 (CA)

6.49 **Facts:** Levolux was engaged under a GC/Works subcontract by Ferson to install curtain wall and louvre systems. A dispute arose concerning the amount due to Levolux on an

[64] See further Ch. 13.
[65] [2002] BLR 334 (CA).
[66] [2003] BLR 118.
[67] [2003] EWHC 1229 (TCC).
[68] [2004] BLR 212, [2004] EWHC 899.
[69] [2012] BLR 373, [2012] EWHC 1254 (TCC).
[70] [2012] EWHC 1958 (TCC).
[71] [2012] EWHC 1956 (TCC).

interim payment, as a result of which Levolux suspended performance and commenced adjudication to recover the monies it claimed was due. Ferson replied by determining the contract in consequence of which clause 29.8.1 provided that 'all sums of money that may be due or accruing due from the contractor to the subcontractor shall cease to be due or accrue due'. Accordingly, Ferson contended that it did not have to make any further payment to Levolux until after the completion of the works and the making good of defects.

6.50 The adjudicator agreed with Levolux and ordered that Levolux should be paid the entire amount of its claim, together with interest. Ferson refused to pay and Levolux brought enforcement proceedings. At first instance, Judge Wilcox rejected Ferson's argument that no further payment was due as a result of clause 29.8.1 since it was a necessary implication of the adjudicator's decision for the payment of money that Levolux had been entitled to suspend the works. Thus Ferson's attempted determination of the subcontract had no contractual effect. Furthermore, the judge held that the contractual scheme which assumed that no monies would be paid until completion of the works did not apply to an adjudicator's award.

6.51 **Held:** The position as stated by Judge Wilcox was correct. The intended purpose of s. 108 was plain. The contract must be construed so as to give effect to the intention of Parliament rather than to defeat it. If that could not be achieved by way of construction, then the offending clause must be struck down. In the words of Lord Mantell:

> 27. [Ferson]... submits that there are three main exceptions to the principle that an adjudicator's decision is binding and enforceable pending final resolution by arbitration or litigation. These he identifies as (1) where the adjudicator did not have jurisdiction or failed to act fairly or in conformity with the applicable procedures, (2) the terms of the contract override the apparent obligation to make payment in accordance with the adjudicator's decision and (3) where the decision is overridden by another applicable adjudication. The first of these is uncontroversial and already referred to at an earlier place in this judgment. Exceptions (2) and (3) are derived from the judgment of His Honour Judge Thornton QC in *Bovis Lend Lease v. Triangle Developments Ltd* 2nd November 2002, a transcript of the judgment of which has been provided. In that case Judge Thornton QC conducted a thorough review of a number of cases at first instance where this point has arisen and having noted the exceptions to the general position, to which I have referred, finally concluded that:
>
> > '(1) The decision of an adjudicator that money must be paid gives rise to a second contractual obligation on the paying party to comply with that decision within the stipulated period. This obligation will usually preclude the paying party from making withholdings, deductions, set-offs or cross-claims against that sum. (2) For a withholding to be made against an adjudicator's decision, an effective notice to withhold payment must usually have been given prior to the adjudication notice being given, or possibly the decision being given, and which was ruled upon and made part of the subject matter of that decision. (3) However, where other contractual terms clearly have the effect of superseding, or provide for an entitlement to avoid or deduct from, a payment directed to be paid by an adjudicator's decision, those terms will prevail. (4) Equally where a paying party is given an entitlement to deduct from or cross-claim against the sum directed to be paid as a result of the same, or another, adjudication decision, the first decision will not be enforced or, alternatively, judgment will be stayed.'
>
> It is, of course, upon the third conclusion that Mr Collings relies in support of his submissions and that which gives rise to the obvious question; was Judge Thornton right in reaching that particular conclusion?
>
> 28. His Honour Judge Thornton was much influenced by a judgment of His Honour Judge LLoyd in *KNS Industrial Services (Birmingham) Ltd v. Sindall Ltd* (2001) 75 Con. LR 71. In that case Judge LLoyd was faced with an argument that a clause in the contract,

which allowed for the determination of the contractor's employment, was capable of over-riding the decision of the adjudicator. At para. 28 Judge LLoyd concluded:

> 'Other rights under the contract which were not the subject of the decision remain available to the relevant party. If, therefore, by the time an adjudicator makes a decision requiring payment by a party to the contract, the contract has been lawfully terminated by that party (or that party has real prospects of success in supporting that termination) or some other event has occurred which under the contract entitles a party not to pay, then the amount required to be paid by the decision does not have to be paid.'

Judge Thornton also had recourse to a decision of this court in *Parsons Plastics (Research and Development) Ltd v. Purac Ltd* (2002) BLR 334. In that case the court was not concerned with a construction contract as such, but with a contract which nevertheless contained what may be termed the s.108 provisions. It follows, that the point of construction was much the same as here. At para. 15 Pill LJ stated:

> 'It is open to the respondents (employment paying party) to set-off against the adjudicator's decision any other claim they have against the appellants (contractor receiving party) which had not been determined by the adjudicator. The adjudicator's decision cannot be re-litigated in other proceedings but, on the wording of this sub-contract, can be made the subject of set-off and counter-claim.'

However, having referred to that passage, Judge Thornton went on to say that the adjudication was a purely contractual adjudication without the statutory backing of the HGCRA and that 'it is only clear words that can trump the payment decision'.

29. It is, I think, important to note that in *Parsons Plastics (Research and Development) Ltd v. Purac Ltd* the court did not have to consider what impact s.108 might have on the construction of the relevant clause and, moreover, was concerned with a set-off and counter-claim upon which there had been no adjudication. On that, if no other basis, it can clearly be distinguished from the instant case. Here what was claimed, in opposition to the application for summary judgment, was a right to withhold payment following a valid determination of the contract. Rightly or wrongly the adjudicator held that there had been no valid determination. So, even accepting that the logic of *Parsons Plastics* applies in the circumstances of the present case, its application would not result in clauses 38A.7 and 38A.9 being overridden by clauses 29.8 and 29.9. And in any event this logic is insufficient to support Judge Thornton's conclusion expressed, as it is, in such broad terms.

30. But to my mind the answer to this appeal is the straight forward one provided by Judge Wilcox. The intended purpose of s.108 is plain. It is explained in those cases to which I have referred in an earlier part of this judgment. If Mr Collings and His Honour Judge Thornton are right, that purpose would be defeated. The contract must be construed so as to give effect to the intention of Parliament rather than to defeat it. If that cannot be achieved by way of construction, then the offending clause must be struck down. I would suggest that it can be done without the need to strike out any particular clause and that is by the means adopted by Judge Wilcox. Clauses 29.8 and 29.9 must be read as not applying to monies due by reason of an adjudicator's decision.

31. For those reasons I would dismiss this appeal.

Balfour Beatty Construction Ltd v Serco Ltd [2004] EWHC 3336 (TCC)

6.52 **Facts:** Balfour Beatty was engaged by Serco to design, supply, install, and test 104 variable message signs at locations on motorways throughout England. The bespoke form of contract included provisions for extensions of time and liquidated damages. Delays occurred which led to Balfour Beatty lodging a claim submission comprising 18 heads of claim many of which related to extensions of time, loss, and expense.

6.53 In his decision, the adjudicator revised the completion date to 'at least' 7 June 2004 in accordance with one of the heads of claim, and allowed financial claims under a number of the other heads. He directed that Serco should pay to Balfour Beatty the sum of £620,664, together with VAT.

6.54 Serco refused to pay the sums due under the adjudicator's decision. By a letter dated 6 December 2004, Serco purported to withhold payment on the basis that practical completion under the contract still had not occurred. The adjudicator had extended time only until 7 June 2004 and therefore, Serco was entitled to set off liquidated and ascertained damages for the period after 7 June 2004. According to Serco's calculations, these damages would exceed the sum payable to Balfour Beatty under the adjudicator's decision.

6.55 **Held:** By the use of the words 'at least', and the fact that the contract was still ongoing at the date of the adjudication, the adjudicator appeared to leave the matter open that there would be further extensions of time. No specific entitlement to liquidated and ascertained damages followed logically from the adjudicator's decision and it was strongly disputed between the parties whether any liquidated and ascertained damages were due and payable at all. The contract required both parties to give effect forthwith to the adjudicator's decision and accordingly Serco was obliged to pay the sum awarded by the adjudicator without setting off the liquidated and ascertained damages which it claimed:

> 48. I must begin this part of the judgment by reference to the authorities. In *VHE Construction PLC v RBSTB Trust Co. Limited* [2000] BLR 187, the parties contracted on the JCT Standard Form With Contractor's Design (1981 edition). The employer sought to set off liquidated damages against monies payable to the contractor under an adjudicator's award. His Honour Judge Hicks QC held that such a set-off was not permissible. Judge Hicks had regard to the overall purpose of Part 2 of the Construction Act. He concluded that the employer's obligation to comply with the adjudicator's decision meant:
>
> '… comply, without recourse to defences or crossclaims not raised in the adjudication.'
>
> 49. In *David McLean Housing Contractors Limited v Swansea Housing Association Limited* [2002] BLR 125 the parties contracted on the same JCT form of contract as was used in *VHE Construction.* Following practical completion, a variety of claims by the contractor were referred to adjudication. The adjudicator awarded certain sums to the contractor. He also determined that the contractor was entitled to an extension of time which fell short of the actual delay in achieving practical completion. On the day after the adjudicator's corrected decision was published, the employer wrote to the contractor stating that it would deduct liquidated and ascertained damages. Thereafter, the employer made only a partial payment to the contractor of the sums awarded by the adjudicator. The employer withheld liquidated and ascertained damages in respect of the period of delay for which there was no extension of time. In subsequent enforcement proceedings, His Honour Judge LLoyd QC held that the employer was entitled to set off the liquidated and ascertained damages which were due. At paragraph 18 of his judgment Judge LLoyd said this:
>
> > i. 'The next point is the real issue: is the claimant entitled to all the money the subject of the adjudicator's decision. All the money that was certified in certificate 20, bar the amount in dispute on liquidated damages, has in fact been paid. Is the claimant entitled to the amount for liquidated damages? That amount now reflects the adjudicator's view about the extension of time that was sought by the claimant so the claimant is bound to accept that conclusion in these proceedings since it was part of the dispute which it referred.'
>
> 50. There is, in my judgment, no inconsistency between the reasoning in *VHE Construction* and *David McLean.* In each case the decision flows from an analysis of what the adjudicator had decided and from the particular circumstances of the case.
>
> …

> 54. In the present case, for the reasons set out in paragraph 5 of this judgment, the adjudicator has not reached any definitive conclusion as to the total extension of time which is due to Balfour Beatty. No specific entitlement to liquidated and ascertained damages follows logically from the adjudicator's decision. It is strongly disputed between the parties whether any liquidated and ascertained damages are due and payable. Paragraph 10 of Appendix A to Schedule 23 of the Contract requires both parties to give effect forthwith to the adjudicator's decision. The effect of paragraph 13 of Appendix A is that Balfour Beatty is entitled to the relief and remedies set out in the adjudicator's decision and, moreover, is entitled to summary enforcement of such relief and remedies. These contractual provisions are consistent with the provisions of Part 2 of the Construction Act and with the Parliamentary intention referred to in the authorities.

William Verry Ltd v London Borough of Camden [2006] All ER (D) 292 (Mar), [2006] EWHC 761 (TCC)

6.56 **Facts:** Camden sought to resist enforcement of an adjudicator's decision on an interim certificate that required payment from Camden to Verry. Camden relied on the final certificate issued 10 days before the decision which, following the deduction of Camden's entitlement to LADs as found by the adjudicator, resulted in a net payment being due to Camden. It also relied on its entitlement to set off a counterclaim for defects which was the subject of a further adjudication that was in progress.

6.57 **Held:** In Ramsey J's view, this approach was incorrect. If payment of an adjudication decision on the sum due on an interim certificate had to be subject to the view of the contract administrator in a subsequent certificate, then the intention of Parliament and the purpose of adjudication would be defeated. Each successive certificate would then defeat the decision by an adjudicator on the previous certificate. The final certificate was, in any event, subject to a number of further impending adjudications on the issues of defects and could not be regarded as conclusive. Camden should therefore comply with the adjudicator's decision and make payment of the sums awarded. The outcome of future adjudications would determine what sums would eventually be payable and by whom under the true final account:

> 47. As I have said, in my judgment, the effect of the statutory provisions is generally to exclude a right of set-off from an adjudicator's decision. In this case, the adjudicator expressly dealt with the question of extension of time and liquidated damages. These matters were raised by Camden, but issues of defects were not raised in the adjudication. If they had been raised, then evidently they could have been taken into account to the extent possible, and Camden could not now raise a set-off. By failing to raise them, I do not consider that Camden can be in a better position by setting off a disputed, unliquidated counterclaim against the adjudicator's decision.

Balfour Beatty Construction Northern Ltd v Modus Corovest (Blackpool) Ltd [2008] EWHC 3029 (TCC)

6.58 **Facts:** Balfour Beatty applied for enforcement of an adjudicator's award together with summary judgment in relation to an earlier payment certificate which remained unpaid. Modus had not served any withholding notice relating to the certificate, but had, after the final date for payment, issued notice under clause 24 of the JCT 1998 conditions that delay damages were payable. Modus applied to set off its counterclaim for delay damages on the basis that:

1. The sums awarded by the adjudicator and certified by the employer's agent fell to be reduced by the later claim for delay damages and were not therefore 'properly due' under the contract.
2. There could be no defence to Modus' counterclaim since (a) Balfour Beatty had itself failed to serve a withholding notice in response to the claim for delay damages, and (b) since Modus' agent was entitled to fix the completion dates, the sums calculated by reference to those dates were due as a debt.

6.59 **Held:** Coulson J rejected each of these arguments. The failure to serve a withholding notice was fatal to any set-off against the earlier payment certificate, and there was a general exclusion of the right to set-off from an adjudicator's decision which certainly applied in the case of the JCT conditions considered. (*Rupert Morgan Building Services (LLC) Ltd v David Jervis, Harriet Jervis* [2004][72] and *William Verry Ltd v London Borough of Camden* [2006] (Key Case)[73] applied.)

6.60 Further, Modus was not entitled to summary judgment for payment of its counterclaim as a debt since:

1. The requirement to serve a withholding notice applied only to stage payments by the employer and not to payments in the other direction.
2. There was no authority to support the proposition that the employer's agent's fixture of the completion dates should automatically be regarded as binding for the purposes of CPR Part 24. This would be contrary to the ordinary rules applicable to CPR Part 24 and would depend on whether or not the contractor had a genuine claim for an extension of time which had at least a realistic prospect of success.

> 86. . . . [the] authorities were considered by Ramsey J in *William Verry Limited v London Borough of Camden* [2006] All ER (D) 292 (Mar), [2006] EWHC 761 (TCC). In that case, the contractual provisions as to compliance with the adjudicator's decision were precisely the same as they are here, although the numbering was different. At paragraph 28 of his judgment, Ramsey J concluded that there was generally an exclusion of the right to set-off from an adjudicator's decision and that, certainly on the terms of the contract in that case, there was no such right.
>
> 87. I respectfully agree with and adopt the approach of Ramsey J in *Verry*. Therefore, there could be no set-off against the Adjudicator's decision in the present case.
>
> 88. In addition, I consider that, as a matter of construction of the contract, there can be no set-off against the sum due by way of clause 30.3, for the reasons which I have previously noted. The absence of a withholding notice is fatal to the alleged set-off against that amount.
>
> 89. Therefore, I conclude that Modus cannot set off their counterclaim against the sums otherwise due to Balfour Beatty. But whether or not Modus are entitled to summary judgment on the counterclaim is a different question and the final issue for me to determine on these applications.

Ledwood Mechanical Engineering Ltd v Whessoe Oil and Gas Ltd [2008] BLR 198, [2007] EWHC 2743 (TCC)

6.61 **Facts:** Whessoe sought to avoid payment of an adjudicator's decision by the application of contractual provisions for a pain/gain share arrangement. In Whessoe's view, the adjudicator's award fell to be reduced according to a formula for deducting 'pain'.

72 [2004] 1 All ER 529, [2003] EWCA Civ 1563 (CA).

73 [2006] All ER (D) 292 (Mar), [2006] EWHC 761 (TCC).

6.62 **Held:** Where it followed logically from an adjudicator's decision that the employer could recover an ascertainable sum for delay, then an employer might set off that sum against amounts due pursuant to an adjudicator's decision. However, the ability to deduct such undisputed or indisputable amounts as liquidated damages was a particular exception to the general rule and did not give a wider power to set off other sums against an adjudicator's decision.

> 37. . . . whilst the natural corollary of the Adjudicator's decision is that it increases the number of expended hours in the 'pain/gain' calculation, the calculation of that effect is not undisputed or indisputable. This is not a case like a calculation of liquidated damages which can be made using the number of weeks decided by the Adjudicator and applying the agreed rate. The issue of deduction of liquidated damages arises in a case where the Adjudicator deals with the extensions of time but does not deal with the consequential effect on an undisputed or indisputable claim for liquidated damages. It is a particular exception which relates to the manner and extent of compliance with the Adjudicator's decision. It does not, in my judgment, give a wider power to set off sums generally against an Adjudicator's decision.
>
> 38. In this case, as can be seen from the disputed evidence, it is necessary to consider both the overall man hours expended and the revised target man hours before the agreed formula can be applied. The evidence shows that both the overall man hours expended and the revised target man hours are disputed. I therefore do not consider that the Joint Venture can set off a sum for the risk/reward adjustment against the Adjudicator's decision relating to Application 19.

Table 6.3 Table of Cases: Set-off

Title	Citation	Issue
VHE Construction v RBSTB Trust	[2000] BLR 187 (TCC), Judge Hicks	**Set-off against an adjudicator's decision** There was no residual right to set off liquidated damages not claimed in the adjudication against sums that the adjudicator had decided were due.
Edmund Nuttall Ltd v Sevenoaks District Council	[2000] Adj LR 04/14 (TCC), Dyson J	**Set-off in respect of claims which had not been raised at adjudication** If an employer has a right to claim liquidated and ascertained damages, then it should raise those damages as a set-off in the adjudication proceedings and not later in enforcement proceedings.
KNS Industrial Services (Birmingham) Ltd v Sindall Ltd	75 Con LR 71, Judge LLoyd	**Set-off against an adjudicator's decision after termination** *Obiter*: other rights under the contract that were not the subject of the adjudication decision remain available to the relevant party. If, therefore, by the time an adjudicator makes a decision requiring payment by a party to the contract, the contract has been lawfully terminated by that party (or that party has real prospects of success in supporting that termination) or some other event has occurred which under the contract entitles a party not to pay, then the amount required to be paid by the decision does not have to be paid. (cf. *Ferson Contractors Ltd v Levolux AT Ltd* [2003] BLR 118, [2003] EWCA Civ 11 (CA).)
David McLean Housing Ltd v Swansea Housing Association Ltd	[2002] BLR 125 (TCC), Judge LLoyd	**Set-off in relation to other matters considered by an adjudicator** Where the adjudicator had awarded an extension of time which fell short of the actual delay in achieving

Title	Citation	Issue
		practical completion, and where it was possible to calculate the residual liquidated damages payable, the employer would be entitled to set those amounts off against the adjudicator's award (cf. *Balfour Beatty Construction Ltd v Serco Ltd* [2004] EWHC 3336 (TCC) L.
Parsons Plastics (Research and Development) Ltd v Purac Ltd	[2002] BLR 334, [2002] EWHC Civ 459 Pill LJ, Mummery LJ, Latham LJ, (CA)	**Set-off against an adjudicator's decision in a contract to which the 1996 Act did not apply** On the express wording of the contract (to which the 1996 Act did not apply) it was open to the employer to set off against the adjudicator's decision any other claim it had against the contractor and which had not been determined by the adjudicator.
Bovis Lend Lease Ltd v Triangle Ltd	[2003] BLR 31, [2002] EWHC 3123 (TCC), Judge Thornton	**Set-off against an adjudicator's decision after termination** Ordinarily, a decision of an adjudicator will give rise to a contractual entitlement to immediate payment without deduction, set-off, withholding, reliance on a cross-claim, abatement, or stay of execution. However, contractual terms might prevail in circumstances where the contract had been validly determined and the contract provided that no further sums were due. NB found to be wrongly decided in *Ferson Contractors Ltd v Levolux AT Ltd* [2003] BLR 118, [2003] EWCA Civ 11 (CA).
Ferson Contractors Ltd v Levolux AT Ltd	[2002] Adj LR 06/26 (TCC), Judge Wilcox; [2003] 1 All ER (Comm) 385, [2003] EWCA Civ 11 (CA), Ward LJ, Mantell LJ, Longmore LJ	**Set-off against an adjudicator's decision after termination** Notwithstanding the existence of contractual terms that provided that no further sums were due following termination, an adjudicator's award should still be paid without any withholding. The contractual scheme which assumed that no monies would be paid until completion of the works did not apply to an adjudicator's award.
Construction Centre Group Ltd v Highland Council	2003 SLT 623, [2003] Scot CS 114 (Inner House, Court of Session), Lord Osborne, Lord Hamilton, Lord Carloway	**Set-off in respect of claims which had not been raised in the adjudication** Having failed to raise a defence of set-off in respect of delay damages during adjudication proceedings, the employer was not entitled to rely upon them as a reason to resist enforcement. To allow a set-off would be to fail to recognize: (i) the nature of the adjudicator's order as being a resolution, albeit provisional, of a dispute between the parties; and (ii) the nature of the present action as being an enforcement mechanism for that order rather than proceedings concerned with any underlying question of the true and ultimate indebtedness (if any) of the employer to the contractor.
Shimizu Europe Ltd v LBJ Fabrications Ltd	[2003] BLR 381, [2003] EWHC 1229 (TCC), Judge Kirkham	**Set-off against an adjudicator's decision** Since the adjudicator had decided that payment only became due on the service of the subcontractor's VAT invoice and therefore had not been due when the

(Continued)

Table 6.3 *Continued*

Title	Citation	Issue
		adjudication commenced, it was open to the main contractor to serve notice of withholding once the VAT invoice was issued. If the subcontractor had issued its VAT invoice prior to commencing its adjudication, then the sum (as confirmed by the adjudicator) would have been due and payable without set-off.
MJ Gleeson Group plc v Devonshire Green Holdings Ltd	[2004] Adj LR 03/19 (TCC), Judge Gilliland	**Set-off against an adjudicator's decision in respect of issues arising after service of the adjudication notice** The attempt to defeat an adjudicator's decision by serving a withholding notice in respect of events which occurred subsequent to the commencement of the adjudication was entirely inconsistent with the statutory purpose of providing a quick and effective remedy on an interim basis. An adjudicator's decision was meant to be enforced and complied with, without regard to other provisions of the contract which might otherwise have allowed sums to be withheld.
Balfour Beatty Construction Ltd v Serco Ltd	[2004] EWHC 3336 (TCC), Jackson J	**Set-off against an adjudicator's decision** There might be limited circumstances where a party might be able to set off against an adjudicator's decision where either: (i) it follows logically from an adjudicator's decision that the employer is entitled to recover a specific sum by way of liquidated and ascertained damages or (ii) the terms of the contract or the circumstances of the case so allow.
Castle Inns (Sterling) v Clark Contracts	[2005] Adj LR 12/29, (Outer House, Court of Session), Lord Drummond Young	**Set-off against later payment certificates** Where the final certificate was issued after the adjudicator's decision, the architect had no power to undo the decisions of an adjudicator. Accordingly, he had to comply with those decisions so far as they were relevant to any of his tasks, including making his valuation in the final certificate. However, if new material had emerged since the date of the adjudicator's decision, the architect was entitled to take that into account in preparing the final certificate, or indeed any interim certificate, and to make any appropriate modification to the adjudicator's decision. What he could not do was challenge an issue of principle in the adjudicator's decision.
David McLean Contractors Ltd v The Albany Building Ltd	[2005] Adj LR 11/10 (TCC), Judge Gilliland	**Set-off in respect of claims which had not been raised in the adjudication** Even where a party had a valid claim to delay damages which would normally entitle him to a legal or equitable right of set-off, that reasoning and principle did not apply where the parties had agreed to comply with the adjudicator's award.
Geris Handelsgesellschaft GmbH v Les Constructions Industrielles de la Mediterrannee S A	[2005] EWHC 499 (TCC), Judge LLoyd	**Set-off against an adjudicator's decision** The adjudicator clearly expected the employer to have an immediate right of set-off and no direction had been given as to the balance that should be paid. Therefore, the employer had reasonable prospects of success in maintaining that the contractor had no immediate right to payment of the sums that

Title	Citation	Issue
		the adjudicator had agreed were due on the certificates. Accordingly, the decision would not be enforced.
Interserve Industrial Services Ltd v Cleveland Bridge UK Ltd	[2006] All ER (D) 49 (Feb.), [2006] EWHC 741 (TCC), Jackson J	**Set-off under successive adjudicators' decisions** Where the parties to a construction contract engage in successive adjudications, each adjudication should focus upon the parties' current rights and remedies. At the end of each adjudication, absent special circumstances, the losing party must comply with the adjudicator's decision.
Hillview Industrial Developments (UK) Ltd v Botes Building Ltd	[2006] All ER (D) 280 (Jun.), [2006] EWHC 1365 (TCC) Judge Toulmin	**Set-off under successive adjudications** The contractor was not entitled to resist summary judgment of an adjudicator's award of liquidated damages on the grounds that it had itself applied for summary judgment for payment on the final account.
William Verry Ltd v London Borough of Camden	[2006] All ER (D) 292 (Mar), [2006] EWHC 761 (TCC), Ramsey J	**Set-off against later payment certificates** An adjudicator decided sums due under an interim certificate, at a time when that certificate had already been superseded by a final payment certificate. It was not permissible to set one off against the other to determine a net sum due. Camden had to comply with the adjudicator's decision and make payment of the sums awarded. The outcome of future adjudications would determine what sums would eventually be payable and by whom under the true final account.
R J Knapman Ltd v Richards	[2006] 108 Con LR 64, [2006] EWHC 2518 (TCC), Judge Coulson	**Set-off in relation to other matters considered by an adjudicator** If it follows logically from the adjudicator's decision that a sum is due to the defendants by way of liquidated damages then the defendants can set off that sum against the adjudicator's award. In this case there was no such logical consequence and therefore no entitlement to set-off. (*Balfour Beatty v Serco* and *William Verry Limited v The Mayor and Burgesses of the London Borough of Camden* applied.)
Avoncroft Construction Ltd v Sharba Homes (CN) Ltd	119 Con LR 130, [2008] EWHC 933 (TCC), Judge Kirkham	**Set-off in relation to other matters considered by an adjudicator** Despite references by the adjudicator to continuing delays and the employer's remedy of liquidated and ascertained damages, these were not issues which he had determined and no damages flowed from his decision. It was not appropriate to construe decisions of an adjudicator as closely as one might a judgment of the court. (*Balfour Beatty Construction Ltd v Serco Ltd* considered.)
Ledwood Mechanical Engineering Ltd v Whessoe Oil and Gas Ltd	[2008] BLR 198, [2007] EWHC 2743 (TCC), Ramsey J	**Set-off in relation to other matters considered by an adjudicator** Whilst the natural corollary of the adjudicator's decision was that it increased the number of expended hours in the 'pain/gain' calculation, the calculation of that effect was not undisputed or indisputable. Consequently, the employer was not entitled to set off sums calculated to give effect to the arrangement against the adjudicator's award. (*Balfour Beatty Construction Ltd v Serco Ltd* considered.)

(Continued)

Table 6.3 *Continued*

Title	Citation	Issue
YCMS Ltd v Grabiner	[2009] BLR 211, [2009] EWHC 127 (TCC), Akenhead J	**Set-off under successive adjudications** The fact that a third decision had been reached which on its face allowed the defendants a net recovery was not a special circumstance which justified departing from the general rule that valid adjudicators' decisions should be enforced promptly. Things might be different if there were effectively simultaneous adjudications and decisions.
Rok Building Ltd v Celtic Composting Systems Ltd	130 Con LR 74, [2010] EWHC 66 (TCC), Akenhead J	**Set-off against an adjudicator's decision** Celtic declined to pay the sums due on the basis that the decision did not require payment to be made within a set period of time and that the sums awarded could be taken into account in subsequent certificates. The adjudicator's decision was enforced as it was clear that the adjudicator was directing Celtic to pay Rok, and simply because he had not specified a period in which payment should be made did not suggest that payment should not be made. In fact it would require clear wording to suggest that the matter should be dealt with in later certificates.
Straw Realisations (No. 1) Ltd v Shaftsbury House (Developments) Ltd	[2011] BLR 47, [2010] EWHC 2597 (TCC), Edwards-Stuart J	**Express clause for no further payment and setting off of accounts on insolvency** The contract included provisions that in the event of the contractor's insolvency the contractual provisions that required further payment shall cease to apply and there was to be a mutual setting off of accounts. It was held that such a clause cannot prevail over an obligation to comply with the decision of an adjudicator.
R&C Electrical Engineers Ltd v Shaylor Construction Ltd	[2012] BLR 373, [2012] EWHC 1254 (TCC), Edwards-Stuart J	**Set-off against an adjudicator's decision** Having determined that payment of an adjudicator's award relating to a sub-contract would only become due once a certificate under the main contract was issued, it was open to the main contractor to serve notice of withholding against the adjudicator's award in respect of delay damages. The contractual right to set off sums against an adjudicator's award was preserved by the fact that the payment had not yet become due. (*Shimizu Europe Ltd v LBJ Fabrications Ltd* applied.)
Beck Interiors Ltd v Classic Decorative Finishing Ltd	[2012] EWHC 1956 (TCC), Coulson J	**Cross-contract set-off** Leaving aside the usual rule that an adjudicator's decision must be enforced, there was no contractual or equitable right of set-off. Equity could not possibly require the court to take account of a claim on a separate contract when addressing the right to enforce an adjudicator's decision under the contract in question.
Squibb Group Ltd v Vertase FLI Ltd	[2012] BLR 408, [2012] EWHC 1958 (TCC), Coulson J	**Set-off against an adjudicator's decision** Neither of the exceptions cited in *Balfour Beatty v Serco* applied. The adjudicator had clearly decided that an amount was to be paid within an extended period of

Title	Citation	Issue
		14 days, but that did not bring the payment within the normal payment cycle so as to allow a withholding notice to be served. The contractual rights of set-off did not apply in the case of an adjudicator's award. Nor could the adjudicator's decision be construed so that liquidated damages in respect of any other period of delay flowed from the decision.

(4) Double Jeopardy

6.63 In summary, the principle of double jeopardy applies to adjudications in the following ways:

1. An adjudicator's decision is binding until finally determined: s. 108(3).
2. Successive adjudication of 'substantially the same' dispute as determined by a previous adjudicator is prohibited.
3. Whether a dispute is substantially the same as a dispute previously determined by an adjudicator is a question of fact and degree in every case and may turn on whether the supporting documentation for the issue is the same in the subsequent adjudication.
4. The eight principles set out by Ramsey J in *HG Construction Ltd v Ashwell Homes (East Anglia) Ltd* (2007) (at 6.66) should be considered for all claims of double jeopardy.
5. The four principles set out by Jackson J in *Quietfield v Vascroft* (2006) (Key Case) (at 6.70) should be considered where double jeopardy is raised in the context of extensions of time.

Introduction

6.64 As discussed above, s. 108(3) requires a construction contract to provide that an adjudicator's decision is binding until the dispute is finally determined. Therefore, if an adjudicator makes a decision in relation to a dispute, that decision is binding on the parties, and the subject matter of the decision from that dispute cannot be redetermined at a subsequent adjudication. The Court of Appeal in *Quietfield Ltd v Vascroft Construction Ltd* (2006) (Key Case)[74] confirmed that a party should not be twice vexed in the same matter, and that the common law concept of *res judicata* (more often referred to in the context of adjudications as 'double jeopardy' or occasionally '*res adjudicata*') applies to adjudications as equally as it does to litigation. This is reinforced by paragraph 9 of the Scheme, which provides that an adjudicator must resign if the dispute the adjudicator is asked to determine has already been the subject of an adjudicator's decision.[75]

6.65 The principle of double jeopardy applies where a determination has already been made on substantially the same dispute. In such circumstances, there is no dispute or difference in existence to refer to the next adjudication.[76] There are two (perhaps obvious) circumstances in which double jeopardy will not arise:

---

[74] [2007] BLR 67(CA), [2006] EWCA Civ 1737 (CA) (Key Case).
[75] See Ch. 4 at 4.44 for discussion of para. 9 of the Scheme.
[76] See e.g. *Watkin Jones & Son Ltd v Lidl UK GMBH*, unreported, 27 Dec. 2001 (TCC).

1. if an adjudication has been commenced but has not resulted in a decision[77]
2. if an adjudicator's decision is a nullity, no dispute will have been determined and the same dispute may be referred again.[78]

Is the Dispute 'Substantially the Same'?

6.66 In *HG Construction Ltd v Ashwell Homes (East Anglia) Ltd* (2007)[79] Ramsey J summarized the applicable considerations for double jeopardy, whether arising under the Scheme or under a JCT Standard Form of Building Contract (with Contractor's Design) 1998 at paragraph 38:

> 38. ... (1) the parties are bound by the decision of an adjudicator on a dispute or difference until it is finally determined by court or adjudication proceedings or by an agreement made subsequently by the parties.
>
> (2) The parties cannot seek a further decision by an adjudicator on a dispute or difference if that dispute or difference has already been the subject of a decision by an Adjudicator.
>
> (3) As a matter of practice, an adjudicator should consider (based either on an objection raised by one of the parties or on his own volition) whether he is being asked to decide a matter on which there is already a binding decision by another Adjudicator. If so he should decline to decide that matter or, if that is the only matter which he is asked to decide, he should resign.
>
> (4) The extent to which a decision or a dispute is binding will depend on an analysis of
>
> (a) the terms, scope and extent of the dispute or difference referred to adjudication and
>
> (b) the terms, scope and extent of the decision made by the adjudicator.
>
> (5) In considering the terms, scope and extent of the dispute or difference the approach has to be to ask whether the dispute or difference is the same or substantially the same as the relevant dispute or difference.
>
> (6) In considering the terms, scope and extent of the decision, the approach has to be to ask whether the Adjudicator has decided a dispute or difference which is the same or fundamentally the same as the relevant dispute or difference.
>
> (7) ... the approach must involve not only the 'same' but also 'substantially the same' dispute or difference. The reason for this, in my judgement, is that disputes or differences encompass a wide range of factual and legal issues. If there had to be complete identity of factual and legal issues then the ability to 're-adjudicate' what was in substance the same dispute or difference would deprive Clause 39A.7.1 of its intended purpose. As Dyson LJ pointed out in *Quietfield* at para 44: 'The cost of a referral can be substantial. No doubt that is one of the reasons why the statutory scheme protects respondents form [*sic*] successive referrals to adjudication of what is substantially the same dispute.' The expense and trouble of successive adjudications on the same or substantially the same dispute or difference in relation to adjudication which provides a temporarily binding decision is something which is to be discouraged and is the purpose behind the provisions of Clause 39A.7.1.

---

[77] *Vision Homes Ltd v Lancsville Construction Ltd* [2009] BLR 525, [2009] EWHC 2042 (TCC) at para. 70, in which Clarke J observed that the Scheme makes no provision for the resignation of an adjudicator in such a situation. The Court of Appeal has also confirmed in *Lanes Group Plc v Galliford Try Infrastructure Limited* [2012] BLR 121, [2011] EWCA 1617 (CA) that the Scheme and several standard forms recognize the right to restart an abortive adjudication (see para. 38).

[78] *Amec Capital Projects Ltd v Whitefriars City Estates Ltd* [2005] 1 All ER 723, [2004] EWCA Civ 1418 (CA); see also *Jacques (t/a C&E Jacques Partnership) v Ensign Contractors Ltd* [2009] EWHC 3383 (TCC) in which the parties agreed an earlier adjudicator's decision was null and void.

[79] [2007] BLR 175, [2007] EWHC 144 (TCC). The clause in question was 39A.7.1 which provided: 'The decision of the Adjudicator shall be binding on the Parties until the dispute or difference is finally determined by arbitration or by legal proceedings or by an agreement in writing between the parties made after the decision of the Adjudicator has been given.'

(8) Whether one dispute is substantially the same as another dispute is a question of fact and degree: see para 46[80] of *Quietfield* per Dyson LJ.

6.67 Where an allegation is made during an adjudication that the dispute referred or an issue within that dispute is substantially the same as a dispute or issue previously determined by an adjudicator, the adjudicator needs to tread carefully. If this allegation is correct, the adjudicator will not have jurisdiction to hear the referral or determine the particular issue, as there is no dispute to determine. Where there is only a sub-issue that is alleged to have been determined in a previous adjudication, if the allegation is incorrect but the adjudicator considers the issue has been previously determined and therefore does not consider the submissions on the point there is likely to be a breach of natural justice.[81]

6.68 Whether a dispute or difference is 'substantially the same' is a question of fact and degree in every case. The challenge will be assessed on whether the subsequent adjudicator was invited to, or in fact did, trespass upon the jurisdiction of the first adjudicator.[82]

6.69 There have been several decisions, particularly since the decision in *Quietfield Ltd v Vascroft Construction Ltd* (2006) (Key Case),[83] in which a subsequent adjudicator's decision was found to be unenforceable because the dispute was substantially the same as a dispute previously determined. Table 6.4 identifies (with shaded entries) eight examples of successful claims that the dispute was substantially the same as one previously determined. A key distinction between the unsuccessful and successful claims of double jeopardy has been identified by Coulson J:

> 22. . . . If there is any sort of common thread in those [unsuccessful] cases it is that the material relied on in the second adjudication was, as a matter of fact, very different to that relied on in the first, giving rise to what was described as 'a separate and distinct factual enquiry' second time round.[84]

Key Case: Double Jeopardy

Quietfield Ltd v Vascroft Contractors Ltd (2006) 109 Con LR 29, [2006] EWHC 174 (TCC); upheld by the Court of Appeal [2007] BLR 67, [2006] EWCA Civ 1737

6.70 **Facts:** There had been a first adjudication (under the Scheme) in which the adjudicator found that the contractor (Vascroft) had failed to demonstrate its entitlement to an extension of time for part of the period between planned and actual completion under the JCT 1998 standard form. A second adjudication started by the employer (Quietfield) claimed LADs of £588,000, and relied upon the adjudicator's first decision that the contractor was not entitled to any extension of time (EOT). The contractor's response included a detailed and lengthy total EOT claim which set out numerous causes of delay, traced the dominant critical path, compared the as-planned and as-built programmes, and traced the interrelationships between different activities on site and various causes of delay. This was new information. The adjudicator declined to consider the contractor's submissions on the grounds that this matter had been determined in the first adjudication, and found for the employer for the full sum claimed. The contractor declined to pay the sums awarded on

[80] This should perhaps be a reference to para. 47.

[81] *Quietfield Ltd v Vascroft Contractors Ltd* (2006) 109 Con LR 29, [2006] EWHC 174 (TCC) (Key Case); upheld by the Court of Appeal [2007] BLR 67, [2006] EWCA Civ 1737.

[82] *Emcor Drake & Skull Ltd v Costain Construction Ltd* (2004) 97 Con LR 142, [2004] EWHC 2439 (TCC).

[83] [2007] BLR 67, [2006] EWCA Civ 1737 (CA).

[84] *Benfield Construction Ltd v Trudson (Hatton) Ltd* [2008] CILL 2633, [2008] EWHC 2333 (TCC).

the ground that the adjudicator's decision was unlawful, and the employer commenced an application for summary judgment for enforcement of the decision.

6.71 **Held:** Jackson J found the effect of s.108 of the 1996 Act to be that:

> 34. . . . once a dispute has been determined by adjudication, there cannot be another adjudication about that same dispute. The adjudicator's decision will remain binding upon the parties unless and until that decision is overtaken by a judgment of the court, an arbitral award or a settlement agreement between the parties.

The judge held that the contractor's alleged entitlement to an EOT in the final adjudication was 'substantially different from the claims for extension of time which were advanced, considered and rejected in the first adjudication'; therefore the adjudicator ought to have considered the contractor's substantive defence. As the adjudicator failed to do so, the decision could not be enforced because he failed to abide by the rules of natural justice.[85] Jackson J distilled four principles applicable to consideration of subsequent EOTs, later endorsed by the Court of Appeal:

> 42. . . . (i) Where the contract permits the contractor to make successive applications for extension of time on different grounds, either party, if dissatisfied with the decisions made, can refer those matters to successive adjudications. In each case the difference between the contentions of the aggrieved party and the decision of the architect or contract administrator will constitute the 'dispute' within the meaning of section 108 of the 1996 Act.
>
> (ii) If the contractor makes successive applications for extension of time on the same grounds, the architect or contract administrator will, no doubt, reiterate his original decision. The aggrieved party cannot refer this matter to successive adjudications. He is debarred from doing so by paragraphs 9 and 23 of the Scheme and section 108(3) of the 1996 Act.
>
> (iii) Subject to paragraph (iv) below, where the contractor is resisting a claim for liquidated and ascertained damages in respect of delay, pursued in adjudication proceedings, the contractor may rely by way of defence upon his entitlement to an extension of time.
>
> (iv) However, the contractor cannot rely by way of defence in adjudication proceedings upon an alleged entitlement to extension of time which has been considered and rejected in a previous adjudication.

6.72 Per Dyson LJ in the judgment of the Court of Appeal:

> 42. In my judgment, therefore, the contractor must present some new material which could reasonably lead the architect to reach a different conclusion from that on which he based his earlier decision or decisions. The judge did not explain what he meant by 'different grounds' in his first principle. I can see no reason to construe clause 25 so as to prohibit the contractor from relying on the same Relevant Event as he relied on in support of a previous application for extension of time, giving materially different particulars of the expected effects and/or a different estimate of the extent of the expected delay to the completion of the Works. If the position were otherwise, the contractor could not make good shortcomings of one application by a later application, and would be obliged to refer the matter to arbitration. That cannot have been intended by the contract. There is nothing in the express language which prevents the contractor from making good the deficiencies of an earlier application in a later application.
>
> 43. So much for the position under clause 25. The judge's first principle may appear to suggest that every dispute arising from the rejection of an application for an extension of time may be referred to adjudication. I do not consider that that is necessarily the case. The question whether a contractor may make successive applications for extensions of time depends on the true construction of clause 25 and any term necessarily to be implied.

[85] This case is also discussed in Ch. 9 (Jurisdictional challenges) and Ch. 10 (Natural Justice).

The question whether disputes arising from the rejection of successive applications for an extension of time may be referred to adjudication depends on the effect of section 108(3) of the 1996 Act and paragraph 9(2) of the Scheme.

44. There are obvious differences between successive applications for extensions of time under the contract and successive referrals of disputes to adjudication. In the real world, there is often a regular dialogue between contractor and architect in relation to issues arising from clause 25. If an architect rejects an application for an extension of time pointing out a deficiency in the application which the contractor subsequently makes good, it would be absurd if the architect could not grant the application if he now thought that it was justified. To do so would be part of the architect's ordinary function of administering the contract. But referrals to adjudication raise different considerations. The cost of a referral can be substantial. No doubt that is one of the reasons why the statutory scheme protects respondents from successive referrals to adjudication of what is substantially the same dispute.

45. Paragraph 9(2) provides that an adjudicator must resign where the dispute is the same or substantially the same as one which has previously been referred to adjudication and a decision has been taken in that adjudication. It must necessarily follow that the parties may not refer a dispute to adjudication in such circumstances.

46. This is the mechanism that has been adopted to protect respondents from having to face the expense and trouble of successive adjudications on the same or substantially the same dispute. There is an analogy here, albeit an imperfect one, with the rules developed by the common law to prevent successive litigation over the same matter...

47. Whether dispute A is substantially the same as dispute B is a question of fact and degree. If the contractor identifies the same Relevant Event in successive applications for extensions of time, but gives different particulars of its expected effects, the differences may or may not be sufficient to lead to the conclusion that the two disputes are not substantially the same. All the more so if the particulars of expected effects are the same, but the evidence by which the contractor seeks to prove them is different.

48. Where the only difference between disputes arising from the rejection of two successive applications for an extension of time is that the later application makes good shortcomings of the earlier application, an adjudicator will usually have little difficulty in deciding that the two disputes are substantially the same.

49. In the present case, I am in no doubt that the judge reached the right conclusion. The first disputed claim which was the subject of the first adjudication was substantially different from the second disputed claim. The written notices which formed the basis of the second claim identified Relevant Events which were substantially more extensive than those which formed the basis of the first claim. The particulars of expected effects were very different too. There will be some borderline cases where it is a matter of judgment whether the two claims are substantially the same and where there may be room for more than one view. In my view, this is not a borderline case.

Table 6.4 Table of Cases: Double Jeopardy (shaded entries indicate the dispute was substantially the same as a previous dispute)

Title	Citation	Issue
Sherwood Casson Ltd v MacKenzie	(2000) 2 TCLR 418 (TCC), Judge Thornton	**Dispute not substantially the same: interim and final accounts** A dispute concerning the settlement of a final account (which involved some revaluation of work done) was not the same dispute as an earlier dispute arising out of claims for interim payments.

(Continued)

Table 6.4 *Continued*

Title	Citation	Issue
Mivan Ltd v Lighting Technology Projects Ltd	[2001] Adj CS 04/09 (TCC), Judge Seymour	**Dispute not substantially the same: withholding notice** The first adjudication decided that, in the absence of a valid withholding notice, the contractor was entitled to be paid a sum. The sum was paid and a second adjudication determined that the contractor had to repay the sum. The second adjudication dealt with a different issue (repayment of monies) from the first (efficacy of withholding notice).
Watkin Jones & Son Ltd v Lidl UK GMBH	27 Dec. 2001 (TCC), Judge LLoyd (unreported)	**Dispute substantially the same: contractor's entitlement to payment in absence of employer payment notices** The first adjudication determined that the contractor's payment application was subject to the notice provisions (under a JCT with Contractor's Design 1998). As the employer had failed to issue those notices, the full amount sought by the contractor was due. The employer then sought determination of the properly calculated amount for which the contractor should have applied. A second adjudicator declined to act. A third adjudicator accepted the nomination, but on a CPR Part 8 application, the court held that the contractor's entitlement had been determined in the first adjudication and the third adjudicator had no jurisdiction.
Naylor (t/a Powerfloated Concrete Floors) v Greenacres Curling Ltd	[2001] Scot CS 163, (Outer House, Court of Session), Lord Bonomy	**Dispute substantially the same: compliance with specification** The first adjudicator considered whether the contractor had fulfilled his contract and was entitled to payment or whether payment should not be ordered because the work failed to comply with the specification for the slab. The fact that, in the second adjudication, the employer sought damages for breach of contract or alternatively rectification of the slab at the expense of the contractor did not alter the fact that the dispute was substantially the same.
Holt Insulation Ltd v Colt International Ltd	[2001] EWHC 451 (TCC), Judge Mackay	**Dispute not substantially the same: or such other sum as the adjudicator sees fit to award** A first adjudication concerned the contractor's entitlement to an interim payment of £110,587.56. The claim had been unsuccessful. A second adjudication was commenced, giving the adjudicator jurisdiction to decide a sum different to the sum claimed in the first adjudication. The notices were crucially different and did not relate to the same dispute because the first adjudication was limited to deciding entitlement to exactly the sum sought or no sum at all.

Title	Citation	Issue
Skanska Construction UK Ltd v The ERDC Group Ltd	[2002] Scot CS 307, (Outer House, Court of Session), Lady Paton	**Dispute not substantially the same: interim and final payment of direct loss and expense** The first adjudication involved the unsuccessful claim for an interim payment of direct loss and expense. Considerable accumulation and exchange of information between the parties followed, and a second adjudication commenced on entitlement to final payment. The dispute was not substantially the same. A different stage in the contract had been reached; different contractual provisions applied; considerably more information may be available by the date of issue of the final account; and different considerations and perspectives may apply.
David McLean Housing Contractors Limited v Swansea Housing Association Limited	[2002] BLR 125 (TCC), Judge LLoyd	**Dispute not substantially the same: notices and LADs** In the first adjudication, the point decided by the adjudicator was the narrow issue of the validity of the withholding notice. In the second adjudication he had dealt with the issue of whether the employer had been entitled to deduct liquidated and ascertained damages at all, having regard to the notices. That was a different issue and therefore there was no duplication.
Emcor Drake & Skull Ltd v Costain Construction Ltd	97 Con LR 142, [2004] EWHC 2439 (TCC), Judge Havery	**Dispute not substantially the same: different claimed critical path** The adjudicator in the first adjudication rejected the contractor's entitlement to an EOT based on the critical path running through the construction of the bedrooms. A second adjudicator considered he had jurisdiction to determine the wider dispute concerning the contractor's entitlement to an EOT for a critical path that was not limited to the issue of the bedrooms. The court agreed with the second adjudicator and enforced his decision.
Amec Capital Projects Ltd v Whitefriars City Estates Ltd	[2005] 1 All ER 723, [2004] EWCA Civ 1418 (CA), Kennedy LJ, Chadwick LJ, Dyson LJ	**First decision a nullity** Clause 9(2) of the Scheme prevents contradictory decisions by two separate adjudicators, but it does not apply where the first adjudicator's decision has been held to be a nullity.
Quietfield Ltd v Vascroft Construction Ltd	[2006] EWHC 174 (TCC), Jackson J; [2007] BLR 67, [2006] EWCA Civ 1737 (May, Dyson, Smith LJJ)	**Not substantially the same dispute: extension of time** In the first adjudication, the claimant relied on matters set out in two letters to support a claim for EOT. In a later adjudication, the contractor was entitled to rely on materials 'substantially different from the claims for extension of time which were advanced, considered and rejected in the first adjudication'. Jackson J's decision at first instance was upheld by the Court of Appeal.

(Continued)

Table 6.4 *Continued*

Title	Citation	Issue
HG Construction Ltd v Ashwell Homes (East Anglia) Ltd	[2007] BLR 175, [2007] EWHC 144 (TCC), Ramsey J	**Substantially the same dispute: LADs** A first adjudication decided that LAD provisions were valid and enforceable. A third adjudication decided there was no basis on which the LAD provisions operated. The third adjudicator's decision was not binding on the parties as it was based on the same or substantially the same dispute as raised by the applicant in the previous adjudication, even where there was a new argument challenging the enforceability of the LADs.
Benfield Construction Ltd v Trudson (Hatton) Ltd	[2008] CILL 2633, [2008] EWHC 2333 (TCC), Coulson J	**Substantially the same dispute: LADs, practical completion, and partial possession** The first and second adjudication decisions dealt with the question of whether practical completion had been achieved and if not, whether LADs were applicable. The third adjudication decided that on partial possession, practical completion had occurred and so LADs should be repaid. The third decision was unenforceable as there were no different material facts presented, or detailed issues considered therein.
Birmingham City Council v Paddison Construction Ltd	[2008] BLR 622, [2008] EWHC 2254 (TCC), Judge Kirkham	**Substantially the same dispute: EOT** The first adjudication decision concerned claims for an EOT, return of LADs and a claim for loss and expense arising out of the EOT. The loss and expense claim was rejected as exaggerated in the first adjudication, but was brought again in the later adjudication. The disputes were substantially the same. Although they were for different amounts, there was no difference in the supporting material, only in the analysis of it. The existence of a claim for damages for breach of contract did not render the dispute in the second adjudication materially different from that in the first.
Barr v Klin Investment UK Ltd	[2009] CILL 2787, [2009] CSOH 104 (Outer House, Court of Session), Lord Glennie	**Dispute not substantially the same: withholding notice, LADs, enforceability, and payment of sum** In the first adjudication, the referring party sought a declaration that the responding party had failed to serve either a payment or withholding notice timeously. The second adjudication concerned whether the LAD provisions were enforceable. The dispute referred in the third adjudication, whether the £375,600 referred to in the withholding notice was payable, was not substantially the same as the disputes determined in the previous adjudications. If unsuccessful in the referral of a point of construction, a party cannot be barred from referring the specific disputed factual basis for payment in question.

Title	Citation	Issue
Vision Homes Ltd v Lancsville Construction Ltd	[2009] BLR 525, [2009] EWHC 2042 (TCC), Clarke J	**Dispute cannot be substantially the same if no decision reached in other adjudication** Two adjudications were commenced and abandoned between the parties. The third and fourth adjudications were both commenced on the same day and both concerned the employer's entitlement to levy LADs. The adjudicator in the third adjudication published his decision before the adjudicator in the fourth and the unsuccessful party sought to argue he was prohibited from so doing because of the fourth adjudication concerning the same dispute. As no decision had been taken in the fourth adjudication, this ground of objection failed. NB this decision was not enforced on different grounds.
Jacques (t/a C&E Jacques Partnership) v Ensign Contractors Ltd	[2009] EWHC 3383 (TCC), Akenhead J	**Dispute not substantially the same where the parties have agreed the earlier decision is null and void** A fourth adjudication between the parties resulted in a decision which the parties agreed was 'null and void' and not binding upon the parties. The parties agreed an order in the TCC to reflect this, and to record that the claimant was entitled to adjudicate the same dispute at any time.
Redwing Construction Ltd v Charles Wishart	[2010] EWHC 3366 (TCC), Akenhead J	**When an issue in adjudication 1 impinges on adjudication 2** Where the dispute is materially different in both adjudications but an issue in the first adjudication impinges on the issues to be decided in the second adjudication, the parties may still be bound by the finding made in the first decision. It was held that it depended on whether the issue was part of the dispute actually referred for decision, but if not then it would not be binding in subsequent adjudications.
Carillion Construction Ltd v Smith	141 Con LR 117, [2011] EWHC 2910 (TCC), Akenhead J	**Substantially the same dispute: delay, disruption, loss, and expense** The court set out a number of factors to be considered in determining whether a dispute that had been referred to adjudication pursuant to an adjudication clause in a contract was the same or substantially the same as one which had been the subject of a previous adjudication between the parties (see 54–7 of judgment). In the instant case the third adjudicator had no jurisdiction to resolve the dispute referred to him, that dispute being the same or substantially the same as that in the second adjudication. Each involved a claim for loss and expense arising out of delays and disruption said to have been caused by breaches of contract and default on Carillon's part. The period up to which delay and disruption ran was the same and the financial heads of claim were

(Continued)

Table 6.4 *Continued*

Title	Citation	Issue
		the same, albeit the figures were different. The fact that the third adjudication claim was expressed to be largely based on an analysis of documentation said to have been produced for the first time during the second adjudication, did not convert it into a different dispute. It was apposite to note that Smith had been in possession of the documents for a substantial time before the conclusion of the second adjudication, and it was clear that Smith had relied on it extensively in that adjudication. It could not therefore be said that he had not had the opportunity to challenge and react to it during the second adjudication. Essentially Smith had, in the second adjudication, exercised the right under the sub-contract to adjudicate the claim for delay and disruption, and he had lost. That decision was binding (paras. 58–9, 67–8).
Jerram Falkus Construction Ltd v Fenice Investments (No. 4)	[2011] BLR 644, [2011] EWHC 1935 (TCC), Coulson J	**Dispute substantially the same: final account conclusive** The contractor (Jerram Falkus) commenced an adjudication claiming an EOT and that the employer (Fenice) had committed an act of prevention by the employer and time was at large. The adjudicator found against the contractor who later commenced proceedings seeking identical declarations from the court. These declarations were refused because they were the same as the matters determined in the adjudication, which had become binding through the conclusivity clause of the contract (discussed further at 6.34-6.35).
Vertase F.L.I Ltd v Squibb Group Ltd	[2012] EWHC 3194 (TCC), Edwards Stuart J	**Dispute substantially the same: change of mind by adjudicator in second adjudication** In a second adjudication the adjudicator appeared to change his mind about a legal issue on which he opined in a prior adjudication. In the first adjudication he had said no LADs were due because no withholding notice had been served, and had also said 'Vertase do not have a right to claim LDs under the subcontract unless they can demonstrate an equivalent loss under the main contract.' The debate concerned whether that view expressed in the first adjudication was a finding on which he based his decision to reject LADs, or whether it was a summary of an assertion by one party that was not ultimately the basis on which the adjudicator reached his decision (and hence would be allowed to express a different view on this in the second adjudication). On close analysis of the decision, the judge decided that it was the adjudicator's own conclusion. The judge rejected the argument that the dispute in the two adjudications was different and that the adjudicator was entitled to reach a different conclusion on this issue (see para. 44). In so doing he relied on Akenhead J in *Redwing*, para. 27(d).

7

ENFORCEMENT

(1) Overview

7.01 This chapter considers the enforcement of an adjudicator's decision by way of court proceedings, usually by following the special procedures developed by the Technology and Construction Court (TCC). It also considers other potential enforcement procedures such as injunctions, declaratory relief, winding-up orders, and final charging orders. It deals with enforcement of part but not all of an adjudicator's decision and reviews the law of severability in this regard; discusses approbation and reprobation, costs and interest, and enforcing the judgment of the court.

7.02 This chapter does not attempt to deal with every case on enforcement, but rather the key cases or cases that illustrate the courts' application of the principles of enforcement. The specific challenges that may be made to resist enforcement of a decision are considered in Chapters 9 to 11.

(2) General Principles of Enforcement

Valid Decisions Should Be Summarily Enforced

7.03 Since the Housing Grants Construction and Regeneration Act 1996 ('the 1996 Act') came into effect in 1998, the High Court and Court of Appeal have repeatedly emphasized the special character of adjudicators' decisions and the policy considerations underpinning s. 108 of the 1996 Act. The adjudication process is recognized as a 'speedy mechanism for settling disputes in construction contracts on a provisional interim basis, and requiring the decisions of adjudicators to be enforced pending the final determination'.[1] Summary judgment should be entered to enforce adjudicators' decisions, and if the decision is wrong, it can be corrected in subsequent court or arbitral proceedings.[2] Adjudicators' decisions are therefore generally summarily enforced without deduction, set-off, withholding, reliance on a cross-claim, abatement, or stay of execution.[3]

7.04 An adjudicator's decision does not involve the final determination of the parties' rights[4] and as a result the courts seek to enforce decisions wherever possible. However, only valid decisions must be enforced. Thus, while some may refer to the 'right of enforcement', this is qualified or contingent upon the validity of the decision itself.[5]

Errors of Fact and/or Law: Answering the Right Question the Wrong Way

7.05 It is now a well-established principle that an adjudicator's decision should be enforced in all but the most exceptional circumstances.[6] In particular, it will be no basis for resisting enforcement of a statutory decision[7] to contend that the adjudicator reached conclusions that were wrong as a question of fact or a matter of law. The adjudicator has jurisdiction to decide the dispute referred. As long as the adjudicator answers the question referred for resolution ('the right question') that is referred to him the decision will be enforced even if the decision is wrong as a matter of fact or law. In *Bouygues v Dahl-Jenson* (2000),[8] the adjudicator had wrongly taken account of a 5 per cent retention figure, which was not yet due for payment. The result of the error was a decision that a repayment of £140,000 was due instead of a payment of £200,000. The Court of Appeal stated that 'unfairness in a specific case cannot be determinative of the true construction or effect of the scheme in general'.[9] In other words the fact that the error resulted in unfairness was not a ground to challenge the validity of the decision.

---

[1] *Macob Civil Engineering Ltd v Morrison Construction Ltd* [1999] BLR 93 (TCC) at para. 14 (Key Case).

[2] *C&B Scene Concept Design v Isobars* [2002] 1 BLR 93, [2002] EWCA 46 (Key Case).

[3] *Bovis Lend Lease Ltd v Triangle Development Ltd* [2003] BLR 31, [2002] EWHC 3123 (TCC). For a discussion of set-off against an adjudication decision see Ch. 6 at 6.36–6.62 and for a discussion of when a stay will be ordered see Ch. 8.

[4] *Carillion Construction Ltd v Devonport Royal Dockyard Ltd* [2006] BLR 15, [2005] EWCA Civ 1358 (Key Case). The parties may agree to treat the adjudication decision as final and binding by a term to this effect in the contract (or after the decision is issued) (see paras. 6.16–6.23).

[5] *Alstom Signalling Ltd v Jarvis Facilities Ltd* [2004] EWHC 1285 (TCC) (Key Case).

[6] See Table 7.1 and the limited successful challenges to enforcement which are shaded.

[7] It is important to note that the doctrine of unreviewable error does not attach to non-statutory or ad hoc adjudication decisions: *Steve Domsalla (t/a Domsalla Building Services) v Kenneth Dyason* (2008) BLR 348, [2007] EWHC 1174 (TCC).

[8] [2000] All ER (D) 1132, 73 Con LR 135 (CA). See also *C&B Scene Concept Design Ltd v Isobars Ltd* [2002] BLR 93 [2002] EWCA 46 (Key Case).

[9] Per Buxton LJ in *Bouygues (UK) Ltd v Dahl-Jensen (UK) Ltd* [2000] All ER (D) 1132, 73 Con LR 135 (CA).

7.06 The short time frame for adjudication means that 'mistakes will inevitably occur' but unless the mistake can be characterized as an excess of jurisdiction that would make the decision unenforceable it should be enforced.[10] This is particularly so because the decision can ultimately be corrected during final determination of the dispute by court proceedings or arbitration.[11] These principles have been confirmed in numerous Court of Appeal decisions since 2001.[12]

7.07 The exception to this principle (discussed in detail in Chapter 9) is if an error goes to the adjudicator's jurisdiction—that is, where the adjudicator is either not empowered to consider the dispute at all or the adjudicator answers a question that is different from that referred. However, the courts will be slow to conclude that an error of fact or law is one that goes to jurisdiction and will examine any such error critically and with a degree of scepticism before they will accept that there has been an error of jurisdiction (or indeed a breach of the rules of natural justice).[13] It is not for the court to examine the correctness of a decision which does not go to jurisdiction, even if it is glaring or serious, but rather to enforce the decision whether it was right or wrong.[14]

Challenges Based on Excess Jurisdiction or Breach of Natural Justice

7.08 There are two main bases for challenging the enforceability of an adjudicator's decision: excess of jurisdiction or a breach of natural justice (including bias). Whilst adjudicators' decisions should generally be enforced, if the adjudicator was not properly vested with jurisdiction or exceeded it, the decision will not be enforced. Similarly, if the adjudicator materially breached the rules of natural justice, or there was a reasonable apprehension of bias, actual bias or pre-determination. the decision will not be enforceable. These issues are addressed in Chapters 9 to 11, which identify the attempts—both successful and unsuccessful—that parties have made to avoid being bound by adjudicators' decisions.

Fraud

7.09 It may also be possible to resist enforcement of an adjudicator's decision on the grounds of fraud, but only in limited circumstances. The principles were addressed by the Court of Appeal in *Speymill Contracts Ltd v Baskind* (2010),[15] which confirmed the approach taken in two previous TCC decisions.[16] In *Speymill* it was held that if an allegation of fraud was or could have been raised before the adjudicator then it would be considered to have been adjudicated upon with the result that it could not be raised to resist enforcement. However, where

---

[10] Rule 4 from *Sherwood & Casson v MacKenzie* (2000) CILL 1577, as endorsed by the Court of Appeal in *C&B Scene Concept Design v Isobars* [2002] 1 BLR 93, [2002] EWCA 46.

[11] *Shimizu Europe Ltd v Automajor Ltd* [2002] BLR 113 (TCC) at para. 23, per Judge Seymour.

[12] *Bouygues (UK) Ltd v Dahl-Jensen (UK) Ltd* [2001] All ER Comm 1041, [2000] BLR 522; *C&B Scene Concept Design Ltd v Isobars Ltd* [2002] BLR 93 [2002] EWCA 46; *Levolux AT Ltd v Ferson Contractors Ltd* [2003] EWCA Civ 11; *Pegram Shopfitters Ltd v Tally Weijl (UK) Ltd* [2003] EWCA 1750, [2004] 1 All ER 818; *Amec Capital Projects Ltd v Whitefriars City Estates Ltd* [2004] EWCA 1418, [2005] BLR 1; *Carillion Construction Ltd v Devonport Royal Dockyard Ltd* [2006] BLR 15 (CA), [2005] EWCA 1358.

[13] *Pegram Shopfitters Ltd v Tally Weijl (UK) Ltd* [2003] EWCA 1750; *Amec Capital Projects Ltd v Whitefriars City Estates Ltd* [2004] EWCA 1418, [2005] BLR 1; *Carillion Construction Ltd v Devonport Royal Dockyard Ltd* (2005) BLR 310, [2005] EWHC 778 (TCC), [2006] BLR 15 (CA), [2005] EWCA 1358 at para. 52.

[14] *Tim Butler Contractors Ltd v Merewood Homes Ltd*, 12 Apr. [2002] 18 Const LJ 74; *Rok Building Ltd v Celtic Composting Systems Ltd (No. 2)* (2010) 130 Con LR 74, [2010] EWHC 66 (TCC).

[15] [2010] BLR 257, [2010] EWCA Civ 120 (CA).

[16] *S. G. South Ltd v King's Head Cirencester LLP* [2010] BLR 47, [2009] EWHC 2645 (TCC); *GPS Marine Contractors Ltd v Ringway Infrastructure Services Ltd* [2010] BLR 377, [2010] EWHC 283 (TCC) (Key Case).

the matters giving rise to the allegation of fraud had only emerged after the adjudication, then it may be possible to rely on them to resist enforcement, but only if the fraud directly impacts on the matters decided by the adjudicator (for example a certificate upon which the decision was based is later found to have been obtained by fraud).

Reservation of Position

7.10 In order to maintain a challenge to the enforceability of an adjudicator's decision, a party should ensure that it adequately reserves its position at all stages during the course of the adjudication, or it may otherwise be deemed to have waived its right to object. Reservation of position is more fully discussed at 5.09–5.16.

Enforcement of Non-pecuniary Decisions

7.11 Where a party disagrees with an adjudicator's decision and does not comply with it (usually by making the required payment within the timescale provided by the adjudicator) the successful party may apply to the courts to enforce the adjudicator's decision, usually by way of summary judgment for payment of the sum in question. However, some decisions do not concern payment of an amount but award some other type of relief, for example a declaration or (where the adjudicator is so empowered) an order for specific performance. An adjudicator may give a declaratory decision about any issue under a contract, for example a declaration as to an entitlement to extension of time,[17] or a declaration that determination for cause was valid.[18] Section 108(3) and, if applicable, paragraph 23(2) of the Scheme for Construction Contracts ('the Scheme') require parties to comply with the decisions of adjudicators and the courts have confirmed that compliance is necessary even where those decisions are declaratory.[19]

7.12 In *Macob Civil Engineering Ltd v Morrison Construction Ltd* [1999] (Key Case)[20] Dyson J said that the appropriate method of enforcement of declarations may be by way of declaration or mandatory injunction. The judge was specifically contemplating adjudicators' decisions concerning the parties' performance or non-performance of the contract such as returning to site, providing access or inspection facilities, opening up work, or carrying out specific work.[21] In *Multiplex Constructions (UK) Ltd v Mott MacDonald Ltd* (2007) (Key Case),[22] the claimant sought a declaration, an order for specific performance, and an injunction for the enforcement of the adjudicator's decision that the defendant provide 'all pertinent records'. The court found that it would be appropriate to enforce that decision by granting a declaration that the adjudicator's decision was enforceable and the parties should comply with it. However the court rejected the applications for an order for specific performance and an injunction stating that neither was an appropriate route to enforcement in this case.

Enforcement where there is an Arbitration Agreement

7.13 An adjudicator's decision can be enforced by the courts even where there is an arbitration agreement in the construction contract[23] as discussed more fully at 8.43–8.48. However, any final determination of the dispute between the parties cannot be by the court.

---

[17] *Sindall Ltd v (1) Abner Solland (2) Grazyna Solland (3) Solland Interiors (a firm) (4) Solland Interiors Ltd* [2001] All ER (D) 370 (TCC).

[18] *Banner Holdings Ltd v Colchester Borough Council* [2010] All ER (D) 226, [2010] EWHC 139 (TCC).

[19] *HS Works Ltd v Enterprise Managed Services Ltd* [2009] BLR 378; [2009], EWHC 729 (TCC) at para. 61 in particular.

[20] *Macob Civil Engineering Ltd v Morrison Construction Ltd* [1999] All ER (D) 143 (TCC) (Key Case).

[21] *Macob Civil Engineering Ltd v Morrison Construction Ltd* [1999] All ER (D) 143.

[22] [2007] All ER (D) 133, [2007] EWHC 20 (TCC) (Key Case).

[23] *Clarke Quarries Ltd v PT McWilliams Ltd* [2009] IEHC 403, [2010] BLR 520 (High Court of Ireland). See also *Macob Civil Engineering Ltd v Morrison Construction Ltd* [1999] BLR 93 (TCC) (Key Case).

Staying Enforcement

7.14 Whilst the general rule is that adjudicators' decisions should be enforced without a stay of execution,[24] there are limited grounds for requesting that the court exercises its discretion to stay the execution of enforcement of a decision, most frequently based on the impecuniosity of the successful party. Stay of execution of enforcement orders and stay of enforcement proceedings are discussed in Chapter 8.

Key Cases: General Principles of Enforcement

C&B Scene Concept Design v Isobars [2002] 1 BLR 94 [2002] EWCA 46

7.15 **Facts:** The adjudicator had been asked to determine whether the contractor was entitled to payment under a JCT 1998 with Contractor's Design standard form. The parties had not elected Alternative A or B and therefore the Scheme applied by virtue of s. 109 of the Act. The adjudicator allowed clause 30.3.5 to operate and decided that the full amount applied for by the contractor was payable. The contractor sought summary judgment to enforce the decision, but the employer argued that the Scheme replaced the whole of clause 30, and that in failing to appreciate that, the adjudicator had addressed himself to the wrong question and in so doing had exceeded his jurisdiction. The TCC declined to enforce the adjudicator's decision and the contractor appealed to the Court of Appeal.

7.16 **Held:** The Court of Appeal allowed the appeal, finding that the decision of the adjudicator should be enforced regardless of any errors of procedure, fact, or law he made. The adjudicator had asked himself the correct question: what was the contractor's entitlement to payment for application six? Any error he made in answering that question, by applying the contract instead of the Scheme, or by finding that the contractor was entitled to be paid £1,500 for claims which did not in fact arise under the contract, were errors of law which did not affect the enforceability of that decision:

> 22. The real question is whether this error on the part of the Adjudicator went to his jurisdiction, or was merely an erroneous decision of law on a matter within his jurisdiction. If it was the former the Recorder was right to hold that summary judgment should not be entered. If it was the latter, then in my judgment the proper course, subject to any question of stay of execution, is that the Claimant is entitled to summary judgment.
>
> 23. The whole purpose of s. 108 of the Act, which imports into construction contracts the right to refer disputes to adjudication, is that it provides a swift and effective means of resolution of disputes which is binding during the currency of the contract and until final determination by litigation or arbitration: s. 108(3). The provisions of s. 109–111 are designed to enable the contractor to obtain payment of interim payments. Any dispute can be quickly resolved by the Adjudicator and enforced through the courts. If he is wrong, the matter can be corrected in subsequent litigation or arbitration...
>
> 26. Errors of procedure, fact or law are not sufficient to prevent enforcement of an adjudicator's decision by summary judgment...
>
> 29. But the Adjudicator's jurisdiction is determined by and derives from the dispute that is referred to him. If he determines matters over and beyond the dispute, he has no jurisdiction. But the scope of the dispute was agreed, namely as to the Employer's obligation to make payment and the Contractor's entitlement to receive payment following receipt by the Employer of the Contractor's Applications for interim payment Nos 4, 5 and 6 (see paragraph 12 above). In order to determine this dispute the Adjudicator had to resolve as

[24] *Bovis Lend Lease Ltd v Triangle Development Ltd* [2003] BLR 31; [2002] EWHC 3123 (TCC).

a matter of law whether Clauses 30.3.3–6 applied or not, and if they did, what was the effect of failure to serve a timeous notice by the Employer. Even if he was wrong on both these points that did not affect his jurisdiction.

30. It is important that the enforcement of an adjudicator's decision by summary judgment should not be prevented by arguments that the adjudicator has made errors of law in reaching his decision, unless the adjudicator has purported to decide matters that are not referred to him. He must decide as a matter of construction of the referral, and therefore as a matter of law, what the dispute is that he has to decide. If he erroneously decides that the dispute referred to him is wider than it is, then, in so far as he has exceeded his jurisdiction, his decision cannot be enforced. But in the present case there was entire agreement as to the scope of the dispute, and the Adjudicator's decision, albeit he may have made errors of law as to the relevant contractual provisions, is still binding and enforceable until the matter is corrected in the final determination.

Carillion Construction Ltd v Devonport Royal Dockyard Ltd [2006] BLR 15, [2005] EWCA Civ 1358 (CA)

7.17 **Facts:** Devonport was employed by the Secretary of State for Defence to carry out works to upgrade the facilities at a dockyard. The contractual relationship between Devonport (the owner of the dockyard) and Carillion (the subcontractor to Devonport) was governed by a subcontract and an Alliance Agreement, which contained a target cost and pain/gain share mechanism. Certain disputes arose between the parties and the adjudicator decided that Carillion was entitled to be paid sums including £1,167,436 (excluding VAT) for the adjusted target cost. Devonport challenged the award on the grounds that the adjudicator lacked jurisdiction and/or had breached natural justice in reaching his decision.[25]

7.18 **Held:** At first instance, Jackson J granted Carillion's application for summary judgment enforcing the adjudicator's decision (reported at [2005] BLR 310) stating:

80. ...

1. The adjudication procedure does not involve the final determination of anybody's rights (unless all the parties so wish).
2. The Court of Appeal has repeatedly emphasised that adjudicators' decisions must be enforced, even if they result from errors of procedure, fact or law: see *Bouygues, C&B Scene* and *Levolux;*
3. Where an adjudicator has acted in excess of his jurisdiction or in serious breach of the rules of natural justice, the court will not enforce his decision: see *Discain, Balfour Beatty* and *Pegram Shopfitters.*
4. Judges must be astute to examine technical defences with a degree of scepticism consonant with the policy of the 1996 Act. Errors of law, fact or procedure by an adjudicator must be examined critically before the Court accepts that such errors constitute excess of jurisdiction or serious breaches of the rules of natural justice: see *Pegram Shopfitters* and *Amec.*

...

87. ... The adjudicator's assessment of target cost is highly likely to be revised either upwards or downwards, if and when an arbitrator or this court comes to determine the matters in issue between the parties. The adjudicator's approach to or assessment of target cost may well embody errors of both fact and law. This would be unsurprising in view of

[25] The natural justice element of this case is discussed at 10.10

the statutory constraints under which he was operating and the sheer volume of evidence and intricate submissions which were thrust upon him. Nevertheless, any such errors of law and fact cannot be characterised as excess of jurisdiction.

7.19 The Court of Appeal upheld the decision, finding no reason to disagree with Jackson J's conclusions that in determining the primary sum due to Carillion, this inevitably involved determination of the target cost under the Alliance Agreement.

Table 7.1 Table of Cases: General Principles of Enforcement (shaded entries indicate the decision was not enforced)

Title	Citation	Issue
Project Consultancy Group v The Trustees of the Gray Trust	(1999) BLR 377 (TCC), Dyson J	**Error of fact and/or law: adjudicator not empowered to make the decision** It was unclear that any construction contract was ever concluded. The adjudicator had wrongly proceeded as if he had jurisdiction. This was not an error within his jurisdiction. Not enforced.
Macob Civil Engineering Ltd v Morrison Construction Ltd	(1999) BLR 93 (TCC), Dyson J	**Error of fact and/or law: enforceable decision** Parliament intended that decisions of adjudicators should be enforced pending the final determination of disputes. Enforced.
Tim Butler Contractors Ltd v Merewood Homes Ltd	(2002) 18 Const LJ 74, Judge Gilliland	**Error of law: existence of a contract entitling stage payments** An adjudicator found that there was no evidence as to duration of the works and therefore no assumption that they would be less than 45 days. Accordingly, the contractor was entitled to stage payments under s. 109 of the 1996 Act. The court did not express a view as to whether this was right or wrong, as it was a decision which went to the terms of the contract rather than the adjudicator's jurisdiction. Enforced.
Sherwood & Casson v MacKenzie	(2000) CILL 15711 (TCC), Judge Thornton	**Error of law: alleged mistake by adjudicator in relation to substantially same dispute** The two disputes (interim and final) were 'clearly different' and the adjudicator's decision to that effect 'clearly correct'. The overall decision was one taken within his jurisdiction. Enforced.
Pro-Design Ltd v New Millennium Experience Company Ltd	unreported, 26 Sep. 2001, TCC	**Fraud: claimant company was a fraudulent vehicle** During enforcement proceedings, the defendant main contractor alleged the claimant lighting subcontractor was a fraudulent vehicle, being a company owned and operated by an employee of the defendant. The fraud issue arose after the conclusion of the adjudication. Not enforced.
Farebrother Building Services Ltd v Frogmore Investments Ltd	(2001) CILL 1762 (TCC), Judge Gilliland	**Error of law: set-off not included in sum awarded** The sum awarded by the adjudicator showed he had not allowed for a set-off for which the responding party had contended. The adjudicator may have been wrong or he may

(Continued)

Table 7.1 *Continued*

Title	Citation	Issue
		have erred in what he did, but it was an error that was, in principle, within his jurisdiction. Enforced.
Bouygues v Dahl-Jensen)	[2001] 1 All ER (Comm) 1041, [2000] BLR 522 (CA), Gibson, Chadwick and Buxton LJJ	**Error of fact: calculation of contract sum incorrect** The adjudicator's calculation was based on a gross sum that wrongly included retention. The mistake of fact was made within the adjudicator's jurisdiction. If an adjudicator answered the right question in the wrong way, his decision is enforceable. If he answered the wrong question, his decision is a nullity. Provided adjudicators act within their jurisdiction, their awards stand and are enforceable. Enforced, but stayed due to insolvency. Affirmed earlier decision of Dyson J (reported at [2000] BLR 49 (TCC).
Northern Developments (Cumbria) Ltd v J. & J. Nichols	[2000] All ER (D) 68, [2000] EWHC 176 (TCC), Judge Bowsher	**Error of fact: possible mistake or decision that delays caused no loss** Whether the adjudicator had mistakenly forgotten to put a figure in for delays or whether he had decided that the delays caused no loss, the decision was 'beyond question'. Enforced.
SL Timber Systems Ltd v Carillion Construction Ltd	[2000] BLR 516, 2002 SLT 997, (Outer House), Lord Macfadyen	**Error of law: sum due** The adjudicator made an error in relation to the effect of s. 110(2) and s. 111 of the 1996 Act. Failure to serve a s. 111 notice did not preclude argument as to the sum due. The adjudicator answered the right question the wrong way. Enforced.
Nolan Davies Ltd v Catton	[2001] All ER(D) 232 (Mar) (TCC), Judge Wilcox	**Error of law: allegation of no concluded contract** It was open to the adjudicator to decide there had been concluded contracts and, in any event, the parties had agreed that they would be bound by the adjudicator's decision on jurisdiction. Enforced.
A. J. Brenton v Jack Palmer	Claim HT 00/436 2001 (TCC), Judge Havery	**Error as to the identity of contracting party** A decision of an adjudicator as to the identity of the contracting party, decided as part of his substantive decision, could not be questioned in enforcement proceedings. Enforced.
Karl Construction (Scotland) Ltd v Sweeney Civil Engineering (Scotland) Ltd	[2002] SCLR 766[26] (Outer House, SCOS), Lord Caplan	**Error of law: application of payment mechanism** The adjudicator decided that the Scheme applied instead of the subcontract payment mechanism. The unsuccessful party claimed that it had not been given adequate opportunity to address the adjudicator on the adequacy of the subcontract payment mechanism. Any mistake

[26] On appeal from [2001] SCLR 95 (Inner House).

Title	Citation	Issue
		that the adjudicator had made was procedural and was not a venture beyond her jurisdiction. Enforced.
C&B Scene Concept Design v Isobars	[2002] All ER (D) 301, [2002] EWCA 46, Potter, Rix, Sir Murray Stuart- LJJ	**Error of law: application of the contract instead of the Scheme** Where an error goes to an adjudicator's jurisdiction, summary judgment should not be entered. However where it does not, it is merely an erroneous decision of law on a matter within his jurisdiction. Even if wrong in relation to questions of law, these points did not affect his jurisdiction. Enforced.
Shimizu Europe Ltd v Automajor Ltd	[2002] All ER (D) 80 (TCC), Judge Seymour	**Error of fact: mistaken belief that the employer would not challenge a sum** The adjudicator included an amount in his decision for payment to the contractor on the mistaken belief that the employer had agreed during the hearing not to challenge the sum. The adjudicator had asked himself the correct question: was any sum due to the contractor, and if so, what amount? If he got the answer wrong because he misunderstood the submissions, this was an error within his jurisdiction to make. Enforced.
Joinery Plus Ltd v Laing Ltd	[2003] All ER (D) 201, [2003] EWHC 439 (TCC)	**Error of law: answering a different question** Where the adjudicator decided the dispute by reference to entirely the wrong contractual terms, that was an error that went to his jurisdiction. He only had jurisdiction to decide the dispute under the terms of the actual construction contract between the parties. Not enforced.
Pegram Shopfitters Ltd v Tally Wiejl (Uk) Ltd	[2004] 1 All ER 818, Hale, Hooper LJJ; [2003] EWCA 1750, Judge Thornton	**Adjudicator has no jurisdiction to decide his own jurisdiction** An adjudicator has no jurisdiction to decide his own jurisdiction unless there is a submission to his jurisdiction in that regard.
Thomas-Frederic (Construction) Ltd v Keith Wilson	[2003] All ER (D) 341, [2003] EWCA, Brown, Judge, Parker LJJ	**Error as to the identity of contracting party** There was no evidence that the responding party had contracted in person. Not enforced.
William Verry Ltd v North West London Communal Mikvah	[2004] All ER (D) 80, [2004] EWHC 1300 (TCC), Judge Thornton	**Error of law: alleged defects in work not considered** The adjudicator failed to consider the value of alleged defects in the works. However, whilst shutting out some of the important issues, it did not affect the overall result. The decision was valid and fair. Enforced.
London & Amsterdam Properties v Waterman Partnerships	[2003] All ER (D) 391, [2003] EWHC 3059 (TCC), Judge Wilcox	**Errors of fact and/or law: non-compliant referral and failure to consider professional negligence** The responding party claimed that no compliant referral had been served because

(Continued)

Table 7.1 *Continued*

Title	Citation	Issue
		it was more than 20 pages and also that the adjudicator had made an error of law concerning professional negligence. The court confirmed the adjudicator's decision to accept the 17-page referral and 1,000+ pages of supporting documentation as compliant with the adjudication procedure in question. If there was an error of law as to the finding of professional negligence, it was not within the power of the court to interfere with that finding in a summary judgment application. Enforced.
Alstom Signalling Ltd v Jarvis Facilities Ltd	[2004] All ER (D) 02, [2004] EWHC 1232 (TCC), Judge LLoyd	**Error of fact and/or law: withholding notice** The adjudicator was wrong to decide that Alstom had failed to serve a withholding notice. Not enforced.
Allen Wilson Shopfitters and Builders Ltd v Buckingham	[2005] All ER (D) 109, [2005] EWHC 1165 (TCC), Judge Coulson	**Error of law: application of Scheme** Adjudicator applied the Scheme rather than the contract payment mechanism. This decision was within his jurisdiction. Enforced.
Andrew Wallace Ltd v Artisan Regeneration Ltd	[2006] EWHC 15 (TCC), Judge Kirkham	**Fraud: no real prospect of establishing fraud** After the adjudication decision, the defendant alleged fraud against the claimant who had been found by the adjudicator to be due payment of architectural fees. While the court's enforcement of a decision would not constitute assistance in the perpetration of a fraud, in this instance, the defendant had no real prospect of establishing its allegation of fraud on the evidence it had adduced. Enforced.
Carillion Construction Ltd v Devonport Royal Dockyard Ltd	[2006] BLR 15, [2005] EWCA Civ 1358 (CA), Sir Anthony Clarke MR, Chadwick LJ, Moore-Bick LJ	**Errors of fact and/or law: assessment of target cost** The need to have the 'right' answer has been subordinated to the need to have an answer quickly. The adjudicator's assessment of target cost may well have embodied errors of both fact and law, nevertheless such errors cannot be characterized as excess of jurisdiction. Enforced, affirming earlier decision of Jackson J (reported at [2005] BLR 310).
Kier Regional Ltd (t/a Wallis) v City & General (Holborn) Ltd	[2006] EWHC 848 (TCC), [2006] BLR 315, Jackson J	**Error of law: refusal to consider 'new' evidence** The adjudicator refused to consider two reports produced by the responding party after the adjudication commenced because they were 'new' evidence not seen before the valuation in dispute was produced. At worst, the adjudicator made an error of law that caused him to disregard the evidence. However, this error was not one of the plainest cases of breach of natural justice and therefore the decision was enforced. Enforced.

Title	Citation	Issue
McConnell Dowell Constructors (Aust) Pty Ltd v National Grid Gas plc	[2007] BLR 92, [2006] EWHC 2551 (TCC), Jackson J	**Error of law: applicable interest rate** Errors of fact or law by an adjudicator do not affect his decision. Therefore, the arguments of the unsuccessful party that the adjudicator had taken too high an interest rate (under the Late Payment of Commercial Debts (Interest) Act 1998) was irrelevant to an enforcement hearing.
Steve Domsalla (t/a Domsalla Building Services) v Kenneth Dyason	[2007] BLR 348, [2007] EWHC 1174 (TCC), Judge Thornton	**The doctrine of unreviewable error does not apply to non-statutory adjudications** The parties had agreed to an ad hoc adjudication and the adjudicator had failed to consider the defences in the absence of withholding notices. To do so in circumstances where those defences could possibly succeed amounted to procedural unfairness and failure to answer the dispute referred. Not enforced.
Able Construction (UK) Ltd v Forest Property Development Ltd	[2009] All ER (D) 176 (Feb), [2009] EWHC 159 (TCC), Coulson J	**Settlement agreement after adjudication did not preclude enforcement** Following an adjudicator's award of some £166,000, a settlement agreement was agreed for £150,000 to be paid in stages, in lieu of which the claimant would be entitled to enforce the adjudicator's decision. There was default of the agreement, and the decision was enforced—this was precisely what the agreement had anticipated.
Rok Building Ltd v Celtic Composting Systems Ltd (No. 2)	[2010] All ER (D) 107, [2010] EWHC 66 (TCC), Akenhead J	**Error of fact: errors of calculation** Celtic refused to comply wholly with the adjudicator's decision which had been corrected in part by application of the slip rule. The court should not conduct a review of the relative correctness of an adjudicator's decision to determine whether the adjudicator 'got it wrong'. Adjudicators must do the best they can in the relatively short period of time available. Akenhead J declined to decide whether there was an error. Even if an error was glaring and serious, the decision should still be enforced.
GPS Marine Contractors Ltd v Ringway Infrastructure Services Ltd	[2010] All ER (D) 232, [2010] EWHC 283 (TCC), Ramsey J	**Fraud: inconsistencies in evidence in adjudication** Where matters of inconsistency of evidence could have been raised in the course of the adjudication, those same matters could not be formulated as an allegation of fraud to resist enforcement. Not enforced on other grounds (arguable that dispute compromised).
Banner Holdings Ltd v Colchester Borough Council	[2010] All ER (D) 226, [2010] EWHC 139 (TCC)	**Declaration that determination was valid** A local authority commenced an adjudication seeking a declaration that it had validly determined the contract for cause. Banner sought a Part 8 declaration that the adjudicator did not have jurisdiction to grant the declaration. The Part 8 declaration was denied.

(Continued)

Table 7.1 *Continued*

Title	Citation	Issue
SG South Ltd v Swan Yard (Cirencester) Ltd	[2010] All ER (D) 04, [2010] EWHC 376 (TCC), Coulson J	**Error of fact: sum incorrectly calculated** The adjudicator had not misread the evidence in relation to the preliminaries amount he had found to be due, and even if he had, it was an error within his jurisdiction to make. Enforced.
SG South Ltd v King's Head Cirencester LLP	[2019] All ER (D) 120, [2009] EWHC 2645 (TCC), Akenhead J	**Fraud: allegations of dishonest record keeping and removal of items from site** Fraud may only be relied on to resist enforcement if it was not or could not have been raised in the course of the adjudication and directly impacts on the matters decided in the adjudication. Enforced.
Geoffrey Osborne Ltd v Atkins Rail Ltd	[2010] BLR 363, [2009] EWHC 2425 (TCC), Edwards-Stuart J	**Error of fact: computation wrong** An adjudicator was invited to correct a significant arithmetical error in his decision but refused to do so. The party the error benefited sought to summarily enforce the decision and the court permitted the challenge to the arithmetic. *Bouygues UK Ltd v Dahl-Jensen UK Ltd* distinguished on the basis that there was no agreement to arbitrate between the parties in the current action. Not enforced in full.
Speymill Contracts Ltd v Baskind	[2010] BLR 257, [2010] EWCA Civ 120, Waller, Jackson, Sir David Keene LJJ	**Allegations of fraud decided by adjudicator** Where the adjudicator had made a decision in respect of the allegations of fraud, those allegations could not be used as a basis for resisting enforcement. *Obiter*: it may be possible to resist enforcement on grounds of fraud where the matters giving rise to the allegation only emerged after the adjudication and were not or could not have been raised as a defence in the adjudication. Enforced.
Urang Commercial Ltd v Century Investments Ltd	[2011] All ER (D) 138, [2011] EWHC 1561 (TCC), Edwards-Stuart J	**Error of law: adjudicator incorrectly considered a withholding notice was required if defence was to succeed** Withholding notices were only contractually required for sums which were stated as due in interim valuations, and therefore the adjudicator wrongly failed to consider the counterclaims. This error was within his jurisdiction to make. Enforced.

(3) Methods of Enforcement

7.20 Methods of enforcement include:

1. Summary judgment pursuant to part 24 of the CPR.
2. Enforcement by identical means to s. 42 of the Arbitration Act 1996 (as provided in paragraph 24 of the Scheme) where the adjudication was conducted under the Scheme and the adjudicator made a peremptory order (however this avenue was not encouraged Dyson J in *Macob v Morrison*).

3. Enforcement by mandatory injunction/specific performance (although recent case law indicates this is not permitted in Part 24 proceedings).
4. Winding up or bankruptcy proceedings where the sums are acknowledged as due and payable.

Summary Judgment

7.21 While it is now widely accepted that summary judgment is the appropriate course for enforcement of adjudicators' decisions, this was not immediately apparent when the 1996 Act was introduced. In *Macob Civil Engineering Ltd v Morrison Construction Ltd* (1999) (Key Case),[27] Dyson J held that the fact that an adjudicator's decision may later be revised was not a good reason for making summary judgment an inappropriate remedy. The grant of summary judgment did not pre-empt any later decision of an arbitrator or court—it simply reflected the fact that at the time the summary judgment was made, there was no defence to the payment of sums awarded by the adjudicator.

7.22 The last 14 years have confirmed Dyson J's approach and successful parties now most commonly use the Special Procedure developed by the TCC for expedited summary judgment of enforcement claims under CPR 24 (discussed more fully at 7.41–7.67). Whilst there remain alternative methods of seeking enforcement of decisions of adjudicators,[28] in most cases the judges of the TCC have given clear guidance that the appropriate course of action is to follow the TCC Special Procedure:

> 18. The... TCC Guide... makes it clear that the TCC will deal with all applications to enforce the decisions of adjudicators, regardless of the value of the decision, and will do so quickly and efficiently in accordance with a procedure worked out in consultation with the construction industry and the users of the court. ... It is important that all parties to adjudication realise that, save in exceptional circumstances, the most efficient way of enforcing the adjudicator's decision is by enforcement proceedings in the TCC. Other ways of enforcing such decisions (such as, for instance, bankruptcy proceedings) are something of a blunt instrument and raise potential issues which have little or nothing to do with the decision which is at the heart of any enforcement application.[29]

7.23 The TCC has recently reinforced that general financial limits for issuing claims in the TCC do not apply to adjudication claims.[30]

Alternative Methods of Enforcement

7.24 Notwithstanding the TCC's clear preference for adjudication decisions to be enforced using its own bespoke procedure for enforcement, there are alternative methods that have been attempted by parties, with varying degrees of success.

Enforcement by Peremptory Order

7.25 The original Scheme contained a potential route to enforcement: paragraph 23(1) of the Scheme provides that the adjudicator may order compliance with his decision peremptorily, that is forthwith and without further investigation. Paragraph 24 says that s. 42 of the Arbitration Act 1996 (Arbitration Act) applies meaning the court may make an order that the parties comply with the adjudicator's peremptory decision.

---

[27] [1999] All ER (D) 143 (TCC) (Key Case).
[28] Discussed at 7.24–7.32.
[29] *Harlow & Milner Ltd v Teasdale* [2006] EWHC 54 (TCC).
[30] *West Country Renovations Ltd v McDowell* [2013] WLR 416, [2012] EWHC 307 (TCC).

7.26 Notwithstanding the inclusion of this peremptory enforcement under the Scheme, there has been judicial reluctance towards this approach. For example, in *Macob Civil Engineering Ltd v Morrison Construction Ltd* (1999) (Key Case),[31] Dyson J confirmed that a court was empowered to enforce the decision pursuant to s. 42 of the Arbitration Act, but considered it was 'not at all clear' why s. 42 had been incorporated into the Scheme. The more appropriate remedy before the courts was summary judgment on the grounds that, at the time of judgment, there was no defence to enforcement of the adjudicator's decision. Shortly after this decision, the TCC was asked to enforce an adjudicator's decision, given as a peremptory order, in *Outwing Construction Ltd v H. Randell & Son Ltd* (1999).[32] The court noted that there was no recognized procedure for enforcement of peremptory decisions but agreed that the claimant was entitled to seek enforcement by issuing a claim.[33]

7.27 Paragraph 24 of the Scheme has been removed following the introduction of the Local Democracy, Economic Development and Construction Act 2009. Therefore, adjudicators appointed to consider disputes arising under construction contracts entered into after 1 October 2011 will no longer have to consider whether to make their decisions take effect as peremptory orders.

Injunction/Specific Performance

7.28 In *Macob Civil Engineering Ltd v Morrison Construction Ltd* (1999) (Key Case)[34] Dyson J considered it was inappropriate to enforce adjudicators' decisions for payment of sums due by means of a claim for a mandatory injunction.[35] It would rarely be appropriate to grant injunctive relief to enforce an obligation on one contracting party to pay the other. Nevertheless the case leaves open the question of whether it might be appropriate to seek an injunction on different facts. Dyson J said the court had jurisdiction to grant a mandatory injunction to enforce an adjudicator's decision which might be appropriate, for example, where a party was ordered by an adjudicator to make payment to a third party,[36] or where the adjudicator's decision contained declarations.[37]

7.29 However, in the case of *Multiplex Constructions (UK) Ltd v Mott MacDonald Ltd* (2007) (Key Case)[38] Jackson J said that there was no basis on which the court in Part 24 proceedings could make an order for specific performance or grant an injunction.

Winding Up/Bankruptcy

7.30 A further alternative to issuing a claim for summary judgment to enforce an adjudicator's decision is the use of winding up or bankruptcy proceedings. These will only be appropriate in limited circumstances, and may carry more inherent risks than the TCC Special Procedure. It has been successfully used in situations where sums were acknowledged as due and payable in accordance with an adjudicator's decision, and no withholding notices had been issued.[39]

---

[31] *Macob Civil Engineering Ltd v Morrison Construction Ltd* [1999] All ER (D) 143 (TCC) (Key Case), and in particular paras. 31, 32, and 38.

[32] [1999] BLR 156, [1999] EWHC 100 (TCC).

[33] In the event the court was not required to order enforcement as the payment had been made in full following the issue of the claim and the case only related the question of costs.

[34] [1999] All ER (D) 143 (TCC).

[35] *Ibid.*, paras. 31–7.

[36] *Ibid.*, para. 35.

[37] *Ibid.*, para. 35 and see the discussion at 7.57–7.64.

[38] [2007] All ER (D) 133, [2007] EWHC 20 (TCC), para. 47 (Key Case).

[39] *Re a Company (No. 1299 of 2001)* [2001] CILL 1745; *Guardi Shoes Ltd v Datum Contracts* [2002] CILL 1934.

7.31 However, the insolvency proceedings approach was unsuccessful in *George Parke v Fenton Gretton Partnership* (2001)[40] in which a judge in the Chancery Division considered Mr Parke had a bona fide counterclaim and set aside the statutory demand—one of the key reasons being that Mr Parke had commenced proceedings in the TCC on the basis that the final account showed a balance in his favour:

> In my judgment it cannot be right that an employer or main contractor can be made bankrupt when it is known that he has proper proceedings on foot which, if successful, will result in a payment to him. I do not accept that the scheme of the 1996 Act is that an adjudication can be pursued to bankruptcy no matter the underlying state of account. The court would be required to close its eyes to the overall position, which in the context of bankruptcy is in my judgment wrong in principle.

7.32 Ultimately, whether an adjudicator's decision can be pursued to insolvency will come down to consideration of the particular circumstances in each case and there is no overriding principle that such an approach is or is not valid in all cases. A key consideration in the exercise of discretion in restraining insolvency proceedings is the impression formed by the court of the strength of any counterclaim relied on.[41]

Key Cases: Methods of Enforcement

Macob Civil Engineering Ltd v Morrison Construction Ltd [1999] All ER (D) 143 (TCC)

7.33 **Facts:** Macob was engaged as a subcontractor to Morrison to carry out groundworks at a retail development in Wales. A dispute arose concerning payment application number six, which was referred to adjudication. The adjudicator directed that Morrison pay forthwith £302,366.34 plus interest together with the adjudicator's fees. The adjudicator's decision:

1. Was purported to be issued in accordance with paragraph 23(1) of the Scheme which provides that an adjudicator may order the parties to comply peremptorily with his decision.
2. Stated that in the event of non-compliance by Morrison, he gave permission under s. 42 of the Arbitration Act 1996 (as modified by paragraph 24 of the Scheme) for either party to apply to the court for an order requiring such compliance.

7.34 Morrison did not comply with the decision, which it claimed was invalid and unenforceable due to breach of natural justice, and referred to arbitration the disputes arising out of or in connection with the adjudicator's decision. Macob sought to enforce the adjudicator's decision and Morrison issued a summons to stay the proceedings under s. 9 of the Arbitration Act because the contract contained a valid arbitration clause which it argued applied to all disputes raised concerning the adjudicator's decision. Morrison argued that disputes concerning the validity of the decision had to be decided by an arbitrator and therefore rendered the adjudicator's decision unenforceable unless it was confirmed by an arbitrator.

7.35 **Held:** Even if there was a challenge to the validity of an adjudicator's decision, it was nevertheless a decision within the meaning of the Act, the Scheme, and the contract in question. The decision was binding and enforceable until the challenge was finally determined by arbitration, legal proceedings or agreement. Dyson J found that the usual course of

[40] [2001] CILL 1712.

[41] *Shaw v MFP Foundations & Piling Ltd* [2010] CILL 2831, [2010] EWHC 9 (Ch) at para. 50. Judge Davies also reconciled the decisions in *Guardi* and *Parke*.

enforcement would be to issue proceedings claiming the sum due, followed by an application for summary judgment:

> 31. Mr Furst submits that there is no power in the court to make an order under section 42 of the Arbitration Act 1996 (as modified by paragraph 24 of Part 1 of the Scheme). His argument is that the power given by section 42 is exercisable 'unless otherwise agreed by the parties'. He says that the parties have otherwise agreed in the present case by agreeing to refer to arbitration disputes arising out of the decision of an adjudicator. In my view, the arbitration clause is not an agreement of the kind envisaged by section 42(1). What that subsection contemplates is an agreement expressly directed to the section 42 power. Ordinarily, this would be an agreement expressly excluding that power, although I accept that there may be other ways of achieving the same object. A general reference of disputes to arbitration is surely insufficient.
>
> 32. In my view, therefore, the court can enforce this decision under section 42.
>
> 33. There was some limited discussion as to whether, section 42 apart, the appropriate procedure was by way of writ and an application for summary judgment, or by way of a claim for a mandatory injunction. . . .
>
> 34. I do not consider that the mere fact that the decision may later be revised is a good reason for saying that summary judgment is inappropriate. The grant of summary judgment does not pre-empt any later decision that an arbitrator may make. It merely reflects the fact that there is no defence to a claim to enforce the decision of the adjudicator at the time of judgment.
>
> 35. I am in no doubt that the court has jurisdiction to grant a mandatory injunction to enforce an adjudicator's decision, but it would rarely be appropriate to grant injunctive relief to enforce an obligation on one contracting party to pay the other. Clearly, different considerations apply where the adjudicator decides that a party should perform some other obligation, e.g. return to site, provide access or inspection facilities, open up work, carry out specified work etc. Nor do I intend to cast any doubt on decisions where mandatory judgments have been ordered requiring payment of money to a third party, e.g. to a trustee stakeholder as in *Drake & Scull Engineering Ltd v McLaughlin & Harvey Plc* [1992] CILL 768.
>
> 36. The words of section 37 of the Supreme Court Act 1981 are widely expressed viz: 'the High Court may by order (whether interlocutory or final) grant an injunction . . . in all cases in which it appears just and convenient to do so'. But a mandatory injunction to enforce a payment obligation carries with it the potential for contempt proceedings. It is difficult to see why the sanction for failure to pay in accordance with an adjudicator's decision should be more draconian than for failure to honour a money judgment entered by the court.
>
> 37. Thus, section 42 apart, the usual remedy for failure to pay in accordance with an adjudicator's decision will be to issue proceedings claiming the sum due, followed by an application for summary judgment.
>
> 38. It is not at all clear why section 42 of the Arbitration Act 1996 was incorporated into the Scheme. It may be that Parliament intended that the court should be more willing to grant a mandatory injunction in cases where the adjudicator has made a peremptory order than where he has not. Where an adjudicator has made a peremptory order, this is a factor that should be taken into account by the court in deciding whether to grant an injunction. But it seems to me that it is for the court to decide whether to grant a mandatory injunction, and, for the reasons already given, the court should be slow to grant a mandatory injunction to enforce a decision requiring the payment of money by one contracting party to another.
>
> 39. The adjudicator did not explain why he thought it appropriate to make a peremptory order. [Counsel for the claimant] was unable to suggest any reason why an injunction

should be granted (other than the one which I have already referred to and rejected). In these circumstances, I am not persuaded that I ought to exercise my discretion in favour of granting an injunction.

40. The plaintiff has not claimed a money judgment in these proceedings. In the result, I think that the relief that I ought to grant is a declaration that (i) the decision of the adjudicator is binding on the defendant until the dispute arising from the decision is finally determined by arbitration, legal proceedings or agreement; and (ii) the defendant was required by the decision to pay the sums identified by the adjudicator forthwith in accordance with the Scheme, and is now in default.

Multiplex Constructions (UK) Ltd v Mott MacDonald Ltd [2007] All ER (D) 133, [2007] EWHC 20 (TCC)

7.36 **Facts:** Multiplex was the main contractor for the construction of the stadium at Wembley. The employer was Wembley National Stadium Ltd (WNSL). Mott was originally engaged as a consultant structural engineer by WNSL but after Multiplex was taken on, WSNL novated its agreement for civil and structural engineering services with Mott to Multiplex. Clause 13.1 of Schedule 4B of the novation agreement required Mott to retain 'all pertinent records' including records of costs; permit the client or its representatives, at all reasonable times, to have access to such records; and deliver copies of the records free of charge at the time and in the manner directed by the client.

7.37 Multiplex sought to exercise its right of access and identified that whilst no definition of 'records' existed in the novation agreement, it sought to broadly define 'records'. Mott considered Multiplex's definition too wide (without providing an alternative definition), but agreed it would provide access to 'appropriate records' which did not include records which it was contractually obliged to keep confidential to WNSL, unless WNSL gave its permission. It also proposed providing the records to a programme which did not interfere with critical elements of the works.

7.38 A dispute was referred to adjudication concerning the meaning of the words 'all records pertinent to the Services' in the novation agreement. Multiplex sought a declaration and/or a decision that it was correct in its definition and/or a declaration as to the true meaning of the words. The adjudicator rejected both the broad definition advanced by Multiplex and the narrow definition advanced by Mott and found that there was no relevant confidentiality obligations and that Mott must provide access to the records within seven days of the decision and then deliver copies as directed by Multiplex.

7.39 Mott was unhappy with the decision but advised Multiplex it would be making a substantial amount of documentation available. Multiplex formed the view that the documents made available were insufficient and that Mott had failed to comply with the adjudicator's decision. Accordingly, Multiplex commenced proceedings in the TCC seeking:

1. A declaration that the decision was binding and that Mott was contractually obliged to render full performance forthwith; and
2. Specific performance ordering compliance with the decision; and/or
3. Alternatively an injunction ordering and/or requiring Mott to comply; and/or
4. Alternatively damages together with interest under s. 35A of the Supreme Court Act 1981.

7.40 **Held:** Jackson J considered the proper course was to grant the declaration sought:

> 44. ... Mott has spent approximately 1000 man hours in identifying and retrieving the documents which have been disclosed. The documentation disclosed amounts to well over 100,000 pages. Multiplex, on the other hand, has only spent two days inspecting that material. Indeed, only one day before the commencement of this litigation. On the first inspection day the representative sent by Multiplex to inspect, namely Mr Edwards, was a relatively junior person.
>
> 45. I am bound to say that I am not favourably impressed by the extent of the inspection carried out by Multiplex before embarking upon this litigation. Nevertheless, it remains the case that there are factual issues between the parties in relation to compliance which this court is not in a position to resolve.
>
> ...
>
> 47. ... On the present evidence I cannot say whether or not there are gaps in the material disclosed by Mott. There is no basis on which this court in Part 24 proceedings could make an order for specific performance or grant an injunction, as sought in Multiplex's application dated 11th December. Nor is there any basis for giving a toned down version of those remedies, as suggested by [Multiplex] in oral argument.
>
> 48. ... I cannot at the moment reach any conclusion on the question whether Mott has complied with the adjudicator's decision. Accordingly, Multiplex fails in its application for summary judgment on its claims for specific performance, an injunction and/or damages...
>
> 49. ... the only issue upon which Multiplex has succeeded is the jurisdiction issue. A question then arises as to whether this court should grant a declaration...
>
> ...
>
> 52. The position therefore remains as it was at the outset of this hearing. There is a dispute on the facts, as to whether Mott has complied with the adjudicator's decision. That factual dispute awaits resolution at a later date. There is also a fully articulated dispute on jurisdiction. Both parties' cases are set out in writing. In addition, both counsel have developed their submissions on jurisdiction in oral argument. Mott maintains its contention that the adjudicator reached a decision which was in excess of its jurisdiction.
>
> 53. ... I consider that the proper course is for this court to grant the declaration sought. I reach this conclusion for three reasons. These are as follows:
>
> (i) The jurisdiction issue has been fully argued. ... Multiplex's case on this issue is of sufficient strength to warrant summary judgment.
> (ii) The question of the adjudicator's jurisdiction remains a live issue between the parties.
> (iii) As a matter of policy the Technology and Construction Court should at each stage of litigation resolve every live issue which is then capable of resolution. It is a tenet of case management that this court is constantly seeking to narrow the issues between the parties.
>
> ...
>
> 55. Let me now draw the threads together. For the reasons set out above, Multiplex is entitled to a declaration along the lines set out in the particulars of claim, although the precise wording will require some modification. Multiplex's application for summary judgment on the balance of its claim is dismissed.

Table 7.2 Table of Cases: Methods of Enforcement

Title	Citation	Issue
Macob Civil Engineering Ltd v Morrison Construction Ltd	[1999] BLR 93 (TCC), Dyson J	**Appropriateness of summary judgment** The fact that an adjudicator's decision may later be revised is not a good reason for making summary judgment an inappropriate remedy. **Peremptory order** Whilst the courts are empowered to enforce decisions peremptorily pursuant to s. 42 of the Arbitration Act, the appropriate remedy was summary judgment. **Mandatory injunction** Summary judgment was also the preferred course over mandatory injunction. It would rarely be appropriate to grant injunctive relief to enforce an obligation on one party to pay the other.
Outwing Construction Ltd v H. Randell & Son Ltd	[1999] BLR 156, [1999] EWHC 100 (TCC), Judge LLoyd	**Peremptory order** An adjudicator directed by peremptory order that Randell pay Outwing. Randell failed to do so within the time period and Outwing sought enforcement by issue of a writ, which it was at liberty to do. Shortly thereafter Randell paid the sum sought.
Re a Company (No. 1299 of 2000)	[2001] CILL 1745, Judge Donaldson	**Winding up/bankruptcy** Where sums are due and payable in accordance with an adjudicator's decision and no withholding notices have been issued, winding up/bankruptcy proceedings may be successful.
George Parke v Fenton Gretton Partnership	[2001] CILL 1712 (Ch D), Judge Boggis	**Winding up/bankruptcy** In this instance, Mr Parke had already commenced proceedings in the TCC dealing with the final account (the subject of the adjudication) and claiming a balance in his favour was due. This was a genuine or substantial cross-claim which made it unjust for the creditor to be permitted to pursue bankruptcy proceedings against Mr Parke. The 1996 Act does not permit an adjudication to be pursued to bankruptcy no matter the underlying state of the account. Application to set aside statutory demand successful.
Sindall Ltd v (1) Abner Solland (2) Grazyna Solland (3) Solland Interiors (a firm) (4) Solland Interiors Ltd	[2001] All ER (D) 370 (TCC), Judge LLoyd	**Enforcement of non-pecuniary decision that determination was wrongful** An adjudicator had decided that Sindall was entitled to a declaration that the determination was wrongful and that Sindall was entitled to an extension of time. Sindall sought to enforce and the defendant sought a declaration that the adjudicator acted without jurisdiction. The court denied that declaration and enforced the decision.

(Continued)

Table 7.2 *Continued*

Title	Citation	Issue
Guardi Shoes Ltd v Datum Contracts	[2002] CILL 1934, Ferris J	**Winding up/bankruptcy** Where sums are due and payable in accordance with an adjudicator's decision and no withholding notices had been issued, winding up/bankruptcy proceedings may be successful. The TCC had earlier entered judgment against Guardi, who defaulted on payment. The judge ordered an injunction (prohibiting the winding-up of Guardi) should be lifted. The approach in this case was considered reasonable in *Shaw v MFP* (2010) because there had been (i) a failure to serve a withholding notice as required by s. 111, (ii) a failure to commence arbitration or litigation within a short compass of the adjudicator's decision, and (iii) no convincing evidence supporting the cross-claim relied on.
Multiplex Constructions (UK) Ltd v Mott MacDonald Ltd	[2007] All ER (D) 133, [2007] EWHC 20 (TCC), Jackson J	**Enforcement of non-pecuniary decision to provide all pertinent records** The court had no power to order specific performance or grant an injunction, but exercised its discretion to grant a declaration that the decision of the adjudicator was binding on Mott who was contractually obliged to render full performance forthwith.
Shaw v MFP Foundations & Piling Ltd	[2010] CILL 2831, [2010] EWHC 9 (Ch), Judge Davies	**Setting aside a statutory demand following an adjudicator's decision** The decision of a district judge refusing to set aside a statutory demand served following non-compliance with an adjudicator's decision was successfully appealed. Judge Davies found that there was nothing in the Act or the Scheme that indicates it was intended that the 'pay now litigate later' philosophy should displace the position that applied to personal insolvency by r. 6.5(4) of the Insolvency Rules (which includes that a statutory demand be set aside if there is a counterclaim which equals or exceeds the amount of the debt), or to corporate insolvency by case law.
MBE Electrical Contractors Ltd v Honeywell Control Systems Ltd	[2010] BLR 561, [2010] EWHC 2244 (TCC), Judge Langan	**Peremptory order** An adjudicator's decision must expressly state that it is to be enforced as a peremptory order for it to be so enforced.

(4) Summary Enforcement Using CPR Part 24

7.41 The following is a summary of enforcement by summary judgment:

1. The appropriate procedure for the enforcement of adjudicators' decisions is to follow the TCC's Special Procedure from the TCC Guide.

2. A claim is issued under either Part 7 (Proceedings) or Part 8 (Alternative Procedure) and an application for summary judgment of that claim is made at the same time under Part 24 (Summary Judgment).
3. Most claims are conducted using Part 8 as they are unlikely to involve substantial disputes of fact. Applications for summary judgment should not involve 'mini trials' of facts which should be left for the substantive determination of the dispute.
4. A valid challenge to enforcement involving a detailed factual enquiry may not be capable of being determined summarily and will need to be tried at the final hearing of the Part 7/Part 8 claim.
5. Applications for summary judgment which involve a point of law (and no dispute of fact) may be determined during the course of the summary judgment itself so as to avoid time and cost implications of a trial on the same point of law.

7.42 Ordinarily, a decision of an adjudicator will give rise to a separate contractual entitlement to immediate payment[42] and it is that right which is being upheld when an adjudicator's decision is being enforced. This section sets out the TCC Special Procedure and CPR procedural issues involved in enforcing that right, but is not intended to be a full commentary on summary judgment procedure or the Civil Procedure Rules, which is beyond the scope of this book.

TCC Special Procedure

7.43 One of the first enforcement applications heard in early 1999 was *Outwing Construction Ltd v H. Randell & Son Ltd* (1999).[43] In this case, the court agreed to abridge time for the summary enforcement proceedings, as this was in line with the strict timetables imposed by s. 108 of the 1996 Act. This case was a precursor to the development by the TCC of a special procedure in the TCC Guide to deal with adjudication enforcement applications.[44] The TCC has also created an express exemption from the need for adjudication enforcement proceedings to comply with the Pre-Action Protocol for Construction and Engineering Disputes.[45] The TCC Guide is '*emphatically* not designed to define' the court's jurisdiction in every case but to set out in simple terms how the court could assist the parties to resolve their disputes.[46]

The whole procedure can take as little as 32 days. The main steps of the TCC's adjudication enforcement procedure, as set out in the TCC Guide, are summarized below in Table 7.3.

7.44 **Table 7.3 TCC's Adjudication Enforcement Procedure**

Claim Form Day 1	Identify the construction contract, the jurisdiction of the adjudicator, the procedural rules under which the adjudication was conducted, the adjudicator's decision, the relief sought, and the grounds for seeking that relief. The claim form should be accompanied by: (i) an application notice setting out procedural directions sought (e.g. abridgment of time and summary judgment under CPR Part 24); (ii) witness statement(s) setting out evidence relied on in support of the adjudication enforcement claim and any associated procedural application and attaching the adjudicator's decision; and

(Continued)

[42] *Bovis Lend Lease Ltd v Triangle Development Ltd* [2003] BLR 31; [2002] EWHC 3123 (TCC).
[43] [1999] BLR 156, [1999] EWHC 100 (TCC).
[44] Found at para. 9.2 to the second revision of the 2nd edn of the TCC Guide which came into effect in Oct. 2010.
[45] Para. 1.2(i) of the Pre-Action Protocol. Harlow & Milner Ltd v Teasdale [2006] All ER (D) 382, [2006] EWHC 535 (TCC).
[46] *Vitpol Building Service v Samen* [2008] EWHC 2283 (TCC), paras. 13–17.

Table 7.3 *Continued*

	(iii) an estimate of the length of time required for the hearing (e.g. one day or half a day). The claim form should be lodged at the appropriate registry or court and marked as being a 'paper without notice adjudication enforcement claim and application for the urgent attention of a TCC judge'.
Judge's Appointment	The parties will be informed of the named judge by the TCC.
Judge's Directions Day 4	The judge will deal with the procedural application on paper and will ordinarily provide directions within three working days of the receipt of the claim form. The directions will deal with the abridged filing of acknowledgement of service and witness statements by the defendant, early return date for the hearing of the application, and identification of the relief being sought; the defendant will be given liberty to apply.
Defendant's Submissions Day 15	Generally, the defendant will be directed to serve its submissions at least 14 days from the date of service of the claimant's submissions.
Submission of Court Bundle Day 30/31	By 4.00 pm one clear working day before the hearing, the parties should lodge a bundle containing: (i) short skeleton arguments summarizing respective contentions as to why the decision is or is not enforceable and explaining the grounds for any other relief sought; (ii) the adjudicator's decision; (iii) documents that will be relied upon; (iv) copies of authorities relied upon; (v) copies of the relevant sections of the 1996 Act; (vi) the adjudication procedural rules under which the adjudication was conducted; and (vii) copies of any adjudication provisions in the contract underlying the adjudication. If the hearing is going to last a half day or less, the skeleton arguments may be submitted no later than 1.00 pm the day before the hearing.
Enforcement Hearing Day 32	This will ordinarily be within 28 days of the directions being made.

CPR Part 24

7.45 An application to enforce an adjudicator's decision is commenced using the claim forms and procedures in either Part 7 (Proceedings) or Part 8 (Alternative Procedure) of the CPR at the same time as an application for summary judgment in accordance with Part 24 (Summary Judgment). If the Part 24 application fails (on the basis that there is an arguable defence that cannot be determined summarily), then the claimant proceeds under the Part 7 or 8 procedure for determination of its claim. However, in certain circumstances, such as where the application to enforce will involve a jurisdictional challenge involving no questions of fact, the claimant may prefer to issue a claim using Part 8 and ask for it to be determined quickly, but not as a Part 24 summary judgment.

7.46 Rule 24.2 provides that the court may give summary judgment against a claimant or defendant on the whole of a claim or on a particular issue if:

(a) it considers that –

(i) that claimant has no real prospect of succeeding on a claim or issue; or

(ii) that defendant has no real prospect of successfully defending the claim or issue;

and

(b) there is no other reason why the case or issue should be disposed of at a trial.

7.47 In *Swain v Hillman* (2001)[47] the words 'no real prospect of succeeding' within r. 24.2(a)(i) were considered as not needing any amplification because they speak for themselves:

> The word 'real' distinguishes fanciful prospects of success, ... they direct the court to the need to see whether there is a 'realistic' as opposed to a 'fanciful' prospect of success.

7.48 In *Glencot Development and Design Co Ltd v Ben Barrett & Son (Contractors) Ltd* [2001] BLR 207 (TCC)[48] the TCC considered the rights of the defendant under r. 24.2(a)(ii) as being largely the same in adjudication enforcement proceedings as in any other applications for summary judgment:

> 13. ... An application under Part 24.2 to enforce payment of an amount ordered by a decision of an adjudicator does not mean that the rights of a defendant are less than if the claim had been for another contractual debt that was apparently due, except that the room to question the amount decided to be due is very limited.

7.49 At the summary judgment application, if the court decides that in relation to the challenge to enforcement the defendant has no real prospect of showing that there was a jurisdictional error or breach of natural justice, then the decision is enforced. If instead the court concludes that the defendant has a real prospect of showing there was a jurisdictional or natural justice error the decision will not be enforced on the summary judgment application. Often the court will be able to decide on the material before it on hearing that application that there was a clear jurisdictional error or breach of natural justice so that the decision is a nullity. However, there will be occasions where (whilst there is a real prospect of success for the defendant) the question of jurisdiction or breach of natural justice cannot be finally decided without determining questions of fact which cannot be done on a summary judgment application. In such cases directions will be given for trial of those factual matters, pending which the decision will not be enforced. The outcome of that trial will determine whether the decision is enforceable or a nullity.

7.50 Where the court is unable to grant summary judgment, it has no power on a Part 24 application to make an order for interim payment under Part 25 (Interim Remedies and Security for Costs) of its own initiative and the party asking for the interim payment must have issued a formal application under Part 25 if it wishes to makes an alternative submission that an interim payment should be made.[49] The court may, however, give conditional leave to defend, for example ordering that a sum be paid into court.

Judgment in Default

7.51 If it becomes clear that a defendant is unlikely to participate in the enforcement proceedings, the claimant could consider proceeding by way of obtaining a judgment in default.[50]

Arguments of Fact at Part 24 Enforcement Hearings

7.52 As stated above at 7.05 it is not a valid challenge to the enforceability of a decision to allege that the adjudicator was wrong in his conclusions as a matter of fact.[51] However, some valid challenges to enforcement, such as jurisdiction challenges, may involve questions of fact that

---

[47] [2001] 1 All ER 91 at 92, per Lord Woolf MR.

[48] [2001] All ER (D) 384 (TCC).

[49] *Glencot Development v Ben Barrett* [2001] BLR 207 at para. 34.

[50] *Coventry Scaffolding Co. (London) Ltd v Lancsville Construction Ltd* [2009] All ER (D) 93, [2009] EWHC 2995 (TCC).

[51] See also 9.47.

cannot be decided summarily, for example whether the contract was formed at all and on what terms.

7.53 On an application for summary judgment it is inappropriate for the court to conduct a 'mini trial' to reach its conclusion.[52] Where there are disputes of fact, the Part 24 test requires a very different approach from that at a trial.[53] In other words, because Part 24 is a summary procedure, it is not appropriate, or possible, for the determination of disputes of fact which require the hearing of oral evidence. Issues of fact that can be determined on the documents are frequently decided by the Part 24 judge. However, if the challenge to enforcement requires the judge to consider an issue that requires a detailed factual analysis and oral evidence then it is likely that summary enforcement will not succeed if that question has a real prospect of success. The court will give permission to defend and give directions for a trial of the enforcement claim.[54]

7.54 For example, in *Beck Interiors Ltd v Russo* 132 Con LR 56, (2009)[55] the successful party in an adjudication sought to enforce the decision against the unsuccessful party's guarantor. Whether the guarantor was bound by the adjudicator's decision depended upon whether the guarantee was conditional and whether the guarantor was bound in his personal capacity. It was not possible to determine such issues in a Part 24 application, and accordingly the summary application was refused.

Arguments of Law at Part 24 Enforcement Hearings

7.55 It is not usually open to an unsuccessful party to challenge enforcement of the decision on the grounds that it contains errors of law.[56] However, if an error of law has been made which results in the adjudicator making a decision which was outside his jurisdiction, that may be a valid ground for challenging the enforceability of the decision. Where such a valid challenge is made, the alleged error of law will be argued at the enforcement hearing.

7.56 Where such a defence to enforcement is raised it should be decided on its merits at the Part 24 hearing and not listed for separate trial.[57]

Declaratory or Injunctive Relief in Relation to an Ongoing Adjudication

7.57 During the course of an adjudication a party may make an application to the court for a declaration on an issue that affects the adjudication. Most commonly this arises where there is a jurisdictional dispute. The party disputing jurisdiction may raise the issue in the adjudication[58] and/or may issue proceedings in court to have the jurisdiction point decided. Usually the application will be made by way of a Part 8 claim for a declaration and/or for an order restraining the adjudicator from continuing with the adjudication.

7.58 *Birmingham City Council v Paddison Construction Ltd* (2008)[59] concerned a claim for a declaration that an adjudicator had no jurisdiction because the dispute had been decided by a previous adjudicator. The TCC judge granted a declaration that the adjudicator had no jurisdiction and must resign and/or that any decision reached would be a nullity and

---

[52] *Swain v Hillman* [2001] 1 All ER 91; *Debeck Ductwork Installation Ltd v T&E Engineering Ltd*, unreported decision of Judge Kirkham in the TCC, Birmingham Registry on 14 Oct. 2002.

[53] *VHE Construction plc v RBSTB Trust Co. Ltd* [2000] All ER (D) 23.

[54] *Multiplex Constructions (UK) Ltd v Mott MacDonald Ltd* (2007) 110 Con LR 63, [2007] EWHC 20 (TCC) (Key Case).

[55] [2009] EWHC 3861 (TCC).

[56] See 7.05–7.07 and 9.47.

[57] *VHE Construction plc v RBSTB Trust Co. Ltd* [2000] All ER (D) 23.

[58] For a discussion of how jurisdiction challenges should be made see Ch. 5 at 5.09–5.16 and Ch. 9 at 9.11–15.

[59] [2008] BLR 622, [2008] EWHC 2254 (TCC).

unenforceable. In *Dalkia Energy & Technical Services Ltd v Bell Group UK Ltd* (2009)[60] the adjudicator had determined that he had jurisdiction to decide the dispute referred and whilst the adjudication was in progress the responding party applied to court for final determination of the question of jurisdiction. The adjudicator's decision that Bell's standard terms had been incorporated into the contract would not ordinarily be a matter with which the court could interfere on enforcement. However, the court could decide the point in Part 8 proceedings, where the court was being asked to give a final and binding decision on the incorporation issue. In this case the court decided the terms had been incorporated and agreed with the adjudicator's finding that he had jurisdiction.

7.59 In *Workplace Technologies v E Squared Ltd* (2000)[61] the responding party applied for an injunction to restrain the adjudication on the ground that the relevant contract was not caught by the 1996 Act. In the meantime an adjudicator had been appointed. On the facts the contract was not finally concluded until, at the earliest, 20 May 1998, thus s.108 of the Act applied to it. The application was therefore dismissed. Judge Wilcox stated that it will be rare for a court to restrain an ongoing adjudication. That the High Court may grant declaratory (or possibly injunctive) relief in relation to an ongoing adjudication in relation to jurisdiction or natural justice issues was clarified in *The Dorchester Hotel Ltd v Vivid Interiors Ltd* (2009).[62] In that case the responding party in the adjudication (Dorchester) applied for declarations to the effect that the timetable fixed in the adjudication was too short to allow Dorchester to deal with the material provided. Dorchester claimed that if additional time was not granted it would amount to a breach of natural justice.

> 12.... If an ongoing adjudication is fundamentally flawed in some way, or may be just about to go off the rails irretrievably, then it seems to me that it must be sensible and appropriate for the parties to be able to have recourse to the TCC: otherwise a good deal of time and money will be spent on an adjudication which will ultimately be wasted. That was recognised in the early cases involving a challenge to the adjudicator's jurisdiction. For example, in *ABB Zantingh Ltd v. Zedal Building Services Ltd* [2001] BLR 66, HHJ Bowsher QC said that it was 'an entirely proper course' for the jurisdiction dispute to be referred to the Court during the adjudication itself, in order to prevent wasted effort and costs being expended on an adjudication which the adjudicator may not have had the jurisdiction to determine.
>
> 13. It seems to me that, if the Court has the power to grant a declaration in respect of an adjudicator's jurisdiction in an ongoing adjudication, it also has the power to grant a declaration if it considers that there has been or will be a breach of natural justice which will have a significantly prejudicial effect on the responding party, in this case the Claimant.

7.60 The judge in *Dorchester* declined to grant the declarations sought for four reasons:

1. The adjudicator thought he could fairly deal with the case within the timetable.
2. The judge could not say prospectively that Dorchester did not have enough time to submit its response (it had been given between 19 December 2008 and 28 January 2009).
3. The judge was unable to determine whether the new material contained new disputes that had not yet crystallized and whether Dorchester would be unable to deal with it in the time available.

---

[60] 122 Con LR 66, [2009] EWHC 73 (TCC).

[61] Unreported (TCC), Judge Wilcox.

[62] [2009] BLR 135, [2009] EWHC 70 (TCC) and the cases cited by the judge in para. 12 of the judgment including *Vitpol Building Service v Samen* [2008] EWHC 2283 (TCC) and *ABB Zantingh Ltd v Zedal Building Services Ltd* [2001] BLR 66 (TCC).

4. The refusal to grant the declarations did not leave Dorchester without a remedy as it would be able to argue that there had ultimately been a breach of natural justice at an enforcement hearing.

7.61 The court will only grant a declaration regarding procedural matters where it is a very clear breach of the rules of natural justice. Coulson J in *Dorchester* considered that the facts of *CJP Builders Ltd v William Verry Ltd* (2008)[63] were an example of the type of clear case that would justify the grant of a declaration if brought during the adjudication. *CJP Builders* was a case where the adjudicator had wrongly found that he could not extend time for service of a defence and had consequently disregarded the defence completely. At enforcement the TCC judge decided this was breach of natural justice and declined to enforce the adjudicator's decision. However, other than in very clear cases the court would be very reluctant to interfere with the adjudicator's discretion to dictate the procedure of the adjudication.

7.62 Applications for an injunction to restrain an adjudication may succeed in rare circumstances. In *Mentmore Towers Ltd v Packman Lucas Ltd* (2010)[64] Edwards-Stuart J granted an injunction restraining adjudications from proceeding where he found it was unreasonable and oppressive for the contractor to be subjected to further adjudication proceedings where the employer had failed to pay sums ordered by previous adjudication decisions or to comply with orders made by the court enforcing those decisions.

7.63 In *Lanes Group Plc v Galliford Try Infrastructure Ltd* [2011][65] requested the ICE appoint an adjudicator and the ICE appointed Mr Klein. However GTI considered Mr Klein was not an appropriate adjudicator and declined to serve the referral on him. Instead it asked the ICE for a replacement adjudicator but this was initially refused. GTI then asserted that the time had passed within which a referral could validly be served and that it was permitted to start again with a fresh notice. The fresh notice was served and pursuant to a fresh request to appoint an adjudicator the ICE appointed Mr Atkinson. Akenhead J was asked to order an injunction restraining GTI from pursuing the Atkinson adjudication on the ground that the referral had been served out of time. That application failed as the judge found the allegation was not made out. The matter came back before the same court a little over a week later when Akenhead J was again asked to restrain the Atkinson adjudication on the ground that GTI had by its actions committed a repudiatory breach of the adjudication agreement. The judge rejected the application because in a statutory adjudication the concept of repudiatory breach of the adjudication agreement did not apply. [66] This was upheld subsequently by the Court of Appeal.[67]

7.64 In *WW Gear Construction Ltd v McGee Group Ltd* (2012)[68] the court was asked to grant a declaration about the construction of a clause in the underlying contract during an ongoing adjudication. The judge agreed the court had power to make such an order but said it would be rare for the court to make such a declaration where there were practical difficulties and a risk of injustice. The judge thought that because the declaration would be given very shortly before the adjudication decision was due it would be an unacceptable imposition on an adjudicator and would probably cause misunderstandings or mistakes because the parties might not be able to make submissions to the adjudicator in light of the court judgment (paras. 25–8).

---

[63] (2008) EWHC 2025 (TCC), [2008] BLR 545.

[64] [2010] BLR 393, [2010] EWHC 457 (TCC).

[65] [2011] EWHC 1234 (TCC)

[66] Paras. 30-1. The same was not true for purely contractual adjudications, as discussed at Ch. 5 at 5.31.

[67] [2012] BLR 121.

[68] [2012] BLR 355, [2012] EWHC 1509.

Final Determination of Substantive Issue at the Enforcement Hearing

7.65 As the decision of an adjudicator is only binding until the dispute is finally determined, it is open to either party to go to court for a final decision on the dispute considered by the adjudicator. Providing that the nature and scope of the dispute permits it to be dealt with within the relatively tight confines of a Part 8 application, a party may seek a Part 8 final declaration from the court on the dispute or a point decided by the adjudicator. That final determination would effectively replace the adjudicator's decision.[69] Where an application to summarily enforce a decision is made, the court may be able to decide such a Part 8 application at the same time. However it will usually only be possible to do this where the Part 8 application does not involve a substantial dispute of fact and so can be finally determined on the material before the court. Further, the application for a final determination can only be heard if there is no arbitration agreement between the parties requiring that arbitration be the form of final resolution.[70] Such an approach is only possible where the issue for final determination is relatively limited. It would suit a point of law or contract interpretation and it may be used where any factual disputes can be determined during a short hearing with narrow focus and without disclosure.[71] In other words where 'any necessary oral evidence can be accommodated within the final hearing'.[72] Using Part 8 inappropriately, however, could 'prove a treacherous shortcut'.[73]

7.66 In *Alstom Signalling Ltd v Jarvis Facilities Ltd* (2004) (Key Case),[74] Judge LLoyd decided that he could consider the substantive disputes relating to whether a valid withholding notice had been issued and Jarvis's entitlement to payment at the same time as considering enforcement of the adjudicator's decision.

7.67 Therefore, there may be certain circumstances in which a judge is open to considering the underlying issues of substance at the same time as an application for enforcement. This gives the parties a more immediate rehearing of the issues upon which the adjudicator had based his decision. However, this approach will most likely be limited to questions of law. If there is a part of the decision that can be isolated and determined by the court, if considered just and expedient to do so, that would give effect to the overriding objective of the CPR.[75] In *Geoffrey Osborne Ltd v Atkins Rail Ltd* (2010),[76] the adjudicator had made a significant arithmetic error in calculating the balance due following his decisions on particular items in a certificate (by not allowing for the sum already paid in respect of those items in the previous certificate). The judge in effect made a final determination of the balance due to the claimant in light of the adjudicator's decisions on the valuation of those items.

Key Cases: Enforcement by Summary Judgment

VHE Construction plc v RBSTB Trust Co. Ltd [2000] BLR 187 All ER (D) 23 (TCC)

7.68 **Facts:** The unsuccessful employer (RBSTB) challenged enforcement of two adjudication decisions by raising issues of law concerning whether the effect of the decisions was to order a sum of money to be paid and whether a valid withholding notice had been served after

[69] *Walter Lilly & Co. Ltd v DMW Developments Ltd* [2008] All ER (D) 214 (Dec), [2008] EWHC 3139 (TCC).
[70] *Geoffrey Osborne Ltd v Atkins Rail Ltd* [2010] BLR 363, [2009] EWHC 2425 (TCC).
[71] *Forest Heath District Council v ISG Jackson Ltd* [2010] All ER (D) 16 (Nov), [2010] EWHC 322 (TCC).
[72] *Vitpol Building Service v Samen* [2008] EWHC 2283 (TCC) at paras. 18(b) and 19.
[73] *Ibid.*, para. 47.
[74] [2004] All ER (D) 02, [2004] EWHC 1285 (TCC).
[75] *Geoffrey Osborne Ltd v Atkins Rail* [2010] BLR 363, [2009] EWHC 2425 (TCC).
[76] *Ibid.*

the decisions justifying withholding of payment. RBSTB argued that the sum decided did not become payable until a VAT invoice was submitted for the amount of the decision, by which time it was said a valid withholding notice had been served. The first adjudicator decided that on a true construction of the contract terms,[77] in the absence of a payment notice or withholding notice, the employer was bound to pay the full amount of the contractor's application in the sum of £1,037,898.[78]

7.69 RBSTB started a second adjudication asking for the application to be reviewed and revised and asking for a determination of the true value of the application. The second adjudicator revised the value of the application down to £254,831.83 and ordered that payment made in excess of that sum be repaid immediately. RBSTB sought to deduct a new claim for LADs and paid only £46,974.69. VHE issued proceedings to enforce both adjudication decisions and claimed £207,857.14. The only challenges to the enforcement application were questions of law raised by RBSTB concerning whether withholding notices had been served (RBSTB said the adjudication submissions could stand as withholding notices) and the legal status of the two decisions (as RBSTB submitted they did not give rise to an order to pay the sums in question).

7.70 **Held:** Judge Hicks determined these points against RBSTB and enforced the decisions such that the balance of £207,857.14 was ordered to be paid. He found that these were issues of law that the court could determine summarily at the Part 24 hearing. The judge observed that whilst the terms of r. 24.2 are different from its predecessors (and are subject to the overriding objective), the following considerations were relevant:

1. Part 24 of the CPR replaced Order 14 and Order 14A of the Rules of the Supreme Court. Order 14A provided, inter alia, a means for the summary disposal of 'any question of law or construction of any document' which would be finally determinative of the proceedings or of any claim or issue in them. Therefore, Part 24 is not intended to exclude these issues.[79]
2. In relation to Order 14, even before Order 14A was introduced, it was recognized that:

   Where the court is satisfied that there are no issues of fact between the parties, it would be pointless to give leave to defend on the basis that there is a triable issue of law, and this is so even if the issue of law is complex and highly arguable.[80]

3. It would be a futile waste of time and costs, particularly in the light of the legislative purpose of s. 108 of the Act, to order a trial at which precisely the same arguments would be repeated as were deployed before the judge on summary application.[81]

Alstom Signalling Ltd v Jarvis Facilities Ltd [2004] EWHC 1285 (TCC)

7.71 **Facts:** Alstom was the main contractor engaged to design, manufacture and install plant for a project to extend the Tyne and Wear Metro. Alstom engaged Jarvis to design, supply, install, test, and commission signalling and telecommunications equipment for the

[77] JCT Standard From of Building Contract (with Contractor's Design) 1981.

[78] It should be noted that the first adjudicator considered the referral was not drafted widely enough for him to also consider whether the amount of the contractor's application was justified (i.e. to open up, review, and revise the amount of the interim payment). He was simply deciding that the interim payment was payable immediately because no notices had been served by the employer.

[79] Para. 71.

[80] Para. 72, quoting from Supreme Court Practice, 1999, para. 14/4/2, citing *RG Carter Ltd v Clarke* [1990] 1WLR 578.

[81] Para. 74.

project. The adjudicator decided that Alstom had not served a valid withholding notice and Jarvis was entitled to the sum in its payment application. Jarvis sought to enforce the decision and Alstom applied for declarations that it had served a valid withholding notice and that the sum due was the amount in Alstom's certificate and not that in Jarvis' payment application.

7.72 **Held:** Judge LLoyd decided that he could expeditiously consider the substantive disputes at the same time as considering enforcing the adjudicator's decision and granted the declarations sought:

> 20. ... If, however, before an application to enforce an adjudicator's decision is heard, the point decided by it is finally determined adversely to the party who is relying on the decision then that application and the action will fail. That might be so if the point related to a standard form of contract and the point was determined in proceedings between other parties. Any other conclusion would be verging on the absurd: to allow the application to enforce the decision and then to set it aside (assuming the defendant had its tackle in order to do so). The decision is binding only in so far as the dispute has not been finally determined. The Act does not say when the final determination may take place. In my judgment the Act does not lead to any such technical absurdity, nor is it permissible under the Civil Procedure Rules as it is directly contrary to the overriding objective and other provisions of Part 1.Once the court is seized of the case it has to take a course which saves expense and is expeditious. To proceed first to deal with the application for summary judgment, to allow it and then to track back and to determine the dispute that gave rise to it is not consistent with the principles of Part 1 of the CPR and it is not in the interests of both parties, when they can be satisfied in an expeditious and less expensive way. Similarly it may be prudent to defer an application to enforce or to stay a judgment if the point in dispute is to be decided soon. Transferring money for a limited period of time may not be sensible. Mr Bowdery suggested that to consider the point in question would effectively destroy the efficacy of adjudication. I disagree. Most adjudications are about issues of fact. In ordinary course of events, they will not be capable of being finally determined, even in this court or in a swift arbitration, before the application for summary judgment is normally heard. It is possible that, particularly where the point is one of law or otherwise capable of being tried early, a party might move with determination and speed and get in first, as it were (as Alstom has done). I do not believe that the court's powers are so circumscribed by the Act that, in an appropriate case, it cannot order that the dispute should be determined prior to or at the same time the application for enforcement is determined. It has happened before in this court. The interests of the parties are surely best served by such a determination and not by uncertainty. Alstom has a right to a determination of the points that it has raised, just as Jarvis has a right to have its application heard and to know if the decision is enforceable. The two can be decided at the same time.

Table 7.4 Table of Cases: Enforcement by Summary Judgment

Title	Citation	Issue
Outwing Construction Ltd v H. Randell & Son Ltd	[1999] BLR 156, [1999] EWHC 100 (TCC), Judge LLoyd	**Abridged time** The court agreed to an abridgment of period for acknowledging service and adducing affidavit evidence. Shortly thereafter Randell paid the sum sought.
VHE Construction plc v RBSTB Trust Co. Ltd	[2000] All ER (D) 23 (TCC), Judge Hicks	**Determination of issues of law** Where there is no dispute of fact, but rather questions of law or construction of documents which would be finally determinative of the

(Continued)

Table 7.4 *Continued*

Title	Citation	Issue
		proceedings, these should be considered by the court. It would be a futile waste of time and costs to order a trial which would consider the same arguments as considered at summary judgment.
Glencot Development v Ben Barrett	[2001] All ER (D) 384 (TCC), Judge LLoyd	**Application of CPR Part 24** The rights of the defendant are largely the same as in any other application for summary judgment except that the room to question the amount decided to be due is very limited. The court has no inherent power in a Part 24 application to make an order for interim payment under Part 25.
Debeck Ductwork Installation Ltd v T&E Engineering Ltd	14 Oct. 2002 (TCC), Judge Kirkham (unreported)	**Mini trial inappropriate** It is inappropriate for the court to conduct a 'mini trial' to reach their conclusion.
Alstom Signalling Ltd v Jarvis Facilities Ltd	[2004] All ER (D), [2004] EWHC 1285 (TCC), Judge LLoyd	**Consideration of substantive dispute during summary judgment** Where the point is one of law (or otherwise capable of being tried early), the interests of the parties (and Part 1 of the CPR) will be best served by determination of the issue during summary proceedings.
Gray & Sons Builders (Bedford) Ltd v Essnetial Box Company Ltd	[2006] 108 Con LR 49, [2006] EWHC 2520 (TCC), Judge Coulson	**Preparation for the TCC Special Procedure** Documents must be put in order properly at the outset; speed is of the essence; careful work is necessary.
Harlow & Milner Ltd v Teasdale	[2006] All ER (D) 382, [2006] EWHC 535 (TCC), Judge Coulson	**TCC Special Procedure** The TCC Guide makes it clear that, save in exceptional circumstances, the TCC will deal with all applications to enforce decisions of adjudicators quickly and efficiently and in accordance with the procedure developed with the construction industry.
Multiplex Constructions (UK) Ltd v Mott MacDonald Ltd	[2007] All ER (D) 133, EWHC 20 (TCC), Jackson J	**CPR 24: evidence a matter of controversy** The TCC should at each stage of litigation resolve every live issue capable of resolution. The question of whether the documents provided by Mott amounted to full compliance was a matter of controversy on the evidence before the court and was one which could not be resolved on the basis of written evidence at a hearing under CPR 24. The court granted the declaration for Mott to provide all pertinent records.
Leading Rule v Phoenix Interiors Ltd	[2007] EWHC 2293 (TCC) Akenhead J	**Part 8 declarations** Phoenix sought declarations (i) to overturn that part of the adjudicator's decision with which it disagreed, and (ii) that the contract did not have the effect of modifying the seven-day period in s. 112 of the 1996 Act. Leading Rule sought, amongst other things, to enforce the adjudicator's decision. The declarations were denied as the court was not satisfied that the preliminary issues would substantially save

Title	Citation	Issue
		time or money and thus they would all be heard together.
Vitpol Building Service v Samen	[2008] EWHC 2283 (TCC)	**Part 7 and Part 8 hybrid** A hybrid Part 7/8 procedure is possible to serve the individual needs of a case. The TCC Guide does not define the court's jurisdiction.
Walter Lilly & Co. Ltd v DMW Developments Ltd	[2008] All ER (D) 214 (Dec) [2008] EWHC 3139 (TCC), Coulson J	**Part 8 declaration** The adjudicator decided that the 'loss of identity' of American black walnut veneers was a breach of contract by Walter Lilly but neither the adjudicator nor the defendant's new contract administrator identified the term of the contract which was said to have been breached. The court refused to grant the declaration sought but did grant one which stated that if the only cause of the fading was natural light (as found by the adjudicator) then such condition, on its own, could not render the claimant in breach of contract.
Coventry Scaffolding Co. (London) Ltd v Lancsville Construction Ltd	[2009] All ER (D) 93, [2009] EWHC 2995 (TCC), Akenhead J	**Judgment in default** Once it is clear that there has been service of the claim form, and it is likely that a defendant is not going to participate, there is no procedural reason why judgment in default should not be obtained. It saves the court's time and claimant's costs. If there is a good reason why the acknowledgement of service has not been filed, then a defendant would also be protected by such rights as it has to apply to the court to have the judgment in default set aside. It is also possible for a claimant to apply to bring forward the hearing, which would also reduce time and costs.
Beck Interiors Ltd v Russo 132 Con LR 56,	[2009] EWHC 3861 (TCC), Ramsey J	**Impossible to resolve questions of guarantee in Part 24 proceedings** Beck had been successful in an adjudication and sought enforcement against the guarantor who maintained the guarantee was conditional and that he was not bound in his personal capacity. Russo had real prospects of successfully defending the claim on both grounds so the decision was not enforced. Not enforced.
Geoffrey Osborne Ltd v Atkins Rail Ltd	[2010] BLR 363, [2009] EWHC 2425 (TCC), Edwards-Stuart J	**Consideration of the substantive dispute: use of Part 8 proceedings** An adjudicator was invited to correct a significant arithmetical error in his decision but refused to do so. The party the error benefited sought to summarily enforce the decision and the court permitted the challenge to the arithmetic as long as the matter was one which could be finally determined on the material before the court. *Bouygues UK Ltd v Dahl-Jensen UK Ltd* distinguished on the basis that in that case the defendant had not sought a declaration to finally determine the issue on which the adjudicator had erred but was simply using the error to found an argument as to excess of jurisdiction.

(Continued)

Table 7.4 *Continued*

Title	Citation	Issue
Forest Heath District Council v ISG Jackson Ltd	[2010] All ER (D) 16 (Nov), [2010] EWHC 322 (TCC), Ramsey J	**Part 8 not suitable for declaration concerning entitlement to EOT** Part 8 was not an appropriate route as it was not a case where there was no substantial dispute of fact or (as in a 'hybrid Part 8 procedure') where the disputed issue of fact could be determined by a short hearing with a narrow focus without disclosure. Declaration not granted.

(5) Enforcing Part of an Adjudicator's Decision

7.73 There have been occasions when it has been found on enforcement that only part of a decision was reached in excess of jurisdiction or in breach of the rules of natural justice. Consistent with the policy of the courts to enforce the decisions of adjudicators whenever possible, rather than allow the whole decision to fall in such circumstances, the decision has been enforced in part, with only the aspects of the decision affected by the errors being unenforceable. This approach has become known as severance of the decision, with the 'bad' part of the decision being severed from the 'good', the latter standing to be enforced.

7.74 Several early cases established that a decision could be split, leaving some of it enforceable and some of it unenforceable,[82] but severance may not always be possible in practice.

7.75 The following general points of principle are discussed in the sections below:

1. Where more than one dispute is referred and decided, a part may be severed (where legally and practically possible) so as to salvage the enforceable parts which should be immediately enforced.
2. The decision may not be severable when one part cannot practically be severed from another, for example (i) when the facts or parts of the decision are so interlinked that the decision cannot be severed, or (ii) when the reasoning in the bad part is integral to the good part, or (iiii) when a breach of natural justice or excess of jurisdiction taints the whole decision and means that no part is enforceable.
3. Where only a single dispute is referred for adjudication severance will usually not be available.[83]

The Test for Severability

7.76 A six-stage test for severability was set out by Akenhead J in *Cantillon Ltd v Urvasco Ltd* (2008):[84]

[82] But see 7.78–7.84 below.

[83] These cases include *Fastrack Contractors Ltd v Morrison Construction* [2000] All ER (D) 11 (TCC); and *Homer Burgess v Chirex* (2000) [2000] BLR 124 (Scottish Outer House Court of Sessions).

[84] [2008] EWHC 282 (TCC).

(a) The first step must be to ascertain what dispute or disputes has or have been referred to adjudication. One needs to see whether in fact or in effect there is in substance only one dispute or two and what any such dispute comprises.
(b) It is open to a party to an adjudication agreement as here to seek to refer more than one dispute or difference to an adjudicator. If there is no objection to that by the other party or if the contract permits it, the adjudicator will have to resolve all referred disputes and differences. If there is objection, the adjudicator can only proceed with resolving more than one dispute or difference if the contract permits him to do so.
(c) If the decision properly addresses more than one dispute or difference, a successful jurisdictional challenge on that part of the decision which deals with one such dispute or difference will not undermine the validity and enforceability of that part of the decision which deals with the other(s).
(d) The same in logic must apply to the case where there is a non-compliance with the rules of natural justice which only affects the disposal of one dispute or difference.
(e) There is a proviso to (c) and (d) above which is that, if the decision as drafted is simply not severable in practice, for instance on the wording, or if the breach of the rules of natural justice is so severe or all pervading that the remainder of the decision is tainted, the decision will not be enforced.
(f) In all cases where there is a decision on one dispute or difference, and the adjudicator acts, materially, in excess of jurisdiction or in breach of the rules of natural justice, the decision will not be enforced by the Court.

7.77 The courts adopt a wide interpretation of 'dispute' when considering if a single dispute or more than one dispute has been referred to adjudication. If the same test is to apply to severance it significantly restricts the opportunities for severing an adjudicator's decision. For example, a final-account dispute which includes matters relating to variations, loss and expense, extension of time, prolongation, and LADs would be regarded as a single dispute (namely what sum is due on the final account) for the purposes of determining whether 'a dispute' had been referred. A strict application of the *Cantillon* test would therefore mean a decision on the value of a final account containing multiple elements would not be severable in circumstances where there was no jurisdiction in relation to even a small part of the single dispute referred. The decision on the whole of the final account should therefore be unenforceable. This result may give rise to unfairness and has been queried in some judgments.[85]

7.78 In *Adonis Construction v O'Keefe Soil Remediation* (2009),[86] Clarke J expressed doubt (*obiter*) that a single dispute could never be severed: 'I entertain some doubt as to whether the adjudicator's entire decision, if otherwise within jurisdiction, is to be regarded as wholly without jurisdiction because of his inapposite application of the costs clause.' That there may need to be a different approach to severance when a decision deals with a single dispute was expressed in *Pilon Ltd v Beyer Group plc* (2010).[87] Coulson J stated as follows:

> 40. … I acknowledge that it may soon be time for the TCC to review whether, where there is a single dispute, if it can be shown that a jurisdiction/natural justice point is worth a fixed amount which is significantly less than the overall sum awarded by the adjudicator, severance could properly be considered. That was, after all, the basis on which summary judgment applications were routinely decided before the *HGCRA*. However, as a result of my other findings, this is not the place to consider that issue further.

---

[85] As noted by the judge in *Quartzelec Ltd v Honeywell Control System Ltd* [2008] EWHC 3315 (TCC).
[86] [2009] CILL 2784, [2009] EWHC 2047 (TCC).
[87] [2010] All ER (D) 197, [2010] EWHC 837 (TCC).

7.79 A possible alternative approach seems to have been considered in *Cleveland Bridge (UK) Ltd v Whessoe-Volker Stein Joint Venture* (2010).[88] That concerned a contract which provided for works some of which were construction operations under the Act, and some were not. The notice of adjudication referred one dispute to the adjudicator who only had jurisdiction to deal with that dispute in so far as it arose under the part of the subcontract which related to construction operations. The notice asked for one decision but also asked in the alternative for separate decisions on the separate parts, in case severance was necessary. The judge said that where a single decision is made on a single dispute, then it cannot be severed, but appeared to say that severance may have been available if separate decisions had been made on the separate parts:

> 108. ... The Adjudicator held that she had jurisdiction over the whole dispute and her Decision is a single decision in relation to that whole dispute. She did not provide a decision which dealt with Cleveland Bridge's alternative position that she had jurisdiction as to part of but not the whole of the dispute.
>
> 109. Having found that the Adjudicator had jurisdiction only in respect of part of the dispute it follows that her Decision on the whole of the dispute, to the effect that £317,500 plus VAT was due and payable to Cleveland Bridge was not a decision for which she had jurisdiction. In general where a single decision is made relating to one dispute and the adjudicator does not have jurisdiction for some element of that dispute, the decision will not be valid and enforceable.
>
> 110. If the Adjudicator had made a decision on the whole dispute but had also made a decision which dealt only with the part of the dispute which was within her jurisdiction then, in my judgment, the decision on the whole dispute would not be enforceable or valid but there would be a valid decision on the part of the dispute which was within her jurisdiction.
>
> 111. Absent such a position is a decision made in respect of a single dispute, part of which is within jurisdiction and part outside jurisdiction, severable so that the part within the jurisdiction can be enforced?

7.80 Notwithstanding that the *Cantillon* 'single dispute' test was strictly applied in *Quartzelec Ltd v Honeywell Control Systems Ltd* (2008)[89] and formed part of the *ratio* of the decision in *Working Environments Ltd v Greencoat Construction Ltd* (2012),[90] in *Beck Interiors Ltd v UK Flooring Contractors Ltd* (2012) (Key Case)[91] Akenhead J said that 'different considerations as to severability may arise in relation to different jurisdictional challenges'. In that case the adjudication notice contained one claim for the extra cost of recarpeting and one claim for LADs. The notice identified separate sums relating to each claim. The LAD claim had not been made prior to the notice of adjudication with the result that the adjudicator had no jurisdiction over that part of the claim (for want of a dispute having crystallized). The judge severed the decision regarding the LAD claim and enforced the decision regarding the cost of carpeting. The judgment does not say that this was possible because two separate disputes were referred.

Other Examples of Adjudicators' Decisions which Have Been Severed

7.81 Examples where the courts have permitted severance of an adjudicator's decision include where:

1. The decision was enforceable only as to two of the four invoices referred, as no dispute had crystallized in relation to the later invoices, served without supporting back-up less

[88] [2010] EWHC 1076 (TCC).
[89] [2009] BLR 328, [2008] EWHC 3315 (TCC).
[90] [2012] All ER (D) 23, [2012] EWHC 1039 (TCC).
[91] [2012] All ER (D) 31, [2012] EWHC 1808 (TCC).

than two weeks before the notice of adjudication: *Ken Griffin & John Tomlinson (t/a K&D Contractors) v Midas Homes Ltd* (2000).[92]

2. The decision was enforceable except that part which related to rescission, as the adjudicator had fallen into error, and the defendant accepted that the other amounts were unaffected by the invalidity of the adjudicator's findings in relation to rescission: *Barr Ltd v Law Mining Ltd* (2001).[93]
3. One of the three components of the decision reached was tainted by breaches of natural justice, but the court salvaged the other two decisions and enforced them immediately: *AWG Construction Services Ltd v Rockingham Motor Speedway Ltd* (2004).[94]
4. The decision was enforced with the exception of the fees order, which was inconsistent with the parties' agreement to pay 50 per cent each: *Interserve Industrial Services Ltd v Cleveland Bridge UK Ltd* (2006).[95]

Key Case: Severability

Beck Interiors Ltd v UK Flooring Contractors Ltd [2012] EWHC 1808 (TCC)

7.82 **Facts:** The claimant referred its financial claim to adjudication following termination of its contract with the defendant. That claim included a sum for liquidated damages in respect of which the adjudicator made an award.

7.83 **Held:** The liquidated damages dispute had not crystallized at the time of referral and the adjudicator therefore had no jurisdiction to award a sum in respect of that head of claim. Since there was a clearly identifiable sum relating to that claim, that part of the decision could be severed. After citing the six-part test in *Cantillon*, Akenhead J stated:

> 21. The Court needs to bear in mind that there are many different types of jurisdictional challenge. They include issues as to whether a dispute has crystallised, whether two disputes have been referred to adjudication, whether the adjudicator has been properly appointed, whether there is an effective adjudication provision, whether there is a contract at all between the parties, whether the subject matter of the contract is exempt from the statutory provisions for adjudication and numerous others. Different considerations as to severability may arise in relation to different jurisdictional challenges.
>
> ...
>
> 32. The next question which arises is therefore whether it is legitimate for the Court to sever the decision and in effect enforce that part of the decision which relates to that which was generally in dispute before the Notice of Adjudication was dispatched. In my judgement, this is a case in which the Court can and should do that. In reality, there was only one crystallised dispute which was referable to adjudication and that related only to the claim (as slightly adjusted) for £31,148.97 for the increased costs of completing the carpeting work. It is clear from the body of the Notice of Adjudication that the presented claim is made up essentially of two parts, £31,148.97 and the £36,000 for the new liquidated damages claim. They are presented in effect as two separate arguments with separable evidence supporting them, albeit that the losses flow from the failure to complete on time or indeed at all.
>
> 33. There is no difficulty in identifying clearly what the adjudicator decided in relation to each claim: £19,763.41 for the increased costs of completion and £33,600 for liquidated damages. It was rightly accepted that the claim for the fixed sum of £100 under the

[92] 78 Con LR 52, [2000] EWHC 182 (TCC).
[93] 80 Con LR 134 [2001], (Scottish Outer House).
[94] [2006] All ER (D) 68, [2004] EWHC 888 (TCC).
[95] [2006] All ER (D) 49, [2006] EWHC 741.

Commercial Late Payment of Debts (Interest) Act could not be enforced because it related most obviously to the liquidated damages claim which is the only element which could be said to give rise to a debt, the other claim relating simply to common law damages. It is also difficult for the Court to apportion the adjudicator's fee which he ordered (primarily) UKFCL to pay; this is because, although one could arithmetically apportion it in relation to sums recovered and others not jurisdictionally recoverable, one can not second guess what the adjudicator would have done. For instance, he might have said that each party should pay half or that Beck should pay the costs of his time relating to the jurisdictional issue and some different proportion of his costs for the balance.

Table 7.5 Table of Cases: Severability (the shaded entries indicate a decision was severed)

Title	Citation	Issue
Homer Burgess v Chirex	[2000] BLR 124; (Scottish Outer House Court of Sessions), Lord Macfadyen	**Severability: part of claim related to work that was not construction operations** The referring party claimed sums relating to its works carried out at a pharmaceutical plant, including the installation of pipework. The adjudicator fell into error in his construction of the expression 'plant' in s. 105(2)(c) and the installation of pipework was held to be an operation which fell within the exception in s. 105(2)(c)(ii). He therefore had no jurisdiction to award sums in respect of that pipework installation. The decision could not stand to its full extent. The consequences for the part of his decision that was within his jurisdiction was reserved for further argument which was not reported (see below).
Homer Burgess v Chirex (no. 2)	Unreported. This decision is referred to in *RSL (South West) Limited v Stansell Limited*	**Severability: either whole decision fails or sever and enforce the part made with jurisdiction** Lord MacFadyen said 'It would, in my view, be open to me to regard the adjudicator's error as to the scope of his jurisdiction as undermining the validity of his decision as a whole, despite there being parts of it that might have been made to the same effect if he had not erred as to his jurisdiction. It would therefore be open to me to reduce the whole of the adjudicator's decision. Alternatively, it would in my view be open to me to approach the matter from the pursuers' rather than the defenders' point of view, ask myself to what extent the decision was intra vires, and grant decree for payment enforcing that part of the decision that was valid and could properly be given the statutory temporary effect'. (But see Ramsey J in *Cleveland Bridge v Whessoe* below, who decided that only the first option was available where one dispute has been referred.)
KNS Industrial Services (Birmingham) Ltd v Sindall Ltd	[2000] All ER (D) 1153 (TCC), Judge LLoyd	**Severablity: where only one dispute, accept whole of decision or seek a final determination of the issue** The unsuccessful party, KNS, commenced the action, arguing that it was due more than the amount which had been determined by the adjudicator. Sindall cross-claimed for enforcement. The court enforced the decision, finding that the figures could not be isolated or severed from the adjudicator's decision. There was only one dispute, even though it embraced a number of claims or issues. It was not right to try to dismantle the decision and the parties had to accept it 'warts and all'. Adjudicators' decisions are not to be used as a launching pad for satellite litigation designed

Title	Citation	Issue
		to obtain what is to be obtained by litigation or arbitration. The parties cannot come to the court to have a decision revised to excise what was unwanted and to replace it with what was or was thought to be right, unless the court is the ultimate tribunal. Followed in *Farebrother v Frogmore.*
Ken Griffin & John Tomlinson (t/a K&D Contractors) v Midas Homes Ltd	78 Con LR 152, [2000] EWHC 182 (TCC), Judge LLoyd	**Severability: the adjudicator's finding in relation to only two of four invoices was enforceable** A dispute concerning payment of four invoices was referred and an adjudicator determined that £52,493.91 was owed to the claimant. The court found that it was difficult to conclude there existed any dispute referable to adjudication on 3 May 2000 about the claims which were unsupported by any back up and in any event only reached the defendant on 19/20 April 2000. Adjudication was not a substitute for discussion and negotiation nor an agenda for such where no dispute had truly existed. Accordingly, the decision was only enforceable in relation to two invoices served in February 2000 amounting to £11,917.
Barr Ltd v Law Mining Ltd	(2001) 80 Con LR 134, (Scottish Outer House), Lord Macfadyen	**Severability: defendant agreed to sever the invalid part of the adjudicator's decision on rescission** The court found that the adjudicator would only have had jurisdiction to consider the claimant's entitlement to certain payment once he had formed the view that there had been no rescission. However, the adjudicator made no such decision and as such had no jurisdiction to determine that part of the dispute affected by that issue. The defendant did not dispute the competency in principle of partial enforcement. Accordingly, the claimant was entitled to a summary decree for those amounts unaffected by the invalidity of the adjudicator's decision on the rescission issue.
Farebrother Building Services Ltd v Frogmore Investments Ltd	(2001) CILL 1762, 20 April 2001 (TCC), Judge Gilliland	**Severability: not right for a court to try to dismantle and reconstruct a decision** In a decision under the TeCSA Rules, the adjudicator forms a binding view as to his jurisdiction. The TeCSA Rules do not permit a party to pick and choose amongst the decisions given, assert or characterize part as unjustified and then allege that the part objected to has been made without jurisdiction. Either the adjudicator has jurisdiction or he does not. If he has jurisdiction, the decision is binding. Followed *KNS v Sindall.*
Shimizu Europe Ltd v Automajor Ltd	[2002] All ER (D) 80 (TCC), Judge Seymour	**Severability: cannot simultaneously approbate and reprobate a decision on a single dispute** Where there is one decision that a sum is payable or that an EOT is due, that decision must either be accepted in its entirety or not at all. Where there is more than one dispute referred to the adjudicator (e.g. a payment claim and an EOT), a valid objection to one dispute will not affect the enforceability of the other dispute.
Hitec Power Protection BV v MCI Worldcom Ltd	[2002] EWHC 1953 (TCC), Judge Seymour	**Severability: an award made in excess of jurisdiction should not be severed** Once a court finds an award has been made in excess of jurisdiction, it is not a part of the court's function to seek to salvage from the debris of an unenforcable award, or an

(Continued)

Table 7.5 *Continued*

Title	Citation	Issue
		award unenforcable in total, something for the benefit of the party seeking to enforce.
RSL (South West) Ltd v Stansell Ltd	[2003] EWHC 1390 (TCC), Judge Seymour	**Severability: decision on EOT not severable** An adjudicator's decision in relation to an EOT was a nullity due to breaches of natural justice. The court was asked to 'salvage from the wreckage' those other elements from the final account that were not tainted by the breaches. The court found that the decision was wholly unenforceable.
AWG Construction Services Ltd v Rockingham Motor Speedway Ltd	[2004] EWHC 888 (TCC), Judge Toulmin	**Severability: three disputes issued to adjudication** The parties agreed that three separate disputes were referred to a single adjudicator (the Oval, grandstand, and tunnel disputes). There had been a breach of natural justice in relation to the Oval dispute. However, the court immediately enforced the other two decisions which were not tainted by the breach of natural justice.
Carillion Construction Ltd v Devonport Royal Dockyard Ltd	[2005] EWCA [2005] All ER (D) 202 (CA), Chadwick, Moore-Bick LJJ	**Severability: cherry-picking from decisions permitted** The unsuccessful party challenged enforcement on four grounds. If the attack on the adjudicator's decision were to succeed under any one of the first three grounds alone (breach of natural justice and exceeding his jurisdiction on issues of target cost and defects), the decision would be unenforceable. However, if the attack was successful on the fourth ground alone (no jurisdiction to award interest), then the award of interest would be severable and the balance of the decision could be enforced. By agreeing that the adjudicator should decide whether interest should be paid the parties conferred on him a jurisdiction to award interest which he would not otherwise have had. The decision was enforced in whole.
Interserve Industrial Services Ltd v Cleveland Bridge UK Ltd	[2006] All ER (D) 49 (Feb); [2006] EWHC 741 (TCC), Jackson J	**Severability: fees** The adjudicator did not have jurisdiction to make an order requiring Cleveland to pay 80 per cent of his fees and expenses, as the parties had agreed in their contract to split the fees and expenses equally. The parties accepted (and the court agreed) that the part of the decision dealing with fees was severable from the rest of the adjudicator's decision.
Cantillon Ltd v Urvasco Ltd	[2008] EWHC 282, [2008] BLR 250 All ER (D) 406 (TCC), Akenhead J	**Severability: six principles** Although the decision did not require a finding on this issue, Akenhead J set out, *obiter*, a six-stage test for severability.
Quartzelec Ltd v Honeywell Control Systems Ltd	[2009] BLR 328, [2008] EWHC 3315 (TCC), Judge Davies	**Severability: one dispute with many issues** The court found that, in refusing to consider a defence raised by the responding party, the adjudicator committed a significant jurisdictional error and failed to act in accordance with natural justice. There was one single dispute and as such the whole decision was unenforceable.
Bovis Lend Least Ltd v Trustees of the London Clinic	[2009] All ER (D) 240, [2009] EWHC 64 (TCC), Akenhead J	**Severability: whether legally and practically possible** The court did not have to determine the issue, due to its finding that the adjudicator had jurisdiction to consider the dispute relating to delay, EOT, the recovery

Title	Citation	Issue
		of liquidated damages, and loss and expense. However, Akenhead J indicated that, had the crystallized dispute not included the loss and expense claim, the decision could have been severed to enforce the EOT and recovery of LADs. The only difficulty would have come with regard to the question of the adjudicator's fees and expenses that were not apportioned as between the two parts of the claim. If that had proved an insuperable difficulty, leave to enforce that part of the decision would simply not have been granted.
Adonis Construction v O'Keefe Soil Remediation	[2009] CILL 2784, [2009] EWHC 2047 (TCC), Clarke J	**Severability: costs** At paras. 49–50 the judge expressed doubt (*obiter*) that where a single dispute was referred the valid part could never be severed and enforced, for example, where the adjudicator's error was only in relation to costs.
Pilon Ltd v Beyer Group plc	[2010] All ER (D) 197, [2010] EWHC 837 (TCC), Coulson J	**Severability: failure to consider defence** Pilon's application for enforcement of an adjudicator's decision was challenged by Beyer who argued that the adjudicator had failed to consider its overpayment defence, worth around £147,000. The dispute determined was 'what, if anything, was due as a result of the interim application of September 2009'. The decision was not severable. Even if the decision was severable in principle, the adjudicator's failure to address the defence may have tainted the decision as a whole with this material error.
Cleveland Bridge (UK) Ltd v Whessoe-Volker Stein Joint Venture	[2010] All ER (D) 206, [2010] EWHC 1076 (TCC), Ramsey J	**Severability: decision made without jurisdiction** An adjudicator made a decision on works under a subcontract, some of which were found by the TCC not to be 'construction operations' under s. 105(2)(c) of the 1996 Act. Where the adjudicator made a single decision on a single dispute referred, the whole of the decision is not enforceable. A temporarily binding decision cannot be dissected to impose a separate and severable obligation to be bound by the adjudicator's decision on each of the component issues on which the adjudicator based that decision. To do otherwise would produce a decision partly made by the adjudicator and partly made by the court. That is not the role of the court.
Carillion Utility Services Ltd v SP Power Systems Ltd	[2011] CSOH 139, Lord Hodge	**Severability: breach of natural justice** The adjudicator breached natural justice by applying commercial rates that, from his experience, he saw as reasonable, but on which he had not given the parties an opportunity to comment. He also held that SP was to pay 60 per cent of his fees and expenses. However, there was no way to sever the decision on expenses from the decision on the merits or alter the decision on the merits while leaving the expenses decision in place. The whole decision was not enforced.

(Continued)

Table 7.5 *Continued*

Title	Citation	Issue
Working Environments Ltd v Greencoat Construction Ltd	[2012] EWHC 1039 (TCC), Akenhead J	**Severability: adjudicator decided matters that were not part of referral and not raised in defence** An adjudication concerning the proper value of an interim application, comprised issues about the value of variations, whether an abatement was valid and whether a set-off was valid. A valid withholding notice served after the adjudication started was relied on and raised further issues. The judge dismissed an argument that the withholding notice was outside the jurisdiction of the adjudicator. He was entitled to consider a legally available defence to the dispute referred (but could not take defence points of his own initiative). The decision on two issues had been made without jurisdiction and was severed. They related to LADs and had not been raised by the referring party until 22 days into the adjudication.
Beck Interiors Ltd v UK Flooring Contractors Ltd	[2012] All ER (D) 31, [2012] EWHC 1808 (TCC), Akenhead J	**Severability: LAD element of claim for costs following termination not previously asserted** In an adjudication of a dispute relating to a financial claim following termination of a contract, the notice of adjudication included one claim for the increased cost of carpeting work and one claim for LADs. The adjudicator's decision included a clearly identified sum in respect of a claim for LADs and a separate sum for the carpeting cost claim. It was found that the LAD claim had not been asserted prior to the notice of adjudication with the result that the adjudicator had no jurisdiction to award a sum in respect of LADs. The court severed that part of the decision relating to LADs whilst enforcing the balance of the decision.
Lidl UK GmbH v RG Carter Colchester Ltd	[2013] CILL 3276, [2012] EWHC 3138 (TCC), Edwards-Stuart J	**Severability: reasoning in bad part integral to whole** As a matter of principle, where a single dispute was referred, it is not possible to sever part of the decision made without jurisdiction from the rest of the decision if the reasoning in the 'bad' part was integral to the reasoning in the 'good' part. But even if a single dispute was referred the court could sever decisions on additional issues brought into the adjudication (whether in oversight or error) if the reasoning in that part was not integral to the whole decision. Failing that, the entire decision would be unenforceable.

(6) Approbation and Reprobation

7.84 The principle of 'approbating and reprobating' or 'blowing hot and blowing cold' may arise where the party in question is to be treated as having made an election from which he cannot resile, but he will not be regarded as having so elected unless he has taken a benefit under or arising out of the course of conduct which he first pursued and with which his subsequent action is inconsistent.[96]

[96] Evershed MR, with whom Singleton and Jenkins LJJ agreed, in *Banque des Marchands de Moscou (Koupetschesky) v Kindersley* [1950] 2 All ER 549, [1951] 1 Ch 112, 119; endorsed in *Redworth Construction Ltd v Brookdale Healthcare Ltd* [2006] BLR 366, [2006] EWHC 1994 (TCC).

7.85 That principle applies in the context of adjudication so that, in summary:

1. A party who has accepted and/or relied on the decision of an adjudicator at one point is prevented, at a later stage, from objecting to the adjudicator's jurisdiction.
2. A party may not be permitted to take a conflicting position in relation to jurisdiction, in subsequent enforcement proceedings.
3. A party may be deemed to have taken a conflicting position through acquiescence.
4. The principle of election applies equally to elections following the issuing of an adjudicator's decision, for example by part payment, invitation to use the slip rule, or reference of a dispute which had been the subject of an adjudication to arbitration.

The Principle of Election

7.86 Dyson J confirmed the applicability of the doctrine of approbation and reprobation in enforcement proceedings and applications for stays under s. 9 of the Arbitration Act in *Macob Civil Engineering Ltd v Morrison Construction Ltd* [1999] (Key Case).[97] Since that case, it has become well established that a party cannot both assert that an adjudicator's decision is valid for certain purposes and at the same time seek to challenge the validity of the decision. The party must elect to take one course or the other. By taking a benefit under an adjudicator's decision, the party will generally be taken to have elected a particular course and will be precluded from challenging the adjudicator's decision.[98] The principle has been described as 'essentially one of estoppel'.[99] An election may arise either in the course of an adjudication or after the decision has been delivered.

7.87 However, there have been differences in approach in the decided cases as to what represents the taking of a benefit, or indeed if taking of a benefit is needed at all in the context of an adjudication, with the result that the question in each case is likely to be highly fact sensitive. In *R. Durtnell & Sons Ltd v Kaduna Ltd* (2003)[100] Judge Seymour took a very broad view as to what constitutes the taking of a benefit but this was subsequently doubted by Coulson J in *Amec Group Ltd v Thames Water Utilities* (2010).[101] Meanwhile, in *PT Building Services Ltd v ROK Build Ltd* (2008)[102] Ramsey J identified that taking a benefit was sufficient but not necessary to establish an election.

Elections Made During an Adjudication

7.88 A number of authorities suggest that a party cannot adopt one position during an adjudication and a contrary position in the course of enforcement proceedings. The question has most often arisen where a party to an adjudication takes a position in relation to the existence, or non-existence, of a construction contract and then takes the opposite position at the enforcement hearing. The defendant may be estopped from resiling from its earlier position.

7.89 Perhaps the most obvious example is where a referring party had a dispute determined partially in its favour by reference to a contract containing one set of terms, and was not permitted to change its ground and abandon its contention that the contract upon which it succeeded in the adjudication was ever made: *Galliford Try Construction Ltd v Michael Heal Associates Ltd* (2003).[103] In that case, Judge Seymour stated that, in seeking to resist an

---

[97] [1999] All ER (D) 143 (TCC).
[98] *PT Building Services Ltd v ROK Build Ltd* [2008] EWHC 3434 (TCC) at para. 26, per Ramsey J.
[99] *RJ Knapman Ltd v Richards* [2006] 108 Con LR 64; [2006], [2006] EWHC 2518 (TCC).
[100] [2003] All ER (D) 281, [2003] EWHC 517 (TCC).
[101] [2010] All ER (D) 267, [2010] EWHC 419 (TCC).
[102] [2008] EWHC 3434 (TCC).
[103] [2003] All ER (D) 07, [2003] EWHC 2886 (TCC).

objection to enforcement of the adjudicator's decision on the grounds that no construction contract existed, Galliford Try was:

> 52. ... playing fast and loose with the process of adjudication, shifting its ground opportunistically to meet the challenge of the moment. No Court can be expected to treat phlegmatically a case in which a successful party to an adjudication comes before it saying, '*I know that I succeeded in the adjudication on a basis which I now recognise was wrong in law, but the adjudicator decided what he was asked to decide and it is just tough luck for the Defendant*'. That attitude seems to come very close to an abuse of the process of adjudication.

7.90 However, what may be more difficult is the position in enforcement proceedings where a party seeks to rely on evidence and/or arguments which it elected not to advance during the adjudication. In *Redworth Construction Ltd v Brookdale Healthcare Ltd* (2006),[104] a party elected to put its argument in a particular way, based on a particular document, arguably in order to obtain the benefit of a decision in its favour without introducing unfavourable evidence in other documents as to sectional completion dates. Upon enforcement, when a challenge was made as to the existence of a construction contract, the court decided that an election had been made and that party could not subsequently rely on a different document.

7.91 *Redworth v Brookdale* has been relied upon in subsequent cases for the proposition that: 'in enforcement proceedings a claimant cannot go beyond the matters on which it relied in the adjudication in support of its argument that the adjudicator had the necessary jurisdiction'.[105] However, it has recently been distinguished in *Nickleby FM Ltd v Somerfield Stores Ltd* (2010),[106] where Akenhead J found himself in 'some disagreement' with the principles expressed in *Redworth v Brookdale* and commented:

> 28. ... an adjudicator, who reaches what is expressed and accepted by him and the parties as a non-binding decision, has only enquired into his jurisdiction as he was entitled to do and it is primarily in the court that a binding decision can be given as to jurisdiction. I can not see that principles of election apply in these circumstances. Of course, if a respondent to an adjudication does not challenge the jurisdiction of the adjudicator during the adjudication when it knows of the grounds of challenge, it will generally be deemed to have waived or abandoned any rights to challenge the jurisdiction on those grounds. That however is not in strict terms election. Whether the Redworth decision was rightly decided or not on this point, one needs to examine in any event with care whether a materially different case on jurisdiction is being mounted in the court proceedings compared with that raised before the adjudicator. It must also be relevant to consider whether at least in a clear case the adjudicator with the correct and full information before him would have reached the same conclusion that he did. It will also be relevant to consider whether the adjudicator in fact and in reality actually did have jurisdiction. If he or she did have jurisdiction to decide the dispute referred to adjudication, and if he or she with the full information available would have inevitably concluded that there was jurisdiction, I can not see why the adjudication decision should not be enforced in those circumstances.

7.92 During the adjudication in question, Nickleby had argued there was a construction contract on the basis of offer accepted by conduct, whereas during the enforcement, its particulars of claim additionally argued on the basis of an oral agreement. Akenhead J found that Nickleby had not advanced a case on jurisdiction in enforcement proceedings that was materially or prejudicially different from that which it advanced in the adjudication. Akenhead J also found that as a matter of fact and law the adjudicator actually did have jurisdiction

[104] [2006] BLR 366, [2006] EWHC 1994 (TCC) (Key Case).

[105] *Durham County Council v Jeremy Kendall (t/a HLB Architects)* [2011] All ER (D) 351, [2011] EWHC 780 (TCC)—on the facts of this case, it was considered a difficult argument to maintain and the matter was not considered in detail (see para. 37 of the decision).

[106] [2010] 131 Con LR 203; [2010] EWHC 1976 (TCC).

at the time he was appointed. There was no reason to believe the adjudicator would have reached any non-binding decision other than that he had jurisdiction if he had been given the additional evidence that was available at enforcement but, by oversight from both parties, had been withheld from him during the adjudication. The clear factual differences between *Redworth* and *Nickleby* are that in *Nickleby* both parties overlooked the additional evidence and ultimately agreed that there was in fact a contract in writing, whereas in *Redworth* it arguably suited the successful party's case to bring only certain documents to the adjudicator's attention on jurisdiction and the parties remained in dispute as to the basis on which the contract had been formed.

7.93 It is important to distinguish the principle of election from a reservation of rights. Whilst a party generally has to reserve its rights to bring a jurisdictional objection, that is only necessary where the party has knowledge of the potential objection. A party is not prohibited from relying upon an argument that an adjudicator has decided something not referred to him or not in dispute at the time of the notice of referral unless, with knowledge of the availability of the point, he has elected not to raise it.[107]

Elections Made After an Adjudication

7.94 The principle of election applies equally to decisions following an adjudication. If a party has treated a decision as a valid decision, then that party may have elected to forgo any opportunity to object to the decision.[108] If a party considers that an adjudicator's decision was not a valid decision, for whatever reason, then that party's actions must consistently show that it believes that, for example, as a result of that invalid decision, there was nothing to pay, nothing to correct, and nothing from which a dispute may be referred to arbitration. Courts may refuse to allow parties to challenge the enforceability of adjudicators' decisions if, through their actions, and without adequately reserving their rights, they have acknowledged the validity of the decision by electing to:

1. Make part payment of the amounts ordered by the adjudicator.[109]
2. In the absence of any circumstances indicating to the contrary, pay the adjudicator's fees and expenses.[110]
3. Use part of the adjudicator's decision as the basis of subsequent payments and a fresh withholding notice.[111]
4. Apply to the adjudicator to correct his decision under the slip rule.[112]
5. Rely on the adjudicator's decision as a defence in a second adjudication.[113]
6. Ask the court to enforce an adjudicator's decision, but simultaneously requesting a Part 8 declaration that part of the adjudicator's decision was wrong.[114]
7. Starting an arbitration disputing the decision of the adjudicator where the contract provided a dispute process that required a final decision be achieved by a challenge to the substance of the adjudicator's decision (as opposed to separate proceedings for final determination of the matter anew).[115]

---

[107] *R. Durtnell & Sons Ltd v Kaduna Ltd* [2003] All ER (D) 281, [2003] EWHC 517 (TCC). See further Ch. 5 at 5.09–5.16 for a discussion of adequate reservation of rights.

[108] *Shimizu Europe Ltd v Automajor Ltd* [2002] All ER (D) 80 (TCC).

[109] *Shimizu Europe Ltd v Automajor Ltd* [2002] All ER (D) 80 (TCC).

[110] *PT Building Services Ltd v ROK Build Ltd* [2008] EWHC 3434 (TCC) at para. 29.

[111] *Amec Group Ltd v Thames Water Utilities Ltd* [2010] EWHC 419 (TCC).

[112] *Shimizu Europe Ltd v Automajor Ltd* [2002] All ER (D) 80 (TCC).

[113] *PT Building Services Ltd v ROK Build Ltd* [2008] EWHC 3434 (TCC); *Linnett v Halliwells LLP* [2009] All ER (D) 36, [2009] EWHC 319 (TCC).

[114] *Pilon Ltd v Beyer Group PLC* [2010] All ER (D) 197, [2010] EWHC 837 (TCC).

[115] *Macob Civil Engineering Ltd v Morrison Construction Ltd* [1999] BLR 93 (TCC); *MBE Electrical Contractors Ltd v Honeywell Control Systems Ltd* [2010] BLR 561, [2010] EWHC 2244 (TCC).

7.95 Therefore, parties considering challenging an adjudicator's decision should be careful not to acknowledge the decision's validity by behaving as if they consider it to be valid.

Key Cases: Approbation and Reprobation

Macob Civil Engineering Ltd v Morrison Construction Ltd [1999] BLR 93 (TCC)

7.96 **Facts:** Macob was engaged as a subcontractor to Morrison to carry out groundworks at a retail development in Wales. A dispute arose concerning payment application 6, which was referred to adjudication. The adjudicator directed that Morrison pay forthwith £302,366.34 plus interest together with the adjudicator's fees. The adjudicator's decision stated that, inter alia, in the event of non-compliance by Morrison, he gave permission under s. 42 of the Arbitration Act (as modified by paragraph 24 of the Scheme) for either party to apply to the court for an order requiring such compliance. Morrison did not comply with the decision, which it claimed was invalid and unenforceable due to breach of natural justice, and referred to arbitration the disputes arising out of or in connection with the adjudicator's decision.

7.97 **Held:** Dyson J found that Morrison had elected to treat the decision as valid and refer the dispute to arbitration and was not permitted to 'blow hot and cold' by then contesting its enforceability:

> 28. . . . In my view, if the defendant wished to challenge the validity of the decision, it had an election. One course open to it was (as it did) to treat it as a decision within the meaning of clause 27, and refer the dispute to arbitration. The other was to contend that it was not a decision at all within the meaning of clause 27, and to seek to defend the enforcement proceedings on the basis that the purported decision was not binding or enforceable because it was a nullity. For the reasons stated earlier in this judgment, this second course would not have availed the defendant.
>
> 29. But what the defendant could not do was to assert that the decision was a decision for the purposes of being the subject of a reference to arbitration, but was not a decision for the purposes of being binding and enforceable pending any revision by the arbitrator. In so holding, I am doing no more than applying the doctrine of approbation and reprobation, or election. A person cannot blow hot and cold: see *Lissenden v CAV Bosch Ltd* [1940] AC 412, and Halsbury's Laws 4th Edition Volume 16, paragraphs 957 and 958. Once the defendant elected to treat the decision as one capable of being referred to arbitration, he was bound also to treat it as a decision which was binding and enforceable unless revised by the arbitrator.

Pegram Shopfitters Ltd v Tally Wiejl (UK) Ltd [2004] 1 All ER 818, [2003] EWCA 1750 (CA)

7.98 **Facts:** The adjudicator had been appointed pursuant to the provisions of the Scheme. In the adjudication the defendant had asserted there was a contract on the basis of the JCT Prime Cost form, or alternatively if that was incorrect then there was no contract at all. In either case the adjudicator would not have jurisdiction, the incorrect appointment process having been followed on its primary case, or the Act not applying on its alternative case. The adjudicator accepted the claimant's case on contract formation, which was that there was a construction contract in writing to which the scheme was implied. On enforcement, the defendant maintained its alternative case that there was no contract. At first instance it was held that the defendant was precluded from contending there was no contract at all, having contended in the adjudication that there was a contract in writing.

Held: **7.99**

29. The judge was, in my judgment, wrong (as was the adjudicator) to proceed on the unquestionable premise that both parties agreed that their relationship was governed by a construction contract. He was wrong to preclude the defendants from contending in the alternative that there was no contract. Their written submissions, witness statements and skeleton arguments before the adjudicator and the judge make it perfectly clear that their first case was that there was a written contract in JCT Prime Cost Terms. Their alternative was that if that were wrong, there was no contract, certainly no contract in writing. They were not contending or accepting that there was a written construction contract come what may. Nor were they accepting that, if their own first case failed, there was a written construction contract on the claimants' conditions. It was not diametrically opposite to the approach of the defendants before the adjudicator to contend in the alternative that there was no written construction contract. The judge incidentally at the outset of his judgment recorded the defendants' contentions in the wrong order. It was not their first contention that there was no contract at all. In my judgment, the judge was wrong to conclude that it was not open to the defendants to advance their alternative contention in the enforcement proceedings. His decision was wrongly premised on the assumption that the defendants had accepted that there was a written construction contract, however the contractual dispute was resolved, and that the only dispute was as to its terms. The judge was also, I think, wrong to suppose that because (as he thought) there was a construction contract but the parties were not able clearly to identify its terms, the Scheme applied because the parties had not produced a construction contract which complied with section 108 of the 1996 Act. This was simply ducking the critical question.

Table 7.6 Table of Cases: Approbation and Reprobation (shaded entries indicate a finding of election by the court)

Title	Citation	Issue
Macob Civil Engineering Ltd v Morrison Construction Ltd	[1999] BLR 93 (TCC), Dyson J	**Election: reference of adjudicator's decision to arbitration** The parties had entered a contract in which adjudication was a mandatory step prior to arbitration and any disputes arising out of an adjudication were to be referred to arbitration. Morrison sought to resist the enforcement of an adjudicator's decision but had already referred the disputes arising from the decision to arbitration. In light of the contractual provisions, the referral to arbitration was found to be an election to treat the decision of the adjudicator as valid which precluded Morrison from contending that it was not valid in the course of enforcement proceedings. If the decision was not valid then there was no dispute arising from the decision that could be referred to arbitration. Morrison should simply have sought to defend the enforcement proceedings.
Shimizu Europe Ltd v Automajor Ltd	[2002] All ER (D) 80 (TCC), Judge Seymour	**Election: slip rule and part payment** The unsuccessful party invited the adjudicator to correct his decision under the slip rule and also made part payment against the decision. On seeking to resist enforcement it was held, *obiter*, that by those actions it had elected to treat the decision of the adjudicator as valid and could not deny there was a valid decision to be enforced.

(Continued)

Table 7.6 *Continued*

Title	Citation	Issue
Pegram Shopfitters Ltd v Tally Wiejl (UK) Ltd	[2004] 1 All ER 818 (TCC), Judge Thornton QC; [2003] EWCA 1750 (CA), May, Hale LJJ, Hooper J	**Election: defendant maintaining alternative case on contract during enforcement** In the adjudication the defendant had asserted there was a contract on the basis of the JCT Prime Cost form, or alternatively if that was incorrect then there was no contract at all. In either case the adjudicator would not have jurisdiction, having been appointed pursuant to the provisions of the Scheme. The adjudicator accepted the claimant's case on contract formation. Held on appeal, overturning the decision at first instance, that the defendant was not precluded from relying on its alternative case that it had raised in the course of the adjudication (i.e. that there was no contract) in resisting enforcement.
Galliford Try Construction Ltd v Michael Heal Associates Ltd	[2003] All ER (D) 07, [2003] EWHC 2886 (TCC), Judge Seymour	**Election: adoption of new position on contract during enforcement** *Obiter*: Where a party argues in an adjudication that one set of contractual terms apply, it will not be permitted to abandon its contention and maintain that no contract on those terms had ever been made.
R. Durtnell & Sons Ltd v Kaduna Ltd	[2003] All ER (D) 281, [2003] EWHC 517 (TCC), Judge Seymour	**Election: where adjudicator exceeds jurisdiction, election does not apply** *Obiter*: for the doctrine of approbation and reprobation to apply to an adjudication, a party with knowledge that it was open to him to challenge the decision must instead elect to take the benefit of part of the decision. Benefit was not restricted to obtaining a cash sum or an entitlement to payment and could include a finding that crystallized, on an interim basis, a liability to pay an identified sum.[116]
RJ Knapman Ltd v Richards	[2006] 108 Con LR 64, [2006] EWHC 2518 (TCC), Coulson J	**Election: failure to comply with part of a decision does not preclude enforcement** Where an adjudicator's decision included matters that required compliance by both parties (in this case, payment of an identified sum by the defendant and a declaration that the claimant was contractually responsible for the installation of doors and windows which were unsatisfactory and incomplete) a failure of one party to comply with its obligations did not affect the other party's obligation to comply with the decision. It was a matter of enforcing the respective obligations and not a case of approbation and reprobation. If the claimant refused to comply with the contractual responsibility as found by the adjudicator, the defendant could seek enforcement of that part of the decision.
Redworth Construction Ltd v Brookdale Healthcare Ltd	[2006] BLR 366, [2006] EWHC 1994 (TCC), Judge Havery	**Election: adoption of new position on contract by referring party on enforcement** In enforcement proceedings, a referring party cannot rely on a basis for the formation of the contract that

[116] This definition of benefit has been doubted by Coulson J in *Amec Group Ltd v Thames Water Utilities* [2010] EWHC 419 (TCC).

Title	Citation	Issue
		it did not rely on in the course of the adjudication. In this case, during the adjudcation Redworth only relied on a particular document to establish that there was a contract on a JCT standard form, which was accepted by the adjudicator. When the formation of the contract was challenged on enforcement, Redworth submitted a different basis for the contract relying on a later document to establish that there was a contract in writing. When it was prevented from relying on that document it could not show there was a contract in writing with the result that the adjudicator had no jurisdiction and the decision was not enforced.
PT Building Services Ltd v ROK Build Ltd	[2008] EWHC 3434 (TCC), Ramsey J	**Election: payment of adjudicator's fees and expenses and relying on first adjudication to stop a subsequent adjudication** A party that had relied on the adjudicator's decision to encourage a second adjudicator to resign on the basis that the matter had already been determined and had paid the adjudicator's fees and expenses was found to have elected to treat the decision as valid with the result that it could not argue that the adjudicator's decision was unenforceable. Whilst there was no benefit to ROK in paying the adjudicator's fees and expenses, the taking of a benefit is sufficient but not necessary to establish an election. Payment of a sum awarded against a party is made in reliance on the decision being valid and, in the absence of circumstances indicating to the contrary, by making a payment a party will be held to have elected to treat the decision as valid, at least to the extent of the matters in respect of which payment was made.
Linnett v Halliwells LLP	[2009] All ER (D) 36, [2009] EWHC 319 (TCC), Ramsey J	**Election: relying on an adjudicator's decision as a defence in a subsequent adjudication** An adjudicator sought payment of fees and expenses from Halliwells who argued that the adjudicator had no jurisdiction and they had not agreed to pay. However, Halliwells' position that the adjudicator had no jurisdiction was inconsistent with its defence in a subsequent adjudication that the subsequent adjudicator could not order payment of a particular sum because it had already been the subject of the first adjudication. *Obiter*: Halliwells had approbated and reprobated by relying on the decision for the purpose of the second adjudication but then asserting in the present proceedings that the first decision was made without jurisdiction.
RWE NPower v Alstom Power Ltd	[2010] 133 Con LR 155, [2010] EWHC 3061 (TCC), Judge Havelock-Allan	**Election: alleged concession during adjudication** Alstom's notice of intention to refer included four heads of financial claim for prolongation under the terms of a single contract. On reading RWE's response it realized that it had no entitlement to three of the heads claimed under the contract in question; those claims arose under a related contract. It therefore wrote to the adjudicator to withdraw those three

(Continued)

Table 7.6 *Continued*

Title	Citation	Issue
		heads of claim. It was held that Alstom's concession in respect of those heads of claim did not preclude it, either by election or estoppel, from asserting that only a single dispute had been referred.
Amec Group Ltd v Thames Water Utilities	[2010] All ER (D) 267, [2010] EWHC 419 (TCC), Coulson J	**Election: making subsequent payments and withholdings based on adjudicator's decision** The employer paid part, but not all, of the sum awarded by the adjudicator and had used the adjudicator's decision as the basis for subsequent payments and, more importantly, a fresh withholding notice. *Obiter*: that conduct arguably amounted to approbation and reprobation, but the point was not decided in the absence of argument on the nature of the benefit required for an election.
Pilon Ltd v Breyer Group PLC	[2010] All ER (D) 197, [2010] EWHC 837 (TCC), Coulson J	**Election: application for a declaration that part of the adjudicator's decision is wrong by party seeking enforcement** In order to address the argument raised by the defendant in resisting enforcement that the adjudicator's failure to consider the defendant's overpayment defence was a breach of natural justice, the claimant sought a declaration that the contract required payment and/or withholding notices before an overpayment defence could even be considered (contrary to the finding of the adjudicator). The court could not make such a declaration because there was an arbitration provision but it held, *obiter*, that asking the court to substitute its own view for that of the adjudicator (whether in the enforcement proceedings or separate Part 8 proceedings) amounted to the clearest possible case of approbation and reprobation.
Nickleby FM Ltd v Somerfield Stores Ltd	[2010] 131 Con LR 203, [2010] EWHC 1976 (TCC), Akenhead J	**Election: jurisdictional basis argued upon enforcement not materially or prejudicially different from that argued during adjudication** The defendant resisting enforcement proceedings on the grounds that there was no contract in writing sought to prevent the claimant from introducing arguments and documents that had not been raised in the adjudication. The defendant, however, accepted that there was in fact a contract in writing. It was held that where an adjudicator does in fact and in reality have jurisdiction and would have inevitably concluded that he had jurisdiction had all the material before the court been available to him then there was no reason not to enforce the decision. The court did not need to consider the question of whether there was an oral agreement confirmed in writing (which had not been argued in the adjudication) as the claimant's alternative case of an offer accepted by conduct was not materially or prejudicially different from the argument before the adjudicator on which he decided he had jurisdiction. He would inevitably have reached the same decision had he been provided with the additional material.

Title	Citation	Issue
		The question of approbation and reprobation did not therefore arise. It was doubted whether the principle of election applied to a non-binding decision on jurisdiction. *Redworth Construction Ltd v Brookdale Healthcare Ltd* [2006] EWHC 1994 (TCC) distinguished.

(7) Costs and Interest

7.100 In proceedings to enforce an adjudicator's decision:

1. Costs are usually awarded on a standard basis.
2. However, costs will be awarded on an indemnity basis if the court wishes to express its disapproval of the way the case was conducted, for example where there was no defence or the claim for staying enforcement was without merit or if contest to the enforcement was not advised in a timely manner.
3. The court may exercise its discretion to award simple interest on all or some of the debt for the period between the date when the defendant failed to comply with the adjudicator's decision, until the date that the sum was actually paid.

Costs

7.101 Part 44 of the CPR identifies the court's discretion as to whether costs are payable, the amount of those costs, and when they are paid.[117] The starting point in court proceedings is that the successful party is to be awarded its costs on a standard basis.[118] The court can make a summary assessment of costs rather than requiring them to go to detailed assessment.[119] The court also has a discretion for costs to be awarded on the indemnity basis which will usually result in a higher rate of recovery than an award on the standard basis:[120]

> 44.4(1) Where the court is to assess the amount of costs (whether by summary or detailed assessment) it will assess the costs –
> (a) on the standard basis; or
> (b) on the indemnity basis,
> the court will not in either case allow costs which had been unreasonably incurred or are unreasonable in amount...
> (2) Where the amount of costs is to be assessed on the standard basis, the court will –
> (a) only allow costs which are proportionate to the matters in issue; and
> (b) resolve any doubt which it may have as to whether costs were reasonably incurred or reasonable and proportionate in amount in favour of the paying party.

117 There are almost as many decisions about costs as there are decisions about enforcement. Accordingly, this chapter only deals with matters of principle and examples of these principles. It does not try to set out each and every case on costs or interest.

118 As a very rough rule of thumb, and pending change as a result of the Jackson reforms on costs, when awarded on the standard basis around 60–70 per cent of the costs incurred will be recovered.

119 This is very common, see e.g. *Balfour Beatty Engineering Services (HY) Ltd v Shepherd Construction Ltd* (2009) 127 Con LR 110, [2009] EWHC 2218 (TCC).

120 As a rule of thumb, when awarded on the indemnity basis around 80–90 per cent of the costs incurred will be recovered.

7.102 These basic rules and practices of reasonableness and proportionality apply also in relation to the assessment of costs on a standard basis where 'conditional fee arrangements' or 'after the event' insurance is involved.[121] The onus is on the party seeking its costs.[122]

7.103 The circumstances in which an award may be made on the indemnity basis was considered by the Court of Appealin *Reid Minty v Taylor* (2002).[123] Whilst the Court of Appeal did not specifically consider costs in an enforcement of an adjudicator's decision, the guidance as to the judicial approach to considering awarding costs on an indemnity basis has been considered by the High Court in adjudication cases,[124] many of which apply May LJ's explanation of the requirement for 'unreasonable' conduct in awarding indemnity costs:

> 28. If costs are awarded on an indemnity basis in many cases there will be some implicit expression of disapproval of the way in which the litigation has been conducted, but I do not think that this will necessarily be so in every case. What is, however, relevant, at the present appeal, is that litigation can readily be conducted in a way which is unreasonable and which justifies an award of costs on an indemnity basis, where the conduct could not properly be regarded as lacking moral probity, or deserving moral condemnation...
>
> 32. There will be many cases in which, although the defendant asserts a strong case throughout and eventually wins, the Court will not regard the claimant's conduct of the litigation as unreasonable and will not be persuaded to award the defendant indemnity costs. There may be others where the conduct of a losing claimant will be regarded, in all the circumstances, as meriting an order in favour of the defendant of indemnity costs. Offers to settle and their terms will be relevant, and if they come within Part 36 may, subject to the Court's discretion, be determinative.

7.104 Coulson J summarized the applicable principles in relation to indemnity costs in *Fitzpatrick Contractors Ltd v Tyco Fire and Integrated Solutions (UK) Ltd* (2008):[125]

> (i) Indemnity costs are no longer limited to cases where the court wishes to express disapproval of the way in which the litigation has been conducted. An order for indemnity costs can be made even where the conduct could not properly be regarded as lacking in moral probity or deserving of moral condemnation (see *Reid Minty v. Taylor* [2002] 1 WLR 2800, [2001] EWCA Civ 1723).
>
> (ii) However, such conduct would need to be unreasonable 'to a high degree. 'Unreasonable' in this context certainly does not mean merely wrong or misguided in hindsight' (see Simon Brown LJ (as he then was) in *Kiam v. MGN Ltd No. 2* [2002] 1 WLR 2810), [2002] EWCA Civ 66.
>
> (iii) It is always important for the court to consider each case on its facts and to decide whether there is something in the conduct of the action or the circumstances of the case in question which takes it out of the norm in a way which justifies an order for indemnity costs (see Waller LJ in *Excelsior Commercial & Industrial Holdings Ltd v. Salisbury Hamer Aspden & Johnson* [2002] EWCA Civ 879).
>
> (iv) Examples of conduct that has led to such an order for indemnity costs include the use of litigation for ulterior commercial purposes (see *Amoco (UK) Exploration Co v. British American Offshore Ltd* [2002] BLR 135, [2001] EWHC 484 (Comm)) and the making of an unjustified and personal attack on one party by the other (see *Clark v. Associated Newspapers* [1998] WLR 1558.
>
> (v) There are a number of decisions, both of the TCC and of other courts, which make plain that the pursuit of a weak claim will not usually, on its own, justify an order for indemnity costs, whereas the pursuit of a hopeless claim (or a claim which the party pursuing it should

[121] *Redwing Construction Ltd v Charles Wishart* [2011] All ER (D) 101, [2011] EWHC 19 (TCC) (Key Case). See also paras. 11.4–11.10 of the Costs Practice Direction.

[122] *Redwing Construction Ltd v Charles Wishart* [2011] All ER (D) 101, [2011] EWHC 19 (TCC) (Key Case), para. 17.

[123] [2002] 1 WLR 2800, [2001] EWCA Civ 1723.

[124] *Harlow & Milner Ltd v Teasdale* [2006] All ER (D) 382, [2006] EWHC 54 (TCC); *Gray & Sons Builders (Bedford) Ltd v Essential Box Company Ltd* (2006) 108 Con LR 49, [2006] EWHC 2520 (TCC).

[125] [2008] 119 Con LR 155, [2008] EWHC 1301 (TCC). Whilst not an adjudication case, it was a construction case decided in the TCC.

have realised was hopeless) will lead to such an order. In both *Wates Construction Ltd v. HGP Greentree Allchurch Evans Ltd* [2006] BLR 45, [2005] EWHC 2174 and *EQ Projects Ltd v Alavi (t/a Merc London)* [2006] BLR 130, [2006] EWHC 29 (TCC) this court was persuaded that, in the circumstances of those cases, an order for indemnity costs was appropriate because the claimants should have realised that their claim was hopeless and should not have taken the matter on to trial. However, in *Healy-Upright v. Bradley* [2007] All ER (D) 29 (Nov), [2007] EWHC 3161 (Ch), the court reiterated that an order for indemnity costs was not justified by the mere fact that the paying party had been found to be wrong, either in fact or in law or both, or by the fact that in hindsight, the result of the case now being known, the position adopted by that party may be thought to have been unreasonable.

7.105 Whilst each case will, of course, turn on its own facts, the courts have found that indemnity costs were appropriate to be awarded in the following situations:

1. Where a party does not comply with the adjudicator's decision, it has also failed to comply with the agreed contractual regime, (and should expect to be penalized for its default by way of both costs and interest).[126]
2. Where the defendant had, and must have known that it had, no defence, and yet forced the claimant to incur costs of enforcement proceedings.[127]
3. Where the claim for staying the enforcement was made without merit.[128]
4. Where the court was not advised until shortly before the proceedings that the defendant did not in fact contest the enforcement.[129]

7.106 Even though indemnity costs may be awarded against the defendant in contested enforcement proceedings, they will not necessarily follow if the court considers that the defences raised were not without merit. The actions of the claimant are also relevant, and may mean that indemnity costs will not be granted.[130] The effect of 'without prejudice save as to costs' offers are also considered by the courts when deciding whether to award on a standard or indemnity basis,[131] or indeed to reduce costs.[132] Other reasons to reduce costs may be to reflect a court's disapproval of lies told by a party to the court or otherwise[133] or to reflect that the successful party on summary enforcement was not successful in all its claims.[134]

Interest

7.107 The majority of enforcement decisions do not interfere with the decision of the adjudicator in relation to interest, even if it is wrong in fact or law. Therefore, for example, an argument that the adjudicator awarded too high an interest rate under the Late Payment of Commercial Debts (Interest) Act 1998 is irrelevant to an enforcement hearing.[135]

---

126 *Fenice Investments Inc. v Jerram Falkus Construction Ltd* (2009) 128 Con LR 124, [2009] EWHC 3272 (TCC).

127 *Harlow & Milner Ltd v Teasdale* [2006] All ER (D) 382, [2006] EWHC 535, (TCC); *Harris Calnan Construction Co Ltd v Ridewood (Kensington) Ltd* [2007] All ER (D) 384, [2007] EWHC 2738 (TCC); *Able Construction (UK) Ltd v Forest Property Development Ltd* [2009] All ER (D) 176 (Feb), [2009] EWHC 159 (TCC); *O'Donnell Developments Ltd v Build Ability* [2009] EWHC 3388 (TCC).

128 *ART Consultancy v Naveda Trading Ltd* [2007] All ER (D) 157 (Jul), [2007] EWHC 1375 (TCC).

129 *Gray & Sons Builders (Bedford) Ltd v Essential Box Company Ltd* (2006) 108 Con LR 49, [2006] EWHC 2520 (TCC) (Key Case).

130 *Gipping Construction v Eaves Ltd* [2008] EWHC 3134 (TCC).

131 *Camillin Denny Architects Ltd v Adelaide Jones & Company Ltd* [2009] All ER (D) 117, [2009] EWHC 2110 (TCC).

132 *Jacques (t/a C&E Jacques Partnership) v Ensign Contractors Ltd* [2009] EWHC 3383 (TCC).

133 *Sughra Sulaman v AXA Insurance plc* [2009] All ER (D) 116 (Dec), [2009] EWCA Civ 1331, as considered in *Cynthia Jacques & Elise Jacques Grombach v Ensign Contractors Ltd* [2009] EWHC 3383 (TCC).

134 *Cynthia Jacques & Elise Jacques Grombach v Ensign Contractors Ltd* [2009] EWHC 3383 (TCC); *Supablast (Nationwide) Ltd v Story Rail Ltd* [2010] BLR 211, [2010] EWHC 56 (TCC).

135 *McConnell Dowell Constructors (Aust) Pty Ltd v National Grid Gas plc* [2007] BLR 92, [2006] EWHC 2551 (TCC).

7.108 The court has a discretion under s. 35A of the Supreme Court Act to order interest upon any sum adjudicated as due which was not paid, from the date it should have been paid until it is paid.[136]

7.109 The court may also choose to impose a punitive rate of interest in accordance with the Late Payment of Commercial Debts (Interest) Act 1998.[137] This is likely to be awarded where there is a debt that is agreed to be overdue and no reason for its non-payment; in such circumstances where 'cash flow is so important. It is necessary for the courts to utilise the 1998 Act, wherever appropriate, to arrive at a significant rate of interest.'[138] Where a party fails to comply with the adjudicator's decision, then whatever the result of the final determination, that party should expect to be penalized for its default by way of both costs (discussed above) and interest.[139]

Key Cases: Costs and Interest

Gray & Sons Builders (Bedford) Ltd v Essential Box Company Ltd (2006) 108 Con LR 49, [2006] EWHC 2520 (TCC)

7.110 **Facts:** The defendant resisted the enforcement of an adjudication decision up until the day before the enforcement hearing.

7.111 **Held:** Judge Coulson reviewed the authorities[140] and concluded that the defendant knew, or ought to have known, that it had no defence to the claim to enforce the adjudicator's decision and yet unreasonably continued to give the impression throughout that the application was resisted, thereby allowing the claimant to incur costs. Judge Coulson warned that in circumstances where the court was only told shortly before the hearing that the application is not in fact contested, that would prima facie result in an order for indemnity costs being appropriate.[141] The judge commented:

> 12. Defendants who avoid paying up in accordance with an Adjudicator's decision until the last moment or beyond are, so it seems to me, seeking to frustrate the adjudication provisions within the Housing Grants Construction and Regeneration Act 1996, or, if it is appropriate, the adjudication scheme that might have taken its place in any given construction contract. In those circumstances, for the reasons that I have given, it seems to me that, as a matter of principle, indemnity costs are appropriate.

7.112 The judge rejected the defendant's submissions in relation to cash flow which he deemed were ultimately irrelevant as the sums were due to the claimant under a contract freely entered into by the defendant. Lord Denning MR had described cash flow as 'the lifeblood of the enterprise of any builder' in *Dawnays Ltd v FG Ltd Minter* [1971][142] and the 1996

---

[136] *Ringway Infrastructure Services Ltd v Vauxhall Motors Ltd* (2007) 115 Con LR 149, [2007] EWHC 2507 (TCC) (Key Case).

[137] *Ruttle Plant Hire Ltd v Secretary of State for Environment Food & Rural Affairs (No. 3)* [2009] 1 All ER 448, [2008] EWHC 238 (TCC) – while not an adjudication case, it is referred to in *Fenice Investments Inc. v Jerram Falkus Construction Ltd* (2009) 128 Con LR 124, [2009] EWHC 3272 (TCC) at para. 49.

[138] *Able Construction (UK) Ltd v Forest Property Development Ltd* [2009] All ER (D) 176 (Feb), [2009] EWHC 159 (TCC), per Coulson J at para. 20.

[139] *Fenice Investments Inc. v Jerram Falkus Construction Ltd* [2009] All ER (D) 176, [2009] EWHC 3272 (TCC). In that case, there had been no claim under the Late Payments of Commercial Debts (Interest) Act 1998 and therefore the adjudicator's calculation of interest was applied for the period after the adjudicator's decision.

[140] *Reid Minty v Taylor* [2002]1 WLR 2800, [2001] EWCA Civ 1723; *Wates Construction Ltd v HGP Greentree Allchurch Evans Ltd* [2006] BLR 45, [2005] EWHC 2174 (TCC). The latter case involved a substantive trial in the TCC which, in the opinion of Judge Coulson, should have been abandoned two months before trial.

[141] Para. 12

[142] [1971] 1 WLR 1205.

Act was 'designed to preserve that cash flow'. As a consequence, indemnity costs were found to be the appropriate basis upon which to assess costs in that case.[143]

7.113 The judge also rejected the defendant's submissions that because it had made an offer which was reasonable, the claimant should be deprived of part of its costs. The claimant had 'done better' in the hearing than the offer. Therefore, there was no ground for saying that the claimant should or could have done something different which would have put it in the same position that it was in following the hearing.

Redwing Construction Ltd v Charles Wishart [2011] BLR 186, [2011] EWHC 19 (TCC)

7.114 **Facts:** Redwing decided to commence enforcement proceedings when the full amount awarded by the adjudicator was not paid. On 3 November 2010, Redwing entered into a conditional fee agreement (CFA) with its solicitors with the effect that, if it lost the enforcement proceedings, it would only be liable to pay the other side's costs; if it won, it paid the solicitor's basic charges, disbursements, the success fee of 100 per cent of the solicitor's basic charges and premium for after the event (ATE) insurance. Proceedings were issued on 4 November 2010. However Mr Wishart was not formally notified of the CFA until 19 November 2010. On 16 November 2010, Redwing took out ATE insurance with a premium of £8,480 for cover of £20,000. The total costs claimed by Redwing were over £40,000, which included the premium and £13,282.50 of basic fees, which was doubled by the 100 per cent success fee under the CFA.

7.115 CPR 44.15.1 requires a party with a CFA or ATE insurance to provide information in accordance with the Costs Practice Direction, paragraph 19 which requires a claimant with a funding arrangement in place before starting proceedings to file a notice of that arrangement. CPR 44.3B(1) provides that, unless the court orders otherwise, a party may not recover any additional liability for any period during which that party failed to provide information about a funding arrangement in accordance with a rule, practice direction, or court order.

7.116 **Held:** Akenhead J did not permit Redwing to recover its total claimed costs. Whilst it was not unreasonable of Redwing to enter into the CFA and obtain ATE insurance, there was no good reason as to why notification of the CFA did not take place, as is usual, with the issuing of the claim form. The 100 per cent success fee should have been based on a risk assessment, and no such assessment was provided to the court. Further, on the face of it, the ATE premium was very high. Accordingly, Redwing was awarded 20 per cent of the CFA uplift and of the ATE insurance premium, together with the other allowances made on summary assessment. Akenhead J set out the general principles in relation to CFAs and ATE insurance at paragraph 15 and went on:

> 16. It is also necessary to consider whether and to what extent CFAs and ATE Insurance have any part to play in adjudication enforcement cases, particularly in the TCC. There is no exemption, as such, in the Rules for these cases. It must follow that parties are entitled to enter into such funding arrangements in such types of case. However, it needs to be borne in mind that the large majority of reported cases on adjudication enforcements are successful and indeed in almost every case the claimants are sufficiently confident to pursue summary judgement applications on the basis that there is no realistic defence. It

[143] See in particular para. 13 of the decision.

must follow that courts, particularly the TCC which deals with virtually all such cases, will think long and hard about allowing substantial CFA mark-ups, particularly when there is a summary judgement application by the party with the CFA. It is important that claimants do not use CFAs and ATE insurance primarily as a commercial threat to defendants. It is legitimate for the Court to ask itself whether, in any particular case, a CFA or ATE Insurance was a reasonable and proportionate arrangement to make.

...

21. ...

(a) A 20 per cent uplift of basic solicitor's charges is reasonable in circumstances in which the Claimant was virtually bound substantially to 'win' its Claim, judged at the time when the CFA was entered into, with the only real risk being that the Defendant might do what it did, namely pay out what he was virtually bound to lose and fight the balance, the risk of losing of which was somewhere between 30 and 40 per cent.
(b) In relation to the ATE Insurance premium, there being no presumption that it was reasonable and bearing in mind that this is a summary assessment of costs on a standard basis, whilst one can understand why a claimant might well want the safety net of such insurance, the risk of losing, judged at the time when it was entered into, was sufficiently low to undermine the reasonableness of imposing anything near 100 per cent of it on the paying party in this case. In the absence of any evidence from the Claimant as to the reasonableness of the premium but without deciding that as such the premium is itself unreasonable, I have formed the view that it would only be reasonable to make Mr Wishart pay 20 per cent of the premium. I must and do presume that a wholly unrealistic assessment of risk was made to justify the imposition of a premium of some 42 per cent of the insured amount. I have a very real doubt that anything more is reasonable.
(c) Particularly in relation to the allowance for the CFA mark-up, I have taken into account in a broad brush manner the fact that part of Redwing 's costs were incurred before it formally notified Mr Wishart that a CFA was in place and that it would not be entitled to any mark-up on cost incurred before 19 November 2010.

Ringway Infrastructure Services Ltd v Vauxhall Motors Ltd (2007) 115 Con LR 149, [2007] EWHC 2507 (TCC)

7.117 **Facts:** The adjudicator had decided that a sum was owed to Ringway in accordance with clause 30.3.5 of a JCT 1998 with Contractor's Design contract. The net amount became payable on 24 May 2007, and Ringway claimed it should have been paid interest from that date. However, the adjudicator awarded interest only from 2 July 2007 to the date that compliance with his decision was expected. Section 35A of the Supreme Court Act states that, in proceedings for recovery of a debt, the court may include simple interest on any sum at such rate as it thinks fit on all or any part of the debt and that the period is that between the date when the cause of action arose and the date of the judgment in the case of the sum for which judgment is given.

7.118 **Held:** Whilst Akenhead J decided that he was unable to interfere with the adjudicator's direction that interest should be payable from 2 July 2007 (rather than 24 May 2007), the court did have discretion to award interest from the date that the defendant failed to comply with the adjudicator's decision. In exercising the court's discretion, Akenhead J found that the sum to be allowable as interest should be based on the daily rate determined by the adjudicator, from the date that Vauxhall failed to honour the adjudicator's decision (22 August 2007) until the date that the underlying sum was actually paid:

> 14. . . . The nature of enforcement of adjudicators' decisions is contractual. Clause 38A.7.2 here requires the parties to 'comply with the decision of the Adjudicator'. The Adjudicator's decision may be right or wrong but, whether right or wrong, it is to be complied with. The failure (in this case by Vauxhall) to comply with the decision of the Adjudicator and pay the sum ordered was a breach of Clause 39A.7.2.
>
> 15. Thus, the cause of action upon which Ringway had to rely and indeed did rely in their Particulars of Claim is the breach of Clause 39A.7. They did not as such sue Vauxhall for a debt or damages said to have arisen as a result of Vauxhall's failure to pay the sum due under Clause 30.3.3.5.
>
> 16. Thus, to relate what has happened to Section 35A(1) (of the Supreme Court Act 1981), the date when the cause of action arose was the date when Vauxhall failed to honour the Adjudicator's decision. The Adjudicator ordered Vauxhall to pay the sums which he decided were due no later than 21 August 2007. Accordingly, by 22 August 2007, the cause of action had arisen upon which Ringway not only did rely but had to rely.
>
> 17. The Court does have a discretion. However, the goalposts or limits of that discretion are the date when the cause of action arose and the date of the judgment. However, because in this case the Adjudicator, sensibly, quantified interest up to 21 August 2007 and then ordered interest to be payable at a daily rate of £375.04 until the sum due under the decision was paid, it is that sum which will be allowable as interest.

Table 7.7 Table of Cases: Costs and Interest

Title	Citation	Issue
Reid Minty v Taylor	[2001] 1 WLR 2800, [2001] EWCA Civ 1723, Ward, Kay, May LJJ	**Costs on an indemnity basis: disapproval of conduct** In many cases, costs on an indemnity basis will be awarded as an 'implicit expression of disapproval' in the way litigation has been conducted. However, litigation can be conducted in an unreasonable way justifying costs on an indemnity basis without conduct lacking moral probity or deserving moral condemnation. There may be cases where conduct of a losing claimant will merit an order of indemnity costs for the defendant. Offers to settle and their terms will be relevant and, if they come within Part 36 (subject to the court's discretion) may be determinative. In this case (concerning libel), the defendant had, and must have known that it had, no defence and yet forced the claimant to incur costs of proceedings. Costs on an indemnity basis were ordered.
Harlow & Milner Ltd v Teasdale	[2006] BLR 359, [2006] EWHC 54 (TCC), Judge Coulson	**Costs on an indemnity basis: no defence** The defendant had, and must have known that it had, no defence and yet forced the claimant to incur costs of enforcement proceedings.
Wates Construction Ltd v HGP Greentree Allchurch Evans Ltd	[2006] BLR 45, [2005] EWHC 2174, Judge Coulson	**Costs on an indemnity basis: claim without merit** In a substantive trial before the TCC, the claimants should have known, two months before the trial commenced, that their claim was entirely without merit. However they continued and indemnity costs were awarded against them. There was no need to prove an ulterior motive. To maintain a claim that the party knows, or ought to know, is doomed to fail on the facts and on the law, is conduct that is so unreasonable as to justify an order for indemnity costs.

(Continued)

Table 7.7 *Continued*

Title	Citation	Issue
Gray & Sons Builders (Bedford) Ltd. v Essential Box Company Ltd	[2006] 108 Con LR 49, [2006] EWHC 2520 (TCC), Judge Coulson	**Costs on an indemnity basis: late 'no contest'** The defendant did not confirm that it did not contest the enforcement proceedings until the day before the hearing. The defendant's claim that it was suffering cash-flow problems was irrelevant. The claimant had done better in the hearing than the defendant's offer.
ART Consultancy v Naveda Trading Ltd	[2007] All ER (D) 157 (Jul), [2007] EWHC 1375 (TCC), Judge Coulson	**Costs on an indemnity basis: no defence or grounds for stay** The defendant had no basis for challenging an adjudicator's decision or seeking to stay the enforcement due to the claimant's financial situation. The judge commented that the 'obviously unjustified allegation' in relation to the stay resulted in an 'entirely negative' impression.
Harris Calnan Construction Co. Ltd v Ridewood (Kensington) Ltd	[2007] All ER (D) 384, [2007] EWHC 2738 (TCC), Judge Coulson	**Costs on an indemnity basis: no defence** There was no substantive basis for challenging the adjudicator's decision. The court rejected the defendant's request that such costs be reduced because (i) the claimant did not need to be legally represented at the enforcement proceedings; and (ii) the claimant produced a skeleton argument. The claimant had the right to be represented and had complied with the court order to produce the skeleton argument.
McConnell Dowell Constructors (Aust) Pty Ltd v National Grid Gas plc	[2007] BLR 92, [2006] EWHC 2551 (TCC), Jackson J	**Adjudicator's decision on interest irrelevant to enforcement hearing** The argument of the unsuccessful party that the adjudicator had taken too high an interest rate (under the Late Payment of Commercial Debts (Interest) Act 1998) was irrelevant to an enforcement hearing.
Ringway Infrastructure Services Ltd v Vauxhall Motors Ltd	115 Con LR 149, [2007] EWHC 2507 (TCC), Akenhead J	**Court's discretion to order interest on any sum adjudicated as due** The court may not interfere with the adjudicator's decision as to the award of interest, it has discretion under s. 35A of the Supreme Court Act to award interest.
Gipping Construction v Eaves Ltd	[2008] EWHC 3134 (TCC), Akenhead J	**No indemnity costs warranted: successful party's behaviour relevant** The defendant had contested the enforcement of the adjudicator's decision. However, the behaviour of the claimant, particularly in filing its documents late, meant that indemnity costs were not awarded against the defendant.
Camillin Denny Architects Ltd v Adelaide Jones & Company Ltd	[2009] All ER (D) 176, [2009] EWHC 2110 (TCC), Akenhead J	**Effect of 'without prejudice save as to costs' offer** The claimant, who was wholly successful in its application for summary judgment, was found to be due £112,347.50 including VAT but excluding costs. While the claimant had 'beaten' the final offer, justice was 'best done by awarding the Claimant simply its costs on a standard basis'. The claimant had only beaten the latest offer by a few thousand pounds and had been prepared to settle for less than the defendant's latest offer.

Title	Citation	Issue
		Interest at 4 per cent as matter of discretion The claimant was entitled to interest at the same rate as ordered by the adjudicator. In relation to legal costs and expenses, interest was allowable as a matter of discretion at a rate of 4 per cent per annum from the date that payment should have been made in accordance with the adjudicator's decision. The same rate of interest was applied to that part of the adjudicator's fees, which the defendant should have paid but failed to pay.
Able Construction (UK) Ltd v Forest Property Development Ltd	[2009] All ER (D) 176 (Feb), [2009] EWHC 159 (TCC), Coulson J	**Costs on an indemnity basis and punitive interest: breach of settlement agreement** The losing party defaulted on its stage payment obligations of a sum awarded by an adjudicator. Enforcement proceedings were commenced and the losing party did not attend. Indemnity costs were awarded and interest in accordance with the Late Payments of Commercial Debts (Interest) Act 1998 permitted.
Balfour Beatty Engineering Services (HY) Ltd v Shepherd Construction Ltd	(2009) 127 Con LR 110, [2009] EWHC 2218 (TCC), Akenhead J	**Costs significantly reduced in summary assessment** The claimant sought costs of £78k (exclusive of VAT). In a summary assessment, the court found only £45k should be paid because: (i) it was unlikely that the claimant was not registered for VAT and therefore it should not be recoverable as part of the costs; (ii) the costs were disproportionate for proceedings which took little more than a month to pursue; (iii) the hours booked for attendances and work on documents seemed exceptionally and unnecessarily high, particularly where there had been very detailed involvement by Counsel; and (iv) a large amount of unnecessary documentation was put before the court, including a whole file of coloured programmes to which reference was never made.
		Simple interest awarded in accordance with the Supreme Court Act 1981 The claimant sought interest either at the default interest rate of 2 per cent above LIBOR for sums due under the subcontract or simple interest under s. 35A of the Supreme Court Act 1981. The appropriate interest rate was the latter as the subcontract related the default interest rate only to interim payments not made by the final date for payment rather than the late payment of a sum awarded under an adjudicator's decision.
Fenice Investments Inc. v Jerram Falkus Construction Ltd	Fenice 128 Con LR 124, [2009] EWHC 3272 (TCC), Coulson J	**Indemnity costs awarded for Part 7 proceedings** Non-compliance with an adjudicator's decision (a contractual breach) can lead to penalty by both interest and costs. The greater costs of the Part 8 proceedings were assessed on a standard basis because the point of law was bona fide, even if wrong.

(Continued)

Table 7.7 *Continued*

Title	Citation	Issue
Jacques (t/a C&E Jacques Partnership) v Ensign Contractors Ltd	[2009] EWHC 3383 (TCC), Akenhead J	**Parties to bear their own costs for certain applications** The parties were ordered to each bear their own costs in relation to further applications which they both lost (rearguing enforcement and payment by instalments). It was unnecessary to punish the claimant (who was successful in summary judgment but lost on the issue of the stay) because of the 'misleading' evidence one of the sisters submitted. Settlement discussions prior to the first hearing were taken into account and the court ordered that it would be appropriate and fair to allow the claimant 40 per cent of its costs up to and including the date of the first hearing.
O'Donnell Developments Ltd v Build Ability	128 Con LR 141, [2009] EWHC 3388 (TCC), Ramsey J	**Indemnity costs up until concession** Build Ability challenged summary enforcement of an adjudicator's decision (which had corrected a 'slip') and sought a stay, which it later withdrew. Build Ability started with several grounds on which they resisted enforcement. However, as proceedings progressed, these were dropped down to one. However, O'Donnell had been put to the cost of putting in evidence on matters which were not reasonably raised. Indemnity costs were awarded up to the point that Build Ability abandoned its arguments. Such conduct is unreasonable and needs to be marked by indemnity costs to prevent parties from seeking to defend on grounds that have no merit.
Supablast (Nationwide) Ltd v Story Rail Ltd	[2010] BLR 211, [2010] EWHC 56 (TCC), Akenhead J	**Indemnity costs not appropriate but costs reduced** While there was an absence of reality about the argument put forward by the defendant, it was not put forward in bad faith, unprofessionally, or wholly unreasonably. Whilst the claimant had the decision enforced, it effectively lost 'the variation issue' and it was clear that time and cost were spent pursuing this less arguable point, leading to a reduction of 20 per cent to reflect both parties' costs spent addressing this issue.
Redwing Construction Ltd v Charles Wishart	[2011] All ER (D) 101, [2011] EWHC 19 (TCC), Akenhead J	**Standard costs: conditional fee agreements and after the event insurance** Sets out the relevant principles for CFAs and ATE insurance.

(8) Enforcing the Judgment of the Court

7.119 Once an adjudicator's decision has been enforced by the courts, the defendant should pay the judgment sum, usually within 14 days in the absence of any express order as to time for payment.[144] Where an adjudicator's decision has been enforced by the courts and still not respected by the unsuccessful party, the following options may be available:

1. Application for an interim charging order, which if not complied with may be followed by an application for a final charging order, and then an application for order of sale;

[144] CPR 40.11.

2. Application for a third party creditor to pay the unsuccessful party's debt to the successful party; and/or
3. Application for the TCC judgment to be enforced in other jurisdictions.

Charging Orders

7.120 If the defendant fails to comply with the court's judgment, the claimant may make an application for an interim charging order. This may be converted into a final charging order upon application of the claimant. A charging order gives the claimant a charge over the assets of the defendant.

7.121 Such orders and their procedures were considered in a series of decisions concerning *Harlow & Milner Ltd v Teasdale* (2006).[145] In the first decision, the defendant requested the court not to make the interim charging order final on the merits; alternatively that it should be stayed in some way pending resolution of an arbitration. Both applications were denied. In relation to the merits point, Judge Coulson stated:

> 3. ... It seems to me that this argument...is quite hopeless. It is a wholly insufficient ground, under CPR 73.8, on which to oppose a Final Charging Order. What the Defendant (and her solicitors) continue to fail to appreciate is that the adjudication process is designed to give rise to a prompt (albeit temporary) result, with which the parties are obliged to comply in full: see the Housing Grants, Constuction and Regeneration Act 1996, and the string of decisions by the Court of Appeal in which they have made plain that Adjudicators' decisions are to be peremptorily enforced, starting with *Macob Civil Engineering v Morrison Construction* [1999] BLR 93 and *Bouygues (UK) v Dahl-Jensen (UK)* [2000] BLR 522. In this case the Defendant was ordered by the adjudicator 9 months ago to pay to the Claimant a sum of about £90,000. The Defendant, in breach of her contractual obligations, continues to refuse to do that, despite the judgment of this court of 16.1.06 which expressly required her to pay the sum awarded by the adjudicator.

7.122 The second decision (*Harlow & Milner Ltd v Teasdale* (2006))[146] followed continued non-payment after the final charging order was in place. The claimant sought an order for sale which the defendant again sought to resist on the basis of an impending arbitration. In refusing to exercise his discretion to avert the sale, Judge Coulson pointed to the fact that the property was not the defendant's home, but an investment property, and that she was in contumelious default because she had refused to pay a judgment sum for reasons which did not, in law, justify such non-payment:

> 20. Standing back from the authorities for a moment, it is worth considering what the effect would be if I acceded to the Defendant's request not to make the order for sale because of the on-going arbitration. It would mean that any unsuccessful party in adjudication would know that, if they refused to pay up for long enough, and started their own arbitration, they could eventually render the adjudicator's decision of no effect. It would be condoning, in clear terms, a judgment debtor's persistent default, and its complete refusal to comply with the earlier judgment of the Court. For those reasons, it is a position which I am simply unable to adopt.

Third Party Creditor

7.123 A further option for the claimant is to seek to enforce the award against a third party who owes a debt to the defendant. Rule 72 of the CPR, allows the court to make an order now known as a 'third party debt order' (formerly a 'garnishee' order) which requires a third party, instead of making payment to the judgment debtor in respect of its own debt, to pay

[145] [2006] BLR 359, [2006] EWHC 1708 (TCC).
[146] [2006] BLR 359, [2006] EWHC 1708 (TCC).

a sum to the judgment creditor. The application of this rule was considered by Coulson J in *Kier Regional Ltd (t/a Wallis) v City & General (Holborn) Ltd and ors* [2008] All ER (D) 189, (2008) (Key Case)[147] and is set out in detail below.

Enforcing a TCC Judgment in Other Jurisdictions

7.124 In the event that parties wish to enforce a judgment in foreign jurisdictions, this can be done in accordance with CPR 74 Part II which provides that the claimant must apply for a certified copy of the judgment and that such application may be made without notice. Different procedures and legislation will then apply subject to the jurisdiction in which it is sought to enforce the judgment.

Key Cases: Enforcing the Judgment of the Court

Kier Regional Ltd (t/a Wallis) v City & General (Holborn) Ltd and ors [2008] EWHC 2454 (TCC) Judge Coulson

7.125 **Facts:** Holborn and its co-defendants entered into a profit/loss sharing arrangement in relation to a property. Following this, Holborn entered into a building contract with Kier for construction works to that property. Many adjudications ensued, including one which resulted in a decision that Kier was entitled to approximately £719,000 of loss and expense for delay caused by Holborn. This decision was enforced by Jackson J in January 2006[148] and led to Kier obtaining a charging order against Holborn's interest in the property in April 2006. Holborn did not object to this, as it pointed out at the time that there was insufficient equity in the property for the charging order to be enforceable. When this proved to be the case, Kier applied again, in 2008, for an interim third party debt order (garnishee order) against Holborn's co-defendants on the basis that they both owed large sums of money to Holborn. However, all of the defendants claimed that these were contingent liabilities and effectively admitted that their accounts were incorrect. The defendant also made a cross-application for a stay of execution of the judgment given against them approximately two and a half years earlier.

7.126 **Held:** The facts of this case presented an 'unusual application' under RSC Order 47 for a stay of execution. Both the imminence of the arbitration and the prejudice that the co-defendants might suffer were considered by Coulson J to be pertinent considerations:

> 25. The fundamental requirement, before any final third party debt order can be made, is that the relationship of creditor and debtor must exist between the judgment debtor and the third party respectively. There must be money due to the judgment debtor from the third party. In particular:
>
> a) There must be a present debt. 'If they [the debts] may hereafter arise, it is possible also they may not hereafter arise, and it would require explicit words to include such future possible debts': see *Fry LJ in Webb v Stenton* (1883) 11 QBD 518 at 529.
> b) Thus, under a building contract, money in the hands of the employer cannot be attached until a certificate is issued by the architect, because it is only then that the employer is liable to pay the contractor: see *Dunlop & Ranken Limited v Hendall Steel Structures* [1957] 1WLR 1102.

[147] [2008] All ER (D) 189, [2008] EWHC 2454 (TCC) (Key Case).
[148] [2008] All ER (D) 189, [2006] EWHC 848 (TCC).

c) A judgment creditor cannot, by means of a third party debt order, stand in a better position as regards the third party than did the judgment debtor: see *Re General Horticultural Co. ex parte Whitehouse* [1886] 32 Ch. D 512.

7.127 CPR 72 provides that a third party debt order may be made interim or final, at the court's discretion. General guidance to the exercising of the discretion to make a final order is found in *Roberts Petroleum Ltd v Bernard Kenny Ltd* (1983)[149] which confirms, inter alia:

1. The burden of showing cause why an interim order shall not be made final is on the judgment debtor.
2. In exercising its discretion, the court must take into account all the relevant circumstances whether they arose before or after the interim order.[150]

7.128 Judge Coulson construed the agreements between the parties (because the parties' Companies House accounts were incorrect) and found that no current debt was in fact owed by the third parties to the defendant. Accordingly, Judge Coulson found that it was not right to say that either co-defendant owed any debt to Holborn at the time of the hearing and the extent of that debt (if any) would not be known until the conclusion of the arbitration. In those circumstances, there was no debt due and he considered it 'quite inappropriate' to make any final third party debt order. The interim third party debt orders were discharged. However, even if there was a debt, it would have been relevant to consider the prejudice which the third parties would have suffered by such an order, including the fact that they had not necessarily signed up for the 'pay now, arbitrate later' philosophy of the 1996 Act.[151] This case was the first consideration of a third party debt order under CPR Part 72 in adjudication enforcement proceedings. Although Judge Coulson concluded that no debt was owed by the co-defendants to Holborn (and therefore that the third party debt order could not succeed) he thought it worthwhile to consider the principles in the case to determine whether a third party debt order and stay would otherwise have been granted. The relevant criteria in both cases were:

1. The prejudice that such orders will cause to the third parties.
2. The imminence of the arbitration (and the issues in that arbitration as they now stand).
3. The delay on the part of Kier.
4. The fact that the original adjudicator's decision was based on what Jackson J himself thought might well be an error.[152]

> 67. Accordingly, I make plain that, had I found there to be a debt due and owing from Cambridge and/or Temple to Holborn, I would have refused to exercise my discretion in favour of making final third party debt orders. The two principal grounds for that conclusion would have been the prejudice to the third parties if such an order was made and the imminence of the arbitration (and a consideration of the issues involved in that arbitration). Neither the delays nor the original nature of the adjudicator's decision would have been of any real significance in the exercise of my discretion.

[149] [1983] 1 WLR 301.

[150] As highlighted by Coulson J in *Kier Regional Ltd (t/a Wallis) v City & General (Holborn) Ltd and ors* [2008] All ER (D) 189, [2008] EWHC 2454 (TCC) (Key Case) at para. 45. See in particular paras. 51–64 of that decision for consideration of the discretion under r. 72.

[151] At para. 51.

[152] As set out by Judge Coulson at para. 56 of the decision.

Table 7.8 Table of Cases: Enforcing the Court's Judgment

Title	Citation	Issue
Harlow & Milner Ltd v Teasdale	[2006] All ER (D) 382, [2006] EWHC 535 (TCC), Judge Coulson	**Interim and final charging order** In breach of the adjudicator's decision and interim charging order issued by the TCC, the defendant sought to avoid payment based on the merits of the case and stay pending arbitral resolution. Both grounds were rejected; the defendant was obliged to comply in full with the adjudicator's decision.
Harlow & Milner Ltd v Teasdale	[2006] BLR 359 [2006] EWHC 1708 (TCC), Judge Coulson	**Order of sale** The TCC exercised its discretion to grant the order of sale because: (i) it did not relate to a matrimonial home but investment properties, (ii) the defendant was in contumelious default for refusing to pay a judgment sum for reasons which do not in law justify non-payment, and (iii) the financial reality was that the judgment debt would not be paid without a sale, and such sale would cover the increasing judgment debt.
Kier Regional Ltd (t/a Wallis) v City & General (Holborn) Ltd and ors	[2008] All ER (D) 189, [2008] EWHC 2454 (TCC), Coulson J	**Third party creditor** A CPR Part 72 application, requires a 'present debt' to be owed from the third party to the unsuccessful party in an adjudication. It was inappropriate to permit the application when the financial balance would not be clear until the outcome of an arbitration.

8

STAYING ENFORCEMENT

(1) Introduction

8.01 The substantive grounds on which the enforcement of an adjudicator's decision may be resisted are discussed in Chapters 9, 10, and 11. However, in circumstances where such grounds do not exist and the decision is therefore enforecable, a party may still be able to avoid the obligation to make a payment in respect of the decision by seeking either a stay of execution of the order enforcing the award, or by seeking a stay of the enforcement proceedings themselves. These approaches, whilst having the same practical effect, are procedurally distinct and are dealt with separately in each section in this chapter.

8.02 RSC Order 47[1] grants the court a wide discretion to stay the execution of a judgment or order where there are 'special circumstances' rendering it 'inexpedient' to enforce a judgment:

> (1) Where a judgment is given or an order made for the payment by any person of money and the court is satisfied on an application made at the time of the judgment, or order, or at any time thereafter by the judgment debtor or other party liable to execution –
>
> (a) that there are special circumstances which render it inexpedient to enforce the judgment or order...
>
> ... the court may by order stay the execution of the judgment or order... either absolutely or for such period and subject to such conditions as the court thinks fit.

8.03 Where there is an enforceable adjudicator's decision the court will order summary judgment. However, a stay of execution of that order prevents the successful party from doing anything with that order until some further order is made lifting the stay (or any conditions attached to the stay cease to apply). The most frequent basis for applying for a stay of execution is

[1] Preserved in Section A of the CPR by operation of Part 50.

the impecuniosity of the successful party.[2] Even though they are entitled to summary judgment in the amount sought, where there is a sufficiently real risk that they may be unable to repay the sum awarded in the event that the temporarily binding adjudicator's decision is later overturned in court or arbitral proceedings, then it is considered inexpedient to require payment to be made.

8.04 As to staying the enforcement proceedings themselves, if parties have agreed a particular method by which disputes are to be resolved, then a court has an inherent jurisdiction to stay court proceedings brought in breach of that agreement,[3] even where that agreement is a general agreement to refer disputes to alternative dispute resolution.[4] However, in most adjudication enforcement cases, the courts have refused to stay, or even adjourn, summary enforcement proceedings pending other proceedings. The extent to which this principle may apply to court proceedings to enforce an adjudicator's award and whether such proceedings should be stayed for the enforcement to be decided by alternative means is discussed in the final section of this chapter.

(2) Stay of Execution of Enforcement Order on Grounds of Impecuniosity

8.05 Stays may be granted where the unsuccessful party convinces the court that the successful party's impecuniosity is such that payment should not be made. The key principles in such a consideration are as follows:

1. The degree of impecuniosity is all important. If the successful party is in liquidation or does not dispute its insolvency, a stay will usually be granted. If there is no proper evidence of financial vulnerability, a stay will be denied. There are many shades of impecuniosity between these two extremes.
2. A stay on grounds of impecuniosity may not be granted where:
   (a) the successful party is not in a significantly worse financial position than at the time the contract was entered into; or
   (b) the successful party's financial condition was due to the failure to be paid the money awarded in the adjudicator's decision.
3. The onus is on the unsuccessful party seeking the stay to adduce evidence of a very real risk of future non-payment.
4. There may be cost ramifications if a party wrongly challenges the enforceability of a decision and seeks a stay in circumstances where no such position is tenable.

Insolvency Rules

8.06 Rule 4.90 of the Insolvency Rules provides:

> (1) This rule applies where, before the company goes into liquidation there have been mutual credits, mutual debts or other mutual dealings between the company and any creditor of the company proving or claiming to prove for a debt in the liquidation.

[2] In most instances, the claimant is the successful party, seeking to enforce the adjudicator's decision and an application for a stay of execution is made in response by the defendant. However, in *J. W. Hughes Building Contractors Ltd v GB Metalwork Ltd* [2003] EWHC 2421 (TCC), it was the unsuccessful party who commenced the proceedings, seeking an order from the court staying the enforcement.

[3] *Channel Tunnel Group Ltd & France Manche SA v Balfour Beatty Construction Ltd* [1993] AC 334; [1993] 2 WLR 262 as mentioned in *Balfour Beatty Construction Northern Ltd v Modus Corovest (Blackpool) Ltd* [2009] CILL 2660, [2008] EWHC 3029 (TCC).

[4] *Cable & Wireless plc v IBM United Kingdom Ltd* [2003] BLR 89, [2002] EWHC 2059 (Comm) as mentioned in *Balfour Beatty Construction Northern Ltd v Modus Corovest (Blackpool) Ltd* [2009] CILL 2660, [2008] EWHC 3029 (TCC).

(2) An account shall be taken of what is due from each party to the other in respect of the mutual dealings and the sums due from one party shall be set off against the sums due from the other.

...

(4) Only the balance (if any) of the account is provable in the liquidation. Alternatively (as the case may be) the amount shall be paid to the liquidator as part of assets.

8.07 The obligation to implement an adjudicator's decision without delay does not prevail over a party's entitlements upon the insolvency of another party, such as mutual setting off of accounts under 4.90 of the Insolvency Rules.[5]

8.08 Accordingly, whilst the general principle is that adjudicators' decisions are to be enforced without cross-claim or set-off, where the judgment creditor (or the successful party from the adjudication) is insolvent, that rule is displaced. If this were not so there would be a real risk that payment in respect of an adjudicatior's decision, which is intended only to be temporarily binding, would lead to unfairness. In particular, if the adjudicator's decision were to be reversed in a subsequent final determination of the dispute in litigation or arbitration, it would be very unlikely that the money paid over pursuant to the adjudicator's decision could be recovered from the insolvent party. This may amount to 'special circumstances' under Order 47(1)(a) justifying a stay of enforcement of the decision. If no stay was granted, the judgment debtor (or unsuccessful party from the adjudication) would be denied its rights to a full reckoning under the Insolvency Rules and would run the real risk of receiving only a limited dividend, usually as an unsecured creditor—all as a consequence of a temporarily binding decision. Therefore, where there is sufficient evidence to establish impecuniosity, summary judgment enforcing the adjudicator's decision is likely to be entered but a stay of execution of that judgment will be ordered.

8.09 The Court of Appeal explained the reasoning for this approach in its decision in *Bouygues (UK) Ltd v Dahl-Jensen (UK) Ltd* (2000)[6] in which Dahl-Jensen was in liquidation:

> 33. ... If Bouygues is obliged to pay to Dahl-Jensen the amount awarded by the adjudicator, those monies, when received by the liquidator of Dahl-Jensen, will form part of the fund applicable for distribution amongst Dahl-Jensen's creditors. If Bouygues itself has a claim under the construction contract, as it currently asserts, and is required to prove for that claim in the liquidation of Dahl-Jensen, it will receive only a dividend pro rata to the amount of its claim. It will be deprived of the benefit of treating Dahl-Jensen's claim under the adjudicator's determination as security for its own cross-claim. ...
>
> 35. ... In circumstances such as the present where there are latent claims and cross-claims between parties, one of which is in liquidation, it seems to me that there is a compelling reason to refuse summary judgment on a claim arising out of an adjudication which is necessarily provisional. All claims and cross-claims should be resolved in the liquidation in which full account can be taken and a balance struck. That is what r.490 of the Insolvency Rules 1986 requires.
>
> 36. It seems to me that those matters ought to have been considered on the application for summary judgment. But the point was not taken before the judge and his attention was not, it seems, drawn to the provisions of the Insolvency Rules 1986. Nor was the point taken in the notice of appeal. Nor was it embraced by counsel for the appellant with any enthusiasm when it was drawn to his attention by this Court. In those circumstances—and in the

[5] *William Verry Ltd v London Borough of Camden* [2006] All ER (D) 292 (Mar), [2006] EWHC 761 (TCC); *Levolux AT Ltd* v *Ferson Contractors* [2003] BLR 118, [2003] EWCA Civ 11; *Integrated Building Services Engineering Consultants Ltd (t/a Operon) v Pihl UK Ltd* [2010] BLR 622; [2010] CSOH 80.

[6] [2001] 1 All ER (Comm) 1041. This case is discussed as one of the Key Cases in Ch. 9 at 9.54–9.55, see in particular paras. 29–36.

circumstances that the effect of the summary judgment is substantially negated by the stay of execution which this court will impose—I do not think it right to set aside an order made by the judge in the exercise of his discretion. I too would dismiss this appeal.

Principles of Stay of Execution

8.10 In *Wimbledon Construction Company 2000 Ltd v Vago* (2005) (Key Case),[7] Judge Coulson set out the following principles as relevant to an application for staying enforcement on the grounds of impecuniosity:

> 26. ... (a) Adjudication ... is designed to be a quick and inexpensive method of arriving at a temporary result in a construction dispute.
>
> (b) In consequence, adjudicators' decisions are intended to be enforced summarily and the claimant (being the successful party in the adjudication) should not generally be kept out of his money.
>
> (c) In an application to stay the execution of summary judgment arising out of an Adjudicator's decision the court must exercise its discretion under Order 47 with considerations (a) and (b) firmly in mind (see *AWG*).
>
> (d) The probable inability of the claimant to repay the judgment sum (awarded by the Adjudicator and enforced by way of summary judgment) at the end of the substantive trial, or arbitration hearing, may constitute special circumstances within the meaning of Order 47 rule 1(1)(a) rendering it appropriate to grant a stay (see *Herschel*).
>
> (e) If the claimant is in insolvent liquidation, or there is no dispute on the evidence that the claimant is insolvent, then a stay of execution will usually be granted (see *Bouygues* and *Rainford House*).
>
> (f) If the evidence of the claimant's present financial position suggested that it is probable that it would be unable to pay the judgment sum when it fell due, that would not usually justify the grant of a stay if:
>
> (i) the claimant's financial position is the same or similar to its financial position at the time the relevant contract was made (see *Herschel*); or
> (ii) the claimant's financial position is due, either wholly or in significant part, to the defendant's failure to pay those sums which were awarded by the Adjudicator (see *Absolute Rentals*).

8.11 These principles have been endorsed in many cases and help to ensure that stays are granted in circumstances 'consistent with the overriding objective, [where] the justice of the case demands it'.[8] The application of these principles will depend on the facts of each case. There will be some cases in which it would be 'wholly unjust and inequitable if the judgment sum was not the subject of a stay of execution'.[9] There are some instances where justice is better served when the stay is made over only part of the judgment sum,[10] or where it is made conditional upon monies being paid into court rather than allowing the defendant the benefit of the sums during the period of the stay.[11]

---

[7] [2005] BLR 374, [2005] EWHC 1086 (TCC).

[8] *AWG Construction Services Ltd v Rockingham Motor Speedway Ltd* [2004] CILL 2154, [2004] EWHC 888 (TCC).

[9] *JPA Design & Build Ltd v Sentosa (UK) Ltd* [2009] All ER (D) 06 (Oct), [2009] EWHC 2312 (TCC).

[10] *Jacques (t/a C&E Jacques Partnership) v Ensign Contractors Ltd* [2009] EWHC 3383 (TCC).

[11] *Rainford House Ltd (in administrative receivership) v Cadogan Ltd* [2001] All ER (D) 144 , [2001] EWHC 18 (TCC); *Baldwins Industrial Services plc v Barr Ltd* [2003] BLR 176, [2002] EWHC 2915 (TCC); *Ashley House Plc v Galliers Southern Ltd* [2002] Adj LR 02/15, [2002] EWHC 274 (TCC).

Types of Financial Hardship

8.12 The first question to be considered on an application for a stay of execution based on impecuniosity is whether the type and/or level of financial hardship is such that the claimant will be unable to repay any sums paid to it pursuant to the adjudication decision. As described above, where the judgment creditor is in liquidation, the Court of Appeal has confirmed there are grounds to either refuse summary judgment or to stay execution.[12] However, there are many different types of financial hardship along the continuum between solvency and liquidation, and this provides scope for the court to exercise its discretion. A stay application must be supported by evidence of impecuniosity, usually in the form of up-to-date financial information relating to the company seeking enforcement.[13] However, it is relevant to consider not just the latest accounts but also the current trading position and future trading prospects of the successful party.[14]

8.13 As can be seen in Table 8.1, the courts have considered many types of financial hardship, from company voluntary arrangements[15] to floating charges over assets,[16] credit references showing 'maximum risk' and county court judgments,[17] or accounts and credit rating indicating a high risk of insolvency.[18] Applications for stays were denied in each of these circumstances.

8.14 In fact, there have only been a handful of successful applications for stays, including where the judgment creditor:

1. Was in administrative receivership, and monies were paid into court.[19]
2. Had gone through incomplete winding-up procedures as at the date of the application for summary judgment (when the court stayed the enforcement pending the hearing of a winding-up petition 14 days later).[20]
3. Had significant debts, an unsatisfied county court judgment, and the judgment debtor had a significant adjudication decision which it could set off against the decision in question.[21]
4. Was in administration by order of the court[22] or in accordance with Schedule B1 of the Insolvency Act 1986.[23]

---

[12] *Bouygues (UK) Ltd v Dahl-Jensen (UK) Ltd* [2001] 1 All ER (Comm) 1041.

[13] *AWG Construction Services Ltd v Rockingham Motor Speedway Ltd* [2004] CILL 2154, [2004] EWHC 888 (TCC).

[14] *Berry Piling Systems Ltd v Sheer Projects Ltd* (2012) 141 Con LR 225, [2012] EWHC 241 (TCC).

[15] *Mead General Building Ltd v Dartmoor Properties Ltd* [2009] BLR 225, [2009] EWHC 200 (TCC). Although see *Pilon Ltd v Breyer Group PLC* [2010] BLR 452, [2010] EWHC 837 (TCC), in which, had the court found that there was an enforceable decision, it would have stayed enforcement due to the CVA and the up-to-date credit rating which indicated 'a high risk of business failure'.

[16] *Total M&E Services Ltd v ABB Building Technologies Ltd (formerly ABB Steward Ltd)* (2002) 87 Con LR 154, [2002] EWHC 248 (TCC) (Key Case).

[17] *Avoncroft Construction Ltd v Sharba Homes (CN) Ltd* (2008) 119 Con LR 130, [2008] EWHC 933 (TCC) (Key Case).

[18] *JW Hughes Building Contractors Ltd v GB Metalwork Ltd* [2003] Adj LR 10/03, [2003] EWHC 2421.

[19] *Rainford House Ltd (in administrative receivership) v Cadogan Ltd* [2001] All ER (D) 144, [2001] EWHC 18 (TCC); *Baldwins Industrial Services PLC v Barr Ltd* [2003] EWHC 2915 (TCC).

[20] *Harwood Construction Ltd v Lantrode Ltd*, 24 Nov. 2000 (TCC) (unreported) (Key Case).

[21] *JPA Design & Build Ltd v Sentosa (UK) Ltd* [2009] 50 EG 68, [2009] EWHC 2312 (TCC).

[22] *Straw Realisations (No. 1) Ltd v Shaftsbury House (Developments) Ltd* [2011] BLR 47, [2010] EWHC 2597 (TCC).

[23] *Integrated Building Services Engineering Consultants Ltd (t/a Operon) v Pihl UK Ltd* [2010] BLR 622, [2010] CSOH 80.

8.15 In *Straw Realisations (No. 1) Ltd v Shaftsbury House (Developments) Ltd* (2010)[24] Edwards-Stuart J set out the applicable principles to the different types of financial difficulties:

> (1) A clause in a contract that purports to supersede the obligation to comply with an adjudicator's decision, in this case the provision for a mutual setting off of the accounts between the parties on the happening of certain events as set out in clause 8.5 and an obligation only to pay the balance and the restriction on any further payments, cannot prevail over an obligation to comply with the decision of an adjudicator: see *Ferson v Levolux* and *Verry v Camden*.
>
> (2) If, at the date of the hearing of the application to enforce an adjudicator's decision, the successful party is in liquidation, then the adjudicator's decision will not be enforced by way of summary judgment: see *Bouygues v Dahl Jensen* and *Melville Dundas*. The same result follows if a party is the subject of the appointment of administrative receivers: see *Melville Dundas*.
>
> (3) For the same reasons, I consider that if a party is in administration and a notice of distribution has been given, an adjudicator's decision will not be enforced.
>
> (4) If a party is in administration, but no notice of distribution has been given, an adjudicator's decision which has not become final will not be enforced by way of summary judgment. In my view, this follows from the decision in *Melville Dundas* as well as being consistent with the reasoning in *Integrated Building Services v PIHL*.
>
> (5) If the circumstances are as in paragraph (4) above but the adjudicator's decision has, by agreement of the parties or operation of the contract, become final, the decision may be enforced by way of summary judgment (subject to the imposition of a stay). I reach this conclusion because I do not consider that the reasoning of the majority in *Melville Dundas* extends to this situation.
>
> (6) There is no rule of English law that the fact that a party is on the verge of insolvency ('*vergens ad inopiam*') triggers the operation of bankruptcy set-off: see *Melville Dundas*, per Lord Hope at paragraph 33. However, the law in Scotland appears to be different on this point (perhaps because the Scottish courts do not enjoy the power to grant a stay in such circumstances).
>
> (7) If a party is insolvent in a real sense, or its financial circumstances are such that if an adjudicator's decision is complied with the paying party is unlikely to recover its money, or at least a substantial part of it, the court may grant summary judgment but stay the enforcement of that judgment.

Evidencing the Impecuniosity

8.16 The burden of proving the impecuniosity is on the party alleging it, who must adduce evidence of 'a very real risk of future non-payment'.[25] The assessment is to be carried out at the likely future date when repayment may be required which necessarily requires a degree of speculation.[26] Notwithstanding the duty to cooperate, there is no obligation on the claimant to 'give widespread disclosure' of all of its financial information 'so that the other party can see whether there is something which gives grounds for an application to stay'.[27] Where there is no 'direct evidence' as to the claimant's financial status, a court may conclude that if it acceded to a stay application, it would 'drive a coach and horses through the adjudication scheme' and 'frustrate Parliament's intention'.[28]

---

[24] [2011] BLR 47, [2010] EWHC 2597 (TCC).

[25] *Total M&E Services Ltd v ABB Building Technologies Ltd* (formerly ABB Steward Ltd) (2002) 87 Con LR 154, [2002] EWHC 248 (TCC) (Key Case).

[26] *Berry Piling Systems Ltd v Sheer Projects Ltd* (2012) 141 Con LR 225, [2012] EWHC 241 (TCC)

[27] *O'Donnell Developments Ltd v Build Ability* (2009) 128 Con LR 141, [2009] EWHC 3388 (TCC).

[28] *Nolan Davis Ltd v Catton* [2001] All ER (D) 232 (Mar).

8.17 Judge Seymour considered the evidentiary aspects of an application for a stay in *Rainford House Ltd v Cadogan Ltd (in administrative receivership)* (2001) (Key Case)[29] in which he stated:

> 11. So far as the question whether to grant a stay of execution is concerned, each case must depend upon its own facts. I agree ... that it is for the applicant for any stay to put before the court credible material which, unless contradicted, demonstrates that the claimant is insolvent. However, in my judgment it is not necessary for the applicant for the stay to go further than to put before the court evidence as to the present financial position of the claimant, so that he does not need to shoulder some additional burden of predicting when any challenge to the correctness in fact of the determination of the adjudicator will be heard or of putting before the court positive evidence as to what the financial position of the claimant will then be. Further, I do not consider that the burden which the applicant for a stay bears is that of demonstrating beyond the possibility of error that the claimant will, come the time when the correctness or otherwise of the decision of the adjudicator is determined, be unable to repay the amount determined by the adjudicator to be payable. ... [I]t is, in my judgment, appropriate when considering an application for a stay of execution in a case such as the present, to proceed on the basis that once the applicant for a stay has adduced apparently credible evidence which, if uncontradicted, shows that the claimant in the action is then insolvent,
>
> (a) it is for the claimant, if it wishes the court not to draw the inference for which the applicant for the stay contends, to seek to contradict the evidence adduced on behalf of the applicant;
> (b) in the absence of evidence to suggest that the position as it appears at the time the application is before the court is likely to alter the inference which should be drawn is that it will not.

8.18 It is clearly important to produce the right amount of evidence to persuade a court that a stay is warranted. In an attempt to overcome this hurdle, in *Treasure & Son Ltd v Martin Dawes* (2007) (Key Case),[30] two forensic independent reports were obtained in support of Mr Dawes' attempts to stay enforcement. Notwithstanding these creative (and expensive) avenues, the court refused to grant the stay as the claimant was neither insolvent nor bankrupt and there was no persuasive evidence that it would not be in a position to repay the monies in question in the event that an arbitrator came to a conclusion different from the adjudicator's decision.

8.19 Depending on the allegations of impecuniosity levelled at the claimant in enforcement proceedings, it may wish to seek to defend its financial health through the introduction of its own evidence. Thus, in *Avoncroft Construction Ltd v Sharba Homes (CN) Ltd* (2008) (Key Case),[31] the claimant introduced proof of a parent company guarantee, and the court declined to grant the stay.

Circumstances in which a Stay Is Unlikely to Be Granted

8.20 Even if the court considers that the impecuniosity threshold has been reached, it may not exercise its discretion to order a stay of execution. As identified in paragraph 26(f)(i) of *Wimbledon v Vago* (Key Case) (see 8.10 above), the courts will not grant a stay where the claimant company is not in a significantly worse financial position at the time it seeks summary judgment than at the time that the contract was made.[32] This is because the defendants 'got the result they contracted for and cannot now use the claimant's ill-health to avoid

---

[29] [2001] All ER (D) 144 [2001] EWHC 18 (TCC).
[30] [2008] BLR 24, [2007] EWHC 2420 (TCC).
[31] 119 Con LR 130, [2008] EWHC 933 (TCC).
[32] *Michael John Construction v Golledge and ors* [2006] TCLR 3, [2006] EWHC 71 (TCC); *Shaw v Massey Foundation & Pilings Ltd* [2009] EWHC 493 (TCC); *Air Design (Kent) Ltd v Deerglen (Jersey) Ltd* [2009] CILL 2657, [2008] EWHC 3047 (TCC); *Pilon Ltd v Breyer Group PLC* [2010] BLR 452, [2010] EWHC 837 (TCC).

judgment'.[33] For example, where the claimant company was not even incorporated when the defendant contracted with it, the defendant was not granted a stay of execution, as it contracted 'with [its] eyes open' to the claimant's 'lack of value and creditworthiness'.[34]

8.21 The courts have also followed paragraph 26(f)(ii) in *Wimbledon v Vago* and have refused to grant a stay where non-payment of the sum awarded by the adjudicator materially caused or contributed to the claimant's financial difficulties.[35] Refusal to pay in accordance with an adjudicator's decision can cause very real hardship for small contractors and may contribute to their problems in paying suppliers, subcontractors, and legal representatives.[36]

8.22 In *Pilon Ltd v Breyer Group PLC* (2010),[37] Coulson J confirmed:

1. The question is whether the defendant's non-payment of the sum found due by the adjudicator has caused the claimant's poor financial position.
2. Allegations of the defendant's non-payment of sums[38] arising under other contracts between the parties was irrelevant.
3. Where the debts of the claimant are not linked to the defendant in any meaningful way, a stay may be granted.

Key Cases: Stay of Execution of Enforcement Order on Grounds of Impecuniosity

Bouygues (UK) Ltd v Dahl-Jensen (UK) Ltd [2000] BLR 522

This case is discussed at 8.09.

Harwood Construction Ltd v Lantrode Ltd, 24 November 2000 (TCC), Judge Seymour (unreported)

8.23 **Facts:** At the date of the application for summary judgment, the claimant was not in liquidation. Therefore, the judge had to weigh up the competing considerations:

1. In favour of the stay: a winding-up petition had been presented and the hearing date was within 14 days of the court's decision on summary judgment. Section 29 of the Insolvency Act had the effect that if a winding-up order was made, the liquidation would be deemed to have commenced at the date of the presentation of the petition.
2. Against the stay: a witness statement explaining the claimant's perception that the likely outcome of the petition was that it would fail. The mere presentation of a petition should not prevent the enforcement of an adjudication award.

8.24 **Held:** Judge Seymour commented that the evidence was insufficient to enable him to be confident that the petition (brought by someone other than the defendant) would succeed or fail. He also considered that (subject to the petition hearing) there was no reason why the claimant should not be awarded summary judgment. The judge ordered a stay until

---

[33] *Michael John Construction v Golledge and ors* [2006] TCLR 3, [2006] EWHC 71 (TCC) at para. 79.

[34] *S. G. South Ltd v (1) King's Head Cirencenster LLP and (2) Corn Hall Arcade Ltd* [2010] BLR 47, [2009] EWHC 2645 (TCC)

[35] *Absolute Rentals Ltd v Glencor Enterprises Ltd* (2000) CILL 1637; *Michael John Construction v Golledge and ors* [2006] TCLR 3, [2006] EWHC 71 (TCC); *Treasure & Son Ltd v Martin Dawes* (2008) BLR 24, [2007] EWHC 2420 (TCC) (Key Case); *S. G. South Ltd v (1) King's Head Cirencenster LLP and (2) Corn Hall Arcade Ltd* [2010] BLR 47, [2009] EWHC 2645 (TCC) at paras. 41 and 42 in particular.

[36] *S. G. South Ltd v (1) King's Head Cirencenster LLP and (2) Corn Hall Arcade Ltd* [2010] BLR 47, [2009] EWHC 2645 (TCC).

[37] [2010] BLR 452, [2010] EWHC 837 (TCC).

[38] In this instance, more than £800,000.

the hearing of the petition for winding up, and that it should be continued thereafter if a winding-up order was made, but otherwise would cease if the petition was dismissed.

Rainford House Ltd (in administrative receivership) v Cadogan Ltd [2001] All ER (D) 144, [2001] EWHC 18 (TCC)

8.25 **Facts:** The adjudicator issued a decision in Rainford's favour but Cadogan refused to pay the amount. Rainford applied to the court for summary judgment seeking payment of the money due. Cadogan did not object to the enforceability of the decision itself, but argued that since Rainford had been in administrative receivership since December 2000, a stay should be granted.

8.26 **Held:** Judge Seymour concluded there was a strong prima facie case that Rainford was insolvent, and therefore issued a stay of execution pending the trial of the counterclaim or further order. The stay was conditional upon Cadogan paying the judgment sum into court. Judge Seymour considered the policy behind the Act as reflected in the prohibition of pay when paid clauses:

> 9. ... I do not consider that the policy of the statute is to transfer as between the parties to construction contracts the risk of insolvency of one of the parties. That this understanding is correct seems to me to be clear from the terms of section 113 of the Act ... Thus, if there is a substantial chance, demonstrated by objective evidence, such as the making of a winding-up order, or the appointment of a receiver, that money the obligation to pay which is actually disputed, notwithstanding that the notice contemplated by section 110 of the 1996 Act has not been given, will, if paid, for practical purposes be lost, it seems to me that that is a circumstance which, as Chadwick LJ indicated in his judgment in *Bouygues (UK) Ltd. v. Dahl-Jensen (UK) Ltd.*, ought to be considered on any application for summary judgment. That is not to say that vague fears or unsubstantiated rumours of insolvency will merit much attention, but evidence that some third party has taken action which puts the continued financial viability of the claimant at hazard must, I think, be evaluated seriously.
>
> 10. Whereas in the case of a company in liquidation it is inevitable that the process contemplated by Rule 4.90 of Insolvency Rules 1986 will be undertaken, that is not the position in a case in which the claimant is a company in administrative receivership. In the latter case one cannot tell what the outcome of the receivership will be. In a case in which there is not, inevitably, a need for a determination more or less as matters then stand between the parties of the net state of accounts, in which process the correctness of the decision of the adjudicator must be evaluated, and there is not otherwise any defence to a claim to enforce the award of an adjudicator, it seems to me that the factors which led Chadwick LJ in *Bouygues (UK) Ltd. v. Dahl-Jensen (UK) Ltd.* to consider that it would not be appropriate to give summary judgment at all are not present. However, if there is credible evidence that the claimant is insolvent, in my judgment that is a highly material matter for the court to consider in relation to any application for a stay of execution of the judgment in favour of the claimant.

Total M&E Services Ltd v ABB Building Technologies Ltd (2002) 87 Con LR 154, [2002] EWHC 248 (TCC), Judge Wilcox

8.27 **Facts:** An adjudicator had found that Total was entitled to be paid £462,788 for additional works. ABB refused to comply with the decision, challenging its enforceability, and applied for a stay. Total feared that if the claimant was paid it would 'cut and run disposing of the monies and liquidating the company'. The claimant company's circumstances were that:

1. It was a labour-only subcontractor and not a fixed-asset risk company and did not own quantities of expensive plant and equipment.

2. A credit report in January 2000 showed that the company had £1,000 paid-up capital by its directors and no accounts filed since 1998.
3. There was a floating charge over the assets of the company in relation to some indebtedness at the time of the hearing.
4. The claimant's credit was spent and no further credit would be given until monies were paid to its own labour supplier.

8.28 **Held:** Judge Wilcox concluded that Total's capacity to pay in the future was directly limited to its present entitlement and use of the adjudicator's award. If Total was deprived of its benefit:

> 53. . . . its subcontractor would be in exactly the same position he would have been in had [the Act] never been enacted. They would be starved of the funds to which present entitlement has been shown and upon which they are dependent for the future progress of their business, and, which governs their earning capacity as to future obligations.

8.29 There had been no real change in the claimant's financial status since January 2002, whereas the defendant had enjoyed the substantial benefit of the claimant's labour. He found the evidence as to the risk of future non-payment was not based on compelling or uncontradicted evidence. Accordingly, Judge Wilcox was satisfied there were no special circumstances which rendered it inexpedient to enforce the judgment and refused the stay:

> 52. Where a stay is sought the Court must consider all the circumstances. It must consider whether there are special circumstances which render it inexpedient to enforce the judgment. The risk of an inability to repay on due time is one of a number of factors to be taken account of in the balancing exercise. Where the risk is high as where there is strong uncontradicted evidence of a present inability to pay or a company is in administration a stay may be appropriate on terms safeguarding the disputed monies. The burden is clearly upon the party seeking a stay to adduce evidence of a very real risk of future non-payment. The balancing exercise is of course subject to the overriding considerations of Part 1 of the CPR ensuring justice and fairness between the parties. In considering what is just and fair in an application for a stay of execution of a summary judgment under Part 24 in circumstances such as these the Court must be careful not to reallocate the commercial risks accepted by the parties who engage in a construction contract mindful of the provisions of the Housing Grants Construction Regeneration Act 1996 and subject to the general safeguards of insolvency law.

Wimbledon Construction Company 2000 Ltd v Vago [2005] BLR 374, [2005] EWHC 1086 (TCC) Coulson J

This case is discussed at 8.10.

Treasure & Son Ltd v Martin Dawes [2008] BLR 24, [2007] EWHC 2420 (TCC) Coulson J

8.30 **Facts:** Mr Dawes applied for a stay of the enforcement of an adjudicator's decision, primarily on the basis of two reports:

1. A preliminary opinion commissioned by an expert quantity surveyor as to whether all or part of the £1m awarded against Mr Dawes in the adjudication might ultimately be recovered by Mr Dawes. The expert formed a preliminary view that Mr Dawes ought

to recover more than £650,000 in an arbitration which had been commenced but not yet concluded.

2. A report prepared by a forensic accountant who had examined the last three years of Treasure's financial accounts.

8.31 **Held:** Akenhead J considered the relevant principles from *Wimbledon Construction v Derek Vago* (2005)[39] and concluded that he was 'wholly satisfied' that there could be no stay of execution. In reaching that conclusion, Akenhead J set out the reasons behind his refusal:

> 54. ... I am wholly satisfied that there can be no stay of execution. My reasons are as follows:
>
> (a) Treasure is certainly not in insolvent or other liquidation;
> (b) Treasure is a very long-established company (over 200 years) and has clearly been trading successfully over many years as a builder;
> (c) Treasure's last three filed financial accounts show a reasonably substantial turnover and a reasonably comfortably gross profit margin;
> (d) Its Balance Sheets over the years 2003, 2004 and 2005 show net assets of £843,000, £1,148,000 and £1,273,000 respectively. Although it is not a very large company, it is certainly not insubstantial;
> (e) The evidence in Mr. Davis's report is the product of an understandably 'brief review'. It confirms in effect both on his limited research to date that a significant element of the sum which is the subject matter of the adjudication decision is or is likely to be owed to Treasure. A sum of something less than £500,000 of the £1,018,821.12 ordered by the Adjudicator to be paid will on his brief analysis effectively remain with Treasure;
> (f) I do not find Mr Davis' evidence convincing although, given his brief involvement, I attach no criticism at all of the efforts which he has made;
> (g) I do not consider that Chadwick's report, limited as it is, gets anywhere near establishing that there will be an inability on the part of Treasure to repay all or some £600,000 of the sum to be paid to Treasure pursuant to the Adjudicator's decision. His thesis is predicated upon the basis that, lawfully, the directors of Treasure could distribute company reserves and/or pay substantial dividends out so that what is left in the company would be insufficient to repay all or any significant part of the sum ordered to be paid by the Adjudicator;
> (h) This risk or possibility does not begin to give rise to some probable inability to repay. There is no evidence of any sort from which it would be proper to infer that the Directors would go down this route. Chadwicks cannot point to anything in the filed financial accounts which demonstrates that there is a past history of the Directors doing any such thing. Secondly, it might well be difficult for the Directors to divest the company of substantial sums lawfully, certainly, if the primary intent was to evade any liability to repay. Thirdly, based on the most recently filed company accounts, there is no reason to suppose that there is any probability that Treasure would not be in a position to repay such sums as an arbitrator might order them to repay to Mr Dawes. Fourthly, the filed company accounts do not take into account the sums that may have been payable by Mr Dawes in 2005 but which were not paid. As Mr Davis infers some £500,000 was payable at least for the period 2005 to early 2007 which Mr. Dawes has not paid. If one takes that into account, and it is legitimate to do so, one finds a better financial position than the most recent accounts indicate.

[39] [2005] BLR 374, [2005] EWHC 1086 (TCC), see para. 8.10 above.

Avoncroft Construction Ltd v Sharba Homes (CN) Ltd (2008) 119 Con LR 130, [2008] EWHC 933 (TCC) Kirkham J

8.32 **Facts:** The claimant sought enforcement of an adjudicator's decision, which the defendant resisted by reference to the validity of LADs and a withholding notice. In addition to its substantive challenge to the adjudicator's decision, the defendant also sought to stay enforcement on the grounds that the claimant's finances were a cause for concern and that there was evidence of a 'grave worsening' of the claimant's financial circumstances.

8.33 **Held:** Judge Kirkham considered the evidence raised by the defendant and, following *Wimbledon v Vago,* concluded that she was not persuaded that the claimant would probably be unable to repay the judgment sum. Nor was she persuaded that there were any special circumstances pursuant to RSC Order 47, r. 1. Therefore, she ordered that the defendant pay the judgment sum into court.

8.34 In reaching the conclusion that no stay was warranted, Judge Kirkham considered the following evidence:

1. Two reports from a credit reference agency in March 2008, which rated the claimant at 39/100 (or 'above average risk') in October 2007 and at 7/100 (or 'maximum risk') in March 2008 (Judge Kirkham approached these references with caution as the criterion applied might be its willingness to lend money—an irrelevant factor in such proceedings).
2. More than £10,000 of county court judgments registered against the claimant (a factor which was taken into account).
3. Letters from suppliers of the claimant contending that they had claims for substantial sums (these were considered as 'largely self-serving documents' and of little assistance).
4. A parent company guarantee from a company with assets of more than £300,000 (it was not significant that this parent company guarantee contained a 'step in' clause allowing the guarantor to run any defence which would be open to that subsidiary).

Table 8.1 Table of Cases: Stay of Execution of Enforcement Order on Grounds of Impecuniosity (shaded entries indicate a stay was ordered)

Title	Citation	Issue
Harwood Construction Ltd v Lantrode Ltd	24 Nov. 2000 (TCC), Judge Seymour (unreported)	**Winding-up procedures incomplete at date of enforcement hearing** The court balanced factors in favour of and against a stay and decided to stay the matter for 14 days when the winding-up petition would be heard and let the outcome of that determine the outcome of the application for a stay. No decision pending winding-up procedures.
Absolute Rentals Ltd v Gencor Enterprises Ltd	[2000] CILL 1637 (TCC), Judge Wilcox	**Irregularity in company returns** The defendant served late statements alleging irregularity in the company returns, which was admitted by the claimant. The court found it was not in a position to decide upon the financial standing of either company. It was not desirable to make a decision on such limited evidence, particularly where any impecuniosity could derive from the defendant's default. The robust and summary procedure of enforcement may create casualties although the determinations are provisional and not final. Stay declined.

(2) Stay of Execution of Enforcement Order on Grounds of Impecuniosity

Title	Citation	Issue
Bouygues (UK) Ltd v Dahl-Jensen (UK) Ltd	[2001] 1 All ER (Comm) 1041, (CA), Gibson, Chadwick, Buxton LJJ	**Stay on grounds of impecuniosity** All claims and cross-claims should be resolved in the liquidation, in which full account could be taken and a balance struck, in accordance with r. 4.90 of the Insolvency Rules. Stay granted.
Nolan Davis Ltd v Catton	[2001] All ER (D) 232 (Mar), Judge Wilcox	**No direct evidence of financial hardship** Where no direct evidence adduced, the stay must be refused. Stay declined.
Rainford House Ltd (in administrative receivership) v Cadogan Ltd	[2001] All ER (D) 144, [2001] EWHC 18 (TCC), Judge Seymour	**Administrative receivership** There was a strong prima facie case that the claimant was insolvent. A stay of execution pending the trial of the counterclaim (or further order) was granted, conditional upon the defendant paying the judgment sum into court. Stay granted.
Baldwins Industrial Services plc v Barr Ltd	[2003] BLR 176, [2002] EWHC 2915 (TCC) Kirkham J	**Joint administrative receivers** Baldwins' financial position and the consequent potential injustice to Barr, together with Barr's stated intention to begin proceedings within a month, constituted special circumstances. The principal sum awarded by the adjudicator was ordered to be paid into court. Stay granted.
Total M&E Services Ltd v ABB Building Technologies Ltd (formerly ABB Steward Ltd)	(2002) 87 Con LR 154, [2002] EWHC 248 (TCC), Judge Wilcox	**Floating charge over assets and no further credit available** The claimant was a labour-only subcontractor with a floating charge over its assets and no further credit. The risk of future non-payment was not based on compelling or uncontradicted evidence. Stay declined.
JW Hughes Building Contractors Ltd v GB Metalwork Ltd	[2003] Adj LR 10/03, [2003] EWHC 2421, Judge Forbes	**High risk of insolvency alleged** An unusual case in which the unsuccessful party in the adjudication brought its own proceedings requesting a stay because of GB Metalwork's accounts and credit rating, combined with a retained loss, suggested a high risk of insolvency. The judge considered evidence confirming the improvement in the company's financial health, the expectation from its bank that its overdraft facility would be extended, and that its situation was not too dissimilar from that at the date of the subcontract. Once the monies were paid, its position would improve even more. Stay declined.
AWG Construction Services Ltd v Rockingham Motor Speedway Ltd	[2004] CILL 2154, [2004] EWHC 888 (TCC), Judge Toulmin	**Principles of granting stay** Up-to-date financial information should be considered. A stay should not be granted unless, consistent with the overriding objective, the justice of the case demands it. *Obiter*, since the decision was not enforceable in any event.
Wimbledon Construction Company 2000 Ltd v Vago	[2005] BLR 374, [2005] EWHC 1086 (TCC), Coulson J	**Principles of granting stay** Six principles for granting a stay based on impecuniosity, as set out verbatim at 8.10 above. Stay refused because, inter alia, the claimant was in a similar financial position as at the date of contract, and therefore the defendant had 'contracted for this result'; further the financial hardship was caused by the actions of the defendant. Stay declined.

(Continued)

Table 8.1 *Continued*

Title	Citation	Issue
ALE Heavy Lift v MSD (Darlington) Ltd	[2006] EWHC 2080 (TCC), Judge Toulmin	**No change in financial status** The claimant was in an uncertain position financially, but was not in insolvent liquidation or in administrative receivership. Rather, its financial position was the same as, or similar to, its financial position at the time the relevant contract was made. Stay declined.
Michael John Construction v Golledge and ors	[2006] TCLR 3, [2006] EWHC 71 (TCC), Judge Coulson	**Claimant financially vulnerable and deteriorating** The financial accounts showed the claimant was always vulnerable to cash-flow problems, even at the time that the contract was entered into; the deterioration of the claimant's financial condition was due, at least in large part, to the defendant's failure to pay the adjudication award. The court was not persuaded that there was a potential counterclaim against the claimants which could cancel out the adjudication amount. Stay declined.
Avoncroft Construction Ltd v Sharba Homes (CN) Ltd	(2008) 119 Con LR 130, [2008] EWHC 933 (TCC), Judge Kirkham	**Evidence of grave worsening of the claimant's financial position** The court was not persuaded by the defendant's evidence of credit ratings, county court judgments against the claimant, and letters from suppliers alleging substantial sums were due. Stay declined.
Treasure & Son Ltd v Martin Dawes	(2008) BLR 24, [2007] EWHC 2420 (TCC), Akenhead J	**Two expert reports adduced to support claim of impecuniosity** Two expert reports by a QS and a forensic accountant were not considered convincing against evidence of filed accounts, balance sheets showing not insubstantial net assets and an unpaid amount of some £500,000, which the defendant should have made in previous years. Stay declined.
Air Design (Kent) Ltd v Deerglen (Jersey) Ltd	[2009] CILL 2657, [2008] EWHC 3047(TCC), Akenhead J	**No evidence of insolvency** Following *Wimbledon v Vago*, the court rejected the application for a stay based on reduction in shareholders' funds or net assets, as the defendant had not discharged its onus proving the claimant's alleged insolvency. Further, Air Design was in no worse financial position than it was when it entered into the contract. Stay declined.
Mead General Building Ltd v Dartmoor Properties Ltd	[2009] BLR 225, [2009] EWHC 200 (TCC), Coulson J	**Company voluntary arrangements** The existence of a CVA is relevant but not determinative of the issue of whether a stay should be granted. The claimant's financial troubles had been directly caused by the defendant's failure to pay. Stay declined.
Shaw v Massey Foundation & Pilings Ltd	[2009] EWHC 493 (TCC), Coulson J	**Financial uncertainty** Although there was a point raised in relation to the uncertain financial position of the respondents, it was clear that its financial position was no worse than at the time the contract was made. Stay declined.
JPA Design & Build Ltd v Sentosa (UK) Ltd	[2009] 50 EG 68, [2009] EWHC 2312 (TCC), Coulson J	**Significant debts and losses** JPA had not ceased trading, but had no ongoing work, had significant debts to third parties and an unsatisfied county court judgment. At the time of contract, JPA was making a modest profit, but at the time of the hearing had significant losses.

Title	Citation	Issue
		Failure to pay the £300,000 would not have prevented JPA's current financial situation. Further, a second adjudicator's decision awarded Sentosa £180,000 against JPA; and a contractual provision made the £300,000 repayable to Sentosa as part of the Final Account and therefore was 'imminently repayable'. Stay granted.
S. G. South Ltd v (1) King's Head Cirencenster LLP and (2) Corn Hall Arcade Ltd	[2010] BLR 47, [2009] EWHC 2645 (TCC), Akenhead J	**Financial difficulties caused by non-payment** The accounts filed by the claimant showed a deficit of nearly £70,000, cash at the bank of only £2,000 and more than £90,000 owed in unpaid tax; failure to pay the adjudicator's fees and expenses promptly; a new company had been set up, which may mean the claimant company would 'slip into dissolution'; and two supposedly unsatisfied county court judgments for £7,000 in total. Whilst undoubtedly in financial difficulty, both provisos to granting a stay in *Wimbeldon v Vago* were made out. Stay declined.
Jacques (t/a C&E Jacques Partnership) v Ensign Contractors Ltd	[2009] EWHC 3383 (TCC), Akenhead J	**Stay imposed on part of the judgment** The inability to repay may go only to part of the enforcement judgment sum. The court did not accept the evidence from the claimant sisters and questioned the accuracy of the purported property and rental valuations. The total amount of the judgment with interest and costs was £130,000. The court imposed a stay on £60,000 including interest, and decided costs should be dealt with separately.
O'Donnell Developments Ltd v Build Ability Ltd	(2009) 128 Con LR 141, [2009] EWHC 3388 (TCC), Ramsey J	**No obligation on the claimant to give widespread disclosure** Build Ability challenged summary enforcement and sought a stay, which it later withdrew. Build Ability had requested information from O'Donnell, who had refused to provide it beyond a letter from its accountants. It was for Build Ability to adduce the evidence and decide whether there were grounds for a stay. Application for stay withdrawn.
Integrated Building Services Engineering Consultants Ltd (t/a Operon) v Pihl UK Ltd	[2010] BLR 622, [2010] CSOH 80, Lord Hodge	**Administration (Schedule B1)** The claimant had administrators appointed to it after it commenced proceedings to enforce an adjudicator's decision. PIHL produced documents indicating unsecured creditors would receive no more than 3p in the pound. Stay granted.
Pilon Ltd v Breyer Group plc	[2010] BLR 452, [2010] EWHC 837 (TCC), Coulson J	**CVA, poor credit rating and serious debts** The claimant company was subject to a CVA and had a very poor credit rating and significant and serious debts (particularly to HMRC). The claimant's financial position was significantly worse than it was at the time of entering into the contract, and was not due to the defendant's failure to pay sums awarded by the adjudicator. Therefore, although the adjudicator's decision was unenforceable on the facts, had it been enforceable, a stay of execution would have been granted.

(Continued)

Table 8.1 *Continued*

Title	Citation	Issue
Straw Realisations (No. 1) Ltd v Shaftsbury House (Developments) Ltd	[2011] BLR 47, [2010] EWHC 2597 (TCC), Edwards-Stuart J	**Administration and no guarantee** The claimant was put into administration only three days after the adjudicator's decision became payable and was 'so heavily insolvent that prior payment would have been neither here nor there'. Whilst earlier reference had been made to an offer of guarantee from Vinci, that offer was not maintained by the claimant. Stay granted.
NAP Anglia Ltd v Sun-Land Development Co. Ltd	[2012] BLR 195, [2012] EWHC 51 (TCC), Edwards-Stuart J	**County court proceedings in progress and accounts showing net liability** The mere existence of court proceedings is insufficient for a stay. The successful party was in a less healthy financial position than at the time of contract formation but was not so bad (trading at a profit) that it would be unable to repay a significant part of the sum awarded. Partial stay granted.
Partner Projects Ltd v Corinthian Nominees Ltd	[2012] BLR 97, [2011] EWHC 2989 (TCC), Edwards-Stuart J	**Company dormant for four years and substantial director's loan outstanding** Even though it was likely that the successful company would not be able to repay the whole of the adjudication award at a later date, the position was brought about to a significant extent by the defendant. It was also relevant to the exercise of the court's discretion that the defendant appeared to have been trading whilst insolvent for around four years. Stay declined.
Berry Piling Systems Ltd v Sheer Projects Ltd	(2012) 141 Con LR 225, [2012] EWHC 241 (TCC), Edwards-Stuart J	**Latest accounts showing a loss and credit rating agency assess as high risk** For the purposes of determining an application for a stay, the likely ability of a company to repay a sum awarded by an adjudicator is to be assessed at the time any repayment may have to be made which necessarily involves a level of speculation. It is relevant to consider not just the latest accounts but also the current trading position and future trading prospects. Stay declined.

(3) Stay of Execution of Enforcement on Grounds Other than Impecuniosity

8.35 Whilst the majority of applications to stay the execution of an order enforcing an adjudicator's decision appear to be based on the financial status of the successful party (as discussed in the preceding section), there are other grounds on which applications for stays have been sought, which are discussed below. It should be noted that in most instances applications for stays based on these other grounds have been rejected.

8.36 In summary:

1. Stays are generally not granted pending other adjudications, as each interim binding decision should be enforced following each fast track resolution.
2. Stays are generally not granted pending substantive determination.
3. Stays or adjournments of enforcements may be granted where there are special circumstances, but each case will turn on its own facts.

Pending Other Adjudications

8.37 A stay of execution is unlikely to be granted pending further adjudications. The courts have repeatedly confirmed that the correct approach, absent special circumstances, is to comply with the adjudicator's decision at the end of each adjudication.[40] The intention of Parliament is that adjudication decisions should be enforced summarily and the successful party should not be prejudiced by being kept out of its money: these factors militate strongly against the grant of a stay for an uncertain outcome of a future adjudication.[41] Accordingly, there is not usually any good reason to await the outcome of another adjudication, even where the result in the second adjudication was expected in a matter of two to three weeks.[42]

Pending Final Determination

8.38 A stay of execution of the enforcement order is equally unlikely to be granted pending the outcome of the final determination of the matters that are the subject of the adjudicator's decision, whether by litigation or arbitration. It should be noted that this is a different principle to that of staying the enforcement proceedings themselves on the basis that there is an arbitration agreement (discussed in the next section). The question in the context of a stay of execution is whether it would be inexpedient to require payment to be made where the sum may be required to be repaid on the (usually imminent) decision in the final determination.

8.39 One of the first decisions was *Herschel Engineering Ltd v Breen Property Ltd* (2000).[43] Dyson J found that it did not matter in an adjudication enforcement application whether there were concurrent litigation or arbitration proceedings; the successful party was entitled to enforce that decision, whatever the state of other concurrent proceedings may be, whether litigation proceedings[44] or mediation.[45] The one case where a stay of execution was ordered pending the outcome of arbitration was explained as being decided on its own very particular facts[46] and was described by the same judge in a subsequent decision[47] as a case where there were 'exceptional circumstances'.

8.40 It is worthy of note in this context that where the unsuccessful party to an adjudication has failed to comply with the decision and court judgments enforcing the same, and then seeks to assert rights under the same contract in separate court proceedings, this may amount to unreasonable and oppressive behaviour that will justify a stay of those proceedings.[48]

Other Reasons for Seeking Stay/Adjournment of Enforcement

8.41 In addition to the reasons discussed previously in this chapter, there have been several (mostly unsuccessful) attempts to persuade the courts not to enforce an adjudicator's decision, for

[40] E.g. in *Interserve Industrial Services Ltd v Cleveland Bridge UK Ltd* [2006] All ER (D) 49 (Feb); [2006] EWHC 741 (TCC); *William Verry Ltd v The Mayor and Burgesses of the London Borough of Camden* [2006] EWHC 761 (TCC).

[41] *William Verry Ltd v London Borough of Camden* [2006] All ER (D) 292 (Mar), [2006] EWHC 761 (TCC).

[42] *Avoncroft Construction Ltd v Sharba Homes (CN) Ltd* (2008) 119 Con LR 130, [2008] EWHC 933 (TCC); *William Verry Ltd v The Mayor and Burgesses of the London Borough of Camden* [2006] EWHC 761 (TCC). This approach appears to have superseded an earlier approach in *William Verry Ltd v North West London Communal Mikvah* [2004] 1 BLR 308, [2004] EWHC 1300 (TCC) where a stay was granted pending an adjudication decision expected six weeks later.

[43] [2000] BLR 272 (TCC).

[44] *Hillview Industrial Developoments (UK) Ltd v Botes Building Ltd* [2006] All ER (D) 280 (Jun); [2006] EWHC 1365 (TCC) (Key Case).

[45] *Balfour Beatty Construction Northern Ltd v Modus Corovest (Blackpool) Ltd* [2009] CILL 2660, [2008] EWHC 3029 (TCC).

[46] *Kier Regional Ltd (t/a Wallis) v City & General (Holborn) Ltd and ors* [2009] BLR 90, [2008] EWHC 2454 (TCC).

[47] *Shaw v Massey Foundation & Pilings Ltd* [2009] EWHC 493 (TCC).

[48] *Mentmore Towers Ltd and others v Packman Lucas Ltd* [2010] BLR 393, [2010] EWHC 457 (TCC). In this case, where there was unreasonable and oppressive behaviour with some elements of bad faith, the court found that the claimant was seeking to avoid the 'pay now, argue later' policy of the 1996 Act. A stay was permitted until the claimant did as contractually required and paid on the adjudicator's decisions and argued later.

example, based on issues of international enforcement[49] or requests for extensions due to the bad health of the unsuccessful party.[50]

Extended Time to Pay

8.42 There may also be grounds upon which the court decides to extend the time to pay, in accordance with a judgment of the court, from the usual 14 days set out in CPR 40.11. This may have the same effect as a stay of execution. The court does have a discretion to do so—indeed it has a discretion to extend or reduce the period of time to pay, or order that payments can be made by instalments.[51] However, it declined to do so where it could not see any difficulty in the defendant having to pay the sums ordered.[52] The relevant principles were set out in *Gipping Construction Ltd v Eaves Ltd* (2008):[53]

> 11. ... First, if a party wishes to persuade the court that a period greater than 14 days should be allowed for payment, it is necessary that that application is supported by proper evidence. Secondly, it is much better generally that, if there is a genuine problem about the defendant paying, or being able to pay, that that is a matter first fully discussed on a 'without prejudice' or even open basis between the parties. Ultimately, of course, the court can be asked to rule upon it, but it is much better if commercial parties meet and discuss the issue between themselves and it would only be if they were unable to agree that the court should consider an alternative longer period. It is unlikely that mere inability to pay will suffice to justify the extension of the normal fourteen day period; usually, inability to pay is no defence and an insolvent debtor must take the usual consequences of its insolvency.
>
> 12. I have formed the view in this case that since there is little or no evidence before me about the inability to pay, and because I do feel that this is a case in which the parties should talk to discuss times for payment, I am not going to alter the usual order. What I have done in other cases is to give permission to the defendant to apply to the court at a later stage to extend the 14-day period. That seems to me to be a sensible course in cases where there have not yet been discussions but any discussions may break down. So in this case I would give such permission to apply, which will be incorporated in the judgment, provided that an application, supported by written evidence, is filed before the end of next week, that is within the fourteen day period so that the Court may reach a decision on the application within that period. What I would very much hope is that the parties, as those best qualified to form a judgment on the matter, would talk first in some detail about what is to be done about payment. If the matter comes back in front of me, I cannot in any way undertake to tie the court's hands but the court will probably not be sympathetic to any such application unless at least some money has been paid on account of this judgment before such an application is made.

Table 8.2 Table of Cases: Staying Enforcement on Grounds Other than Impecuniosity (shaded entries indicate where a stay was ordered)

Title	Citation	Issue
Harwood Construction Ltd v Lantrode Ltd	24 Nov. 2000 (TCC), Judge Seymour (unreported)	**Pending court proceedings** It was unsustainable to claim a stay based on pending court proceedings. However, no decision would be given for 14 days, pending winding-up procedures.

[49] *McConnell Dowell Constructors (Aust) Pty Ltd v National Grid Gas plc (formerly Transco Plc)* [2007] BLR 92, [2006] EWHC 2551 (TCC).

[50] *Brenton (t/a Manton Electrical Components) v Palmer* [2001] EWHC 436 (TCC).

[51] *Yoram Ansalem v Raivid* [2009] EWHC 3226 (TCC) (not a case concerning adjudication).

[52] *Jacques (t/a C&E Jacques Partnership) v Ensign Contractors Ltd* [2009] EWHC 3383 (TCC).

[53] [2008] EWHC 3134 (TCC). Followed in *Gulf International Bank v Al Ittefaq Steel Products Co* [2010] EWHC 2601 (QB).

Title	Citation	Issue
Herschel Engineering Ltd v Breen Property Ltd	[2000] BLR 272 (TCC), Dyson J	**Concurrent proceedings** The successful party in an adjudication is entitled to enforce it, regardless of whether there were concurrent litigation or arbitration proceedings. Stay declined.
Absolute Rentals Ltd v Gencor Enterprises Ltd	[2000] CILL 1637 (TCC), Judge Wilcox	**Arbitration pending** The enforcement of an adjudicator's decision is entirely without prejudice to the determination of the final merits by arbitration. To allow a stay of enforcement to arbitration would frustrate the 1996 Act. The purpose of the Scheme is to provide a speedy mechanism for settling disputes on a provisional interim basis. The robust and summary procedure may create casualties although the determinations are provisional and not final. Stay declined.
Brenton (t/a Manton Electrical Components) v Palmer	[2001] EWHC 436 (TCC), Havery J	**Defendant's health problems** No adjournment was possible due to the defendant's serious health problems. Granting the adjournment would simply have the effect of keeping the claimant out of his money for even longer, which was not the purpose of the Act. Adjournment declined.
Alstom Signalling Ltd (t/a Alstom Transport Information Solutions) v Jarvis Facilities Ltd	(2004) 95 Con LR 55, [2004] EWHC 1232 (TCC), Judge LLoyd	**Matter substantively determined during summary enforcement** It may be prudent to defer an application to enforce or to stay a judgment if the point in dispute is to be decided soon. Transferring money for a limited period of time may not be sensible.
William Verry Ltd v North West London Communal Mikvah	[2004] 1 BLR 308, [2004] EWHC 1300 (TCC), Judge Thornton	**Decision not drawn up for six weeks** The procedural realities were that, if the defendant believed it had real defect claims against the claimant, it could commence an adjudication and have a decision within six weeks. If not, it would suggest its claimed abatement has 'little merit'. Order not drawn up for six weeks so as to allow adjudication or commercial discussions.
Harlow & Milner Ltd v Teasdale	[2006] EWHC 535 (TCC), Judge Coulson	**Pending arbitration** An arbitration had been commenced (although a hearing was not scheduled for many months). The court refused to stay the final charging order until the resolution of the arbitration. The defendant could not ignore the judgment of the court. Otherwise, parties could commence arbitral proceedings just to delay enforcement, which would be contrary to the purpose of adjudication. Stay declined.
Michale John Construction v Golledge and ors	[2006] TCLR 3, [2006] EWHC 71 (TCC)	**Arbitration notice not pursued** Whist the defendant had served an arbitration notice, it had subsequently done nothing to pursue the claim further. Stay declined.
Interserve Industrial Services Ltd v Cleveland Bridge UK Ltd	[2006] All ER (D) 49 (Feb); [2006] EWHC 741 (TCC), Jackson J	**Pending further adjudications** Where there are successive adjudications, unless there are special circumstances, each adjudicator's decision should be enforced. Stay declined.

(Continued)

Table 8.2 *Continued*

Title	Citation	Issue
William Verry Ltd v London Borough of Camden	[2006] All ER (D) 292 (Mar), [2006] EWHC 761 (TCC), Ramsey J	**Pending further adjudication** William Verry sought enforcement of an adjudicator's decision that Camden should pay more than £530,000 as a result of an EOT for 27 weeks and 5 days. A further adjudication decision in relation to Camden's defects counterclaim was due about three weeks after the enforcement hearing. The outcome of that adjudication was uncertain and not a matter on which the judge considered it appropriate to speculate. Stay declined.
Hillview Industrial Developments (UK) Ltd v Botes Building Ltd	[2006] All ER (D) 280 (Jun); [2006] EWHC 1365 (TCC), Judge Toulmin	**Pending litigation** Adjudication decisions should be enforced at the end of each adjudication, regardless of whether legal proceedings were pending, unless there is a risk of manifest injustice. Stay declined.
Gray & Sons Builders (Bedford) Ltd v Essential Box Company Ltd	(2006) 108 Con LR 49, [2006] EWHC 1365 (TCC), Judge Toulmin	**Pending arbitration** The court rejected a claim that because the defendant had commenced arbitration proceedings, the claimant could not and should not be allowed to enforce the adjudication and/or that the adjudicator had no jurisdiction to consider that adjudication. Stay declined.
McConnell Dowell Constructors (Aust) Pty Ltd v National Grid Gas plc (formerly Transco Plc)	[2007] BLR 92, [2006] EWHC 2551 (TCC), Jackson J	**International enforcement: bond provided** National Grid argued that it had 'special circumstances' in accordance with RSC Order 47 because: (i) it had good prospects in future proceedings for 'clawing back' much of the money awarded by the adjudicator; and (ii) whilst at the time of contracting, McConnell had a registered branch in the UK, that office had subsequently closed down. Therefore, any judgment or arbitral award obtained by National Grid would have to be enforced in Victoria, Australia. Jackson J saw much force in the argument that it would be more time consuming and expensive for National Grid to enforce outside of the UK. However, McConnell offered to provide a bond and Jackson J considered this to be sufficient to meet the specific concerns raised by National Grid. Stay declined.
Avoncroft Construction Ltd v Sharba Homes (CN) Ltd	(2008) 119 Con LR 130, [2008] EWHC 933 (TCC), Judge Kirkham	**Pending further adjudications** A second adjudication decision was due in two weeks in which the defendant may have recovered significantly more than it was ordered to pay in the first adjudication. Stay declined.
Kier Regional Ltd (t/a Wallis) v City & General (Holborn) Ltd and ors	[2009] BLR 90, [2008] EWHC 2454 (TCC), Coulson J	**Final charging order and arbitration award imminent** Where the successful party in the adjudication already had a final charging order, its instant application for a third party debt order had failed and the arbitration to finally determine the matter was only eight weeks away (and was proceeding on the basis of the delay, loss, and expense claims being put on a very different basis from that on which the adjudication was decided), it was appropriate to restrain any further enforcement applications. Stay granted.
Balfour Beatty Construction Northern Ltd v Modus Corovest (Blackpool) Ltd	[2009] CILL 2660, [2008] EWHC 3029 (TCC), Coulson J	**Pending agreement to mediate** The construction contract contained a clause permitting the parties to suggest mediation of disputes. It was no more than an agreement to agree. The court could not

Title	Citation	Issue
		conclude that there was an arguable defence on which the other party had a realistic prospect of success. Stay declined.
Gipping Construction Ltd v Eaves Ltd	[2008] EWHC 3134 (TCC), Akenhead J	**Time to pay** Sets out principles when considering whether to extend usual 14-day period for payment. The court did not extend in this instance, but encouraged the parties to have sensible discussions and revert with an application at a later stage.
SG South Ltd v (1) King's Head Cirencenster LLP and (2) Corn Hall Arcade Ltd	[2010] BLR 47, [2009] EWHC 2645 (TCC), Akenhead J	**Final account pending** There is no authority which permits the court to defer enforcement because there is a possibility of an imminent resolution of other disputes between the parties which might alter the overall balance between the parties. Stay declined.
Jacques (t/a C&E Jacques Partnership) v Ensign Contractors Ltd	[2009] EWHC 3383 (TCC), Akenhead J	**Time to pay** The effect of the court's decision was that the defendant was to pay £54,000 by 8 January 2010. It sought to extend that time as it claimed it needed to obtain personal loans. The court did not accept this, and rejected the application to extend the time to pay.

(4) Stay of Enforcement Proceedings

8.43 Parties have generally not succeeded where they have sought to stay enforcement proceedings where there is an arbitration agreement, or where legal proceedings for final determination are underway.

Section 9 of the Arbitration Act

8.44 Section 9 of the Arbitration Act 1996 relevantly provides:

> (1) A party to an arbitration agreement against whom legal proceedings are brought (whether by way of claim or counterclaim) in respect of a matter which under the agreement is to be referred to arbitration may (upon notice to the other parties to the proceedings) apply to the court in which the proceedings have been brought to stay the proceedings so far as they concern that matter.
>
> (2) …
>
> (3) An application may not be made by a person before taking the appropriate procedural step (if any) to acknowledge the legal proceedings against him or after he has taken any step in those proceedings to answer the substantive claim.
>
> (4) On an application under this section the court shall grant a stay unless satisfied that the arbitration agreement is null and void, inoperative, or incapable of being performed.

8.45 It is an obvious point, but this section will only apply where the parties have properly entered into an arbitration agreement.[54] Arbitration agreements in construction contracts often now

[54] In *(1) Walter Llewellyn & Sons Ltd v Excel Brickwork Ltd* [2011] CILL 2978, [2010] EWHC 3415 (TCC) the parties had failed to properly execute the arbitration agreement in an NEC contract.

expressly state that the enforcement of an adjudicator's decision is not subject to the arbitration agreement. Where there is no such express provision, the question has arisen whether enforcement proceedings should be stayed with the question of enforcement to be decided by arbitration. Adjudication enforcement proceedings are 'legal proceedings' and therefore potentially subject to s. 9 of the Arbitration Act 1996 where the underlying construction contract contains an arbitration agreement.

Stay of Execution Unlikely

8.46 However, since Dyson J's judgment in *Macob Civil Engineering Ltd v Morrison Construction Ltd* (1999) (Key Case),[55] the courts have repeated that in adjudication enforcement proceedings, a stay under s. 9 is unlikely. The reason for this is that in most instances, the court will prefer to enforce any valid decision by an adjudicator, and then let the parties deal with the merits of the dispute in the arbitration. To do otherwise would permit parties to commence arbitral proceedings just to delay enforcement,[56] and this would make a nonsense of the adjudication process.[57] This policy is applied in most instances, even if there are 'casualties'[58] along the way:

> 30. ... [The] parties had made a contract which, by an express provision, contained an arbitration clause, and, by the statutory implication of terms, incorporated the Scheme. The relationship between arbitration and adjudication is dealt with in paragraphs 21 and 23(2) of the Scheme. These provide, respectively, that the decision of the adjudicator is to be complied with immediately, and that the decision is binding and is to be complied with 'until the dispute is finally determined by legal proceedings, by arbitration or by agreement'.
>
> 31. ... [This is] a clear articulation of the 'pay now, argue later' policy which underlies Part II of the Construction Act and the Scheme itself. That policy would be stultified if a reference to arbitration under clause 30.3 were to put a brake, whether permanently or otherwise, on the carrying through of the adjudication process to enforcement. [The unsuccessful party in an adjudication] is free to take any points which are open to it in arbitration, but this does not entitle it to set on one side the Scheme which is part and parcel of the agreement into which it entered. Objections as to the adjudicator's jurisdiction, if they are to bar enforcement of his award, will have to be made in the enforcement proceedings. Questions which relate to the merits of the dispute must be left to the arbitration. In that way, proper weight is given both to the arbitration clause and to the importation of the Scheme into the contract.[59]

8.47 As can be seen from Table 8.3 below, the court has almost always allowed the enforcement proceedings to continue notwithstanding the existence of an arbitration agreement. The only successful application has been in special circumstances where, before the adjudication had commenced, the parties agreed to refer the dispute as to whether a construction contract existed to arbitration.[60]

8.48 Where the contract provided that disputes arising out of an adjudication were to be referred to arbitration (in contrast to the matter being finally determined anew) it was held that the unsuccessful party must elect either to refer the dispute to arbitration or contend that the decision was not a binding and enforceable decision—it could not do both.[61]

[55] *Macob Civil Engineering Ltd v Morrison Construction Ltd* [1999] BLR 93 (TCC).

[56] *Harlow & Milner Ltd v Teasdale* [2006] EWHC 535 (TCC).

[57] *MBE Electrical Contractors Ltd v Honeywell Control Systems Ltd* [2010] BLR 561, [2010] EWHC 2244 (TCC).

[58] *(1) Walter Llewellyn & Sons Ltd v Excel Brickwork Ltd* [2011] CILL 2978, [2010] EWHC 3415 (TCC).

[59] *MBE Electrical Contractors Ltd v Honeywell Control Systems Ltd* [2010] BLR 561, [2010] EWHC 2244 (TCC).

[60] *Cygnet Heathcare v Higgins City Ltd* (2000) 16 Const LJ 394 (TCC) (Key Case).

[61] *Macob Civil Engineering Ltd v Morrison Construction Ltd* [1999] BLR 93 (TCC).

The unsuccessful party may in any event be barred from applying for a stay of the enforcement proceedings as a result of s. 9(3). An application to invoke the assistance of the court to dispose of a claim or counterclaim (in the absence of a simultaneous application for a stay or an effective reservation of the right to apply for a stay to arbitration) must be 'a step';[62] and filing an affidavit in reply to an application for summary judgment is also 'a step'.[63] It is, however, no answer to an application for a stay under s. 9 that the defendant has no arguable defence to the claimant's claim.[64]

Key Cases: Staying Enforcement

Macob Civil Engineering Ltd v Morrison Construction Ltd [1999] BLR 93 (TCC)

8.49 **Facts:** Macob was engaged as a subcontractor to Morrison to carry out groundworks at a retail development in Wales. Macob sought to enforce an adjudicator's decision and Morrison referred to arbitration the disputes arising out of or in connection with the adjudicator's decision and issued a summons to stay the proceedings under s. 9 of the Arbitration Act 1996.

8.50 **Held:** Even if there was a challenge to the validity of an adjudicator's decision, it was nevertheless a decision within the meaning of the Act, the Scheme, and the contract in question. The decision was binding and enforceable until the challenge was finally determined by arbitration, legal proceedings or agreement. Justice Dyson rejected the argument that there should have been a stay under the Arbitration Act 1996.

> 27. [The defendant's counsel] submits that even if (as I have held) the adjudicator's decision was a decision within the meaning of clause 27, the defendant is entitled to a stay of these enforcement proceedings under section 9 of the Arbitration Act 1996 because there is a dispute as to whether it was a decision. He argues that this is the conclusion to which I am driven by Halki. He relies on the fact that by letter dated 13 January 1999 the defendant gave the plaintiff notice of arbitration in respect of 7 disputes relating to the adjudicator's decision. These included: 'Was the purported Adjudicator's Decision dated 6 January 1999 of any force or effect?' Thus, he submits that the dispute as to the validity of the decision has been the subject of a notice of arbitration, and the current proceedings must be stayed. Mr Furst accepts that, where there is a dispute as to the merits of a decision, the effect of section 108(3) of the Act, paragraph 23(1) of the Scheme and clause 27(i) of the contract is that it is binding and enforceable pending the final resolution of the dispute by arbitration or otherwise. But if the dispute is as to the validity of the decision, the position is otherwise where the defendant to the enforcement proceedings has referred that dispute to arbitration and seeks a stay under section 9.
>
> 28. This is an ingenious argument, but I cannot accept it. In my view, if the defendant wished to challenge the validity of the decision, it had an election. One course open to it was (as it did) to treat it as a decision within the meaning of clause 27, and refer the dispute to arbitration. The other was to contend that it was not a decision at all within the meaning of clause 27, and to seek to defend the enforcement proceedings on the basis that the purported decision was not binding or enforceable because it was a nullity. For the reasons stated earlier in this judgment, this second course would not have availed the defendant.
>
> 29. But what the defendant could not do was to assert that the decision was a decision for the purposes of being the subject of a reference to arbitration, but was not a decision for

[62] *David McLean Housing Contractors Ltd v Swansea Housing Association Ltd* [2002] BLR 125 (TCC).

[63] *Turner & Goudy v McConnell* [1985] 1 WLR 898, referred to in *David McLean Housing Contractors Ltd v Swansea Housing Association Ltd* [2002] BLR 125 (TCC).

[64] *Halki Shipping Corporation v Sopex Oils Ltd* [1998] 1 WLR 726 (CA); *Collins (Contractors) Ltd v Baltic Quay Management* (1994) Ltd [2005] BLR 63, [2004] EWCA Civ 1757 (CA).

> the purposes of being binding and enforceable pending any revision by the arbitrator. In so holding, I am doing no more than applying the doctrine of approbation and reprobation, or election. A person cannot blow hot and cold: ... Once the defendant elected to treat the decision as one capable of being referred to arbitration, he was bound also to treat it as a decision which was binding and enforceable unless revised by the arbitrator.

Hillview Industrial Developments (UK) Ltd v Botes Building Ltd [2006] All ER (D) 280 (Jun); [2006] EWHC 1365 (TCC)

8.51 **Facts:** Botes did not comply with an adjudicator's decision that almost £300,000 in liquidated damages be paid within seven days of the decision. The employer commenced enforcement proceedings and the contractor conceded it had no defence to the enforcement claim. The contractor requested agreement to adjourn the enforcement application because it was about to issue legal proceedings for summary judgment for the final account resulting in payment of less than £200,000. It was argued on behalf of the contractor (Botes) that:

1. Although in normal circumstances, Hillview would be entitled to summary judgment, under CPR 24.2 it would be open to the court to conclude that the circumstances in this case provided a compelling reason why the case should be disposed of at trial rather than by summary judgment.
2. The application should be adjourned in order that both applications for summary judgment could be heard at the same time; or by ordering immediate payment of the difference between the adjudicator's award and the sums claimed by Botes and adjourning the application.
3. The contractor was entitled to set off the balance due to it against the larger sum awarded by the adjudicator.
4. If judgment was given against Botes for the full amount, the court should grant a stay for the short time until Botes' summary judgment application could be heard.

8.52 **Held:** Judge Toulmin rejected each of Botes' arguments and gave judgment for Hillview for the full sum awarded by the adjudicator plus statutory interest. Judge Toulmin found that the decision in an adjudication must be enforced even where court proceedings are involved:

> 29. ... It would, in my view, be an abuse of the process of the court to allow the case to proceed to trial where Botes concedes that it has no defence.
>
> 30. I can also see no justification for adjourning the summary judgment hearing. The purpose of the 1996 Act is to provide a means of resolving construction disputes on a provisional basis and providing that sums awarded should be paid promptly. This is set out clearly in the contract. I can see no reason why this application should be adjourned so that both applications can be heard at the same time.
>
> 31. Put bluntly, Hillview was entitled to be paid within seven days of the Adjudicator's award on 27th March 2006 and has not yet been paid.
>
> ...
>
> 33. ... [T]he purpose of the 1996 Act is to provide a statutory framework which would enable justice to be done between parties to a dispute. It was not intended to cause injustice. ... This can, in appropriate cases, be dealt with by the grant of a stay. I am satisfied that the jurisdiction in adjudication enforcement cases to grant a stay under the CPR must be limited to cases where there is a risk of manifest injustice.
>
> 34. I see no injustice in ordering Botes to pay the sum awarded by the adjudicator forthwith, nor do I find it a curious result if Botes is required to pay the sum awarded by the Adjudicator in full only (if it be the case) for a substantial part of it to have to be repaid by Hillview in a short time from now. Such a solution is entirely consistent with the legislation.

Table 8.3 Table of Cases: Staying Enforcement Where There Is an Arbitration Agreement (the shaded entry indicates a stay was granted)

Title	Citation	Issue
Macob Civil Engineering Ltd v Morrison Construction Ltd	[1999] BLR 93 (TCC), Dyson J	**Effect of arbitration agreement** An adjudicator's decision is binding and enforceable until finally determined. Stay inappropriate and declined.
Cygnet Healthcare v Higgins City Ltd	(2000) 16 Const LJ 394 (TCC), Judge Thornton	**Arbitration agreed on question of law** The claimant sought to enforce an adjudicator's decision that it was entitled to LADs and damages for defective works. Prior to the adjudication commencing, the parties agreed to refer the issue of whether a construction contract was in existence to arbitration. There was a statutory entitlement to adjudicate at any time, whether or not parallel court or arbitral proceedings had been commenced (see *Herschel Engineering Ltd v Breen Property Ltd*). However, the parties had entered into an ad hoc arbitration agreement covering the same disputes and there was no practical purpose achieved by allowing the adjudication enforcement proceedings to occur in advance of the arbitration. Therefore, the judge adjourned the application for summary judgment, the application to amend the arbitration application, and the underlying arbitration itself. Stay effectively granted until the arbitrator's award in relation to the underlying contract.
Amec Capital Projects Ltd v Whitefriars City Estate Ltd	[2003] EWHC 2443 (TCC), Judge LLoyd	**Dispute as to enforceability of decision was not subject to an arbitration agreement** Works were performed to a letter of intent and 'amended standard form' which named a particular adjudicator. A different person was appointed and Amec sought to enforce that adjudicator's decision. Whitefriars applied for a stay, arguing that the real dispute between the parties was the subject of an arbitration agreement. The policy of the Act is that substantive disputes are to be resolved by litigation or, if selected, arbitration. Matters preliminary to that substantive dispute (such as whether the adjudicator made an enforceable decision) are not matters for the arbitrator but for the courts on enforcement. Stay declined.
Shaw v Massey Foundation & Pilings Ltd	[2009] EWHC 493 (TCC), Coulson J	**Delay in applying for summary judgment** A party is not entitled to a stay of adjudication enforcement proceedings to arbitrate. The judge did not accept the defendant's argument that the claimant's failure to apply instantly for summary judgment meant that a stay for arbitration should be granted. Stay declined.
MBE Electrical Contractors Ltd v Honeywell Control Systems Ltd	[2010] BLR 561, [2010] EWHC 2244, Judge Langan	**Relationship between existence of arbitration agreement and the Scheme** The contract provided that any dispute arising would be finally resolved by a panel of three arbitrators. The Scheme was implied into the contract. The interaction between these two provisions meant that

(Continued)

Table 8.3 *Continued*

Title	Citation	Issue
		the parties had to comply with the adjudicator's decision. Any points of merit were open to Honeywell in the arbitration. It would make a nonsense of the adjudication process if the losing party could avoid the consequences of an adjudicator's decision by claiming that the dispute should be referred to arbitration. Stay declined.
(1) Walter Llewellyn & Sons Ltd v Excel Brickwork Ltd	[2011] CILL 2978, [2010] EWHC 3415 (TCC), Akenhead J	**Parties failed to execute arbitration agreement** The parties used an NEC2 contract but left blank the part of the form where the tribunal's identity could be entered. The standard conditions only provided for arbitration if the parties expressly agreed to it. Stay declined.

9

JURISDICTIONAL CHALLENGES

(1) General Principles

Source and Nature of the Adjudicator's Jurisdiction

9.01 In the context of adjudication, to refer to the jurisdiction of the adjudicator is to refer to the power of the adjudicator or the scope of the adjudicator's authority. Any decision reached by an adjudicator who had no jurisdiction to make that decision will not be enforceable. The adjudicator may lack jurisdiction to make any decision at all or, for a variety of reasons that are discussed below, may lack the jurisdiction to make a particular decision. Equally, the adjudicator may have jurisdiction in relation to part of the decision but not in relation to another part. In the latter case if the decision is severable then it may be that only the part of the decision that was made without jurisdiction will be unenforceable.[1]

9.02 Unlike the courts, an adjudicator has no inherent power to make any binding determination in respect of the rights or obligations of any party. An adjudicator gains such power by virtue of his appointment to adjudicate an identified dispute and the referral of that dispute for determination, but only if the party purporting to appoint the adjudicator is entitled to seek the appointment of an adjudicator to determine that dispute.

9.03 A party has no inherent entitlement to appoint an adjudicator to determine a dispute; any such entitlement only arises by virtue of the appointing party's contractual arrangements with the party with whom it is in dispute. The entitlement to appoint an adjudicator may arise as the result of an express provision in the contract, or be implied by the operation of s.

[1] See Ch. 7 at 7.73–7.81 for discussion of when an adjudicator's decision may be severed.

108 of the Act if the contract is a 'construction contract' within the meaning of s.104.[2] The right to appoint an adjudicator may also arise as a result of an ad hoc agreement between parties to resolve that dispute by adjudication.[3]

9.04 The first step in identifying the jurisdiction of an adjudicator is therefore an examination of the contract between the parties to the dispute. If there is no express adjudication provision and the requirements of s. 104 of the Act are not satisfied then, in the absence of an ad hoc agreement to adjudicate, any adjudicator appointed will have no jurisdiction.

9.05 The second step is to identify the limits of the adjudicator's jurisdiction, in particular the matters on which the adjudicator has the power to reach a decision and the relief which may be granted. The scope of the adjudicator's jurisdiction is initially circumscribed by the content of the notice of adjudication.[4] In particular the adjudicator may only determine the dispute identified in that notice and may only award the particular relief requested in that notice.[5]

9.06 It is important to note that, unlike natural justice challenges, if the excess of jurisdiction is made out, it is irrelevant whether or not the other party has suffered prejudice.[6]

Jurisdiction to Consider any Available Defences Raised

9.07 It is now well settled that the responding party is entitled to raise any defence that is a proper defence in law to the claim being made and the adjudicator is required to consider all defences properly raised.[7] It is irrelevant that the circumstances giving rise to the defence are not set out in the adjudication notice or were not raised or 'in dispute' prior to the reference to adjudication. It would be 'absurd' if the referring party could, through 'some devious bit of drafting, put beyond the scope of the adjudication' the responding party's otherwise legitimate defence to the claim.[8] The failure of an adjudicator to consider a valid defence is usually the result of the adjudicator misunderstanding the scope of his or her jurisdiction which may result in a breach of natural justice.[9] This topic is discussed in more detail in Chapter 10 at 10.36–10.53.

9.08 The adjudicator's jurisdiction may therefore be extended as a result of any matters raised by the responding party that necessarily require the adjudicator to determine those matters. For example, a contractor's notice may be limited to a claim for payment of outstanding sums; that claim may be met with a set-off for the cost of defects or liquidated damages. The adjudicator has jurisdiction to deal with the matters raised in defence,[10] even though there was no reference to those defences in the notice of adjudication.[11]

---

[2] See Ch. 1 at 1.03 onwards.

[3] See Ch. 5.

[4] *KNS Industrial Services (Birmingham) Ltd v Sindall Ltd* (2000) 75 Con LR 71; *Mecright v TA Morris Developments Ltd* [2001] Adj LR 06/22. See also Ch. 4 at 4.09–4.15

[5] *Ken Griffin and John Tomlinson (t/a K&D Contractors) v Midas Homes Ltd* (2000) 78 Con LR 152 (TCC). The position may be different where the parties have incorporated express adjudication rules that allow for further specifics of the relief sought to be provided in the referral, as was the case in *Jerome Engineering Ltd v Lloyd Morris Electrical Ltd* [2002] CILL 1827.

[6] *IDE Contracting Ltd v R. G. Carter Cambridge Ltd* [2004] BLR 172, [2004] EWHC 36; *Vision Homes Ltd v Lancsville Construction Ltd* [2009] BLR 525, [2009] EWHC 2042 (TCC) at para. 56.

[7] *KNS Industrial Services (Birmingham) Ltd v Sindall Ltd* (2000) 75 Con LR 71 (TCC); *Cantillon Ltd v Urvasco Ltd* BLR 250, [2008] EWHC 282 (TCC) at para. 54; *Quartzelec Ltd v Honeywell Control Systems Ltd* [2009] BLR 328, [2008] EWHC 3315 (TCC).

[8] *Pilon Ltd v Breyer Group plc* 2010 BLR 452, [2010] EWHC 837 (TCC).

[9] It may also be characterized as a failure to exercise his or her jurisdiction, see *Ballast plc v Burrell Company (Construction Management) Ltd* [2001] BLR 529 and the cases discussed on this topic in Ch. 10.

[10] Subject always to the proviso that the defence must be one that is legally available to the responding party. It may be the defence raised is not legally available because a withholding notice was required but had not been served. For a discussion about this see Ch. 10 at 10.39–10.40.

[11] The adjudicator is not however entitled to raise the defence himself or herself, it must be one which is raised by the responding party: *Working Environments Ltd v Greencoat Construction Ltd* [2012] EWHC 1039 (TCC) at para. 24.

Expanding the Adjudicator's Jurisdiction During the Adjudication

9.09 It is also possible for the parties to extend the adjudicator's jurisdiction by express agreement in the course of the adjudication.

9.10 The adjudicator has no jurisdiction until the dispute has been validly referred.[12] Before that time the adjudicator has no power to decide the dispute even though he has been appointed.

Adjudicator's Investigation of Jurisdiction

9.11 A responding party will on occasion seek to prevent an adjudication proceeding by identifying jurisdictional issues at the outset in the hope that the adjudicator will resign due to not having jurisdiction. Whilst there is no obligation on an adjudicator to enquire into his own jurisdiction[13] it may assist the parties if the adjudicator does so to prevent costs being wasted on the adjudication, only for the referring party to discover that it has an unenforceable decision.

9.12 However, unless the adjudication agreement provides[14] or the parties agree otherwise, any conclusion the adjudicator reaches in relation to his or her own jurisdiction will not bind the parties. The responding party will therefore be entitled, subject to the question of reservation of rights discussed below, to raise the jurisdictional objection in the courts, either during the adjudication (through a Part 8 application)[15] or in defence of enforcement proceedings. However, one exception to this principle was identified in *Air Design (Kent) Ltd v Deerglen (Jersey) Ltd* (2008)[16] where it was found that an adjudicator's decision on a matter that went to his jurisdiction was binding on the parties where the adjudicator was also required to answer the same question to determine the substantive dispute he had been asked to decide.

9.13 Otherwise, there are three ways in which an adjudicator's decision on jurisdiction will bind the parties:

1. If the adjudication rules provide that such decisions are binding, then even if the adjudicator's decision on jurisdiction was wrong in fact or law it will not provide a basis for resisting enforcement.
2. A party may not raise a jurisdictional challenge if, by its communications or conduct, that party conferred power on the adjudicator to make a decision as to jurisdiction.
3. After having raised a jurisdictional objection, or if a potential jurisdictional objection was or should have been reasonably apparent in the course of the adjudication but was not raised, then if the party continues to participate in the adjudication it will only be able to rely on the objection when resisting enforcement if it clearly stated that it reserved its

---

[12] *Lanes Group Plc v Galliford Try Infrastructure Ltd* (2011) 137 Con LR 1, [2011] EWHC 1679 (TCC).

[13] *AMEC Capital Projects Ltd v Whitefriars City Estates Ltd* [2005] 1 All ER 723, [2004] EWCA Civ 1418 (CA).

[14] Such as para. 12 of the TeCSA Rules which provided 'The Adjudicator may rule upon his own substantive jurisdiction, and as to the scope of the Adjudication' as considered in *Farebrother Building Services Ltd v Frogmore Investments Ltd*, 20 Apr. 2001, [2001] CILL 1762 (TCC)

[15] See Ch. 7 at 7.57–7.64.

[16] [2009] CILL 2657, [2008] EWHC 3047 (TCC). Akenhead J has applied his decision in subsequent cases, including *Camillin Denny Architects Ltd v Adelaide Jones & Company Ltd* [2009] BLR 606, [2009] EWHC 2110 (TCC) and *Supablast (Nationwide) Ltd v Story Rail Ltd* [2010] BLR 211, [2010] EWHC 56 (TCC).

rights to raise the objection later and that its participation in the adjudication was under protest.

Reservation of Rights

9.14 Parties to adjudications must be careful not to inadvertently bestow jurisdiction upon an adjudicator. This may occur when a dispute or an issue is referred to the adjudicator over which he or she does not have jurisdiction, but no adequate objection is taken to it being decided by the adjudicator.[17] If a party wants to raise the jurisdictional objection at a later date it may be held to have acquiesced in the adjudicator deciding the point, or to have waived its right to object to an increase in jurisdiction or a particular procedural course. The best way to avoid losing such rights of challenge is via an adequate reservation of rights. A reservation of rights may be either general (e.g. 'the responding party reserves the right to raise any jurisdictional issues in due course whether previously raised or not') or specific. Whilst a general reservation of rights is undesirable, it is nevertheless effective.[18] However, it is important that any objection and reservation of rights (general or specific) is expressly and consistently maintained throughout the adjudication if a party is to be able to rely on the objection(s) as a challenge to enforcement. If a responding party initially raises both specific and general challenges to the jurisdiction, but later only refers to the specific challenges, it may well be found to have abandoned any general objection and only be permitted to rely on the specific objections that had been maintained.[19] Similarly, if a responding party only raises a specific ground that proves unsuccessful, it cannot raise different grounds of objection when seeking to resist enforcement.[20]

9.15 In order to avoid these difficulties when seeking to resist enforcement on jurisdictional grounds it is suggested that the following steps are taken:

1. Make and maintain a general reservation of rights at each stage of the adjudication and on each occasion that a specific challenge is raised or repeated.
2. Any specific jurisdictional challenge that is identified during the adjudication should be raised with the adjudicator and the referring party.
3. Where a specific challenge is made, the party should expressly state that in making the challenge it does not confer on the adjudicator jurisdiction to make a binding jurisdictional determination and that it reserves its right to raise the specific challenge in any enforcement proceedings that may follow.
4. Maintain any specific objection and reservation of rights at each stage of the adjudication.

Table 9.1 Table of Cases: General Principles

Title	Citation	Issue
KNS Industrial Services (Birmingham) Ltd v Sindall	(2000) 75 Con LR 71 (TCC), Judge LLoyd	**Limits of jurisdiction and possible defences** *Obiter*: the notice of adjudication determines an adjudicator's jurisdiction. Whilst the adjudicator's jurisdiction does not derive from other documents, they are likely to help the adjudicator determine what is to be decided. A possible consequence of a

[17] For a fuller discussion of when there may be an 'ad hoc' conferral of jurisdiction on an adjudicator and the related table of cases, see Ch. 5.

[18] *GPS Marine Contractors Ltd v Ringway Infrastructure Services Ltd* [2010] BLR 377, [2010] EWHC 283 (TCC).

[19] As happened in *Durham County Council v Jeremy Kendall* (2011) BLR 425 (TCC).

[20] *Allied P&L v Paradigm Housing Group Ltd* [2010] BLR 59, [2009] EWHC 2890 (TCC).

Title	Citation	Issue
		dispute referred to adjudication is the deployment of any defence to the claim made.
Farebrother Building Services Ltd v Frogmore Investments Ltd	20 April 2001, [2001] CILL 1762 (TCC), Judge Gilliland	**The adjudicator's view of his jurisdiction was binding** Para. 12 of the TeCSA Rules granted the adjudicator the power to rule on his own substantive jurisdiction.
Pegram Shopfitters Ltd v Tally Wiejl (UK) Ltd	[2004] 1 All ER 818, [2003] EWCA Civ 1750 (CA), May, Hale LJJ, Hooper J[21]	**Adjudicator has no jurisdiction to decide own jurisdiction** An adjudicator has no jurisdiction to decide his or her own jurisdiction unless there is a submission to jurisdiction in that regard.
Cantillon Ltd v Urvasco Ltd	[2008] BLR 250, [2008] EWHC 282 (TCC), Akenhead J	**Defence not aired before adjudication** The responding party can put forward any arguable defence in adjudication, whether propounded before the adjudication or not. It must follow that the adjudicator can rule not only on that defence but also upon the ramifications of that defence to the extent that it is successful in so far as it impacts upon the fundamental dispute. The defendant's defence, that loss and expense claimed related to a period later than the specific weeks identified in the claim, introduced the issue for the adjudicator to decide on that basis in the claimant's favour. Decision enforced.
Air Design (Kent) Ltd v Deerglen (Jersey) Ltd	[2009] CILL 2657, [2008] EWHC 3047 (TCC), Akenhead J	**Overlap of jurisdiction and substance** Where a substantive matter within the adjudicator's jurisdiction overlaps with a question of jurisdiction, an error in the decision relating to the substantive matter cannot result in a challenge to jurisdiction.
Quartzelec Ltd v Honeywell Control Systems Ltd	[2009] BLR 328, [2008] EWHC 3315 (TCC), Judge Davies	**Failure to consider defence to payment claim** A responding party is entitled to raise any defence open to it. In refusing to consider the defence, the adjudicator committed a significant jurisdictional error and failed to act in accordance with natural justice.
Bovis Lend Lease Ltd v Cofely Engineering Services	[2009] EWHC 1120 (TCC), Coulson J	**The right to object is not lost simply because it is not raised in previous adjudications** Where a party fails to raise an objection in an adjudication on a particular point (e.g. the identity of the adjudicator), the party is not barred from raising that objection in subsequent adjudications.
Camillin Denny Architects Ltd v Adelaide Jones & Co. Ltd	[2009] BLR 606, [2009] EWHC 2110 (TCC), Akenhead J	**Overlap of jurisdiction and substance** *Obiter*: an adjudicator may have jurisdiction to resolve a jurisdictional issue if and to the extent that the issues that go to jurisdiction are part of the substantive dispute referred to adjudication.
Supablast (Nationwide) Ltd v Story Rail Ltd	[2010] BLR 211, [2010] EWHC 56 (TCC), Akenhead J	**Overlap of jurisdiction and substance** *Obiter*: the decision whether steel works were instructed as a variation to the first contract or were carried out under a second contract was within the adjudicator's substantive jurisdiction and the decision on the point would be binding.

[21] See also first-instance decision: [2003] 3 All ER 98, [2003] EWHC 984 (TCC), Judge Thornton.

(2) Matters Giving Rise to Jurisdictional Challenge

9.16 In some cases a jurisdictional issue arises because there was no jurisdiction at the outset, such as where there is no right to adjudicate because the contract is outside the ambit of the Act. Equally, where the matter referred has not crystallized into a 'dispute' there will be no jurisdiction to adjudicate upon it. Jurisdictional challenges that fall into this category are considered at 9.18–9.25 below.

9.17 In other situations the jurisdiction issue arises because of something that happens during the adjudication. This is the case where there are errors in the appointment of the adjudicator, where the referral is served outside the strict time limits or where the decision is late. Such situations may result in the adjudicator proceeding with the adjudication when he has no power to do so or making a decision out of time which he has no power to do. This category also includes errors made in the decision itself. Whilst the majority of errors of fact and/or law do not affect the enforceability of an adjudicator's decision,[22] there are certain errors which the courts have deemed are such that 'they cannot be permitted to stand'[23] or that go to the 'heart' of the adjudicator's jurisdiction.[24] So, if the adjudicator decides the wrong question, instead of the one actually referred, the resulting decision will be one made without jurisdiction and so unenforceable. Furthermore a procedural mistake by the adjudicator can result in the making of a decision that he or she is not empowered to make, such as making the decision out of time or revising the decision after it is issued in an unauthorized manner.[25] Such decisions made without jurisdiction will not be enforceable and will be a nullity.[26]

No Jurisdiction at the Outset

Not a Construction Contract in Writing

9.18 Where reliance is placed on the Act to establish the initial threshold right to adjudicate, an adjudicator will not have jurisdiction if the requirements of the Act are not met. Consequently, challenges have been made on the basis that the contract in question is not a construction contract within the meaning of ss. 104–6 or is not an agreement in writing within the meaning of s. 107. The reader is referred to Chapters 1 and 2 for a detailed consideration of the relevant principles. However, it should be noted that courts generally adopt a robust approach to challenges based on there being no contract in writing.[27]

Not a Party

9.19 The responding party may contend that the adjudication was brought by or against the wrong party, with the result that one of the parties to the adjudication was not a party to the underlying contract. Pursuant to s. 108 of the Act, the adjudicator is only empowered to decide disputes arising under the construction contract and that means between the parties to that contract. This is clear from the wording of the Act which says that 'A party

---

[22] See below at 9.47.

[23] *The Project Consultancy Group v The Trustees of the Gray Trust* [1999] BLR 377, in which the TCC refused to summarily enforce an adjudicator's decision because it was unclear whether any contract was ever concluded and the adjudicator was not empowered to give a binding decision that there was a contract.

[24] *Joinery Plus Ltd v Laing Ltd* [2003] BLR 184, [2003] EWHC 3513 (TCC); *Hart Investments Ltd v Fidler and anor* [2007] BLR 30, [2006] EWHC 2857 (TCC).

[25] For a discussion of the power to correct slips or clerical errors, see Ch. 4 at 4.109–4.117.

[26] *Joinery Plus Ltd v Laing Ltd* [2003] BLR 184, [2003] EWHC 3513 (TCC).

[27] *Pegram Shopfitters Ltd v Tally Wiejl (UK) Ltd* [2004] 1 All ER 818, [2003] EWCA Civ 1750 (CA).

to a construction contract has the right to refer a dispute . . .'.[28] There is no jurisdiction to decide a dispute between different parties unless that jurisdiction is otherwise conferred on the adjudicator.

9.20 Such cases turn on the evidence of who the contracting parties were. If that is a question that cannot be decided either way on a summary basis then at the enforcement hearing directions will be given for a full trial of that question.[29] If the adjudication is started against a party that is 'plainly wrong' then the Court of Appeal has confirmed that the decision will not be enforced.[30] However, as can be seen from Table 9.2 below, the majority of decisions have been enforced, with the courts applying the principles developed in relation to arbitration, so that where the party has simply been misnamed in the adjudication and there has been no confusion as to who the parties to the adjudication were, then a 'wrong party' challenge will fail. See further at 3.20.

Table 9.2 Table of Cases: Not a Party (shaded entries indicate a successful challenge to enforcement)

Title	Citation	Issue
A. J. Brenton v Jack Palmer	19 Jan. 2001 (TCC) Judge Havery (unreported)	**Individual or company** Mr Palmer claimed his company should have been the responding party. A decision of an adjudicator as to the identity of the contracting party, decided as part of his substantive decision, could not be questioned in enforcement proceedings. Enforced.
Total M & E Services Ltd v ABB Building Technologies Ltd (formerly ABB Steward Ltd)	(2002) 87 Con LR 154, [2002] EWHC 248 (TCC), Judge Wilcox	**Misnomer** The referring party was named as 'Total Mechanical and Engineering Services'. In a case of clear misdescription where no one was misled as to the identity of the referring party, the court made a declaration that the referring party was the contracting party. *Obiter*: where there are similar company names, as for instance in a group of companies or where there are subsidiaries with overlapping management systems and some common directors, a precise description of the referring party could be critical. Enforced.
Gibson v Imperial Homes	[2002] All ER (D) 367 (Feb), Judge Toulmin	**Company not incorporated at time of contract** The responding party argued that the contract was made with Chinadome Ltd. One of the responding party's representatives had ostensible authority to make the contract on behalf of either company and the documents made it clear that the contract was made on behalf of Imperial Homes and Development Ltd. The referring party was in any event entitled to sue in his own name whether: (i) in his own right if he was the contracting party; or (ii) as agent of a limited company that was pre-incorporation at the time of the contract formation if that was the contracting party. Enforced.

(Continued)

[28] s. 108(1).

[29] As in *Estor Ltd v Multifit (UK) Ltd* [2009] EWHC 2108 (TCC).

[30] *Thomas-Frederic's (Construction) Ltd v Keith Wilson* [2004] BLR 23, [2003] EWCA Civ 1494 (CA).

Table 9.2 *Continued*

Title	Citation	Issue
Thomas-Frederic's (Construction) Ltd v Keith Wilson	[2004] BLR 23, [2003] EWCA Civ 1494 (CA), Brown, Judge, Parker LJJ	**Identity of contracting party** There was no evidence that the responding party had contracted in person. It was plainly wrong that Mr Wilson was a party to the relevant contract. Not enforced.
Redworth Construction Ltd v Brookdale Healthcare Ltd	[2006] BLR 366, [2006] EWHC 1994 (TCC), Judge Harvey	**Novation** There had been no novation of the contract with the result that the parties to the adjudication were the contracting parties. Not enforced on other grounds (no contract in writing).
ROK Build Ltd v Harris Wharf Development Company Ltd	[2006] EWHC 3573 (TCC), Wilcox J	**Associated companies** It was arguable that it was not the referring party but an associated company that was a party to the contract. Not enforced.
Williams (t/a Sanclair Construction) v Noor (t/a India Kitchen)	[2007] All ER (D) 51 (Dec), [2007] EWHC 3467 (TCC), Judge Hickinbottom	**Identity of contracting party** On an objective view, the responding party had always intended to deal with whoever was the legal person trading as Sanclair Construction. There was only ever a misdescription of the party in the adjudication. The court made a declaration that the parties to the adjudication were the contracting parties. Enforced.
Camillin Denny Architects Ltd v Adelaide Jones & Company	[2009] BLR 606, [2009] EWHC 2110 (TCC), Akenhead J	**Novation** There was no realistic prospect of establishing that there had been a clear, unqualified and fully retrospective novation of the contract. If there had, then the decision would not have been enforceable. Enforced.
Estor Ltd v Multifit (UK) Ltd	(2009) 126 Con LR 40, [2009] EWHC 2108 (TCC), Akenhead J	**Identity of original contracting party** There was no summary enforcement as the question of the identity of the contracting party could not be determined on a summary basis. There was 'just what can be described as a realistic prospect of Estor establishing that it was not the company which entered into the contract with Multifit'. However, at trial of the issue two months later,[31] it was later found that the referring party was the contracting party and the decision was then enforced.
Durham County Council v Jeremy Kendall (t/a HLB Architects)	[2011] BLR 425, [2011] EWHC 780 (TCC), Akenhead J	**Sole trader** A sole trader's trading personality has no legal existence independent from that of the individual who is trading. Where the trading name sues or is sued, it is the trader in person that is suing or being sued. Similarly, a contract made under the trading name is a contract made by the trader. Enforced

No Dispute Crystallized

9.21 To invoke the right to adjudicate under s. 108 of the Act there must be a 'dispute' which is capable of being referred to adjudication. An adjudicator has no jurisdiction to determine a claim that has not yet crystallized into a 'dispute' unless that power is otherwise

[31] See associated proceedings at [2010] CILL 2800, [2009] EWHC 2565 (TCC) Akenhead J.

conferred.[32] Enforcement of a decision may be challenged on the grounds that the matter decided, or part of it, was referred prematurely and the decision was made without jurisdiction. Although once a popular ground of challenge, the guidance given by the courts over the years has narrowed the circumstances within which absence of a dispute may be argued and challenges on this basis are now rarely successful. For a full discussion of this topic see Chapter 3 at 3.29–3.38. Equally there is no jurisdiction to adjudicate a dispute that has previously been compromised by the parties.

More than One Dispute Referred

9.22 As discussed in Chapter 3 at 3.49–3.52, the Act permits the referral of a single dispute to adjudication, although the Scheme for Construction Contracts (the Scheme) says the parties are free to agree to refer more than one dispute as long as both parties consent. Absent specific agreement to confer more than one dispute[33] the decision may be challenged on the basis that more than one dispute was referred. To avoid this danger, the adjudication notice should be worded so that what is being referred is a single dispute. Consistent with the courts' general approach of supporting the intention of Parliament by enforcing adjudication decisions wherever possible, the courts have given a broad interpretation to what comprises a single dispute and have usually construed adjudication notices as containing one dispute even though a number of different issues, or component parts of that dispute, are identified in the notice. Where more than one dispute is decided, it may be possible to sever the decision and enforce only the part made with jurisdiction. The subject of severance is discussed in Chapter 7 at 7.73–7.81.

Dispute Does Not Arise under the Contract

9.23 The wording of s. 108(1) of the Act is clear: it permits the referral to adjudication of disputes 'arising under the contract'. The Scheme includes the same wording. It is therefore legitimate to challenge enforcement of a decision on the ground that the dispute decided did not 'arise under' the contract. As to the meaning of this expression and a discussion of cases in which it has been considered, see Chapter 3 at 3.10–3.14. It has also been held that there is no jurisdiction for an adjudicator to decide a dispute which concerns the taking of a single account as anticipated by r. 4.90 of the Insolvency Rules.[34]

Dispute under Multiple Contracts or Side Agreements

9.24 Where the dispute decided by an adjudicator arose under more than one contract the decision may be vulnerable to a jurisdictional challenge. Equally if the dispute arises wholly or in part under a side agreement it may not be a dispute that arises under the contract, which contains the adjudication agreement. This subject is discussed in Chapter 3 at 3.15–3.19.

Dispute Decided in Previous Adjudication

9.25 An adjudicator's decision on a dispute is binding until finally determined in legal or arbitration proceedings or by agreement of the parties.[35] The effect of this is to prevent successive adjudications on the same or substantially the same dispute. Thus an adjudicator has no jurisdiction to determine a dispute that has previously been decided in a different adjudication

---

32 For instance, an adjudication agreement may be drafted so widely that it permits the adjudicator to decide both disputes and claims made even before they crystallize into a dispute.

33 Which may be contained in the wording of the contract specific adjudication agreement.

34 *Enterprise Managed Services Ltd v Tony McFadden Utilities Ltd* [2010] BLR 89, [2009] EWHC 3222 (TCC).

35 s. 108(3).

(whether by the same or a different adjudicator). This is often referred to as the rule against 'double jeopardy' and is discussed in Chapter 6 at 6.63-6.69.

Jurisdiction Issues Arising During Adjudication

9.26 In general, as discussed in Chapter 10, procedural errors in a validly constituted adjudication will not result in an unenforceable decision. However, certain procedural errors have been found to deprive the adjudicator of jurisdiction. In particular, an adjudicator will only have jurisdiction to decide the dispute referred if the adjudicator has been properly appointed in accordance with either any express adjudication rules or the Scheme, the dispute has been properly referred within the relevant time limits, and the decision has been provided in accordance with the applicable rules. Each of these aspects is considered in the sections below.

Errors in Appointment of Adjudicator

9.27 If an adjudicator has been appointed other than in accordance with an express contractual provision (if such exists), or the Scheme, then the appointment will be invalid and the adjudicator will have no jurisdiction. Each case will turn on its own particular facts, but as can be seen from the case summaries in Table 9.3, appointments have been found to be invalid as a result of the referring party applying to the wrong nominating body[36] or an adjudicator being appointed who is not the named adjudicator.[37]

9.28 The timing of the request for appointment may also be critical. Under the Scheme, a request to a named adjudicator or nominating body is to be made 'following the giving of a notice of adjudication…'.[38] In both *IDE Contracting v R. G. Carter Cambridge Ltd* (2004)[39] and *Vision Homes Ltd v Lancsville Construction Ltd* (2009),[40] that Scheme provision was found to be strict, so that an approach to a nominating body or a named adjudicator (even to check availability), cannot be made prior to the issue of the notice of adjudication, however close in time the notice may have followed. Whether the same is true under a contractual appointment provision will turn on the interpretation of the contract in question. For example, it is not essential under the JCT adjudication provisions that the notice of adjudication is issued before a request is made to the nominating body.[41]

Table 9.3 Table of Cases: Failure in Appointment (shaded entries indicate a successful challenge based on appointment failure)

Title	Citation	Issue
Watson Building Services Ltd v Harrison	[2001] SLT 846 (Outer House, Court of Session), Lady Paton	**Correct appointment procedure followed** The court found no adjudication provisions had been incorporated into the contract and so the adjudicator appointed under the Scheme was correctly appointed.

[36] *IDE Contracting Ltd v R. G. Carter Cambridge Ltd* [2004] BLR 172, [2004] EWHC 36 (TCC).

[37] *AMEC Capital Projects Ltd v Whitefriars City Estates Ltd* [2005] 1 All ER 723, [2004] EWCA Civ 1418 (CA).

[38] Para. 2(1).

[39] [2004] BLR 172, [2004] EWHC 36 (TCC).

[40] [2009] BLR 525, [2009] EWHC 2042 (TCC).

[41] *Palmac Contracting Ltd v Park Lane Estate Ltd* [2005] BLR 301, [2005] EWHC 919 (TCC) and the decision in *Dalkia Energy & Technical Services Ltd v Bell Group UK Ltd* [2009] EWHC 73 (TCC) concerning a provision that was equivalent to the JCT provision.

Title	Citation	Issue
David Mclean Housing Contractors Ltd v Swansea Housing Association Ltd	[2002] BLR 125 TCC, Judge LLoyd	**Correct appointment procedure followed** On a proper construction of the contract, no express appointment provision had been incorporated and the Scheme applied with the result that the adjudicator was correctly appointed. Enforced (but counterclaim for LADs flowing from EOT award allowed).
Pegram Shopfitters Ltd v Tally Weijl (UK) Ltd	[2004] 1 All ER 818, [2003] EWCA Civ 1750, May, Hale LJJ, Hooper J	**Invalid appointment** The adjudicator was appointed under the Scheme, which the referring party said applied to the contract. The Court of Appeal upheld the responding party's challenges that the appointment was invalid because: (i) the contract included the JCT Prime Cost Contract 1998 and the adjudicator should have been appointed under the procedure in that contract; or alternatively (ii) there was no contract at all. The CA held this was a valid challenge which should have prevented the decision from being summarily enforceable under Part 24: 'The fact that adjudication under the Scheme and adjudication under the JCT Prime Cost Contract would be similar procedures does not overcome the twin difficulties that Mr Morris was appointed under the Scheme, and that a sufficiently secure identification of the contractual terms was intrinsically necessary to the proper performance of his adjudication task.' NB see cf *Bovis Lend Lease Ltd v Cofely Engineering Services* [2009] EWHC 1120 (TCC).
London & Amsterdam Properties Ltd v Waterman Partnership Ltd	[2004] BLR 179, [2003] EWHC 3059 (TCC), Judge Wilcox	**Consistency of adjudicator's fee proposal with scheme** The adjudicator's fee proposal made after his appointment, whereby he would charge for his time actually spent, was consistent with para. 25 of the Scheme and did not invalidate his appointment. Not enforced on other grounds (bias).
IDE Contracting Ltd v R. G. Carter Cambridge Ltd	[2004] BLR 172, [2004] EWHC 36 (TCC), Judge Havery	**Early approach to nominating body** The issue of the notice of adjudication must precede an approach to a named adjudicator to check availability. Not enforced.
Bennett v FMK Construction Ltd	(2005) 101 Con LR 92, [2005] EWHC 1268 (TCC), Judge Havery	**Request to nominating body made on afternoon of seventh day after notice** Clause 41A of the 1998 JCT Standard Form Private without Quantities relating to the time for appointment of the adjudicator is directory and not mandatory. Part 8 declaration granted (appointment was valid).
AMEC Capital Projects Ltd v Whitefriars City Estates Ltd	[2005] 1 All ER 723, [2004] EWCA Civ 1418 (CA), Dyson, Kennedy, Chadwick LJJ	**Death of named adjudicator** The contractual appointment mechanism failed on the death of the named adjudicator and the Scheme applied. The adjudicator had been correctly appointed under the Scheme. Enforced.

(Continued)

Table 9.3 *Continued*

Title	Citation	Issue
Palmac Contracting Ltd v *Park Lane Estate Ltd*	[2005] BLR 301, [2005] EWHC 919 (TCC), Judge Kirkham	**Contractual appointment provision did not require notice to precede request for appointment** Under clause 39A of the JCT Standard Form of Building Contract (with Contractor's Design) 1998, the notice of adjudication does not need to precede the request for nomination. Enforced. Clause 41A.5.6 provides: 'Any failure... to comply with any requirement of the adjudicator under 41A.5.5 or with any provision in or requirement under clause 41A shall not invalidate the decision of the adjudicator.' This clause is limited to procedural steps in a validly constituted adjudication. This could not validate the appointment of an adjudicator invalidly appointed (*obiter*).
Lead Technical Services Ltd v CMS Medical Ltd	[2007] BLR 251, [2007] EWCA Civ 316 (CA), Buxton, Rix, Moses LJJ	**Contractual appointment provisions not followed** There was a real prospect that the deed of appointment containing express adjudication provisions (nomination by TECSA) was effective. The adjudicator had been nominated by the ICE and had therefore not been appointed in accordance with those provisions. Not enforced.
Makers UK Ltd v Camden London Borough Council	[2008] BLR 470, [2008] EWHC 1836 (TCC), Akenhead J	**Request for particular adjudicator** A request to the nominating body to appoint a particular adjudicator did not invalidate the appointment. Enforced.
Dalkia Energy & Technical Services Ltd v Bell Group UK Ltd	(2009) 122 Con LR 66, [2009] EWHC 73 (TCC), Coulson J	**Timing of approach to nominating body** The contractual adjudication scheme did not require there to be an attempt to agree an adjudicator before an approach was made to the nominating body, nor did it require the notice of adjudication to be served before the approach was made to the nominating body. Part 8 declaration rejected.
Bovis Lend Lease Ltd v Cofely Engineering Services	[2009] EWHC 1120 (TCC), Coulson J	**Part 8 application to determine correct contractual appointment mechanism** On the proper construction of the contract, the adjudicator had been properly appointed. *Obiter*: the adjudicator's appointment was not invalid when appointed pursuant to a RICS nomination provision that was arguably invalid, because the fall-back was that the Scheme would have applied and the application to the RICS would have been valid under the Scheme. So on these facts, whichever procedure applied, the actual appointment made by the RICS would have been valid. No declaration granted.
Vision Homes Ltd v Lancsville Construction Ltd	[2009] BLR 525, [2009] EWHC 2042 (TCC), Clarke J	**Request for appointment preceded notice of adjudication** The requirement in the Scheme that the request to the nominating body should follow the giving of the notice of adjudication is strict. Where the notice followed the request by just 18 minutes the appointment was invalid. Not enforced.

(Continued)

Title	Citation	Issue
Profile Projects Ltd v Elmwood (Glasgow) Ltd	[2011] CSOH 64 (Outer House, Court of Session), Lord Menzies	**Adjudicator was not appointed pursuant to contractually identified appointing body** The contractual adjudication scheme was not Act-compliant so was replaced by the Scheme. However, that did not have the effect of nullifying the contractual provision identifying the agreed appointing body, with the result that the referring party did not have a choice over the appointing body and the adjudicator appointed by another body did not have jurisdiction. Not enforced.
Sprunt Ltd v London Borough of Camden	[2012] BLR 83, [2011] EWHC 3191 (TCC), Akenhead J	**Nomination provisions allowing one party to select adjudicator not Act-compliant** The contract provision allowing the employer to appoint an adjudicator of its choosing was found to be non-compliant with the Act. It offended the policy of having actually and ostensibly impartial adjudicators. The adjudicator was properly appointed according to the Scheme. Enforced.

Late Service of Referral

9.29 Section 108(2)(b) of the Act requires that any construction contract:

> Shall provide a timetable with the object of securing the appointment of the adjudicator and referral of the dispute to him within 7 days of such notice.

9.30 Paragraph 7 of the Scheme provides that:

> (1) ... the referring party shall, not later than seven days from the date of the notice of adjudication, refer the dispute in writing (the 'referral notice') to the adjudicator.
> (2) A referral notice shall be accompanied by copies of, or relevant extracts from, the construction contract and such other documents as the referring party intends to rely upon.

9.31 The time for referral runs from the date the notice of adjudication is sent and not the date of receipt by the responding party.[42] A referral notice received by fax after 4.00 pm is received that day and will not be deemed to have been received on the following day, as the Civil Procedure Rules do not apply to adjudications.[43]

9.32 In *Hart Investments Ltd v Fidler and anor* (2006) (Key Case),[44] after considering the then conflicting authorities in relation to late decisions (discussed in at 9.37–9.44), the court found that the requirement for service within seven days in the Scheme is strict and cannot be extended; and if the referral notice is served outside that time then it is not a valid referral notice. The effect is that the dispute has never been referred, with the result that the adjudicator has no jurisdiction to decide that dispute. Shortly after this decision, the same principles were held to apply to the adjudication provisions of the JCT 1998 Standard Form (clause 41A.4.1) in *Cubitt Building & Interiors Ltd v Fleetglade Ltd* (2006).[45] However, in that

[42] *William Verry Ltd v North West London Communal Mikvah* [2004] BLR 308, [2004] EWHC 1300 (TCC).

[43] *Cubitt Building & Interiors Ltd v Fleetglade Ltd* (2006) 110 Con LR 36, [2006] EWHC 3413 (TCC).

[44] [2007] BLR 30, [2006] EWHC 2857 (TCC).

[45] In this case, the decision (to the extent that it found the time limits were not strict) in *William Verry Ltd v North West London Communal Mikvah* [2004] BLR 308, [2004] EWHC 1300 (TCC) was held to have been decided on its own particular facts.

case, while the time limits were found to be mandatory, the service of the referral notice on the eighth day was permitted as being in accordance with a common sense interpretation of clause 41A.4.1.

9.33 In *Lanes Group Plc v Galliford Try Infrastructure Ltd* (2011)[46] the court considered the question of whether a failure to refer in time affects the referring party's right to start again with another notice, appointment, and referral. An earlier application for an injunction to restrain the referring party from making a further referral of a particular dispute was unsuccessful,[47] although it was specifically only argued on the basis of repudiatory breach of the agreement to adjudicate. When the question was reconsidered at the enforcement proceedings it was held that there was no basis for implying an absolute or qualified bar to restarting an adjudication where there had been a previous failure to refer that same dispute in time and there was no prejudice arising from such conduct. This was upheld by the Court of Appeal.[48] Akenhead J also said that failure to refer in time is, however, a breach (albeit non-repudiatory) of the implied adjudication agreement between the parties, which may sound in damages and could potentially lead to the court awarding injunctive relief.[49]

9.34 An element of flexibility has, however, been allowed in relation to paragraph 7(2) of the Scheme where there has been late service of documents supporting the referral notice, although it will always be a question of fact and degree whether the late documents make the referral so deficient that it affects the validity of service of the referral altogether.[50]

Key Case: Late Referral

Hart Investments Ltd v Fidler and anor [2007] BLR 30, [2006] EWHC 2857 (TCC)

9.35 **Facts:** The referring party served its referral eight days after the notice of intention to refer had been served.

9.36 **Held:** The adjudicator has no jurisdiction until he is properly seized of the dispute, which requires that the matter has been properly referred. The adjudicator cannot 'make good' a bad referral through directions for extension of time. The seven-day time limit in the Scheme is a strict requirement and any referral made after that time is invalid with the result that the adjudicator has no jurisdiction:

> 49 My initial reaction to this point was to consider that, in the overall scheme of things, it might be difficult to say that the delay of one day in the provision of the referral notice should be accorded great significance, and that it would be harsh to rule that the whole adjudication was a nullity because of that one day's delay. But, on a more detailed analysis, I do not consider this reaction to be so easy to justify. Indeed, all kinds of difficult questions arise if the failure to comply with the time period is ignored: What if the delay was not one day, but one month? What if important events occurred during the period of any delay in the provision of a referral notice which put the responding party in a much worse position as against the referring party than it would have been if there had been no delay? If the words 'not later than seven days' are to be qualified in some way, then how is such a qualification to be formulated, let alone assessed? 'Not later than seven days and perhaps one or two more'? 'Not later than a period that seems just and equitable in the circumstances'?

[46] 137 Con LR 1, [2011] EWHC 1679 (TCC).
[47] *Lanes Group Plc v Galliford Try Infrastructure Ltd* [2011] BLR 438, [2011] EWHC 1035 (TCC).
[48] [2012] BLR 121, [2011] EWCA Civ 1617.
[49] [2011] BLR 438, [2011] EWHC 1035 (TCC) at para. 29, [2011] EWHC 1679 (TCC) at para. 42.
[50] *PT Building Services Ltd v ROK Build Ltd* [2008] EWHC 3434 (TCC).

50 The whole point of adjudication is that speed is given precedence over accuracy. What matters is a quick decision, not necessarily a correct one. There is a summary timetable with which both the parties and the adjudicator must comply. If the swift timetable is kept to, the vast majority of adjudicators' decisions are then enforced by this court in accordance with the 1996 Act. If the timetable can be extended without consent either, as here, at the beginning of the process or, as in Simons, at the end of the 28 days, there is a great danger of uncertainty and of a watering-down of the critical importance of the tight timetable on which the entire adjudication process is based. In other words, if, as I consider it to be, Ritchie is a correct statement of the position at the conclusion of the 28 days, it seems to me that the same principle must also apply to the event which signals the commencement of the same 28 day period, namely the provision of the referral notice within 7 days of the intention to refer.

51 I agree with Mr. Quiney that the provisions of the 1996 Act at ss.108(2)(b) and (c) address the 28 day period for the decision in different, and possibly stronger, language than the seven days for the referral notice. But, even then, the Act requires the appointment and the referral notice to be 'secured' within seven days. Moreover, the Scheme is, I think, entirely clear on this point. The referral notice must be provided by a date which is not later than seven days after the notification of the notice of intention to refer. If it is not, it cannot be a referral notice in accordance with the Scheme. In that event, of course, the responding party may consent, expressly or by implication, to waive the irregularity. There was no such waiver here. If the responding party does not waive the irregularity the referring party must start again, which is precisely the same course of action envisaged in Ritchie. Larchpark had that choice to make. They decided not to start again, and it seems to me that they are, therefore, obliged to accept the consequences of that decision.

52 At one point Mr. Quiney, with customary acuity, suggested that the adjudicator could extend without consent the seven day time limit as part of his general powers under para.13 of the scheme. That was a typically ingenious argument, but I do not believe that it can be right. Everything done pursuant to the Scheme, including the 28 day period for the adjudication itself, flows from the date of the referral notice. The adjudicator is not seized of the adjudication until the referral notice is provided and the 28 day period starts to run. He therefore has no power until he gets the referral notice; thus he has no power to extend the seven day period which occurs before his jurisdiction begins. In any event, I do not consider that para.13 of the Scheme permits the adjudicator to disregard the time limits set out in the Scheme if the relevant extension is not agreed to. He certainly could not do so retrospectively, which is what I consider the adjudicator purported to do here.

53 This leads me to a related aspect of Mr. Quiney's submissions, namely the argument that the adjudicator's decision retrospectively to grant an extension of the seven day period (which extension was expressly not agreed) was a matter of law which, rightly or wrongly, the adjudicator was entitled to make. I reject that contention. If, as I have found, the adjudicator had no jurisdiction to consider the adjudication, because the referral notice was invalid, and that invalidity was not waived, then the fact that he went on to consider the issue and concluded (wrongly) that he did have jurisdiction is ultimately irrelevant to the powers of this court. The validity of the referral notice went to the heart of the adjudicator's jurisdiction and was not an issue on which he could bind the parties. The line of authority, starting with C & B Scene, is therefore of no application in this case.

54 Accordingly, I have concluded that the referral notice was irregular/invalid because it was not served in accordance with the 1996 Act or para.7 of the Scheme. Hart were entitled to refuse to waive that irregularity, which they did. The adjudicator, therefore, had no jurisdiction to enter on the reference and the award was a nullity. I therefore decline to enforce it.

Table 9.4 Table of Cases: Late Referral (successful challenges are indicated by shaded entries)

Title	Citation	Issue
Costain Ltd v Wescol Steel Ltd	[2003] EWHC 312 (TCC)	**Service by wrong method** JCT DOM/1 Clause 38A.5.6: 'Shall not invalidate decision of adjudicator.' Breach of a procedural requirement to serve notices and documents by fax and first class post did not invalidate the decision of the adjudicator. Clause 38.5.6 renders 38A requirements non mandatory. (See also *Palmac* and *Cubitt.*)
William Verry Ltd v North West London Communal Mikvah	(2004) BLR 308, [2004] EWHC 1300 (TCC), Judge Thornton	**JCT 1998 condition 41A.1.4** Section 108(2)(b) did not preclude a contract from being drafted to allow a longer period for service than seven days. Enforced (but subsequently identified as a case decided on its own facts).
Hart Investments Ltd v Fidler and anor	(2007) BLR 30, [2006] EWHC 2857 (TCC), Coulson J	**Referral served eight days after notice of intention to refer** The time limit in the Scheme is strict and any referral more than seven days after the notice of intention to refer has been served is not a valid referral. Not enforced.
Cubitt Building & Interiors Ltd v Fleetglade Ltd	(2006) 110 Con LR 36, [2006] EWHC 3413 (TCC), Coulson J	**Referral served eight days after notice of intention to refer** A referral served eight days after notice of intention to refer was valid. The late appointment of the adjudicator at 5.35 pm on the seventh day was due to the nominating body's delay. The appointment was immediately followed by an offer to fax the referral notice with supporting documents to follow by courier the next day. This offer was rejected. Whilst JCT 1998 condition 41A.1.4 contains mandatory time frames, it must be interpreted in a sensible and businesslike way. Enforced. Agreed with *obiter* in *Palmac* (above at Table 9.3). If a late referral notice would render an adjudication invalid, clause 41A.5.6 cannot rescue that default.
PT Building Services Ltd v ROK Build Ltd	[2008] EWHC 3434 (TCC), Ramsey J	**Late service of supporting documents** A delay of one day in providing a copy of the construction contract did not invalidate the referral. Validity of the referral in such circumstances will depend on the degree of the deficiency in providing documents.
Linnett v Halliwells LLP	[2009] BLR 312, [2009] EWHC 319 (TCC), Ramsey J	**Late service of supporting documents** Failure to serve supporting documents within the time limit for service of the referral is a procedural non-compliance that can be saved by condition 41A.5.6 of the 1998 JCT Standard Form. The court should be slow to find that a failure to comply with a detailed procedural aspect of the adjudication provisions deprives the adjudicator of jurisdiction. *Obiter*: condition 41A.5.6 cannot save an invalidly constituted adjudication. The referral was valid.

Title	Citation	Issue
Lanes Group Plc v Galliford Try Infrastructure Ltd	[2011] BLR 438, [2011] EWHC 1035 (TCC), Akenhead J	**Second adjudication after failure to refer first adjudication in time** A failure to refer in time is not a repudiatory breach of the adjudication agreement. No injunction granted.
Lanes Group Plc v Galliford Try Infrastructure Ltd	[2011] EWHC 1234 (TCC), Akenhead J	**Late service of supporting documents** The parties expressly incorporated the ICE adjudication procedure which (in addition to requiring the adjudicator to be appointed to a timetable with the object of securing the referral of the dispute to him within seven days of the notice of adjudication) required the referral to be sent to the adjudicator and copied to the responding party within two days of the adjudicator confirming his appointment. Some supporting documents were sent to the responding party a few hours after the two-day limit, but the referral was served within the seven-day time period. It was held that service was in time, the two-day limit referred to sending, not receipt, and was not an 'unless' provision. Application for declaration dismissed.
Lanes Group Plc v Galliford Try Infrastructure Ltd	[2012] BLR 121, [2011] EWCA Civ 1617 (CA), Richards LJ, Stanley Burnton LJ, Jackson LJ	**Second adjudication after failure to refer in time** The Court of Appeal affirmed the finding that there is no implicit bar to restarting an adjudication after a failure to refer in time. Enforced (finding of Judge Waksman on bias (2011) 137 Con LR 1, [2011] EWHC 1679 (TCC) overturned by Court of Appeal).

Late Decision

9.37 Section 108 (2) of the 1996 Act requires that a construction contract must contain certain provisions so as to:

(c) require the adjudicator to reach a decision within 28 days of referral or such longer period as is agreed by the parties after the dispute has been referred;
(d) allow the adjudicator to extend the period of 28 days by up to 14 days, with the consent of the party by whom the dispute was referred;

9.38 These requirements are reflected at paragraphs 8(3) and 19(1) of the Scheme. There were a number of early conflicting first instance decisions as to whether the time limit imposed by the Act and the Scheme was mandatory or simply directory. Those decisions were considered, *obiter*, in *Hart Investments Ltd v Fidler and anor* (2006) (Key Case),[51] with the preferred approach being that it is a mandatory requirement that the adjudicator reaches a decision within a 28-day period (or longer if agreed by the parties within the terms of the Scheme or other rules).

[51] [2007] BLR 30, [2006] EWHC 2857 (TCC).

9.39 Time starts running from the date of receipt of the referral and not the date on which it was sent, and in calculating whether it is late, no account is to be taken of fractions of a day.[52] As most recently confirmed in *Lee v Chartered Properties (Building) Ltd* (2010),[53] there is a two-stage test to determine if a decision is in time. First, the adjudicator must reach the decision within the time limits imposed by the Act and Scheme or any expressly agreed adjudication rules, subject to any agreed extensions of time. Secondly, having reached the decision within the time limits, it must be communicated to the parties as soon as possible thereafter. Some cases have described this communication obligation as 'forthwith' or within 'a short additional period'.[54] Once the relevant period of time for reaching and/or communicating the decision has expired, in the absence of any agreed extension of time properly granted in accordance with the Scheme or rules, the adjudicator will no longer have jurisdiction to decide the dispute and any purported decision thereafter will be a nullity.

9.40 Thus a term that the adjudicator's decision 'shall nevertheless be valid if issued after the time allowed' does not contravene the mandatory provisions of s. 108(2)(c) of the 1996 Act, but delay in issuing may still invalidate the decision subject to the particular circumstances, including the length of the delay.[55] Such a term is still to be found in the standard form GC/Works/1 (1998) at clause 59(5).

9.41 However a term which goes further than this and allows the adjudicator not only to issue his decision after the 28 days but also to reach his decision after that time has expired does contravene s. 108(2)(c) of the 1996 Act.[56]

9.42 Such a clause was under consideration in *Epping Electrical Company Ltd v Briggs and Forrester* (2007).[57] In that case the judge was considering paragraph 25 of the CIC Model Adjudication procedure, which purported to provide that any decision reached after the expiry of the 42 days would still be valid, provided it was reached before the appointment of a replacement adjudicator. Judge Havery concluded that such a provision did not comply with the 1996 Act, holding that 'the apparent effect of paragraph 25 of the CIC Procedure is inconsistent with the mandatory nature of section 108(2)'. He went on to conclude that as a result, in accordance with s. 108(5) of the 1996 Act, the Scheme applied in place of the adjudication provisions of the contract and that, on the facts of that case, the decision was not enforceable because it had been reached outside the statutory time limits. The CIC amended paragraph 25 of its Adjudication Procedure following this decision.

9.43 A clause in these terms is also to be found in the ICE Adjudication Procedure 1997 at Clause 6.4 which provides that 'If the Adjudicator fails to reach and notify his decision in due time but does so before the dispute has been referred to a replacement adjudicator under paragraph 6.3 his decision shall still be effective.' The effect of *Epping* and *Dalkia* is that any party wishing to use this form ought to strike through these words in clause 6.4 to avoid the whole procedure being replaced by the Scheme pursuant to s. 108(5).[58]

---

[52] *Aveat Heating Ltd v Jerram Falkus Construction Ltd* (2007) 113 Con LR 13, [2007] EWHC 131 (TCC).

[53] [2010] BLR 500, [2010] EWHC 1540 (TCC).

[54] *Barnes & Elliott Ltd v Taylor Woodrow Holdings Ltd* [2004] BLR 111, [2003] EWHC 3100 (TCC); *Cubitt Building & Interiors Ltd v Fleetglade Ltd* (2006) 110 Con LR 36, [2006] EWHC 3413 (TCC); *Dalkia Energy & Technical Services Ltd v Bell Group Ltd* (2009) 122 Con LR 66, [2009] EWHC 73 (TCC).

[55] *Dalkia Energy & Technical Services Ltd v Bell Group Ltd* (2009) 122 Con LR 66, [2009] EWHC 73 (TCC) at paras. 72 and 73 of the judgment (*Aveat Heating Ltd v Jerram Falkus Construction Ltd* (2007) 113 Con. LR 13, [2007] EWHC 131 (TCC) doubted).

[56] *Dalkia Energy & Technical Services Ltd v Bell Group Ltd*, approving *Epping Electrical Company Ltd v Briggs & Forrester* (Plumbing Services) Ltd [2007] BLR 126, [2007] EWHC 4 (TCC).

[57] [2007] BLR 126, [2007] EWHC 4 (TCC).

[58] This clause has been removed in the 2012 edn of the ICE Adjudication Procedures.

9.44 There has been a tendency for some adjudicators to include a 'lien' condition in their terms of appointment that provides that the decision will only be released on payment of the adjudicator's fees. However, the adjudicator's prevailing obligation is to deliver the decision to the parties as soon as possible after it has been reached, and if the insistence on compliance with such a term results in a delay to the delivery of the decision, then it may well be held unenforceable.[59] Similarly, the existence of CPR Part 8 proceedings cannot excuse non-compliance with the timetable for reaching and communicating a decision, unless both parties agree otherwise.[60]

Key Case: Late Decision

Mott MacDonald Ltd v London & Regional Properties Ltd (2007) 113 Con LR 33, [2007] EWHC 1055 (TCC)

9.45 **Facts:** The adjudicator had sought to impose a term in his acceptance of appointment that 'Prior to releasing my Decision I will require payment of my fees and expenses by the Referring Party.' The referring party agreed to a 14-day extension of time and the decision was reached in time but the adjudicator did not deliver his decision for a further five days until his fees had been paid.

9.46 **Held:** The decision had not been delivered as soon as possible after it was reached with the result that it was out of time and unenforceable. An adjudicator may not impose a lien on his decision. Judge Thornton said:

> Issue 7 – Was the adjudicator entitled to impose a precondition on the delivery of his decision to the parties that his fees should first be paid by the referring party?
>
> 75. The adjudicator is obliged to comply with the scheme rules because he is required to comply with any relevant terms of the construction contract and the scheme rules have effect as implied terms of that contract by virtue of section 114(4) of the HGCRA. In complying with the scheme rules, the adjudicator must act impartially, rule 12(a) of the scheme. The adjudicator must reach a decision which must be accompanied by reasons if these are requested, rules 19 and 22. This decision, with the reasons if these are requested, must be delivered to the parties as soon as possible after he has reached his decision, rule 19(3).
>
> 76. It follows from these provisions that the adjudicator may not impose a lien on his decision or reasons and not deliver it pending the payment of his fees. This is because he is restricting himself from complying with his obligation to deliver these documents as soon as possible after he has reached his decision which must be reached within 28 days or, if the referring party agrees, within 42 days.
>
> 77. Moreover, the adjudicator appeared to lack impartiality in making it a condition of his appointment that his fees would first have to be paid by the referring party before he delivered his decision to the parties and by then appearing to enforce that pre-condition. An adjudicator appointed under a construction contract to which Part II of the HGCRA is applicable, particularly where the agreement does not contain an overriding contractual adjudication clause. His appointment is not consensual in the same way as an arbitrator's appointment is consensual and he has a quasi-judicial function since he is imposed unilaterally by the state onto one of the parties to reach a binding, albeit temporary, decision about their dispute. The adjudicator may not, therefore, be or appear to be financially

[59] *Mott MacDonald Ltd v London & Regional Properties Ltd* (2007) 113 Con LR 33, [2007] EWHC 1055 (TCC) (Key Case).

[60] *Vision Homes Ltd v Lancsville Construction Ltd* [2009] BLR 525, [2009] EWHC 2042 (TCC) at paras. 72–4.

beholden to one party, particularly the referring party, or place himself in the position in which he might appear to be more partial to one side than the other. The imposition of a lien on his decision which has to be lifted by the referring party in order to obtain his decision gives an appearance of partiality and amounts to a breach of rule 12(a) of the scheme.

78. Thus, the adjudicator was in breach of his contractual obligations imposed by rules 12(a) and 19(3) of the scheme in imposing this condition and, subsequently, in implementing it.

...

Issue 9 – Was a copy of the decision delivered to each of the parties as soon as it was reached?

81. The decision was not delivered to each of the parties as soon as it was reached. There were three reasons why the decision was not delivered to the parties on the day it was reached on Friday 8 December 2006 but was instead received on Thursday 14 December 2006. Firstly, the adjudicator imposed a pre-condition that the decision would not be released until MM paid his fees; secondly, the adjudicator implemented that condition and did not release the decision for 5 days whilst awaiting payment; and thirdly, the adjudicator failed to send the decision by fax, despite his direction that all communications in the adjudication should be sent in this way, but only sent it by first class post so that it arrived one day after it had been sent. In the context of the scheme rules, 'delivery to each of the parties' means getting the decision into their hands rather than dispatching it to them.

82. It follows that the decision was not delivered in compliance with rule 19(3) since it was not delivered as soon as possible after it had been reached nor was it delivered prior to the end of the 42-day period whose last day was 13 December 2006. There was a delay of five days or three working days in delivering it. There was no reason in principle for the adjudicator to delay delivering his decision as soon as he has reached it even if the time for delivery has not passed. The delay was caused by his breach of rule 12(a) in imposing a pre-condition of the release of his decision that MM should first pay his fees and then enforcing this pre-condition and by his failure to comply with his own stipulated procedure whereby all communications to and from the parties should initially be by fax. However, when the decision has been reached within the relevant 28-day or 42-day period, it is incumbent on the adjudicator to deliver it as soon as it has been finished and certainly to deliver it before the relevant period of 28 or 42 days has expired.

Issue 10–What is the effect on the validity and enforceability of the decision of the answers given to issues (7)–(9)?

83. There are now a long line of decisions in the Technology and Construction Court that have held that a decision that is not delivered promptly by the most rapid available means of delivery is invalid. These decisions include *Bloor Construction (UK) Ltd v Bowmer & Kirland (London) Ltd, St Andrew's Bay Developments Ltd v HBG Management Ltd, Barnes & Elliott Ltd v Taylor Woodrow Holdings Ltd, Ritchie Brothers (PWC) Ltd v David Philip (Commercials), Hart Investments Ltd v Fidler & Others and Cubitt Building & Interiors Ltd v Fleetglade Ltd.*

84. The rationale for the principle I have already summarised and which is derived from these authorities is as follows:

(1) Adjudication is intended to be a rapid and informal means of resolving disputes on a temporary basis.
(2) To that end, the scheme rules, and all other adjudication rules, provide that the adjudicator must deliver his decision promptly.
(3) Given the rationale for adjudication in its present rapid form, the rules are to be construed as being mandatory. They are rules which the adjudicator is obliged to comply with.

(4) So as to comply with this rationale, the adjudicator should use the most rapid means of delivery that are reasonably available. This will ordinarily involve use of email or facsimile facilities.
(5) Any delay after the end of the relevant adjudication period in delivering the decision must be minimal and, if the decision has been reached before the end of that period, it should be delivered within that period.
(6) Any failure to comply with the requirement of prompt and rapid delivery will render the decision unenforceable and, probably, a nullity.

85. There was no good reason for the adjudicator to have delayed in providing his decision to the parties after Friday 8 December 2006. It follows that since I agree with, and adopt, these principles, the decision of the adjudicator, even if it had been reached within the adjudicator's jurisdiction, is unenforceable and, probably, a nullity. An additional reason for reaching this conclusion is that the adjudicator failed to act impartially in imposing a pre-condition that MM should pay his fees prior to the his providing a copy of his decision to the parties.

Table 9.5 Table of Cases: Late Decision (successful challenges are indicated by shaded entries)

Title	Citation	Issue
Bloor Construction (UK) Ltd v Bowmer & Kirkland (London) Ltd	[2000] BLR 314 (TCC), Judge Toulmin	**Late communication of decision** *Obiter*: the provision in JCT condition 41A.5.3 (that the adjudicator should send his decision 'forthwith' once it had been reached) required that the process of communication be commenced immediately after the decision was reached. Enforced.
St Andrews Bay Development Ltd v HBG Management Ltd	2003 SLT 740 (Outer House Court of Session), Lord Wheatley	**Late communication of decision** The adjudicator delayed delivery of the decision until her fees had been paid. A failure by an adjudicator to produce a decision within the time limits is a serious matter but is not of sufficient significance to render the decision a nullity. Enforced.
Barnes & Elliott Ltd v Taylor Woodrow Holdings Ltd and George Wimpey Southern Ltd	[2004] BLR 111, [2003] EWHC 3100 (TCC), Judge LLoyd	**Late communication of decision** The decision was made before the time limit expired but was communicated one day late. To be enforceable the decision must be reached within the time limit. Communication of the decision outside the time limit by a day—or possibly two days—does not lead to the decision being unenforceable. Enforced.
Simons Construction Ltd v Aardvark Developments Ltd	[2004] BLR 117, [2003] EWHC 2474 (TCC), Judge Seymour	**Decision reached out of time** A decision reached out of time is enforceable. Enforced.
Ritchie Brothers (PWC) Ltd v David Philp (Commercials) Ltd	[2005] BLR 384, [2005] CSIH 32, (Inner House Court of Session), Lord Gill LJC, Lord Abernethy, Lord Nimmo Smith	**Decision reached after 28 days had expired** Subject to any agreed extension of time, an adjudicator's jurisdiction expires on the expiry of the 28-day time limit. A decision reached out of time is therefore a nullity. Not enforced.

(Continued)

Table 9.5 *Continued*

Title	Citation	Issue
Cubitt Building & Interiors Ltd v Fleetglade Ltd	(2006) 110 Con LR 36, [2006] EWHC 3413 (TCC), Coulson J	**Decision delayed by request for payment of lien** An adjudicator's decision had been reached within an agreed extended time. The adjudicator delayed release by a matter of hours in an attempt to exercise a lien, which on an analysis of the law and the contract, he was not entitled to do. However, the decision was communicated on the day following the agreed extension. Enforced.
Epping Electrical Co. Ltd v Briggs & Forrester (Plumbing Services) Ltd	[2007] BLR 126, [2007] EWHC 4 (TCC), Judge Havery	**Late decision cannot be saved by rules** Consent to an extension of time for an adjudicator to reach his decision had been conditional on the decision being issued by a particular date. When that did not happen and no valid further extension was granted, the decision was made out of time and was unenforceable. Adjudication rules (in this case the CIC rules) that seek to validate a decision delivered out of time are not Act-compliant and will be replaced with the Scheme. Not enforced.
Aveat Heating Ltd v Jerram Falkus Construction Ltd	(2007) 113 Con LR 13, [2007] EWHC 131 (TCC), Judge Havery	**Late decision cannot be saved by rules** Adjudication rules (in this case the rules in the GC/Works contract) that seek to validate a decision delivered out of time are not Act-compliant and will be replaced in their entirety with the Scheme. The date of the referral is the date on which it is received by the adjudicator; no account is to be taken of fractions of a day. Not enforced.[61]
Mott MacDonald Ltd v London & Regional Properties Ltd	(2007) 113 Con LR 33, [2007] EWHC 1055 (TCC), Judge Thornton	**Decision not delivered pending payment of adjudicator's fees** A decision that was not sent for five days pending payment of fees was not communicated as soon as possible after it was reached. Not enforced.
A. C. Yule & Son Ltd v Speedwell Roofing & Cladding Ltd	[2007] BLR 499, [2007] EWHC 1360 (TCC), Coulson J	**Extension of time agreed by silence** The 28 days was extended to 42. On day 42, the adjudicator asked for a further extension of two days to which the referring party consented but the responding party remained silent. The decision was handed down on day 43. The responding party had made it clear by its conduct it had accepted the extension and was estopped from denying the decision was on time. Enforced.

[61] In *Dalkia Energy & Technical Services Ltd v Bell Group UK Ltd* (2009) 122 Con LR 66, [2009] EWHC 73 (TCC) at para. 75, Coulson J stated that Judge Havery may have been in error in expressing the view that he did, that the clauses in *Epping* and *Aveat* were very similar, when in truth they referred to different aspects of the adjudication process.

Title	Citation	Issue
Treasure & Son Ltd v Martin Dawes (2007)	[2008] BLR 24, [2007] EWHC 2420 (TCC), Akenhead J	**Unsigned decision** There is no implied requirement that a decision needs to be signed by the adjudicator to be valid. Enforced.
Letchworth Roofing Co. v Sterling Building Co.	[2009] CILL 2717, [2009] EWHC 1119 (TCC), Coulson J	**Extension of time agreed in referral** There was found to be an agreement to extend time based on the referral and the failure of the parties to raise any objection to timetables notified by the adjudicator. *Obiter*: if the adjudicator fails to complete his decision within the time limit (as agreed) the decision is a nullity. Enforced.
Dalkia Energy & Technical Services Ltd v Bell Group UK Ltd	(2009) 122 Con LR 66, [2009] EWHC 73 (TCC), Coulson J	**Time for communicating decision** The standard terms and conditions for the Bell Group provided that the adjudicator's decision would 'nevertheless be valid if issued after the time allowed'. This was found merely to reflect the state of the law, which allows a short additional period after the decision has been reached within which it may be issued. Whether or not the actual delay will invalidate the decision will depend on the circumstances of each case. Part 8 declaration rejected.
Lee v Chartered Properties (Building) Ltd	[2010] BLR 500, [2010] EWHC 1540 (TCC), Akenhead J	**Decision communicated after expiry of time limit** A decision delivered three days after it was reached was not delivered as soon as possible. There was no good reason for the delay to delivery. Not enforced.

Errors of Fact or Law

9.47 It is now a well-established principle that an adjudicator's decision should be enforced in all but the most exceptional circumstances.[62] In particular, it will be no basis for resisting enforcement of a statutory decision[63] to contend that the adjudicator reached conclusions that were wrong as a question of fact or a matter of law. The adjudicator has jurisdiction to decide the dispute referred. As long as the adjudicator answers 'the right question' that is referred, the decision will be enforced even if the decision is wrong as a matter of fact or law. The courts will be slow to conclude that an error of fact or law is one that goes to jurisdiction and will examine any such error critically and with a degree of scepticism before they will accept that there has been an error of jurisdiction (or indeed a breach of the rules of natural justice).[64]

---

[62] See 7.05–7.07.

[63] It is important to note that the doctrine of unreviewable error does not attach to non-statutory or ad hoc adjudication decisions: *Steve Domsalla (t/a Domsalla Building Services) v Kenneth Dyason* (2008) BLR 348, [2007] EWHC 1174 (TCC).

[64] *Pegram Shopfitters Ltd v Tally Weijl (UK) Ltd* [2003] EWCA Civ 1750, [2004] 1 All ER 818; *Amec Capital Projects Ltd v Whitefriars City Estates Ltd* [2004] EWCA Civ 1418, [2005] BLR 1; *Carillion Construction Ltd v Devonport Royal Dockyard Ltd* [2005] BLR 310, [2005] EWHC 778 (TCC), [2006] BLR 15 (CA), [2005] EWCA Civ 1358 at para. 52.

Answering the Wrong Question

9.48 In *Bouygues (UK) Ltd v Dahl-Jensen (UK) Ltd* (2000) (Key Case),[65] the Court of Appeal held that if an adjudicator has answered the right question in the wrong way, the decision will be binding, but if an adjudicator answers the wrong question the decision will be a nullity.

9.49 In *Joinery Plus Ltd v Laing Ltd* (2003),[66] Judge Thornton set out the following seven principles which provide guidance to the issue of whether the adjudicator answered the wrong question:[67]

> 51. . . . 1. The precise question giving rise to the dispute that has been referred to the adjudicator must be identified.
>
> 2. If the adjudicator has answered that referred question, even if erroneously or in the wrong way, the resulting decision is both valid and enforceable. If, on the other hand, the adjudicator has answered the wrong question, the resulting decision is a nullity.
>
> 3. In determining whether the error is within jurisdiction or is so great that it led to the wrong question being asked and to the decision being a nullity, the court should give a fair, natural and sensible interpretation to the decision and, where there are reasons, to the reasons in the light of the disputes that are the subject of the reference. The court should bear in mind the speedy nature of the adjudication process which means that mistakes will inevitably occur. Overall, the court should guard against characterising a mistaken answer to an issue that lies within the scope of the reference as an excess of jurisdiction.
>
> 4. A mistake which amounts to a slip in the drafting of the reasons may be corrected by the adjudicator within a reasonable time but this is a limited power that does not extend to jurisdictional errors or errors of law.[68]
>
> 5. In deciding whether an error goes to jurisdiction, it is pertinent to ask whether the error was relevant to the decision and whether it caused any prejudice to either party.
>
> 6. A wrong decision as to whether certain contract clauses applied; or whether they had been superseded by the statutory Scheme for Adjudication; or as to whether a particular sum should be evaluated as part of, or should be included in the arithmetical computation of, the Final Contract Sum in a dispute as to what the Final Contract Sum was do not go to jurisdiction.
>
> 7. However, where the claim that was considered by the adjudicator was significantly different in its factual detail from the claim previously disputed and referred, the resulting decision was one made by reference to something not referred, was without jurisdiction and was unenforceable since the adjudicator had asked and answered the wrong question.

9.50 As explained in the General Principles section of this chapter at 9.01–9.15, the jurisdiction of the adjudicator is circumscribed by the applicable adjudication rules, the notice of adjudication, and any defences that are properly raised in the course of the adjudication. The adjudicator must decide, as a question of construction of the referral notice, what matter has been referred.[69] If an adjudicator erroneously decides that the dispute referred is wider than

---

[65] [2001] 1 All ER (Comm) 1041 (CA).

[66] [2003] BLR 184, [2003] EWHC 3513 (TCC).

[67] These principles are based on the following cases: *Nikko Hotels (UK) Ltd v MEPC Plc* [1991] 2 EGLR 103 (Ch D); *Bouygues (UK) Ltd v Dahl-Jensen (UK) Ltd* [2000] 1 All ER (Comm) 1041 (CA) (Key Case) at paras. 12–14 affirming [2000] BLR 49 (TCC) paras. 19 and 35–6; *Bloor Construction (UK) Ltd v Bowmer & Kirkland (London) Ltd* [2000] BLR 314 (TCC); *Discain Project Services Ltd v Opecprime Development Ltd* (No. 2) [2001] BLR 285; *Shimizu Europe Ltd v Automajor Ltd* [2002] BLR 113 (TCC); *C&B Scene Concept Design Ltd v Isobars Ltd* [2002] BLR 93 (CA); *Edmund Nuttall Ltd v R. G. Carter Ltd* [2002] BLR 312, [2002] EWCA Civ 46, (TCC); and *Balfour Beatty Construction Ltd v The London Borough of Lambeth* [2002] BLR 288; [2002] EWHC 597 (TCC).

[68] The issue of slips in decisions is discussed in detail in Ch. 4 at 4.109–4.177.

[69] *Ballast plc v The Burrell Company (Construction Management) Ltd* [2001] BLR 529 (Outer House, Court of Session).

it is, the Court of Appeal has stated that, in so far as the adjudicator has exceeded his or her jurisdiction, that decision cannot be enforced.[70]

9.51 In order to determine if the adjudicator has answered the wrong question and therefore acted without jurisdiction, it is necessary to construe the documents produced in the course of the adjudication. That will include the adjudication notice, the referral and submissions of both parties (to determine the extent of the jurisdiction) and the decision of the adjudicator (to determine what questions he actually decided). It is also necessary to identify what sub-issues the adjudicator needed to determine in order to decide the question referred; such necessary sub-issues, including defences, will also be within the adjudicator's jurisdiction.[71] Otherwise, save where there is express or implicit agreement between the parties, the adjudicator's jurisdiction cannot be expanded in the course of the adjudication.

9.52 Each challenge on this ground, including whether a matter raised in reply is beyond the jurisdiction of the adjudicator, will be determined in light of these principles and on its own particular facts and circumstances. Whilst each case will be different, the cases in Table 9.6 show that, generally, an adjudicator will have jurisdiction to make a decision awarding different sums[72] or periods of time[73] from that claimed in the referral, or a decision considering the effect or validity of a withholding notice.[74] However, an adjudicator will not be permitted to reach a decision:

1. On a different basis from that advanced in the notice and referral,[75] for example where the referring party puts forward a materially different case in its reply to address shortcomings in its original referral identified by the responding party.
2. On the wrong contractual terms.[76]
3. In reliance on evidence the parties agreed was irrelevant.[77]
4. That does not answer the question at all.[78]

9.53 The question of the scope of the adjudicator's jurisdiction arises not only where the adjudicator takes an overly wide view of jurisdiction but also where an erroneously narrow view of jurisdiction is adopted with the result that the adjudicator fails to decide a matter that has been properly referred. The principles applicable to such circumstances are addressed in Chapter 10 in the context of breach of the rules of natural justice and the reader is referred to 10.36–10.41 for a discussion of the applicable principles and tables of cases. In short, where the adjudicator fails to consider a defence on its merits as a result of a mistaken conclusion that there was no jurisdiction to consider the defence at all, the decision will be unenforceable.

---

[70] *C&B Scene Concept Design Ltd v Isobars Ltd* [2002] BLR 93, [2002] EWCA Civ 46 (CA).

[71] *KNS Industrial Services (Birmingham) Ltd v Sindall Ltd* (2000) 75 Con LR 71 (TCC).

[72] *OSC Building Services Ltd v Interior Dimensions Contracts Ltd* [2009] CILL 2688, [2009] EWHC 248 (TCC); *Workspace Management Ltd v YJL London Ltd* [2009] BLR 497, [2009] EWHC 2017 (TCC); *Volker Stevin Ltd v Holystone Contracts Ltd* [2010] EWHC 2344 (TCC).

[73] *Cantillon Ltd v Urvasco Ltd* [2008] BLR 250, [2008] EWHC 282 (TCC).

[74] *SL Timber Systems Ltd v Carillion Construction Ltd* [2001] BLR 516, 2002 SLT 997 (Outer House, Court of Session); *HS Works Ltd v Enterprise Managed Services Ltd* [2009] BLR 378, [2009] EWHC 729 (TCC); *Windglass Windows Ltd v Capital Skyline Construction Ltd* (2009) 126 Con LR 118, [2009] EWHC 2022 (TCC); *Barr Ltd v Klin Investment UK Ltd* [2009] CSOH 104 (Outer House, Court of Session).

[75] *(1) Ken Griffin and (2) John Tomlinson (t/a K&D Contractors) v Midas Homes Ltd* (2000) 78 Con LR 152 (TCC); *Edmund Nuttall Ltd v R. G. Carter Ltd* [2002] BLR 312, [2002] EWHC 400 (TCC); *McAlpine PPS Pipeline Systems Ltd v Transco Plc* [2004] BLR 352, [2004] EWHC 2030 (TCC); *AWG Construction Ltd v Rockingham Motor Speedway Ltd* [2004] EWHC 888 (TCC).

[76] *Shimizu Europe Ltd v LBJ Fabrications Ltd* [2003] BLR 381, [2003] EWHC 1229 (TCC); *Joinery Plus Ltd v Laing Ltd* [2003] BLR 184, [2003] EWHC 3513 (TCC).

[77] *Shimizu Europe Ltd v LBJ Fabrications Ltd* [2003] BLR 381, [2003] EWHC 1229 (TCC); *Primus Build Ltd v Pompey Centre Ltd* [2009] BLR 437, [2009] EWHC 1487 (TCC).

[78] *Ballast plc v The Burrell Company (Construction Management) Ltd* [2001] BLR 529 (Outer House, Court of Session).

However, an error of fact or law that leads the adjudicator to reject the defence because he decided there was a requirement for a withholding notice that had not been met is an error within jurisdiction that cannot found a challenge to enforcement.

Key Cases: Answering the Wrong Question

Bouygues (UK) Ltd v Dahl-Jensen (UK) Ltd [2001] 1 All ER (Comm) 1041 (CA)

9.54 **Facts:** The adjudicator issued a decision in which he had made an arithmetical error when he calculated the sum due using a gross sum which included retention whilst deducting the amount paid to date which did not include retention. The effect of that error was to release the sums retained as part of the sum awarded. The magnitude of the error was in the region of £350,000 with the result that instead of a decision that a payment of £200,000 was due, there was around £140,000 repayment due.

9.55 **Held:** The decision contained an error, but it was an error made within the adjudicator's jurisdiction and was therefore enforceable. Per Chadwick LJ:

> 27. The first question raised by this appeal is whether the adjudicator's determination in the present case is binding on the parties—subject always to the limitation contained in section 108(3) and in paragraphs 4 and 31 of the Model Adjudication Procedure to which I have referred. The answer to that question turns on whether the adjudicator confined himself to a determination of the issues that were put before him by the parties. If he did so, then the parties are bound by his determination, notwithstanding that he may have fallen into error. As Knox J put it in *Nikko Hotels (UK) Ltd v MEPC plc* [1991] 2 EGLR 103 at page 108, letter B, in the passage cited by Buxton LJ, if the adjudicator has answered the right question in the wrong way, his decision will be binding. If he has answered the wrong question, his decision will be a nullity.
>
> 28. I am satisfied, for the reasons given by Buxton LJ, that in the present case the adjudicator did confine himself to the determination of the issues put to him. This is not a case in which he can be said to have answered the wrong question. He answered the right question. But, as is accepted by both parties, he answered that question in the wrong way. That being so, notwithstanding that he appears to have made an error that is manifest on the face of his calculations, it is accepted that, subject to the limitation to which I have already referred, his determination is binding upon the parties.

McAlpine PPS Pipeline Systems Joint Venture v Transco Plc [2004] BLR 352, [2004] EWHC 2030, (2004) BLR 352 (TCC)

9.56 **Facts:** An adjudicator's decision considered the relevant compensation events under an NEC 2 contract when the only issue referred to the adjudicator was one that concerned unpaid interest on those compensation events. The responding party argued that there had been a change in the basis of the dispute. The adjudicator had decided that he could serve the parties better by deciding the dispute which he believed ought to have been referred, namely the dispute which required an investigation of the underlying facts concerning entitlement to compensation events.

9.57 **Held:** The court rejected the application for summary judgment and found that the defendant had a real prospect of contending that the adjudicator had acted beyond his jurisdiction and decided a dispute not referred to him. Judge Toulmin stated that the problem in such cases was not whether there was a dispute but rather the defining of the nature of the dispute or difference that has been referred to the adjudicator. After reviewing the relevant authorities, the judge set out the following nine questions that 'may provide pointers to adjudicators as to the questions which they should have in mind when their jurisdiction is challenged or where further evidence is tendered in the course of the adjudication':

1. What issues were discussed at any meetings between the parties before service of the referral to adjudication and the notice of dissatisfaction?
2. What is the dispute that was referred to the adjudicator after the defendant had had an opportunity to respond to the claims put forward by the claimant?
3. What was the basis on which the dispute was referred? Was it (a) general and/or (b) by reference to specific issues?
4. Was the adjudicator's decision responsive to the issues referred?
5. Were new issues raised in the course of the adjudication?
6. If so, were the new issues objected to by the other party?
7. Was any such objection one which goes to the fundamental nature of the dispute referred?
8. If so, does the objection go to the fairness of the procedure?
9. If there was a breach of the procedure, does it significantly affect the fairness of the decision?

9.58 Judge Toulmin emphasized the importance of the originating referral as it defines the extent of the adjudicator's jurisdiction:

> 145. It seems to me that it is clear from the Act that it is for the party who refers the dispute to adjudication to define the issues which are referred. In the absence of agreement between the parties to vary the terms on which the dispute is referred, the adjudicator has no jurisdiction to vary the basis on which the reference has been made. He can, of course, take the initiative in ascertaining the facts and the law in relation to the dispute referred to him. (See section 108(2)(f) of the Act and paragraph 13 of the scheme.)
>
> 146. Unfortunately, it is not enough for the adjudicator to say that he was sure that both parties would want to conclude the matter without recourse to further proceedings. If the existing referral does not enable him to deal with the dispute in the way in which he wishes, he is powerless to alter the terms of the referral in the absence of the agreement of both parties. So long as the dispute remains before him, he must decide only the issues referred to him.

Table 9.6 Table of Cases: Answering the Wrong Question (successful challenges are indicated by shaded entries)

Title	Citation	Issue
Northern Developments (Cumbria) Ltd v J&J Nichol	[2000] BLR 158 (TCC), Judge Bowsher	**Jurisdiction extended by implied agreement** The adjudicator's jurisdiction under the Scheme was extended by an implied agreement that the adjudicator had jurisdiction to make an award of costs. That agreement arose from both parties having requested in writing that they be awarded their costs without raising any question of the adjudicator's jurisdiction to do so. Enforced.
(1) Ken Griffin and (2) John Tomlinson (t/a K&D Contractors) v Midas Homes Ltd	(2000) 78 Con LR 152 (TCC), Judge LLoyd	**Scope of notice of adjudication** The notice of adjudication defines the dispute to be referred, specifies precisely the redress sought and the parties so that the adjudicator knows the ambit of his jurisdiction. On its proper construction, the notice of adjudication only gave the adjudicator jurisdiction to consider

(Continued)

Table 9.6 *Continued*

Title	Citation	Issue
		the questions of whether there had been a wrongful determination of the contract, and payments prior to that determination; he had no jurisdiction to decide the consequences of the determination. Not enforced in part (decision severed).
Bouygues (UK) Ltd v Dahl-Jensen (UK) Ltd	[2001] 1 All ER (Comm) 1041 (CA), Gibson, Chadwick, Buxton LJJ	**Answering a different question** An adjudicator must confine himself to the determination of the issues that were put before him. If an adjudicator has answered the right question in the wrong way, the decision will be binding. If an adjudicator answers the wrong question, his decision will be a nullity. Enforced.
KNS Industrial Services (Birmingham) Ltd v Sindall Ltd	(2000) 75 Con LR 71 (TCC), Judge LLoyd	**Defence of lawful determination of contract within jurisdiction** Where a referring party sought a decision on the amount that it was to be paid, the adjudicator's jurisdiction extended to consideration of any defence that justified non-payment. In this case, that included whether the employment had been lawfully determined as such a determination would affect the entitlement to payment. Enforced.
LPL Electrical Services Ltd v Kershaw Mechanical Services Ltd	(unreported) 2 Feb. 2001 (TCC), Judge Havery	**Interpretation of notice of adjudication** Whilst the notice of adjudication referred to a particular interim application for payment, it also sought payment of a particular sum with the result that the adjudicator had jurisdiction to award that sum even though it did not technically arise under the interim application identified in the notice of adjudication. Enforced.
Sindall Ltd v Abner Solland and ors	(2001) 80 Con LR 152 (TCC), Judge LLoyd	**Necessary part of decision on matter within jurisdiction** Whilst there was no pre-existing dispute relating to a claim for an extension of time, a decision on entitlement to extension of time was a necessary part of determining the wider question of whether termination had been unlawful which was within the jurisdiction of the adjudicator. Enforced.
Ballast plc v The Burrell Company (Construction Management) Ltd	[2002] Scot CS 324, (Extra Division, Inner House, Court of Session), Lord Cullen LP, Lord Johnston, Lord Weir	**Not reaching a decision** In his reasons, the adjudicator stated that he had 'found it impossible to reach what I can substantiate as a reasonably legally based decision'. An adjudicator cannot narrow or enlarge his jurisdiction by misconstruing the limits of his jurisdiction. An adjudicator must reach a decision on the matters referred; if he does not, any decision is a nullity. The adjudicator has a duty to consider the validity of each and all claims. They must either succeed or fail either in whole or in part. Not enforced.

Title	Citation	Issue
SL Timber Systems Ltd v Carillion Construction Ltd	[2001] BLR 516, 2002 SLT 997, (Outer House, Court of Session), Lord Macfadyen	**Error in decision relating to the effect of a failure to issue a withholding notice** Errors in a decision are not to be too readily characterized as jurisdictional errors. A decision on the effect of a failure to issue a withholding notice was not a matter that went to jurisdiction even though it led the adjudicator not to consider whether the sums claimed were sums due under the contract. He answered the question referred, namely whether the respondent was required to pay the sums claimed. Enforced.
Shimizu Europe Ltd v Automajor Ltd	[2002] BLR 113 (TCC), Judge Seymour	**Inclusion of sums in respect of variations** The notice of adjudication required the adjudicator to decide the sums due under the contract. That is what he decided and any erroneous inclusion in that sum of amounts in respect of alleged variations was merely an error in the calculation of the sum and not an error that affected his jurisdiction. Enforced.
C&B Scene Concept Design Ltd v Isobars Ltd	[2002] BLR 93, [2002] EWCA Civ 46 (CA), Potter, Rix LJJ, Sir Murray Stuart-Smith	**Mistake as to contractual terms** The scope of an adjudicator's jurisdiction is determined by and derives from the dispute that is referred to him. To resolve a payment dispute the adjudicator had to determine the relevant contractual terms and their effect. Errors in that determination were not jurisdictional errors. Enforced.
Edmund Nuttall Ltd v R. G. Carter Ltd	[2002] BLR 312, [2002] EWHC 400 (TCC), Judge Seymour	**Dispute determined on different grounds from those claimed** The jurisdiction of the adjudicator derives from the notice of adjudication which is to be construed in accordance with the normal principles of contractual construction. The dispute referred was the referring party's entitlement to an extension of time and loss and expense as defined by its 'May Claim'. The adjudicator did not have jurisdiction to decide the entitlements on different grounds from those originally claimed. Not enforced.
Cowlin Construction Ltd v CFW Architects (a firm)	[2003] BLR 241 (TCC), Kirkham J	**Jurisdiction to determine VAT** Where the notice to adjudicate expressly referred to VAT, that was a matter within the jurisdiction of the adjudicator. A mistake as to whether VAT was payable was not an error that went to jurisdiction. Enforced.
Joinery Plus Ltd v Laing Ltd	[2003] BLR 184, [2003] EWHC 3513 (TCC), Judge Thornton	**Answering a different question** Where the adjudicator decided the dispute by reference to entirely the wrong contractual terms, that was an error which went to his jurisdiction. He only had jurisdiction to decide the dispute under the terms of the actual construction contract between the parties. Not enforced.

(Continued)

Table 9.6 *Continued*

Title	Citation	Issue
Shimizu Europe Ltd v LBJ Fabrications Ltd	[2003] BLR 381, [2003] EWHC 1229 (TCC), Judge Kirkham	**Diverging from agreed contractual basis** The parties had agreed the position as to the contractual relationship between them (that their agreement was based on a letter of intent not on JCT DOM/1). Accordingly, the adjudicator did not have jurisdiction to decide the dispute on a different contractual basis. Not enforced.
AWG Construction Services Ltd v Rockingham Motor Speedway Ltd	[2004] EWHC 888 (TCC), Judge Toulmin	**General negligence was not part of dispute referred** In deciding the respondent had been negligent on a basis that had not been set out in the dispute that had been referred, but rather was based on the reply, the adjudicator had exceeded his jurisdiction. Not enforced.
McAlpine PPS Pipeline Systems Joint Venture v Transco Plc	[2004] BLR 352, [2004] EWHC 2030 (TCC), Judge Toulmin	**Different basis of claim introduced in reply** It is important to consider precisely what matters were referred to the adjudicator and in what terms. Where the question referred is a specific one, in the absence of agreement between the parties, the adjudicator only has jurisdiction to answer the question referred. The referring party cannot change the basis on which it puts its claim by way of reply. Not enforced.
Carillion Construction Ltd v Devonport Royal Dockyard Ltd	[2006] BLR 15, [2005] EWCA Civ 1358 (CA), Sir Anthony Clarke MR, Chadwick, Moore-Bick LJJ	**Award of interest** The determination of the sum due (which was within the adjudicator's jurisdiction) necessarily involved making an assessment of the target cost, the decision in respect of which was therefore also within the adjudicator's jurisdiction. There is no freestanding jurisdiction to award interest under para. 20(c) of the Scheme. The adjudicator only has jurisdiction to award interest where a claim for interest is made and there is a contractual right to interest, or the parties have implicitly or expressly agreed that the adjudicator should have jurisdiction to award interest. Enforced.
John Roberts Architects Ltd v Parkcare Homes (No. 2) Ltd	[2006] BLR 106, [2006] EWCA Civ 64 (CA), May, Keene, Scott Baker LJJ	**Jurisdiction to award costs** Where the adjudication rules gave the adjudicator jurisdiction to award costs, on a proper construction of the rules he had that power even where no substantive decision was made after the referral was withdrawn. Enforced.
Cantillon Ltd v Urvasco Ltd	[2008] BLR 250, [2008] EWHC 282 (TCC), Akenhead J	**Decision on a basis different from that set out in the referral** The adjudicator had jurisdiction to determine that the relevant delay occurred during a later and shorter period than that claimed, as contended for by the respondent. Enforced.

Title	Citation	Issue
YCMS Ltd (t/a Young Construction Management Services) v (1) Stephen Grabiner and (2) Miriam Grabiner	[2009] BLR 211, [2009] EWHC 127 (TCC), Akenhead J	**Reference to matters raised in defence** Where a claim was made for sums certified but unpaid up to a specified date, the adjudicator had jurisdiction to refer to a later certificate where that certificate had been raised in defence of the claim. Not enforced on other grounds (exceeding power to correct a slip).
OSC Building Services Ltd v Interior Dimensions Contracts Ltd	[2009] CILL 2688, [2009] EWHC 248 (TCC), Ramsey J	**Final account or interim payments claim** Challenge based on the referral being wider than the notice of adjudication. On a proper interpretation, the notice of adjudication was referring a claim in respect of the last interim assessment even though the words 'final account' had been used. That is what the adjudicator decided. Where the precise sum to be awarded was left open to the adjudicator by use of the words 'or such other sum as the adjudicator may decide' in the notice of adjudication, an increase in the sum claimed in the referral did not expand the claim. The adjudicator had jurisdiction to determine the sums due when the notice of adjudication was issued. Enforced.
HS Works Ltd v Enterprise Managed Services Ltd	[2009] BLR 378, [2009] EWHC 729 (TCC), Akenhead J	**Consideration of the relevant defence of withholding notice** The dispute contained in the notice involved a primary assertion that there was no effective withholding notice as required by the contract. In accepting that case, it was unnecessary for the adjudicator to consider the responding party's case on the merits of the cross-claims. The challenge that the adjudicator had failed to answer the dispute referred was not successful. Further, the adjudicator could not be criticized for not addressing every difference in the accounting position between the parties—a spot check was adequate. Enforced.
Letchworth Roofing Co. v Sterling Building Co.	[2009] CILL 2717, [2009] EWHC 1119 (TCC), Coulson J	**Dispute limited to question of validity of withholding notice** On a proper interpretation of the notice of adjudication, the adjudicator had only been asked to determine the validity of the withholding notice and not the merit of any cross-claim.
Workspace Management Ltd v YJL London Ltd	[2009] BLR 497, [2009] EWHC 2017 (TCC), Coulson J	**Answering the question of a proper valuation** Where an adjudicator had been asked to determine the proper valuation of a particular certificate, he had jurisdiction to decide the proper valuation was substantially lower than the sum certified and that the responding party had overpaid. He also had jurisdiction to award payment of the sum overpaid but declined to do so, considering he did not have the jurisdiction. Enforced.

(Continued)

Table 9.6 *Continued*

Title	Citation	Issue
Windglass Windows Ltd v Capital Skyline Construction Ltd	(2009) 126 Con LR 118, [2009] EWHC 2022 (TCC), Coulson J	**Decision on validity of withholding notice** A decision on the validity of the purported withholding notices was a necessary part of determining the claim for sums due and was squarely within the adjudicator's jurisdiction. Enforced.
Vision Homes Ltd v Lancsville Construction Ltd	[2009] BLR 525, [2009] EWHC 2042 (TCC), Clarke J	**Decision on applicability of contract terms within jurisdiction** Where the question of entitlement to levy LADs had been referred to the adjudicator, who had been asked specifically 'to investigate all the issues of the agreement', it was open to the adjudicator to determine what parts of the original contract remained in operation and with what effect. Not enforced on other grounds (failure in appointment).
Primus Build Ltd v (1) Pompey Centre Ltd and (2) Slidesilver Ltd	[2009] BLR 437, [2009] EWHC 1487 (TCC), Coulson J	**Consideration of matters the parties agreed were irrelevant** The adjudicator did not have jurisdiction to consider evidence that the parties had agreed was irrelevant, in this case certain accounts. Not enforced.
Banner Holdings Ltd v Colchester Borough Council	(2010) 131 Con LR 77, [2010] EWHC 139 (TCC), Coulson J	**Contractual limit to adjudicator's jurisdiction** The contract between the parties based on the GC Works/1 form expressly provided that an adjudicator had no jurisdiction to vary or overrule a decision made by the employer in respect of determination of the contract. The adjudicator had jurisdiction because he had been asked to confirm the validity of a determination decision and not to overrule or vary it. *Obiter*: the provision seeking to restrict the jurisdiction of the adjudicator was not Act-compliant and would be replaced by the Scheme. Enforced.
Volker Stevin Ltd v Holystone Contracts Ltd	[2010] EWHC 2344 (TCC), Coulson J	**Reduction of sum claimed in course of adjudication** The adjudicator had jurisdiction to consider the reduced claim advanced in the reply, not least because the notice did not confine the dispute to a particular sum. Enforced.
Urang Commercial Ltd v Century Investments Ltd and anor	[2011] EWHC 1561 (TCC), Edwards-Stuart J	**Rejection of defence of set-off was a mistake of law within jurisdiction** Where an adjudicator wrongly thought he could not consider a set-off defence because he wrongly concluded that a withholding notice was required and was not provided, the judge found this was a mistake of law which did not render the decision unenforceable. Enforced.

Title	Citation	Issue
Herbosh-Kiere Marine Contractors Ltd v Dover Harbour Board	[2012] EWHC 84 (TCC), Akenhead J	**No dispute over approach to assessment of delay claim** The dispute referred related to the determination of the periods of delay for which the employer was responsible. The parties agreed on the approach to valuation, namely that any delay should be valued using specific resource rates and specific periods of delay. The adjudicator's decision calculated an award based on the overall period of delay multiplied by a composite rate, which was not the dispute referred. Not enforced (also breach of natural justice).

(3) When Can a Claim Be Withdrawn or an Adjudication Restarted?

9.59 The Court of Appeal has confirmed that an individual claim can be withdrawn from an adjudicator, subject to potential cost penalties, unless there is something in the relevant adjudication agreement or rules which says otherwise: *Lanes Group Plc v Galliford Try Infrastructure Ltd* (2011)[79] approving *Midland Expressway Ltd and ors v Carillion Construction Ltd and ors* (2006).[80] In *Midland* Jackson J pointed out that whilst the 1996 Act is silent about any entitlement of a party to withdraw a claim advanced in adjudication, there is also no express prohibition on a party withdrawing a disputed claim. If there were such a restriction it would lead to the bizarre consequence that the parties would be forced to press on with a claim that the referring party no longer wished to pursue leading to wastage of costs and resources on the part of all parties.

9.60 Attention must of course be paid to the terms of any contractual adjudication agreement, or any adjudication rules incorporated by reference. Either may suggest (expressly or impliedly) that claims, once commenced, cannot be withdrawn from an adjudicator although such a provision would be unusual and there is currently no reported decision concerning such a term.

9.61 The Court of Appeal has also confirmed that a party may abandon an adjudication before one adjudicator and start again before a second adjudicator: *Lanes Group Plc v Galliford Try Infrastructure Ltd* (2011).[81] In that case, following an application by Galliford for the appointment of an adjudicator, the ICE appointed Mr Klein who confirmed his appointment in writing. Galliford immediately wrote to point out that Mr Klein had acted against Galliford in a series of acrimonious adjudications recently that, they submitted, would make it difficult for him to be impartial. They served a fresh notice and asked for a new adjudicator but the ICE refused at this point. Galliford did not serve the referral within the time stipulated by the rules and thereby allowed that reference to 'lapse'. Mr Klein did not resign. However Galliford then served another fresh notice and the ICE appointed Mr Atkinson

[79] [2011] EWCA Civ 1617, [2012] BLR 121 at paras. 35–42 of the judgment.
[80] [2006] EWHC 1505 (TCC), [2006] BLR 325 at paras. 100–6.
[81] [2011] EWCA Civ 1617, [2012] BLR 121 at para. 40 of the judgment.

who accepted his appointment. Lanes applied to the TCC for an injunction to restrain the Atkinson adjudication on the grounds that it was unfair and oppressive, an abuse of process, and Galliford had committed a repudiatory breach of the adjudication agreement.

9.62 Akenhead J held that the doctrine of repudiation is not applicable to adjudication procedures included in contracts governed by the 1996 Act,[82] although it may well be applicable to adjudication agreements that were purely contractual.[83] Whilst refusing to serve a referral notice on the adjudicator could be, and was, a breach of the adjudication agreement—which could in some circumstances justify injunctive relief—where there were justifiable reasons such as an honest belief that there was apparent bias, the breach and damages would be nominal.[84] The reasoning of Akenhead J was approved in the Court of Appeal.[85]

9.63 A few months after Akenhead J's judgment, the case came back before the High Court when Lanes sought to resist enforcement of the Atkinson adjudication decision on two grounds, one of which was that, because an adjudication of the the same dispute had been started and Mr Klein nominated, Mr Atkinson had no jurisdiction.[86] On the question of whether Galliford had been entitled to withdraw the claim from Mr Klein and start again, Judge Waksman decided there could be no absolute nor any qualified bar on starting again for six reasons (upheld by the Court of Appeal[87]):

1. The scheme at paragraph 9(2) prohibits adjudication on a matter already decided by an adjudicator. It could have but didn't include a prohibition on withdrawing a dispute from an adjudicator and starting again.
2. Until the referral was received by an adjudicator he or she had no jurisdiction over the dispute.
3. The existing authority of *Hart v Fidler* and *Vision Homes v Lancsville* suggests the referring party may start again where the referral has been served out of time.
4. An absolute bar would be draconian penalty where the late service of the referral was due to inadvertence such as the drafter falling sick.
5. A qualified bar was unworkable as there was no clear definable scope for its operation.
6. No prejudice was suffered by starting again.

[82] [2011] EWHC 1035 (TCC), Akenhead J, at paras. 24, 28, 30, and 31 of the judgment.
[83] Para. 24 of the judgment.
[84] Judgment para. 29.
[85] [2011] EWCA Civ 1617, [2012] BLR 121 at para. 35 of the judgment.
[86] *Lanes Group Plc v Galliford Try Infrastructure Ltd* [2011] EWHC 1679 (TCC), Judge Waksman.
[87] [2011] EWCA Civ 1617, [2012] BLR 121 at para. 42.

10

CHALLENGES BASED ON BREACH OF NATURAL JUSTICE

(1) Summary of Challenges Based on Breach of Natural Justice

10.01 The following points of general application have emerged from the cases on natural justice:

1. An adjudicator is required to act impartially under s. 108(2)(e) of the Housing Grants Construction and Regeneration Act 1996 ('the 1996 Act'), and under the common law. A party has the right to present its case before an impartial tribunal where justice must be seen to be done.
2. An adjudicator must conduct proceedings in accordance with the rules of natural justice, or as fairly as the time constraints permit. However the speed with which adjudications are conducted means that some breaches of natural justice which have no demonstrable consequences may be disregarded.
3. The adjudicator's conduct is key, and this may in some cases require asking whether the adjudicator himself considered he had sufficient time to conduct proceedings fairly. If he considers that justice cannot be done then he may request an extension of time or, if necessary, resign.
4. Parties can, by their conduct, agree to waive or acquiesce in a breach of natural justice. If so, a subsequent challenge to enforcement proceedings is unlikely to be successful.

(2) General Principles of Natural Justice

10.02 Section 108(2)(e) of the 1996 Act imposes a duty of impartiality upon the adjudicator. Where an adjudication is being conducted under the Scheme for Construction Contracts ('the Scheme'), this duty to act impartially is also imposed by paragraph 12.

10.03 In addition, the common law principles of natural justice require that those affected by decision-makers are dealt with in a fair manner. The House of Lords have said, in another context, that quite what 'fairness' requires will depend on the character of the decision-making body, the questions it is being asked to answer and the framework (statutory or otherwise) in which it is operating.[1] In the most general terms, natural justice requires that a party should have the right to prior notice of the case against it and a fair opportunity to make representations before a decision is made; the specific requirements will depend on the circumstances of each particular case.[2]

10.04 Absent agreement of the parties, adjudication under the Act is subject to strict time limits. That necessarily limits the time for preparation and presentation of evidence and argument and subsequent consideration by the adjudicator. Questions of unfairness in the course of adjudication therefore need to be considered against that background. An adjudicator is required to conduct proceedings in accordance with broad concepts of procedural fairness.[3] Thus whilst the adjudicator must be impartial and allow the parties an opportunity to present their cases, the nature of the process will introduce limitations as to what is required. As observed in *Discain Project Services Ltd v Opecprime Development Ltd* (2000) BLR 402,[4] adjudication is inherently an imperfect but not necessarily final process[5] which makes regard for the rules of natural justice 'more rather than less important', particularly where there is no appeal on fact or law from the adjudicator's decision.

10.05 There is therefore something of a balancing act to be performed. It is now well established that adjudicators must conduct proceedings 'in accordance with natural justice, or as fairly as the limitations imposed by Parliament permit'.[6] However, the speed of adjudication makes it inevitable that natural justice might sometimes be compromised. Indeed, the courts have recognized that adjudication is 'not a finely tuned instrument' but rather a 'summary and at times blunt instrument for the resolution of disputes'.[7]

10.06 Thus, the principles of procedural fairness (or the need to observe the rules of natural justice) are not to be regarded as diluted for the purposes of the adjudication process, but are judged in the light of time restraints, the provisional nature of the decision, and any concessions or agreements made by the parties as to the nature of the process in a particular case.[8]

---

[1] *Lloyd v McMahon* [1987] 2 WLR 821, [1987] 2 WLR 821, [1987] AC 625.

[2] *Dean & Dyball Construction Ltd v Kenneth Grubb Associates Ltd* (2003) 100 Con LR 92, [2003] EWHC 2465 (TCC) (Key Case).

[3] *Austin Hall Building Ltd v Buckland Securities Ltd* [2001] BLR 272, [2001] EWHC 434 (TCC).

[4] [2000] BLR 402.

[5] Although it has been noted that the practical (if not strictly legal) effect of very many, if not most, adjudicators' decisions is final: *Balfour Beatty Construction Ltd v The Mayor and Burgesses of the London Borough of Lambeth* [2002] BLR 288; [2002] EWHC 597 (TCC) at para. 29, Judge LLoyd (Key Case).

[6] *Glencot Development and Design Co. Ltd v Ben Barrett & Son (Contractors) Ltd* [2001] BLR 207 (TCC); *Austin Hall Building Ltd v Buckland Securities Ltd* [2001] BLR 272, [2001] EWHC 434 (TCC); *Discain Project Services v Opecprime Development Ltd* [2000] BLR 402.

[7] *Try Construction Ltd v Eton Town House Group Ltd* [2003] BLR 286, [2003] EWHC 60 (TCC) at para. 29; *London and Amsterdam Properties Ltd v Waterman Partnerships Ltd* [2004] BLR 179, [2003] EWHC 3059 (TCC).

[8] *Ibid.*

Materiality of Breach

10.07 Not every breach of natural justice necessarily leads to an invalidation of the decision.[9] The breach must be a material breach, which will be a question of fact and degree in each case. The courts will ask whether the issue at the heart of the natural justice objection had any 'demonstrable consequence' and if not, 'repugnant' as it may be, such breaches are disregarded.[10] Thus, an unsuccessful party must not merely assert a breach of natural justice, 'any breach proved must be substantial and relevant'.[11]

10.08 Guidance for applying the test of materiality was provided in *Cantillon Ltd v Urvasco Ltd* (2008).[12] In that case, the adjudicator awarded prolongation costs for a nine-week period; however this was a different period from that claimed in the dispute referred to him. Urvasco sought to resist enforcement on the grounds that it had not been given an opportunity to address the adjudicator on the prolongation costs which applied to this alternative period, even though the contending periods had clearly been discussed during the adjudication. In refusing the challenge, Akenhead J made the following points:

> 57. ... (a) It must first be established that the Adjudicator failed to apply the rules of natural justice;
>
> (b) Any breach of the rules must be more than peripheral; they must be material breaches;
>
> (c) Breaches of the rules will be material in cases where the adjudicator has failed to bring to the attention of the parties a point or issue which they ought to be given the opportunity to comment upon if it is one which is either decisive or of considerable potential importance to the outcome of the resolution of the dispute and is not peripheral or irrelevant.
>
> (d) Whether the issue is decisive or of considerable potential importance or is peripheral or irrelevant obviously involves a question of degree which must be assessed by any judge in a case such as this.

10.09 Provided that the procedure is fair, then the rules of natural justice cannot be used to protect a party against an outcome that may be considered unfair. As discussed in Chapter 7, adjudication decisions may be enforced even where they contain manifest errors of fact or law as long as they were made within the adjudicator's jurisdiction. Considerations of natural justice, therefore, concern whether the process has been conducted fairly. They are not concerned with the question of whether the adjudicator got the decision wrong unless it is alleged that was the result of a breach of natural justice in the way the procedure was conducted. However, the terms of the decision are often relied upon to support a challenge based on breach of natural justice.

Adjudicators' Decisions Should Generally Be Enforced

10.10 Parties cannot treat every perceived procedural unfairness as grounds for challenging enforcement. The observations of Chadwick LJ in *Carillion Construction Ltd v Devonport Royal Dockyard Ltd* (2005) (Key Case)[13] apply equally to challenges based on natural justice as challenges based on excess jurisdiction:

---

[9] *Macob Civil Engineering Ltd v Morrison Construction Ltd* [1999] BLR 93, per Dyson J.

[10] *Discain Project Services Ltd v Opecprime Development Ltd* [2000] BLR 402 at 405.

[11] *Discain Project Services Ltd* v *Opecprime Development Ltd (No. 2)* [2001] BLR 285 at para. 68. In Scotland, the test may be slightly easier to meet in that a challenge may be made on the possibility that the breach produced an injustice, as there is no need at Scottish law to demonstrate 'actual injustice': *Costain Ltd v Strathclyde Builders Ltd* [2003] Scot CS 316 at paras. 23 and 24 in particular, but cf. *Highlands and Islands Airports Ltd v Shetland Islands Council* [2012] CSOH 12 at 33 which suggests the approaches in the jurisdictions may be consistent.

[12] [2008] BLR 250, [2008] EWHC 282 (TCC), (Key Case).

[13] [2006] BLR 15, [2005] EWCA Civ 1358 (CA) (Key Case).

1. It is only too easy in a complex case for a party dissatisfied with the decision of an adjudicator to comb through the adjudicator's reasons and identify points upon which to present a challenge under the labels 'excess of jurisdiction' or 'breach of natural justice'.
2. The objective behind the 1996 Act requires the courts to respect and to enforce an adjudicator's decision unless it is plain that the manner in which it was reached was obviously unfair.

10.11 The need to have the 'right' answer has been subordinated to the need to have an answer quickly, and, as the cases discussed below show, it will only be in the clearest of circumstances that enforcement of a decision may be resisted on these grounds. As the Court of Appeal affirmed in *AMEC Capital Projects Ltd v Whitefriars City Estates Ltd* (2004),[14] allegations of breach of natural justice are 'easy enough' to make, however only those challenges based on a 'properly arguable objection' should be allowed to succeed:

> 22. It is easy enough to make challenges of breach of natural justice against an adjudicator. The purpose of the scheme of the 1996 Act is now well known. It is to provide a speedy mechanism for settling disputes in construction contracts on a provisional interim basis, and requiring the decisions of adjudicators to be enforced pending final determination of disputes by arbitration, litigation or agreement. The intention of Parliament to achieve this purpose will be undermined if allegations of breach of natural justice are not examined critically when they are raised by parties who are seeking to avoid complying with adjudicators' decisions.

The Adjudicator's Conduct Is Key

10.12 The manner in which the adjudicator has conducted himself in order to perform the balancing act between natural justice and tight time constraints is the prime consideration. In particular, a Court will ask whether he has complied with the adjudication process established by the parties' agreement or the Act and/or the Scheme. The case law confirms that adjudicators must apply natural justice principles 'without fear or favour' and that it is the adjudicator's conduct of the adjudication which is key, rather than the action of the parties:

> 31. … The introduction of systems of adjudication has undoubtedly brought many benefits to the construction industry in this country, but at a price. The price, which Parliament, and to a large extent the industry, has considered justified, is that the procedure adopted in the interests of speed is inevitably somewhat rough and ready and carries with it the risk of significant injustice. That risk can be minimised by adjudicators maintaining a firm grasp upon the principles of natural justice and applying them without fear or favour. The risk is increased if attempts are made to explore the boundaries of the proper scope and function of adjudication with a view to commercial advantage.[15]

Impact of the Human Rights Act 1998

10.13 Early discussions after the 1996 Act came into force considered whether the process of adjudication would stand up to the scrutiny of the Human Rights Act 1998 ('the Human Rights Act'). The Human Rights Act reflects the rights under the European Convention on Human Rights ('the Convention') including Article 6(1) of the Convention whereby, in the determination of its civil rights and obligations, a party is entitled to a fair and public hearing by an independent and impartial tribunal with public announcement of the judgment.

10.14 In 2001, the TCC confirmed in *Elanay Contracts Ltd v The Vestry* (2001)[16] that the Convention did not apply to adjudications, which were 'not in any sense a final determination', but could be reopened. This decision was criticized by the commentators in the Building Law Reports

---

[14] [2005] 1 All ER 723, [2004] EWCA Civ 1418 (CA) (Key Case).
[15] *RSL (South West) Ltd v Stansell Ltd* [2003] EWHC 1390 (TCC) (Key Case).
[16] [2001] BLR 33 (TCC).

but was quickly followed by the case of *Austin Hall Building Ltd v Buckland Securities Ltd* (2001).[17] Despite approaching 'the matter from a different direction', Judge Bowsher concluded that the process of adjudication could not be described as legal proceedings by a public authority (and thus caught by the Convention and the Human Rights Act) as adjudications were designed to avoid the need for legal proceedings.[18]

10.15 No further challenge to the inherent fairness of the adjudication procedure based on the Convention or the Human Rights Act has been made since 2001. Accordingly, whilst the adjudication procedure might be considered by some to be inherently unfair, an adjudicator is bound by the time limits that parliament prescribed[19] and there is no legal right to a public hearing or public pronouncement of an adjudicator's decision. On the contrary, adjudication is generally approached as though it is a confidential process.[20]

(3) Possible Grounds for a Natural Justice Challenge

Inadequate Time or Opportunity to Respond

10.16 Absent agreed extension of time, in an adjudication under the Act the adjudicator's decision must be reached within 28 days of the referral of the dispute. Even with agreed extensions of time adjudication is a rapid procedure by comparison with litigation or arbitration. Parties have sought to use the rapid nature of adjudication as the basis for resisting enforcement of a decision by asserting that the dispute referred was simply too complex for adjudication, that it had insufficient time to respond to the referral, or that it had insufficient time to deal with specific material, particularly where such material was provided late in the timetable.

Dispute Too Complex

10.17 In *London & Amsterdam Properties Ltd v Waterman Partnership Ltd* (2003) (Key Case)[21] there was a suggestion (*obiter*) that there may be some disputes, in particular final-account disputes, that may be too complex to be decided fairly in the time limits imposed by adjudication. A similar comment was made in *AWG Construction Services Ltd v Rockingham Motor Speedway Ltd* (2004).[22] That argument has subsequently been raised by parties seeking to resist enforcement—but without success to date.

10.18 The Act does not impose any restriction on the scale of dispute that may be referred to adjudication[23] and it provides a mechanism for extension of time either by the referring party or by agreement between the parties (such extensions therefore being reliant, at least, on the cooperation of the referring party). There is no power for the adjudicator to extend time unilaterally. However, the courts have made clear the approach that should be taken by adjudicators: if the adjudicator considers that the dispute is too complex to allow a fair procedure and decision within the time limits, then the parties should be invited to agree an extension of time, failing which the adjudicator should resign.[24] The courts have taken the view that

---

[17] [2001] BLR 272, [2001] EWHC 434 (TCC).

[18] *Ibid.* see paras. 19 and 20 in particular.

[19] See the decision of Judge LLoyd in *Glencot Development and Design Co Ltd v Ben Barratt & Sons (Contractors) Ltd* [2001] BLR 207 (TCC).

[20] See 6.03-6.04.

[21] [2004] BLR 179, [2003] EWHC 3059 (TCC).

[22] [2004] EWHC 888 (TCC).

[23] *CIB Properties Ltd v Birse Construction Ltd* [2005] 1 WLR 2252, [2004] EWHC 2365 (TCC).

[24] *William Verry (Glazing Systems) Ltd v Furlong Homes Ltd* [2005] EWHC 138 (TCC); *CIB Properties Ltd v Birse Construction Ltd* [2005] 1 WLR 2252, [2004] EWHC 2365 (TCC) (Key Case); *Balfour Beatty Construction v London Borough of Lambeth* [2002] BLR 288 (Key Case), per Judge LLoyd at para. 36, [2002] EWHC 597 (TCC).

the adjudicator is best placed to make the assessment as to whether there is sufficient time for submissions by the parties and consideration by the adjudicator. If it can be shown that the adjudicator properly considered whether a fair decision could be reached in the time allowed, then such challenges are unlikely to succeed. That is likely to be discernable from the adjudicator's reaction to requests for more time during the adjudication. Absent such contemporaneous requests, protestations after the event that there was insufficient time will lack credibility.

Key Case: Dispute Too Complex

AMEC Group Ltd v Thames Water Utilities Ltd [2010] EWHC 419 (TCC)

10.19 **Facts:** Thames Water had let over 300,000 packages of work to AMEC under a framework agreement. AMEC referred to adjudication its entitlement to payment under the framework agreement and was awarded just under £1 million by the adjudicator. Thames Water had served a response with supporting information running to 50 lever-arch files. Following an extension of time, AMEC in turn provided an extensive reply. The adjudicator's decision expressly stated as follows:

> Whilst I have not carried out a forensic analysis of all documents submitted by the Parties, which are numerous, I have spent considerable time and I believe sufficient time reviewing the documents in order to appreciate the nature of the issues presented to me and to understand the case of each party in relation to the principal issues. In respect to quantum, I am satisfied that I am able to do justice between the Parties and arrive at an overall figure which properly reflects the merits of the case as I find them ... I do not believe the dispute is so complex that I am unable to give a proper and considered decision within the time constraints of this adjudication.

10.20 **Held:** Coulson J held that size or complexity of the dispute would not in themselves be sufficient to ground a complaint based on breach of natural justice. The relevant question is whether the adjudicator considered he had sufficient time to deal with the matter fairly. In this case the adjudicator had regard to all the material provided and there was no breach of natural justice.

> 60. In my judgment, therefore, the law on this subject can be summarised as follows:
>
> (a) The mere fact that an adjudication is concerned with a large or complex dispute does not of itself make it unsuitable for adjudication: see *CIB v Birse*.
> (b) What matters is whether, notwithstanding the size or complexity of the dispute, the adjudicator had: (i) sufficiently appreciated the nature of any issue referred to him before giving a decision on that issue, including the submissions of each party; and (ii) was satisfied that he could do broad justice between the parties (see *CIB v Birse*).
> (c) If the adjudicator felt able to reach a decision within the time limit then a court, when considering whether or not that conclusion was outside the rules of natural justice, would consider the basis on which the adjudicator reached that conclusion (*HS Properties*). In practical terms, that consideration is likely to amount to no more than a scrutiny of the particular allegations as to why the Defendant claims that the adjudicator acted in breach of natural justice.
> (d) If the allegation is, as here, that the adjudicator failed to have sufficient regard to the material provided by one party, the court will consider that by reference to the nature of the material; the timing of the provision of that material; and the opportunities available to the parties, both before and during the adjudication, to address the subject matter of that material.

Table 10.1 Table of Cases: Dispute Too Complex

Title	Citation	Issue
CIB Properties Ltd v Birse Construction Ltd	[2005] 1 WLR 2252, [2004] EWHC 2365 (TCC), Judge Toulmin	**Dispute too complex** There is no limit to the complexity of dispute that can be referred to adjudication. The adjudicator's decision clearly stated that if he had felt unable to reach a fair decision in the time available, he would not have provided a decision. Enforced.
William Verry (Glazing Systems) Ltd v Furlong Homes Ltd	[2005] EWHC 138 (TCC), Judge Coulson	**Dispute too complex** If an adjudicator considers he does not have time to produce a fair decision he should inform the parties and should decline to produce an unfair decision. Enforced.
HS Works Ltd v Enterprise Managed Services Ltd	[2009] BLR 378, [2009] EWHC 729 (TCC), Akenhead J	**Dispute too complex** A most important factor is whether, and if so upon what basis, the adjudicator felt able to reach his decision in the time available. It is relevant to consider the opportunity the responding party has had to consider the material prior to the adjudication. Enforced.
AMEC Group Ltd v Thames Water Utilities Ltd	[2010] EWHC 419 (TCC), Coulson J	**Dispute too complex** Size and complexity alone are insufficient to establish a breach of natural justice. Enforced.

Ambush and Inadequate Opportunity to Respond

10.21 Statutory adjudication (and often contractual adjudication) is by its nature a rapid process which necessarily abbreviates the time available for the responding party to deal with, and the adjudicator to consider, the matters in issue. On enforcement proceedings there is therefore often a plea by the responding party that it was not granted a fair time to deal with the referral with the result that there was a breach of natural justice. Such challenges generally do not succeed.

10.22 The time limits imposed on adjudication and the inherent nature of the procedure undoubtedly place the referring party at an advantage. Since the timetable does not start until the notice of adjudication is served, the referring party has the opportunity to prepare a detailed case over an extended period of time, only giving notice of adjudication when it is satisfied that it has its case in order. By comparison, the response and decision will be required in a very limited period of time, subject to any extensions of time that may be granted. The referring party's exploitation of this process, often referred to as an 'ambush', has been acknowledged as a 'real public concern'[25] and 'a matter of regret'[26] and is considered by some to be an abuse of the adjudication process. However, because the Act permits a referring party to refer a dispute 'at any time', it is not uncommon for referring parties to do so 'in order to obtain the greatest possible advantage from the summary adjudication procedure'.[27] Other tactics that could

[25] *Austin Hall Building Ltd v Buckland Securities Ltd* [2001] BLR 272, [2001] EWHC 434 (TCC) at para. 12.

[26] *Dorchester Hotel Ltd v Vivid Interiors Ltd* [2009] BLR 135, [2009] EWHC 70 (TCC) at para. 23 (Key Case).

[27] *Dorchester Hotel Ltd v Vivid Interiors Ltd* [2009] BLR 135, [2009] EWHC 70 (TCC) at para. 23 (Key Case).

also be categorized as ambush include deliberately referring the matter during holiday periods or holding back material until the adjudication, or even until late in the timetable.

10.23 However, there is at least a degree of protection conferred on responding parties by the requirement that the dispute must have crystallized before it can be referred to adjudication. It is therefore likely that much of the material and arguments will not be new to the responding party. *CIB Properties Ltd v Birse Construction Ltd* (2005)[28] demonstrates that the courts will have little sympathy for a party who fails to use the time following the initial notification of a claim productively. The result is that ambush on its own will not necessarily be enough to found a breach of the rules of natural justice, as confirmed in *London & Amsterdam Properties Ltd v Waterman Partnership Ltd* (2004) (Key Case):[29]

> 179. ... mere ambush however unattractive does not necessarily amount to procedural unfairness. It depends upon the case. It may be an important part of the context in which the Adjudicator is required to operate and in which his conduct may fall to be judged in the light of the fundamental common law requirements statutorily underpinned in Section 108 (2)(e) of the Act.

10.24 The courts have enforced decisions where parties have alleged that they did not have enough time to respond;[30] or that submission of compendious documents from an earlier adjudication deprived them of a fair opportunity to respond;[31] or where a reasonable extension was given to the referring party to deal with a refined and enhanced extension of time claim submitted by the responding party.[32]

10.25 The only examples of successful challenges have been the result of either not receiving the referral documents at all,[33] a late change in the referring party's case, or late additional evidence that should have been provided with the original referral rather than by way of reply.[34]

10.26 As with the question of the complexity of the matter as a whole, the primary consideration is whether the adjudicator considers he has time to deal with the matter fairly.[35] If the adjudicator considers that there is insufficient time in the adjudication timetable for a party to respond adequately to new issues arising during the course of an adjudication, then the adjudicator should seek the parties' consent to extend the timetable, or should resign.[36] If the adjudicator does neither and then bases the decision on that evidence, then there may be grounds for the decision to be challenged. Akenhead J in *Bovis Lend Lease Ltd v Trustees of the London Clinic* (2009)[37] set out the reasons for this and highlighted the course that should be taken if an adjudicator believes the timetable is insufficient:

> 51. ... [T]he statutory framework of the [1996 Act] is one which enables a party to a construction contract to refer anything, which might be classified as a dispute to adjudication, in

---

28 [2005] 1 WLR 2252, [2004] EWHC 2365 (TCC).

29 [2004] BLR 179 [2003], EWHC 3059 (TCC).

30 See e.g. *Edenbooth Ltd v Cre8 Developments Ltd* [2008] CILL 2592, [2008] EWHC 570 (TCC).

31 *Emcor Drake & Scull Ltd v Costain Construction Ltd* (2004) *97 Con LR 142*, [2004] EWHC 2439 (TCC).

32 *William Verry (Glazing Systems) Ltd v Furlong Homes Ltd* [2005] EWHC 138 (TCC).

33 *Rohde (t/a M. Rhode Construction) v Markham-David* [2006] BLR 291, [2006] EWHC 814.

34 *London & Amsterdam Properties Ltd v Waterman Partnerships Ltd* [2004] BLR 179, [2003] EWHC 3059 (TCC) (Key Case); *AWG Construction Services Ltd v Rockingham Motor Speedway Ltd* [2004] EWHC 888; *McAlpine PPS Pipeline Systems Joint Venture v Transco Plc* [2004] BLR 352, [2004] EWHC 2030 (TCC) (Key Case).

35 *Dorchester Hotel Ltd v Vivid Interiors Ltd* [2009] BLR 135, [2009] EWHC 70 (TCC) (Key Case) at paras. 26–7; *CIB Properties Ltd v Birse Construction Ltd* [2005] 1 WLR 2252, [2004] EWHC 2365 (TCC).

36 *William Verry (Glazing Systems) Ltd v Furlong Homes Ltd* [2005] EWHC 138 (TCC); *Balfour Beatty Construction Ltd v London Borough of Lambeth* [2002] BLR 288 (TCC), per Judge LLoyd at para. 36.

37 123 Con LR 15, [2009] EWHC 64 (TCC).

the ordinary course of events for a decision to be provided by the adjudicator within 28 days of the reference. Therefore, the threshold to a reference to adjudication is simply and only that there is a crystallised dispute. Thus, if a dispute has arisen by 23rd December in a given year, the referring party may refer that dispute to adjudication on 24 December. That might give rise to an assertion that there has been an 'ambush' because the defending party may well have insufficient time, given the Christmas break common in the construction industry, to prepare its defence. It is not uncommon, similarly, for claiming parties to refer matters to adjudication during the summer holidays when it is known that key personnel of the defending party are away. Again, this might be said to be an 'ambush'. However, for better or for worse, Parliament does not expressly give an adjudicator the power to extend the 28 days by reason of that fact. However, there is a sensible school of thought which suggests that in those circumstances an adjudicator can in effect decline to accept the appointment on the grounds that justice cannot be done or the adjudicator can simply say to the claiming party words to the effect: 'Unless you agree to an extension of time I will not be able to produce my decision within 28 days.' Indeed, that is commonly what adjudicators will do and it is a very rare case when the claiming party does not accede to some extension of time accordingly.

10.27 A similar approach was identified in *London & Amsterdam Properties Ltd v Waterman Partnership Ltd* (2004) (Key Case)[38] where Judge Wilcox suggested that the appropriate course of action for an adjudicator to take, where there was late submission of substantial new material but the referring party refused to extend time, was to exclude the late material.

10.28 Whilst some breaches of natural justice may not be apparent until the decision is communicated, a party should be acutely aware at the time it is responding in the adjudication that it faces real difficulties responding in time. Any failure to raise such concerns contemporaneously is likely to be fatal to any subsequent claim that it had inadequate time. For instance, a complaint that an opportunity to make further submissions was denied would appear to be doomed to failure unless such an opportunity was requested at the time.[39]

10.29 Whether the adjudicator allowing late provision of material will amount to a breach of natural justice will be a question of fact and degree considering the particular circumstances of each case. In *Balfour Beatty Construction Northern Ltd v Modus Corovest (Blackpool) Ltd* (2008),[40] Judge Coulson identified that such a finding would depend on the information concerned, the lateness of that material, whether it could properly be described as an 'ambush', the surrounding facts and, most importantly, the adjudicator's obligation to comply with the timetable.[41]

10.30 In *AWG Construction Services Ltd v Rockingham Motor Speedway Ltd* (2004)[42] the referring party made out a completely new case in its reply and it was held that the responding party was not given adequate opportunity to deal with the new case. AWG had only five days to respond before the adjudicator issued his decision in which he found in favour of Rockingham on the grounds set out in Rockingham's late report. Whilst the case was primarily decided on the basis of lack of jurisdiction, it was determined, *obiter*, that the adjudicator's failure to afford AWG a proper opportunity to give a fully considered response to the additional material that had been served at such a late stage had clearly prejudiced AWG.

---

38 [2004] BLR 179, [2003], EWHC 3059 (TCC).

39 *Bovis Lend Lease Ltd v The Trustees of the London Clinic* (2009) 123 Con LR 15, [2009] EWHC 64 (TCC).

40 [2009] CILL 2660, [2008] EWHC 3029 (TCC).

41 *Ibid.* at para. 53.

42 *AWG Construction Services Ltd v Rockingham Motor Speedway Ltd* [2004] EWHC 888 (TCC).

Key Cases: Ambush and Inadequate Opportunity to Respond

London & Amsterdam Properties Ltd v Waterman Partnership Ltd [2004] BLR 179, [2003] EWHC 3059 (TCC)

10.31 **Facts:** Prior to referring the dispute to adjudication the referring party had refused, without any apparent reason, to provide further information and substantiation for its claim that late design information had led to critical delays to the steelwork package. Many months later, the dispute was referred to adjudication without the responding party ever having seen the build-up to the referring party's claim. After Waterman's response had been served, LAP served its reply containing a further witness statement exhibiting considerable further evidence in support of LAP's quantum claim. It was evidence that was available at the time of the referral notice but was not served at that time. It had not been made available to Waterman before service of a reply. Waterman requested an extension of time to deal with the evidence on the basis that its quantum expert was away and would not be back until after the decision was due to be made.

10.32 **Held:** The adjudicator had wrongly admitted evidence, and had not given the responding party the opportunity to comment on the voluminous information submitted (some of it for the first time) in the referral; additional quantum evidence was served for the first time in the reply; and the responding party's quantum expert had been unavailable to deal with the reply. This combination of circumstances, and in particular the adjudicator's acceptance of the late quantum evidence served with the reply, amounted to a 'substantial and relevant' breach of natural justice. Judge Wilcox commented: 'A referring party who is permitted to ambush a respondent by deploying fresh arguments and using documentation held on to until the eve of the reference may have an expensive and hollow victory in the event.' For the purposes of the Part 24 application, the defendant had demonstrated a substantial live and triable issue. In the circumstances, that decision was not permitted to stand as it was reached by breach of natural justice.

> 164. ... LAP chose not to reveal its case as to causation and quantum until the adjudication commenced and even then the evidence although approaching 1000 A4 pages was incomplete thus giving rise to the necessity to support its case by seeking to adduce the additional evidence introduced very late in the adjudication process. It could not therefore have been taken account of in Waterman's response. It was not made available until after Waterman's response when their quantum expert drew attention to the lack of substantiation of the payments to William Hare. There clearly was an evidential ambush. The decision to withhold the quantum evidence requested in July and August was clearly deliberate. The decision to serve the considerable body of detailed evidence at the time of the referral was deliberate. The omission to serve the necessary additional evidence may have been merely oversight or neglect.
>
> 165. Mr Akenhead submits '*that even if Waterman was ambushed that is of no relevance: the Adjudicator made his decision and both the HGCRA, the contract and case law make it clear that it must be complied with*'. Mr Bartlett contends that this was no 'mere' ambush. The reception of the additional evidence and the failure to give Waterman the opportunity to deal with it amounted to a breach of natural justice, because the Adjudicator as to vital issues based his decision on matters that Waterman could not properly deal with. He was not therefore impartial ...
>
> 180. The Adjudicator ... nowhere dealt with the question as to whether Waterman had sufficient time to answer and rebut Mr Baker's evidence, nor did he find that Waterman

did not need an extension; he merely pointed out ... that he did not have power to grant the extension that was sought. At paragraph 1.23 he referred to the further evidence as additional information. In my judgment that was new evidence supporting LAP's existing case on quantum which could and should have been adduced much earlier. ... [T]he Adjudicator expressly found that there was an absence of evidence that Mr Baker's settlement recommendations and advice were unreasonable. The additional material belatedly produced in the final stages of the adjudication was part of that advice. The Adjudicator found that the settlement was reasonable on the basis of the very evidence about which Waterman complained and in the absence of the expert quantum evidence which Waterman did not have a fair opportunity to adduce ...

182. Had the claimant supplied the quantum information when it was first requested in July both parties would have been able to consider their differences in a sensible commercial way reflecting the legal strength and weaknesses of their respective positions before adjudication commenced. The claimant chose not to. Where, as in this case the dispute is complex, involving the evaluation of the activities of a number of parties over a long period of time and issues of professional negligence and where the project is substantially complete the post mortem is best suited to arbitration or litigation.

Dorchester Hotel Ltd v Vivid Interiors Ltd [2009] BLR 135, [2009] EWHC 70 (TCC)

10.33 **Facts:** Vivid served a notice of intention to refer on 19 December 2008 with 37 lever-arch files supporting its final account claim, which was a different amount from its previous claim. It included five entirely new files which had not been seen by Dorchester before the adjudication. Dorchester objected to the pre-Christmas service and the date, albeit extended, for its response of 28 January 2009. It applied to the court in a Part 8 application, claiming that the timetable was too tight and that there was a very real risk of a breach of natural justice.

10.34 **Held:** Coulson J refused to grant the declaration sought by the claimant for four main reasons:

1. The adjudicator was of the view that he could fairly determine the adjudication within the agreed timetable—this was in many ways the most important factor. The adjudicator's duty to determine the adjudication fairly is a continuing one.
2. The timetable had been extended by agreement and the responding party not only had from 19 December 2008 until 28 January 2009 to submit its response (which could have included working over the Christmas period if it so wished), but was also timetabled to have the 'final say' by way of scheduled rejoinder.
3. It was not clear whether there might be future difficulties in relation to challenges that there was no crystallized dispute because of so much new material or the majority of the claim figures having changed. Such points would have to await analysis of the adjudicator's decision once it was issued.
4. Refusal to grant the declarations at that stage did not leave the claimant without a remedy. The claimant was entitled to rely on any breaches of natural justice which could be made out to resist enforcement.

10.35 Coulson J reserved costs until after the adjudicator had handed down his decision. If the parties complied with the decision then the costs would follow the event (against the claimant). However, if the claimant challenged the decision on natural justice grounds, the judge determining the challenge would also determine the costs of the Part 8 proceedings.

Table 10.2 Table of Cases: Ambush and Insufficient Opportunity to Respond (successful challenges are indicated by shaded entries)

Title	Citation	Issue
Austin Hall Building Ltd v Buckland Securities Ltd	[2001] BLR 272, [2001] EWHC 434 (TCC), Judge Bowsher	**Alleged ambush not accepted** No breach of natural justice where the responding party had the draft final account in its possession for nine months prior to it being served with the referral. In the circumstances the responding party could not be said to have been ambushed by the referral and had a fair opportunity to respond. Enforced.
JW Hughes Building Contractors Ltd v GB Metalwork Ltd	[2003] Adj LR 10/03, [2003] EWHC 2421, Judge Forbes	**Referral documents not received during adjudication** A failure by the respondent's solicitors to pass on the referral and evidence to the respondent did not disadvantage the respondent in circumstances where all but one of the documents in evidence (which was not of great significance) were common to the parties and the respondent had a sufficient understanding of the case it had to answer to allow it to serve a detailed response. The respondent had not taken up the adjudicator's invitation to raise the matter of the missing paperwork further. Enforced.
London & Amsterdam Properties Ltd v Waterman Partnerships Ltd	[2004] BLR 179, [2003] EWHC 3059 (TCC), Judge Wilcox	**Late quantum evidence** The mere fact that the referring party has 'ambushed' the responding party does not of itself give rise to a breach of natural justice. The referring party's substantial quantum evidence was only provided with its reply late in the adjudication timetable and there was insufficient time to allow the responding party a fair opportunity to deal with the material. That evidence could and should have been provided much earlier and a central issue in the decision was decided on the basis of that material. Natural justice required that the adjudicator should have either sought an extension of time or, if refused, excluded the late material from his consideration. Not enforced.
AWG Construction Services Ltd v Rockingham Motor Speedway Ltd	[2004] EWHC 888 (TCC), Judge Toulmin	**New case put forward in reply** The referring party set out an entirely new basis for its claim in its reply a few days before the end of the adjudication. The adjudicator's decision was based on the new case to which the responding party had not had a proper opportunity to respond. Not enforced.
McAlpine PPS Pipeline Systems Joint Venture v Transco Plc	[2004] BLR 352, [2004] EWHC 2030 (TCC), Judge Toulmin	**Further substantial evidence provided at request of adjudicator** The referring party served substantial evidence (a further 1,500 pages) throughout the adjudication, up until five days before the responding party's response was due. The decision was based to a significant extent on the late material, which had been provided

Title	Citation	Issue
		at the adjudicator's request. This was new evidence in support of a new case. *Obiter*: there was a realistic prospect that the responding party had been denied a fair opportunity to respond to that material, having only had five days from receipt of the last material when it would have had five weeks to respond had the material been served with the referral, which the adjudicator expressly stated it should have been. Not enforced (primarily on jurisdictional grounds).
CIB Properties Ltd v Birse Construction Ltd	[2005] 1 WLR 2252, [2004] EWHC 2365 (TCC), Judge Toulmin	**Responding party failed to act constructively in time available before adjudication** Following a substantial referral (comprising around 100 files including 24 experts' reports with £12m in issue and a further 55 files served during the adjudication) there was no ambush or breach of natural justice where any time constraints were of the responding party's own making. The responding party had failed to provide a constructive response to the original claim in the 15 weeks between first notification of the claim and referral to adjudication but rather used that time to manoeuvre tactically to set up an argument that no dispute had crystallized. The decision expressly stated that the adjudicator was satisfied that he had sufficient time to reach a fair decision. Enforced.
Emcor Drake & Scull Ltd v Costain Construction Ltd	(2004) 97 Con LR 142, [2004] EWHC 2439 (TCC), Judge Havery	**Voluminous submissions** An adjudication requires the responding party to make a quick response to extensive paperwork. This is well known and does not of itself give rise to a breach of natural justice. The fact that the same documentation (more than 4,000 pages) appeared in two successive adjudications was wholly insufficient as a ground for claiming abuse of process. Enforced.
William Verry (Glazing Systems) Ltd v Furlong Homes Ltd	[2005] EWHC 138 (TCC), Judge Coulson	**Party claimed insufficient time to deal with matters raised in response** If an adjudicator considers he does not have time to produce a fair decision he should inform the parties and should decline to produce an unfair decision. On the evidence there was adequate time to deal with the material. Where a party refers the entirety of its final account to adjudication it cannot later complain of unfairness as a result of the nature of the process. Enforced.
Rohde (t/a M. Rhode Construction) v Nicholas Markham-David	[2006] BLR 291, [2006] EWHC 814 (TCC), Jackson J	**Adjudication documents not received** If a referring party takes a deliberate decision in relation to the choice of address for service of documents in order to deny the responding party the opportunity to participate in an

(Continued)

Table 10.2 *Continued*

Title	Citation	Issue
		adjudication that may give rise to a breach of natural justice. Not enforced/default judgment set aside.
Edenbooth Ltd v Cre8 Developments Ltd	[2008] CILL 2592, [2008] EWHC 570 (TCC), Coulson J	**Not all documents received: claim of insufficient time to respond** The timetable was extended to allow for communication difficulties which were in part self-inflicted and the responding party had been able to serve a response. Time pressures are an inevitable consequence of the adjudication process. Enforced.
Balfour Beatty Construction Northern Ltd v Modus Corovest (Blackpool) Ltd	[2009] CILL 2660, [2008] EWHC 3029 (TCC), Coulson J	**Late material in reply** Findings of breach of natural justice as a result of late material depend on the nature of the information, the lateness of service, whether it can properly be regarded as an 'ambush', the surrounding facts and the adjudicator's obligation to comply with the timetable. The timetable did not allow for a rejoinder and the responding party did not request such an opportunity. There were no new points in the reply. Enforced.
Dorchester Hotel Ltd v Vivid Interiors Ltd	[2009] BLR 135, [2009] EWHC 70 (TCC), Coulson J	**Claim of inadequate time to deal with material: Part 8 declaration sought** The adjudicator's view as to whether he can fairly conduct the adjudication within the agreed timetable is the most important factor. Part 8 claim dismissed.
Bovis Lend Lease Ltd v The Trustees of the London Clinic	(2009) 123 Con LR 15, [2009] EWHC 64 (TCC), Akenhead J	**No contemporaneous complaint about insufficient time** It is the conduct of the tribunal and not the parties that is relevant in determining if there has been a breach of natural justice. The absence of contemporaneous complaint to the tribunal is a most important factor. There was no suggestion that the referring party knew about the maternity leave of one of the key personnel. On the facts there was no breach of natural justice. On the one occasion the responding party requested an extension of time for its response it was granted. Otherwise there was no complaint of insufficient time during the adjudication. Enforced.
NAP Anglia Ltd v Sun-Land Development Co. Ltd	[2012] BLR 110, [2011 EWHC 2846 (TCC), Edwards-Stuart J	**A procedure that allows the referring party the final submission is not unusual and is fair** It is neither unusual for the referring party to be allowed the final submission, nor for the period of time allowed for each round of submissions to be shorter than the previous round. Where that was the approach to timetable and submissions adopted by the adjudicator, there was no breach of natural justice. Enforced.

Failure to Consider Defence/Counterclaim

10.36 It is frequently the case in construction disputes that both parties consider they have valid claims; for example, a contractor's claim for payment is often met with a counterclaim for defective work. It is now well settled, as discussed in Chapter 9 on jurisdiction, that the responding party is entitled to raise any defence that is a proper defence in law to the claim being made. It matters not that the matters giving rise to the defence are not set out in the adjudication notice or were not raised or 'in dispute' prior to the reference to adjudication.[43]

10.37 Whilst there is no doubt that a defendant can raise whatever matters it wishes by way of defence for the adjudicator to consider, that general principle does not permit a defendant to rely on a cross-claim which should have been the subject of a withholding notice, but was not. In other words, a defendant cannot avoid the absence of a valid withholding notice if, by reference to the contract and on the facts of the particular dispute, the raising of the cross-claim in question required such a notice. To hold otherwise would be to obviate the need for withholding notices at all.

10.38 The decided cases generally fall into three categories, although some of the earlier decisions in this area have subsequently been doubted. First, where the adjudicator erroneously determines as a matter of jurisdiction alone (where such jurisdictional decisions do not bind the parties) that he cannot consider a defence and therefore ignores it, there will have been a breach of natural justice.[44] There is clear injustice where a party is denied having its defence considered on its merits when it should have been, and such errors will generally be considered to be a material breach, as long as the defence has a real, as opposed to fanciful, prospect of success. However, the court will not investigate the facts to determine whether the adjudicator would have reached a different decision had the evidence and argument in question been considered.[45]

10.39 Secondly, cases where the adjudicator determines on the merits that the defence does not represent a proper legal defence. This often arises as a result of the adjudicator's determination that the circumstances are such that a withholding notice is required before any sum can be deducted from an interim payment. Such determinations are not determinations as to jurisdiction but are a consideration of the merits of the defence. They will usually require an investigation of the true construction of the payment and withholding notice provisions and a factual investigation of any documents that purport to be notices. If the adjudicator errs in any part of those investigations, that is an error of fact or law made whilst determining a matter within the adjudicator's jurisdiction and is generally unimpeachable.[46]

10.40 An adjudicator therefore has to decide whether or not a withholding notice is required to permit a cross-claim to be raised as a defence, and if so, whether or not there has been a valid notice. If he concludes that no notice was required, or that a notice was required and that

---

[43] See 9.06–9.07.

[44] *Buxton Building Contractors v Governors of Durand Primary School* [2004] BLR 374; [2004] EWHC 733 (TCC); *Boardwell v k3D Property Partnership Ltd* [2006] Adj CS 04/21; *Quietfield Ltd v Vascroft Contractors Ltd* [2006] EWHC 174 (TCC), [2007] BLR 67; [2006] EWCA Civ 1737 (Key Case); *CJP Builders Ltd v William Verry Ltd* [2008] BLR 545, [2008] EWHC 2025 (TCC) (Key Case); *Quartzelec Ltd v Honeywell Control Systems Ltd* [2009] BLR 328, [2008] EWHC 3315 (TCC) (Key Case); *Thermal Energy Construction Ltd v AE & ELentjes UK Ltd* [2009] All ER (D) 271 (Jan), [2009] EWHC 408 (TCC) (Key Case); *Pilon Ltd v Breyer Group Plc* [2010] BLR 452, [2010] EWHC 837 (TCC) (Key Case); *RBG Ltd v SGL Carbon Fibers Ltd* [2010] BLR 631, [2010] CSOH 77; *PC Harrington Contractors Ltd v Tyroddy Construction Ltd* [2011] All ER (D) 162 (Apr), [2011] EWHC 813 (TCC).

[45] *CJP Builders Ltd v William Verry Ltd* [2008] BLR 545, [2008] EWHC 2025 (TCC) (Key Case).

[46] *Humes Building Contracts Ltd v Charlotte Homes (Surrey) Ltd*, unreported, 4 January 2007 (TCC); *HS Works Ltd v Enterprise Managed Services Ltd* [2009] BLR 378, [2009] EWHC 729 (TCC); *Urang Commercial Ltd v Century Investments Ltd* (2011) 138 Con LR 233, [2011] EWHC 1561 (TCC).

there was a valid notice, then he must take the cross-claim into account in arriving at his decision. If he concludes that a notice was required, and that either there was no notice or that the notice that has been served was invalid for any reason, then he is entitled to disregard the cross-claim, and the court cannot interfere with that decision: *Letchworth Roofing Co. v Sterling Building Co.* (2009).[47] If the adjudicator wrongly decides that a withholding notice was required and because there was no notice he rejects a defence of set-off, then he has made an error of law rather than a breach of natural justice.[48]

10.41 Thirdly, where the adjudicator's decision does not expressly refer to a defence having been considered. Such cases turn on their own particular facts, including the nature of the defence that it is said was ignored and whether it can be said from the terms of the decision that the defence was considered and dismissed. What is clear is that an adjudicator's decision is not required to expressly refer to each and every defence or alternative defence raised and to recite and address the arguments and why they were considered unpersuasive.[49] However, where the terms of the decision suggest that the adjudicator has entirely ignored a fundamental defence, that may well found a material breach of the rules of natural justice.[50] An entire defence was expressly ignored in *CJP Builders Ltd v William Verry Ltd* (2008) (Key Case),[51] the adjudicator considering that the applicable adjudication rules did not allow him any discretion to consider a response that was served late or grant an extension of time for service. As a result the adjudication was in effect not contested and the court found that the responding party's right to be heard had been infringed, albeit entirely innocently and as a result of an erroneous interpretation of the adjudication rules. Whilst such a procedural failure is not strictly an error as to jurisdiction, it is broadly analogous with the cases in the first category. It certainly could not be categorized as a case where the adjudicator had made an error of fact or law whilst considering the merits of the defence.

Key Cases: Failure to Consider Defence

William Verry (Glazing Systems) Ltd v Furlong Homes Ltd [2005] EWHC 138 (TCC)

10.42 **Facts:** Furlong Homes referred a dispute arising out of a final account to adjudication. The notice of adjudication was widely drafted and stated that, inter alia, the adjudicator would be requested to decide that the extension of time granted by Furlong to 2 February 2004 was correct. Verry's response sought an extension of time until 27 July 2004. Furlong Homes said the new extension of time had not previously been seen and argued that the adjudicator's jurisdiction was restricted to the crystallized dispute. The adjudicator decided that Verry was entitled to an extension of time to 27 July 2004. Furlong contended that the adjudicator did not have the jurisdiction to consider Verry's 'new claim' for an extension of time and/or Verry's new claim led to unfairness and a substantial risk of injustice.

[47] [2009] CILL 2717 (TCC), [2009] EWHC 1119 (TCC).

[48] *Urang Commercial Ltd v Century Investments Ltd* [2011] EWHC 1561 (TCC).

[49] *Carillion Construction Ltd v Devonport Royal Dockyard Ltd* [2005] BLR 310, [2005] EWHC 778 (TCC) at para. 102. Jackson J's decision was confirmed by the Court of Appeal [2006] BLR 15, [2005] EWCA Civ 1358 (Key Case). A similar approach was taken in *South West Contractors Ltd v Birakos Enterprises Ltd* [2006] All ER (D) 63 (Nov), [2006] EWHC 2794 (TCC). For a different approach, see *Thermal Energy Construction Ltd v AE & E Lentjes UK Ltd* [2009] All ER (D) 271 (Jan), [2009] EWHC 408 (TCC) (Key Case), discussed below at 10.68.

[50] As was the case in *Thermal Energy Construction Ltd v A. E. & E. Lentjes UK Ltd* [2009] EWHC 408 (TCC) (Key Case). Further consideration is given to such cases, where it was contended that failures to address particular arguments represented breaches of natural justice, in the next section.

[51] [2008] BLR 545, [2008] EWHC 2025 (TCC).

10.43 **Held:** It would be excessively legalistic to classify Section D of Verry's response as a new claim. Verry was not restricted to the points in its letter of 2 July 2004; there was nothing in the documents, and in particular the notice of adjudication and the referral notice, that could lead to any such restriction. Even if it could be classified as a new claim for an extension of time, it clearly formed part of the dispute which was referred by Furlong to adjudication:

> 47. Verry were responding to this claim and were not themselves the referring party. They did not start the adjudication. They had to defend themselves as best they could against the allegation that their only entitlement to an extension of time was down to the 2nd February 2004 and that liquidated damages should be deducted for the period of delay thereafter. In my judgment, they were not to be taken as having agreed that they could only defend themselves in the adjudication with half a shield, able to rely on some matters of fact but being artificially prevented from referring to others. It seems to me that that would be an absurd result. In my judgment, Verry were entitled to take whatever points they liked to defend themselves against the assertion that their extension entitlement was limited in the way advanced by Furlong and the adjudicator was obliged to consider all the points which they raised.

Quietfield Ltd v Vascroft Contractors Ltd [2007] BLR 67, [2006] EWHC 174 (TCC); [2006] EWCA Civ 1737 (CA)

10.44 **Facts:** There had been a first adjudication (under the Scheme) in which the adjudicator found that Vascroft had failed to demonstrate its entitlement to an extension of time under the JCT 1998 Standard Form. There was another adjudication started by Quietfield who claimed liquidated and ascertained damages ('LADs') of £588,000, and relied upon the adjudicator's first decision that Vascroft was not entitled to any extension of time. Vascroft's response included a detailed and lengthy total extension of time claim which set out numerous causes of delay, tracked the dominant critical path, compared the as-planned and as-built programmes, and traced the interrelationships between different activities on site and various causes of delay. This was new information. The adjudicator declined to consider Vascroft's submissions on the basis that the extension of time had been determined in the first adjudication, and found that Quietfield was entitled to levy the full LADs claimed. Vascroft refused to pay the sums awarded on the ground that the adjudicator's decision was unlawful, and Quietfield commenced an application for summary judgment for enforcement of the decision. The question for determination by the court was:

> 29. … was the adjudicator correct in treating his own decision in the first adjudication as conclusive in relation to extension of time? If the answer to this question is 'Yes', then the adjudicator's decision … must be enforced. If the answer to this question is 'No', then it follows that the adjudicator has expressly refused to consider both the written submissions and the evidence which constitute Vascroft's only substantive defence in the adjudication. In that event there has been a breach of the rules of natural justice and the adjudicator's decision cannot be enforced.

10.45 **Held:** At first instance, Jackson J found that the answer to the above question was 'no' and there had been a breach of natural justice in not considering the new information, which supported a dispute which was not 'substantially the same' as that information put forward in the first adjudication. Jackson J found (at paragraphs 49–51) that Vascroft was entitled to advance any available defence, irrespective of whether the defence had been notified when the relevant dispute arose. Jackson J also held that Vascroft's alleged entitlement to an extension of time in the final adjudication was 'substantially different from the claims for extension of time which were advanced, considered and rejected in the first adjudication.' Therefore the adjudicator ought to have considered Vascroft's substantive defence. As the adjudicator failed to do so, the decision could not be enforced because he failed to abide by the rules of natural justice. Had it been enforced, Quietfield may have received a substantial sum of money to which it was not entitled, even on an interim basis. It should be noted that the claimant conceded that if the adjudicator was wrong on the threshold point, then the adjudicator's decision could not stand by reason of breach of natural justice. The court acted on the concession, without considering if the adjudicator's error was in any event one made within jurisdiction and not subject to challenge.

10.46 The Court of Appeal (May, Dyson, and Smith LJJ) found that Jackson J had come to the correct conclusion for the correct reasons. The responding party was not restricted to defences of which it had previously given notice and which had thereby generated the 'dispute' referable to adjudication.

CJP Builders Ltd v William Verry Ltd [2008] BLR 545, [2008] EWHC 2025 (TCC)

10.47 **Facts:** William Verry engaged CJP to perform brickwork, blockwork, and stonework under the DOM/2 1981 edition incorporating Amendments 1–8. Clause 38A required a response to be served within seven days of the referral. The adjudicator refused William Verry's request for an extension for its response on the basis that he had no power to go behind clause 38A and that an extension could only be agreed between the parties. An extension was in fact agreed between CJP and William Verry to midday on 14 May 2008. The response was not served until 5.30 pm on 14 May 2008. CJP submitted that the adjudicator could not consider the late response. The adjudicator did not consider the late response, advising that he had no discretion under clause 38A to extend time for service. William Verry argued that, as one of the subcontract documents referred to the TeCSA adjudication rules, these should be the applicable rules, allowing the adjudicator discretion for the procedure and timescale.

10.48 CJP sought to enforce the adjudicator's decision and William Verry challenged the enforcement on the basis that the decision had been reached in a breach of natural justice when the adjudicator refused to consider its response. CJP's counter-argument was that even if there was a breach of natural justice (which was denied), it was immaterial because a second adjudication had already determined that William Verry's defence in the first adjudication would have been unsuccessful.

10.49 **Held:** Akenhead J refused to enforce the adjudicator's decision because it was reached through a breach of natural justice. Akenhead J found:

1. The adjudicator had discretion to extend time for service of the response. There was nothing in clause 38A which, in any express prescriptive language, barred the adjudicator from doing so. In fact, clause 38A.2.5.5 stated that the adjudicator shall 'set his own procedure' and gave him an 'absolute discretion' in taking the initiative in ascertaining laws and facts as deemed necessary as long as it was within the 28 day deadline. Very clear words would be required to ensure that the right to be heard was to be denied.[52]
2. The adjudicator decided for wholly honest but ultimately wrong reasons to exclude from consideration the substantial response in terms of argument and evidence. Once

[52] Para. 79.

he did so, it was inevitable that he would come to the conclusion that he did. It was as if the adjudication was an uncontested claim.[53] Even though it was honest and open, the adjudicator's express and conscious decision not to consider William Verry's response was a breach of natural justice.[54]

3. The failure to regard the whole of William Verry's response both as to argument and evidence was and must have been material: 'There comes a point when a breach of the rules of natural justice is so pervasive that the only proper conclusion to come to is that the breach was material.' The court need not determine whether the adjudicator would have reached a different decision if he had considered the response. The test is whether there was a real (rather than fanciful) possibility that the adjudicator could have reached a different decision.[55]

Quartzelec Ltd v Honeywell Control Systems Ltd [2009] BLR 328, [2008] EWHC 3315 (TCC)

10.50 **Facts:** A scope change dispute was referred to adjudication and the responding party (Honeywell) raised an 'omission' defence (based on deduction for cost saving of a separate variation in the interim certificates) for the first time. The adjudicator accepted the referring party's position that he should refuse to consider the omission defence and awarded £135,000 to the referring party for the scope change. Honeywell argued that the decision was unenforceable for the following reasons:

> 25. … (a) the adjudicator was wrong to decide that he did not have jurisdiction to consider the omissions defence; (b) the adjudicator was obliged to consider the omissions defence in order that he could fulfil his obligation under paragraph 20 of the Scheme to decide the matters in dispute; (c) his failure to consider the omissions defence was a significant jurisdictional error and was also a serious breach of natural justice

10.51 **Held:** Judge Davies refused to enforce the adjudicator's decision. The judge followed the decision in *Cantillon v Urvasco* (Key Case) that it is open to a respondent to an adjudication to raise any ground which would amount in law or in fact to a defence of the claim. The adjudicator made a significant jurisdictional error and did not act in accordance with the requirements of natural justice in refusing to consider the omissions defence. It was a defence which was open to Honeywell to advance as a defence to Quartzelec's money claim, and it should have been considered by the adjudicator on its merits.

> 31. … if the adjudicator had considered the defence and decided, even if wrongly, that it could not succeed in the absence of a withholding notice, that would be a decision within his jurisdiction and would not be one which this court could review on an enforcement hearing. This is consistent with the judgment of Lord MacFadyen in *SL Timber Systems Limited v Carillion Construction Limited* [2001] BLR 516 … at paragraph 23. However the corollary of that, in my judgment, is that since the adjudicator has jurisdiction to consider such defences, he ought to do so, and if he does not do so then he does not properly perform the task which he has been appointed to do. In those circumstances, he also does not in my judgment act in accordance with natural justice, because he has not heard the respondent on all of the defences which he seeks and is entitled to put forward.

[53] Para. 82.
[54] Para. 84.
[55] Para. 84.

Pilon Ltd v Breyer Group plc [2010] BLR 452, [2010] EWHC 837 (TCC)

10.52 **Facts:** A dispute relating to payment in respect of interim certificates 26–62 was referred to adjudication. In defence the responding party claimed there had been an overpayment in relation to interim certificates 1–25 and sought to set off the amount of the overpayment in defence of the claim. The adjudicator considered that he did not have jurisdiction to consider the set-off as the dispute referred related only to certificates 26–62.

10.53 **Held:** Coulson J determined that the adjudicator's decision as to jurisdiction was not binding on the parties and that it had been made in error. The dispute referred related to payment and not just valuation, with the result that the adjudicator was obliged to consider the overpayment in respect of the earlier certificates as a defence. His failure to consider the defence was a material breach of the rules of natural justice:

> 17. An adjudicator can make an inadvertent mistake when answering the question put to him, and that mistake will not ordinarily affect the enforcement of his decision: see *Bouygues (UK) Ltd v Dahl-Jensen (UK) Ltd* (1999) 70 ConLR 41, [2000] BLR 49. If, on the other hand, he considers and purports to decide an issue which is outside his jurisdiction, then his decision will not be enforced: see the discussion in *Sindall Ltd v Solland* [2001] 3 TCLR 712. But there is a third category, which is where the adjudicator takes an erroneously restrictive view of his own jurisdiction, with the result that he decides not to consider an important element of the dispute that has been referred to him. This failure is usually categorised as a breach of natural justice
>
> ...
>
> 22. As a matter of principle, therefore, it seems to me that the law on this topic can be summarised as follows:
>
> 22.1. The adjudicator must attempt to answer the question referred to him. The question may consist of a number of separate sub-issues. If the adjudicator has endeavoured generally to address those issues in order to answer the question then, whether right or wrong, his decision is enforceable: see *Carillion v Devonport.*
>
> 22.2. If the adjudicator fails to address the question referred to him because he has taken an erroneously restrictive view of his jurisdiction (and has, for example, failed even to consider the defence to the claim or some fundamental element of it), then that may make his decision unenforceable, either on grounds of jurisdiction or natural justice: see *Ballast, Broadwell,* and *Thermal Energy.*
>
> 22.3. However, for that result to obtain, the adjudicator's failure must be deliberate. If there has simply been an inadvertent failure to consider one of a number of issues embraced by the single dispute that the adjudicator has to decide, then such a failure will not ordinarily render the decision unenforceable: see *Bouygues* and *Amec v TWUL.*
>
> 22.4. It goes without saying that any such failure must also be material: see *Cantillon v Urvasco* and *CJP Builders Ltd v William Verry Ltd* [2008] EWHC 2025 (TCC), [2008] BLR 545. In other words, the error must be shown to have had a potentially significant effect on the overall result of the adjudication: see *Keir Regional Ltd v City and General (Holborn) Ltd* [2006] EWHC 848 (TCC), [2006] BLR 315.
>
> 22.5. A factor which may be relevant to the court's consideration of this topic in any given case is whether or not the claiming party has brought about the adjudicator's error by a misguided attempt to seek a tactical advantage. That was plainly a factor which, in my view rightly, Judge Davies took into account in *Quartzelec* when finding against the claiming party.

...

25. It is not uncommon for adjudicators to decide the scope of their jurisdiction solely by reference to the words used in the notice of adjudication, without having regard to the necessary implications of those words: that was, for example, what went wrong in *Broadwell*. Adjudicators should be aware that the notice of adjudication will ordinarily be confined to the claim being advanced; it will rarely refer to the points that might be raised by way of a defence to that claim. But, subject to questions of withholding notices and the like, a responding party is entitled to defend himself against a claim for money due by reference to any legitimate available defence (including set-off), and thus such defences will ordinarily be encompassed within the notice of adjudication.

26. As a result, an adjudicator should think very carefully before ruling out a defence merely because there was no mention of it in the claiming party's notice of adjudication. That is only common sense: it would be absurd if the claiming party could, through some devious bit of drafting, put beyond the scope of the adjudication the defending party's otherwise legitimate defence to the claim.

27. I understand that it may be tempting for a claiming party in an adjudication to seek to limit the adjudicator's jurisdiction in a way in which that party believes to be to its advantage. I am in no doubt that is what happened here: Pilon did not wish the adjudicator to have any regard to batches 1–25, and therefore deliberately limited the scope of the adjudication notice to batches 26–62. It was their case that the over-payment claim was outside the adjudicator's jurisdiction, and that is what they (successfully) urged on the adjudicator. Thus, this is a case where Pilon sought a tactical advantage by putting forward an erroneous statement of the adjudicator's jurisdiction and, as the decision in *Quartzelec* shows, that can be a dangerous tactic to adopt.

Table 10.3 Table of Cases: Failure to Consider Defence (successful challenges are indicated by shaded entries)

Title	Citation	Issue
KNS Industrial Services (Birmingham) Ltd v Sindall Ltd	[2000] All ER (D) 1153 (TCC), Judge LLoyd	**The responding party's defence is part of the dispute referred** A party to a dispute who identifies the dispute in simple general terms has to accept that any ground that exists which might justify the action complained of is comprehended within the dispute for which adjudication is sought.
Farebrother Building Services Ltd v Frogmore Investments Ltd	20 Apr. 2001 (TCC), (2001) CILL 1762, Judge Gilliland	**Failure to consider defence of set-off** An adjudication was conducted pursuant to TeCSA Adjudication Rules v1.3 which gave power to the adjudicator to rule on his own substantive jurisdiction. Frogmore argued the adjudicator failed to consider its claim to set off £300,000 arising from Farebrother's alleged failure to proceed regularly and diligently. The adjudicator's decision stated that he did not have jurisdiction to consider Frogmore's counterclaim. Any error was an error within jurisdiction and could not be challenged. Enforced.

(Continued)

Table 10.3 *Continued*

Title	Citation	Issue
Mecright Ltd v T. A. Morris Developments	[2001] Adj LR 06/22 (TCC), Judge Seymour	**Cross-claim on termination** The adjudicator is allowed under the Scheme to take into account 'any other matters which the parties to the dispute agree should be within the scope of the adjudication or which are matters under the contract which he considers are necessarily connected with the dispute'. This meant where the referring party sought a decision that it had validly terminated the contract and was entitled to damages for that cancellation, the adjudicator was entitled to find for the responding party that the contract had been repudiated. However, the adjudicator did not have jurisdiction to decide the responding party was entitled to be paid the value of work performed and damages for repudiation. Those were not matters necessarily connected with the dispute referred.
Ballast plc v The Burrell Company (Construction Management) Ltd	[2001] BLR 529 (OHSCOS), Lord Reid	**Failure to reach a decision on the matter referred rendered the decision a nullity** The adjudicator erroneously misconstrued his powers such that he considered he could not determine the dispute referred to him. Decision a nullity.
Buxton Building Contractors Ltd v Governors of Durand Primary School	[2004] BLR 347; [2004] EWHC 733 (TCC), Judge Thornton	**Failure to consider cross-claim** Decision showed that the adjudicator did not consider the merits of the cross-claim, erroneously assuming that the cross-claim must have been taken into account in the valuation of the certificate that was in dispute. Not enforced. NB: this case has not been followed by several cases.
William Verry Ltd v North West London Communal Mikvah	[2004] BLR 308 [2004] EWHC 1300 (TCC), Judge Thornton	**Failure to consider defect defence** There had been an adjudication in respect of an earlier interim valuation of the works in question and there had been no further work carried out. Defects had since been identified. The adjudicator erred in law as to whether defects could result in a reduction of a later interim valuation. However, that was an error within his jurisdiction with the result that there was no breach of natural justice. Additionally, the adjudicator erroneously found that he had not been asked to decide the defects issue. Enforced, but no order drawn up for six weeks to give the responding party the opportunity to adjudicate the defects issue.
William Verry (Glazing Systems) Ltd v Furlong Homes Ltd	[2005] EWHC 138 (TCC), Judge Coulson	**Defence not limited to half shield** The dispute referred asked the adjudicator to decide that the existing extension of time was adequate. The defendant was entitled in response to ask for a greater extension of time, even if that formulation had not been aired previously in that way. A declaration was granted that the adjudicator's decision was binding until determined by the Court.

Title	Citation	Issue
Boardwell v k3D Property Partnership Ltd	[2006] Adj CS 04/21 (TCC Salford), Judge Raynor	**Defence not considered as a result of jurisdictional error** Adjudicator's erroneous conclusion that he did not have jurisdiction to consider counterclaims because they were not contained in the notice of adjudication led to a material breach of natural justice. Not enforced.
Quietfield Ltd v Vascroft Contractors Ltd	(2006) 109 Con LR 29, [2006] EWHC 174 (TCC); [2007] BLR 67, [2006] EWCA Civ 1737 (CA), May LJ, Dyson LJ, Smith LJ DBE	**Failure to consider extension of time defence: prior adjudication** The adjudicator refused to consider an EOT defence to an LAD claim on the basis that he was bound by the decision in an earlier adjudication that there was no entitlement to an EOT. The adjudicator erred (as a question of fact) as to the scope of what had been decided in the first adjudication and should have considered the defence. There was a concession that if the adjudicator was wrong as to the effect of the first adjudication then there was a breach of natural justice. Not enforced. Appeal dismissed.
Humes Building Contracts Ltd v Charlotte Homes (Surrey) Ltd	unreported, 4 January 2007 (TCC), Judge Gilliland	**Failure to consider defence following erroneous finding that withholding notice was required** The adjudicator erroneously determined he could not consider a defence based on defective works in the absence of a withholding notice. That alone was insufficient to found a breach of natural justice as it was an error of law within his jurisdiction. This decision was not enforced on other grounds (neither party had suggested that a withholding notice was required and the adjudicator did not invite submissions on the point).
Steve Domsalla v Kenneth Dyason	[2007] BLR 348, [2007] EWHC 1174 (TCC), Judge Thornton	**Failure to consider defence following erroneous finding that withholding notice was required** This was a contract with a residential occupier with an express adjudication provision. The adjudicator erred in finding that the withholding notice requirement in a consumer contract was fair under the Unfair Terms in Consumer Contracts Regulations. That led to him not considering a defective-works defence. The policy considerations underpinning enforcement of statutory adjudications irrespective of errors of fact or law do not apply to contractual adjudication.[56] Not enforced.

(Continued)

[56] This case was not followed in *AMEC Group Ltd v Thames Water Utilities* [2010] EWHC 419 (TCC) at para. 24.

Table 10.3 *Continued*

Title	Citation	Issue
CJP Builders Ltd v William Verry Ltd	[2008] BLR 545, [2008] EWHC 2025 (TCC), Akenhead J	**Failure to consider response served 'out of time'** The adjudicator erred as to the effect of the applicable adjudication rules and considered he did not have discretion to consider a response that was served late, with the result that the defence was not considered. DOM/2 Clause 38A.5.1.2 provides that the responding party *may* send the adjudicator a response within seven days of the referral. Akenhead J said this was not the latest date a defence could be considered. This was particularly so where the contract provided that the adjudicator *shall* set his own procedure and at his absolute discretion may take the initiative in ascertaining the facts and the law as he considers necessary. Not enforced.
Quartzelec Ltd v Honeywell Control Systems Ltd	[2009] BLR 238, [2008] EWHC 3315 (TCC), Judge Davies	**Failure to consider defence to payment claim** The adjudicator erroneously considered he did not have jurisdiction to consider a defence that had not been 'in play' prior to the adjudication. The defence was not dismissed on the basis that there was no withholding notice—that would have been an error within jurisdiction and would have been unimpeachable. A responding party is entitled to raise any defence open to defend itself against a claim for payment, regardless of whether or not it was raised as a discrete ground of defence in the run-up to the adjudication. Not enforced.
Letchworth Roofing Company v Sterling Building Company	[2009] EWHC 1119 (TCC), Coulson J	**Cross-claim not permitted when it should have been subject of witholding notice** Whilst the general principle is that a responding party can raise defences in an adjudication, that does not permit the responding party to run counterclaims that should have been the subject of a withholding notice (DOM/1 Clause 38A considered). Enforced.
Thermal Energy Construction Ltd v AE & E Lentjes UK Ltd	[2009] All ER (D) 271 (Jan), [2009] EWHC 408 (TCC), Judge Davies	**Failure to consider defence at all** The decision did not make any reference to the set-off defence that had been raised. On the evidence, the adjudicator had failed to consider the defence at all. Injustice arose as the responding party was deprived from having its defence considered. Not enforced.
HS Works Ltd v Enterprise Managed Services Ltd	[2009] EWHC 729 (TCC), Akenhead J	**Consideration of the relevant defence of withholding notice** The dispute contained in the notice involved a primary assertion that there was no effective withholding notice as required by the contract. In accepting that case, it was unnecessary for the adjudicator to consider the responding

Title	Citation	Issue
		party's case on the merits of the cross-claims. The challenge that the adjudicator had failed to answer the dispute referred was not successful. Further, the adjudicator could not be criticized for not addressing every difference in the accounting position between the parties—a spot check was adequate. Enforced.
Pilon Ltd v Breyer Group plc	[2010] BLR 452, [2010] EWHC 837 (TCC), Coulson J	**Defence not considered as a result of jurisdictional error** The adjudicator erroneously formed a restrictive view of his jurisdiction which led him to ignore entirely the overpayment defence that had been raised. Not enforced.
RBG Ltd v SGL Carbon Fibers Ltd	[2010] BLR 631, [2010] CSOH 77, Lord Menzies	**Defence not considered as a result of jurisdictional error** Decided on the Scottish principle of failure to exhaust jurisdiction. Not enforced.
PC Harrington Contractors Ltd v Tyroddy Construction Ltd	[2011] All ER (D) 162 (Apr), [2011] EWHC 813 (TCC), Akenhead J	**Erroneous conclusion on jurisdiction led to adjudicator not considering defence** A jurisdictional ruling by which an adjudicator denied himself the opportunity to consider the merits of a defence resulted in a breach of natural justice. Further, natural justice required the adjudicator to invite submissions as to his jurisdiction where he declined to consider a defence on jurisdictional grounds. Not enforced.
Urang Commercial Ltd v Century Investments Ltd	(2011) 138 Con LR 23, [2011] EWHC 1561 (TCC), Edwards-Stuart J	**Adjudicator considered a withholding notice was required if defence was to succeed** The adjudicator had jurisdiction to decide if a withholding notice was required. Even if his decision on that point was wrong, it was unimpeachable as it would simply be an error of fact or law that did not go to jurisdiction. Enforced.

Failure to Consider Evidence/Arguments

10.54 The preceding section discussed cases where a failure to consider a defence in its entirety, as long as it is not the result of having considered the substance of the defence on its merits, will invariably lead to the decision not being enforced. This can be contrasted with cases where some lesser failure is asserted by the party resisting enforcement. Such matters include alleged failures to consider particular aspects of the evidence or particular arguments raised. Such challenges have not typically met with success.

10.55 The Court of Appeal's decision in *Carillion Construction Ltd v Devonport Royal Dockyard Ltd* (2005) (Key Case)[57] endorsed the following propositions laid down by Jackson J at first instance:

[57] [2006] BLR 15, [2005] EWCA Civ 1358 (CA). The Court of Appeal's decision confirmed that to the extent that an earlier decision, *Buxton Building Contractors v Governors of Durand Primary School* [2004] BLR 374; [2004] EWHC 733 (TCC), was inconsistent with the proposition, it was wrongly decided. This was reinforced in *Kier Regional Ltd (t/a Wallis) v City & General (Holborn) Ltd* [2006] BLR 215, [2006] EWHC 848 (TCC) (Key Case). In *Buxton*, Judge Thornton had decided that the adjudicator's failure to consider a cross-claim

> 53. ... If an adjudicator declines to consider evidence which, on his analysis of the facts or the law, is irrelevant, that is neither (a) a breach of the rules of natural justice nor (b) a failure to consider relevant material which undermines his decision on Wednesbury grounds or for breach of paragraph 17 of the Scheme. If the adjudicator's analysis of the facts or the law was erroneous, it may follow that he ought to have considered the evidence in question. The possibility of such error is inherent in the adjudication system. It is not a ground for refusing to enforce the adjudicator's decision.

10.56 In *Kier Regional Ltd (t/a Wallis) v City & General (Holborn) Ltd* (2005) (Key Case),[58] the adjudicator's decision not to consider two expert reports prepared after an adjudication commenced was viewed as a matter of law within the adjudicator's jurisdiction to determine, and 'certainly not one of "the plainest cases" of breach of natural justice' envisaged by the Court of Appeal in *Carillion Construction Ltd v Devonport Royal Dockyard Ltd* (2005) (Key Case).[59] As such, the adjudicator's decision was enforced.[60] *Kier* seems to have turned on the reasons which the adjudicator gave for rejecting the reports (see Key Case extract below at 10.59–10.60). In particular, the adjudicator had thought he was bound to consider whether Valuation 32 was right based only on the information available to the contract administrator at the time. The two reports in question had not been in existence at the time. The judgment does not reveal whether it was argued that the adjudicator asked himself the right or the wrong question.

10.57 Whilst there appears to be conflicting *obiter* comment in a recent TCC decision,[61] it is suggested that, in light of the Court of Appeal decision in *AMEC Capital Projects Ltd v Whitefriars City Estates Ltd* (2004) (Key Case),[62] an adjudicator is not required to consider jurisdictional submissions where any decision on jurisdiction will not bind the parties. However, if a party is positively asserting that an adjudicator has jurisdiction, for example to deal with its defence, then a failure to consider that submission may lead to a breach of natural justice if the adjudicator wrongly declines to address the defence.

Key Cases: Failure to Consider Evidence/Arguments

Carillion Construction Ltd v Devonport Royal Dockyard [2006] BLR 15 [2005] EWCA Civ 1358 (CA)

10.58 See 10.55 for discussion and case extract.

Kier Regional Ltd v City & General (Holborn) Ltd [2006] BLR 315, [2006] EWHC 848 (TCC)

10.59 **Facts:** The relevant adjudication scheme was clause 41A of the JCT Standard Form 1998. Kier's claim was for loss and expense. City & General served an expert report which had not been seen before and Kier argued that it should be ignored. The adjudicator decided that

raised by the responding party was fatal to the enforceability of the decision. The cross-claim had concerned the school's refusal to pay based on defects in the works performed by Buxton, and the adjudicator had refused to consider the correspondence provided by the school to support its assertions. Buxton had convinced the adjudicator that the absence of a valid withholding notice meant that there was no defence available to the school.

[58] [2006] EWHC 848 (TCC).

[59] [2006] BLR 15, [2005] EWCA Civ 1358 (CA).

[60] Not to enforce the decision would require 'following that part of the reasoning in *Buxton* which has been disapproved by the Court of Appeal'. *Ibid.*, para. 42.

[61] *PC Harrington Contractors Ltd v Tyroddy Construction Ltd* [2011] EWHC 813 (TCC).

[62] [2005] 1 All ER 723, [2005] EWCA Civ 1418 (CA) (see Key Case at 11.32).

the reports 'were not before the Contract Administrator when he produced his Valuation no. 32 and they are not therefore relevant to the way in which he prepared his valuation. I am required to decide whether the CA was right in all of the circumstances known to him at the time to reject, in whole or in part, Wallis' claim for £1,330,012 based on a pro rata calculation using the contract preliminary rates.' He decided that he should not consider the expert reports. City & General challenged the validity of the decision on the ground that the reports should have been taken into account and the process leading to the decision was manifestly unfair.

10.60 **Held:** The error made by the adjudicator was not one that could invalidate the decision and was at worst an error of law:

> 42. I must say that, despite Mr Williamson's persuasive submissions, I see considerable force in Mr Blackburn's contention that the Adjudicator ought to have taken the two expert reports into account. However, it is not necessary finally to decide this point for one simple reason: this is that the error allegedly made by the Adjudicator is not one which could invalidate his decision. It can be seen from the decision as a whole that the Adjudicator considered each of the arguments advanced by CG in its written response. At worst, the Adjudicator made an error of law which caused him to disregard two pieces of relevant evidence, namely the expert reports of Driver Consult and Precept. In the light of the Court of Appeal's decision in *Carillion*, that error would not render the Adjudicator's decision invalid: see in particular paragraph 84 of *Carillion*. If I refuse to enforce the Adjudicator's decision on this ground, I should be following that part of the reasoning in *Buxton* which has been disapproved by the Court of Appeal.

Jacques (t/a C&E Jacques Partnership) v Ensign Contractors Ltd [2009] EWHC 3383 (TCC)

10.61 **Facts:** The adjudicator decided not to consider the decision of an earlier adjudicator which had been deemed null and void. The contractor argued that in doing so, the adjudicator had failed to understand its defences that included the employer's wrongful allegations of omissions and defects which did not exist, the implementation of various variations and the quantum of actual omissions and defects being significantly exaggerated.

10.62 **Held:** Akenhead J followed the approach in *Carillion Construction Ltd v Devonport Royal Dockyard Ltd* (2005) (Key Case)[63] and enforced the adjudicator's decision, finding that even if the adjudicator had been wrong to conclude that the earlier decision was irrelevant and/or inadmissible, that was a decision within the adjudicator's jurisdiction with which the court would not interfere. Akenhead J summarized the law as follows:

> 26. In the context of this case, I draw the following conclusions:
>
> (a) The Adjudicator must consider defences properly put forward by a defending party in adjudication.
> (b) However, it is within an adjudicator's jurisdiction to decide what evidence is admissible and, indeed, what evidence is helpful and unhelpful in the determination of the dispute or disputes referred to that adjudicator. If, within jurisdiction, the adjudicator decides that certain evidence is inadmissible, that will rarely (if ever) amount to a breach of the rules of natural justice. The position is analogous to a court case in which the Court decides

[63] [2006] BLR 15, [2005] EWCA Civ 1358 (CA).

that certain evidence is either inadmissible or of such little weight and value that it can effectively be ignored: it would be difficult for a challenge to such a decision on fairness grounds to be mounted.
(c) Even if the adjudicator's decision (within jurisdiction) to disregard evidence as inadmissible or of little or no weight was wrong in fact or in law, that decision is not in consequence impugnable as a breach of the rules of natural justice.
(d) One will need in most and possibly all 'natural justice' cases to distinguish between a failure by an adjudicator in the decision to consider and address a substantive (factual or legal) defence and an actual or apparent failure or omission to address all aspects of the evidence which go to support that defence. It is necessary to bear in mind that adjudication involves, usually, the exchange of evidence and argument over a short period of time and the production of a decision within a short time span thereafter. It is simply not practicable, usually, for every aspect of the evidence to be meticulously considered, weighed up and rejected or accepted in whole or in part. Primarily, the adjudicator, needs to address the substantive issues, whether factual or legal, but does not need (as a matter of fairness) to address each and every aspect of the evidence. The adjudicator should not be considered to be in breach of the rules of natural justice if the decision does not address each aspect of the evidence adduced by the parties.

Table 10.4 Table of Cases: Failure to Consider Evidence/Arguments

Title	Citation	Issue
Gillies Ramsay Diamond v PJW Enterprised Ltd	[2004] BLR 131 (SIHCOS), Clerk LJ, Lord Macfadyen, and Lord Caplan	**Decision did not discuss legal authorities referred to by responding party** An adjudicator is under a duty to consider any relevant information submitted by either party. It should be presumed that the adjudicator complied with that obligation unless his decision and reasons suggest otherwise.[64] Enforced.
Carillion Construction Ltd v Devonport Royal Dockyard Ltd	[2006] BLR 15, [2005] EWCA Civ 1358 (CA), Chadwick, Moore-Bick LJJ	**Non-consideration of irrelevant evidence** If an adjudicator determines on analysis of the facts or law, rightly or wrongly, that evidence is irrelevant and declines to consider it, that is not a breach of natural justice. An adjudicator's decision is not required to expressly recite and reject every argument that is not accepted. Enforced.
All in One Building & Refurbishments Ltd v Makers UK Ltd	[2006] CILL 2321, [2005] EWHC 2943 (TCC), Judge Wilcox	**Adjudicator's approach to evidence** The court will not conduct an analysis of the adjudicator's reasoning as to how he arrived at the primary factual position and the application of the law to those facts. The court will not deal with detailed criticisms of how the adjudicator dealt with the evidence. Enforced.

(Continued)

[64] *Balfour Beatty Construction Northern Ltd v Modus Corovest (Blackpool) Ltd* [2008] EWHC 3029 (TCC) endorsed this proposition.

Title	Citation	Issue
AMEC Capital Projects Ltd v Whitefriars City Estates Ltd	[2005] 1 All ER 723, [2004] EWCA Civ 1418 (CA), Dyson, Kennedy, Chadwick LJJ	**No obligation to consider arguments relating to jurisdiction** Procedural fairness does not require an adjudicator to consider arguments in respect of his jurisdiction where any jurisdictional decision would not bind the parties. Enforced, overturning the decision at first instance ([2003] EWHC 2443 (TCC)).
Kier Regional Ltd (t/a Wallis) v City & General (Holborn) Ltd	[2009] CILL 2660, [2006] EWHC 848 TCC, Jackson J	**Failure to consider expert reports** At worst, the adjudicator made an error of law that caused him to disregard the experts' reports. That was not an error of law with which the court could interfere. Enforced.
Balfour Beatty Construction Northern Ltd v Modus Corovest (Blackpool) Ltd	[2009] CILL 2660, [2008] EWHC 3029 (TCC), Coulson J	**Decision failed to deal expressly with a secondary defence** It is wrong to infer that because a decision does not expressly refer to a matter that the adjudicator did not deal with the matter. There is a presumption that the adjudicator has complied with the obligation to consider the information submitted by both parties unless the decision and reasons suggest otherwise. Enforced.
Jacques (t/a C&E Jacques Partnership) v Ensign Contractors Ltd	[2009] EWHC 3383 (TCC), Akenhead J	**Failure to consider defence set out in previous adjudicator's decision which was null and void** It is within an adjudicator's decision to decide what evidence is admissible and helpful or unhelpful in determining the dispute. A decision that evidence is inadmissible will rarely if ever be a breach of natural justice, even if it is wrong in fact or law. Enforced.
GPS Marine Contractors Ltd v Ringway Infrastructure Services Ltd	[2010] BLR 377, [2010] EWHC 283 (TCC), Ramsey J	**Failure to take into account submissions in a rejoinder** A procedure that allows for referral, response and reply submissions is conventional and does not fail to comply with the rules of natural justice where the reply was confined only to matters raised in the response. An adjudicator is entitled to limit the number of rounds of submissions. Enforced.
AMEC Group Ltd v Thames Water Utilities Ltd	[2010] EWHC 419 (TCC), Coulson J	**Adjudicator did not deal in detail with further submission; failure to deal with specific defences** Where an adjudicator invites further submissions late in the process, he should consider any submissions made. An adjudicator is not obliged to consider in detail a second round submission served late in the process. Unless the adjudication rules provide, a party does not have a right to respond to every submission made by the other party. The adjudicator is not obliged to respond to each and every issue raised in the parties' submissions. Enforced.

(Continued)

Table 10.4 *Continued*

Title	Citation	Issue
Lee v Chartered Properties (Building) Ltd	[2010] BLR 500, [2010] EWHC 1540 (TCC), Akenhead J	**Failure to consider jurisdictional challenges** Where an adjudicator has no authority to determine questions of his own jurisdiction, there is no obligation on the adjudicator to consider or rule on his jurisdiction. Not enforced on other grounds (decision not delivered as soon as possible after it was reached).

Failure to Give Reasons

10.63 Closely related to the question of whether the adjudicator failed to consider particular evidence or arguments in reaching the decision are the cases where challenges have been mounted in relation to adjudicators' decisions that are unreasoned or unintelligible. Natural justice does not require that a reasoned decision be given;[65] any requirement for reasons is the result of the specific rules under which the adjudication is conducted. Paragraph 22 of the Scheme provides that, if requested by one of the parties, the adjudicator shall provide reasons for his decision.[66] Other adjudication rules may provide that a reasoned decision is required, or shall be provided if so requested.

10.64 Unsuccessful parties have attempted to argue that the provision of an unreasoned decision does not satisfy the adjudicator's obligation to provide a decision. In *Gillies Ramsay Diamond and ors v PJW Enterprises Ltd* (2003),[67] Clerk LJ considered such a challenge and confirmed that it is not enough that an adjudicator has merely failed to set out relevant issues, and that the courts should assume that the adjudicator did consider the relevant information unless the decision and reasons suggest otherwise. He concluded that a challenge based on intelligibility could succeed 'only if the reasons are so incoherent that it is impossible for the reasonable reader to make sense of them. In such a case, the decision is not supported by any reasons at all and on that account is invalid'.

10.65 This approach of the Scottish courts was endorsed as part of the five principles laid down in *Carillion Construction Ltd v Devonport Royal Dockyard Ltd* (2005) (Key Case):[68]

> 81. … (5) If an adjudicator is requested to give reasons pursuant to paragraph 22 of the Scheme, in my view a brief statement of those reasons will suffice. The reasons should be sufficient to show that the adjudicator has dealt with the issues remitted to him and what his conclusions are on those issues. It will only be in extreme circumstances, such as those described by Lord Justice Clerk in *Gillies Ramsay*, that the court will decline to enforce an otherwise valid adjudicator's decision because of the inadequacy of the reasons given. The complainant would need to show that the reasons were absent or unintelligible and that, as a result, he had suffered substantial prejudice.

[65] *Multiplex Construction (UK) Ltd v West India Quay Development Company Eastern Ltd* (2006) 11 Con LR 33, [2006] EWHC 1569 (TCC); *Balfour Beatty Engineering Services (HY) Ltd v Shepherd Construction Ltd* (2009) 127 Con LR 110, [2009] EWHC 2218 (TCC) (Key Case).

[66] See 4.105–4.108.

[67] [2004] BLR 131, [2003] Scot CS 343 (24 Dec. 2003), Second Division, Inner House, Court of Session.

[68] [2005] BLR 310, [2005] EWHC 778 (TCC). On appeal, the Court of Appeal expressed itself as being in broad agreement with the five propositions: [2006] BLR 15, [2005] EWCA Civ 1358. Those propositions have also been followed by Coulson J in *Balfour Beatty Construction (Northern) Ltd v Modus Corovest (Blackpool) Ltd* [2009] CILL 2660, [2008] EWHC 3029 (TCC) and Judge Davies in *Thermal Energy Construction Ltd v AE & E Lentjes UK Ltd* [2009] All ER (D) 271 (Jan), [2009] EWHC 408 (TCC) (Key Case).

In *Balfour Beatty Construction Northern Ltd v Modus Corovest (Blackpool) Ltd* (2008),[69] Coulson J concluded by analogy to be drawn with the arbitration authorities that an adjudicator is not obliged to decide each and every point argued, but only those that are genuinely 'en route' to deciding the underlying dispute between the parties.[70] **10.66**

In the one challenge of this type to have succeeded, *Thermal Energy Construction Ltd v AE & E Lentjes UK Ltd* (2009) (Key Case),[71] the absence of reasons evidenced the adjudicator's failure to consider the merits of a defence altogether and the case can therefore better be categorized as such and not as a failure to provide reasons per se. **10.67**

Key Cases: Failure to Give Reasons

Thermal Energy Construction Ltd v AE & E Lentjes UK Ltd [2009] EWHC 408 (TCC)

Facts: The claimant was the defendant's subcontractor. The parties conducted an adjudication under the TeCSA Adjudication Rules, paragraph 31 of which obliged the adjudicator to provide written reasons, if so requested by any or all of the parties within seven days of the referral. The referral contained such a request. The response raised a defence by way of set-off and counterclaim of liquidated damages under the main contract and this issue was the subject of further submissions by both parties. The adjudicator's 23-page decision contained no express reference to the defence. **10.68**

Held: Judge Davies refused to enforce the adjudicator's decision for three main reasons. First, an adjudicator is obliged to give reasons so as to make it clear that he has decided all of the essential issues which he must decide as being issues properly put before him by the parties, and so that the parties can understand, in the context of the adjudication procedure, what it is that the adjudicator has decided and why. In this case, there was no express reference to the set-off defence as being one that the adjudicator had to decide or did in fact decide. Judge Davies explained the confusion caused by the decision: **10.69**

> 22. So, for example, in this case it would be important for any reader of the decision to know whether or not firstly the Adjudicator had purported to decide the set-off and counter-claim, and secondly, if so, on what grounds. By way of practical illustration, there is clearly a significant difference between a decision to the effect that the Adjudicator did not have jurisdiction to decide the set-off and counter-claim, which in principle would be subject to consideration by the Courts in the event of an adjudication enforcement application such as the present, and a decision within his jurisdiction that having considered the defence, he rejected it on the merits. In the latter case, in accordance with established principles, a party seeking to resist enforcement would not be entitled to challenge the correctness of that decision if made within his jurisdiction.

Secondly, the defendant suffered two types of substantial injustice because it is unclear whether or not the adjudicator in fact considered the set-off and counterclaim on the merits: **10.70**

1. The defendant lost the opportunity of having the adjudicator deal with that defence, and therefore lost the prospect of the adjudicator deciding that point in his favour.

[69] [2009] CILL 2660, [2008] EWHC 3029 (TCC).

[70] See para. 45. The arbitration cases include *Checkpoint Ltd v Strathclyde Pension Fund* [2003] All ER (D) 56 (Feb), [2003] EWCA (Civ) 84 and *World Trade Corporation Ltd v C Czarnikow Sugar Ltd* [2004] 2 All ER (Comm) 813, [2004] EWHC 2332 (Comm).

[71] Energy [2009] All ER (D) 271 (Jan), [2009] EWHC 408 (TCC).

2. If the defendant was to seek to launch a further adjudication to seek to recover these losses, then first of all it would have to comply with the adjudicator's decision and pay up in the meantime if the adjudicator's decision was enforced.

10.71 Thirdly, the defendant's case was not affected by its failure to request the adjudicator, in accordance with paragraph 32 of the TeCSA Rules, to correct his decision by operation of the slip rule. There is no warrant for the submission that it is a precondition for resisting enforcement on the grounds of a failure to deal with all matters and/or a failure to give reasons, that the losing party must first exercise the right conferred by paragraph 32.

Balfour Beatty Engineering Services (HY) Ltd v Shepherd Construction Ltd (2009) 127 Con LR 110, [2009] EWHC 2218 (TCC)

10.72 **Facts:** Balfour Beatty referred to adjudication its claim for payment in respect of mechanical and electrical works carried out under subcontract to Shepherd. The adjudicator found that Balfour Beatty was entitled to an EOT. Shepherd asserted that the adjudicator's decision was insufficiently clear, failing to explain the reasoning relating to the identification of delay events under the contract and therefore the entitlement to EOT.

10.73 **Held:** The court found that the decision was no work of art with poor grammar and unconventional sentence structure that was in places confusing and repetitive. Akenhead J considered *Gillies Ramsay*, *Balfour Beatty v Modus Corovest* and *Thermal Energy v A. E. &E. Lentjes* and concluded:

> 48. … (a) The decision needs to be intelligible so that the parties, objectively, can know what the adjudicator has decided and why.
>
> (b) A decision which is wholly unreasoned but which is required to be reasoned is not a decision for the purposes of the Scheme or under contractual machinery which requires a reasoned decision. It would therefore not be enforceable as such.
>
> (c) Because the Courts have said time and again that the decision cannot be challenged on the grounds that the adjudicator answered the questions, which he or she was required to address wrongly, the fact that the reasons given are, demonstrably or otherwise, wrong in fact or in law or even in terms of emphasis will not give rise to any effective challenge.
>
> (d) The fact that the adjudicator does not deal with every single argument of fact or law will not mean that the decision is necessarily unreasoned. He or she should deal with those arguments which are sufficient to establish the route by which the decision is reached.
>
> (e) The failure to give reasons is not a breach of natural justice.
>
> (f) The reasons can be expressed simply. If the reasons are so incoherent that it is impossible for the reasonable reader to make sense of them, it will not be a reasoned decision.
>
> (g) Adjudicators are not to be judged too strictly, for instance by the standards of judges or arbitrators, in terms of the reasoning. This reflects the fact that decisions often have to be reached in a short period of time and adjudicators are often not legally qualified. It certainly reflects the fact that there has not been a full judicial or arbitral type process.
>
> (h) The fact that reasoning in a decision is repetitive, diffuse or even ambiguous does not mean that the decision is unreasoned.

10.74 Akenhead J also confirmed that the following facts do not make an adjudicator's decision unreasoned: (i) the decision does not reach the standard of an average court judgment; (ii) it contains reasoning which is wrong in fact or law; or (iii) judges would give more reasons than the adjudicator gave.[72] Akenhead J described the decision as being 'idiosyncratic', in part in shorthand and 'to a marked extent ungrammatical',[73] and as 'no work of art': 'The grammar is poor, the sentence structure is unconventional to say the least and it is in places confusing and repetitive.'[74] However, he found that it was necessary to 'see through all that' and determined that, on the key issues, there was some intelligible reasoning. Accordingly, the court rejected the challenge that the decision was unreasoned and also rejected the assertion that there was a critical gap in the adjudicator's reasoning because it failed to properly identify a delay event, as required by the contract. In relation to this second challenge, the court was 'unconvinced ... that challenges on this type of basis are appropriate':[75]

> 79. ... I am sure that the legislators would not have had in mind a challenge on the basis that some arguably logical link in the chain of reasoning was missing from the decision. Provided that the broad thrust of the reasoning is provided, the Court should enforce. Otherwise, many adjudicators' decisions would be open to an effective challenge.

Table 10.5 Table of Cases: Failure to Provide Reasons (successful challenges are indicated by shaded entries)

Title	Citation	Issue
Gillies Ramsay Diamond v PJW Enterprises Ltd	[2004] BLR 131, [2003] Scot CS 343, Clerk LJ, Lord Macfadyen, and Lord Caplan	**Party claimed that reasons given were unintelligible** Where an adjudicator is required to give a reasoned decision, a challenge to the intelligibility of the reasons can only stand where the reasons are so incoherent that it is impossible for the reasonable reader to make sense of them. The reasons in the instant case were sufficient to show that the adjudicator had dealt with the issues remitted and showed his conclusions on each issue. Enforced.
Carillion Construction Ltd v Devonport Royal Dockyard Ltd	[2005] BLR 310 (TCC), Jackson J	**Sufficient reasons provided** The defendant raised a counterclaim in respect of defective works, which prior to the adjudication it had quantified at £2.9 million. The adjudicator made an allowance in respect of the first batch of defects of around £2.3 million, having applied a reduction of 20 per cent. The adjudicator found that some later defects had not properly been notified to Carillion and that there was no satisfactory evidence substantiating those defects. Jackson J held that sufficient reasons had been given; the adjudicator had explained why he rejected the claim for the later defects and explained the basis of the reduction in respect of the claim for the original defects. Accordingly, a brief statement of reasons will suffice. Only if those reasons were absent or unintelligible and substantial prejudice had been suffered because of that would a decision be a nullity. Enforced.

(Continued)

[72] *Ibid.*, para. 75.
[73] *Ibid.*, para. 29.
[74] *Ibid.*, para. 76.
[75] *Ibid.*, para. 79.

Table 10.5 *Continued*

Title	Citation	Issue
Multiplex Construction (UK) Ltd v West India Quay Development Company Eastern Ltd	(2006) 111 Con LR 33, [2006] EWHC 1569 (TCC), Ramsey J	**Failure to give adequate reasons** Natural justice does not require that reasons be given. In the absence of a specific provision in the rules or a request that reasons be given, an adjudicator is not required to give reasons. Enforced.
Balfour Beatty Construction Northern Ltd v Modus Corovest (Blackpool) Ltd	[2009] CILL 2660, [2008] EWHC 3029 (TCC), Coulson J	**Unreasoned decision** The adjudicator described his own decision as 'not reasoned'. An adjudicator is not required to set out extensively a response to every element of the responding party's case or give detailed reasons for every part of his conclusion. There is a presumption that the adjudicator has complied with his obligation to consider the information submitted by both parties unless his decision and reasons suggest otherwise. The adjudicator had however provided adequate reasons and the challenger could not show prejudice. Enforced.
Thermal Energy Construction Ltd v AE & E Lentjes UK Ltd	[2009] All ER (D) 271 (Jan), [2009] EWHC 408 (TCC), Judge Davies	**Unreasoned decision** The adjudicator's failure to refer expressly to a defence of set-off and counterclaim in the reasons caused significant prejudice to the responding party. The decision was not enforced, notwithstanding the fact that the responding party did not request the adjudicator to amend his decision, using para. 32 of the TeCSA Rules (slip rule). Not enforced (see also failure to consider defence).
Vision Homes Ltd v Lancsville Construction Ltd	[2009] BLR 525, [2009] EWHC 2042 (TCC), Clarke J	**Unreasoned decision** The decision contained idiosyncratic expressions such as 'dumped the Rule book' and 'out went the LADs'. However, it was not so unclear on its face that no effect could be given to it. Not enforced on other grounds (improper appointment).
Balfour Beatty Engineering Services (HY) Ltd v Shepherd Construction Ltd	(2009) 127 Con LR 110, [2009] EWHC 2218 (TCC), Akenhead J	**Unreasoned decision** Adjudicators' decisions are not required to be to the same standard as an average court decision. Despite a decision being 'idiosyncratic' and 'ungrammatical', on the key issues, there was intelligible reasoning. Enforced.
NAP Anglia Ltd v Sun-Land Development Co. Ltd	[2012] BLR 110, [2012] EWHC 2846 (TCC), Edwards-Stuart J	**Alleged failure to address issues** An adjudicator is not required to give reasons for not accepting particular submissions as long as it is clear that the submissions in question have been considered. Enforced.
Pihl UK Ltd v Ramboll UK Ltd	[2012] CSOH 139, Lord Malcolm	**Unreasoned decision** The Court rejected the allegations that the adjudicator had failed to explain (i) the legal source of the obligation to repay the sum ordered and (ii) the basis upon which the sum

Title	Citation	Issue
		was calculated. The adjudicator's reasoning was not open to challenge. *Gillies Ramsay Diamond v PJW Enterprised Ltd* followed.

Denial of Opportunity to Respond to New Material

10.75 Parties in an adjudication have a legitimate expectation that they will be afforded an opportunity to present their case.[76] A party is entitled to know the case against it in order that it can have a fair chance to respond. This necessarily requires that a party has knowledge of all the material on which the adjudicator may rely in reaching a decision. That requirement is expressly reflected at paragraph 17 of the Scheme which requires the adjudicator to:

> consider any relevant information submitted to him by any of the parties to the dispute and shall make available to them any information to be taken into account in reaching his decision.

10.76 The material before the adjudicator is likely to come from several sources including the parties' submissions and information gained through meetings with the parties or their witnesses or experts. Whilst it is normal that in the course of an adjudication all information and correspondence is copied to the opposing party, on occasion such procedures can break down with the result that a party is ignorant of certain information.[77] Equally, the general safeguards normally employed to eliminate the possibility of oral contact between the adjudicator and one party in the absence of the other party may not be followed, with the result that the adjudicator receives information unknown to the other party.[78] In exceptional circumstances, the adjudicator's involvement in a related adjudication may mean the adjudicator has knowledge from that involvement on which he may rely.[79]

10.77 Judge Seymour in *Dean and Dyball Construction Ltd v Kenneth Grubb Associates Ltd* (2003) (Key Case)[80] summarized the requirement of natural justice and dangers inherent in an adjudicator relying on material without submission:

> 52. ... natural justice requires that, if an adjudicator receives a communication about a matter of significance to the substance of the adjudication from one party alone in the absence of the other, he should inform the absent party of the substance of the communication so as to give that party an opportunity to deal with it if it wishes. ... The absent party must be told the substance of what has been said and afforded an opportunity to comment upon it. The reason is fairly obvious, namely that it would be grossly unfair for the person charged with making a decision to rely upon material provided by one party to the dispute of which the other party was unaware and with which it had not been given a chance to deal. ... It is in justice to the provider of the material that it is desirable to confirm with it that the tribunal has understood correctly what it has been told. Obviously if the tribunal has misunderstood what it has been told, and that is pointed out by the party providing the information, the corrected information should then be put to the other party.

10.78 If the adjudicator, in reaching a decision, relies on information of which one or both parties are unaware and/or has not invited submissions on that information, that will often lead to a challenge to the enforceability of the decision. Such challenges are often successful, as long as

[76] *RSL (South West) Ltd v Stansell Ltd* [2003] EWHC 1390 (TCC).
[77] *Woods Hardwick Ltd v Chiltern Air-Conditioning Ltd* [2001] BLR 23 (TCC).
[78] *Discain Project Services Ltd v Opecprime Development Ltd* (No. 2) [2001] BLR 285.
[79] *Pring & St Hill v C. J. Hafner* (t/a Southern Erectors) [2002] EWHC 1775 (TCC).
[80] 100 Con LR 92, [2003] EWHC 2465 (TCC).

it can be shown that it is probable that the information materially affected the adjudicator's decision. Such challenges have not succeeded where it has been found on enforcement that the procedure followed was agreed by the parties,[81] or that the relevant information was in fact provided with the parties being given an adequate opportunity to respond.[82] The same principle applies when an adjudicator decides a point on a basis for which neither party has contended as discussed below.

Wrongful Reliance on Third Party Advice

10.79 As well as information submitted by the parties, information can also be provided by any third parties appointed by the adjudicator to provide legal or expert advice. An adjudicator is positively required to consider the information provided by the parties—and may be assisted by the involvement of an expert. However, any appointment of a third party to assist should be done within the ambit of any power conferred on him to do so, either according to the Scheme or the applicable contractual adjudication rules. Under the Scheme, for instance, an adjudicator cannot appoint an expert without advising the parties; nor can the adjudicator receive the benefit of an expert's opinion without sharing that with the parties.[83]

10.80 Adjudicators' decisions have not been enforced where advice from legal experts has not been provided to the parties,[84] even if informally received,[85] or where a planning expert's final report was not circulated to the parties.[86] It is important to give the parties an opportunity to respond to any expert opinion.[87] Where an opportunity to review and respond to an expert's report or advice is given, it is likely to remove any scope for challenge.[88]

Key Cases: Wrongful Reliance on Third Party Advice

Balfour Beatty Construction Ltd v The Mayor and Burgesses of the London Borough of Lambeth [2002] BLR 288, [2002] EWHC 597 (TCC)

10.81 **Facts:** Balfour Beatty referred a dispute to adjudication concerning the amount of liquidated damages payable by it. This inevitably required a consideration of the extension of time to which Balfour Beatty was entitled. Lambeth argued that Balfour Beatty had failed to establish its case on any of the four different programme analyses submitted by it. Balfour Beatty was requested by the adjudicator to produce a schedule setting out the claimed relevant events, the date of those relevant events, the activity directly affected by them, and the timing of the event(s). The adjudicator then decided to verify Balfour Beatty's as built programmes to divine a critical path. The adjudicator did not invite comment from either party as to either his course of action or the documents he produced.

10.82 **Held:** The decision was not enforced. Lambeth had shown there was a real danger that the adjudicator had not complied with the principles of natural justice because he had

[81] *Try Construction v Eton Town House Group Ltd* [2003] BLR 286, [2003] EWHC 60 (TCC).

[82] *Dean and Dyball Construction Ltd v Kenneth Grubb Associates Ltd* (2003) 100 Con LR 92, [2003] EWHC 2465 (TCC) (Key Case).

[83] See paras. 13(f) and 17 of the Scheme set out above at 4.57 and 4.80-4.82 respectively.

[84] *Costain Ltd v Strathclyde Builders Ltd* [2003] Scot CS 316; *BAL (1996) Ltd v Taylor Woodrow Construction Ltd* [2004] All ER (D) 218 (Feb).

[85] *Highlands and Islands Airports Ltd v Shetland Islands Council* [2012] CSOH 12 (Scottish OH).

[86] *RSL (South West) Ltd v Stansell Ltd* [2003] EWHC 1390 (TCC).

[87] *Balfour Beatty Construction Ltd v The Mayor and Burgesses of the London Borough of Lambeth* [2002] BLR 288, [2002] EWHC 597 (TCC).

[88] *Try Construction Ltd v Eton Town House Group Ltd* [2003] BLR 286, [2003] EWHC 60 (TCC).

drawn on his own knowledge and expertise and not given Lambeth (the responding party) an opportunity to deal with arguments which had not been advanced by either party but which were included in the decision. Judge LLoyd found that the adjudicator failed to inform the parties of the 'as built' programme he had drawn up which formed the basis of his divination of the retrospective critical path, and of the methodology he intended to adopt in considering the extension of time claimed. That methodology was different from those put forward by either party. The responding party had objected that the referring party had failed to make its case and should therefore lose its claim. The adjudicator, however, used his own knowledge and experience to 'make good fundamental deficiencies in the material presented' by the referring party. The court viewed this as going beyond taking the initiative and found that the adjudicator effectively 'did BB's work for it' and made out the referring party's case. This was such a potentially serious breach of either impartiality or fairness that the decision was invalid because it was not a decision which the adjudicator was authorized to make:

> 33. ... If an adjudicator intends to use a method which was not agreed and has not been put forward as appropriate by either party he ought to inform the parties and to obtain their views as it is his choice of how the dispute might be decided. An adjudicator is of course entitled to use the powers available [to] him but he may not of his own volition use them to make good fundamental deficiencies in the material presented by one party without first giving the other party an proper [*sic*] opportunity of dealing both with that intention and with the results. The principles of natural justice applied to an adjudication may not require a party to be aware of 'the case that it has to meet' in the fullest sense since adjudication may be 'inquisitorial' or investigative rather than 'adversarial'. That does not however mean that each party need not be confronted with the main points relevant to the dispute and to the decision.
>
> ...
>
> 36. Furthermore the adjudicator ought to have given Lambeth the opportunity of commenting on the use of the analysis chosen by the adjudicator. ... An adjudicator does not act impartially or fairly if he arrives at a decision without having given a party a reasonable opportunity of commenting upon the case that it has to meet (whether presented by the other party or thought to be important by the adjudicator) simply because there is not enough time available. An adjudicator, acting impartially and in accordance with the principles of natural justice, ought in such circumstances to inform the parties that a decision could not properly reasonably and fairly be arrived at within the time and invite the parties to agree further time. If the parties were not able to agree more time then an adjudicator ought not to make a decision at all and should resign.

RSL (South West) Ltd v Stansell Ltd [2003] EWHC 1390 (TCC)

10.83 **Facts:** Stansell (a building contractor) engaged RSL (structural steel fabricators and erectors) at a site in Bristol for certain structural steelwork and staircases to be fabricated and erected. The form of contract was JCT Domestic Subcontract DOM/2 1981 edition, incorporating Amendments 1–8. Stansell accepted responsibility for a delay to commencement of works but there was a dispute between the parties as to the causes of the delay to completion. RSL commenced an adjudication concerning the EOT to which it was entitled and the adjudicator sought assistance on programming issues from a specialist planner.

10.84 Upon learning of the appointment, Stansell stated that it required sight of any report prepared and reasonable time to comment. The adjudicator provided the 'initial findings' of the planner and invited the parties to make their observations. Stansell did not respond, and the reason subsequently given was that the expert had concluded that RSL had failed to prove its case. The adjudicator requested an extension of time but without explaining why it was required and the parties did not agree to it. The adjudicator then handed down his decision which was '[b]ased upon the responses received and subsequent findings' of the planning expert. However he did not circulate any further report from the expert beyond the first initial report. The adjudicator awarded a 55 working-day EOT, apparently after considering a further report from the expert which had not been circulated to the parties for comment.

10.85 **Held:** Upon application for summary enforcement, Judge Seymour refused to enforce the adjudicator's decision for four main reasons:

1. The adjudicator breached natural justice for failing to provide a copy of the planning expert's report to the parties before basing his decision upon it.
2. Lack of time in the adjudication process was no excuse. If the adjudicator had explained that he required an extension to allow the parties to comment on the final report, and such extension been granted, the parties would have waived their right to object – however here, that did not occur.
3. A party to a dispute resolution procedure has a legitimate expectation that he will be afforded opportunities promised to him to present his case.
4. Justice must be seen to be done.

> 32. It is elementary that the rules of natural justice require that a party to a dispute resolution procedure should know what is the case against him and should have an opportunity to meet it ... It is absolutely essential, in my judgment, for an adjudicator, if he is to observe the rules of natural justice, to give the parties to the adjudication the chance to comment upon any material, from whatever source, including the knowledge or experience of the adjudicator himself, to which the adjudicator is minded to attribute significance in reaching his decision. Thus in the present case it was plain, in my judgment, that [the adjudicator] should not have had any regard to the final report of [the expert programmer] without giving both RSL and Stansell the chance to consider the contents of that report and to comment upon it ...
>
> 33. A further aspect of the requirements of natural justice is that a party to a dispute resolution procedure has a legitimate expectation that he will be afforded opportunities promised to him to present his case ... It is immaterial whether, strictly, [the adjudicator] needed the agreement of the parties to the employment of [the expert] to assist him, because he chose to seek that agreement, whether strictly he needed it or not ... In informing both parties of the terms of the further communication which he had received from [the expert] he created the impression that he would share with them, and give them an opportunity to comment upon, any further communications from [the expert]. That [the adjudicator] did not do ... The need to provide an opportunity to Stansell to comment upon the final report was, as it seems to me, self-evidently the greater, if, in the final report, [the expert] significantly altered the position adopted in the initial report that the evidence was insufficient to make out any claim by one party against the other in respect of delay to the completion of the Sub-Contract Works.
>
> 34. It is, of course, correct that one does not know whether [the expert] in fact modified in his final report the position adopted in his initial report concerning which the parties were afforded the chance to comment. The probability must be that he did, because otherwise paragraphs 72 and 77 of the Decision would not have been phrased as they were. However, insofar as there may be any doubt as to the terms of [the] final report, far from that being a reason to disregard the objection taken on behalf of Stansell, it brings into play another aspect of the rules of natural justice, justice being seen to be done. In my judgment the mere fact that the material before the Court indicates that Mr. Hinchcliffe

took into account in reaching his decision in relation to extensions of time for completion of the Sub-Contract Works a report which may have been of importance in the eventual decision which was not disclosed to the parties to the adjudication is sufficient in itself to mean that Mr. Hinchcliffe's decision was reached in breach of the rules of natural justice and should not be enforced.

Dean and Dyball Construction Ltd v Kenneth Grubb Associates Ltd (2003) 100 Con LR 92, [2003] EWHC 2465 (TCC)

10.86 **Facts:** The adjudication was conducted under the CIC Model Adjudication Procedure which expressly included the power for the adjudicator to meet the parties separately (r. 17(v)). The adjudicator did precisely this, and upon enforcement, the unsuccessful party challenged the decision on the grounds of breach of natural justice.

10.87 **Held:** Judge Seymour upheld the adjudicator's decision, finding that the adjudicator had acted in accordance with the applicable adjudication procedure. The court was of the opinion that the objection was more of form than substance, given that there was no allegation that the adjudicator had regard to evidence which the other party was unaware:

> 53. Rule 17 of the Procedure gave Mr. Lester a discretion as to how to conduct the adjudication in this case. That discretion, although described as *'complete'*, was obviously not totally unfettered—he had still to observe the fundamental requirements of natural justice. It does not seem to me that natural justice necessarily requires that a party have an opportunity to cross-examine witnesses for the opposite party, although it may do, especially if the credibility of a witness is in issue. Again, it does not seem to me that natural justice necessarily requires that evidence from witnesses for one party be taken in the presence of the opposite party or its representatives. It may be sufficient for a tribunal taking evidence from the witnesses for one party simply to indicate to the opposite party, to give it an opportunity to deal with it, what those witnesses have said to which the tribunal is minded to attribute importance. That is essentially the course which Mr. Lester took in the present case. As to whether, absent an express provision in the relevant adjudication procedure permitting such a course to be taken, it would be appropriate to hear witnesses from one party in the absence of the other, I have some doubts. Certainly I have grave difficulty in seeing that adopting such a course could ever be appropriate without the tribunal indicating to the absent party what has been said to which the tribunal is minded to have regard and providing an opportunity for a response. The practice of submissions on behalf of a party to legal proceedings or arbitration proceedings being made in writing, and in that sense in the absence of the opposite party, is now well-established. It is accepted that in a case in which submissions are made in writing all that natural justice and fairness require is that a party should have an opportunity to deal with the primary submissions of the opposite party (rather than necessarily the answer to its own primary submissions). In other words, knowing what is said for the case of the opposite party or against its own case, a party should have an opportunity to answer. There is no obvious reason, at least in a case in which the credibility of witnesses is not in issue, why a similar procedure should not suffice to satisfy the requirements of natural justice and fairness in relation to evidence if parties have agreed that a tribunal may, if it wishes, hear evidence from witnesses for the opposite party in its absence.
>
> 54. There was no suggestion in the present case on behalf of Grubb that Mr. Lester did in fact have regard to evidence given on behalf of D and D of which it was unaware or which it did not have an opportunity to answer. The objection was simply one of form. It involved the somewhat unpromising proposition that the Procedure, if operated in

accordance with its express terms, could be operated unfairly. Perhaps that could happen, but I am satisfied that it did not happen in the present case and that no dispassionate observer, aware of the circumstances, would consider that there was a risk of actual unfairness on the part of Mr. Lester.

Table 10.6 Table of Cases: Reliance on Material without Submission (successful challenges are indicated by shaded entries)

Title	Citation	Issue
Woods Hardwick Ltd v Chiltern Air-Conditioning Ltd	[2001] BLR 23 (TCC), Judge Thornton	**Reliance on material not provided to responding party** The adjudicator relied on information from the referring party and third parties (including subcontractors) that was not made available to the responding party. Not enforced.
Discain Project Services Ltd v Opecprime Development Ltd	[2001] BLR 285, [2001] EWHC 435 (TCC), Judge Bowsher	**Telephone communications with one party** On three occasions the adjudicator communicated with the referring party's representative by telephone in relation to the substance of the matters in dispute. The fact of the communications was notifed by fax to the other party but materially important aspects of the discussions were not reported. Not enforced.
Pring & St Hill v C. J. Hafner (t/a Southern Erectors)	[2002] EWHC 1775 (TCC), Judge LLoyd	**Adjudicator had previously decided related disputes** The decision in an adjudication between subcontractor and sub-subcontractor (down the contractual chain) was not enforced in circumstances where the adjudicator had also adjudicated the same dispute between main contractor and subcontractor (up the contractual chain). Decided on lack of jurisdiction. *Obiter*: The involvement in an earlier adjudication with one party without making available all relevant materials from that adjudication in the present adjudication meant that the adjudicator had information from that adjudication that he was unable to provide to the responding party to allow it to make submissions. Not enforced.
Baune v Zduc Ltd	[2002] All ER (D) 55 (TCC), Judge Seymour	**Reliance on advice from a quantity surveyor** Advice not provided to the parties for comment. *Obiter*: there will be no breach of natural justice where the adjudicator relied on the advice of a quantity surveyor without providing that advice to the parties for comment. JCT adjudication rules applied, which rules allowed the adjudicator to obtain advice. Not enforced on other grounds (jurisdiction).
Try Construction v Eton Town House Group Ltd	[2003] BLR 286, [2003] EWHC 60 (TCC), Judge Wilcox	**Adjudicator relied on programming expert** The parties were found to have agreed the procedure by which a programming expert would be appointed and the nature and method of his analysis. There could be no complaint when that procedure was followed. The decision did not depart from the evidence put forward and arguments developed by the parties. Enforced.

Title	Citation	Issue
RSL (South West) Ltd v Stansell Ltd	[2003] EWHC 1390 (TCC), Judge Seymour	**Adjudicator relied on specialist planner** The adjudicator provided the planner's initial findings that the referring party had not proved its entitlement to an EOT. However, a further report was not shown to the parties and the adjudicator relied on that further report in awarding a 55-day EOT. It is absolutely essential for an adjudicator, if he is to observe the rules of natural justice, to give the parties to the adjudication the chance to comment upon any material, from whatever source, including the knowledge or experience of the adjudicator himself, to which the adjudicator is minded to attribute significance in reaching his decision. Not enforced.
Dean and Dyball Construction Ltd v Kenneth Grubb Associates Ltd	(2003) 100 Con LR 92, [2003] EWHC 2465 (TCC), Judge Seymour	**Oral evidence received without other party present** Natural justice requires that, if an adjudicator receives a communication about a matter of significance to the substance of the adjudication from one party alone in the absence of the other, he should inform the absent party of the substance of the communication so as to give that party an opportunity to deal with it if it wishes. Such a procedure was followed in the instant case. Enforced.
Costain Ltd v Strathclyde Builders Ltd	[2003] Scot CS 316 (Outer House Court of Session), Lord Drummond Young	**Legal advice obtained by adjudicator** An adjudicator sought and received an extension of time to discuss a particular point with a legal adviser he had appointed. Neither party was made aware of the terms or results of the discussion and there was therefore no opportunity to make submissions in respect of the advice. The decision was not enforced notwithstanding the fact that neither party requested any opportunity to know the advice or make submissions thereon.
BAL (1996) Ltd v Taylor Woodrow Construction Ltd	[2004] All ER (D) 218 (Feb), Judge Wilcox	**Legal advice obtained by adjudicator** The parties had not been advised of the details of the legal appointment or the advice received in relation to a cap on payment. Adjudicators should clearly be seen to give the parties a fair opportunity to deal with the material matters upon which an adjudicator may be basing his decision. A party cannot be considered to have acquiesced to a procedure that on its face was in breach of natural justice unless the acquiescence is clear, informed and unambiguous. Not enforced.
AMEC Capital Projects Ltd v Whitefriars City Estates Ltd	[2005] 1 All ER 723, [2004] EWCA Civ 1418, (CA), Dyson, Kennedy, Chadwick LJJ	**Legal advice on previous adjudication** The mere fact that legal advice had been obtained in the first adjudication would not have led the fair-minded informed observer to conclude that there was a real possibility that the adjudicator would have approached the point with a closed

(Continued)

Table 10.6 *Continued*

Title	Citation	Issue
		mind. Leave to appeal from earlier decision ([2004] EWHC 393 (TCC), Judge Toulmin) granted.
PC Harrington Contractors Ltd v Tyroddy Construction Ltd	[2011] All ER (D) 162 (Apr), [2011] EWHC 813 (TCC), Akenhead J	**Adjudicator did not invite submission when deciding a question of jurisdiction** Natural justice required the adjudicator to invite submissions as to his jurisdiction where he declined to consider a defence on jurisdictional grounds. Not enforced (primarily on grounds of failure to consider a defence).
CRJ Services Ltd v Lanstar Ltd (t/a CSG Lanstar)	[2011] EWHC 972 (TCC), Akenhead J	**Witness statement supporting jurisdictional argument not received by opposing party** No material breach of natural justice where, in reaching his conclusion on jurisdiction, the adjudicator had limited regard to material in the statement and the jurisdictional decision was not substantively part of the dispute referred. *Obiter*: conduct in investigating jurisdiction may result in a conclusion of unfairness if the adjudicator's disregard for rules of natural justice is sufficiently serious. Enforced.
Highlands and Islands Airports Ltd v Shetland Islands Council	[2012] CSOH 12 (Scottish Court of Session, Outer House), Lord Menzies	**Adjudicator sought informal legal advice** The adjudicator had formed his own view of the meaning of a contractual provision, then sought informal free verbal legal advice to confirm that view (which it did). He neither informed the parties that he had sought that advice nor sought their submissions on the particular provision in question. That represented a breach of natural justice as, however brief and informal, it was still legal advice on a matter that the adjudicator considered important enough to seek third party confirmation. Not enforced.

Deciding on a Basis Not Argued

10.88 The cases in the preceding section dealt with circumstances where the adjudicator makes use of specific material in reaching a decision where that material has not been provided to a party (or both parties). A related category includes the cases where the adjudicator does not make use of additional material as such, but decides the case on a basis that has not been put forward by either party or in direct opposition to a position that is agreed between the parties on a particular point.

10.89 Jurisdictional challenges are often made against such decisions because the adjudicator decided matters that went beyond the dispute referred to him (discussed at 9.48–9.53). However, overstepping jurisdiction can often overlap with a breach of natural justice, as the adjudicator should obtain the parties' submissions on any novel approach, argument, defence, or evidence[89] so that the parties do not effectively see it for the first time in the decision.

[89] *Primus Build Ltd v Pompey Centre Ltd* [2009] BLR 437, [2009] EWHC 1487 (TCC).

10.90 The early approach to such issues was generally to enforce the adjudicators' decisions. Thus, in *Macob Civil Engineering Ltd v Morrison Construction Ltd* (1999),[90] the unsuccessful party claimed not to have been given an opportunity to provide comment on whether the contract provided an adequate payment mechanism but the decision was still enforced. A further example concerning adequate payment mechanisms is *Karl Construction (Scotland) Ltd v Sweeney Civil Engineering (Scotland) Ltd* (2002) 85 Con LR 59, [2002] SCLR 766:[91]

> 8. ... it is clear that the adjudication process as envisaged by the 1996 Act, and in this case as incorporated in the parties' written contract, is a process far removed from the traditional adversarial format adopted in the courts. In the second place, it follows from that and from what we have already said that the parties had no good reason to think that the adjudicator would be circumscribed, in reaching her decision, by the terms of any written representations made to her, let alone representations that should somehow circumscribe her in her appraisal of the relevant law. This is particularly so bearing in mind that, if she had wished to do so, the adjudicator could have taken her own independent legal advice ... And, lastly, it has to be borne in mind that the adjudicator was under compulsion to issue her decision within the very short time scale of, at most, six weeks.

10.91 A party cannot complain if it has failed to make an argument that it should have presented itself on the material before the adjudicator. However, if the adjudicator intends to take account of new issues in his or her decision, then it is clear that the parties must be given the opportunity to comment or put forward further evidence.[92] It may be difficult for an adjudicator who sees that the case put by the referring party must fail, but that it may have a successful claim based on a different argument. However, the TCC confirmed in *Primus Build Ltd v Pompey Centre Ltd* (2009)[93] that an adjudicator must avoid 'filling the gaps' in such instances, and ask himself the question 'do I need to give notice of and obtain submissions about, that alternative approach?' Coulson J explained:

> 40. ... these things are always a matter of fact and degree. An adjudicator cannot, and is not required to, consult with the parties on every element of his thinking leading up to a decision, even if some elements of his reasoning may be deprived from, rather than expressly set out in, the parties' submissions. But where, as here, an adjudicator considers that the referring party's claims as made cannot be sustained, yet he himself identifies a possible alternative way in which a claim of some sort could be advanced, he will normally be obliged to raise that point in advance with the parties.

10.92 However, there is no general requirement for the adjudicator to warn the parties of any likely conclusions and invite them to specifically address these matters before making a final determination. As Jackson J commented in the first instance decision in *Carillion Construction Ltd v Devonport Royal Dockyard Ltd* (2005) (Key Case):[94]

> 81 (3) ... It is often not practicable for an adjudicator to put to the parties his provisional conclusions for comment. Very often those provisional conclusions will represent some intermediate position, for which neither party was contending. It will only be in an exceptional case such as *Balfour Beatty v London Borough of Lambeth* that an adjudicator's failure to put his provisional conclusions to the parties will constitute such a serious breach of the rules of natural justice that the Court will decline to enforce his decision.

[90] [1999] BLR 93 (TCC) (Key Case in Ch. 7).
[91] [2002] SCLR 766 (confirming the earlier decision which is reported at [2001] SCLR 95).
[92] *Cantillon Ltd v Urvasco Ltd* [2008] BLR 250, [2008] EWHC 282 (TCC) (Key Case).
[93] [2009] BLR 437, [2009] EWHC 1487 (TCC).
[94] [2005] BLR 310, [2005] EWHC 778 (TCC).

10.93 Challenges of this type have succeeded where it has been found that the adjudicator's decision was contrary to the clearly agreed position of the parties,[95] where the adjudicator has decided the matter on some materially different basis that cannot be said to be derived from the cases presented by the parties,[96] or rejected a defence without warning.[97] However, it is most frequently found that the adjudicator has applied personal experience to assess the material put forward with the result that some of it is either accepted or rejected, whether wholly or in part and the decision is found to have been properly derived from the evidence and arguments before the adjudicator. This process has also been described as 'development and exposition'[98] of the contentions of the parties. There will always be something of a grey area as to whether the adjudicator has departed from the submissions of the parties to such an extent that he can be considered to have gone on a 'frolic of his own' and such cases turn on their own specific facts. However, the courts are slow to find a breach of natural justice in all but the clearest of circumstances.

Improper Use of Own Expertise

10.94 In a number of cases the dispute on enforcement has arisen from the adjudicators making use of their own experience. That is something the adjudicator is entitled to do and is an inherent aspect of the adjudicator assessing the information that has been provided. This is generally the reason why many adjudicators are not formally legally qualified but instead have experience and qualifications in fields such as architecture, engineering, quantity surveying, or programming. It would defeat the object of their appointment if they were not permitted to make use of that expertise in reaching their decisions. However, the principles outlined in the preceding section (namely whether the adjudicator should invite submissions on an intended approach) also apply to the use of the adjudicator's own expertise.

10.95 There is a distinction between the adjudicator using his or her own experience to assess the evidence and submissions of the parties[99] and using a completely different method from that contended by either party. In the latter case, natural justice requires that the parties be given an opportunity to comment,[100] and if no adequate opportunity is provided, there will be grounds for resisting enforcement. The question of what the adjudicator actually did will always be a fact-sensitive issue in light of the material provided by the parties and the contents of the decision, in particular whether the adjudicator has simply assessed the material provided or gone off 'on a frolic of his own'.[101] That a decision only relies on information known to both parties and on which they have made submissions may not be enough if the decision is reached on a legal basis or factual interpretation which has not been argued or

---

[95] *Shimizu Europe Ltd v LBJ Fabrications Ltd* [2003] BLR 381, [2003] EWHC 1229 (TCC) (contractual relationship governed by letter of intent but adjudicator determined a standard form applied); *Primus Build Ltd v Pompey Centre Ltd* [2009] BLR 437, [2009] EWHC 1487 (TCC) (evidence as to a party's accounts was agreed by the parties to be irrelevant but the adjudicator relied on it).

[96] *Ardmore Construction Ltd v Taylor Woodrow Construction Ltd* [2006] Scot CS CSOH 3.

[97] *PC Harrington Contractors Ltd v Tyroddy Construction Ltd* [2011] All ER (D) 162 (Apr), [2011] EWHC 813 (TCC).

[98] *Vision Homes Ltd v Lancsville Construction Ltd* [2009] BLR 525, [2009] EWHC 2042 (TCC).

[99] *All in One Building & Refurbishments Ltd v Makers UK Ltd* [2006] CILL 2321, [2005] EWHC 2943 (TCC); *Multiplex Construction (UK) Ltd v West India Quay Development Company (Eastern) Ltd* [2006] EWHC 1569 (TCC); *Cantillon Ltd v Urvasco Ltd* [2008] BLR 250, [2008] EWHC 282 (TCC) (Key Case); *Vision Homes Ltd v Lancsville Construction Ltd* [2009] BLR 525, [2009] EWHC 2042 (TCC); *Hyder Consulting (UK) Ltd v Carillion Construction Ltd* (2011) 138 Con LR 212, [2011] EWHC 1810 (TCC).

[100] *Balfour Beatty Construction Ltd v The Mayor and Burgesses of the London Borough of Lambeth* [2002] BLR 288; [2002] EWHC 597 (TCC) (Key Case); *Carillion Utility Services Ltd v SP Power Systems Ltd* [2011] CSOH 139; *Herbosh-Kiere Marine Contractors Ltd v Dover Harbour Board* [2012] BLR 177, [2012] EWHC 84 (TCC) (Key Case).

[101] *Cantillon Ltd v Urvasco Ltd* [2008] BLR 250 [2008] EWHC 282 (TCC) (Key Case).

put forward by either side.It may also mean the adjudicator has decided a dispute that was not referred or has answered a question different to that actually put to him (about which see 9.48–9.53).

Key Case: Deciding on a Basis Not Argued

Cantillon Ltd v Urvasco Ltd [2008] EWHC 282, (2008) BLR 250 (TCC)

10.96 **Facts:** Urvasco had engaged Cantillon under a JCT Standard Form Private without Quantities 1998 to carry out demolition, piling, and other works. Cantillon commenced an adjudication claiming an EOT and loss and expense for a 13-week period. Urvasco's response included its defence that the losses claimed by Cantillon did not represent the period in which the actual delay occurred, and in fact related to a different, later period. The adjudicator's decision, given by agreed extension some five months after the dispute was referred, was that Cantillon was entitled to be paid its loss and expense for a later nine-week period, calculated at an average weekly rate. The adjudicator preferred the evidence of Cantillon's programming expert. Urvasco challenged enforcement of the decision, arguing that the adjudicator should not have addressed any issue for any period other than the 13 weeks specified by Cantillon. Urvasco also argued that the adjudicator had failed to give it any or any reasonable opportunity to make submissions and adduce evidence in relation to the amount of prolongation costs being incurred in the later period. Urvasco claimed that it would be £17,000 to £60,000 lower than the sum awarded.

10.97 **Held:** The adjudicator's decision was enforced. The fact that the adjudicator may have made some mistakes in his assessment of the loss and expense did not establish that he failed to have regard to the rules of natural justice. He had to do the best that he could on the available information. The finding was within the scope of the dispute referred to the adjudicator. Urvasco had raised the fact that the delay occurred in a later period in its defence and could have predicted that he might adopt the course he did, putting in submissions accordingly.

> 57. From this and other cases, I conclude as follows in relation to breaches of natural justice in adjudication cases:
>
> (a) It must first be established that the Adjudicator failed to apply the rules of natural justice;
> (b) Any breach of the rules must be more than peripheral; they must be material breaches;
> (c) Breaches of the rules will be material in cases where the adjudicator has failed to bring to the attention of the parties a point or issue which they ought to be given the opportunity to comment upon if it is one which is either decisive or of considerable potential importance to the outcome of the resolution of the dispute and is not peripheral or irrelevant.
> (d) Whether the issue is decisive or of considerable potential importance or is peripheral or irrelevant obviously involves a question of degree which must be assessed by any judge in a case such as this.
> (e) It is only if the adjudicator goes off on a frolic of his own, that is wishing to decide a case upon a factual or legal basis which has not been argued or put forward by either side, without giving the parties an opportunity to comment or, where relevant put in further evidence, that the type of breach of the rules of natural justice with which the case of *Balfour Beatty Construction Company Ltd -v- The Camden Borough of Lambeth* was concerned comes into play. It follows that, if either party has argued a particular point and the other party does not come back on the point, there is no breach of the rules of natural justice in relation thereto.

...

74. ...

(a) If the adjudicator had jurisdiction to address the issue, it was up to the parties to put in such evidence as they thought fit to address the realistic permutations which might well apply. It is clear that Urvasco's team was confident that, if there were any such costs payable, any prolongation costs would apply in the later period.
(b) With the Referral, detailed records and other documentation about all the prolongation heads of cost was provided by Cantillon for the whole or virtually the whole of the contract period up towards the end of 2006. The same sort of exercise could have been done for the later period as it was for the earlier period. It is difficult therefore for Urvasco to say that it did not have the opportunity to address the quantum ramifications of there being a delay finding which reflected their own assertion that any prolongation occurred during the later period. The fact that they did not, and deliberately decided not to, take up that opportunity does not convert what happened in to some breach of the rules of natural justice.
(c) The Adjudicator did not, deliberately or otherwise, mislead the parties as to what he was or was not going to do. If anything, he hinted broadly that he might be finding that any compensable delay could well relate to the period when the piling work was actually being done.
(d) The fact that Cantillon on 23 October 2007 asked the Adjudicator to do something which was within his jurisdiction should have alerted Urvasco to the possibility that the Adjudicator might go down that route.
(e) The remarks made at the time that the Adjudicator should not 'make Cantillon's case' for them were not in point. He was not making Cantillon's case: he was assessing what was due to Cantillon on the evidence and argument. He was entitled to investigate the facts and evidence as presented.
(f) There was time, as the parties well knew, after the 23 October 2007 for Urvasco to put in argument and indeed evidence about the later period. It was known that Dr Mastrandrea would take up to about 4 weeks to produce his decision. Urvasco for their own reasons decided not to put in any such argument or evidence even though, squarely, Cantillon had asked the Adjudicator on 23 and 24 October 2007 to ascertain the loss and expense by reference to the evidence and the prolongation period found by him.

Herbosh-Kiere Marine Contractors Limited v Dover Harbour Board [2012] EWHC 84 (TCC), [2012] BLR 177

10.98 **Facts:** The claimant contractor referred to adjudication the valuation of its final account. The claim was calculated on the basis of identified periods of delay to identified resources. The responding party accepted the basis of calculation but disputing the periods of delay. The adjudicator calculated the claim using a composite rate which was not a basis for which either party had contended.

10.99 **Held:** The decision was not enforced. By using a method of assessment that was no part of the dispute referred the adjudicator had exceeded his jurisdiction. Akenhead J also found, *obiter*, that by not allowing the parties the opportunity to make submissions as to the calculation method he adopted, the adjudicator had breached the rules of natural justice and that breach had a material effect on the outcome of the dispute:

33. In essence, and doubtless for what he believed were good and sensible reasons, the adjudicator has gone off 'on a frolic of his own' in using a method of assessment which neither party argued and which he did not put to the parties. In some cases, this may not be sufficient to prevent enforcement of the decision where the 'frolic' makes no material difference to the outcome of the decision. Thus, an adjudicator who refers to a legal authority which neither party relied upon, may have his or her decision enforced nonetheless if

the application of that legal authority obviously makes no difference to the outcome. The breach of the rules of natural justice has to be material. Here, for the reasons indicated above, the breach is material and has or has apparently led to a very substantial financial difference in favour of HKM but necessarily against the interests of DHB.

34. It follows from the above that the adjudicator's decision can not be enforced because not only has he exceeded his jurisdiction by addressing and finding a method of assessment which formed no part of the dispute referred to him but also he has breached the rules of natural justice, doubtless unwittingly, by deciding the case not only on the basis not argued by either party at any stage but also without giving each party the opportunity to make submissions at least on the method of assessment which the adjudicator considered that he should adopt.

Table 10.7 Table of Cases: Deciding on a Basis Not Put Forward (successful challenges are indicated by shaded entries)

Title	Citation	Issue
Karl Construction (Scotland) Ltd v Sweeney Civil Engineering (Scotland) Ltd	[2002] SCLR 766 (Extra Division of Inner House), Lords Marnock, Dawson, and Clarke	**Adjudicator decided dispute on different legal basis** By their submissions, the parties agreed that the contractual payment mechanism was compliant with the Act. The adjudicator found that the contract was non-compliant and the Scheme provisions applied. An adjudicator is obliged to apply the relevant law as he sees it; the decision on the law cannot be circumscribed by the agreement of the parties. There was no obligation to invite submissions where the adjudicator decided the law in a different way from that put forward by the parties. Enforced.
Balfour Beatty Construction Ltd v The Mayor and Burgesses of the London Borough of Lambeth	[2002] BLR 288; [2002] EWHC 597 (TCC), Judge LLoyd	**Adjudicator's own knowledge and experience** The adjudicator determined his own critical path, in effect making the referring party's case for it, but did not invite comment from either party. This was such a potentially serious breach of fairness that the decision was invalid. Not enforced.
Shimizu Europe Ltd v LBJ Fabrications Ltd	[2003] BLR 381, [2003] EWHC 1229 (TCC), Judge Kirkham	**Adjudicator ignored the parties' agreement as to the basis of their contract** The parties had agreed that their contractual relationship was governed by a letter of intent but the adjudicator found other conditions bound the parties. *Obiter*: an adjudicator acts in breach of natural justice where he decides the dispute on a different contractual basis from that agreed by the parties without inviting submissions. Not enforced (primarily on jurisdictional grounds).
Palmac Contracting Ltd v *Park Lane Estate Ltd*	[2005] BLR 301, [2005] EWHC 919 (TCC), Judge Kirkham	**Issue in decision had been subject of invitation for comment** The court found that there was no agreed position as to the requirements for the service of applications for payment from which the adjudicator departed. In any event there could be no breach of natural justice where the adjudicator received comments from the

(Continued)

Table 10.7 *Continued*

Title	Citation	Issue
		parties, incorporated no new material and made no investigation of his own in relation to the matter about which subsequent complaints were made. Enforced.
All in One Building & Refurbishments Ltd v Makers UK Ltd	[2006] CILL 2321, [2005] EWHC 2943 (TCC), Judge Wilcox	**Conclusion on evidence within professional competence of adjudicator** There was no breach of natural justice where the adjudicator reached a judgment having formed his own view as to the appropriate figure for lost profits and overheads based on the evidence presented by both parties. The resisting party had an opportunity to make submissions on the enforcing party's case. Enforced.
Ardmore Construction Ltd v Taylor Woodrow Construction Ltd	[2006] Scot CS CSOH 3 (Scottish Court of Sessions), Lord Clarke	**Issue raised for first time in decision** The adjudicator decided the basis of liability to be paid overtime on a basis that had never been raised or discussed before the adjudicator or on which the resisting party had been given an opportunity to provide evidence or submissions. Where an adjudicator reaches his decision with an objective disregard for fair play which appears to have had a substantial and material effect on the decision, the courts should be prepared to intervene. Not enforced.
Multiplex Construction (UK) Ltd v West India Quay Development Company Eastern Ltd	(2006) 111 Con LR 33, [2006] EWHC 1569 (TCC), Ramsey J	**Distinction between assessment of evidence and adoption of own method** There was no breach of natural justice where an adjudicator based his decision on his assessment of the programming material provided by the parties after making due allowance for his concerns about aspects of it. This was not the imposition of a new methodology and was not based on materials other than those relied on by the parties. Enforced.
Humes Building Contracts Ltd v Charlotte Homes (Surrey) Ltd	unreported, 4 January 2007 (TCC), Judge Gilliland	**Failure to consider defence following erroneous finding that withholding notice required** The adjudicator erroneously determined he could not consider a defence based on defective works in the absence of a withholding notice. Neither party had suggested that a withholding notice was required and the adjudicator did not invite submissions on the point. Not enforced.
Cantillon Ltd v Urvasco Ltd	[2008] BLR 250, [2008] EWHC 282 (TCC), Akenhead J	**EOT derived from parties' submissions** The adjudicator awarded an EOT based on a different period of time from that claimed by the referring party but it was based on the responding party's own case as to when the delay happened. There will be a material breach of natural justice where the adjudicator fails to bring to the attention of the parties a point or issue they ought to be given the opportunity to comment upon which is either decisive for

Title	Citation	Issue
		or of considerable importance to the outcome. Enforced.
(1) CSC Braehead Leisure Ltd and (2) Capital & Regional (Braehead) Ltd v Laing O'Rourke Scotland Ltd	[2008] CSOH 119 (Scottish Outer House of Sessions), Lord Menzies	**Further submissions on adjudicator's figures not permitted** The adjudicator had indicated he would accept further submissions from the parties on the figures set out in his decision. However, the parties only had four minutes to respond before the adjudicator's jurisdiction expired. The adjudicator had considered the material evidence and formed a view, and both parties had been treated equally in receiving the decision at the same time. Enforced.
Primus Build Ltd v Pompey Centre Ltd	[2009] BLR 437, [2009] EWHC 1487 (TCC), Coulson J	**Alternative calculation of the referring party's claim based on documents the parties had agreed should be ignored** *Obiter*: where an adjudicator considers that the referring party's claims as made cannot be sustained, yet he himself identifies a possible alternative way in which a claim of some sort could be advanced, he will normally be obliged to raise that point with the parties in advance of his decision. It is even more important to do so if the alternative approach is based on material the parties have agreed should be ignored. Not enforced (primarily on the grounds jurisdiction).
Barr Ltd v Klin Investment UK Ltd	[2009] CSOH 104 (Scottish Outer House, Court of Session), Lord Glennie	**Adjudicator did not invite further submissions on findings of fact** Provided the factual conclusions are based on the material provided by the parties there is no obligation on the adjudicator to provide his proposed findings to the parties and seek further submissions—even if they represent a 'middle ground' between the parties' submissions. The dispute concerned entitlement to payment of a sum mentioned in a withholding notice. Therefore it was open to both parties to advance any arguments about entitlement to withhold that sum. Enforced.
Vision Homes Ltd v Lancsville Construction Ltd	[2009] BLR 525, [2009] EWHC 2042 (TCC), Christopher Clarke J	**Decision on applicable contract terms inconsistent with agreement of parties** Following significant omissions from the works, the operability of various contractual provisions was in issue. Where the adjudicator's decision is a development and exposition of the contentions of the parties there is no breach of natural justice. Not enforced on other grounds (timing of service of notice of intention to refer).
Straw Realisations (No. 1) Ltd v Shaftsbury House (Developments) Ltd	[2011] BLR 47, [2010] EWHC 2597 (TCC), Edwards-Stuart J	**Party claimed adjudicator had produced his own programme** Where an adjudicator makes it plain that he decided the dispute only on the basis of the evidence and submissions of the parties there was no breach of natural justice. Enforced.

(Continued)

Table 10.7 *Continued*

Title	Citation	Issue
Ellis Building Contractors Ltd v Vincent Goldstein	[2011] CILL 3049, [2011] EWHC 269 (TCC), Akenhead J	**Adjudicator's findings on basis of contract between the parties** The adjudicator made a finding that it was open to him to make based on the material and submissions that were before him. Enforced.
Hyder Consulting (UK) Ltd v Carillion Construction Ltd	(2011) 138 Con LR 212, [2011] EWHC 1810 (TCC), Edwards-Stuart J	**Party claimed adjudicator applied a different methodology** The adjudicator was not obliged to choose between one set of submissions and the other; it was open to him to adopt his own approach if he had decided to reject the submissions of both parties. There was no breach of natural justice where the adjudicator did not use any information which the responding party had not been able to consider, and based his method on a contractual provision in respect of which both parties had made submissions. Enforced
Carillion Utility Services Ltd v SP Power Systems Ltd	[2011] CSOH 139, Lord Hodge	**Adjudicator applied own view of appropriate rate** There was a breach of natural justice where the adjudicator applied a commercial rate that materially affected his calculation. There was no evidence for the rate in the material before him and he did not invite submissions on the rate used. Not enforced.
Herbosh-Kiere Marine Contractors Ltd v Dover Harbour Board	[2012] BLR 177, [2012] EWHC 84 (TCC), Akenhead J	**No dispute over approach to assessment of delay claim** The dispute referred related to the determination of the periods of delay for which the employer was responsible. The parties agreed on the approach to valuation, namely that any delay should be valued using specific resource rates and specific periods of delay. The adjudicator's decision calculated an award based on the overall period of delay multiplied by a composite rate, which was not the dispute referred and had therefore decided the case on a basis not argued by either party, without inviting submissions on his intended approach. Not enforced (also excess of jurisdiction).
Berry Piling Systems Ltd v Sheer Projects Ltd	(2012) 141 Con LR 225, [2012] EWHC 241 (TCC), Edwards-Stuart J	**Delay claim failed for want of proof** The adjudicator's decision was primarily based on the responding party's (i) failure to provide any proper analysis of delay and (ii) acceptance that it was contractually responsible for the matters that gave rise to delay. The decision was not founded on a point which had not been raised with the parties. Enforced.

11

CHALLENGES BASED ON BIAS AND PREDETERMINATION

(1) The Principles of Bias and Predetermination

Actual and Apparent Bias

11.01 As explained in Chapter 10, an adjudicator is required to act in accordance with the principles of natural justice. That includes an obligation to act fairly and without bias.[1] That obligation is reflected in s. 108(2)(e) of the Housing Grants Construction and Regeneration Act 1996 ('the 1996 Act') which provides that an adjudicator is required to be impartial.[2] The Court of Appeal explained the concept of bias in *In re Medicaments and Related Classes of Goods (No. 2)* (2000):[3]

> 37. Bias is an attitude of mind which prevents the judge from making an objective determination of the issues that he has to resolve. A judge may be biased because he has reason to prefer one outcome of the case to another. He may be biased because he has reason to favour one party rather than another. He may be biased not in favour of one outcome of the dispute but because of a prejudice in favour of or against a particular witness which prevents an impartial assessment of the evidence of that witness. Bias can come in many forms. It may consist of irrational prejudice or it may arise from particular circumstances which, for logical reasons, predispose a judge towards a particular view of the evidence or issues before him.

11.02 There are two types of bias: actual and apparent. However, in practice, findings of actual bias are rare because of the difficulties of proof.[4] Consequently, in order to resist enforcement of an adjudicator's decision, it is more likely that a party will argue that the circumstances are

---

[1] *AMEC Capital Projects Ltd v Whitefriars City Estates Ltd* (2005) 1 All ER 723, [2004] EWCA Civ 1418 (CA) (Key Case) at para.14.

[2] This obligation is reiterated in para. 12(a) of the Scheme for Construction Contracts ('the Scheme') and most adjudication rules.

[3] *Sub. nom. Director General of Fair Trading v Proprietary Association for Great Britain*, [2001] 1 WLR 700 (CA), per Lord Phillips.

[4] *Lanes Group Plc v Galliford Try Infrastructure Ltd (t/a Galliford Try Rail)* [2001] 1 WLR 700 (CA), [2011] EWCA Civ 1617 (CA) (Key Case), per Jackson LJ at para. 46.

such that there is a real possibility or appearance of bias. It is paramount that 'justice should not only be done but should manifestly and undoubtedly be seen to be done'.[5]

The 'Fair-minded and Informed Observer' Test

11.03 The subjective impression formed by the aggrieved party is not enough to establish bias.[6] The test for apparent bias was clarified in *Medicaments*:

> 86. ... The court must first ascertain all the circumstances which have a bearing on the suggestion that the judge was biased. It must then ask whether those circumstances would lead a fair-minded and informed observer to conclude that there was a real possibility, or a real danger, the two being the same, that the tribunal was biased.

11.04 *Medicaments* was applied in the context of adjudication under the 1996 Act in *Glencot Development and Design Co. Ltd v Ben Barrett & Son (Contractors) Ltd* (2001) (Key Case)[7] where particular consideration was given to the summary nature of the statutory procedure:

1. The adjudicator must conduct the proceedings in accordance with the rules of natural justice or as fairly as limitations imposed by Parliament permit.
2. Regardless of which adjudication rules applied, the concept of 'impartiality' in the Act must be given its usual meaning, which is the same meaning as at common law which entitles everyone to a fair hearing by an independent and impartial tribunal.
3. The test is an objective one: whether the circumstances would lead to a fair minded and informed observer, having considered the facts, to conclude that there was a real possibility, or a real danger, that the tribunal was biased.[8]

11.05 *Medicaments* was subsequently referred to by the Court of Appeal in *AMEC Capital Projects Ltd v Whitefriars City Estates Ltd* (2004) (Key Case)[9] which confirmed that the 'fair-minded and informed observer' test applied to adjudication. In 2006, the House of Lords confirmed that this test was applicable when considering an allegation of apparent bias.[10] In 2011 the Court of Appeal in *Lanes Group Plc v Galliford Try Infrastructure Ltd* (2011) (Key Case)[11] discussed some of the difficulties in the application of this test:

> 51. One complication in recent years is the elaboration of the 'fair minded observer' test. ... [T]he fair minded observer must be assumed to know all relevant publicly available facts. He or she must be assumed to be neither complacent nor unduly sensitive or suspicious. He or she must be assumed to be fairly perspicacious, because he or she is able 'to distinguish between what is relevant and what is irrelevant, and when exercising his judgment to decide what weight should be given to the facts that are relevant': see *Gillies* at paragraph 17.
>
> 52. There are conceptual difficulties in creating a fictional character, investing that character with an ever growing list of qualities and then speculating about how such a person would answer the question before the court. The obvious danger is that the judge will simply project onto that fictional character his or her personal opinions. Nevertheless, this approach is established by high authority ...

11.06 There are myriad scenarios in which the apprehension of bias may arise and each case will be decided on its own facts and circumstances by applying the above test. The courts have

[5] *R v Sussex Justices ex p. McCarthy* [1924] 1 KB 256 (KBD).
[6] *Makers UK Ltd v Camden London Borough Council* [2008] BLR 470, [2008] EWHC 1836 (TCC).
[7] [2001] BLR 207 (TCC). This case is discussed at 11.35.
[8] The latter part of this test is laid down in *Magill v Weeks* [2002] 2 AC 357, [2001] UKHL 67 at para. 103.
[9] [2005] BLR 1, [2004] EWCA Civ 1418 (CA).
[10] *Gillies v Secretary of State for Work & Pensions* [2006] 1WLR 781, [2006] UKHL 2.
[11] [2012] BLR 121, [2011] EWCA Civ 1617 (CA), per Jackson LJ.

confirmed that it would be dangerous or futile to attempt to exhaustively define each of the possible forms of bias in which there were real grounds for 'doubting the ability of the judge to ignore extraneous considerations, prejudices and predilections and bring an objective judgment to bear on the issues before him'.[12] However, it is useful to identify some of the scenarios in which a real danger of bias could arise:

1. Personal friendship or animosity between the tribunal and a member of the public involved in the case, particularly if the credibility of that individual could be significant in the decision of the case.[13]
2. Where a tribunal may have previously rejected the evidence of a particular person in such outspoken terms as to throw doubt on the tribunal's ability to approach that person's evidence with an open mind on any later occasion.[14]
3. If the tribunal had expressed views, particularly in the course of the proceedings, on a particular issue in such extreme and unbalanced terms as to throw doubt on his ability to try the issue with an objective judicial mind.[15]

11.07 While the guidelines and tests for establishing bias can be stated relatively simply, their successful application is more difficult and in most cases even apparent bias is difficult to prove. Following the decision of the Court of Appeal in *AMEC v Whitefriars* (2004) (Key Case),[16] the party resisting enforcement has succeeded in only one of the 14 cases where apparent bias was raised as a ground for resisting enforcement.[17] By comparison, in the period before *AMEC v Whitefriars*, enforcement was successfully challenged on the grounds of apparent bias in six of nine reported cases.[18]

Evidence from the Adjudicator

11.08 The explanation of the tribunal under review as to its knowledge or appreciation of the circumstances said to give rise to apparent bias is relevant and the fair-minded observer test should assess whether there was a possibility of bias in light of any such explanation.[19] Evidence from the adjudicator has been considered in a number of cases.[20] However, such

[12] *Locabail (UK) Ltd v Bayfield Properties Ltd* [2000] 2 WLR 870, [2000] QB 451 at para. 25.

[13] *Ibid.*

[14] *Ibid.* For example, in *Ealing London Borough Council v Jan* [2002] EWCA Civ 329, it was decided that a judge should not perform the retrial where he had twice said of the respondent that he could not trust him 'further than he could throw him'; or in *Timmins v Gormley* [2004] EWCA Civ 1761 (CA) in which sufficient danger was found to exist that a personal injuries matter was not returned to a recorder who had published articles expressing 'pronounced pro-claimant anti-insurer views' (as referred to in *AMEC Capital Projects Ltd v Whitefriars City Estates Ltd* [2005] BLR 1, [2004] EWCA Civ 1418 (CA) (Key Case) at para. 21).

[15] *Vakauta v Kelly* (1989) 167 CLR 568, (1988) 13 NSWLR 502, referred to in *Locabail (UK) Ltd v Bayfield Properties Ltd* [2000] 2 WLR 870, [2000] QB 451 (CA) at para. 25.

[16] [2005] BLR 1, [2004] EWCA Civ 1418 (CA).

[17] See Table 11.1. It was *Mott MacDonald Ltd v London & Regional Properties Ltd* (2007) 113 Con LR 33, [2007] EWHC 1055 (TCC) which concerned an apprehension of impartiality due to a lien over the adjudicator's fees. There was a further decision of the TCC refusing to enforce an adjudicator's decision for apparent bias, but it was overturned by the Court of Appeal in *Lanes Group Plc v Galliford Try Infrastructure Ltd (t/a Galliford Try Rail)* [2012] BLR 121, [2011] EWCA Civ 1617 (CA) (Key Case).

[18] See Table 11.1. These cases included *AMEC Capital Projects Ltd v Whitefriars City Estates Ltd* (2005) 1 All ER 723, [2004] EWCA Civ 1418 (CA) (Key Case) at first instance.

[19] *In re Medicaments and Related Classes of Goods (No. 2) sub. nom. Director General of Fair Trading v Proprietary Association for Great Britain and anor* [2001] 1 WLR 700 (CA) at 86.

[20] Including *Discain Project Services v Opecprime Development Ltd* [2000] BLR 402; *Woods Harwick Ltd v Chiltern Air Conditioning Ltd* [2001] BLR 23 (TCC); *A & S Enterprises Ltd v Kema Holdings Ltd* [2005] BLR 76 (TCC), [2004] EWHC 3365 (TCC); *Makers UK Ltd v Camden London Borough Council* [2008] BLR 470, [2008] EWHC 1836 (TCC); *Fileturn Ltd v Royal Garden Hotel Ltd* [2010] BLR 512, [2010] EWHC 1736 (TCC).

evidence should only extend to a neutral factual account of the conduct of the adjudication; if the adjudicator goes any further, it could support the allegations of apparent bias.[21]

11.09 One TCC judge has suggested that it may be appropriate for the court to notify any adjudicator in the event that serious irregularity is asserted, in the same way as occurs with arbitration challenges.[22] In *Discain Project Services Ltd v Opecprime Development Ltd (No. 2)* (2001),[23] the adjudicator was called by the court rather than either party so that the parties would have the same opportunity to cross-examine him.

Bias and Natural Justice

11.10 In addition to the obligation to adopt a fair procedure, the adjudicator must also be free from bias or any reasonable perception of bias. As the Court of Appeal identified in *AMEC Capital Projects Ltd v Whitefriars City Estates Ltd* (2004) (Key Case),[24] a fair procedure and bias are conceptually distinct:

> 14. ... It is quite possible to have a decision from an unbiased tribunal which is unfair because the losing party was denied an effective opportunity of making representations. Conversely, it is possible for a tribunal to allow the losing party an effective opportunity to make representations, but be biased. In either event, the decision will be in breach of natural justice, and be liable to be quashed if susceptible to judicial review, or (in the world of private law) to be held to be invalid and unenforceable.
>
> 22. ... It is only where the defendant has advanced a properly arguable objection based on apparent bias that he should be permitted to resist summary enforcement of the adjudicator's award on that ground.

11.11 Notwithstanding the conceptual difference between natural justice and bias, a breach of natural justice may be such that it gives rise to an appearance of bias on the part of the adjudicator, in turn giving rise to parallel challenges on grounds of both a breach of natural justice and apparent bias.

Predetermination

11.12 Predetermination arises when a judge or other decision-maker, in this instance, an adjudicator, reaches a final conclusion before being in possession of all the relevant evidence and arguments. In *Lanes Group Plc v Galliford Try Infrastructure Ltd* (2011),[25] the Court of Appeal confirmed that predetermination is 'conceptually somewhat different' from bias, although it is often treated as a species of bias. The test of the fair-minded observer (outlined above at 11.03–11.05) is also applicable to allegations of apparent pre-determination by an adjudicator.[26]

(2) Possible Grounds for Challenge Based on Bias or Predetermination

Prior Connection with Parties or Matter

11.13 The Scheme does not say that an adjudicator shall be independent but paragraph 4 requires that a person requested or selected to act as an adjudicator shall not be an employee of any party to the dispute and shall declare 'any interest, financial or otherwise, in any matter

21 As happened in *Woods Harwick Ltd v Chiltern Air Conditioning Ltd* [2001] BLR 23 (TCC).

22 *Makers UK Ltd v Camden London Borough Council* [2008] BLR 470, [2008] EWHC 1836 (TCC) at 35(8).

23 [2001] BLR 285, see para. 13.

24 [2005] BLR 1, [2004] EWCA Civ 1418 (CA).

25 [2012] BLR 121, [2011] EWCA Civ 1617 (CA).

26 *Ibid.*, per Jackson LJ at para. 50.

relating to the dispute'. This means that the appointment of an employee of either party will be an invalid appointment. Such an adjudicator will have no jurisdiction to decide the matter and any decision will not be enforceable. It is less clear what remedy is available to a party if the adjudicator validly selected and appointed by a nominating body declares an interest in the dispute that is objectionable to one party but still will not resign.

11.14 In *Makers (UK) Ltd v London Borough of Camden* (2008)[27] one party suggested to RIBA to appoint a named adjudicator (because he was qualified as an architect and as a lawyer) and RIBA acceded to that requested. The challenge based on bias was rejected and the court found there was no duty to consult the other party where suggestions about nomination were made.

11.15 A claim of bias commonly arises where an adjudicator has been involved in a previous dispute concerning the project or the parties, including in some instances re-adjudication of exactly or essentially the same dispute following an earlier decision being held to be unenforceable. However, the mere fact that a tribunal has previously decided the issue is not of itself sufficient to justify a conclusion of apparent bias.[28] As long as the circumstances are not such as to lead the fair-minded observer to conclude that the adjudicator approached the subsequent adjudication with a closed mind, then there will be no finding of bias. That will be so even if the same decision is reached, particularly if there are no different or further submissions or evidence. That the same decision is reached in such circumstances has been described as 'entirely unsurprising'.[29]

11.16 Equally, the mere fact of a previous relationship or connection between the adjudicator and one of the parties or their representatives will not, without more, lead to a finding of bias. It will have to be shown that other circumstances exist, such as a personal rather than a professional relationship, that work provided by the party or its adviser accounts for a material proportion of the adjudicator's practice or that the connection was such that the adjudicator had particular knowledge of the party that may be relevant to the dispute.[30]

11.17 In *Lanes Group Plc v Galliford Try Infrastructure Ltd* (2011)[31] the adjudicator nominated by the ICE had previously acted against the referring party in a series of adjuications that had become acrimonious. The referring party considered that would make it difficult for the nominated adjudicator to appear impartial so did not make the referral but instead requested a further nomination. In hearing an application by the responding party for an injunction to prevent the referring party continuing the adjudication with the second nominated adjudicator, the judge found that the circumstances did not give rise to an appearance of bias. Further, the failure to pusue an adjudication following the nomination of an adjudicator and instead seeking the appointment of a different adjudicator was not a repudiation of the adjudication agreement and the adjudication by the second nominated adjudicator was allowed to continue.

11.18 The fact that an adjudicator has previously acted as dispute-resolver in proceedings involving one of the parties is also not, of itself, sufficient grounds to allege apparent bias or a lack of impartiality: *Andrew Wallace Ltd v Jeff Noon* (2008).[32] In that case the adjudicator had been a mediator of an unrelated dispute with AWL only two days before being appointed as adjudicator by RIBA.

---

[27] [2008] EWHC 1836, [2008] BLR 470.

[28] *AMEC Capital Projects Ltd v Whitefriars City Estates Ltd* [2005] BLR 1, [2004] EWCA Civ 1418 (CA) (Key Case).

[29] *Michael John Construction v Golledge and ors* [2006] EWHC 71 (TCC).

[30] *Fileturn Ltd v Royal Garden Hotel Ltd* [2010] BLR 512, [2010] EWHC 1736 (TCC).

[31] [2001] BLR 438, [2011] EWHC 1035 (TCC). This case was subject to a number of applications and appeals, discussed more fully in 9.33, 9.59–9.63 and 11.28.

[32] [2009] BLR 158.

The problem partly arose because when RIBA asked if the adjudicator had an existing relationship with either of the parties, the adjudicator said no. Whilst it could be said that the bias accusation might have been avoided had he made the disclosure before being appointed, the editors of the Building Law Reports have noted that the Court of Appeal in *Taylor v Lawrence* (2003)[33] said that 'judges should be circumspect about declaring the existence of a relationship where there is no real possibility of it being regarded by a fair-minded and informed observer as raising the possibility of bias'.[34]

11.19 Ultimately, the TCC Judge in *Jeff Noon* found the allegation of bias unsubstantiated and was led to this conclusion by the fact that the adjudicator: (i) had no personal knowledge of the parties; (ii) was a professionally qualified arbitrator; (iii) was appointed by RIBA as opposed to being a party appointee; and (iv) had no current relationship with either party.

11.20 If the adjudicator is in possession of relevant knowledge as a result of a relationship or prior adjudication, then unless that knowledge is made known to the other party there will be stronger grounds for a finding of bias. That was the case in *Pring & St Hill Limited v C. J. Hafner (t/a Southern Erectors)* (2002)[35] where confidentiality requirements prevented the responding party being informed of the matters raised in the earlier adjudication, leading to a finding of apparent bias:

> 23. ... In my view, in these circumstances, there is a very real risk that an adjudicator in [this adjudicator's] position ... would be carrying forward from an earlier adjudication not merely what he had seen or been told but also the judgments which he had formed, the opinions which he had reached, which led him to conclude that sum was the correct measure of McAlpine's damages recoverable from PSH.

11.21 That same principle emerged in the subsequent case of *London & Amsterdam Properties Ltd v Waterman City Estates Ltd* (2003)[36] (in which *Pring* was not cited) where one party unsuccessfully alleged relevant knowledge which the adjudicator denied. Judge Wilcox stated:

> 93. ... Had he been in possession of relevant information which affected his decision, he was under a duty to tell the parties. If he was bound by confidentiality and unable to do so he should have recused himself.
>
> 94. A defendant seeking to impugn an adjudication cannot merely raise the spectre of bias without founding the allegation upon some credible evidence and demonstrating that the knowledge or information was central to the decision. In this case the Adjudicator addressed his mind to the risk, and stated that he knew of no matters giving rise to risk. His duty to guard against bias on this basis would have been a continuing one. There is no basis upon which it can be said that he was in breach of that obligation.

11.22 Such circumstances are likely to give rise to a parallel objection on the grounds of natural justice, the effect being that the adjudicator reached his decision in reliance on material that was not provided to a party.

Unilateral Communication

11.23 The court has given guidance that unilateral communication between the adjudicator and one party should be avoided but if it is considered absolutely necessary, it should be done in

[33] [2003] QB 528 at para. 64.

[34] However see Lord Phillips in *Smith v Kvaerner Cementation Foundations Ltd* [2007] 1 WLR 370, [2006] EWCA Civ 242 citing Lord Denning's proposition, in *Metropolitan Properties v Lannon* [1968] 3 WLR 694 that no man could act as advocate or adviser for or against a party in one dispute and at the same sit as judge of that party in another proceeding without people inevitably thinking he was biased.

[35] [2004] 20 Const LJ 402, [2002] EWHC 1775 (TCC).

[36] [2004] BLR 179, [2003] EWHC 3059 (TCC).

writing.[37] The reason is plain: where there has been unilateral oral communication it is very likely to raise the suspicions of the other party as to the nature of the discussion and how it may have affected the mind of the adjudicator. However, such subjective concern is insufficient alone to resist enforcement. Where it is apparent that such discussions were of a purely administrative nature then there is unlikely to be a finding of apparent bias.[38] However, where the substance of the dispute has been discussed between the adjudicator and just one party with the other party not being informed of the full detail of what was discussed and/or given an opportunity to make submissions in respect of the matters discussed, then it is more likely that apparent bias will be established.[39]

11.24 The same parallel natural justice objection as set out in relation to prior connections may also arise in such circumstances, but only if it is shown that information imparted during the unilateral communication materially affected the adjudicator's decision.

Approach to Evidence

11.25 Attempts have been made to resist enforcement on the assertion that terms of the decision itself would lead the fair-minded observer to conclude that there was a real risk the adjudicator was biased. Such concerns have been based on the approach to or treatment of expert[40] or factual[41] witnesses or costs[42] or the way the evidence has been used to support a conclusion.[43] Such challenges have only succeeded in exceptional circumstances.

Access to 'Without Prejudice' Material

11.26 Parties have attempted to resist enforcement on the basis that there was apparent bias as a result of the adjudicator being provided with, or reference being made to, 'without prejudice' material. It is generally argued that there is a real risk that such material could be perceived to influence the adjudicator's view of the merits of the position of the party that made the offer or otherwise asserted a position. Such a challenge is very unlikely to succeed where the adjudicator has been made aware of the mere fact of an offer having been made.[44] Negotiations and offers prior to a referral to adjudication are not uncommon and probably the norm. An adjudicator will therefore generally expect there to have been negotiations and offers made and actual knowledge that is the case does not lead the informed fair-minded observer to consider that the adjudicator's mind has been affected.[45]

11.27 Whilst the particular circumstances in the reported cases have not been sufficient to establish apparent bias, it would appear that it remains a possible ground for resisting enforcement. In *Ellis Building Contractors Ltd v Vincent Goldstein* (2011),[46] following a review of the authorities, the judge commented, *obiter*, that 'Where an adjudicator decides a case primarily upon the basis of wrongly received "without prejudice" material, his or her decision may well not be enforced.' However, absent such clear influence on the decision, such challenges will be unlikely to succeed.

---

37 *Makers UK Ltd v Camden London Borough Council* [2008] BLR 470, [2008] EWHC 1836 (TCC) at para. 37.

38 *Ibid.*

39 As in *Discain Project Services Ltd v Opecprime Development Ltd* [2000] BLR 402, [2004] EWHC 3365 (TCC).

40 *Balfour Beatty Engineering Services (HY) Ltd v Shepherd Construction Ltd* (2009) 127 Con LR, [2009] EWHC 2218 (TCC).

41 *A&S Enterprises Ltd v Kema Holdings Ltd* [2005] BLR 76 (TCC), [2004] EWHC 3365 (TCC).

42 *Aveat Heating Ltd v Jerram Falkus Construction Ltd* (2007) 113 Con LR 13, [2007] EWHC 131 (TCC); *Camillin Denny Architects Limited v Adelaide Jones & Co. Limited* [2009] BLR 606, [2009] EWHC 2110 (TCC).

43 *Barr Ltd v Klin Investment UK Ltd* [2009] CILL 2787, [2009] CSOH 104 (OH).

44 *Volker Stevin Ltd v Holystone Contracts Ltd* [2010] EWHC 2344 (TCC).

45 *Ibid.*

46 [2011] CILL 3049, [2011] EWHC 269 (TCC) at para. 29(b).

Predetermination

11.28 Bias towards or against a particular party could lead an adjudicator to predetermine a dispute. However, there may also be circumstances in which no bias exists but an adjudicator makes up his or her mind (or appears to do so) before all evidence or arguments are made and then fails to properly consider such submissions. There is a risk of such an allegation arising where, for example, an adjudicator issues a draft decision or 'preliminary views' document for consideration.[47] Whilst taking such a step may be helpful to avoid allegations of breach of natural justice or excess of jurisdiction (by giving the parties an opportunity to make submissions on the points), an adjudicator must not reach a final decision prematurely:

> 56. There is nothing objectionable in a judge setting out his or her provisional view at an early stage of proceedings, so that the parties have an opportunity to correct any errors in the judge's thinking or to concentrate on matters which appear to be influencing the judge. Of course, it is unacceptable if the judge reaches a final decision before he is in possession of all relevant evidence and arguments which the parties wish to put before him. There is, however, a clear distinction between (a) reaching a final decision prematurely and (b) reaching a provisional view which is disclosed for the assistance of the parties.

Key Cases: Bias and Predetermination

Lanes Group Plc v Galliford Try Infrastructure Ltd (t/a Galliford Try Rail) [2012] BLR 121, [2011] EWCA Civ 1617 (CA)

11.29 **Facts:** Before the response had been served, the adjudicator had communicated to the parties a preliminary view in a document that was similar in form to a final decision which included phrases such as 'I find' and 'I hold'. However, there was a paragraph at the beginning of the preliminary view which expressly said that such statements:

> are not and not intended to be decisions of the adjudicator but preliminary views and findings of fact preparatory to the decision. The preliminary views and findings are a step in making the decision and I am not bound by them. I do not commit myself to communicate nor issue amendments or further Preliminary Views and Findings of Fact.

11.30 Many of the conclusions in the preliminary views document were reflected in the final decision. The TCC judge concluded that there was a real risk that the adjudicator had prejudged the dispute before considering the responding party's submissions with the result that apparent bias was made out and the adjudicator's decision was a nullity.

11.31 **Held:** The Court of Appeal overturned the TCC's decision. Jackson LJ (with whose judgment Richards and Stanley Burnton LJJ agreed) reviewed the relevant legal principles in relation to bias and predetermination and then stated:

> 57. In my view the fair-minded observer, with all the admirable qualities identified above, would have no difficulty in deciding this case. He would characterise the Preliminary View as a provisional view, disclosed for the assistance of the parties, not as a final determination reached before Mr Atkinson had considered Lanes' submissions and evidence ...
>
> 59. On the apparent bias issue I come to a different conclusion from the judge. In my view the adjudicator's decision was not tainted by apparent bias or apparent pre-determination. Therefore Mr Atkinson's award was and is enforceable.

[47] *Lanes Group Plc v Galliford Try Infrastructure Ltd (t/a Galliford Try Rail)* [2011] EWCA Civ 1617 (CA) (Key Case).

> 60. I am re-inforced in this conclusion by the fact that we are dealing with an adjudication decision, not an arbitration award or a judicial decision. Adjudication is a rough and ready process carried out at great speed. Vast masses of submissions and evidence have to be assimilated by the adjudicator in a short space of time. The adjudicator will fashion his procedure in whatever way enables him to discharge his onerous duties most swiftly, effectively and fairly. See clause 5.5 of the ICE Adjudication Procedure and paragraph 13 of the Scheme. An adjudication decision is not final. It is only binding until such time as the parties have concluded their litigation or their arbitration or their settlement negotiations or some other form of ADR.
>
> 61. Because adjudication has all these features, courts are reluctant to strike down adjudication decisions for breach of natural justice or on similar grounds, unless the complainant's case is clearly made out: see the judgment of the Court of Appeal in *Carillion Construction Ltd v Devonport Royal Dockyard Ltd* [2005] EWCA Civ 1358, [2006] BLR 15 at paragraphs 52–53 and 84–87.

AMEC Capital Projects Ltd v Whitefriars City Estates Ltd [2005] BLR 1, [2004] EWCA Civ 1418 (CA)

11.32 **Facts:** AMEC was engaged under the JCT Standard Form of Building Contract (with Contractor's Design) 1998 to carry out pre-construction works. AMEC referred a dispute in relation to the final account to an adjudicator who was not the named adjudicator. That adjudication resulted in a decision in favour of AMEC but it was found to be unenforceable because the adjudicator had not been appointed in accordance with the contractual adjudication rules and therefore lacked jurisdiction. AMEC re-adjudicated the dispute, applying to RIBA to nominate an adjudicator, the contractual appointment mechanism having since failed owing to the death of the named adjudicator. AMEC suggested to RIBA that the same adjudicator be nominated in order to save time and cost. RIBA nominated the same adjudicator who was duly appointed. AMEC spoke to the adjudicator by telephone and informed him that he had been requested because the facts and issues were the same as in the first adjudication. The adjudicator reached the same decision. Whitefriars sought to resist enforcement on grounds which included apparent bias resulting from the fact the adjudicator had previously decided the same dispute. Whitefriars contended that in such circumstances the adjudicator could not be relied on to approach the matter with an open mind. At first instance Judge Toulmin considered that the fact of the re-adjudication alone was insufficient to establish apparent bias, but in the circumstances where the adjudicator had failed to invite submissions on legal advice he had obtained in relation to his jurisdiction and in light of the conversation between AMEC and the adjudicator, apparent bias was made out with the result that the enforcement application was dismissed. AMEC appealed.

11.33 **Held:** The Court of Appeal concluded that whether the matters were taken individually or together there was nothing which would have led a fair-minded and informed observer to conclude that there was a real possibility of bias by the adjudicator. AMEC's appeal was allowed and the adjudicator's decision was enforced.

11.34 Per Dyson LJ (with whom Kennedy and Chadwick LJJ agreed):

> 19. The risk of apparent bias may need to be considered where the decision of a tribunal is allowed on appeal, a rehearing is ordered and the question arises whether the rehearing should

be conducted by the same or a different tribunal ... The question that falls to be decided in all such cases is whether the fair-minded and informed observer would consider that the tribunal could be relied on to approach the issue on the second occasion with an open mind, or whether he or she would conclude that there was a real (as opposed to fanciful) possibility that the tribunal would approach its task with a closed mind, predisposed to reaching the same decision as before, regardless of the evidence and arguments that might be adduced. Usually, the reason for sending a case back for a rehearing will be that there is fresh evidence or a new point, or the appeal court has held that the tribunal made some mistake which, it is to be expected, will not be repeated on the rehearing. The present case is unusual in that the court did not find that Mr Biscoe had made any mistake in arriving at his first decision, and, so far as the RIBA were aware, there was no fresh material.

...

20. In my judgment, the mere fact that the tribunal has previously decided the issue is not of itself sufficient to justify a conclusion of apparent bias. Something more is required. Judges are assumed to be trustworthy and to understand that they should approach every case with an open mind. The same applies to adjudicators, who are almost always professional persons. That is not to say that, if it is asked to redetermine an issue and the evidence and arguments are merely a repeat of what went before, the tribunal will not be likely to reach the same conclusion as before. It would be unrealistic, indeed absurd, to expect the tribunal in such circumstances to ignore its earlier decision and not to be inclined to come to the same conclusion as before, particularly if the previous decision was carefully reasoned. The vice which the law must guard against is that the tribunal may approach the rehearing with a closed mind. If a judge has considered an issue carefully before reaching a decision on the first occasion, it cannot sensibly be said that he has a closed mind if, the evidence and arguments being the same as before, he does not give as careful a consideration on the second occasion as on the first. He will, however, be expected to give such reconsideration of the matter as is reasonably necessary for him to be satisfied that his first decision was correct ...

21. The mere fact that the tribunal has decided the issue before is therefore not enough for apparent bias. There needs to be something of substance to lead the fair-minded and informed observer to conclude that there is a real possibility that the tribunal will not bring an open mind and objective judgment to bear.

...

25. I confess that I fail to understand how the fact that the original decision was made without jurisdiction has any relevance to the issue of apparent bias. The fact that the first decision was a nullity did not make Mr Biscoe any more or less likely to approach the second adjudication with a closed mind than if the first decision had been one which he had jurisdiction to make. In my view, the first point is misconceived.

...

32. ... The mere fact that legal advice had been obtained in the first adjudication would not have led the fair-minded informed observer to conclude that there was a real possibility that Mr Biscoe would have approached the ... point with a closed mind.

...

37. ... I do not see how the position is affected by Mr Cassidy's comment that the reason why the dispute was being referred to Mr Biscoe was that he was familiar with the facts. In particular, I do not accept that this remark amounted to an invitation to Mr Biscoe to reach the same decision as on the previous occasion, still less that it is to be inferred that there was a real possibility that Mr Biscoe would reach the same decision by reason of that remark. I would accept that conversations between one party and the tribunal in the absence of the other party should be avoided. Communications should ordinarily be in writing with copies to all parties. But I see nothing in the circumstances of this conversation, which arose out of an innocuous telephone call to Mr Biscoe's office, which would lead the fair-minded

and informed observer to conclude that what was said would give rise to a real possibility of bias.

...

47. The judge held that there was a real possibility of bias in the present case by reason of the combined effect of the fact that (a) on AMEC 's case, the issues were the same in the two adjudications; (b) on the basis of Mr Biscoe's findings, the issues that he had to decide were the same in both adjudications; (c) the legal advice obtained in the first adjudication may have been 'carried forward' into the second adjudication and influenced the second decision; (d) Mr Biscoe did not give the parties an opportunity to comment on the legal advice obtained on the jurisdiction issue in the second adjudication; and (e) Mr Cassidy had a private conversation with Mr Biscoe. For the reasons that I have given, I do not consider that these factors, whether taken individually or in combination, justify the conclusion that there was apparent bias in this case.

Glencot Development and Design Co. Ltd v Ben Barrett & Son (Contractors) Ltd [2001] BLR 207 (TCC)

11.35 **Facts:** An adjudicator acted as a quasi-mediator during an adjudication, in an attempt to settle the dispute. He met separately with both parties. The matter did not settle and one of the parties objected to his continuing as adjudicator. Notwithstanding that objection, he continued to hand down a decision, after seeking advice from counsel. The claimant applied for summary judgment under CPR Part 24 and the defendant resisted, arguing that the decision was invalid as a consequence of the adjudicator not having withdrawn following the unsuccessful mediation.

11.36 **Held:** Judge LLoyd found that the defendant had real prospects of establishing that the adjudicator was no longer impartial as a result of his involvement in the mediation. Any fair-minded and informed observer would conclude that there was a real possibility of the adjudicator being biased:

24. In this case Mr. Kennedy submitted that there was no evidence that anything emerged in the discussions that might have affected Mr. Talbot's decision or approach. That very submission effectively makes the defendant's case. Whilst in an adjudication it is permissible to make inquiries and receive evidence and submissions from one party alone there is a clear obligation on the adjudicator to give any absent party a complete and accurate account of what has taken place. Mr. Talbot went to and fro between the parties. We do not know what he heard or learned. He was under no obligation to report it, nor given that the content was 'without prejudice' and confidential ought there to be any inquiry as to what happened. Those private discussions could have conveyed material or impressions which subsequently influenced his decision. On the evidence he was or may have been instrumental in resolving the issue about the 3 per cent discount which was one of the matters that he later had to decide (in the event against the defendant). Of much more consequence in my view is the fact that the discussions on 29 September were heated so that it would have been only understandable if some view had been formed about some people or a party. In the adjudication Mr. Talbot was asked to decide certain points about which there was no documentary evidence, in other words to form a view about the credibility of the applicant's case. These are areas where unconscious or insidious bias may well be present. Mr. Talbot's action in writing the letter of 2 October tellingly suggests that he was concerned about [what] an outsider might reasonably think about what had taken place.

25. Accordingly and taking account of Mr. Talbot's commendable openness and explanations which I accept as accurate I have nevertheless reached the conclusion any fair-minded and informed observer would conclude [from] Mr. Talbot's participation in the lengthy discussions on 29 September that there was a real possibility of him being biased.

Table 11.1 Table of Cases: Bias and Predetermination (shaded entries indicate the allegation was upheld)

Title	Citation	Issue
Woods Hardwick Ltd v Chiltern Air Conditioning Ltd	[2001] BLR 23 (TCC), Judge Thornton	**Bias: reliance on material not provided to responding party; adjudicator as witness** The adjudicator relied on information from the referring party and third parties that was not made available to the responding party. The adjudicator provided a witness statement in the enforcement proceedings that went beyond a neutral statement of the facts and sought to justify his actions and decision. Not enforced.
Glencot Development and Design Co. Ltd v Ben Barrett & Son (Contractors) Ltd	[2001] BLR 207 (TCC), Judge LLoyd	**Bias: adjudicator acted as quasi-mediator** The adjudicator attempted to settle the dispute and met separately with both parties. This was not anticipated by the applicable rules. Any fair-minded observer would have perceived unfairness/apprehension of bias because he could have gathered material or formed impressions during those separate meetings, which contributed to his decision. Not enforced.
Discain Project Services Ltd v Opecprime Development Ltd	[2001] BLR 285 (TCC), Judge Bowsher	**Bias: telephone communications with one party** On three occasions the adjudicator communicated with the referring party's representative by telephone in relation to the substance of the matters in dispute. The fact of the communications was notifed by fax to the other party but materially important aspects of the discussions were not reported. Apparent bias made out. Not enforced.
R. G. Carter Ltd v Edmund Nuttall Ltd	[2002] BLR 359 (TCC), Judge Bowsher	**No bias: re-adjudication of similar dispute before same adjudicator** On the same day as an adjudicator's decision was found to be unenforceable for lack of jurisdiction, Carter gave notice of a fifth adjudication, covering some of the same ground as the unenforceable decision. The same adjudicator was appointed again, despite Carter's request that he not be. Carter sought an application to set aside the adjudicator's appointment on the basis of bias. The mere fact that an adjudicator is required to consider the same or similar matters as in a previous adjudication is insufficient to cause an apprehension of bias. Fair and reasonable professional people in the construction industry are perfectly capable of a change of mind based upon different evidence being presented. Application refused.

Title	Citation	Issue
Pring & St Hill Limited v C. J. Hafner (t/a Southern Erectors)	[2002] EWHC 1775 (TCC), Judge LLoyd	**Bias: adjudicator decided related disputes** The decision in an adjudication between subcontractor and sub-subcontractor (down the contractual chain) was not enforced in circumstances where the adjudicator had also adjudicated the same dispute between main contractor and subcontractor (up the contractual chain). Decided on lack of jurisdiction. *Obiter*: An apprehension of bias arose from the fact that there was a very real risk the adjudicator would carry forward from the previous adjudication what he had seen on the site inspection and been told, the judgments he had formed, and the opinions he had reached with the result that he may be predisposed to particular views. There was nothing in the decision to show the adjudicator had considered matters afresh. Further, the involvement in an earlier adjudication with one party without making available all relevant materials from that adjudication in the present adjudication meant that the adjudicator had in effect had private exchanges with that party. Not enforced.
London & Amsterdam Properties Ltd v Waterman Partnership Ltd	[2004] BLR 179, [2003] EWHC 3059 (TCC), Judge Wilcox	**No bias: adjudicator previously decided related disputes** That the adjudicator had decided a previous dispute involving one of the parties to the present adjudication which arose from the same project but involved different matters and facts was, on its own, insufficient to establish bias. Not enforced on other grounds (insufficient time to consider late material).
Specialist Ceiling Services Northern Ltd v ZVI Construction (UK) Ltd	[2004] BLR 403 (TCC), Judge Grenfell	**No bias: disclosure of 'without prejudice' offer** The referral identified the fact a 'without prejudice' offer had been made and the responding party's valuation of the account and assessment of allowable EOT at the time the offer was made. Apparent bias was not made out in circumstances where the adjudicator had applied the correct test in determining whether he could continue and the decision showed that the adjudicator had not been influenced, rejecting as he had the submissions based on the 'without prejudice' material. Enforced.
A&S Enterprises Ltd v Kema Holdings Ltd	[2005] BLR 76 (TCC), [2004] EWHC 3365 (TCC), Judge Seymour	**Bias: adjudicator unfairly citicized a witness for not attending a telephone hearing** The adjudicator had not indicated before the hearing that he considered the witness to be crucial or expected to hear from the witness. The terms of the decision suggested the criticism of the witness had led to the adjudicator placing less weight on the party's evidence and submissions. Not enforced.
AMEC Capital Projects Ltd v Whitefriars City Estates Ltd	[2005] 1 All ER 723, [2004] EWCA Civ 1418 (CA), Dyson, Kennedy, Chadwick LJJ	**No bias: same adjudicator for rehearing** The facts that a tribunal had previously decided the same issue (without jurisdiction), did not invite submissions on legal advice relating to his jurisdiction and had a conversation with one party in relation to the reason for his appointment were insufficient to justify a conclusion of apparent bias. Enforced, overturning the decision at first instance ([2003] EWHC 2443 (TCC) Judge LLoyd).

(*Continued*)

Table 11.1 *Continued*

Title	Citation	Issue
Michael John Construction v Golledge and ors	[2006] EWHC 71 (TCC), Judge Coulson	**No bias: adjudicator had previously decided the same dispute with different responding party** The fact the adjudicator had previously decided the same matters against a different party was insufficient on its own to establish bias. There was nothing in the correspondence or conduct of the second adjudication that could lead to a finding of apparent bias. The adjudicator specifically invited submissions on valuation from the responding party but it failed to address the point and it was therefore unsurprising that the same conclusion was reached. Enforced.
Aveat Heating Ltd v Jerram Falkus Construction Ltd	(2007) 113 Con LR 13, [2007] EWHC 131 (TCC), Judge Havery	**No bias: adjudicator made an award of costs and expenses** There was no bias even though the adjudicator had not received any substantiation for the claimed sums. Enforced (but costs and expenses award not enforced for lack of jurisdiction).
Mott MacDonald Ltd v London & Regional Properties Ltd	(2007) 113 Con LR 33, [2007] EWHC 1055 (TCC), Judge Thornton	**Bias: adjudicator lacked impartiality by demanding payment of his fees before delivering decision** An adjudicator may not be or appear to be 'financially beholden' to one party, particularly the referring party, or place himself in the position in which he might appear to be more partial to one side than the other. The imposition of a lien on his decision which has to be lifted by the referring party in order to obtain his decision gave an appearance of partiality and amounted to a breach of paras. 12(a) and 19(3) of the Scheme. Not enforced on this ground and additionally lack of jurisdiction and procedural irregularity.
Makers UK Ltd v Camden London Borough Council	[2008] BLR 470, [2008] EWHC 1836 (TCC), Akenhead J	**No bias: telephone calls prior to appointment: suggestion to nominating body of appointment of particular adjudicator** Telephone conversations prior to the request to the nominating body between the adjudicator and referring party were insufficient to establish apparent bias. Those conversations were for the purposes of checking availability and conflict and were purely administrative in nature. The adjudicator was selected by the nominating body, not the referring party. Subjective impression of bias on the part of the responding party was insufficient. Enforced.
Andrew Wallace Ltd v Jeff Noon	[2009] BLR 158, Judge Grant	**No bias: adjudicator had recently acted as mediator in an unconnected dispute involving one party** Mediating an unrelated dispute involving one party to the adjudication, even very close in time to the appointment as adjudicator, was insufficient to establish apparent bias. Enforced.

Title	Citation	Issue
Barr Ltd v Klin Investment UK Ltd	[2009] CILL 2787, [2009] CSOH 104 (Scottish Outer House, Court of Session), Lord Glennie	**No bias: attempt to establish apparent bias from terms of decision itself** The adjudicator made findings in relation to the meaning of a withholding notice that the responding party claimed gave the appearance the adjudicator was wishing to help the referring party. Bias was not made out in circumstances where the adjudicator had taken a proper approach to construing the notice. The approach, in any event, was based on the responding party's own arguments and worked in the responding party's favour. Enforced.
Camillin Denny Architects Limited v Adelaide Jones & Company Limited	[2009] BLR 606, [2009] EWHC 2110 (TCC), Akenhead J	**No bias: adjudicator's approach to costs award to successful party** An award of 90 per cent of the referring party's costs in circumstances where it had succeeded on 60 per cent of the value of its claim did not give rise to apparent bias. The adjudicator had acted properly considering all arguments on costs and then exercising his discretion in a fair manner. The court would not interfere, even if the adjudicator's discretion was exercised wrongly. Enforced.
Balfour Beatty Engineering Services (HY) Ltd v Shepherd Construction Ltd	(2009) 127 Con LR 110, [2009] EWHC 2218 (TCC), Akenhead J	**No bias: adjudicator's approach to expert evidence** The adjudicator found that the expert for the responding party had been instructed to ignore early delays to access. The court found that was a fair inference and did not give rise to a finding of apparent bias; an adjudicator's finding which is adverse or even a criticism of the behaviour of one party or its witnesses does not give rise to a valid challenge of either actual or apparent bias. Enforced.
Fileturn Ltd v Royal Garden Hotel Ltd	[2010] BLR 512, [2010] EWHC 1736 (TCC), Edwards-Stuart J	**No bias: adjudicator was previously a co-director of party's representative** Where the adjudicator had previously been a co-director with the claims consultant representing one party there was no apparent bias in the absence of any: (i) relationship outside their professional activities; (ii) significant contact in the six years since the adjudicator left the claims consultant; (iii) financial interest in the claims consultant; (iv) material dependency on the claims consultant for the adjudicator's practice; or (v) particular knowledge of or connection with the parties themselves. Enforced.
Volker Stevin Ltd v Holystone Contracts Ltd	[2010] EWHC 2344 (TCC), Coulson J	**No bias: adjudicator had knowledge of the fact that a 'without prejudice' offer had been made** Knowledge of the fact of an offer would not affect the mind of the adjudicator. An adjudicator would generally expect offers to have been made. Enforced.

(Continued)

Table 11.1 *Continued*

Title	Citation	Issue
Ellis Building Contractors Ltd v Vincent Goldstein	[2011] CILL 3049, [2011] EWHC 269 (TCC), Akenhead J	**No bias: adjudicator was provided with a redacted 'without prejudice' letter to show that a defence raised in the adjudication had not been raised before** The adjudicator did not appear to rely on the fact that the defence had not been raised before. Decision was reached based on analysis of open documents. *Obiter*: where a decision is based primarily upon wrongly received 'without prejudice' material the decision may well not be enforced. Enforced.
Sprunt Limited v London Borough of Camden	[2012] BLR 83, [2011] EWHC 3191 (TCC), Akenhead J	**No bias: nomination provisions allowing one party to select adjudicator offend principles of fairness** A contract provision allowing the employer to appoint an adjudicator of its choosing offended against the policy in the Act of having actually and ostensibly impartial adjudicators. The contract was therefore not Act-compliant and the adjudication scheme (including the appointment mechanism) was replaced by the Scheme. Enforced.
Lanes Group plc v Galliford Try Infrastructure Ltd	[2011] BLR 438, [2011] EWHC 1035 (TCC), Akenhead J	**No bias: nominated adjudicator had been referring party's opponent in a series of previous adjudications that had become acrimonious** The involvement of the nominated adjudicator as an opponent in a series of adjudications that had resulted in a 'spat' some 14 months earlier was not enough to give rise to an appearance of bias. However, the referral had been abandoned and no injunction was granted to stop a referral to a second adjudicator.
Lanes Group Plc v Galliford Try Infrastructure Ltd (t/a Galliford Try Rail)	[2012] BLR 121, [2011] EWCA 1617 (CA), Richards, Stanley Burnton, Jackson LJJ	**Predetermination: communication of preliminary views** The distribution of a document containing preliminary views before receipt of the responding party's submissions did not amount to apparent bias or apparent predetermination. Enforced. Overturned decision of Judge Waksman (sitting as Deputy HCJ) [2011] EWHC 1679 (TCC).

PART IV

PAYMENT UNDER CONSTRUCTION CONTRACTS

12

SECTIONS 109 AND 110(1): INTERIM AND FINAL PAYMENTS

(1) Overview

Introduction and Summary

12.01 This chapter looks at the circumstances in which s. 109 applies to create a right to interim payment in contracts over 45 days' duration. It also considers the requirement under s. 110(1) to include an 'adequate mechanism' to determine the amounts and timings of payments. Both of these sections set minimum requirements which must be observed in the parties' contract, otherwise the provisions of the Scheme for Construction Contracts ('the Scheme') will apply.

12.02 Neither ss. 109 nor 110(1) have been amended substantially under the Local Democracy, Economic Development and Construction Act 2009 ('the 2009 Act'), and those amendments which have been made relate almost entirely to the prohibition on pay when certified clauses which is discussed at Chapter 16 below. The sole exception to this is a minor drafting amendment in s. 109(4) which has been made for the purposes of consistency with the new wording of ss. 110A, 110B, and 111(1).

12.03 The requirements for providing an 'adequate mechanism' to determine the dates and amounts of payments are summarized below:

1. All construction contracts must contain an adequate mechanism for determining what payments become due and when. In the case of contracts over 45 days' duration, this must include an adequate mechanism for determining interim payments as well as a final payment (s. 109).

2. If the parties have not provided for interim payments in contracts of over 45 days' duration, the Scheme will apply a payment interval of 28 days.
3. The question of whether the parties have provided an 'adequate mechanism' is one of fact, not law. What will be adequate in some cases will be inadequate in others; however the requirement of certainty is a central ingredient.
4. If the parties have failed to provide an adequate (or any) mechanism to determine the amounts and dates of payments, the Scheme will apply to 'fill the gaps'.
5. All construction contracts are required to stipulate a final date for payment. If they do not, then the Scheme applies so that the final date for payment is 17 days after the due date.

(2) The Right to Interim Payment under s. 109

Introduction

12.04 Section 109 creates a statutory right for a contractor or subcontractor to be paid before work is complete. Previously, such arrangements were left entirely for the parties to agree, often on uncertain terms. If work continued for prolonged periods, contractors and subcontractors undertook a high risk that they might expend significant sums in employing labour and providing materials with no right (or no enforceable right) to receive payment until many weeks or months afterwards. In such circumstances, cash flow would be severely restrained and if disputes arose, there was a risk that the contractor or subcontractor might face insolvency before they were resolved. Section 109 therefore underpins all of the other payment provisions of ss. 110–13 in creating an enforceable right to receive cash flow throughout a project, albeit that the parties remain free to decide the amounts and timings of such payments.[1]

12.05 Whilst it remains possible (subject to the risk that a payment mechanism may be found to be inadequate and therefore replaced by the Scheme) for a payer to avoid making regular or substantial payments by, for example, setting payment dates at a considerable distance from one another or specifying nominal amounts, the mere existence of the Act, and of s. 109 in particular, has strengthened the position of contractors and subcontractors so that such abuses should now be rare. The fact that the 2009 Act has recommended no change to the existing freedom of the parties to determine the conditions of interim payment, suggests that s. 109 has, by and large, worked as intended.

12.06 If a construction contract fails to allow for interim payments, then the relevant provisions of the Scheme apply to introduce payment intervals of 28 days and a method of valuation based on the actual costs of the work performed during the period.[2] The relevant provisions of the Scheme are the same whether the parties have failed to make *any* allowance for interim payments (s. 109), or whether such provision has been made but it is not 'adequate' (s. 110(1)). The question of how the Scheme applies is considered in sections (4) and (5) below.

Duration of Work More than 45 Days

12.07 Section 109 provides that a party to a construction contract is entitled to payment by instalments, stage payments, or other periodic payments for any work under the contract unless:

1. it is specified in the contract that the duration of the work is to be less than 45 days or
2. it is agreed between the parties that the duration of the work is estimated to be less than 45 days.

---

[1] s. 109(2).
[2] s. 103(3).

12.08 The question of whether the work is of less than 45 days' duration may not always be clear, and it is possible to think of numerous instances where works might extend beyond 45 days even if the parties had not originally intended that this should happen. For example:

1. where no start or finish dates are specified so that it is unclear whether the parties initially envisaged an original duration of less than 45 days
2. where the scope of work is extended, or delays occur, so that the original contract duration is extended beyond 45 days
3. where the contract is for call-off works so that the duration of the contract is longer than 45 days, but works may only be carried out on a few days throughout this period.

12.09 In such cases, it will be a matter of construction of the contract whether the parties have agreed that the duration of the work was estimated to be less than 45 days. The mere fact that works have extended beyond their original intended duration is unlikely to satisfy this requirement, even in cases where additional works have been instructed or extensions of time granted for matters beyond the payee's control. In both cases, unless the additional work was foreseen at the start of the contract, it would not invalidate the parties' original agreement as to the estimated duration. However, in cases where it is envisaged at the start of the works that they will be extended in due course, whether, for example, as a result of scope still being specified or as a result of provisional sum items within the bills of quantities, it would probably be prudent to include a regime for interim payments or risk the Scheme being implied.[3] Equally, framework agreements and term contracts should probably include a provision for interim payment, if not in the main agreement, at least in the terms of any work order instructed of more than 45 days' duration.

12.10 Perhaps unsurprisingly, given the likelihood that the works involved are of small value, there are no reports of the courts having been called upon to decide any cases where it has been argued that the interim payment provisions should not apply because the works were of less than 45 days' duration. However, in *Tim Butler Contractors Ltd v Merewood Homes Ltd* (2002)[4] this question was considered by an adjudicator who decided that the parties had neither specified nor agreed a duration less than 45 days.

12.11 The adjudicator's decision was upheld by Judge Gilliland as being, right or wrong, a decision which was within the adjudicator's jurisdiction to make. Whilst this case provides little guidance in so far as the parties' agreement was evidenced by a chain of correspondence unlikely to be repeated elsewhere, it does at least demonstrate that the courts are likely to view the implication of the payment provisions of the Scheme as matters solely of contract interpretation within an adjudicator's jurisdiction to decide. Thus, if an adjudicator wrongly decides that the Scheme applies because the contract is of duration more than 45 days, this does not open up his decision to a jurisdictional challenge (as it would if he wrongly decided that the contract was a construction contract within the meaning of ss. 104–7).[5]

(3) Section 110(1): Providing an 'Adequate Mechanism'

Introduction

12.12 Whilst s. 109 provides that construction contracts should contain a mechanism for interim payments if they are over 45 days' duration, s. 110(1) contains details as to how and when

[3] Although, even in such cases, it would appear possible that such an implication might be avoided through the use of sufficiently clear language evidencing the parties' *agreement* to a duration of less than 45 days.

[4] [2002] 18 Const LJ 74.

[5] Jurisdictional challenges are discussed in Ch. 9.

both these and final payments should be made. Section 110 is not limited to contracts of longer duration and the provisions therefore apply whether or not the specified or estimated duration of the work is more than 45 days. The only difference between contracts of long and short duration, therefore, is that the payment mechanism will only to apply to a single, final payment in short contracts, whereas it will apply to both interim and final payments in contracts of longer duration.

12.13 The purpose of s. 110, like s. 109, is to ensure that contracts are drafted to include the adequate payment mechanisms. Provided they do, the Act will not interfere to substitute other provisions. However, if the contract does not include the adequate mechanisms, the Scheme will apply.

12.14 Section 110(1) requires every construction contract to provide an adequate mechanism for determining *what* payments become due. Contracts must also contain an adequate mechanism for determining *when* the payments become due and stipulate a final date for payment. This use of terminology is at first sight confusing as, for many people, the fact that a payment is due means that it should be paid immediately and not at some point in the future before the final date for payment. This situation can be particularly confusing when the parties agree a payment mechanism whereby both the due date and the final date for payment are some time after the payment has been invoiced, assessed and certified. However, the purpose of the due date is to provide certainty as to when during the long assessment process the payee's right to be paid a particular sum crystallizes. The purpose of the final date for payment, on the other hand, is to set a long-stop date by which this payment must be made.[6] The due date and the final date for payment thereby form a 'window' within which payment must be made, and there is no need to wait until the final day to make payment (although payers often do).

What Constitutes an 'Adequate Mechanism'?

12.15 The question of what is an 'adequate mechanism' is one of fact. What is adequate in some circumstances will not be adequate in others: *Mair v Arshad* (2007) (Key Case). The issue is therefore open to subjective interpretation and questions were raised during the Committee proceedings on the original Housing Grants, Construction and Regeneration Bill as to whether minimum requirements should be laid down so that the question of adequacy did not become one that was frequently contested in the courts. In the end, this suggestion was rejected on the familiar ground that any further attempts to describe the essential features of the payment mechanism would constitute too great an intrusion on the parties' freedom of contract.

12.16 Nevertheless, it seems that parties rarely have any difficulty in practice in agreeing a mechanism which meets the test of adequacy, or else recognizing where contract terms fall short of this standard and agreeing to substitute the Scheme. Indeed, the issue is not one that has been frequently before the courts, possibly because both parties and the draftsmen of standard-form contracts have used the Scheme to guide them. As Sheriff O'Carroll put it in *Mair v Arshad* (2007):

> The adequacy of the mechanism is a question of fact though, no doubt, some assistance may be gained in the final assessment by reference to the Scheme.[7]

12.17 Taking the elements of the Scheme, therefore, a contract should satisfy the test of adequacy if it contains:

1. a clear payment interval or condition for interim payment in contracts over 45 days' duration[8]

---

[6] *Reinwood Ltd v L. Brown & Sons Ltd* [2008] 1WLR 696, [2008] UKHL 12 (HL) at para. 49 (Key Case at 13.130).

[7] At para. 21 (Key Case).

[8] s. 109.

2. a provision for either the making of a claim by the payee or certification by or on behalf of the payer, in either case specifying the amount of the payment which is considered due and the basis on which it has been calculated[9]
3. in relation to interim payments, a method of valuation to support the assessment of the amount due[10] and
4. a provision that payment should become due within a definite period of the invoice or certification,[11] provided that failure by the payer or certifier to issue a certificate should not be allowed to delay payment indefinitely.[12]

12.18 Of course, this is not the only type of payment mechanism which will satisfy the test, and parties should not rely entirely on differences between the Scheme's provisions and those of the contract in order to demonstrate inadequacy in the latter. The only binding obligation is to comply with the terms of the Act, which is drafted in far less prescriptive terms than the Scheme. The lack of one or all of the above elements will not necessarily be fatal. In simple contracts an agreement to make an unspecified advance without any method of valuation to support it, may be reasonable.[13] Agreed schedules of payments on specific dates or on completion of sections of the work are also common and, provided they create sufficient certainty, are likely to be valid.

12.19 In the cases considered below, the common complaint relates to a lack of certainty as to the valuation and timing of payments. Although there is no standard which can be applied across all contracts, a contract mechanism which makes payment conditional on events which are reasonably certain to happen may be deemed adequate, even when the event is beyond the control of the parties: *Alstom Signalling Ltd v Jarvis Facilities Ltd* (2004) (Key Case). On the other hand, where the payment mechanism depends upon the parties reaching agreement on the amount of payments, without any timetable for agreement to be reached nor any mechanism for resolving any failure to agree other than by reference to adjudication, then the payment mechanism may fail the test: *Maxi Construction Management Ltd v Mortons Rolls Ltd* (2001) (Key Case).

12.20 For contracts entered into after the 2009 Act came into force, new ss. 110(1A) and 110(1D) identify that any provision which makes payment conditional upon the performance of obligations or on a payment notice or certificate having been given under another contract, will be deemed inadequate. Had such provisions been present in the 1996 Act, then some of the cases set out below may have been decided differently. Sections 110(1A)—(1D) are considered in Chapter 16 below.

Final Date for Payment

12.21 In addition to providing an adequate mechanism for determining what payments become due and when, s. 110(1) also requires the parties to a construction contract to specify a final date for payment. A contract payment scheme that does not identify a final date for payment is not an adequate mechanism: *Karl Construction (Scotland) Ltd v Sweeney Civil Engineering (Scotland) Ltd* (2000) (Key Case). The final date for payment is the last date by which the payer must make payment of the amount due, unless it has served a notice of its intention to

---

[9] See definition of 'claim by the payee' under para. II.12 of the Scheme together with the requirement for the payer to give notice of the amount of the payment proposed to be made under s. 110(2) of the 1996 Act.

[10] See definition of 'value of work' contained in para. II.12 of the Scheme.

[11] See paras. II.4, 5, 6, and 7 of the Scheme.

[12] *Ringway Infrastructure Service Ltd v Vauxhall Motors Ltd* [2007] All ER (D) 333, [2007] EWHC 2421 (TCC), Akenhead J (Key Case).

[13] *Peter Mair v Mohammed Arshad* [2007] Sheriffdom of Tayside, Central & Fife at Cupar, A538/04, Sheriff Derek O'Carroll, Oct. 2007 (Key Case).

withhold or pay less. If the payer fails to do so, then the payee may suspend performance of the works in accordance with s. 112.[14]

12.22 The parties are free to decide the final date for payment, and it is common for the date to be later in respect of final payments than it is for interim payments to allow the payer time to check the final account. If the parties do not stipulate a final date for payment in relation to any amounts which may become due (whether interim, final, or any other payments) then the Scheme applies to provide that this shall be 17 days from the date that the payment becomes due. Equally if the parties specify a final date for payment which is conditional on other events occurring, then this must satisfy the test of adequacy in exactly the same way as required for the due date.[15]

12.23 The difference between the due date and final date for payment was considered by the House of Lords in the case of *Reinwood Ltd v L. Brown & Sons Ltd* (2008)[16] in which it was stated that the obligation to pay arose and was crystallized on the due date, whilst the final date for payment was 'akin' to making time of the essence:

> 49. ... A sum becomes due under a certificate when it is issued, and the 'final date for payment' is the date by which failure to pay can have serious consequences for the employer. As my noble and learned friend, Lord Walker of Gestingthorpe, observed during the argument, the function of the 'final date' is akin to making time of the essence of the payment as at that date. [17]

12.24 It is possible that a dispute may arise as to the amount to be paid even before it is required to be paid. In *All in One Building & Refurbishments Ltd v Makers (UK) Ltd* (2005)[18] the contract had been brought to an end with each party blaming the other for the repudiation. Regardless of the employer's position than no sums were yet due to be paid, Judge Wilcox upheld the adjudicator's decision that a dispute had crystallized such that All in One was entitled to apply for determination of the amounts due to be paid to them for the alleged financial consequences of the breach:

> 27. [The employer's] fallback argument that the assessed figures were akin to a draft final account, and that because under the contract two months was allowed for payment of a final account a dispute would not crystallise until the expiration of that period I also reject. The contractual date for payment is prescribed in the contract depending upon whether it is characterised as an interim payment or a final payment, but a distinction must be drawn between the date for payment and an entitlement to payment. It was the entitlement to payment that was being denied in relation to these claims. The contract may prescribe a time when payment becomes due and that may be a material factor amongst many others in arriving at the conclusion as to whether a claim is denied or not. It is not determinative as to whether a dispute has arisen.

The 2009 Act

12.25 The 2009 Act makes significant changes to the notification provisions contained in ss. 110(2) and 111. Although these are considered in detail in the next chapter, they include a right for the payee to issue a payment notice if the payer fails to issue a payment notice when required. Section 110B(3) now provides that where this happens, the final date for payment will be postponed by the same number of days after the date that the payer was to have given notice that the payee gives his notice. This provision has been reflected in some, but not all, of the standard forms considered at section 6 below.

---

[14] Discussed in Ch. 15.
[15] See 12.15 to 12.20.
[16] [2008] 1 WLR 696, [2008] EWCA Civ 1090 (CA).
[17] Per Lord Neuberger of Abbotsbury at para. 49. Key Case at 13.130.
[18] (2005) CILL 2321, [2005] EWHC 2943 (TCC).

Consequences of Missing the Final Date for Payment

12.26 Since s. 111(1) of both the 1996 and 2009 Acts imposes an obligation on the payer to pay the amount due (now the 'notified sum') by the final date for payment, it follows that any attempt to pay less than the amount that is properly due (which may be less than the amount of the payment application) without giving notice constitutes a breach of contract. Stipulations as to time for payment are not generally of the essence in commercial contracts unless a contrary intention appears from the terms of the contract.[19] Their Lordships' judgment in *Reinwood v Brown* indicates that the final date for payment has this effect and that there should be no separate need to serve notice making time of the essence.[20] Even so, the failure is not automatically a repudiation of the contract, giving rise to a right to terminate. The breach must still go to the root of the contract.[21] Whilst it is unlikely that a repudiatory breach would occur in the case of one or more short delays in making payment, it may occur when persistent late or non-payment causes a cumulative effect which may be so serious as to justify the innocent party bringing the contract to a premature end.[22]

12.27 The 1996 Act anticipates these problems and provides its own remedy for breach of contract in failing to pay by the final date for payment. This is contained in s. 112 and provides that a payee may suspend performance provided that certain conditions apply. The right to suspend appears to supplement and not to replace the payee's other rights in such circumstances, including the right to terminate (in cases where this arises) and a right to interest on late payment either under the contract or under the Late Payment of Commercial Debts (Interest) Act 1998. It should be noted, however, that the question of whether a payee was justified in terminating the contract where a statutory right of suspension existed has not yet been tested by the courts.

Key Cases: An 'Adequate Mechanism'

Peter Mair v Mohammed Arshad [2007] Sheriffdom of Tayside, Central & Fife at Cupar, A538/04, Oct. 2007

12.28 **Facts:** The contractual scheme for payment merely required that 'an invoice will be issued at the end of each working week with payment due on presentation'. It did not oblige the payee to specify exactly the basis for the calculation of the payments. Without such an obligation, it was contended, there was potential for abuse. The defending employer relied upon the Scheme as providing 'an analogue' or test for the type of provision that was required in any construction contract in order to meet the requirement of adequacy arguing that the contract did not meet that test and was therefore not adequate.

12.29 **Held:** *Obiter*: Sheriff Derek O'Carroll drew a common-sense distinction between the level of particularity which would be required in large complex construction operations as opposed to simple, straightforward operations such as the one in question. The adequacy of the mechanism was a question of fact, not law, though no doubt some assistance might be gained in the final assessment by reference to the Scheme. In this case, the employer was not entitled to demand further specification of the nature of the claim. The work was

---

[19] Sale of Goods Act 1979, s. 10(1).

[20] See *Behzadi v Shaftesbury Hotels* [1992] Ch 1 (CA).

[21] *Woodar v Wimpey Ltd* [1980] 1 WLR 277 (HL).

[22] *Rice (t/a The Garden Guardian) v Great Yarmouth Borough Council* [2000] All ER (D) 902 (CA), per Hale LJ at para. 35; *Alan Auld Associates Ltd v Rick Pollard Associates and anor* [2008] BLR 419, [2008] EWCA Civ 655 (CA); cf. *Decro-Wall SA v Practitioners in Marketing Ltd* [1971] 1 WLR 361.

carried out on the employer's own property and so he would know what work had been done. He also employed an architect to assist him. There was nothing to suggest that he did not or could not find out what was done on his property. The contract was for a fixed amount and the whole sum claimed was less than that amount. Therefore, the fact that the contract failed to provide for particularity or substantiation of the contractor's invoices did not necessarily mean that the contract was inadequate:

> 21. I was invited by the solicitor for the defender to find that as a matter of law, the mechanism was not adequate (construing the terms of the contractual mechanism with the scheme mechanism). In my view, that is not the correct approach. In my view, whether a contractual mechanism is 'adequate' is a question of fact, not law. What will be adequate in some cases will be inadequate in others. Large complex construction operations will no doubt require complex highly structured mechanisms. Equally, simple, straightforward construction operations may be satisfactorily served by a very simple mechanism. I do not accept that I can determine as a matter of law whether the contractual mechanism in this case is adequate, whether by reference to the scheme or otherwise. The adequacy of the mechanism is a question of fact though, no doubt, some assistance may be gained in the final assessment by reference to the scheme.

Maxi Construction Management Ltd v Mortons Rolls Ltd [2001] Scot HC 78, (2001) CILL 1784 (Outer House, Court of Session)

12.30 **Facts:** The parties entered into a contract based on the Scottish version of the JCT Standard Form of Building Contract with Contractor's Design 1998. The payment clauses had been amended by the incorporation of the Employer's Requirements, such that two conflicting final dates for payment had been stated and there was no longer any provision for the giving of either a payment notice or a withholding notice. Moreover, paragraph 2.5.20 of the Employer's Requirements required the contractor and employer's agent (Bucknall Austin) to agree the amount of each valuation before the contractor could submit its payment application.

12.31 Maxi submitted a valuation in accordance with the contractual scheme, requiring the employer's agent to agree payment. Maxi's valuation included numerous claims for variations and loss and expense, the build up of which was unclear. No payment was forthcoming and so Maxi sought payment in accordance with Clause 30.3.5 of the standard JCT terms which provide that, in the absence of a payment or withholding notice, the amount of the contractor's application becomes due and payable.

12.32 Both parties accepted that the inconsistencies in the contract conditions and lack of provision for payment and withholding notices meant that the Scheme would be implied to deal with these issues. The questions for Lord MacFadyen to decide, therefore, were:

1. whether the contractual machinery for determining what payments became due was otherwise capable of being preserved so that Maxi's valuation still needed to be agreed with Bucknall Austin prior to the application being submitted
2. if not, whether Maxi's valuation was a 'claim by the payee' within the meaning of paragraph II.12 of the Scheme such that a payment-due date had been triggered under Clause 30.3.5.[23]

[23] See 12.69 to 12.70 below for a discussion of para. II.12.

Interestingly, neither party sought to argue that the Scheme should apply to substitute the whole of the payment mechanism so that Clause 30.3.5 would also be rendered invalid.[24]

Held: **12.33**

1. Clause 2.5.20 of the Employer's Requirements was incompatible with s. 110(1)(a) of the Act, because no timetable had been provided for agreement of the sums due, nor was there any means of resolving a failure to reach such agreement other than by reference to adjudication. Accordingly, the payment mechanism was inadequate to determine what payments became due, or when they became due. Paragraphs II.4 and II.12 of the Scheme applied so that payment became due within seven days of a 'claim by the payee'.
2. Maxi's valuation did not amount to a 'claim by the payee' because, on the facts, it was simply a request to agree a sum due, and not an application for payment. *Obiter*: even if Maxi's valuation had been expressed as an application for payment, it did not satisfy the test of paragraph II.12 since it was insufficiently detailed to show the basis on which it had been calculated.

> 28. … A requirement that a valuation be agreed by the employer's agent before a claim for payment can be made is not necessarily, in my view, incompatible with section 110(1)(a), provided a timetable for the process of agreement, and a means of resolving a failure to reach agreement are provided. But paragraph 2.5.20 makes no such provision. Failure on the part of Bucknall Austin to agree a valuation could hold up the making of a claim for payment indefinitely. That, in my view, means that the contract does not provide an adequate mechanism for determining when payments become due under the contract. Although I understand the distinction which Mr McIlvride sought to draw between machinery for determining what payments become due and machinery for determining when those payments become due, I do not consider that paragraph 2.5.20 can be regarded as bearing only on 'what' and not on 'when'. The absence of a timetable and of a means for resolving deadlock has the effect that paragraph 2.5.20 renders inadequate the machinery for determining when payments are due.

Alstom Signalling Ltd v Jarvis Facilities Ltd [2004] EWHC 1285 (TCC)

Facts: Alstom was the main contractor engaged by Railtrack plc to design, manufacture and install plant for a project to extend the Tyne and Wear Metro. Alstom engaged Jarvis to design, install, test, and commission the signalling and telecommunications systems for the project under an amended IChemE Model Form for Process Plants, second edition 1997. The subcontract contained the following bespoke conditions: **12.34**

> The Contractor shall pay to the Subcontractor the amount due on the certificate within seven days of the Railtrack certificate being issued in accordance with Annex F1—the Project Cut-Off Dates (which contains the amount certified against the application aforementioned).

Annex F1 contained calendars for 2000 and 2001 with the relevant dates for invoicing and certification being highlighted.

12.35 Issues arose regarding payment under a certificate and whether Alstom had issued a withholding notice that it was required to have issued.[25] In addition, Jarvis challenged the adequacy of the payment scheme since it contained payment dates that Jarvis argued could be changed unilaterally, robbing the agreement of any certainty.

---

[24] Cf. *C&B Scene Concept Design Ltd v Isobars Ltd* [2001] CILL 1781 (TCC) (Key Case).

[25] See further discussion of this case at 13.105.

12.36 **Held:** Judge LLoyd considered that there was no reason why the contract did not comply with s. 110 in terms of providing an adequate mechanism to determine when a payment was due and when it should be paid. The agreement was predicated on the assumption that Railtrack would comply with its obligations and issue a certificate at the times set out in the Railtrack payment cycle at Annex F1. If it did not, then, on the wording of the subcontract, this would not entitle Alstom to withhold payment from Jarvis since the date for payment remained the same and was independently ascertainable from the subcontract:

> 22. Section 110 of the Act calls for an adequate mechanism to determine the final date for payment. I find myself at somewhat of a loss to understand why Schedule F does not comply with section 110 of the Act in terms of an adequate mechanism to determine when a payment was due for the purposes of section 110(1). The subcontract was made by reference to the main contract, both formally and financially. Clause 1.3 of the Schedule F demonstrated a common approach as regards a 'Neutral Cash Flow' ... The applications for payments were linked to the Railtrack cycle ... The contractor is to issue certificates (see clause 2.5). That establishes when a payment is due for the purposes of the second limb of section 110(1)(a) and the requirements for the content of the certificate comply with section 110(2). (I deal later with the first limb of section 110(1)(a).) Conventionally it seems that Alstom was to issue a certificate within 14 days of the receipt of an application (see, for example, its letter of 13 June 2003). Clause 2.6 said that payment would be made within seven days of the Railtrack certificate being issued in accordance with Annex F1. There was therefore certainty as to the final date for payment—seven days of the Railtrack certificate. That satisfies section 110(1)(b). The fact that Railtrack, probably in breach of its contract with Alstom, might fail to issue its certificate in accordance with Annex F1 does not mean that for the purposes of section 110(1)(b) there was no final date. The final date for payment remains seven days after the issue of the certificate. The fact that a date is set by reference to a future event does not render it any the less a final date. Section 110(1) says, very clearly:
>
> 'The parties are free to agree how long the period is to be between the date on which a sum becomes due and the final date for payment.'
>
> The event could be a stage, or milestone or completion, practical or substantial. It could be the result of action by a third party, such as a certificate under a superior contract or transaction, as is found in financing arrangements. Provided that the event is readily recognisable and will produce a date by reference to which the final date can be set, there is no reason why it cannot be used. Payment of a subcontractor by reference to the date of a main contract certificate accords with industry practice and, on this project, is not at all inconsistent with the aim of Neutral Cash Flow (taking into account clause 2.1(b) of Schedule F). No difficulty could arise after the end of 2001 as the pattern set by the two years could easily be projected beyond the end of 2001 (and, evidently, was so projected). Put another way, if Railtrack did not issue a certificate on time Alstom could hardly use it as a defence since clause 2.6 is written on the assumption of due compliance. I therefore do not understand how it could be said that the date could be changed unilaterally.

Table 12.1 Table of Cases: Providing an 'Adequate Mechanism'

Title	Citation	Issue
Maxi Construction Management Ltd v Mortons Rolls Ltd	[2001] Scot HC 78, (2001) CILL 1784 (Outer House, Court of Session), Lord MacFadyen	**Inadequate mechanism: dates and amounts of payment subject to parties agreement** The contract required agreement before a payment could be made. However, no timetable had been provided for agreement to be reached, nor any means

Title	Citation	Issue
		of resolving any disagreement other than adjudication. The Scheme applied to determine the dates for payment, but the contractor's payment application did not suffice to trigger payment because it was insufficiently detailed.
Tim Butler Contractors Ltd v Merewood Homes Ltd	[2002] 18 Const LJ 74, Judge Gilliland	**Contract over 45 days' duration** The adjudicator considered the chain of correspondence and concluded that there was no evidence of the parties' agreement to a works duration of 4 weeks, notwithstanding the existence of a programme showing this to be the period to completion. Consequently the contract satisfied the test of being over 45 days' duration, and the contractor was entitled to be paid the amount of its interim application with interest.
Alstom Signalling Ltd v Jarvis Facilities Ltd	[2004] EWHC 1285 (TCC), Judge LLoyd	**Adequate mechanism: final date for payment determined by reference to superior contract** The final date for payment under a subcontract was linked to the dates upon which certificates were to be issued under the main contract. The mechanism was adequate since failure by the main client to make payment on the dates specified would not excuse payment under the subcontract.
All in One Building & Refurbishments Ltd v Makers (UK) Ltd (2005)	[2005] EWHC 2943, (2005) CILL 2321 (TCC), Judge Wilcox	**Final date for payment** A dispute may arise between parties regarding payment even before the final date for payment has been passed if it is clear that the employer disputes that an entitlement exists.
Mair v Arshad	[2007] A538/04, October 2007 (Sheriffdom of Tayside, Central and Fife at Cupar), Sheriff O'Carroll	**Adequate mechanism is a question of fact** The adequacy of the mechanism was a question of fact not law, though the provisions of the Scheme could provide guidance. A simple mechanism requiring the contractor to issue an invoice without any level of particularity might be adequate in some circumstances.
Ringway Infrastructure Services Ltd v Vauxhall Motors Ltd	[2007] All ER (D) 333, [2007] EWHC 2421 (TCC), Akenhead J	**Adequate mechanism: payment mechanism adequate even if open to misapplication** A payment mechanism should not be deemed inadequate simply because it might be misapplied. The contract could be construed to avoid a precondition to payment which might otherwise enable the employer to profit from its breach.
Reinwood Ltd v L Brown & Sons Ltd	[2008] 1WLR 696, [2008] BLR 219 (House of Lords)	**Final date for payment** The final date for payment has the effect of making time of the essence, following which failure to pay may result in the payee being able to show a repudiatory breach.

(4) What Happens if a Payment Scheme Is Not 'Adequate'?

Introduction

12.37 If an adjudicator decides that no provision has been made for interim payments where it should have been, or that the contract provisions for determining the amount due or dates for payment are not adequate, then the Scheme applies to substitute the defective machinery.[26] In most cases where there is any doubt as to the effectiveness of the contractual provisions, claimants will seek relief under the Scheme as an alternative. However it may also be open to an adjudicator to take the initiative in awarding payment under the Scheme even where this has not specifically been sought, provided that the terms of the Referral do not restrict the adjudicator from doing so: *Karl Construction (Scotland) Ltd v Sweeney Civil Engineering (Scotland) Ltd* (2000) (Key Case).

12.38 It appears that an adjudicator's decision as to whether or not the Scheme applies is a matter of contractual interpretation solely within an adjudicator's jurisdiction to decide.[27] If an adjudicator wrongly decides that the Scheme applies then this does not open up his decision to a jurisdictional challenge as it would if, for example, he wrongly decided that the contract was a construction contract within the meaning of ss. 104–7: *Tim Butler Contractors Ltd v Merewood Homes Ltd* (2002).[28]

Applying the Scheme in Whole or in Part

12.39 A construction contract may be non-compliant in one respect with the payment regime contemplated by Part II of the Act, but compliant in others. For instance, it may contain an adequate valuation mechanism, but fail to stipulate dates on which a payment becomes due, or a final date for payment.

12.40 This raises the question of whether the Scheme applies in whole to substitute an entire payment regime into the contract, or whether it applies in part so as to 'fill in the gaps'. The position in relation to the adjudication provisions of the Act is that the Scheme applies in whole to replace a non-compliant adjudication agreement.[29] However, the language relating to payment provisions in both s. 110(3) of the Act and paragraph II.1(a) of the Scheme is different, and provides that the 'relevant provisions of paragraphs 2 to 4 shall apply'. This suggests that the Scheme will only apply to 'fill the gaps' left by the parties and will not require a wholesale replacement of terms which have otherwise been agreed.

12.41 Many arguments on the subject start from the point that if the contractual machinery is tainted in one respect, then it should fall by the wayside and be replaced in its entirety.[30] Others have argued by analogy with the Unfair Contract Terms Act 1977, that if a term is unfair, then it should be struck out altogether.[31] As Coulson J put it in *Banner Holdings v Colchester Borough Council* (2010): 'I do not believe that it should be for the court to have to piece together a compliant set of provisions from two different sources.'[32]

12.42 In *C&B Scene Concept Design Ltd v Isobars Ltd* (2001) (Key Case), the failure of the contractual payment scheme was considered fundamental to the operation of the contract, and the employer was able to argue successfully in enforcement proceedings that it fell to be substituted in its entirety by the Scheme.

12.43 However, this position must be contrasted with two cases that permitted only the offending parts of the contact to be substituted with the relevant provisions of the Scheme. In *Maxi*

[26] s. 110(3).

[27] *C&B Scene Concept Design Ltd v Isobars Ltd* [2002] EWCA Civ. 46 at para. 29; *Allen Wilson Shopfitters v Buckingham* [2005] EWHC 1165 (TCC), paras. 28–30.

[28] [2002] 18 Const LJ 74, see 12.10.

[29] *Yuanda (UK) Co Ltd v WW Gear Construction Ltd* [2010] EWHC 720 at para. 62. See also 3.100–3.101 above.

[30] See *dicta* of Judge Toulmin in *John Mowlem v Hydra-Tight Ltd* [2002] 17 Const LJ 358 at 363

[31] See Coulson J, *Construction Adjudication* (Oxford University Press, 2nd edn, 2011), p. 128.

[32] [2010] EWHC 139 (TCC) at para. 43.

Construction Management Ltd v Mortons Rolls Ltd (2001) (Key Case) various offending provisions of the contract were required to be substituted by the Scheme, but clause 30.3.5 of the JCT conditions was preserved allowing the contractor to claim the amount of its interim payment application.

12.44 In the Scottish case of *Hills Electrical & Mechanical plc v Dawn Construction Ltd* (2003) (Key Case) the subcontractor successfully argued before Lord Clarke that the contract did not comply with s. 110(1) of the Act because it failed to provide for dates for applications to be made triggering a payment becoming due. However, it was unsuccessful in showing that the necessary corollary of this failure was that the Scheme would also substitute a final date for payment 17 days after the date on which payment became due, as opposed to 28 days provided by the contract. The timing of the final date for payment was of vital importance for the subcontractor in *Hills* as the main contractor had gone into administration between the two dates contended by the opposing parties.

12.45 The decision in *Hills* was approved by Edwards-Stuart J in *Yuanda (UK) Co. Ltd v WW Gear Construction Ltd* (2010); however, it was also stated that the question would be one of fact.[33] The payment provisions of the Act, unlike the adjudication provisions, do not represent either a single term or a seamless set of provisions. On the contrary, the provisions of ss. 109–13 are drafted so as to leave the maximum amount of freedom to the parties to determine how payments should be made. The language of the Scheme is similar and provides quite clearly that the parties' agreement as to the terms of payment and withholding should take precedence wherever possible. In *Hills*, Lord Clarke saw no reason to substitute provisions of the contractual payment regime which were not in any conflict with the provisions that needed to be imported by the Scheme.

Key Cases: Applying the Scheme in Part

Karl Construction (Scotland) Ltd v Sweeney Civil Engineering (Scotland) Ltd [2000] Scot CS 330 (Outer House, Court of Session)

12.46 **Facts:** The parties entered into a bespoke subcontract that provided for payments to become due, but stated that they would only be payable once the value of the equivalent works was included in a certificate or otherwise approved by the employer under the main contract.

12.47 Although neither party had argued that the terms of the subcontract did not adequately specify a final date for payment, the adjudicator held that this was the case and decided that Sweeney was entitled to immediate payment of its interim application under the Scheme instead. Karl refused to make payment on the grounds that the adjudicator had decided questions she was not asked to decide and was therefore acting outside of her jurisdiction.

12.48 **Held:** Lord Caplan upheld the adjudicator's decision. It was clear that the issue that had been referred to adjudication related to the timing of Sweeney's payment and, even though the parties had confined their arguments to the terms of the contract, the adjudicator was entitled to rely on her own knowledge of the Act to determine that payment was due under the Scheme instead:

> 28. Given the terms of the referral itself it is not surprising that the adjudicator was persuaded that Sweeney were invoking any feature of the Sub-Contract or adjudication legislation which would entitle them to payment of their Application for Payment No 5. It was equally clear that Karl Construction were disputing their obligation to make such a payment. Both parties appear to have got at least their principal contentions in support

[33] [2010] EWHC 720 (TCC), Edwards-Stuart J.

of their respective cases wrong ... However, looking at the issue more broadly, the parties were in dispute as to when was the time for payment of the outstanding works set out in Application for Payment 5. This is certainly what Sweeney, at least, would want to resolve ...

29. The adjudicator is no doubt experienced both in the application of the 1996 Act and of the Scheme. Once she decides that the Sub-Contract in fact contains no adequate mechanism for deciding when instalment payments become due, her knowledge of Part II of the Scheme is no doubt sufficient to permit her to make a finding that in the circumstances Sweeney were entitled to the redress they have claimed in the adjudication.

12.49 The decision was appealed to the Inner House[34] but again upheld, Lord Marnoch commenting:

7. ... What, after all, [Sweeney's] submission amounted to was that the adjudicator's understanding and application of the relevant law could in some way be circumscribed by the agreement of the parties. In our opinion, however, that proposition only has to be stated in order to be discarded. The adjudicator, who was in terms charged with ascertaining, inter alia, the law, had no option but to apply the relevant law as she saw it and, indeed, in the present case, any other conclusion would enable parties effectively to contract out of the mandatory provisions of the 1996 Act.

C&B Scene Concept Design Ltd v Isobars Ltd [2001] CILL 1781 (TCC)

12.50 **Facts:** The parties had failed to complete the Appendix to the JCT Standard Form of Building Contract with Contractor's Design 1998, to specify which of two competing payment regimes should apply. Clause 30.3.5 of this form was retained so that in the absence of the employer's payment certificate the contractor claimed the amounts shown in its payment application. The adjudicator agreed with the contractor's interpretation and awarded these amounts.

12.51 **Held:** The conflict was fundamental to the operation of the contract, and so the payment regime contained in clause 30.3 fell to be substituted in its entirety by the Scheme. Accordingly, the contractor was not entitled to rely on clause 30.3.5, and the adjudicator had been wrong to award these amounts in the absence of a payment notice or withholding notice from the employer. As Mr Recorder Robert Moxon Browne QC explained:

35. ... it seems to me plain that the adjudicator overlooked the fact that clause 30.3.5 of the JCT Form had no application to the contract which, as the claimant itself had asserted before him, lacked agreed provisions in relation to either 'alternatives A or B' in Appendix 2 to the form. As I have explained, clause 30.3 of the contract, relating to applications for interim payment, is predicated on agreement of either alternative A or B, and in the absence of agreement as to either of these alternatives, the Scheme for Construction Contracts specified by the HGCRA applies (see HGCRA s109(3)).

36. Part 2 of the Scheme provides that where the parties have failed to agree the machinery for interim payments, s2 of Part II shall apply. It is unnecessary to recite these provisions. Suffice to say the language employed makes it clear that the amounts due to the contractor will depend on the value of the work done and/or materials in fact supplied. The language is far removed from clause 30.3.5 of the JCT Form.

[34] 85 Con LR 59 [2002].

12.52 When the case went to appeal[35] the contractor relied on the words in s. 110(3) to argue that the Scheme should apply only to fill the gaps in the contract and not to replace the contract scheme in full. Section 110(3) provides 'if or to the extent that a contract does not contain such provision as is mentioned in Subsection (1) and (2), the relevant provision of the Scheme for Construction Contracts apply'. In the event, however, the Court of Appeal declined to decide the point since they had already decided that, if the adjudicator had been wrong in law, this was not a matter that went to his jurisdiction. They therefore overturned the Recorder's decision, but were content to assume that he had been right to apply the Scheme in full:

> 22. For the purposes of this judgment I am content, without deciding, to assume that the Recorder was right ... and that accordingly the Adjudicator was wrong; he ought as a matter of law to have held that Clauses 30.3.3–6 were not part of the contract and that under the scheme the failure to serve a timeous notice did not prevent the Defendants relying on the matters of the defence which it wished to advance, on the basis that all the Claimant was entitled to be paid was what was 'due under the contract'; or at least these points were arguable.[36]

Hills Electrical & Mechanical plc v Dawn Construction Ltd [2003] Scot CS 107, [2004] SLT 477 (Outer House Court of Session)

12.53 **Facts:** Hills entered into a subcontract with Dawn in June 2000. In October 2000 Dawn entered into administration leaving Hills' latest interim payment application unpaid. The contract contained a final date for payment that was 28 days after the due date for the payment application. By virtue of the other provisions of the contract, Dawn was not liable to pay Hills if it was insolvent at the final date for payment. Hills sought to argue that because the contract contained no dates for making interim payment applications and a 'pay-when-paid' clause, both of which were contrary to the 1996 Act, the Scheme should apply in full to substitute the payment mechanism in the contract. Under the Scheme the final date for payment would have been 17 days after the payment application had become due, and accordingly Hills would have been entitled to be paid before the insolvency. Dawn did not accept that the subcontract terms were deficient in any manner, but argued that even if they were, the final date for payment was a matter that had been agreed within the subcontract. This did not fall to be substituted by the provisions in the Scheme.

12.54 **Held:** Parliament could not have intended that terms that had been expressly agreed between the parties should be supplanted by the Scheme simply because they had omitted to deal with one or more of the other requirements of the legislation. On the contrary the language of the Scheme pointed the other way. Accordingly, the final date for payment remained 28 days after the date on which the sums became due, as agreed by the parties under the contract. In Lord Clarke's opinion, the legislature intended to interfere with the parties' freedom of contract only to the extent that this was clearly provided for, either expressly or by clear implication by the terms of the legislation itself'. Sections 109, 110(1), and 111(3) all confirmed that the parties were free to agree certain pertinent matters:

> 18. ... It is only, in the absence of agreement, in relation to any of these things, in the parties' own contract, that the provisions of the Scheme apply. Moreover, as has been

[35] *C&B Scene Concept Design Ltd v Isobars Ltd* [2002] BLR 93 (CA).

[36] Per Sir Murray Stuart-Smith in the Court of Appeal at para. 22.

seen, Section 114(4) provides that where any provision of the Scheme does apply to a construction contract, in default of a contractual provision agreed by the parties, the effect is that the Scheme's provision becomes an implied term of the contract in question. That sub-section begins with the words 'where any provisions of the Scheme'. The emphasised words, in my judgement, clearly envisage that it was not intended by the legislature that expressly agreed terms relating to the matters covered by the Scheme were to be supplanted by the provisions of the Scheme simply because of the fact that the parties had omitted to provide for one or other of the matters desiderated by the legislation or had failed to deal with it adequately, having regard to the statutory provisions. On that contrary, the language of the Scheme itself points the other way. ... The use of the word 'relevant' in my judgment ... makes it clear that simply because the parties have omitted to provide, or have provided inadequately, in their contract, for one matter in relation to payment, does not mean that anything more than the Scheme's provisions in relation to that matter, is imported into the parties' contract as an implied term of that contract. Paragraph 8 of the Part II of the Schedule, in dealing with the final date for payment, stands alone and only comes into effect, in my opinion, when the parties themselves have failed to specify a final date for payment in relation to payments which have become due under the contract. ...

20 On a proper analysis of the legislative provisions on the Scheme, it is therefore, in my judgment, apparent that the approach adopted by the pursuers in this case is unsound. I, accordingly, disagree with the view expressed at first instance in the case of *CB Scene Concept Design Ltd.*[37]

Table 12.2 Table of Cases: Applying the Scheme

Title	Citation	Issue
Karl Construction (Scotland) Ltd v Sweeney Civil Engineering (Scotland) Ltd	[2000] Scot CS 330, [2001] SCLR 95 (Outer House, Court of Session), Lord Caplan; [2002] 85 Con LR 59, [2002] SCLR 766 (Inner House, Court of Session), Lords Marnoch, Dawson, Clarke	**Inadequate mechanism: Scheme applies automatically to substitute inadequate contractual provisions** The final date for payment was conditional upon a certificate being issued, or payment otherwise agreed under a superior contract. Even though neither party had argued that the payment mechanism was inadequate, it was open to the adjudicator to apply her own knowledge of the Act to decide that the Scheme applied instead.
C&B Scene Concept Design Ltd v Isobars Ltd	[2001] CILL 1781 (TCC), Mr Recorder Moxon Brown QC; [2002] BLR 93 (CA)	**Applying the Scheme in full** The parties' failure to specify which of two alternative payment regimes applied was fundamental to the operation of the contract, and so the payment regime contained in clause 30.3 fell to be substituted in its entirety by the Scheme.
Hills Electrical & Mechanical plc v Dawn Construction Ltd	[2003] Scot CS 107, [2004] SLT 477 (Outer House Court of Session), Lord Clarke	**Applying the Scheme in part** Although the parties had failed to provide an adequate mechanism for determining when payments became due, there was no need to apply the Scheme in full to also substitute a final date for payment. Parliament could not have intended that terms that had been expressly agreed between the parties should be supplanted by the Scheme

[37] *C&B Scene Concept Design Ltd* v *Isobars Ltd* [2002] BLR 93 (CA).

Title	Citation	Issue
		simply because they had omitted to deal with one or more of the other requirements of the legislation. On the contrary the language of the Scheme pointed the other way.
Banner Holdings v Colchester Borough Council	[2010] EWHC 139 (TCC), Coulson J	**Applying the Scheme as a whole** In a case concerning the adjudication provisions of the scheme, Coulson J questioned[38] whether 'more generally' it should not be for the court to have to piece together a compliant set of provisions from two different sources. That would not make for certainty.
Yuanda (UK) Co Ltd v WW Gear Construction Ltd	[2010] EWHC 720 (TCC), Edwards-Stuart J	**Applying the Scheme in part** *Obiter*: in a case examining whether the adjudication provisions of the Scheme applied in full or in part, the judge approved the case of *Hills* as the correct approach to dealing with cases in which there was non-compliance with the payment provisions of the Act. However, much would depend on the facts and there might be situations where the whole of the Scheme's provisions should be substituted as no other solution was feasible.

(5) Payments under the Scheme

Introduction

12.55 This section considers the provisions that will be implied into the contract by the Scheme *only* if a contract fails to make adequate (or any) provision for the dates and amounts of payments. If the contract does provide an adequate mechanism, but the paying party does not comply with it, then the payee must enforce the terms of the contract to determine what is due and when.

Interim Payments under the Scheme

12.56 If the contract fails to provide an adequate mechanism for determining the amounts to be paid, then the Scheme will apply to substitute a method of assessment according to whether the payment in question is an interim payment or a final payment. In the case of interim payments paragraphs II.2 to II.4 of the Scheme apply equally to cases where the parties have failed to provide for interim payments to be made at all (s. 109), and where they have made provision but that provision is inadequate (s. 110).[39] In summary:

1. The starting point for valuations under the Scheme is 'the value of any work performed in accordance with the relevant construction contract', plus the value of materials brought to site (where relevant).[40] The 'value of work' performed is defined by paragraph II.12 as the amount 'determined in accordance with the construction contract under which the work is performed or where the contract provides no such provision, the cost of any work performed in accordance with that contract, together with an amount equal to any overhead and profit included in the contract price'.

[38] At paras. 42 and 43.
[39] Para. II.1 of the Scheme.
[40] Para. II.2(2)(a) and (b).

2. Any other sum that the contract specifies shall be payable in the relevant period can also be added.[41] The valuation period is determined by the contract, failing which it is the 28-day period.[42]
3. Each payment is the difference between what has already been paid and the aggregate valuation to the end of the relevant period or stage (Scheme paragraph II.2(1)).
4. There is a ceiling whereby the sum of all interim or stage payments must not exceed the contract price (Scheme II.1(4)). However, see 12.65.

Value of Work Performed

12.57 The Scheme endeavours to preserve any agreement set out in the contract as to the method of valuation provided that one exists. Typically, this might consist of Bills of Quantities which set out rates, or a requirement that amounts are to be calculated by reference to a method of measurement[43] at published or open market rates. The contract might also set out stage payments payable on the completion of given milestones. If the contract does not provide any mechanism for determining the value of work performed, then the Scheme requires payment to be on a 'cost plus' basis, with profit or overhead being payable only where there is evidence of their inclusion within the contract price.

12.58 Although there is no further definition in the Scheme of 'value of work performed', the payee is likely to be entitled to:

1. wages and add-ons for taxes, social costs (holidays, pensions, etc.) and possibly bonuses and allowances if they are directly related to the work
2. accommodation and site facilities; this may depend on whether the breakdown of the contract price indicates that these costs are intended to be included in any overhead or fee
3. payments to subcontractors for work performed, including all of the elements listed in point 1 above
4. materials incorporated into the work; these will also fall within the definition of 'value of work performed' and should therefore be reimbursed on the same basis as labour costs (i.e. according to any method of valuation stipulated in the contract, or else on a 'cost plus' basis).

12.59 Paragraphs II.2(b) and (c) also require payment to be made for materials not yet incorporated in the works (such costs will only be included where the contract so allows and their 'value' will be reimbursed provided that they have been manufactured on site or brought onto site during the relevant period);[44] and any other amount or sum which the contract specifies shall be payable in respect of the relevant period. If the contract is not specific as to the elements included in the contract price, then no amounts will be payable under this category.[45]

12.60 In order to value all of these costs, the payer may require to see the payee's accounts and records, for example in support of the hours worked and payments made. If these have not been kept, then there would appear to be nothing preventing the payer from making its own estimate of the work performed.

---

[41] Para. II.2(2)(c).

[42] See para. II.2 and the definition of 'relevant period' in para. II.12.

[43] See e.g. the Standard Method of Measurement, 7th edn, and CESMM3.

[44] Para. II.2(2)(b) of the Scheme. There is no further definition of the term 'value of materials' and it is therefore not at all clear whether this would include any overheads attributable to the cost of procuring materials, including delivery, samples, etc. The fact that para. II.2(2)(b) does not mention any overheads or profit payable on such materials would tend to suggest, however, that these should not be claimed until the materials are incorporated in the work.

[45] Paragraph II.2(2)(c).

12.61 In addition to the above, the definition of 'value of work performed' allows for overheads and profits to be payable in the following circumstances:

1. Where valuation takes place under the contract valuation mechanism, then overheads and profit will be included in interim payments if the contract states they should be. Often these elements will be factored into the rates and prices set out in the contractor's bills of quantities.
2. Where valuation takes place under the Scheme, overhead and profit will only be payable to the extent that it is possible to break down the contract price in order to isolate any overhead and profit elements within it.

12.62 No further definition of overheads or profit is provided by the Scheme, but it is likely that these are intended to cover all costs which are not directly incurred in performing the work and which therefore cannot be claimed under the cost categories set out above. Examples might include head office and design costs, procurement, insurance, and risk; however site overheads should probably be claimable under the definition of *value of work performed.* Such a breakdown may be difficult in practice, especially where elements such as the main contractor's margin on subcontract costs are not obvious from the contract price.

12.63 Where it is not possible to identify these elements within the contract price, then it is unlikely that they will be recoverable as overhead in interim payments. However, paragraph II.5 of the Scheme should allow for overhead and profit to be captured within the final payment which represents the balance of the contract price payable after all interim payments have been deducted.

12.64 The reference in paragraphs II.2(2) and II.12 of the Scheme to '*work performed in accordance with the contract*' appears to mean that the value of any work not performed in accordance with the contract should be omitted. This should permit the payer to deduct elements of the payee's payment application relating to defects or work which was not required by the contract. Any such withholding should be made in accordance with ss. 110(2) and 111 of the Act, including service of a payment or withholding notice (now a 'pay-less' notice) where necessary. (See further Chapter13 below which describes the circumstances in which a withholding notice may not be necessary because the value of the work itself has been reduced: *SL Timber Systems Ltd v Carillion Construction Ltd* (2001).)[46]

Assessing Final Payments under the Scheme

12.65 Paragraph II.5 applies to final payments under the Scheme and provides that the final payment should be the difference (if any) between the 'contract price', and the aggregate of any instalments or stage or periodic payments which have become due under the contract. The 'contract price' is defined in paragraph II.12 as 'the entire sum payable under the construction contract in respect of the work'. Thus:

1. Where the contract contains a lump sum or fixed price for the work, then 'the entire sum payable' relates to this lump sum adjusted where so provided under the contract.
2. Where the contract provides a mechanism for allowing additional sums to be paid in respect of extended or varied work, then the 'entire sum payable' would include these additional amounts, valued in accordance with the contract. If no such mechanisms exist then, the fallback would be to value changes on the basis of a reasonable sum.[47]

---

[46] [2001] BLR 516 (Outer House Court of Session, Lord MacFadyen).

[47] See *Greenmast Shipping Co SA v Jean Lion & Cie SA, The Saronikos* [1986] 2 Lloyd's Rep 277; *Rover International Ltd v Cannon Film Sales (No. 3)* [1989] 3 All ER 423, [1989] 1 WLR 912 (CA).

3. Where there is no predetermined contract price, a valuation under the Scheme rules would have no predetermined ceiling and the ultimate contract price would be the total value of the work properly performed pursuant to paragraph II.2(2) of the Scheme.
4. In respect of other arrangements, including target or guaranteed maximum price mechanisms, it will be a matter for interpretation of the contract whether these elements survive to limit or moderate the amounts payable under the Scheme.[48]

Timing of Interim Payments

12.66 Paragraph II.12 of the Scheme for Construction Contracts stipulates a payment interval of 28 days if none has been agreed. However the parties are free to adopt a longer or shorter period, or even a schedule of payment dates which may be spaced however the parties choose. In such cases, even where the parties have failed to specify a method of valuation, it seems that the Scheme will apply only to determine the value of the work, and not to substitute the 28-day interval in place of the contract intervals.[49] Indeed, there would appear to be no conflict between applying the Scheme's provisions for the valuation of work, together with dates that the parties have agreed for interim valuations to be made. If the contract stipulates a payment interval of 60 days, there should be no reason why work could not be valued in accordance with the Scheme on 60-day intervals.

Due Dates under the Scheme

12.67 The date upon which a payment becomes due under the Scheme depends upon what type of payment it is as shown in Table 12.3 below. In all cases, the payee's application ('claim by the payee') is a condition of payment which may be delayed indefinitely if the payee fails to put in its invoice.

12.68 By paragraph II.12 of the Scheme, the payee's payment application must be in writing and must specify the amount of any payment which it considers is due and the basis on which it is calculated.

'Claim by the Payee'

12.69 Whether the contractor's payment application constitutes a 'claim by the payee' such as to trigger a payment under the Scheme was considered in the case of *Maxi Construction Management Ltd v Mortons Rolls Ltd* (2001).[50] Here, the contractor had failed to set out any basis of calculation for numerous loss-and-expense items claimed in its interim valuation. Lord Macfadyen decided that the interim valuation failed to comply with paragraph II.12 of the Scheme so as to constitute a 'claim by the payee' when viewed in the context of previous payment applications. It was clear that it contained large numbers of claims within it which failed to specify the basis on which the payment claimed was calculated:

> [29] ... It is not, in my view, appropriate to demand of an application for an *interim* payment that it set out in full detail the basis of calculation of items already paid for under earlier applications. But paragraph 12 does, in my view, require specification of the basis of calculation of the new matter included in the application in question. In my opinion many of the items in the 'Variations/Contract Instructions' part of Interim Valuation No. 10 cannot be regarded as specifying the basis on which they are calculated. ... I therefore take the view that, on that account too, Interim Valuation No. 10 cannot be regarded as a claim by the payee in a form which satisfies paragraph 12 of the Scheme.

---

48 See *Hills Electrical & Mechanical Plc v Dawn Construction* [2004] SLT 477 (Key Case at 12.53.

49 See further section 4 above.

50 [2001] Scot HC 78, (2001) CILL 1784 (Outer House, Court of Session), Lord Macfadyen (Key Case at 12.29 above).

12.70 The case gives little comfort to contractors or subcontractors who might be tempted to put in unspecified claims for loss and expense in the hope that the time limits implied by the Scheme will place pressure on the payer to pay the amounts stated. The appropriate response from payers in such circumstances might be to do nothing on the basis that no payment had become due or, more prudently, to serve a withholding notice within the time limits set out in the contract or the Scheme citing a lack of specification as reason for refusing payment.

Table 12.3 Due Dates under the Scheme

Type of Payment	Due Date under the Scheme	Relevant Paragraphs of the Scheme	Comment
An interim payment under a relevant construction contract (i.e. a contract over 45 days' duration).	The later of: (a) the expiry of 7 days following the end of the payment period specified in the contract, or, if none exists, 7 days following the end of each 28-day period (b) the making of a claim by the payee.	II.2, II.4, II.12	The payee's application is a condition of payment under the Scheme, and the payee should therefore ensure that a substantiated invoice is submitted within seven days of the end of the payment period or risk payment being delayed.
A final payment due under a construction contract over 45 days' duration.	The later of: (a) 30 days following the completion of the work (b) the making of a claim by the payee.	II.5, II.12	The longer period for submission of the payee's invoice allows time for the payee to prepare its final account taking into consideration the need to substantiate all claims for varied and prolonged work. The final date for payment under the Scheme remains the same (17 days) leaving limited time for the payer to challenge the contactor's final valuation. (See further 12.22 above)
The payment of the whole contract price under a construction contract of 45 days or less duration.	The later of: (a) the expiry of 30 days following the completion of the work (b) the making of a claim by the payee.	II.6, II.12	See comments above in relation to final payments under contacts over 45 days' duration.
Any other payment under a construction contract.	The later of: (a) the expiry of 7 days following the completion of the work to which the payment relates (b) the making of a claim by the payee.	II.7, II.12	This provision might cover interim payments agreed under contracts of less than 45 days' duration, or payments to be made other than in relation to work performed or materials supplied (e.g. for insurance or bonds where this is not specified in the contract to be part of the work performed).

Final Payments: Requirement for Completion

12.71 Final payments and 'other' payments under paragraph II.7 of the Scheme require work to have been 'completed' before applications for payment are made. 'Completion' is not defined in either the Act or the Scheme, and relies upon the definition within the contract. If the contract contains no such definition then there is little judicial guidance as to whether completion requires the complete fulfillment by the payee of the works required by the contract.[51] If there is a contractual term requiring completion to be 'practical' or 'substantial' save for *de minimis* defects, then it will be a matter for interpretation whether entitlement to the final payment arises at this time or later when all outstanding defects have been remedied.

Final Date for Payment under the Scheme

12.72 If the parties to a construction contract fail to stipulate a final date for payment, then paragraph II.8(2) provides that this shall be 17 days from the date that payment becomes due. This period remains the same whether the payment is an interim or final instalment.

(6) Payment Provisions under Standard Forms of Contract

12.73 All of the main standard forms of contract have been amended since 1998 in order to comply with the provisions of the Act and the Scheme. They have been amended again for contracts entered into under the 2009 Act although, with one exception, the changes introduced by the 2009 Act do not directly affect the assessment or timings of payments—only the requirement to serve notices as considered in the next chapter. The one exception relates to the final date for payment in the event that the payee serves a payment notice in default of the payer. In this case the final date for payment is postponed by the number of days after the payer should have served a payment notice that the payee serves its notice.

12.74 Nevertheless, the JCT, IChemE, and PPC have each taken the opportunity to make quite extensive amendments to the provisions relating to due dates and final dates for payment. The reason for this appears to be to avoid the dates for payment being conditional upon the issue of a notice or certificate by the payer or third party certifier which is now contrary to the 2009 Act by virtue of s. 110(1D).[52] For example, under the JCT Standard Building Contract 2005, payment-due dates are triggered by the architect's certificate issued on specified dates. If the architect fails to issue a certificate, then it is likely that a payment, valued in accordance with the contractual valuation mechanism, will still become due from the date that the architect should have issued the certificate: *Ringway Infrastructure Service Ltd v Vauxhall Motors Ltd* (2007).[53] Under the 2009 Act, however, it is likely that such provisions will now be judged inadequate to specify a date for payment. This issue is dealt with in the JCT's 2011 edition which now provides that the dates specified in the contract particulars are the due dates. However, the final payment remains dependent upon the conditions for the final certificate being satisfied and the certificate being issued. It remains to be seen whether a payment will become due if these conditions are disputed. See further the discussion of the 2011 provisions on pay when certified clauses in Chapter 16.

12.75 The ways in which some of the Standard Forms of Contract deal with the dates on which payments become due and the final dates for payment are set out and compared in Table 12.4.

---

[51] Although see *Emson Easton Ltd v EME Developments Ltd* [1991] 55 BLR 114 which considered 'completion' for the purposes of the insolvency provisions of a JCT contract to mean the same as 'practical completion' defined elsewhere in the contract.

[52] For discussion of s. 110(1D) see Ch. 16.

[53] [2007] All ER (D) 333, [2007] EWHC 2421 (TCC) (Key Case at 13.21).

Table 12.4 Comparison of Payment Provisions under Standard Forms of Contract

	Assessing the Amount Due	Dates upon which Payments Become Due	Final Dates for Payment	2011 Amendments to Due Dates and Final Dates for Payment
NEC 3	The project manager assesses the amount due at each assessment date and certifies payment within one week (clauses 50.1 and 51.1). The project manager's certificate is the notice of payment on behalf of the employer (clause Y2.2). No distinction is made between interim and final payments under NEC.	7 days after the assessment date (clause Y2.2).	14 days after the due date or a different period if stated in the contract data (clause Y2.2).	None relevant to the timing or assessment of payments.
JCT 2005 Standard Building Contract, JCT 2011 Standard Building Contract	**Interim payments** The architect/contract administrator issues interim certificates on the dates provided for in the contract particulars (clause 4.9). Not later than 5 days after the certificate is issued, the employer gives written notice of payment (clause 4.13.3). **Final payments** The architect/contract administrator issues the final certificate not later than 2 months (now 28 days under the 2011 amendments) after certain events have occurred (clause 4.15.1). Not later than 5 days after this, the employer gives a written payment notice (clause 4.15.3).	**Interim and final payments** Not specified in the 2005 version, but considered to be the date of the architect/contract administrator's certificate or the final certificate.	**Interim payments** 14 days from the date of issue of the interim certificate (clause 4.13.1). **Final payments** 28 days from the date of issue of the final certificate (clause 4.15.4).	**Interim payments** The due dates for interim payments are now specified to be the dates provided for in the contract particulars (clause 4.9.1). The final date for payment is 14 days after the due date and is postponed if the contractor serves a payment notice in default (clauses 4.12.1 and 4.12.4). **Final payments** The due date is the date of the final certificate, with the final date for payment 28 days thereafter (clause 4.15.3). The final date for payment is postponed if the contractor serves a payment notice in default (clause 4.15.6.2).

(Continued)

Table 12.4 *Continued*

	Assessing the Amount Due	Dates upon which Payments Become Due	Final Dates for Payment	2011 Amendments to Due Dates and Final Dates for Payment
JCT 2005 Design and Build Contract, JCT 2011 Design and Build Contract	**Interim payments** The contractor makes an application for interim payment on the dates set out in the contract particulars, (clause 4.9). Not later than 5 days after the application is issued, the employer gives written notice of payment (clause 4.10.3). **Final payments** Within 3 months of practical completion, the contractor issues a final account and final statement which become conclusive on certain events occurring (clause 4.12.1 to 4.12.4). Not later than 5 days after the final statement becomes conclusive, the employer gives written notice of payment (clause 4.12.8). If the contractor does not submit a final account or a final statement, then the employer may do so (clause 4.12.5).	**Interim payments** Not specified, but considered to be the later of the date specified in the contract particulars or the date of receipt of the contractor's application. **Final payments** Not specified but considered to be the date that the contractor's or the employer's final statement becomes conclusive.	**Interim payments** 14 days after receipt of the contactor's application (clause 4.10.1). **Final payments** 28 days after the contractor's or the employer's final statement becomes conclusive (clause 4.12.9).	**Interim payments** The due date is now specified to be the later of the dates set out in the contract particulars and the date of receipt of the contractor's application (clause 4.8.1.3). The final date for payment is 14 days after the due date (clause 4.9.1). **Final payments** The due date is one month from the later to occur of the final statement and the events listed (clause 4.12.5). The final date for payment is 28 days after the due date (clause 4.12.7).
JCT 2011 Standard Building Subcontract	**Interim payments** The subcontractor may make a payment application not later than 7 days prior to the due date. Not later than 5 days after the due date, the contractor gives a payment notice specifying the sum to be paid. (clauses 4.9.3 and 4.10.2). **Final payments** Not later than 5 days after the due date, the contractor sends the final payment notice to the subcontractor. The final payment notice shows the final contract sum less the total amount previously due as interim payments (clause 4.12.2).	**Interim payments** The due date is the date specified in the Subcontract Particulars and at one month intervals thereafter (clause 4.9.1). **Final payments** The due date is 2 months after the later of: (i) the retention release date or (ii) the the contractor's final statement (clause 4.12.1).	**Interim payments** 21 days after the due date (clause 4.10.1). **Final payments** 28 days after the due date (clause 4.12.3). **In both cases** The final date for payment is postponed if the subcontractor serves a payment notice in default.	**N/A**

	Assessing the Amount Due	Dates upon which Payments Become Due	Final Dates for Payment	2011 Amendments to Due Dates and Final Dates for Payment
ICE 1997	**Interim payments** The engineer certifies the sums due within 25 days of receiving the contractor's monthly statement (clause 60(2)). **Final payments** The engineer certifies the final amount due within 3 months of receiving the contactor's final account and all information required to verify it (clause 60(4)). **In both cases** The engineer's certificate is required to be sent to the employer and forwarded by the employer to the contractor as notice of payment (clause 60(9)).	**Interim and final payments** Upon issue of the engineer's certificate (clause 60(2)).	**Interim payments** 28 days after receipt of the contractor's monthly statement (clause 60(2)). **Final payments** 28 days after the due date (clause 60(4)).	**Interim and Final Payments** No amendments made.
I/Chem E (Red Book) 2001	**Interim and final payments** The project manager certifies payment within 14 days of receiving the contractor's monthly statement (56 days in the case of the final certificate) (clause 39.4). Thereafter the contractor submits an invoice for the amount certified (clause 39.5).	**Interim and final payments** The date of receipt by the purchaser of the contractor's invoice in the amount of the sums certified (clause 39.5).	**Interim and final payments** 14 days after receipt by the purchaser of the contractor's invoice (clause 39.5).	**Interim and final payments** The contractor requests payment at monthly intervals with payments falling due 14 days (56 days in the case of the final payment) after the date of receipt by the project manager of the request (clauses 39.3 to 39.5). The final date for payment is 28 days (70 days for the final payment) after the date of receipt by the project manager of the request (clause 39.5).

(Continued)

Table 12.4 *Continued*

	Assessing the Amount Due	Dates upon which Payments Become Due	Final Dates for Payment	2011 Amendments to Due Dates and Final Dates for Payment
PPC2000 (including 2008 and 2011 amendments)	**Interim payments** The constructor submits an application to the client and the client's representative at the intervals stated in the Price Framework and any Pre-Construction Agreement or (if no intervals are stated) at the end of each calendar month (clause 20.2). The client's representative issues a payment notice within 5 days thereafter (clause 20.3). **Final payments** The client's representative issues a final account within 20 working days after notice of rectification of defects. On or after 40 days after the notice, the constructor issues an application in the agreed amount or a different amount. The client issues a payment notice within 5 days thereafter (clause 20.16).	**Interim payments** The date of receipt by the client of the constructor's application (clause 20.2). **Final payments** The date of receipt of the constructor's application (clause 20.16.1).	**Interim and final payments** The later of (i) 20 working days after the due date or (ii) 15 working days after receipt of the constructor's VAT invoice (clause 20.2).	N/A

13

SECTIONS 110(1) AND 111: PAYMENT AND WITHHOLDING NOTICES UNDER THE 1996 ACT

(1) Overview

Introduction and Summary

13.01 Sections 110(2) and 111 of the 1996 Act deal with the notices which the paying party must serve informing the payee of the amount which it proposes to pay, and whether any amounts have been withheld. The straightforward intention behind both of these provisions is to ensure transparency so that the payee is made aware of, and able to challenge if appropriate, any dissatisfaction on the part of the payer as to the work performed. Thus:

1. Section 110(2) requires every construction contract to provide for the giving of a notice by the paying party specifying the amount (if any) of the payment proposed to be made, and the basis on which that amount was calculated.
2. Section 111 provides that a party to a construction contract may not withhold payment of a sum due under the contract after the final date for payment unless it has given effective notice in advance of its intention to do so.

13.02 Unfortunately, ss. 110(2) and 111 have proved problematic in their application for a number of reasons. The right to payment under s. 111 hinges on proving the 'sum due under the contract'. As the phrase suggests, the sum due is thus determined by the terms of the contract itself and not necessarily the amount either applied for by the payee, or set out in the payment notice served by the payer. Compounding this problem is the fact that the 1996 Act provides no fall-back position in the event that the payer does not serve the payment notice that he is supposed to provide.

13.03 The resulting uncertainty as to what payment is due, and therefore whether a withholding notice is required if the payer intends to pay less, is clearly undesirable. Further, since many important matters hinge upon the payee being paid the sum due, not least its ability to suspend work under s. 112, and the right to claim interest, the unhelpful drafting of ss. 110(2) and 111 has become all important. Dissatisfaction with this position has led to both ss. 110(2) and 111 being substantially redrafted in the Local Democracy, Economic Development and Construction Act 2009 ('the 2009 Act'). The position under the 2009 Act affecting construction contracts entered into after 1 October 2011 is considered in Chapter 14. The requirements relating to payment and withholding notices under ss. 110(2) and 111 of the 1996 Act are summarized below:

1. Every construction contract shall provide for the giving of a payment notice not later than five days after the date on which payment becomes due under the contract specifying the amount, if any, of the payment made or proposed to be made and the basis on which the amount was calculated. (s. 110(1)).
2. If the contract does not so provide then the relevant provision of the Scheme for Construction Contracts ('the Scheme') (Part II, paragraph 9) will be implied into the contract to require that a payment notice should be given.
3. The payer may not withhold payment after the final date for payment of a sum due under the contract unless it has given effective notice of its intention to withhold payment (s. 111).
4. To be effective, the notice must comply with the requirements of the Act as to content and timing. A payment notice under s. 110(2) of the Housing Grants Construction and Regeneration Act 1996 ('the 1996 Act') may also give effective notice of withholding providing it also complies with the requirements of s. 111.
5. Notices should be in writing addressed to the other party, and specify the grounds and amounts in sufficient detail for the party receiving the notice to understand any complaints made against it. Correspondence issued prior to the payment application is unlikely to be effective.
6. An effective withholding notice entitles the employer to withhold sums against the amount otherwise due after the final date for payment, but only until the payee successfully challenges (if indeed they mount such a challenge) the grounds for withholding.
7. The notice will generally only be effective to the extent that the amounts or grounds relied upon in any later justification of the withholding are those set out in the notice.

8. If an adjudicator finds that the notice was effective (ie that it was served in time and contained the necessary information), but that the amounts withheld should be paid (the basis for withholding having been found invalid) then payment should be within seven days.
9. If no withholding notice is given, or if it is not effective, then payment should be made immediately.
10. In such circumstances, the amount that should be paid depends on how the contract defines the 'sum due'. This will not necessarily be the same as the amount that has been applied for, or even the amount in the payment notice.

(2) Payment Notices under s. 110(2)

Introduction

13.04 The purpose of s. 110(2) is: to enable both contractor and employer to know the basis on which there is a disagreement between them on any given payment application. It provides an agenda either for further discussion or for a subsequent adjudication.[1]

The notice is also supposed to serve a function of acting as a 'baseline' from which the payee can judge whether any further withholdings have been applied before the final date for payment and, therefore, whether a withholding notice should be served.[2]

13.05 The problem with this mechanism is that s. 110(2) requires simply that the contract must contain a provision for the payer to give notice of payment within five days of it becoming due. However, the1996 Act specifies no statutory penalty where the contract contains such a mechanism but the payer fails to comply with it. Notwithstanding that this will be a breach of contract, the practical effect is that the payer may be free to disregard this requirement.

13.06 Furthermore it is the contract terms which determine what sum is due and therefore the s. 110(2) payment notice may serve no purpose other than to notify the payee in advance of payment how much the payer considers is due. Depending on the terms of the contract, the following situations might arise:

1. The absence of a s. 110(2) notice may not matter much in contracts where the terms provide that the amount due is that contained in a certificate issued by a third party (for example JCT Standard Building Contract 2005 clause 4.9.1). In such a contract the certificate both crystallizes the sum due and notifies the contractor of the sum due. It is against this crystallized sum that the employer may or may not decide to withhold other sums.
2. The same will apply in contracts where the sum due is crystallized by the payee's application followed by the payer's notice (for example the JCT Standard Form of Building Contract (with Contractor's Design) 1998, clause 30.3). In those contracts the sum due is determined by either the unchallenged contractor's application or by the employer's payment notice.
3. The difficulty arises in contracts which define the sum due as 'the value of work performed in accordance with the contract' or similar formulations. In these situations the payee may not know how much the payer considers to be due until payment is made. Depending on the precise terms of the contract, this may be the case regardless of whether

---

[1] *Ringway Infrastructure Service Ltd v Vauxhall Motors Ltd* [2007] All ER (D) 333, [2007] EWHC 2421 (TCC) at para. 68.

[2] See e.g. the Committee Proceedings in the House of Lords leading up to the introduction of the 1996 Act, Hansard, 23 July 1996, col. 1347.

a contract administrator has issued a certificate, and even where the payer himself has issued a payment notice.[3]

13.07 The problems that this third type of contract poses for the payee go further than this. In the absence of a withholding notice, the paying party has potentially far greater opportunities to challenge a contractor's application for payment than it does where the contract provides for interim certificates setting out the sum due for stage payments. This is explained in more detail in 13.28–13.39 below.

13.08 For these reasons, s. 110(2) has been replaced in its entirety under the 2009 Act to create a new fall-back position similar to that described in point 2 above.[4] The detail of this provision, and how it will operate for contracts entered into after the new Act comes into force, is discussed in Chapter 14 below.

Section 110(2): Fall-back Application of the Scheme

13.09 Notwithstanding the above issues, it remains a statutory requirement for construction contracts to make provision for the service of a payment notice in the terms set out in 13.01 above. If the contract does not include such a provision, then the Scheme applies: s. 110(3). The Scheme requires, quite simply, that the party making payment should give such a notice in the identical terms and within the same five-day period as is required by s. 110(2) of the Act.[5] Unlike most of the payment provisions of the Act, it is not permissible for the parties to agree an alternative time for issue of the payment notice, and therefore this notice should be served in all cases within the five-day period set out in the Act.

Section 110(2): Payment Notices Required even where Payment Is Zero

13.10 A payment notice should be given in all circumstances, whether or not an assessment results in a payment being due. Thus, if a periodic or stage payment is due to be made, but rights of set-off or abatement exist so that the amount due is reduced or extinguished by the amounts payable by the payee, then a payment notice should still be given setting out how the zero sum has been determined. Indeed, it is in circumstances such as these that the payment notice is arguably of greatest value, enabling the payee to understand the reasons for non-payment and if necessary giving it valuable time to prepare to challenge the assessment of the sum due.

Can a Third Party Serve a s. 110(2) Notice?

13.11 Section 110(2) provides that the notice should be given by a party to the contract and there is no provision for a notice to be served by a third party such as the contract administrator or project manager. Certain contracts, such as the NEC 3 Option Y2.2, expressly provide that the project manager's certificate constitutes notice from the employer under s. 110(2) of the 1996 Act. In these cases it would seem clear that the project manager has authority as agent to the employer and there should be no question that the certificate stands as the payment notice. In other cases judges have been willing to accept that the certificate of a third party does stand as the employer's payment notice[6] but those were cases where the point was not in dispute. This issue has now been addressed in the 2009 Act which provides that a 'specified person' may serve the payment notice or pay-less notice on behalf of the payer.[7]

---

[3] See *Emcor Drake & Scull Ltd v Sir Robert McAlpine Ltd* [2004] EWHC 1017 (TCC) at para. 71 in which Judge Havery rejected the submission that the amount in the payer's s. 110(2) notice was *ipso facto* the amount due under the contract. For further discussion of this case, see 13.60–13.62 below.

[4] See s. 137 of the 2009 Act which introduces new ss. 110A and 110B into the 1996 Act.

[5] Para. II.9 of the Scheme.

[6] E.g. *Emcor Drake & Scull Ltd v Robert McAlpine Ltd* [2004] EWHC 1017 (TCC) at para. 69.

[7] See Ch. 14 below.

Section 110(2) Notice May Serve as s. 111 Notice

13.12 Where a payment notice has been served setting out the basis for all deductions from sums which would otherwise have been due, there is no further need to serve a withholding notice. Section 111(1) of the 1996 Act provides that the notice mentioned in s. 110(2) may suffice as a notice of intention to withhold payment provided it complies with the requirements of s. 111. This requires that the payment notice should:

1. set out the amount proposed to be withheld and the ground for withholding or, if there is more than one ground, each ground and the amount attributable to it and
2. be given not later than the prescribed period before the final date for payment (i.e. seven days under the Scheme, or such other time as the parties have agreed under the contract).

13.13 Thus, even if the payment notice has been served later than the five-day period prescribed by s. 110(2), it will still be effective to give notice of withholding, provided it has been served no later than the date for withholding notices set out in the contract (or, if no date has been specified, the Scheme).

13.14 There is therefore a large degree of overlap between the payment notice specified by s. 110(2) and withholding notices under s. 111 which has led to some confusion as to why separate notices have been prescribed, and their respective purposes. This is yet another of the issues dealt with by the 2009 Act which brings the content of payment and withholding notices into line so that the withholding notice becomes, in effect, a revision of the payment notice.[8] This appears to confirm that the purpose of both notices is in fact the same in that, although one is expressed as a payment notice and the other a withholding notice, their joint purpose is in informing the payee of the amounts proposed to be paid (if any), and the basis of their calculation.

13.15 Even under the original provisions of the 1996 Act, provided a notice is sufficiently clear as to these points, then it can stand as both a s. 110(2) and s. 111 notice regardless of whether it is referred to as a payment notice or a withholding notice. The payer can serve two separate notices, or combine them into one. The potential for two separate notices effectively gives the payer two bites at the cherry: it may serve notice of withholding in the payment notice, or wait until the last day for serving a withholding notice to give notice of the same or further deductions.

What Happens if a Payment Notice Is Not Served or Is Served Late?

13.16 As has been seen, the consequences of not serving a payment notice may not be severe since it is the contract and not necessarily the payment notice which determines the sum due. Problems only arise when the contract states that it is the payment notice that determines the amount that is due and/or the timing of payments. In such cases the absence of a payment notice is likely to cause real problems for determining the amount of the payment that is due and/or the date that it is due and the final date for payment. Moreover, if the sum due is not capable of being ascertained in the absence of a payment notice, it will be unclear to the payee whether any amounts have been withheld which should have been the subject of a withholding notice. This situation is considered at13.28–13.39.

13.17 In many contracts, however, the sum due and timings of payments are defined not by the payment notice, but by the amounts specified in a third party's payment certificate or else measured by reference to the value of work performed. In these cases the absence of a payer's payment notice may not be material although other issues may arise in terms of ascertaining

[8] s. 111(4) as substituted by s. 139 of the 2009 Act.

the sum due for the purposes of determining the amount to be paid and whether a withholding notice should have been issued. These cases are discussed at 13.40–13.57 below.

Contracts where the Payment Notice Determines the Sum Due

13.18 The parties are free to include terms in the contract whereby the payer's payment notice determines the sum due under the contract. However, in such cases, if no payment notice is served, then one of the first questions that may arise is whether a payment has become due at all. Some standard form and bespoke contracts provide that the payer's payment notice determines not only the amount due, but also the date for payment. If this is the case, then it will be a question for interpretation of the contract whether the payment mechanism is adequate to provide for a due date and final date of payment, or whether the provisions of the Scheme apply instead.

13.19 Such a contract was under consideration in *Ringway Infrastructure Service Ltd v Vauxhall Motors Ltd* (2007)[9] (Key Case). Here, the judge decided that, despite no payment notice having been issued, the date for payment had been triggered on the date on which the employer should have given notice. It was not open to the employer to rely on its own breach of contract to delay payment indefinitely. As to the amount that became due, the fall-back position in the JCT contract applied and the amount of the contractor's application was the sum due.

13.20 It should be noted that the decision in *Ringway* is not directly applicable in other situations since it relied upon the provision of clause 30.3.5 of the JCT Standard Form of Building Contract (with Contractor's Design) 1998 which states that in the absence of a payment or withholding notice the contractor's application becomes due instead. This condition does not appear in other main standard forms of contract issued prior to 2011.[10] Since the 2009 Act came into force, however, a fall-back position similar to that in *Ringway* now has statutory force and will apply in situations where the payer (or specified person) is required to issue a payment notice but fails to do so.

Key Case: Where the Payment Notice Determines the Sum Due

Ringway Infrastructure Service Limited v Vauxhall Motors Limited [2007] All ER (D) 333 (TCC), [2007] EWHC 2421 (TCC)

13.21 **Facts:** The parties contracted under the JCT Standard Form of Building Contract (with Contractor's Design) 1998, clause 30.3.3 of which requires that within five days of the contractor submitting its application for payment, the employer should issue notice of payment specifying the amounts to be paid and the basis of calculation. Following issue of the payment notice, the contractor would be entitled to issue an invoice which would set the clock running for the final date for payment. Clause 30.3.5 applies where the employer has failed to serve both a payment notice under clause 30.3.3 and a withholding notice under clause 30.3.4, and requires the employer, in such circumstances, to pay the contractor the amount stated in the application for interim payment.

13.22 Ringway issued an application for interim payment, to which Vauxhall failed to respond. Vauxhall argued that, since Ringway had not issued an invoice but only an application for payment, no dispute had arisen and no sums were due.

[9] [2007] All ER (D) 333 (TCC), [2007] EWHC 2421 (TCC).

[10] See e.g. the JCT Design and Build Contract 2005 which provides instead that the amount due shall be calculated by reference to the method of valuation set out in the contract (at clauses 4.8.5 and 4.10.5).

13.23 **Held:** Ringway had issued a valid application for payment and Vauxhall's failure to respond within the requisite time limits crystallized the dispute. The issue of a payment notice under clause 30.3.3 was a mandatory obligation on the part of Vauxhall, and therefore Vauxhall were in breach of contract. No invoice was required in circumstances where the employer was itself in breach of clause 30.3.3. The clause 30.3.3 notice was the trigger for the submission of the contractor's invoice and it was not open to an employer by reason of its own breach to defer its obligation to pay indefinitely:

> 67. I consider that it is impossible to construe Interim Application No. 11 as anything other than a claim for payment of money. It cannot be construed either on its face or in context as a request for a consideration of the various claims for variations and other matters on some academic or pure valuation basis. This Application was a commercial document by which payment was sought. It was not an academic exercise.
>
> 68. Thus, Interim Application No. 11 was a claim in fact and under the Contract for payment. Whether or not Ringway was or became entitled to be paid depended upon what Vauxhall and its agent Walfords did or did not do. If Vauxhall believed that the net sum the subject matter of Interim Application No. 11 was overstated to a greater or lesser degree, the contractual machinery enabled them to give an appropriate notice saying so. That is the written notice under Clause 30.3.3. That provision reflects Section 110 of the HGCRA 1996: a primary purpose of Clause 30.3.3 and Section 110(2) is to enable both contractor and employer to know the basis upon which there is disagreement between them on any given payment application. It provides an agenda either for further discussion or for a subsequent adjudication.
>
> 69. Clause 30.3.4 of the Contract Conditions similarly requires the Employer if it is so advised 'not later than five days before the final date for payment of an amount due pursuant to Clause 30.3.3 to give a written notice relating to what amount is proposed to be withheld and/or deducted from' the sum identified as due under Clause 30.3.3. The operation of Clause 30.3.4 is contingent upon a Clause 30.3.3 statement. It is important to note that there is a distinction between Clauses 30.3.3 and 30.3.4: under the former, it is a mandatory obligation on the part of the Employer to give the written notice to the Contractor specifying the amount of payment proposed to be made in respect of any given Interim Application for Interim Payment. Clause 30.3.4 is discretionary; that is not surprising because in any given case the Employer may not wish to deduct from a sum otherwise due to the Contractor.
>
> 70. The contractual machinery for payment thus requires the Employer to give a Clause 30.3.3 written notice. Failure to do so on the part of the Employer is effectively a breach of contract. This applies even more in the case of this particular contract which, by an amended Clause 30.3.6.1, makes the final date of payment of an Interim Payment contingent upon the submission of an invoice; however upon a proper construction of Clause 30.3.6.1, no invoice can be submitted and no final date for payment established unless and until the Employer has served its written notice under Clause 30.3.3. Thus, a breach by the Employer of Clause 30.3.3 would in theory secure that the final date for payment was extended indefinitely. It cannot have been intended that the Employer should be able to defer an obligation to pay as a result of its own breach in failing to do that which the Contract specifically required it to do.
>
> 71. The effect of Clause 30.3.5 is simple: where the Employer has failed to give the requisite written notice under Clause 30.3.3, the Employer must pay the Contractor the amount stated in the requisite Application.
>
> 72. Mr Paul Darling QC, for Vauxhall, seeks to argue that in any event, even if there has been a default by the Employer in failing to serve the requisite Clause 30.3.3 notice, an invoice by necessary implication must be served by the Contractor before there is any

entitlement to payment. That is wrong because Clause 30.3.5 makes it clear that the obligation to pay falls due immediately following the seven-day period after receipt of an Application for Payment where the Employer has not given the requisite written notice.

73. Similarly, for the reasons adumbrated above, the Application for Interim Payment is a claim for payment. It is not necessary for the Contractor, having failed to receive from the Employer the Clause 30.3.3 notice, to submit a further claim or request for immediate payment before it is entitled to be paid.

(3) Withholding Notices under s. 111

When Is a Withholding Notice Required?

13.24 Section 111 prohibits the withholding of payment of a 'sum due under the contract' after the final date for payment unless the payer has given an effective notice of intention to withhold that payment. Thus, when a party to a construction contract covered by the 1996 Act wishes to enforce a claim for payment it must show first that a particular sum was due under the contract and, secondly, that there was no effective notice entitling the payer to withhold payment of that sum due.

13.25 The 1996 Act does not dictate how the sum due under the contract shall be defined, that is a matter left to the terms of the contract. The only requirement of the 1996 Act is that the contract shall contain an adequate mechanism for determining the sum due, the due date, and a final date for payment.[11] Construction contracts determine the sum due under the contract in a variety of ways. Therefore, depending on the terms of the contract, failure to serve a withholding notice may not preclude arguments about the amount actually due under the contract in question.

13.26 This reading of s. 111(1) has left the door open for payers to challenge claims for interim or stage payments, even where there is no withholding notice, on the terms that the sum was never due under the contract in the first place.

Withholding against the Sum Due

13.27 The following sections consider the requirement for a withholding notice to be served in a variety of different contractual situations, and demonstrate the importance of considering each contract carefully on its terms.

When the Sum Due Is the Value of Work Performed

13.28 Some contracts fail to specify that any of the payment notice, certificate, or the payee's application represent the sum due. Often, this is a matter of interpretation and an intent might be discovered that one of these three is in fact the sum due. In other cases, however, the sum due may simply be defined by reference to the valuation mechanism contained in the contract. In these cases the 'sum due' may be disputed between the parties which may cause difficulties for determining whether a withholding notice is required to be issued.

13.29 One of the earliest cases on the subject was *SL Timber Systems Ltd v Carillion Construction Ltd* (2001) (Key Case)[12] a decision of the Scottish Outer House. Despite Carillion having served no payment notice and a withholding notice that was out of time,[13] it was successfully

---

[11] s. 110(1) as discussed in Ch. 12 above. It is only if the mechanism is not 'adequate' as discussed, that the Scheme's provisions for determining the amount due and timing of payment will apply.

[12] [2001] BLR 516.

[13] As to the timing of withholding notices, see 13.116–13.123.

argued before Lord Macfadyen that the sum claimed by SL Timber was not the amount due under the contract because it was not the value of work performed in the relevant period. The adjudicator had erred in awarding the amount of SL Timber's application in lieu of notice; however the decision was still enforced as it was an error of law and was not an error the consequences of which went to the adjudicator's jurisdiction.

13.30 Whilst the judgment does not reveal the precise terms of the contract that applied, it appears that the sum due was determined as the value of work performed. Such terms are similar to those found in the Scheme. Thus, Carillion should have been entitled as a matter of law to raise a dispute about whether work had been done at all, whether it had been properly measured or valued, or whether some other event had occurred on which a contractual liability to make payment depended. These arguments went to the question of whether the sum claimed was due under the contract and did not involve an attempt to 'withhold ... a sum due under the contract'. Therefore there was no requirement to serve a notice of intention to withhold payment.

13.31 The conclusion reached by Lord Macfadyen in *SL Timber* was consistent with the earlier decisions of Judge Thornton in *Woods Hardwick Ltd v Chiltern Air Conditioning Ltd* (2001),[14] and Judge LLoyd in *KNS Industrial Services Birmingham Limited v Sindall Limited* (2000)[15] (Key Cases).

13.32 Whether a withholding notice is required under contracts of this variety depends on whether the arguments for not making payment go to the question of whether the sum claimed is due under the contract, or not. This means that matters which are properly characterized as a defence to the sum due under the contract can still be raised. Matters which are properly characterized as a cross-claim, must be contained in a withholding notice.[16]

13.33 The decision in *SL Timber* has its critics, notably the editors of *Keating on Building Contracts* (9th edn), at paragraph 18–060, who submit that a court will give s. 111 a purposive construction to meet the mischief intended, so that in the absence of a withholding notice a payer may not rely on set-off or abatement as a defence to a payee's claim.

Key Cases: Withholding where the Sum Due Is the Value of Work Performed

Woods Hardwick Ltd v Chiltern Air Conditioning Ltd [2001] BLR 23 (TCC)

13.34 **Facts:** This early case concerned an architect's application for payment and whether sums could be abated against it because of alleged breaches of the contract for services.

13.35 **Held:** Notwithstanding the employer's failure to serve any notice of withholding, it was not obliged to pay the amounts claimed:

> 10. any abatement properly relied upon by Chiltern, would not of course be caught by Section 111 of the Act, so Chiltern's abatement defence could, in principle, defeat or reduce Woods Hardwick's claims.

---

[14] [2001] BLR 23 (TCC).

[15] 75 Con LR 71 (TCC).

[16] For the difference between a defence to a claim for payment and a cross-claim, see the analysis in *Hanak v Green* [1958] 2 QB 9. For an example of contractual set-off see *PC Harrington Contractors Ltd v Multiplex Constructions (UK) Limited* [2008] BLR 16, [2007] EWHC 2833 (TCC) (discussed at 13.63–13.64). A contractual or equitable set-off operates so as to discharge the sum due under the contract, and abatement is a doctrine that denies that monies are due at all. It is said that this means that in such a case no sum would be due and there is no need to serve a withholding notice; see also *Keating on Building Contracts* (9th edn), at 19–060.

KNS Industrial Services Birmingham Limited v Sindall Limited (2000) 75 Con LR 71 (TCC)

13.36 **Facts:** KNS had applied for payment based upon its own valuation of the works under the DOM/1 conditions. KNS relied upon Judge Bowsher's decision in *Northern Developments (Cumbria) Ltd v J. & J. Nicol* (2000)[17] that 'there is no dispute about any matter not raised in the notice of intention to withhold payment'.

13.37 **Held:** Judge LLoyd dismissed this as a misreading of the case. In this instance, it was not safe to conclude that the adjudicator had accepted the subcontractor's gross valuation of the works, and then made deductions from it in the manner of a withholding. It was possible, for instance, that the deduction for 'non-complaint work' might mask matters which would affect the valuation itself. He also referred to the decision in *Fastrack*[18] that 'the "dispute" is whatever claims, heads of claims, issues or contentions or causes of action that are then in dispute which the referring party has chosen to crystallise into an adjudication reference'. Since the issue that had been referred to the adjudicator in *KNS* was the valuation of the works themselves, then it was open to the adjudicator to reduce the subcontractor's valuation for non-compliant work, even in the absence of a withholding notice:

> 17 The term withhold is ... used in section 111 to cover both the situation where in arriving at a valuation the contractor had not taken account of a countervailing factor as well as the situation where there is to be reduction in or deduction from an amount that had been declared or thought to be due. In the former case the word 'withhold' may not always be correct for one cannot withhold what is not due.

SL Timber Systems Ltd v Carillion Construction Ltd [2001] BLR 516 (Outer House Court of Session)

13.38 **Facts:** Disputes arose as to SL Timber's right to stage payments and Carillion challenged the valuations upon which SL Timber's applications were based. Carillion served no s. 110(2) notice and the withholding notice served under s. 111(1) was out of time. The adjudicator decided that Carillion was not entitled to dispute the amount claimed by SL Timber because no withholding notice had been served. Carillion challenged the decision and argued that the sum claimed by SL Timber was not the amount due under the contract because it was not the value of work performed in the relevant period.

13.39 **Held:** Lord Macfayden held that the adjudicator's conclusion was wrong in law and Carillion should have been entitled to run arguments that the sum was not due under the terms of the contract despite the absence of a s. 110(2) notice or a valid s. 111(1) withholding notice. Despite this, the adjudication decision in *SL Timber* was enforced as the error made by the adjudicator was not an error the consequence of which went to his jurisdiction:

> [19] In my opinion the adjudicator fell into error in the first place by conflating his consideration of sections 110 and 111 of the 1996 Act. In my opinion Mr Howie was correct in his submission that these sections have different effects and the notices which they contemplate have different purposes. Section 110(2) prescribes a provision which every

[17] See Table 13.1.
[18] See Ch. 3, Table 3.3.

construction contract must contain. Section 110(3) deals with the case of a construction contract that does not contain the provision required by section 110(2) by making applicable in that case the relevant provision of the Scheme, namely paragraph 9 of Part II. By one or other of these routes every construction contract will require the giving of the sort of notice contemplated in section 110(2). But there the matter stops. Section 110 makes no provision as to the consequence of failure to give the notice it contemplates. For the purposes of the present case, the important point is that there is no provision that failure to give a section 110(2) notice has any effect on the right of the party who has so failed to dispute the claims of the other party. A section 110(2) notice may, if it complies with the requirements of section 111, serve as a section 111 notice (section 111(1)). But that does not alter the fact that failure to give a section 110(2) notice does not, in any way or to any extent, preclude dispute about the sum claimed. In so far, therefore, as the adjudicator lumped together the defenders' failure to give a section 110(2) notice with their failure to give a timeous section 111 notice, I am of opinion that he fell into error. He ought properly to have held that their failure to give a section 110(2) notice was irrelevant to the question of the scope for dispute about the pursuers' claims.

[20] The more significant issue in the present case, in my opinion, is whether the defenders' failure to give a timeous notice under section 111 had the effect that there could be no dispute at all before the adjudicator as to whether the sums claimed by the pursuers were payable. The section provides that a party 'may not withhold payment after the final date for payment of a sum due under the contract unless he has given an effective notice of intention to withhold payment'. In my opinion the words 'sum due under the contract' cannot be equiparated with the words 'sum claimed'. The section is not, in my opinion, concerned with every refusal on the part of one party to pay a sum claimed by the other. It is concerned, rather, with the situation where a sum is due under the contract, and the party by whom that sum is due seeks to withhold payment on some separate ground. Much of the discussion of the section in the cases has been concerned with what circumstances involve 'withholding' payment and therefore require a notice. Without the benefit of authority, I would have been inclined to say that a dispute about whether the work in respect of which the claim was made had been done, or about whether it was properly measured or valued, or about whether some other event on which a contractual liability to make payment depended had occurred, went to the question of whether the sum claimed was due under the contract, therefore did not involve an attempt to 'withhold ... a sum due under the contract', and therefore did not require the giving of a notice of intention to withhold payment. On the other hand, where there was no dispute that the work had been done and was correctly measured and valued, or that the other relevant event had occurred, but the party from whom payment was claimed wished to advance some separate ground for withholding the payment, such as a right of retention in respect of a counterclaim, that would constitute an attempt to 'withhold ... a sum due under the contract', and would require a notice of intention to withhold payment. There are some circumstances in which it is difficult to be clear as to whether the position which the party against whom the claim has been made wishes to adopt is to be analysed as disputing that the sum is due under the contract or as seeking to withhold a sum due under the contract. It occurs to me, too, that Scots law may analyse some cases differently from English law. For present purposes, however, it is in my view unnecessary to draw the borderline between the two categories with precision, because it is clear that the adjudicator has adopted the extreme position that any and every attempt to dispute a claim made under a construction contract is to be regarded for the purposes of section 111 as an attempt to 'withhold' payment, and therefore as requiring a notice of intention to withhold payment. In my opinion, in so holding, and in consequently declining to address whether the sums claimed by the pursuers were due under the contract, the adjudicator fell into error.'

When the Sum Due Is that Stated in a Certificate

13.40 A different approach was taken by the Court of Appeal in *Rupert Morgan Building Services (LLC) Ltd v David Jervis and Harriet Jervis* (2003)[19] (Key Case) where the construction contract determined the sum due as being that contained in a third party certificate.

13.41 It is common for standard-form contracts to say expressly that the sum due to a contractor is the amount contained within an interim valuation certificate issued by a third party such as the project manager or architect. An example of such a provision is found in clause 4.9.1 of the JCT Standard Building Contract 2005 which provides that the architect shall issue interim certificates 'stating the amount due to the Contractor from the Employer'. Whilst the architect is obliged to include within the interim certificate the gross value of the works (assessed in accordance with the contract rules)[20] the amount due to the contractor is defined simply as the amount contained in the certificate.

13.42 In such contracts the Court of Appeal, in *Rupert Morgan*, said that the sum is determined by the certificate. It is not the actual work done which defines the sum due; it is simply defined as the amount in the certificate.[21] The certificate may be wrong and the architect may have missed out work he ought to have included, or may have wrongly measured the work in question but, in the absence of a withholding notice, s. 111(1) operates to prevent the withholding of payment of the sum that has been certified.[22] This means the payee is entitled to the money right away. Arguments about whether the payee should repay some or all of that stage payment on the grounds that it was overvalued can be made in the next payment cycle or when the final account is being negotiated.

13.43 The Court of Appeal in *Rupert Morgan* construed the contract as meaning that when a sum is contained in a certificate, that crystallizes the sum due and no further defences can be raised under the contract to that being the sum due. That interpretation of the contract may be thought to be something of a stretch given that the relevant terms of the contract said 'the employer shall pay to the Contractor the amount certified within 14 days of the date of the certificate, subject to any deductions and set-offs due under the contract'.[23] Nevertheless, the Court of Appeal was clearly minded to give a purposive construction to the terms of the ASI contract in issue, and appear to have interpreted this proviso as relating to deductions and set-offs properly notified in a withholding notice.

13.44 The decision has been applied in both the TCC and the Outer House Court of Session in relation to various contracts issued by the JCT: *Balfour Beatty Construction Northern Ltd v Modus Corovest (Blackpool)* Ltd (2008),[24] *Urang Commercial Ltd v Century Investments Ltd* (2011),[25] *Fleming Buildings Ltd v Forrest or Hives* (2008)[26] (Key Cases).

---

[19] (2004) 1 WLR 1867, [2003] EWCA Civ 1563 (CA).
[20] See clause 30.2 of the JCT Standard Form of Building Contract, 1980 edn.
[21] Para. 11 of the judgment.
[22] *Ibid.*
[23] Emphasis added.
[24] [2009] C.I.L.L. 2660; [2008] EWHC 3029 (TCC).
[25] [2011] C.I.L.L. 3061; [2011] EWHC 1561 (TCC).
[26] [2008] CSOH 103.

Key Cases: Withholding Where the Sum Due Is that Stated in a Certificate

Rupert Morgan Building Services (LLC) Ltd v David Jervis and Harriet Jervis (2004) 1 WLR 1867, [2003] EWCA Civ 1563 (CA)

13.45 **Facts:** The contractor, Rupert Morgan, claimed from the employers, Mr and Mrs Jervis, the sum due under an interim certificate issued by an architect in the course of building works carried out on the employers' property. There was a written contract in a standard form provided by the Architecture and Surveying Institute ('ASI'). The contractor sought summary judgment before the clients filed their defence. The employers appealed on the basis that it was open to them by way of defence to prove that the items of work which went to make up the unpaid balance were not done at all, were duplications of items already paid, were charged as extras when they were within the original contract, or represented 'snagging' for works already done and paid for.

13.46 The contractor argued that once it was shown that there was a certificate and no withholding notice, the certified sum must be paid—it could not be withheld. The clients countered that if work had not been done there could be no 'sum due under contract' and that accordingly s. 111(1) simply did not apply.

13.47 **Held:** Under the contract terms the sum due was determined as the amount in the certificate; it was not the actual work performed which defined the sum due. In the absence of a withholding notice the clients were prevented from withholding the sum due, i.e. the sum in the certificate, and the contractor was entitled to the sum right away. The rival arguments did not arise since they were based on the unspoken but mistaken assumption that the provision was dealing with the ultimate position between the parties. Even if the certificate had been a final certificate (which it was not), this would only have had the effect of debarring the employers from withholding sums until such time as the dispute had been decided by adjudication or litigation. The principle of pay now and argue later applied to ensure that both parties' remedies were preserved.

13.48 Jacob LJ:

> 7. ... Some of the debate seems to have been based upon an unspoken but mistaken assumption, namely that the provision is dealing with the ultimate position between the parties. That is not so as is pointed out by Sheriff J.A. Taylor in *Clark Contracts v The Burrell Co.* [2002] SLT 103. He casts a flood of light on the problem ...
>
> 11. In this ASI contract, the sum is determined by the certificate. Clause 6.1 provides that 'payments shall be made to the Contractor only in accordance with the Architects certificate.' Clause 6.32 defines the sum—essentially the approved gross value of work done less retention and amounts previously paid. Cl. 6.33 says when it is to be paid: 'the employer shall pay to the Contractor the amount certified within 14 days of the date of the certificate, subject to any deductions and set-offs due under the Contract.' So it is not the actual work done which either defines the sum or when it is due. The sum is the amount in the certificate. The due date is 14 days from certificate date. The certificate may be wrong—the architect may (though this is unlikely because he will be working from the builder's bill) have missed out work done (which would operate against the contractor) or he may have included items not in fact done or items already paid for (which would operate

against the client). In the absence of a withholding notice, s.111(1) operates to prevent the client withholding the sum due. The contractor is entitled to the money right away. The fundamental thing to understand is that s.111(1) is a provision about cash-flow. It is not a provision which seeks to make any certificate, interim or final, conclusive. Analysed this way one sees that there is something inconsistent about the clients' argument here. Their duty to pay now and the sum they have to pay arise only because of the certificate. Yet they wish to ignore the certificate to reduce the amount they have to pay.

12. All this becomes blindingly clear following Sheriff Taylor's analysis. He was dealing with a case like this, one involving a system of architect's certificates. This is what he said:

'There was no dispute that the architect had issued an interim certificate. It therefore seems to me that the defenders became entitled to payment of the sum brought out in the interim certificate within 14 days of it being issued. In my opinion that is an entitlement to payment of a sum due under the contract. In order to reach the figure in the interim certificate one has made use of the contractual mechanism. To use the words deployed by Lord Macfadyen [in *SL Timber Systems*] in para. 20, the issue of an interim certificate was the occurrence of "some other event on which a contractual liability to make payment depended". This situation falls to be contrasted with the position in *SL Timber Systems* where, before the adjudicator, there had been no calculation of the sum sued for by reference to a contractual mechanism and which gave rise to an obligation under the contract to make payment. There had been no more than a claim by the pursuers which claim had not been scrutinised by any third party. Thus, in my opinion, if The Burrell Co (Construction Management) Ltd wished to avoid a liability to make such payment because the works did not conform to the contractual standard they would be withholding payment of a sum due under the contract. In order to withhold payment they would require to give notice in terms of s.111(1) of the Act. No such notice was given.

The interim certificate is not conclusive evidence that the works in respect of which the pursuers seek payment were in accordance with the contract (see cl. 30.10). That however does not preclude the sum brought out in an interim certificate being a sum due under the contract. The structure and intent of the Act, as I understand it, and accepted by the solicitor for the defenders, is to pay now and litigate later.'

13. Sheriff Taylor earlier explained why Lord Macfadyen's case, was different. The contract there had no architect or system of certificates. The builder simply presented his bill for payment. The bill in itself did not make any sums due. What, under that contract, would make the sums due is just the fact of the work having been done. So no withholding notice was necessary in respect of works not done—payment was not due in respect of them.

14. Sheriff Taylor's analysis, once articulated, is obviously right. And it has a series of advantages:

(a) It makes irrelevant the problem with the narrow construction—namely that Parliament was setting up a complex and fuzzy line between sums due on the one hand and counterclaims on the other—a line somewhere to be drawn between set-off, claims for breach of contract which do no more than reduce the sum due and claims which go further, abatement and so on.
(b) It provides a fair solution, preserving the builder's cash flow but not preventing the client who has not issued a withholding notice from raising the disputed items in adjudication or even legal proceedings.
(c) It requires the client who is going to withhold to be specific in his notice about how much he is withholding and why, thus limiting the amount of withholding to specific points. And these must be raised early.
(d) It does not preclude the client who has paid from subsequently showing he has overpaid. If he has overpaid on an interim certificate the matter can be put right in subsequent certificates. Otherwise he can raise the matter by way of adjudication or if necessary arbitration or legal proceedings.

(e) It is directed at the mischief which s.111(1) was aimed at. This mischief is mentioned in *Keating*. A report called the Latham report had identified a problem, namely that 'main contractors were abusing their position to wrongfully withhold payment from sub-contractors who were in no position to make any effective protest'. Actually the provision has gone further than just dealing with the position between main and sub-contractors since it covers the position between client and main contractor too—but the main contractor will need paying himself so he can pay the sub-contractor. And he may have his own cash flow needs too. Incidentally s.109 (requiring stage payments for long contracts) is part of the same legislative policy.

15. The principal disadvantage of the statutory scheme from the client's point of view is that if he has overpaid he is at risk from insolvency of the builder. But the risk is one which he can avoid by checking the certificate and giving a timeous withholding notice. No doubt a good architect would inform a lay client about the possibility of serving such a notice—indeed the architect may (I express no opinion) have a duty to do so. Moreover the client may (again I express no opinion) have a remedy against the architect if the latter negligently issues a certificate for too much.

16. Thus I think the appeal should be dismissed. The clients must pay for the present. They are not without remedy if as a result they have overpaid. Preferably their remedy would be by the issue of an accurate final certificate by the architect rather than litigation of any sort. We were surprised to be told that the architect has not done this even though this dispute has been going on for some two years—the sooner he does so the better. He will obviously have to consider the points which the clients were mistakenly seeking to raise in their defence to this claim. The fact that he is the brother of a director of the builders is, of course, irrelevant to his duties to his clients.

Balfour Beatty Construction Northern Ltd v Modus Corovest (Blackpool) Ltd [2009] CILL 2660; [2008] EWHC 3029 (TCC)

13.49 **Facts:** This case involved the JCT Standard Form of Building Contract with Contractors' Design 1998. The employer's agent had issued a payment notice and clause 30.3.3 provided that 'subject to clause 30.3.4, [the Employer] shall pay the amount proposed no later than the final date for payment'.

13.50 **Held:** Clause 30.3.4 clearly defined the amount in the payment notice as the 'amount due' and required the employer to serve a withholding notice if it intended to make any further deductions. The employer had failed to do this and so, the sum was required to be paid by the employer to the contractor without recourse to cross-claims or other argument.

Urang Commercial Ltd v (1) Century Investments Ltd and (2) Eclipse Hotels (Luton) Ltd [2011] 138 Con LR 233, [2011] EWHC 1561 (TCC)

13.51 **Facts:** The contractor, Urang, applied to enforce two separate adjudication decisions in relation to works carried out under the JCT Standard Building Contract 2005 and JCT Design and Build Contract 2005 respectively. The Standard Building Contract provided for the quantity surveyor to carry out the interim valuation whereas the Design and Build Contract appears to have been amended to allow a quantity surveyor to issue the payment notice on behalf of the employer following which the contractor would provide an invoice in the amount of the payment notice.

13.52 The employers failed to pay the amounts that the quantity surveyor had certified/notified without serving any further payment notice or withholding notice. Urang served a notice of adjudication for payment of the amounts certified, and also in relation to a number of other claims concerning prolongation, retention, and interest. The employers counterclaimed costs for remedial works, loss of revenue, and liquidated and ascertained damages ('LADs').

13.53 The adjudicator held that the employers' counterclaims should have been the subject of a withholding notice and awarded Urang the unpaid amounts of the interim certificates, and sums in respect of Urang's other claims. The employers sought to resist enforcement proceedings on the grounds that the sums certified were not the sums due (*SL Timber Systems Ltd v Carillion Construction Ltd* (2002) (Key Case)[27] referred to), and that the adjudicator had been incorrect to dismiss their counterclaims on the basis that they should have been the subject of a withholding notice.

13.54 **Held:** The employers' reliance on *SL Timber* was misplaced. In that case there had been no provision for certification and the interim payment in issue had simply been the subject of an application by a subcontractor which had not been scrutinized by any third party:

> 24 In the case of *Rupert Morgan Building Services (LLC) Ltd v Jervis* [2004] 1 WLR 1867, the Court of Appeal drew a clear distinction between interim payments that have been certified, and were therefore due under the contract, and sums which had not been the subject of any third party scrutiny but were simply claimed as due by the contractor or subcontractor in question.

13.55 The sums stated in the Interim Valuation and payment notice were sums due under the contract and, in the absence of a withholding notice should be paid: *Rupert Morgan Building Services (LLC) Ltd v David Jervis and Harriet Jervis* (2003)[28] applied. However, the adjudicator had been wrong to decide that Urang's other claims required a withholding notice to be served. These were not sums due under interim valuations and therefore the employers would have been entitled to set off their counterclaims against these claims without the need to serve a withholding notice. Nevertheless, the failure to consider the counterclaim was the result of an error of law that did not go to jurisdiction, and so the decision of the adjudicator was unimpeachable on enforcement.

> 28. It is clear from the terms of the contract I have already summarised, that the need to issue a withholding notice applies only to sums stated as due in interim applications. There is no requirement to serve a withholding notice in relation to other claims made by a contractor, whether under another provision in the contract or for damages. The requirement for a withholding notice is confined to the procedure in relation to interim valuations as required by section 110 and 111 of the Housing Grants, Construction and Regeneration Act 1996.

Fleming Buildings Ltd v Forrest or Hives (2008) CSOH 103 (Outer House Court of Session)

13.56 **Facts:** The parties entered into an SBCC Standard Building Contract (with contractor's design portion), the relevant terms of which were contained in clauses 30.1.1.1–30.1.1.5. This scheme provided for the architect to issue interim certificates and the final date for

[27] [2001] BLR 516.

[28] (2004) 1 WLR 1867, [2004] EWCA Civ 1563 (CA).

payment of the sum certified was 14 days from the date of issue. Not later than five days from the issue of the certificate the employer was required to serve a payment notice, and if any withholding notice was to be issued, this should be done not later than five days before the final date for payment. Where the employer does not give either of these notices, the employer was required pay the contractor the amount due under the certificate.

13.57 **Held:** This scheme clearly excluded the right to common law retention or set-off: the only means by which an employer might properly refrain from making payment of a sum due under an architect's certificate was by the contractual mechanism for the provision of payment and withholding notices.

When the Amount Due Is the Payee's Unchallenged Application

13.58 The JCT Standard Form of Building Contract (with Contractor's Design) 1998 provides that the employer must reply to a contractor's interim application within seven days with a written notice stating the amount of payment proposed. The employer is then required to pay that sum under clause 30.3.3, subject to any withholding notice it may provide under clause 30.3.4. The contract goes on to say that where the employer fails to give a notice under 30.3.3 and/or 30.3.4 then the employer shall pay the amount stated in the contractor's application. Under this form of contract, the employer's payment notice plays an essential part in determining the amount due under the contract for interim payments. The failure to provide an employer's payment notice and/or a withholding notice triggers the operation of a fall-back provision in 30.3.5 whereby it is the contractor's application itself which then determines the amount of an interim payment that is due: *Ringway Infrastructure Service Limited v Vauxhall Motors Limited* (2007)[29] (Key Case).

13.59 In a contract such as this, the 'sum due' is determined by either the employer's payment notice or the contractor's application. Contracts of this type will be caught by the rule in *Rupert Morgan*[30] and a withholding notice will be required if the employer intends to pay less than the sum stated in either the employer's payment notice or the contractor's application (whichever is applicable).[31]

Where the Status of the Certificate Is Unclear

13.60 If a contract fails to express the contractual status of a system of certification, in that it fails to say that the certificate represents the sum due to the payee that shall be paid, then it may be that the sum due may not be that in the certificate but may be determined by reference to some other method such as the value of the work performed. In *Emcor Drake & Scull Ltd v Sir Robert McAlpine Ltd* (2004),[32] there was a dispute as to the terms of contract and a formal subcontract was never entered into. The terms under which Emcor Drake & Scull Ltd ('EDS') was working were set out in a purchase order which specified a net monthly account. The parties had adopted an ad hoc system whereby EDS made monthly applications for payment for work done and materials supplied, and Sir Robert McAlpine Ltd ('SRM') certified the gross and net valuations and the amount of the payment due at the end of the month.

13.61 Judge Havery was 'content to assume' that SRM's certificate amounted to the employer's notice of payment under s. 110(2), but disagreed that the amount of the proposed payment

[29] [2007] All ER (D) 333 (TCC), [2007] EWHC 2421(TCC) at para. 71.

[30] (2004) 1 WLR 1867, [2004] EWCA Civ 1563 (CA).

[31] *Balfour Beatty Construction Northern Ltd v Modus Corovest (Blackpool) Ltd* [2009] CILL 2660, [2009] EWHC 3029 at paras. 73 and 74 (Key Case).

[32] [2004] EWHC 1017 (TCC).

stated in a paying party's notice under s. 110 was *ipso facto* the amount due under the contract.[33] There was scarcely any evidence of the contractual status of the system of certification that the parties had adopted and there was a suggestion that the relevant person in SRM certified whatever he thought right as being due to EDS. This was insufficient to show any legal obligation on SRM to pay an amount certified by virtue of its certification. It was the fact of the work having been done (assuming that it was) that made any sum due.

13.62 The situation in *Emcor* was perhaps quite unusual in that the terms of the contract were so unclear. Where standard forms of building contract have been used, there has not yet been any decided case in which the status of a payment certificate in establishing the amount due has been successfully challenged. It is likely that the courts will construe a contract with a certification process as meaning that the amount due is that set out in the third party certificate, even if the express words of the contract do not explicitly say this. However, caution should be used in assuming that this will necessarily be the case, and it will be interesting to see what decision might be reached, for example, in the case of the NEC/ECC contracts where the amount due is defined separately to the project manager's obligation to certify payment.[34]

Where the Contract Permits Cross-claims to Be Deducted

13.63 In *PC Harrington Contractors Ltd v Multiplex Constructions (UK) Limited* (2007)[35] Clarke J considered whether claims made by the contractor against the subcontractor should form part of a separate process of back-charging such that the subcontractor remained entitled to be paid the sum that he had applied for. The contractual mechanism required a payment certificate to be issued certifying the amounts that had been calculated under the contractual valuation mechanism, 'less any amounts that are to be deducted pursuant to clause 21.10'. Clause 21.10 referred to sums which the subcontractor was liable to pay pursuant to the contract, and to any abatements which the contractor was entitled to assess against sums due to the subcontractor. Clause 21.10 went on to state that the contractor could either deduct these sums in computing the amount of any certificate of payment, or else could issue an invoice to the subcontractor for these amounts. Despite the rather 'infelicitous' wording of the clauses which made it unclear whether the sum due was that before or after deduction, the judge decided the latter. The contract provided that in establishing all of the positive amounts due to the subcontractor it was necessary, at the same time, to take into account all of the negatives which went into the contractual computation of the sum due. Consequently, the subcontractor's application was not a reflection of the amounts that were due and the contractor was entitled to assess them within his payment certificate.

13.64 In all cases the answer has been to look to the contract to determine how the amount due is required to be calculated, and if the payee's payment application is correct on every ground then there can be no reason to withhold payment without service of a withholding notice. However, if the contract provides that the sum due takes account of matters which the payee has not included in his payment application, whether these are in the nature of counterclaims or damages as in *PC Harrington v Multiplex*, or relate to more usual questions of valuation and measurement as referred to by Lord Macfadyen in *SL Timber*, then the payer may be well within his rights to assess and pay a different sum even without the service of a withholding notice.

---

[33] At para. 71.

[34] See clauses 50.2 and 51.1 as discussed in *Keating on NEC 3* (Sweet and Maxwell, 2012) Ch. 5. The position might be clearer under Options C–F than it is in Options A and B since the project manager's assessment forms an integral part of establishing the price of work done to date in Options C–F, whereas the amount may be separately ascertainable in Options A and B.

[35] [2008] BLR 16, [2007] EWHC 2833 (TCC).

Withholding Pursuant to Contractual Terms on Insolvency or Termination

13.65 One of the determining factors in *Rupert Morgan*[36] was the Court of Appeal's observation that debate on the requirement to serve withholding notices seemed to have been based upon an unspoken but mistaken assumption, namely that the provision was dealing with the ultimate position between the parties. However, the intention of s. 111 was merely to protect the builder's cash flow; it did nothing to prevent a client who has not issued a withholding notice from later raising the disputed items in adjudication or other legal proceedings. The only real injustice which might occur to a payer was where the payee became insolvent in the period after which a withholding notice could no longer be served. However, as Jacob LJ remarked, the risk was one which it could avoid by checking the certificate and giving a timeous withholding notice.[37]

13.66 In *Melville Dundas Ltd v George Wimpey UK Ltd* (2007)[38] (Key Case), determined by the House of Lords, the majority of their Lordships disagreed. This case involved consideration of clauses 27.3.4 and 27.6.5.1 of the JCT Standard Form of Building Contract (with Contractor's Design) 1998, which together provided that in the case of the contractor's insolvency the employer could terminate the contract, and any provisions which required any further payments to be made to the contractor did not apply. In this case, the final date for payment had already passed when the contractor became insolvent and the employer operated the clauses in order to determine the contract and withhold the payment which had been certified. By then, it was too late to serve a withholding notice. In the view of Lord Hoffman:

> 13 A provision such as clause 27.6.5.1, which gives the employer a limited right to retain funds by way of security for his cross-claims, seems to me a reasonable compromise between discouraging employers from retaining interim payments against the possibility that a contractor who is performing the contract might become insolvent at some future date (which may well be self-fulfilling) and allowing the interim payment system to be used for a purpose for which it was never intended, namely to improve the position of an insolvent contractor's secured or unsecured creditors against the employer.

13.67 In one of the two dissenting speeches, Lord Neuberger pointed out[39] that such a result would cut across the purpose of the Act. His alarm was echoed by numerous commentators who pointed out that clause 27.6.5.1 was not limited to cases of the contractor's insolvency but also applied in cases where the contact had been determined due to the contractor's default. The decision would therefore potentially open a route for unscrupulous employers to avoid making payments without serving notice, simply by terminating the contractor's employment.

13.68 In the event, the uncertainty as to whether the judgment would have wider application was quickly resolved by the case of *Pierce Design International Ltd v Mark Johnson and Deborah Johnson* (2007)[40] (Key Case). In this case, Judge Coulson considered the same provisions of the JCT contract, albeit this time in relation to contractual determination rather than insolvency. He distinguished the decision in *Melville Dundas* on the grounds that clause 27.6.5.1 contained a proviso that the employer might not unreasonably withhold payment or do so after 28 days have passed since the amount fell due. An attempt to set off sums without a withholding notice in circumstances other than the contractor's insolvency would be likely to be unreasonable, and in this case had occurred more than 28 days after the payment had become due.

---

36 See Key Case at 13.45.
37 At para. 15.
38 [2007] 1 WLR 1136, [2007] UKHL18.
39 At para. 77.
40 [2007] BLR 381, [2007] EWHC 1691 (TCC).

13.69 It is suggested that this common-sense approach is more likely to be followed than any attempt to construe a wider interpretation of the *Melville Dundas* decision than appears to have been intended. In any event, the 2009 Act now puts the matter beyond doubt in respect of contracts entered into after the new provisions come into force. The new s. 111(10) provides an exemption from the need to serve a withholding notice only in circumstances where the contract provides that, if the payee becomes insolvent, the payer need not pay any sum due. The exemption is thereby limited to the specific circumstances encountered in *Melville Dundas*, and only applies in cases where it would otherwise have been too late to serve a withholding notice.[41]

Key Cases: Withholding Pursuant to Contractual Terms on Insolvency and Termination

Melville Dundas Ltd v George Wimpey UK Ltd [2007] 1 WLR 1136, [2007] UKHL 18 (HL)

13.70 **Facts:** Wimpey contracted with Melville Dundas Ltd ('MDL') for the construction of a housing development in Whitecraigs, Glasgow for a total sum of £7,088,270. The contract incorporated the conditions of JCT Standard Form of Building Contract (with Contractor's Design) 1998 which provided in clause 30 for monthly applications for interim payments. By clause 30.3.6 the final date for payment of the amount due in an interim payment was 14 days after receipt by the employer of the application.

13.71 On 2 May 2003 the contractor applied for an interim payment of £396,630. There was no dispute that the contractor was entitled to be paid that sum or that the final date for payment was therefore 16 May 2003. Wimpey did not pay on that date and on 22 May 2003 administrative receivers of the contractor were appointed by its bank. Clause 27.3.4 provided that if the contractor had an administrative receiver appointed, the employer could determine the employment of the contractor. Wimpey exercised this right on 30 May 2003. That brought into effect clause 27.6.5.1, which stated:

> Subject to clauses 27.5.3 and 27.6.5.2 the provisions of this contract which require any further payment or any release or further release of retention to the contractor shall not apply; provided that clause 27.6.5.1 shall not be construed so as to prevent the enforcement by the contractor of any rights under this contract in respect of amounts properly due to be paid by the employer to the contractor which the employer has unreasonably not paid and which, where clause 27.3.4 applies, have accrued 28 days or more before the date when under clause 27.3.4 the employer could first give notice to determine the employment of the contractor ...

13.72 In the lower courts it appeared to have been conceded that the effect of this clause was that, upon determination by Wimpey, the interim payment was no longer payable. It had accrued less than 28 days before 22 May 2003, which was the date on which Wimpey could first have given notice of determination. Their Lordships agreed that the clause was intended to have this effect and therefore concerned themselves with determining whether the effect of clause 27.6.5.1 was instead invalidated by the provisions of Part II of the Act.

13.73 **Held:** Section 111 should be construed to give effect to clause 27.6.5.1 so that, if it was not possible for the employer to have complied with the notice requirements of s. 111, this should not defeat the contractual freedom of the parties to agree the circumstances in

[41] See s. 111(10)(b).

which a payment should not be made. Insisting upon strict compliance with s. 111 would nullify the effect of the clause, and could never have been the intention of Parliament.

Lord Hoffman:

13.74

19. What is the purpose of the notice requirement in section 111(1)? Obviously to enable the contractor to know immediately and with clarity why a payment is being withheld. It is primarily part of the machinery of adjudication, so that the contractor can decide whether he should dispute the employer's right to withhold the payment and refer the question to adjudication. But I suppose it also provides the contractor with information for the purpose of any other action which may depend upon knowing the reason why a payment is being withheld.

20. In the case of clause 27.6.5.1 the contractor will have been given notice of why the payment is being withheld because he will have received the notice of determination. But the retrospective operation of the clause means that he will not have received it within the time stipulated in the statute. It seems to me, however, that it would be absurd to impute to Parliament an intention to nullify clauses like 27.6.5.1, not by express provision in the statute, but by the device of providing a notice requirement with which the employer can never comply. Section 111(1) must be construed in a way which is compatible with the operation of clause 27.6.5.1.

...

22. The problem arises because I very much doubt whether Parliament, in enacting section 111(1), took into account that parties would enter into contracts under which the ground for withholding a payment might arise *after* the final date for payment. One cannot therefore find an answer in a close examination of the language of the section. I would prefer simply to say *lex non cogit ad impossibilia* and that on this ground section 111(1) should be construed as not applying to a lawful ground for withholding payment of which it was in the nature of things not possible for notice to have been given within the statutory time frame. That may not be particularly elegant, but the alternative is to hold that the parties' substantive freedom of contract has been indirectly curtailed by a mere piece of machinery, the operation of which would serve no practical purpose. This I find even less attractive. I would therefore allow the appeal and restore the interlocutor of the Lord Ordinary.

Lord Neuberger (dissenting)

13.75

63. As I see it, the respondent's case is very simple, and it proceeds as follows. Section 110(1)(b) requires a construction contract to 'provide a final date for payment in relation to any sum which becomes due'. In this case, the contract complied with this requirement in relation to interim payments through the medium of clause 30.3.6. On the facts of this case, the 'final date' for the payment of the sum was 16 May 2003. Section 111(1) prohibits the appellant from 'withhold[ing] payment ... ' after 'the final date for payment of a sum due under the contract'. In this case, that must mean that the appellant 'may not withhold payment' of the sum after 16 May 2003. Accordingly, in so far as clause 27.6.5.1 has the effect of permitting the appellant to withhold payment of the sum, it is purporting to permit that which section 111(1) prohibits. Therefore, to that extent, it is ineffective. That simple approach commended itself to the Inner House.

64. It is also an approach that commends itself to me, at least as a matter of simple statutory interpretation. On the face of it at any rate, if a statute provides that a person 'may not withhold payment' after a specified date has passed, it appears to me that a contractual provision that he may do so must be ineffective. That conclusion is supported, in my view, by the fact that sections 110 and 111 (and, indeed, sections 108, 109 and 113) appear to have the aims of (a) providing a clear and simple system to ensure that parties to construction contracts know where they are with regard to payments, and (b) ensuring that contractors and sub-contractors can be confident about their cash-flow.

Pierce Design International Ltd v Mark Johnson and Deborah Johnson [2007] BLR 381, [2007] EWHC 1691 (TCC)

13.76 **Facts:** The defendants Mr and Mrs Johnston engaged the claimant, Pierce, to carry out construction works at their property. The contract incorporated the JCT Standard Form of Building Contract (with Contractor's Design) 1998. These were the same conditions (save for certain amendments not in issue) which were considered in the House of Lords' decision in *Melville Dundas* (above).

13.77 Between March 2006 and November 2006 the defendants withheld sums against five interim certificates issued by the employer's agent. No s. 111 notices were served in respect of any of the withheld amounts which together totalled £93,460.33. On 7 March 2007 the defendants served a notice of default pursuant to clause 27.2.1 of the JCT Conditions, notifying Pierce that it was not proceeding regularly and diligently with the work. The defendants alleged that the default was not remedied and, on 30 March 2007 they purported to determine Pierce's employment relying upon clause 27.6.5.1 for avoiding payment of further amounts. Pierce disputed the validity of that determination and applied for summary judgment for payment of the withheld amounts plus interest.

13.78 The issues for Judge Coulson to decide were, first, whether Clause 27.6.5.1 complied with the 1996 Act in allowing sums to be withheld without notice and, if so, whether the terms of the clause had been complied with including the requirement that the withholding had been 'reasonable'.

13.79 **Held:**

1. Clause 27.6.5.1 complied with the Act, even though it allowed sums to be withheld without a withholding notice. This was the House of Lords' finding in *Melville Dundas* and it was not possible to construe this decision so that the clause complied in cases of insolvency, but not in other cases.
2. However, the conditions necessary to apply clause 27.6.5.1 did not apply in this case because the employer's actions were not reasonable, and the withholding did not take place within the 28-day period prescribed by the clause.

E. Does Clause 27.6.5.1 on the Facts of the Present Case Fall Foul of Section 111 of the 1996 Act?

22. Miss Garrett submitted that their Lordships' conclusion as to the proper operation of Clause 27.6.5.1 should be limited to the facts of Melville Dundas and that therefore, absent the insolvency of the contractor and/or the impossibility of serving withholding notices, which she maintained was the situation in the present case, the clause fell foul of Section 111 of the 1996 Act and should therefore be struck down. The principal difficulty with this argument, as Mr. Coplin forcefully pointed out, was that the House of Lords have considered this very clause and concluded, albeit by a majority, that it was not at odds with Section 111. I am bound by that decision. It is not for me to endeavour to restrict the clear consequences of the decision in Melville Dundas.

23. In addition, I should say that I am not attracted to an argument which seeks to suggest that, on one set of facts, a clause in a standard form contract complies with the 1996 Act whilst, on another set of facts, it does not. That seems to me to be a recipe for uncertainty and endless dispute. I consider that a clause of this type either complies with the Act or it does not. If compliance turns on minute factual gradations, then all the commercial certainty that is usually provided by the use of a standard form of contract will be lost. Accordingly, I decline to embark on the voyage of discovery into the factual differences between this case and Melville Dundas, on which Miss Garrett was so keen to send me.

...

25. For these reasons therefore, I consider that, the House of Lords having ruled that Clause 27.6.5.1 complies with Section 111 of the 1996 Act, that is the end of the matter. The first basis for the Claimant's CPR Part 24 application must therefore fail.

...

F. How Should the Proviso to Clause 27.6.5.1 Be Operated?

32. Under the terms of this contract the non-payment of the sums due under Clause 30.3.3 can only be justified if there is a withholding notice under Clause 30.3.4. If there is a withholding notice the sum identified under Clause 30.3.3 no longer becomes due under the contract. It is reduced (or extinguished altogether) by the figure in the withholding notice. Thus, so it seems to me, as a matter of construction of Clause 27.6.5.1 in the context of the contract as a whole, a sum due by way of an interim payment under Clause 30.3.3 would reasonably have not been paid by the employer if there was a valid withholding notice in respect of that sum under Clause 30.3.4. Conversely, if there was no withholding notice, the sum would unreasonably have not been paid by the employer. That is the position here. As a matter of simple interpretation, therefore, I consider that *prima facie* the third and final part of the proviso has also been made out by the Claimant ...

38. ... just stepping back from the detail for a moment, it seems to me that, if the proviso is to be construed in the way that I have set out in paragraph 32 above, it has the additional benefit of meeting head-on many of the concerns which have been expressed about the approach adopted in *Melville Dundas*, to the effect that the decision might allow an unscrupulous employer to use determination as a way of avoiding his responsibility to make interim payments. Indeed, provided that the sum has been due and 'unreasonably not paid' more than twenty-eight days before the determination then, on my interpretation of the proviso, it would satisfy precisely Lord Hoffmann's point, at paragraph 13 of his speech, that employers should be 'discouraged from retaining interim payments against the possibility that a contractor who is performing the contract might become insolvent at some future date (which may well be self-fulfilling)'. My construction would, so it seems to me, also be in accordance with the underlying purpose of the 1996 Act. Furthermore, in circumstances like these, where there is no evidence whatsoever to suggest that the Claimant/contractor is or might be insolvent, my construction of the proviso does not and cannot cause any permanent prejudice to the Defendants. It is not a determination of their rights. All it does is to require them to pay, on an interim basis, the sums which, pursuant to the contract, they ought to have paid months ago.

39. Of course, because the Claimant in this case is not insolvent, I do not need to consider the possible tension between, on the one hand, the clear effect of the proviso as I have construed it, and on the other, the possible risk that the position of the insolvent contractor's secured or unsecured creditors will be improved (as against the employer) if payments are made in accordance with its terms. But, in truth, it seems to me that *Bouyges v Dahl-Jensen* provides a complete answer to that point. As Chadwick LJ explains in that case, an insolvent contractor is not entitled to enforce its right to any further sums following its insolvency. Thus, if the contractor is insolvent, the 'enforcement' identified in the proviso to Clause 27.6.5.1 would be 'prevented' in any event.

40. For all these reasons, despite the clear way in which he put his case, I reject Mr. Coplin's construction of the proviso. In my judgment, for the reason that I have given, the natural meaning of the words used provides that a sum due, which has unreasonably not been paid by the employer, is a sum which the employer's agent has said is due under Clause 30.3.3; which sum has not been the subject of a withholding notice under Clause 30.3.4; and the non-payment of which is therefore a breach of contract which occurred more than 28 days before the determination. Furthermore, I consider that this interpretation brings with it a number of common-sense and commercial benefits noted in paragraph 38 above.

Withholding Amounts against the Final Certificate

13.80 Many contracts make provision for a final certificate which becomes final and binding as to the amounts assessed after a certain period of time. Clause 1.10 of the JCT Standard Building Contract 2005 and 2011 has this effect and provides that the final certificate will be final and binding as to the matters stated unless either party serves notice of adjudication, arbitration, or other legal proceedings within 28 days.

13.81 In such cases it is important that the payer observes any requirements in the contract as to the serving of a withholding notice before payment, or else makes a claim within the required period for repayment of any sums that it considers have been overpaid. If the payer fails to do either and simply refuses to pay the amounts certified, then he may be at risk that the payee will apply for summary judgment for payment of the amount certified as due.

13.82 The situation may also be complicated if, when the final certificate is issued, there are ongoing adjudication proceedings concerning earlier interim payment certificates. If the payee succeeds in recovering amounts in respect of its earlier claims, then such amounts must be paid either immediately or within seven days, depending whether a valid withholding notice had been issued.[42] This might require adjustment of the final account meaning that either party might wish to reserve its rights to challenge the final certificate by serving notice within the time required. Such a situation arose in *William Verry Ltd v London Borough of Camden* (2006)[43] (Key Case).

Key Case: Withholding Amounts against the Final Certificate

William Verry Ltd v London Borough of Camden [2006] EWHC 761 (TCC)

13.83 **Facts:** Camden employed Verry under the terms of a JCT Intermediate Form of Contract 1998. Several adjudications were held in respect of the valuation of interim certificates and Camden sought to resist enforcement of the adjudicator's award in adjudication number 3 on the basis, inter alia, that the final certificate had since been issued showing that no further sums were due.

13.84 **Held:** The adjudicator's decision should be enforced (*Ferson Contractors Ltd v Levolux AT Ltd* (2003)[44] applied). If the adjudicator's decision was subject to the view of a contract administrator in a subsequent certificate, then the intention of Parliament would be defeated. The parties had agreed to be bound by the adjudicator's decision by the terms of the contract, and their reserved rights were not prejudiced as compliance was on a temporary basis only. Since Verry had served notice of adjudication in respect of the final certificate within the required period, the status of the final certificate was not conclusive.

> 43. ... the final certificate in this case has no conclusive effect, given that adjudication number 4 was commenced within 28 days of the final certificate. In addition, clause 4.7.1 provides that it is not conclusive on any matter which is the subject of proceedings, including adjudication commenced before the date of the final certificate. In my judgment, provided that a matter is the subject of adjudication number 3, then the final certificate could not, in any event, be conclusive as to that matter. As a result, the final certificate has only the status, in my judgment, of a payment certificate which is disputed and has no conclusive effect.

[42] See 13.136–13.137.
[43] [2006] EWHC 761 (TCC).
[44] [2003] BLR 118, [2003] EWCA Civ 11 (CA).

Withholding Notices under the Unfair Terms in Consumer Contracts Regulations 1999

13.85 Finally, mention should be given to the unusual circumstances in the case of *Steve Domsalla (t/a Domsalla Building Services) v Kenneth Dyason* (2007)[45] which concerned a contract not caught by the 1996 Act as the employer was in effect a residential occupier. However, the contract incorporated the JCT Minor Works Conditions which contained requirements for payment and withholding notices to be served. Judge Thornton considered that the provision for withholding notices fell foul of the Unfair Terms in Consumer Contracts Regulations 1999 ('UTCCR'), primarily because the contract had been negotiated and was being administered by Mr Dyason's insurers, even though it was Mr Dyason who was named as employer. Under the terms of the arrangement between Mr Dyason and his insurers, Mr Dyason was unable to operate the withholding notice provisions of the contract, and would therefore be severely prejudiced in the event that he became personally liable under the contract. Moreover, the judge considered that certain of the terms set out in Schedule 2 of UTCCR, namely those terms which may be regarded as unfair, were applicable in the circumstances of the case. These were:

(1) The following terms which have the object or effect of–
- (b) inappropriately excluding or limiting the legal rights of the consumer vis-à-vis the supplier in the event of total or partial non-performance or inadequate performance by the supplier of any contractual obligations, including the option of offsetting a debt owed to the supplier against any claim which the consumer may have against him.
- (i) irrevocably binding the consumer to terms which he had no real opportunity of becoming acquainted before the conclusion of the contract;
- (o) obliging the consumer to fulfil all his obligations where the supplier does not perform his;
- (q) excluding or hindering the consumer's right to take legal action or exercise any other legal remedy, ... unduly restricting the evidence available to him ...

13.86 Whilst it is unlikely that the case will be of wide application due to the unusual nature of its facts, *Domsalla* provides a useful illustration of other circumstances in which the courts have found reason to overturn the public policy presumption in favour of the payee, back in favour of protecting an individual payer.

Table 13.1 Table of Cases: Requirement to Serve a Withholding Notice

Title	Citation	Issue
Northern Developments (Cumbria) Ltd v J. & J. Nichol	[2000] BLR 158, [2000] EWHC Technology 176 (TCC), Judge Bowsher	**Withholding against a contractor's valuation** If there was to be a dispute about the amount of a payment, that dispute was to be mentioned in a notice of intention to withhold payment and the adjudicator had no jurisdiction to consider matters which had not been raised in the notice.
VHE Construction plc v RBSTB Trust Co. Ltd	[2000] BLR 187, EWHC Technology 181 (TCC), Judge Hicks	**Withholding against a contractor's invoice** The terms of the JCT Standard Form of Building Contract (with Contractor's Design) 1981 contained identical terms to the 1998 edn as to the determination of the amount due (see 13.20 above). RBSTB failed to serve either a payment or withholding notice.

[45] [2007] BLR 348, [2007] EWHC 1174 (TCC) (Key Case).

Table 13.1 *Continued*

Title	Citation	Issue
		1. The amount due was the amount set out in Contractor's VAT invoice issued under clause 30.6.1. This amount had earlier been approved by an adjudicator's award. 2. The words 'may not withhold payment' were ample in width to have the effect of excluding set-offs and there was no reason why they should not mean what they said.
KNS Industrial Services v Sindall	(2000) 75 Con LR 71, EWHC Technology 75 (TCC), Judge LLoyd	**Withholding against contractor's valuation** Since the issue that had been referred to the adjudicator was the valuation of the works themselves, it was open to the adjudicator to reduce the subcontractor's valuation for non-compliant work, even in the absence of a withholding notice. 'One cannot withhold what is not due.'
Whiteways Contractors (Sussex) Ltd v Impresa Castelli Construction UK Ltd	(2000) 75 Con LR 92 EWHC Technology 61 (TCC), Judge Bowsher	**Withholding against an 'amount due'** 1. In considering the dispute the adjudicator will make his own valuation of the claim before him and in doing so, he may abate the claim in respects not mentioned in the withholding notice. 2. However, he 'ought not to look into abatements outside the four corners of the claim unless they have been mentioned in a notice of intention to withhold payment'.
Millers Specialist Joinery Co. Ltd v Nobles Construction	[2001] CILL 1770, Salford CC 64/00 (TCC), Judge Galliland	**Withholding against contractor's valuation** 1. It did not follow that merely because an abatement may technically reduce the amount due and payable under the contract that Parliament was to be taken to have intended that the amount claimed could not be challenged if no notice was given under s. 111. 2. An abatement normally involved a breach of contract on the part of the contractor and was in the nature of a cross-claim which operated to reduce the amount which could be recovered. It was developed by the common law as a procedural means whereby justice could be done as between the parties without the need for the defendant to bring a cross-action. It was not in substance different from a set-off.
Woods Hardwick Ltd v Chiltern Air Conditioning Ltd	[2001] BLR 23 (TCC), Judge Thornton	**Withholding against a consultant's valuation** 1. Notwithstanding the employer's failure to serve any notice of withholding, sums could be abated against an architect's application for payment because of alleged breaches of the contract. 2. Any abatement properly relied upon by Chiltern, would not be caught by s. 111 of the Act, so Chiltern's abatement defence could, in principle, defeat or reduce Woods Hardwick's claims.
SL Timber Systems Ltd v Carillion Construction Ltd	[2001] BLR 516 (Outer House, Court of Session), Lord Macfadyen	**Withholding against a contractor's valuation** 1. The absence of a timeous notice of intention to withhold payment did not relieve the party making the claim of the ordinary burden of showing that he is entitled under the contract to receive the payment he claims. 2. The words 'sum due under the contract' did not have the same meaning as the words 'sum claimed'. Section 111 was not concerned with every refusal on the part

Title	Citation	Issue
		of one party to pay a sum claimed by the other. It was concerned, rather, with the situation where a sum is due under the contract, and the party by whom that sum is due seeks to withhold payment on some separate ground.
Clark Contracts Ltd v The Burrell Company Construction Management Ltd	[2002] SLT 103 (Glasgow Sheriffs' Court), Sheriff Taylor	**Withholding against a certified amount** 1. There was a distinction between contracts where the contract had no architect or system of certificates and those that did. 2. In the present case the terms of the Standard Form of Building Contract with Quantities 1980 required the employer to pay the amount that had been certified by the architect. The sum that had been certified was the sum due under the contract. If the employer wished to avoid liability to make such payment, it would be required to give a withholding notice. 3. The interim certificate is not conclusive evidence that the works had been carried out in accordance with the contract. That did not, however, mean that the sum stated as due in the interim certificate was not the 'sum due' under the contract.
Rupert Morgan Building Services (LLC) Ltd v (1) David Jervis and (2) Harriet Jervis	(2004) 1 WLR 1867, [2004] EWCA Civ 1563 (CA), Schiemann LJ, Sedley LJ, Jacob LJ	**Withholding against a certified amount** 1. Where an independent certifier had certified an amount due under a contract, there were clear advantages to upholding the requirement for a withholding notice to be served in all cases if the employer intended to pay less than the amount which had been certified. 2. This was a fair solution which preserved the subcontractor's cash flow but did not prevent the paying party from arguing the proper amount in later proceedings.
Emcor Drake and Scull Ltd v Sir Robert McAlpine Ltd	[2004] EWHC 1017 (TCC), Judge Havery	**Withholding against the employer's payment notice** The amount of the proposed payment stated in a paying party's notice under s. 110 is not *ipso facto* the amount due under the contract. There was virtually no evidence as to the contractual status of the system of certification that had been adopted. This was insufficient to show any legal obligation on SRM to pay an amount certified by virtue of its certification. It was the fact of the work having been done (assuming that it was) that made any sum due.
Steve Domsalla (t/a) Domsalla Building Services v Kenneth Dyason	[2007] BLR 348, [2007] EWHC 1174 (TCC), Judge Thornton	**Withholding notices and UTCCR** Where the employer deals as a residential occupier then the provision for withholding notices may fall foul of the UTCCR.
Melville Dundas Ltd v George Wimpey UK Ltd	[2007] 1 WLR 1136, [2007] UKHL 18 (HL), Lord Hoffman, Lord Walker, Lord Mance and Lord Neuberger	**Withholding in the case of a contractor's insolvency** Where the contract contains specific provisions which allow the employer to withhold sums in the event of the contactor's insolvency, then this might be a reasonable

(*Continued*)

Table 13.1 *Continued*

Title	Citation	Issue
		exception to the requirement to serve a withholding notice. Section 111 of the 1996 Act should be construed to give effect to such a clause.
Ringway Infrastructure Services Ltd v Vauxhall Motors Limited	[2007] All ER (D) 333, [2007] EWHC 2421 (TCC)	**Withholding against a contractor's valuation** The JCT Standard Form of Building Contract (with Contractor's Design) 1998 provided that the amount due was determined by either the employer's notice or the contractor's application. A withholding notice will be required if the employer intends to pay less than the sum stated in either the employer's notice or the contractor's application (whichever is applicable).
P. C. Harrington Contractors Ltd v Multiplex Constructions (UK) Limited	[2008] BLR 16, [2007] EWHC 2833 (TCC), Clarke J	**Withholding against a contractor's valuation** 1. The sum applied for by PCH took no account of any abatement or any claim which would operate under the contract to reduce the sums due. It could not therefore be taken to represent the sum that was due. Multiplex were entitled to assess the amount due taking into account these amounts. 2. There was no two-part process specified by the contract for the positive amounts due to PCH to be assessed and paid, and the negative amounts to be claimed separately. The process required that both positive and negative amounts should be entered in the same calculation of the amount due.
Pierce Design International Ltd v Mark Johnson and Deborah Johnson	[2007] BLR 381, [2007] EWHC 1691 (TCC), Judge Coulson	**Withholding after termination** It was not possible to distinguish the decision in *Melville Dundas* to prevent it from also applying in the case of termination not related to any insolvency. However, an attempt to set off sums without a withholding notice in circumstances other than the contractor's insolvency would be likely to be unreasonable, and in this case had occurred more than 28 days after the payment had become due.
Quartzelec Ltd v Honeywell Control Systems Ltd	[2008] EWHC 3315 (TCC), Judge Davies	**Withholding against a contractor's valuation** The contract allowed Honeywell to bring into account, at the interim application stage any savings attributable to a variation omitting part of the works; this was part and parcel of the valuation process. It was therefore not necessary to serve a withholding notice to be entitled to raise the defence in any subsequent adjudication where the question of what the contractor is entitled to be paid under that interim valuation is in issue. *P. C. Harrington Contractors v Multiplex Construction* applied.
Fleming Buildings Ltd v Forrest or Hives	[2008] CSOH 103 (Outer House Court of Session), Lord Menzes	**Withholding against an architect's certificate** The scheme for payment in the SBCC Standard Building Contract (with Contractor's Design Portion), required the architect to issue interim certificates and the final date for payment of the sum certified was 14 days from the date of issue. The amount certified was the sum due and in the absence of either a payment notice or withholding notice the employer was required to pay the contractor the amount due under the certificate.

Title	Citation	Issue
Letchworth Roofing Company v Sterling Building Company	[2009] CILL 2717, [2009] EWHC 1119 (TCC), Coulson J	**Withholding against a contractor's application** The subcontract incorporated the terms of the JCT Standard Form of Subcontract DOM/1. Sterling refused to pay on Letchworth's interim application because of cross-claims for delay. However, the adjudicator decided that Sterling's withholding notice was invalid and awarded the amount claimed without set-off. There was no breach of natural justice nor had the adjudicator acted outside of his jurisdiction. Having decided that a withholding notice was required, but was not given, the adjudicator was entitled to disregard Sterling's defence. A defendant cannot avoid the absence of a valid withholding notice if, by reference to the contract and on the facts of the particular dispute, the raising of the cross-claim in question required such a notice.
Windglass Windows Ltd v Capital Skyline Construction Ltd	(2009) 126 Con LR 118, [2009] EWHC 2022 (TCC), Coulson J	**Withholding against a contractor's application** The terms of the contract were 'inadequate' and the Scheme applied. Windglass applied to an adjudicator for payment of its claims for payment and the adjudicator obliged, holding that Capital's purported withholding notices were ineffective. The point that arose in this case was indistinguishable from *Letchworth*. In the circumstances of the case, the absence of a withholding notice was fatal to bringing a cross-claim in adjudication. The 1996 Act did not permit a party to put in an ineffective withholding notice and then, in the subsequent adjudication, seek to put together an entirely different justification for withholding payment. Such a 'foot in the door' approach is contrary to the 1996 Act, which emphasizes the obligation on the paying party to give good reasons.
Urang Commercial Ltd v (1) Century Investments Ltd and (2) Eclipse Hotels (Luton) Ltd	[2009] CILL 2660, [2011] EWHC 1561 (TCC), Edwards-Stuart J	**Withholding against an interim valuation and payment notice** The amounts stated in the quantity surveyors' interim valuation and payment notice were the amounts due—*Rupert Morgan* applied. However, it was clear from the terms of the contract that the need to serve a withholding notice was restricted to the sums stated as due in the interim valuation and payment notice. The adjudicator had been wrong to decide that a withholding notice was necessary in the case of the contractor's other claims for prolongation costs, release of retention, and interest as these had not been part of the interim valuation process. Accordingly, the employers had been entitled to rely on rights of set-off against these claims.

(4) The Effect, Content, and Timing of Withholding and Pay-less Notices

Introduction

13.87 This section considers the requirements for a withholding notice to be effective in terms of its timing and content. It also looks at what happens if the grounds on which amounts are withheld are later found to be invalid or are superseded by later events. For the reasons stated below, much of the existing law in relation to these matters should continue to apply to pay-less notices issued under the 2009 Act, and this section should therefore also be read in conjunction with Chapter 14.

What Is an Effective Notice?

13.88 Section 111(1) of the 1996 requires the payer to give 'effective notice' of its intention to withhold payment, otherwise the withholding is not valid and the amount due remains payable in full. For a withholding notice to be 'effective' it must simply comply with the requirements as to content and timing set out in s. 111(2) and (3) of the 1996 Act.

13.89 The 2009 Act does not use the same language, but the intent is the same: that in order to be effective both the payment and pay-less notices must comply with the requirements as to content and timing set out in s. 110A(1), (2), and (3) for payment notices, or s. 111(4) and (5) for pay-less notices. If the payment notice does not contain these details or if it is not sent in time, then it would appear that no notified sum becomes due, and the contractor may serve notice in default.[46] In the case of a pay-less notice, this will be ineffective to prevent payment of the whole of the previously notified sum.

13.90 The requirement for detail means that, in the first instance the payee should be able understand the cross-claim made against it. A withholding notice will stand or fall on its contents and cannot subsequently be improved. Whilst the courts have tended not to apply too fine a textual analysis to notices which convey the substance of any cross-claim, notices which fail to demonstrate any contractual basis for withholding amounts may be overturned.[47] Attempts to bring new claims at adjudication that are not set out in the notice will also generally fail: *Letchworth Roofing Company v Sterling Construction Company* (2009).[48]

13.91 The service of an effective withholding or pay-less notice prevents the payee from asserting a right to immediate payment of the sum due or notified sum after the final date for payment; however it may still challenge the validity of the reasons for withholding at any time. The reasons for withholding become the focus of the dispute between the parties and, until this is determined or settled, the payer remains within its rights to continue withholding the sums specified in reliance on the notice.

13.92 The notice is therefore a procedural measure, in that it permits a set-off to be relied on by the paying party as long as a valid notice is served, unless and until the substance of that set-off is successfully challenged in adjudication, arbitration, or court. Section 111 is not dealing with the ultimate position between the parties. As Sedley LJ said in *Rupert Morgan Building Services (LLC) Ltd v David Jervis and Harriet Jervis* (2003):[49]

> There is ... nothing irrevocable about the s.111 process. It is designed simply to ensure that once a certificate is issued, payment follows unless proper notice of withholding is given. It has no legal effect, even presumptively, on the true incidence of liability.

13.93 An adjudicator, arbitrator, or judge may therefore decide that a withholding or pay-less notice complied with the requirements of form and timing in s. 111, but that all or part of the sums withheld should be paid because the grounds for withholding were not justified as a matter of substance. The effect of such a decision is not to overturn the interim or final payment to which the withholding notice relates, nor does it require it to be reopened. Instead, s. 111(4)

---

[46] See further Ch.14.

[47] See *Leander Construction Ltd v Mulalley & Company Ltd* [2011] EWHC 3449, discussed at 13.96.

[48] [2009] CILL 2717, [2009] EWHC 1119 (TCC); see 13.124.

[49] [2004] 1 WLR 1867, [2003] EWCA Civ 1563 (Key Case at 13.45).

of the 1996 Act (now s. 111(8) and (9) of the 2009 Act) provides that the repayment of the withheld sums should be made no later than seven days after the adjudicator's decision, or the date which, apart from the notice, would have been the final date for payment, whichever is later. This contemplates that the final date for payment might still be sometime in the future when the adjudicator's decision is made, although such situations will be rare.

13.94 In most cases, therefore, where the adjudicator decides sums withheld should be paid, the payment will be required to be made outside of the contractual mechanism and after the final date for payment of the interim payment to which the decision relates. This matter has created its own body of case law as to whether the paying party can serve a further withholding notice against payment of the adjudicator's decision (see Chapter 6 above).

13.95 As to whether interest will be payable under the contract or (if none is stipulated) under the Late Payment of Commercial Debts (Interest) Act 1998, this will depend upon the terms of the contract. Equally, the date from which interest flows will also depend upon the contract, and it may be that the payer will be liable for interest from the final date for payment of the sum that the adjudicator decides has been wrongfully withheld.

13.96 A distinction must be made between withholding notices that demonstrate a ground for withholding which is later determined to be invalid, and those that fail to demonstrate *any* grounds for withholding, albeit that sometimes this distinction might be difficult to see. In *Leander Construction Ltd v Mulalley and Company Ltd* (2011),[50] the contractor (Mullalley) sought to set off claims against the subcontractor (Leander) on the basis that it had failed to comply with dates set out in the Activity Schedule. Leander complained that these dates were of no contractual effect and that therefore Mulalley's withholding notice was invalid. At adjudication, Mulalley argued that there was an implied term in the contract that Leander should proceed regularly and diligently towards complying with these dates and that failure to meet them constituted breach of contract. Coulson J disagreed. The original withholding notice proceeded on the false assumption that the dates in the Activity Schedule were contractually binding and there was no implied term that made them so. The withholding notice was ineffective and the sums should be paid forthwith with interest from the date that the original payment should have been made.[51]

How Much Detail Needs to Be Given for a Notice to Be Effective?

13.97 To be effective it appears that a withholding or pay-less notice must be given in writing notwithstanding the provisions of s. 115 of the Act which gives the parties the freedom to agree on the manner of service of any notice under the Act. In *Strathmore Building Services Ltd v Colin Scott Grieg (t/a Hestia Fireside Design)* (2001)[52] Lord Hamilton in the Scottish Court of Session concluded that:

> Although the words 'in writing' are not expressly used, I am satisfied that it unmistakably appears that writing in some form is required. This is so, in my view, having regard not only to the language of section 111 itself, including the use of the indefinite article ('an effective notice', 'a notice') and the requirement to 'specify' particular matters, but also to the language of section 115 and, in particular, section 115(6) which contemplates that a notice under Part II will be in some form of writing. A telephone message, even one referring to a particular letter of earlier date, will not suffice.

---

[50] [2012] BLR 152, [2011] EWHC 3449 (TCC).

[51] See further 13.136–13.137 which deals with circumstances where a withholding notice is ineffective to permit a withholding to be made.

[52] [2001] Const LJ 72 (Outer House, Court of Session), per Lord Hamilton at para. 13 (Key Case).

13.98 Although it seems that the notice need not be particularly formal or bear any particular label, nevertheless it must recognizably answer to the description in the contract or ss. 110 and 111, as the case may be. It should also be addressed to the other party. In *VHE Construction plc v RBSTB Trust Co. Ltd* (2000),[53] RBSTB attempted to argue that its submissions in two adjudications (one of which had already been determined) constituted notice of withholding for the purposes of setting off against VHE's invoice. However, Judge Hicks held that not only was RBSTB debarred from setting off against the invoice because the amount had been awarded by the adjudicator in the first adjudication,[54] but its submissions answered neither of the above criteria.

13.99 In *Buxton Building Contractors v The Governors of Durand Primary School* (2004),[55] on the other hand, the school governors wished to challenge the contract administrator's certificate releasing the retention on the grounds that it had suffered consequential loss by the contractor's delays in repairing defects. It set its arguments out in correspondence sent to the contractors after the certificate had been issued, but before the contractor had issued its invoice. The letters set out the nature of the school's claim, and the amounts that it intended to withhold. Albeit that the letters were not all sent together and 'were not in the form in which a notice of intention to withhold payment is normally drafted', Judge Thornton held that it was at least arguable that a valid withholding notice had been served timeously. The failure of the adjudicator to consider the letters at all before concluding that no valid withholding notice had been served amounted to a breach of natural justice such that his decision should not be enforced.

13.100 The 1996 Act provides that to be effective the notice must specify the amount proposed to be withheld and the ground for withholding, or if there is more than one ground, each ground and the amount attributable to it. The detail required to be given in the notice still requires some clarification from the courts, although, under the 2009 Act, the pay-less notice is only required to provide a revised computation of the notified sum and need not set out any grounds. Nevertheless, in both cases it is obvious that some payers will take a less formal approach than others, and since the purpose of the withholding or pay-less notice is for the payee to understand the complaints against him, it follows that sufficient detail must be given for this to happen.

13.101 In *Thomas Vale Construction plc v Brookside Syston Ltd* (2006),[56] TVC had failed to rectify defects in accordance with a timetable agreed between the parties; however BSL's withholding notice cited the costs of employing others to remedy the defects, which costs had not yet been expended. TVC pointed out that the damages did not flow from the breach alleged which concerned the failure to comply with the defect rectification timetable. Notwithstanding this, the judge considered that the notice was valid since it was clear that BSL was withholding payment because (as TVC did not challenge) TVC had not completed work or remedied defects. In the view of Judge Kirkham:

> it would be inappropriate to apply fine textual analysis to a notice which is intended to communicate to the other party why a payment is not to be made.

13.102 A broad view was also taken by Judge Humphrey LLoyd in *Alstom Signalling Ltd v Jarvis Facilities Ltd* (2004)[57] (Key Case) by which he permitted himself to take account of all the circumstances leading up to and surrounding the employer's payment notice (which also served as notice of withholding under s. 111(1) of the 1996 Act). Whilst it could be said that the

---

[53] [2000] BLR 187, EWHC Technology 181 (TCC).
[54] As to set-off against adjudicators' awards, see Ch. 6
[55] [2004] BLR 374, [2004] EWHC 733 (TCC).
[56] (2009) 25 Const LJ 675, [2006] EWHC 3637 at para. 43.
[57] [2004] EWHC 1285 (TCC) at para. 36.

notice on its face might not precisely meet the requirement of specifying 'each ground and the amount attributable to it', the judge decided it was relevant to consider the form of contract which provided for Alstom to assess the amounts due to Jarvis, and the voluminous nature of Jarvis' application together with its timing. In the circumstances, he had little doubt that the recipient, Jarvis, would understand the grounds and the amounts attributable to them.

13.103 In *Aedas Architects Ltd v Skanska Construction UK Ltd* (2008)[58] (Key Case) the issue considered was whether withholding notices were effective if they failed to apportion the sum withheld against each of the numerous grounds cited, so that in effect a global sum was being claimed. The judge considered that, whilst s. 111 required 'attribution' of an amount to each ground, it did not ask for apportionment. All of the grounds which could have been calculated had been, and a global figure was attributed to the rest. That, in the judge's view, constituted compliance.

13.104 Under the 1996 Act there was no requirement that the withholding notice should set out the amount that the payer actually intended to pay, only those amounts that he intended to withhold. This is an odd omission and has led to some uncertainty as to whether a withholding notice will be valid if no payment notice has been served. In order to address the doubt, some employers have served a payment notice (or a revised payment notice) alongside the withholding notice, even if this means that the payment notice is served late. Alternatively, it is possible to combine the two notices into one as contemplated by s. 111(1) and set out how the amount due has been calculated and any set-offs included in the calculation. Provided that the notice is sufficiently clear as to how the calculation has been carried out, and that it has been served by the latest date for serving a withholding notice, then it is likely that this approach will satisfy the requirements of both ss. 110(2) and 111. See, for example, *Alstom Signalling Ltd v Jarvis Facilities Ltd* (2004)[59] (Key Case) in which Judge LLoyd found that the employer had, in effect, served proper notice of how the amount due had been calculated including contra-charges, which amounted to a notice of withholding under s. 111(1).

Key Cases: Amount of Detail Required in Withholding Notice

Alstom Signalling Ltd v Jarvis Facilities Ltd [2004] EWHC 1285 (TCC)

13.105 **Facts:** See Key Case at 12.33 for the contractual background in this case. In addition to considering whether the contract contained an adequate mechanism for determining when payments were due and the final date for payment, Judge LLoyd was asked to determine whether Alstom's certificate complied with s. 111 in terms of content and timing, so that an effective withholding notice had been given. Alstom countered that Jarvis had abused the payment application process by serving its application too early, failing to send it to the correct place, bombarding Alstom with information, and then by failing to respond to specific questions and queries.

13.106 **Held:**

1. There was no requirement for a separate withholding notice to be served. The contract clearly allowed Alstom to assess Jarvis' application so that the amount due was that which Alstom assessed was due. Section 110(2) allowed that a payment notice could serve as a withholding notice provided that it complied with the provisions of Section 111.
2. Taking account of the circumstances, Alstom's certificate complied with Section 111 as regards both content and timing.

[58] [2008] CSOH 64 (Outer House, Court of Session).
[59] [2006] EWHC 3637 at para. 43.

36. If the relevant period commenced on 18 June 2003 (as I have decided) then and if the application were treated as received on that day the due date for payment would be 25 June and the final date 17 days thereafter, i.e. 11 July. A withholding notice would have to be given by 4 July. In my judgment Alstom's letter of 2 July 2003 therefore not only satisfied [the requirements of the contract], since, read along with Alstom's questions and queries (including 'the lilac file') it set out how Alstom's valuation had been arrived at and the reasons, but it also satisfied section 111 of the Act, were it to be relevant and applicable, both in terms of timing and content. In the case of the latter I take account of all the circumstances leading up to and surrounding the letter, such as the nature of the contract, the application made and its timing, since it could be said that the letter and its attachment on their face might not precisely meet the requirement that a notice should specify 'each ground and the amount attributable to it'. However I have little doubt that the recipient, Jarvis, would know the grounds and the amounts.

Aedas Architects Ltd v Skanska Construction UK Ltd [2008] CSOH 64 (Outer House Court of Session)

13.107 **Facts:** The pursuers sought summary judgment for immediate payment on six invoices on the grounds that ineffective notice of withholding had been given. Although each of the applications for payment had been met by a counter-notice served in time, these simply listed a number of grounds for withholding and either attributed an amount equal or greater to that claimed against all of them globally, or else attributed specific amounts against some but not all.

13.108 Referring to the decision of Judge Kirkham in *Thomas Vale Construction plc v Brookside Syston Ltd* (2006)[60] the defendant argued that none of the counter-notices should be subjected to fine textual analysis. They were not addressed to lawyers but to contract managers and others who were aware of what was happening on site in an ongoing contract concerning several places. The grounds and amounts had been specified and that was enough.

13.109 **Held:** Given that this was an application for summary judgment, it was by no means clear from the counter-notices alone that the defence was bound to fail. Issues of fact could arise which would allow evidence of meetings and conversations to explain the events surrounding the notices. For example, the defenders would be entitled to explain why they were unable to make any financial attribution against particular terms.

13.110 In any event, the documents themselves were effective under s. 111. The Act called for 'attribution' and not 'apportionment' and the counter-notices adopted a competent method of attributing a global figure to some or all of the grounds cited. That, in the judge's view, constituted complicance:

> [17] ... I am also of the opinion that the documents themselves are 'effective' under section 111. I think the matter is clear for items 6/59 to 61. In these, sufficient attribution has been made against five of the enumerated grounds. That in itself is enough to hold that it cannot be said the defence is bound to fail.
>
> [18] As to the last three, 6/62 to 64, what has been done is to debit the whole attribution against any money due. The contract demands attribution to each ground. It does not ask for any apportionments and in my view it is a competent way to proceed by debiting all sums. The Statute speaking of 'each ground' says attribution 'to it' must take place. In my view that also is what the counter-notice has done. All the grounds which can be calculated have so been and a global figure debited. That in my opinion is compliance.

[60] See 13.101.

Must a Contractor Give Notice of Intention to Withhold?

13.111 One question which has arisen is whether a contractor or subcontractor would also be required to serve a withholding notice in circumstances where a payment certificate was negative or where there were otherwise sums owing to the employer? Put another way, if the contractor is able to claim summary judgment for an amount due where no withholding notice has been served, then should the employer not be similarly entitled in situations where he has served notice of his claim?

13.112 This situation was addressed in the case of *Balfour Beatty Construction Northern Ltd v Modus Corovest (Blackpool)* Ltd (2008)[61] (Key Case), in which the employer applied for summary judgment in respect of its claim for delay damages on the grounds (inter alia) that it had served notice under clause 24 of the JCT 1998 Conditions, and the contractor had not served any notice of withholding. The employer said that by operation of the 1996 Act, the final date for payment was 17 days after service of the clause 24 notice, and there could be no defence to the employer's claim.

13.113 In Coulson J's view, however, this argument was without foundation. The provisions of the 1996 Act applied only to stage payments from the employer and not to payments in the other direction:

> 106. The whole purpose of Part II of the 1996 Act and the Scheme for Construction Contracts (set out in Statutory Instrument 1998 No. 649) was to improve cashflow for contractors and subcontractors. That was the principal concern of the Latham Report, which gave rise to both the Act and the Scheme. Not only is there nothing specific in either the Act or the Scheme about payments to the employer, there are many parts of both which I consider to be inconsistent with the construction for which Mr Bowdery contends.

13.114 The position that withholding notices are only required in cases of interim payments and not in the case of other claims between the parties has also been confirmed by Edwards-Stuart J in *Urang Commercial Ltd v (1) Century Investments Ltd and (2) Eclipse Hotels (Luton) Ltd* (2011)[61] (Key Case at para. 13.51 above).

13.115 The 2009 Act is no more specific than the 1996 Act in referring to payments from contractors or subcontractors to their employers and there would appear to be no reason to distinguish between these cases and contracts entered into after 1 October 2011. The amendments published by the JCT to deal with the 2009 Act do however provide, at clause 4.15.4 (Standard Building Contract)[62] that the contractor issues a pay-less notice in circumstances where there is an amount due to the employer under the final certificate. There is no similar provision relating to interim payment certificates showing a negative amount due, and in the absence of any apparent statutory penalty, it would appear that the contractor's withholding may remain valid even in the absence of a pay-less notice.

Timing of the Withholding or Pay-less Notice

13.116 A withholding or pay-less notice must be given not later than the prescribed period before the final date for payment.[63] The parties are free to agree the prescribed period for the giving of the notice in the contract. However, if no period is prescribed in the contract, then the Scheme applies to provide that this period should not be less than seven days before the final date for payment.[64] The final date for payment is also determined either in accordance with

[61] [2009] CILL 2660, [2008] EWHC 3029 (TCC).

[61] 138 Con LR 233, [2011] EWHC 1561 (TCC).

[62] Clause 4.10.2.2 of the Design and Build Contract.

[63] s. 111(2)

[64] Scheme for Construction Contracts SI [1998/649], para. 10.

the contract or, where no such provision is made, 17 days from the date that the payment becomes due (see Chapter 12 above).

13.117 The question has arisen as to whether it is possible to serve a withholding notice too early so as to render it ineffective. This issue arose in *Strathmore Building Services Ltd v Colin Scott Grieg (t/a Hestia Fireside Design)* (2001)[65] (Key Case), in which an employer sought to rely on correspondence in which notice was given of the employer's intention to withhold certain sums, dated prior to the contractor's payment application. Such a situation might be far from uncommon as parties may often be able to show that disputes have been canvassed in correspondence, even if a formal withholding notice has not been served. In this case, the judge found that, where a contract provided for interim payments to be made, a valid withholding notice could not be given before the contractor had made his application for payment. In the judge's view the notice of withholding:

> 13. ... should constitute a considered response to the application for payment, in which response it is specified how much of the sum applied for it is proposed to withhold and the ground or grounds for withholding any amount. Such a response cannot, in my view, effectually be made prior to the application itself being made.

13.118 It remains to be seen whether a similar finding would be made in a case where the contractor's application for payment is not an essential part of the payment process, such as the NEC/ECC contracts where the contractor's application must be considered by the project manager only if one is provided.[66] It is likely, however, that an effective withholding notice cannot be served before an amount has been certified since a certificate issued by a third party on behalf of the employer is likely to crystallize the sum due under the contract which gives rise to the contractor's right to be paid the certified sum. This would then supersede the effect of any earlier withholding notice.

13.119 The 2009 Act puts these matters beyond doubt for contracts entered into after 1 October 2011. Here, s. 111(5)(b) provides that the pay-less notice cannot be given sooner than the payment notice is issued either by the specified person or the payee. It is assumed that any pay-less notice given prior to the payer issuing a payment notice would be superseded by the payment notice in any event.

Key Case: Timing of Withholding Notice

Strathmore Building Services Ltd v Colin Scott Grieg (t/a Hestia Fireside Design) [2001] Const LJ 72 (Outer House Court of Session)

13.120 **Facts:** Strathmore was engaged by HFD to perform work pursuant to a contract in the form of an amended Scottish Building Contract with Contractor's Design (July 1997 Revision). Delays occurred to the works and by August 1999 various interim payments had been made to the pursuers but a balance of about £40,000 of the contract sum remained unpaid. On 17 August 1999 HFD's agent wrote to Strathmore stating that a balancing payment of £40,379.40 was outstanding, but making reference to counterclaims against Strathmore in respect of liquidated damages for delay and 'direct costs incurred'. On a 'without prejudice' basis, however, HFD was prepared to make a payment of £4,971 in full and final settlement of its obligations under the contract.

13.121 On 27 November 1999 Strathmore delivered an invoice for the sum of £41,277.05 (inclusive of VAT) in respect of monies claimed under the contract. HFD did not respond in

[65] [2001] Const LJ 72 (Outer House, Court of Session).
[66] Clause 50.4.

be incorporated.

4. There have been no ~~agreements reached on the purchase of Alan Harris's share~~ of the business. ~~Discussions about Mr Harris's~~ oral or written discussions or agreements concerning the purchase of Alan Harris's share of the business at:

a) a fixed price
b) in accordance with a formulation or
c) at a discount.

5 There have been no discussions or agreements concerning

METROPOLITAN HOTELS

1. Decision to incorporate

To enable the business to grow working capital without the partners/directors incurring a tax liability.

2) This was not a priority at the time and also a cost consideration – The partners believed that the options in the Partnership agreement still applied

3. When a shareholders Agreement is drawn up this Option Agreement &

writing, but claimed that on the day it received the invoice it telephoned Strathmore's office and left a message to the effect that the parties were already in dispute regarding further payment and specifically referring to the agent's letter of 17 August 1999.

13.122 **Held:** A withholding notice could not be served before the contractor had submitted his application for payment. Section 111 required that the notice should present a considered response to the application for payment showing how much of the sum being applied for the employer proposed to withhold:

> [14] The second matter raised was whether a notice effective for the purposes of Section 111 could be a communication in writing sent earlier than the making of the relevant Application. Mr d'Inverno pointed out that, while Section 111(2) provided that any notice must be given not later than a particular time, it did not provide that it required to be given after any particular time, i.e. there was no *terminus a quo*. The letter of 17 August, albeit sent prior to the invoice of 27 November, was (or was arguably) a notice of intention to withhold payment within the meaning of Section 111. I am unable to accept that argument. The purpose of Section 111 is to provide a statutory mechanism on compliance with which, but only on compliance with which, a party otherwise due to make a payment may withhold such payment. It clearly, in my view, envisages a notice given under it being a considered response to the application for payment, in which response it is specified how much of the sum applied for it is proposed to withhold and the ground or grounds for withholding any amount. Such a response cannot, in my view, effectually be made prior to the application itself being made. It may, of course, be that the matter of withholding payment of any sum which might in the future be applied for has previously be raised. In such circumstances a notice in writing given after receipt of the application but which referred to or incorporated some earlier written communication might suffice for the purpose—though I reserve my opinion on that matter. But such an earlier written communication, whether alone or referred to subsequently in an oral communication, cannot, in my view, suffice. This is, as a matter of statutory interpretation, in my view, unmistakeably the case.

Raising New Grounds for Withholding after the Notice Has Been Served

13.123 Service of a withholding or pay-less notice will not generally entitle a payer to raise new grounds of set-off not mentioned in the notice when the matter comes before an adjudicator. The adjudicator may consider the validity of the matters cited in the notice, and justification for the amounts withheld, but he may not consider any new grounds or amounts which the payer may try to introduce at adjudication.

13.124 Often arguments have been made on the grounds of natural justice that an adjudicator failed to consider the payer's counterclaim in awarding amounts due under an interim payment application or certificate. Such arguments were raised in *Letchworth Roofing Company v Sterling Construction Company* (2009)[67] where the subcontractor had asked the adjudicator to consider the contractor's counterclaims, but only for the purposes of assisting in settling the eventual final account. The adjudicator decided that no valid withholding notice had been served, and although he assessed sums under the contractor's counterclaim (as he had been requested) he awarded the amounts stated under the interim application without set-off. Coulson J summarized the position as follows:

> 25. Take the general position first. Whilst there is no doubt that a defendant can raise whatever matters he likes by way of defence for the adjudicator to consider, that general principle

[67] [2009] CILL 2717, [2009] EWHC 1119 (TCC); see also *Urang Commercial Ltd v (1) Century Investments Ltd and (2) Eclipse Hotels (Luton) Ltd* [2011] EWHC 1561; *London & Scottish Properties plc v Riverbrae Construction Ltd* [1999] BLR 346.

does not permit a defendant to rely on a cross-claim which should have been the subject of a withholding notice, but was not. In other words, a defendant cannot avoid the absence of a valid withholding notice if, by reference to the contract and on the facts of the particular dispute, the raising of the cross-claim in question required such a notice. To hold otherwise would be to obviate the need for withholding notices at all: see *Harwood Construction Ltd v Lantrode Ltd* (Unreported, 24.11.00).

33. Sometimes, the interface between the adjudicator's jurisdiction and the scope and validity of withholding notices can be said to give rise to difficulties which, on a proper analysis, are simply not there. In my judgment, the general position is clear. An adjudicator has to decide whether or not a withholding notice is required to permit a cross-claim to be raised as a defence, and if so, whether or not there has been a valid notice. If he concludes that no notice was required, or that a notice was required and that there was a valid notice, then he must take the cross-claim into account in arriving at his decision. If he concludes that a notice was required, and that either there was no notice or that the notice that has been served was invalid for any reason, then he is not entitled to take the cross-claim into account when reaching his conclusion. That is the general position, and it applies here, notwithstanding the slightly unusual words of the referral notice.

13.125 In *Windglass Windows Ltd v Capital Skyline Construction Ltd* (2009),[68] the contractor complained that Windglass, the subcontractor, had failed to submit its payment applications in a particular form with sufficient supporting documentation. It set out these complaints as its reason for withholding payment in two purported withholding notices. At adjudication, however, it relied upon an alleged counterclaim in respect of defects and delay as reason for sums not being due. Coulson J rejected the contractor's position: the withholding notices failed to set out any grounds on which amounts had been withheld, and the counterclaim was so unparticularized that it could not amount to a valid set-off or counterclaim in any event:

27 The 1996 Act does not permit a party to put in an ineffective withholding notice and then, in the subsequent adjudication, seek to put together an entirely different justification for withholding payment. Such a 'foot in the door' approach (if I may call it that) is contrary to the 1996 Act, which emphasises the obligation on the paying party to give good reasons, there and then and in advance of the date for payment, if any part of a sum otherwise due is not going to be paid. If that paying party does not do so, then, in the words of Chadwick LJ, it has to pay now and argue later.

13.126 In order successfully to raise new cross-claims in an adjudication, therefore, the employer needs to show that no withholding notice was in fact necessary in respect of the new grounds relied upon. An example arose in *Urang Commercial Ltd v (1) Century Investments Ltd and (2) Eclipse Hotels (Luton) Ltd* (2011),[69] where the contractor brought adjudication proceedings against the employer which included amounts unpaid against interim valuations, and other claims. The other claims did not require a withholding notice to be served against them and therefore the adjudicator had been wrong not to take them into account. The error was one of law, however, and not one that went to his jurisdiction.

Change in Circumstances Rendering Reasons for Withholding Invalid

13.127 The question of whether a withholding notice remains effective even where the grounds for withholding ceased to be valid before the final date for payment was considered by the House of Lords in *Reinwood Ltd v L. Brown & Sons Ltd* (2008).[70] Their Lordships' concluded that provided a withholding notice was effective when payment was made it should not be

[68] 126 Con LR 118, [2009] EWHC 2022 (TCC).
[69] 138 Con LR 233, [2011] EWHC 1561 (TCC) (Key Case at 13.51).
[70] [2008] 1 WLR 696, [2008] UKHL 12 (HL).

undermined by later events. Thus the grant of the extension of time, after the valid withholding notice was served and payment was made, did not deprive the employer of its right to rely on the withholding notice to deduct liquidated damages that were extinguished by a subsequent extension of time.

13.128 Whether this principle is of general application remains to be seen, but a note of caution needs to be sounded. First, the reasons given by Lord Neuberger in part turned on the wording of the contract extension-of-time clause: the effect of the extension of time was to 'cancel' the existing certificate of non-completion on which the LADs deduction depended. Lord Neuberger said that that the word 'cancel' did not have retrospective effect. Thus the deduction of LADs was valid when it was made and remained so. Secondly, the House of Lords gave no opinion on whether the result would have been the same had the extension of time been granted earlier before payment was made.[71] This leaves open the question of whether the withholding notice could be relied on when making payment if the change of circumstances occurred in the intervening period. However, as Lord Neuberger pointed out, such a construction appeared to be consistent with the purposes of the Act, and an employer might face practical difficulties if an entitlement to additional sums was only realized shortly before payment was to be made.[72]

13.129 The lack of guidance on this question is unfortunate because, as Lord Neuberger also pointed out,[73] it raises questions as to when *any* notice, certificate, or even payment could be relied upon as satisfying the employer's obligations within the payment cycle to which it relates. Payment becomes due on the payment-due date and not on the final date for payment. If events were to be taken into account after a payment became due, then not merely could neither party rely on a valid withholding notice as conclusively determining their rights and obligations with regard to payment on an interim certificate, but neither party could even rely on an actual payment, correct at the time it was made, as being effective. Possibly their Lordships were mindful that any answer to this question would depend inextricably on the facts of the case.

Key Case: Change in Circumstances Rendering Reasons for Withholding Invalid

Reinwood Ltd v L. Brown & Sons Ltd [2008] 1 WLR 696, [2008] UKHL 12 (HL)

13.130 **Facts:** The parties entered into a contract on the terms of JCT Standard Form of Building Contract 1998 for the construction of 59 apartments in Manchester. Clause 24.2.1 of the contract entitled Reinwood Ltd, as employer, to deduct LADs provided that: (1) the architect had issued a certificate of non-completion by the completion date and (2) that a notice of withholding had been served. Clause 24.1 provided that if, after issuing a certificate of non-completion, the completion date was extended, the certificate would be 'cancelled' and the architect would issue any further certificate as required. Clause 30 concerned 'certificates and payments' and provided that the architect would from time to time issue interim certificates stating the amount due to the contractor, and the final date for payment would be 14 days from the date of issue of each interim certificate. If the employer intended to serve notice of withholding, then he should do so not later than five days before the final date for payment.

71 See speech of Lord Neuberger at para. 51.
72 At para. 53.
73 *Ibid.*

13.131 On 14 December 2005, the architect issued a certificate of non-completion ('the December non-completion certificate'). On 11 January 2006, he issued interim certificate number 29 ('the interim certificate'), showing a net amount payable of £187,988. Pursuant to clause 30, the final date for payment of this sum was 25 January.

13.132 On 17 January, the employer served notice of withholding stating that he proposed to withhold £61,629 LADs from the sum due under certificate number 29. Three days later, on 20 January, the employer paid the contractor £126,359. On 23 January, the architect granted an extension of time until 10 January 2006 ('the January extension').

13.133 Reinwood did release payment of the LADs on 1 February; however, Brown claimed to be entitled to determine the contract on, inter alia, the ground that Reinwood had failed to pay the amount due under interim certificate number 29 by the final date for payment. Brown claimed that the extension of time, granted before the final date for payment, disentitled Reinwood from relying on the architect's December non-completion certificate such that the amount due, as certified under interim certificate number 29, should have been paid in full.

13.134 **Held:**

1. The extension of time did not have the effect of denying Reinwood the right to rely on the non-completion certificate. Whilst the effect of the extension of time was to cancel the non-completion certificate, such a cancellation was not retrospective in its effect. Thus the withholding notice was valid when served and the employer was entitled to rely on it when making payment.
2. 25 January was the 'final date for payment' and not the date on which payment became due. The function of the final date 'was akin' to making time of the essence of the payment as at that date following which failure to pay might have serious consequences for the employer.

13.135 Lord Neuberger of Abbotsbury:

> 38. There is no doubt that, if the January extension had been granted before 11 January 2006, the employer would not have been entitled to deduct the LADs resulting from the December non-completion certificate, as that certificate would have been 'cancelled' under clause 24.1 as a result of the January extension, by the time of the issue of the interim certificate. Equally, there is no doubt that, if the January extension had been granted after 25 January 2006, the employer's deduction of the LADs based on the December non-completion certificate would have been unassailable, as that certificate would not have been cancelled under clause 24.1 by the January extension until after the 'final date for payment' under the interim certificate. The difficulty in this case arises from the fact that the January extension was granted after the date of issue of the interim certificate, but before the 'final date for payment' thereunder …
>
> 42. So the question is whether, in accordance with the submission of the contractor, the issue of the January extension, after the employer had both served the withholding notice and paid on the assumption that he had the right to rely on the December non-completion certificate, deprived the employer of that right. In my judgment, in agreement with the Court of Appeal, the issue of the January extension did not have that effect.
>
> 43. It is true that, by virtue of clause 24.1, the effect of the January extension was to 'cancel' the December non-completion certificate, upon which the employer's right to deduct depended. However, such a cancellation was not retrospective in its effect. That is not the normal meaning or effect of the word 'cancel'. Indeed, it would be an absurd meaning to give the word here. It would mean that, even if the January extension had been granted after 25 January, it would have resulted in the employer having underpaid on the interim certificate.

44. Once one accepts that the effect of the January extension was not to cancel the December non-completion certificate retrospectively, it appears to me to follow that, in making any payment before the January extension was granted, the employer was entitled to rely on that certificate, unless the provisions pursuant to which the payment was made provided otherwise …

49. It was argued by the contractor that this conclusion lies uneasily with the fact that the January extension had cancelled the effect of the December non-completion notice by 25 January, which was the date by which the interim certificate had to be paid. In my opinion, 25 January was the 'final date for payment', not the date on which payment became due. A sum becomes due under a certificate when it is issued, and the 'final date for payment' is the date by which failure to pay can have serious consequences for the employer. As my noble and learned friend, Lord Walker of Gestingthorpe, observed during the argument, the function of the 'final date' is akin to making time of the essence of the payment as at that date. It is fair to say that the contract is not completely clear on this issue. However, section 110(1) required this contract to specify both a date 'when' 'payments become due under [it]' and 'a final date for payment', and this requirement is inherent in section 110(2). As the contract has to comply with that requirement, it seems to me that it must be construed as so complying unless it is impossible to do so. I have no difficulty in reading clause 30.1.1.1 as having the effect of rendering a payment under a certificate due as at the date of its issue; indeed, it is otherwise hard to see the purpose of the word 'final' in that clause.

50. It was also argued by the contractor that accepting the employer's argument could lead to abuse by employers. I do not agree. Unlike clause 30.1.4, which appears to contemplate the employer being able to rely on any deduction in his withholding notice, clause 24.2 only covers deductions which are claimed in reliance on a certificate of non-completion issued by the architect. Even where, as may well have been the position in this case, the deduction is made at a time when the employer has reason to believe that an extension of time will be granted in the near future, the contractor will be able to retrieve the monies very quickly after the extension is granted under clause 24.2.2, as already explained.

51. It follows that, for my part, I would dismiss the contractor's appeal. An outstanding question which should be mentioned is whether the employer would still have succeeded if the January extension had been granted after the service of the clause 24.2.1.2 withholding notice, but before the employer actually paid out on the interim certificate (e.g. if the January extension had been granted on 18 or 19 January). It was not a point which was debated (as it does not strictly arise) and therefore it seems to me that we should only express a view on it if the answer is tolerably clear.

52. There is undoubtedly a case for saying that the employer should not have succeeded on those facts. There is plainly a difference between paying in reliance on a withholding notice which is accurate at the time of payment and paying in reliance on such a notice which is no longer accurate at the time of payment. Further, clause 24.2.3 seems to refer to clause 24.2.1[b], but not to clause 24.2.1.2, which tends to suggest that it may not be possible to rely upon a withholding notice under the latter clause once the certificate of non-completion on which it is based is cancelled.

53. However, there are arguments the other way. The principle that a withholding notice, valid when it is served, should be able to be relied on in relation to the payment to which it relates, even after its basis has been undermined, appears at least arguably consistent with the policy of the 1996 Act as discussed above. Further, the points that an employer could face practical difficulties in relation to payment where an extension of time is granted shortly before the final date for payment, and may unfairly lose his right to rely on a new certificate of non-completion because of the time limit in clause 24.2.1, appears to apply to a case where the extension of time is granted before actual payment pursuant to an interim certificate almost as much as it applies where the extension is granted after payment.

> 54. In these circumstances, while it is generally desirable to give as much guidance as possible to the meaning and effect of a provision such as clause 24, I consider that it would be wrong to express a view on this outstanding question

What Happens if No Valid Notice Is Served?

13.136 As seen in 13.93 above, s. 111(4) of the 1996 Act and s. 111(8) and (9) of the 2009 Act, provide a regime for payment of sums wrongly withheld within seven days of an adjudicator's decision to that effect. However, this regime is intended for circumstances when an adjudicator decides that sums withheld in reliance of a complaint payment, or withholding or pay-less notice, should nevertheless be paid because the basis of the payer's calculations or his grounds for withholding were unjustified. In cases where no withholding notice is served, or if it is ineffective for any of the reasons given above, then different considerations apply. Provided that a sum is in fact due under the contract, then the payer will not be entitled to withhold amounts due after the final date for payment. Payment should be made without delay[74] and interest may be payable from the final date for payment either under the contract or the Late Payment of Commercial Debts (Interest) Act 1998.

13.137 It will not generally be appropriate for the payer to wait until the next interim payment is due in order to include the amounts that have been wrongly withheld. It therefore follows that the ordinary contractual process for invoices to be issued, assessed, and paid must be abridged and payment should be made as soon as this can possibly be arranged. Any delay in this process risks the payee applying to court for summary judgment or enforcement of an adjudicator's award.

Cross-contract Set-off

13.138 Set-off between contracts, is clearly contemplated by s. 110(2) of the 1996 Act which refers to 'set-off or abatement by reference to any sum claimed to be due under one or more other contracts'.[75] Under the 1996 legislation, therefore, cross-contract set-offs are amongst the items which should be included in the payment notice. Although the Construction Act Review payment working group recommended that cross-contract set-off should in future be prohibited by legislation, the 2009 Act makes no provision for this.[76] However, all reference to cross-contract set-off has been removed and the 2009 Act is silent on the matter. It is thought that, where the right to set-off exists, the right to do so will remain and that the set-off should continue to be the subject of a valid payment or pay-less notice.

13.139 Cross-contract set-off is, in any event, closely circumscribed by law, and may in fact be prohibited by the terms of the contacts in question. In most circumstances, the issue will not arise. However, in circumstances where cross-contract set-off is possible, either a payment notice or a withholding notice should be issued under the contract under which the payment has become due, specifying details of the amount owed under the associated contract together with the basis of calculation for the balance due. See, for example, *Allied London & Scottish Properties plc v Riverbrae Construction Ltd* (1999)[77] in which the failure by the petitioners to serve a withholding notice was one of the reasons justifying the adjudicator's decision not to delay payment of sums due on account of claims brought under other contracts.

---

[74] *Outwing Construction Ltd v H. Randell & Son Ltd* [1999] BLR 156.

[75] s. 110(2)(b).

[76] Chairman's Final Report of the Deliberations of the Payment Working Group, Section 2.8, Department of Trade and Industry, 6 Sep. 2004.

[77] (1999) BLR 346 (Outer House, Court of Session).

Table 13.2 Table of Cases: The Effect, Content, and Timing of Withholding Notices

Title	Citation	Issue
Allied London & Scottish Properties plc v Riverbrae Construction Ltd	[1999] BLR 346 (Outer House, Court of Session), Lord Kingarth	**Withholding notice required for cross-contract set-off** The employer had engaged the contractor to perform works under a number of different contracts. The contractor made claims for payment in respect of four of these contracts which the employer sought to resist on the grounds that there were liquidated damages owing under three other contracts as well as an alleged over-payment in respect of a fourth. The adjudicator found that there was no right at law for the employer to set off these amounts and also that such claims should have been, but were not, the subject of withholding notices. The court held that, in these circumstances, the adjudicator had acted within his jurisdiction by refusing to take into account or assess the employer's claims.
VHE Construction plc v RBSTB Trust Co. Ltd	[2000] BLR 187, EWHC Technology 181 (TCC), Judge Hicks	**Form of withholding notice** Although notice need not be particularly formal or bear any particular label, nevertheless it must recognizably answer to the description in the contract or s. 111, as the case may be. For that purpose it must be addressed to the other party; s. 111 made no sense without such an implication.
Strathmore Building Services Ltd v Colin Scott Grieg (t/a Hestia Fireside Design)	[2001] Const LJ 72 (Outer House, Court of Session), Lord Hamilton	**Withholding notice must be served after payment application** 1. A withholding notice could not be served before the contractor had submitted his application for payment. Section 111 required that the notice should present a considered response to the application for payment showing how much of the sum being applied for the employer proposed to withhold. 2. A withholding notice must be in writing.
Buxton Building Contractors Ltd v The Governors of Durand Primary School	[2004] Adj LR 03/12, [2004] EWHC 733 (TCC), Judge Thornton	**Form of withholding notice** Letters setting out the nature of the employers' claim and the amounts and grounds for withholding might constitute valid notice of withholding (if served timeously), albeit that they were not all sent together and were not 'in a form in which a notice of intention to withhold payment is normally drafted'.
Rupert Morgan Building Services (LLC) Ltd v (1) David Jervis and (2) Harriet Jervis	(2004) 1 WLR 1867, [2004] EWCA Civ 1563 (CA), Schiemann LJ, Sedley LJ, Jacob LJ	**Effect of withholding notice** The s. 111 notice does not represent the ultimate position between the parties. It is designed simply to ensure that once a certificate is issued, payment follows unless proper notice of withholding is given. It has no legal effect, even presumptively, on the true incidence of liability. The grounds for withholding may be overturned by an adjudicator or other tribunal.
Alstom Signalling Ltd v Jarvis Facilities Ltd	[2004] EWHC 1285 (TCC), Judge LLoyd	**Need for and form of withholding notice** 1. There was no requirement for a separate withholding notice to be served. The contract clearly allowed Alstom to assess Jarvis' application so that the amount due was that which Alstom assessed was due. 2. In any event, Alstom's certificate complied with s. 111 as regards both content and timing. It was

(*Continued*)

Table 13.2 *Continued*

Title	Citation	Issue
		appropriate to take all of the circumstances into account including the nature of the contract, the application made and its timing. Having regard to Alstom's certificate and covering letter, together with the questions that it had previously raised on Jarvis' application, it was clear that Jarvis would know the amounts of the withholding and the grounds.
Thomas Vale Construction plc v Brookside Syston Ltd	(2009) 25 Const. L.J. 675, [2006] EWHC 3637 (TCC), Judge Kirkham	**Detail required by a withholding notice** It is inappropriate to apply fine textual analysis to a notice which is intended to communicate to the other party why a payment is not to be made.
Balfour Beatty Construction Northern Ltd v Modus Corovest (Blackpool) Ltd	[2009] CILL 2660, [2008] EWHC 3029 (TCC), Coulson J	**Withholding notices not required from contractor** The provisions of the 1996 Act apply only to stage payments from the employer and not to amounts due from the contractor to the employer as delay damages or otherwise.
Aedas Architects Ltd v Skanska Construction UK Ltd	[2008] CSOH 64 (Outer House, Court of Session), Lord McEwan	**Attribution of global sum in the withholding notice** Whilst s. 111 required 'attribution' of an amount to each ground, it did not ask for apportionment. All of the grounds which could have been calculated had been, and a global figure was attributed to the rest. That, in the judge's view, constituted compliance.
Reinwood Ltd v L. Brown & Sons Ltd	[2008] 1 WLR 696, [2008] UKHL 12 (HL) Lords Hope of Craighead, Scott of Foscote, Walker of Gestingthorpe, Brown of Eason-under Heywood, and Neuberger of Abbotsbury	**Effect of later events on a withholding notice** On the facts of the case, a withholding notice that was effective when served remained so when payment was made. Events that undermined the grounds for withholding that occurred after payment was made but before the final date for payment did not retrospectively deprive the employer of its right to rely on the withholding notice. This was in part because, whilst the effect of the extension of time was to cancel the non-completion certificate, such a cancellation was not retrospective in its effect.
Letchworth Roofing Company v Sterling Building Company	[2009] CILL 2717, [2009] EWHC 1119 (TCC), Coulson J	**Raising new cross-claims in adjudication** If a withholding notice is required to be served in respect of a cross-claim, then failure to serve it, or failure to serve an effective notice, means that the adjudicator is not entitled to take that cross-claim into account in reaching his decision.
Windglass Windows Ltd v Capital Skyline Construction Ltd	(2009) 126 Con LR 118, [2009] EWHC 2022 (TCC), Coulson J	**Raising new grounds for withholding in an adjudication** In the circumstances of the case, the absence of an effective withholding notice was fatal to bringing a cross-claim in adjudication. The 1996 Act did not permit a party to put in an unparticularized withholding notice and then, in the subsequent adjudication, seek to put together an entirely different justification for withholding payment. Such a 'foot in the door' approach is contrary to the 1996 Act, which emphasizes the obligation on the paying party to give good reasons.

Title	Citation	Issue
Urang Commercial Ltd v (1) Century Investments Ltd and (2) Eclipse Hotels (Luton) Ltd	(2011) 138 Con LR 233, [2011] EWHC 1561 (TCC), Edwards-Stuart J	**Withholding notices only required for interim or final payments** The 1996 Act only requires withholding notices to be served in relation to interim valuations. There was no provision of the contract which required withholding notices to be served in other circumstances where the contractor had submitted claims for prolongation costs, retention, and interest.
Leander Construction Ltd v Mulalley and Company Ltd	[2012] BLR 152, [2011] EWHC 3449 (TCC), Coulson J	**Withholding notice contains no contractual basis of withholding** In order to be effective a withholding notice must demonstrate a ground of withholding. The notice in question relied on a non-existent implied term and was therefore invalid to justify withholding any payment.

14

PAYMENT AND PAY-LESS NOTICES UNDER THE 2009 ACT

(1) Overview

Introduction and Summary

14.01 As a result of the problems that have been encountered in the application of ss. 110(2) and 111, these provisions have now been entirely rewritten by the Local Democracy, Economic Development and Construction Act 2009 ('the 2009 Act'). The changes apply to all construction contracts entered into after 1 October 2011. This chapter considers the changes and how they have been incorporated into the Scheme for Construction Contracts ('the Scheme') and the main standard forms. It should be read in conjunction with the previous chapter insofar as that chapter discusses issues relating to the effect, content, and timing of notices which are considered to be common to notices served under contracts governed by the Housing Grants Construction and Regeneration Act 1996 ('the 1996 Act') and the 2009 Act.

The following summary sets out the changes to ss. 110(2) and 111 under the 2009 Act:

1. References to the 'sum due' and 'withholdings' have been removed: instead the payer is obliged to pay the 'notified sum' which is the sum set out in the payer's, specified person's, or payee's payment notice.
2. If the payer intends to pay less than the notified sum, then it or the specified person must issue a further notice (the 'pay-less notice') which is, in effect, a revision of the payment notice.
3. If the payer or specified person is required to issue the payment notice but fails to do so by the required date, then the notified sum is the amount set out in the payee's application or, if none, a notice given by the payee setting out the amounts to be paid. If the latter, then the final date for payment may be adjusted.
4. The Scheme has been amended to provide for the giving of a payment notice by the payer or specified person not later than five days after the payment-due date, and the pay-less notice not later than seven days before the final date for payment. These provisions remain consistent with the 1996 Act but take account of the new terminology.

5. Further amendments have been made to clarify the position regarding withholdings in the case of the payee's insolvency (*Melville Dundas Ltd v George Wimpey UK Ltd* (2007))[1] and to allow a third party ('specified person') to provide a payment notice in place of the payer.

(2) The 2009 Act Amendments to Notice Requirements

Introduction

14.02 The 2009 Act removes all mention of the terms 'withholding' and 'sum due', and in their place is a new obligation to pay the 'notified sum' which is the sum set out in a payment notice given either by the payer, a 'specified person' (who will normally be a project manager, architect, or other administrator), or by the payee itself. If the payer intends to pay less than the notified sum, then it is obliged to set this out in a revision to the payment notice by the latest date for serving such a notice under the contract or, if no date is specified, the date calculated in accordance with the Scheme.

14.03 Complex though the new provisions appear, their drafting has no doubt been informed by the need to be as precise as possible in order to avoid the ambiguities that have blighted the notice provisions of the 1996 Act. The draftsmen have therefore focused on imposing obligations which are as prescriptive as possible, and on providing remedies where these are not observed. Thus:

1. The parties are required to include provisions within their contract for a payment notice to be given no later than five days after a payment-due date setting out the amount that the notifying party considers due as at the payment-due date, and the basis on which it has been calculated (s. 110A(1)–(3)).
2. The contract may specify that the party giving the notice may be either the employer, a specified person, or the contractor (s. 110A(1)).
3. If the contract fails to provide terms to this effect, then the Scheme will apply (s. 110A(5)) so that the employer gives the notice within five days after the payment-due date.[2]
4. If the contract provides for the employer or a specified person to give the notice, but that party fails to provide it, then the contractor may serve a payment notice instead. In such cases, the final date for payment is postponed by the number of days between the latest date for service of the employer's or specified person's notice and the date that the contractor serves notice in default (s. 110B(1)–(3)).
5. However, if prior to the payer or specified person serving its payment notice, the contract allows for the payee to serve a notice setting out the sum that the payee considers will become due and the basis on which it has been calculated (this will normally be the payee's payment application or invoice), then this notice may stand in place of the payment notice if the payer or specified person defaults. In such cases, there is no need to extend the final date for payment (s. 110B(4)).
6. The payer is obliged to make payment of the notified sum by the final date for payment (s. 111(1)).
7. If the payer intends to pay less then the notified sum, either the payer or the specified person must serve a further notice setting out the amount that the payer or specified

---

[1] (2007) 1 WLR 1136, [2007] UKHL 18 (HL).

[2] Para. 9 of the Scheme, as amended by the 2011 Regulations.

person considers to be due on the date that the notice is served, and the basis on which it has been calculated (s. 111(3) and (4)).

8. The notice setting out the payer's intention to pay less than the notified sum must be served no later than the prescribed period (as stated in the contract or, where none, the seven days specified by the Scheme) before the final date for payment (s. 111(5) and (7)).
9. Where the payer has complied with the requirements set out in 7 and 8 above, the payer is obliged only to pay the amounts set out in the revised notice, and not the original payment notice (s. 111(6)).

14.04 Figure 14.1 sets out the process under the 2009 Act.

14.05 As well as providing a process for the notification and payment of the notified sum, the new provisions have sought to clarify some previous areas of uncertainty and to avoid unnecessary duplication. These areas are considered below.

Payment Certificate as Notice of Payment

14.06 Section 110(2) of the 1996 Act provided that only a party to the contract could serve notice of the payment made or proposed to be made, meaning that a notice served by a third party such as the contract administrator or project manager might not comply unless the contract expressly so provided.[3] Whilst judges have treated a third party's certificate as determining the amount due (*Rupert Morgan Building Services (LLC) Ltd v David Jervis and Harriet Jervis* (2003)),[4] thereby overlooking the requirement for a separate payment notice, the 2009 Act avoids potential duplication by extending the ability to serve payment notices to persons who are named in or identifiable from the contract.[5]

14.07 It appears that the contract need not specifically identify the payment certificate as the payment notice for the purposes of s. 110A(1). Instead, provided that the contractual requirements for the provision of a third party certificate comply with s. 110A(1) and (2) as to timing and content, and provided that the certificate is in fact given in accordance with these requirements, then it seems that the certified amount will be the notified sum for the purposes of requiring the payer to make payment.

14.08 There would therefore appear to be some scope for confusion where a contract provides that in addition to a project manager or contract administrator issuing a certificate, the payer also issues a payment notice complying with s. 110A. If the payment certificate is issued but the payer fails to serve a payment notice, then it would appear that the requirements of s. 110A(1) and (2) would still have been complied with and it will not be open to the payee to serve a payment notice in default.[6]

Contractor May Serve Payment Notice

14.09 A further innovation of s. 110A of the 2009 Act is to allow the parties to agree that the payment notice can be served by the payee instead. Where a payment notice is served by the payee then the amount specified in the notice is the amount that the payer must pay unless it issues a pay-less notice complying with s. 111(3) within the requisite period. If it does not, then the payer retains its right to challenge the amount notified by the payee, by adjudication

---

[3] See further 13.11.

[4] [2004] 1 WLR 1867, [2003] EWCA Civ 1563 (CA).

[5] s. 110A(1)(a) and (2)(b).

[6] It is notable that the JCT's amendments to the Standard Building Contract now remove the requirement for the employer to serve a separate payment notice. See Table 14.1 below.

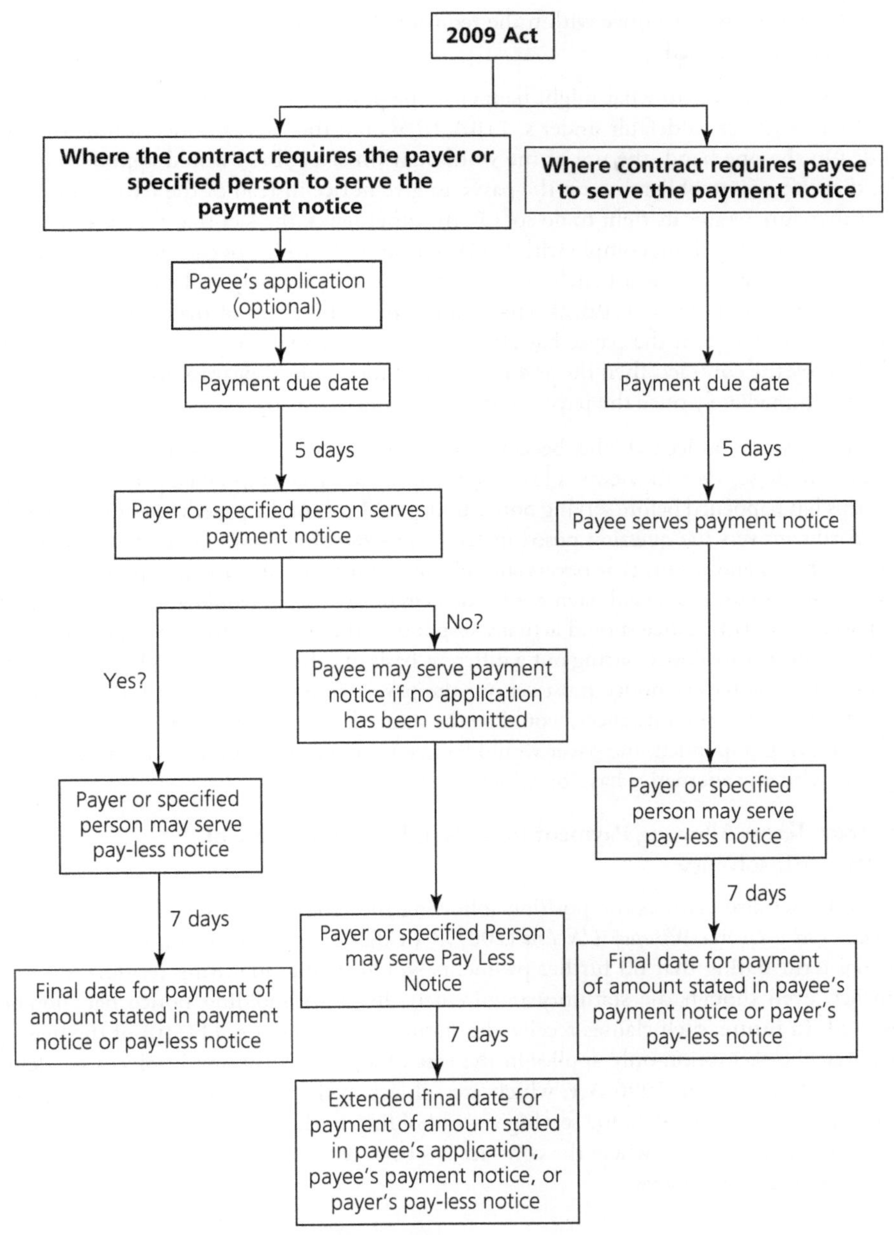

Figure 14.1 Flowchart of Notice Requirements

proceedings if necessary, provided that it pays the notified sum by the final date for payment should the adjudication decision not have been reached by this date. Any later adjustments may be made in a subsequent certificate.

14.10 Section 110B of the 2009 Act provides that, where the contract requires the payer or specified person to serve the payment notice then, if it fails to do so, either the payee's application stands as notice of the payment to be made by the final date for payment or, if there is no payment application, the payee may serve notice in default. In both cases the payer retains its right to serve a pay-less notice within the requisite time, and in the latter case the final date for payment is extended.

14.11 The question arises as to what might happen if the payer serves its notice late but before the payee serves notice in default under s. 110B(2)? Would the payer's notice still be treated as valid? On the one hand, the use of the word 'may' in s. 110B(2) would appear to indicate that that there is no obligation on the payee to give notice instead of the payer and that the payer therefore retains its right to do so. On the other hand, the fact that the payer's notice is late means that it will not comply with s. 110A(1) and (2), which default may not be capable of remedy by late service. In such circumstances, the payee may still be entitled to serve a counter-notice under s. 110B(2) which might have the effect of overturning the payer's notice. Furthermore, if the payee has already served a payment application or invoice provided for by the contract, then the amount applied for would appear to become the 'notified amount' immediately once the payer's notice is overdue.

14.12 If the payee serves notice (whether because this is required by the contract or because the payer has failed to do so) then the new s. 111(5)(b) requires the payer or specified person to wait until after this has happened before serving notice under s. 111(3) of the payer's intention to pay less. This partly answers the question posed in *Strathmore Building Services Ltd v Colin Scott Grieg* (13.117 above) and means that payers should not in future be able to rely upon earlier correspondence as a reason to avoid payment in such circumstances. It would also appear likely that the payer's s. 111(3) notice should actually respond to the payee's notice or application rather than provide a cross-notice setting out a different basis of calculation. This is because s. 111(3) requires that the payer's notice must set out the payer's intention to *pay less than the notified sum* which, in such circumstances, would be the sum calculated in accordance with the payee's notice. Having responded, the payer would be free to set out the sum that it considers is due, together with any alternative basis on which it was calculated.[7]

Contract Terms Allowing Payment to Be Withheld in the Case of the Payee's Insolvency

14.13 The 2009 Act also clarifies the position following the House of Lords' decision in *Melville Dundas Ltd v George Wimpey UK Ltd* (2007).[8] In that case, their Lordships upheld a contractual term stating that no further payments would be due following termination of the contract, even though the statutory requirement for a withholding notice had not been observed. In future, such clauses receive the statutory backing of s. 111(10) of the new Act. However, the exemption only applies in the case of a payee becoming insolvent (as defined in s. 113(2)–(5) of the 1996 Act, which are subject to the limitations set out in Chapter 16 below) after the last date for serving a withholding notice has passed, and has no wider application in other cases where the contract has been terminated. The issues relating to this are discussed at13.68 above.

(3) The 2009 Act and Standard Contract Amendments

14.14 Most of the main providers of standard form contracts have issued updates to coincide with the date that the 2009 Act became effective (1 October 2011). The exception is the Institution of Civil Engineers who, from August 2011, disassociated themselves from the ICE Form of Contract (7th edn), on the basis that they will be promoting the NEC form in future. It follows

[7] s. 111(4).

[8] [2007] 1 WLR 1136, [2007] UKHL 18 (HL).

that the ICE form may still be used for contracts entered into after 1 October 2011, but that bespoke amendments must be made to ensure compliance with the 2009 Act.

14.15 Equally, it is vital that clients using the published forms ensure that their precedents are updated to reflect the changes. Since only the JCT has taken the opportunity to reissue its contracts as new editions, it might be easy to overlook the published amendments. This might result in the contract being non-compliant with the 2009 Act and the Scheme being applied.

14.16 The different approaches taken by the institutions to amending their standard forms is interesting: the JCT, IChemE, and PPC have issued extensive amendments to deal not only with the new terminology, but also with the fall-back position of the contractor issuing the payment notice. The requirement for a payment notice to be issued by the employer under the JCT SBC 2005 has been dropped and the architect/contract administrator's certificate becomes the notified sum (provided that it is issued). The NEC on the other hand appears to have taken the view that the fall-back position applies automatically and there is no need to include extensive provisions simply to repeat the requirements of the law.

14.17 The way in which the various standard forms have been amended to deal with the new notice provisions is set out and contrasted below. For amendments to the dates for payment, see Table 12.3 in Chapter 12.

Table 14.1 Standard Contract Amendments to deal with 2009 Act changes to notice requirements

	Assessing the Notified Sum	Contractor's Notice in Default	The Pay-less Notice
NEC 3	The project manager assesses the amount due at each assessment date and certifies payment within one week (clause 50.1 and 51.1—unamended). The project manager's certificate is the notice of payment specifying the notified sum (clause Y2.2—slightly amended). No distinction is made between interim and final payments under NEC.	No provision made for the contractor's notice in default. The contractor's payment application is optional under NEC3.[9] However, provided that there is one, and that it contains the information required by s. 110(B)(4), then this should become the notified sum by operation of the 2009 Act. Otherwise, the contractor must serve a further payment notice after the default and the final date for payment should be deferred as required by ss. 110(B)(2) and (3).	Seven days before the final date for payment, as before. The pay-less notice sets out the amount considered to be due and the basis on which that sum is calculated (clause Y2.3 amended).
JCT 2011 Standard Building Contract	**Interim payments** The architect/contract administrator issues interim certificates no later than five days after the due date (clause 4.10.1). Subject to any pay-less notice given by the employer, the amount stated in the interim	**Interim and final payments** The contractor's payment application, if issued, is an 'interim payment notice' which, if the architect/contract administrator fails to give a certificate within five days of the payment-due date, becomes the sum to be paid by	**Interim and final payments** Five days before the final date for payment, as before (clause 4.12.5). The pay-less notice sets out the amount considered to be

(Continued)

[9] See clause 50.4.

Table 14.1 *Continued*

	Assessing the Notified Sum	Contractor's Notice in Default	The Pay-less Notice
	certificate is the sum to be paid by the employer (clause 4.12.2). Note that the JCT does not use the term 'notified sum', but it is likely that the terms are sufficiently clear so that the amount in the certificate is the notified sum. **Final payments** The architect/contract administrator issues the final certificate not later than two months after certain events have occurred (clause 4.15.1). The final payment is the amount shown in the final certificate as due to or from the contractor (clause 4.15.2).	the employer. If no payment application is made, then the contractor may issue an 'interim payment notice' at any time after the five-day period and the final date for payment shall be extended accordingly (clauses 4.11.2 and 4.12.4). **Final payments** The contractor can issue a final payment notice if the final certificate is not issued within the two-month period. The final date for payment is adjusted accordingly and the amount must be paid unless the employer issues a pay-less notice (clause 4.15.6).	due and the basis on which that sum is calculated (clause 4.14.1). **Final payments** The pay-less notice may be served by the employer or the contractor depending on which party is required to make the final payment (clause 4.15.4).
JCT 2011 Design and Build Contract	**Interim payments** The contractor makes an interim application on the dates set out in the contract particulars (clause 4.8.1). Not later than five days after the due date, the employer gives written notice of payment (clause 4.10.3). **Final payments** The contractor issues a final statement following practical completion (clause 4.12.1). If the contractor does not submit a final statement within three months of practical completion, then the employer may do so (clause 4.12.3). The party by whom the final payment is payable must give written notice of payment within five days after the due date (clause 4.12.7).	**Interim payments** If no payment notice is given within five days, then the contractor's interim application becomes the sum to be paid by the employer (clause 4.9.3). **Final payments** If no payment notice is given then the balance stated in the final statement is the amount to be paid (clause 4.12.7).	**Interim and final payments** Five days before the final date for payment, as before (clause 4.9.4). The pay-less notice sets out the amount considered to be due and the basis on which that sum is calculated (clause 4.10.2). **Final payments** The pay-less notice may be served by the employer or the contractor depending on which party is required to make the final payment (clause 4.12.8). Except to the extent that notice is given by either party to the other disputing anything in the final statement before the due date, the statement shall, upon the due date, become conclusive as to the sums due (clause 4.12.6).

	Assessing the Notified Sum	Contractor's Notice in Default	The Pay-less Notice
JCT 2011 Standard Building Sub-contract	**Interim payments** The subcontractor may make a payment application not later than seven days prior to the due date. Not later than five days after the due date, the contractor gives a payment notice specifying the sum to be paid (clauses 4.9.3 and 4.10.2). **Final payments** Not later than five days after the due date, the contractor sends the final payment notice to the subcontractor. The final payment notice shall show the final contract sum less the total amount previously due as interim payments (clause 4.12.2).	**Interim and final payments** If the contractor fails to issue a payment notice within five days of the payment-due date, the subcontractor's payment application shall become the sum to be paid by the contractor (clause 4.10.3). If the subcontractor has not issued a payment application, before the due date, then he may do so at any time after the contractor fails to issue the payment notice and the final date for payment is extended (clause 4.9.3). **Final payments** If the final payment notice is not issued by the contractor, then the subcontractor may issue a default payment notice and the final date for payment is extended (clause 4.12.6).	**Interim and final payments** Not later than five days before the final date for payment (clause 4.10.5). The pay-less notice sets out the amount considered to be due and the basis on which that sum is calculated (clause 4.10.8). **Final payments** The pay-less notice may be served by the contractor or the subcontractor depending on which party is required to make the final payment (clause 4.10.8).
PPC2000	**Interim payments** The constructor submits an application to the client and the client's representative at the intervals stated in the price framework and any pre-construction agreement or (if no intervals are stated) at the end of each calendar month (clause 20.2). The client's representative issues a payment notice within five days thereafter (clause 20.3). **Final payments** The client's representative issues a final account within 20 working days after notice of rectification of defects. On or after 40 days after the notice, the payee issues an application in the agreed amount or a different amount. The payer issues a payment notice within five days thereafter (clause 20.16).	**Interim payments** If the client's representative or the client does not issue a payment notice within five days, the constructor's application for payment shall be treated as the payment notice (clause 20.6). **Final payments** If a payment notice is not issued within five days of the due date for payment, the application for payment shall be treated as the payment notice (clause 20.16(iii)).	**Interim and final payments** Not later than two working days before the final date for payment. The pay-less notice may be served by either party depending on which party is required to make the payment (clause 20.7). The pay-less notice sets out the amount considered by the payer to be due on the date the notice is served and the basis on which that sum is calculated.

(Continued)

Table 14.1 *Continued*

	Assessing the Notified Sum	Contractor's Notice in Default	The Pay-less Notice
I/Chem E (Red Book)	The project manager certifies payment within 14 days of receiving the contractor's interim request for payment (56 days in the case of the final payment) (clause 39.4). The amount certified is the amount due for payment (clause 39.5). Note that, like the JCT, the IChemE does not use the term 'notified sum' although it appears sufficiently clear that this is what is intended.	If no certificate is issued within 14 days of receiving the contractor's interim request for payment (or within 56 days for the final payment), the contractor's request for payment becomes due for payment by the final date for payment (clause 39.6).	One day before the final date for payment. The pay-less notice sets out the amount the purchaser considers to be due on the date the notice is given and the basis on which that sum is calculated (clause 39.7).

15

SECTION 112: SUSPENDING PERFORMANCE

(1) Overview

Introduction and Summary

15.01 This chapter considers s. 112, which creates a statutory right for a payee to suspend performance for non-payment. It examines the conditions which must apply for the right to arise, and looks at the consequences of both a proper and improper exercise of the right. The key issues relating to suspending performance are:

1. The right to suspend arises when the sum due or notified sum has not been paid in full by the final date for payment.
2. The payee must give at least seven days' notice of its intention to suspend.
3. Where the right to suspend is operated correctly, payees may be able to apply for loss and expense directly incurred as a result of the suspension as damages for breach of contract. Under the Housing Grants Construction and Regeneration Act 1996 ('the 1996 Act'), the contractor's right to an extension of time is limited to the actual period of suspension. The Local Democracy, Economic Development and Construction Act 2009 ('the 2009 Act') has attempted to clarify the contractor's rights to claim loss and expense and an extension of time.
4. If a payee exercises its suspension rights incorrectly, this may amount to a repudiatory breach of contract entitling the payer to terminate. However, this will depend upon the nature of the breach, and the facts and circumstances of the case.

(2) The Statutory Right to Suspend Performance

The Introduction of a Statutory Right to Suspend Performance

15.02 Section 112 of the 1996 Act created a statutory right for a contractor or subcontractor (a payee) to suspend performance of a contract for non-payment. Before the 1996 Act came into force, the common law did not recognize a remedy of withholding contractual performance whilst keeping the contract alive.[1] A payee who suspended or deliberately slowed down work in response to non-payment ran the risk of being found guilty of a breach of a contractual obligation to proceed regularly and diligently with the works.[2] Whilst delay to completion is not normally treated on its own as repudiatory, especially where the contract allows for liquidated damages,[3] there was a risk that suspension might be seen as sufficiently serious to be judged repudiatory. Unsurprisingly, few payees took this risk.

15.03 Notwithstanding the statutory backing now given to acts of suspension, there have been very few reported instances of payees suspending work under the provisions of the 1996 Act. Respondents to a consultation conducted by the Department of Trade and Industry in June 2007[4] indicated that the right was being exercised in fewer than 1 in 100 cases of non-payment.

15.04 The mere availability of the remedy is likely to have deterred most payers from failing to make timely payment. This, combined with the relatively well-understood payment provisions in ss. 109 and 110(1) and the statutory right to adjudicate, has no doubt contributed to the lack of reported decisions concerning suspension. However, the lack of clarity in the 1996 Act suspension provisions (discussed below) and the wish to avoid a reputation as a (sub)contractor who suspended works, may have also contributed to an under utilization of this provision.

Supplemental to Other Rights

15.05 Section 112(1) provides that the right to suspend work is 'without prejudice to any other rights or remedy'. Thus interest remains payable on the late payment, either under the contract or pursuant to the Late Payment of Commercial Debts (Interest) Act 1998. In certain cases, an unpaid payee may also be entitled to treat the contract as repudiated and terminate its performance of it, provided that the amount or manner of the underpayment is such as to indicate that the payer is unwilling or unable to meet its contractual obligations.[5]

15.06 The right to suspend is therefore supplemental to these rights, and because it is derived from statute, it will override any term of the contract which purports to exclude or limit its application. Accordingly, any contractual provision which purports to alter the statutory conditions under which the payee may suspend, such as the introduction of longer notice periods or a 'cooling off' period after the final date for payment during which the payee cannot suspend, is likely to be ineffective. However, agreed conditions which merely seek to deter the payee from exercising its statutory right to suspend, such as making the payee liable for the costs of de-mobilization and re-mobilization, may be valid under the 1996 Act.

---

[1] *Channel Tunnel Group Ltd v Balfour Beatty Construction Ltd* [1993] 2 WLR 262 (HL).

[2] *Canterbury Pipe Lines Ltd v Christchurch Drainage Board* (1979) 16 BLR 76 (CA); *Lubenham Fidelities and Investments Co. Ltd v South Pembrokeshire District Council* (1986) 33 BLR 39, 6 Con LR 85 (CA).

[3] See *Felton v Wharrie* (1906) 2 Hudson's BC (4th edn) 398, (10th edn) 344, 613, 709 (CA).

[4] 'Improving Payment Practices in the Construction Industry: 2nd Consultation on proposals to amend Part II of the Housing Grants, Construction and Regeneration Act 1996 and the Scheme for Construction Contract (England and Wales) Regulations 1998', DTI, June 2007.

[5] *Moschi v Lep Air Services Ltd* (1972) 2 WLR 1175 (HL); *D. R. Bradley (Cable Jointing) Ltd v Jefco Mechanical Services Ltd* (1988) 6-CLD-07–19. See further 12.26–12.27 regarding the circumstances when late payment may be repudiatory.

The 2009 Act Amendments

15.07 The right to suspend under the 2009 Act has been broadened in order to deal with the issues discussed above. The changes are discussed in the relevant sections below.

(3) Suspending Performance

What Is Required for the Right to Suspend to Arise?

15.08 The conditions necessary for exercising the right to suspend are set out in s. 112(1)–(4) of the 1996 or 2009 Act and may be summarized as follows:

1. A payment has become due.
2. No effective withholding or pay-less notice has been served.
3. The final date for payment has passed.
4. Payment has not been made in full.
5. The payee has given a valid seven-day notice of intention to suspend.

Each of the above conditions is examined in turn in the paragraphs below.

15.09 If in doubt as to whether any of these requirements have been met, a payee may decide to seek a declaration from an adjudicator or the court under CPR Part 8 before suspending performance. There would appear to be no reason why such an application could not be sought during the notice period for suspension, allowing the payee to serve a suspension notice first and stop work once the adjudicator or court has given its decision, if full payment has not been made in the meantime.

A Payment Has Become Due

15.10 Section 112(1) of the 1996 Act provides that the right to suspend arises '[w]here a sum due under a construction contract is not paid in full'. However, the question of what sum is due is not always straightforward. See Chapter 13 for a discussion of the issues and potential difficulties with how to ascertain the sum due.

15.11 For contracts entered into after 1 October 2011, s. 112(1) is now activated if the payer has not paid the 'notified sum'. In theory, this should be more straightforward as the notified sum will be either the sum set out in the payer's or specified person's payment or pay-less notice, or else in a payment notice or application provided by the payee. However, these new provisions will be subject to the inevitable teething problems as the industry comes to grips with the new payment regime and four possible sources for the 'notified sum'.

15.12 Whether under the 1996 Act or the 2009 Act, the right to suspend does not arise where the payee simply disagrees with the amount that has been certified or is included in the payment notice. In such cases, the payee will need to bring adjudication or other proceedings to have the amount adjusted.

No Effective Withholding or Pay-less Notice

15.13 The requirements for a withholding or pay-less notice to be effective are set out at 13.87 and following, and call for the notice to be served within the contractual time limit (or that statutorily implied by the Scheme for Construction Contracts ('the Scheme')) and to contain certain information depending on whether the 1996 or 2009 Act applies.

15.14 If the payee considers a withholding or pay-less notice to be ineffective, either because it is late or because it is insufficiently particularized, then the payee may be entitled to suspend performance. Clearly in such cases the payee is running a risk since, if wrong, it may be held to be in breach of contract (possibly repudiatory breach) and liable for damages by suspending.[6]

The Final Date for Payment Has Passed

15.15 The right to suspend arises under both the 1996 and 2009 Acts if the relevant sum has not been paid in full by 'the final date for payment', which is discussed at 12.26–12.27.

15.16 This suggests that the right to suspend starts from 00.01 hours on the day following the final date for payment. This is important since, if the payee suspends work too soon, it loses the statutory protection of s. 112 and may be in breach of contract. However, since the suspension must in any event be preceded by seven days' notice, any doubt should be capable of being resolved by stating clearly in the notice when the payee considers the right to suspend to have arisen and when it will be exercised.

Payment Not Made in Full

15.17 Provided that a sum due or notified sum can be established, then it should be a simple matter to determine whether or not payment has been made in full. The question of whether a contractor could suspend, and eventually determine, the contract for non-payment of VAT due on a payment ordered by an adjudicator arose in *Leading Rule v Phoenix Interiors Ltd* (2007);[7] however the court declined to answer the issue. Therefore there is no judicial guidance on this point.

A Valid Notice of Intention to Suspend Performance

15.18 Section 112(2) provides that the right to suspend may not be exercised without first giving the party in default 'at least seven days' notice' stating the ground(s) on which it is intended to suspend performance. The drafting of this subsection is imprecise and neither expressly excludes the parties from agreeing a longer notice period nor requires that the final date for payment should first have passed before the payee gives notice.

At Least Seven Days' Notice

15.19 The requirement for 'at least' seven days' notice has been interpreted by some payers as allowing the parties to stipulate a longer notice period in the contract. A report in 2000 of the Construction Liaison Group found that a longer period had been specified in 20 per cent of cases, and in 25 per cent of these the contracts stipulated a notice period in excess of 21 days.[8]

15.20 However, it is notable that s. 112, unlike s. 111(3) for example, does not expressly state that the parties are free to agree an alternative notice period. It is therefore possible that the intention of Parliament was that the seven days' notice period should not be interfered with and that the words 'at least' are intended simply to imply a discretion for the payee to extend the notice period if it appears that payment might be forthcoming.

15.21 The question was almost put to the test in *Leading Rule v Phoenix Interiors Ltd* (2007),[9] in which the lay adjudicator had decided that a contractual notice period of 28 days was valid and that Phoenix had suspended its performance of the work too early. However, Akenhead J declined to rule on that matter and the case proceeded no further in the courts.

[6] As the adjudicator found in *KNS Industrial Services (Birmingham) Ltd v Sindall Ltd* (2001) 75 Con LR 71, [2001] Const LJ 170 (TCC).

[7] [2007] EWHC 2293 (TCC).

[8] Construction Liaison Group Report on the Operation of the Act, June 2000.

[9] [2007] EWHC 2293 (TCC).

15.22 It should be noted that s. 116 provides that Christmas Day, Good Friday, and bank holidays should be excluded for the purpose of calculating time periods under the Act.[10] These days do not count, therefore, as part of the seven days' notice period.

Anticipatory Notice

15.23 The decision in *All in One Building & Refurbishments Ltd v Makers UK Ltd* (2005)[11] suggests that a dispute crystallizes as soon as it is clear that the payer does not intend to fulfil its obligations to make payment on the date required. Therefore, in theory, a payee could give an 'anticipatory' notice before the final date for payment if it considers that payment is unlikely to be forthcoming, and suspend performance as soon as both the final date for payment has passed and the notice period expired. This has not been specifically tested by the courts.

Contents of the Notice

15.24 The question of what information is necessary in the notice was considered in the case of *Palmers Ltd v ABB Power Construction Ltd* (1999)[12] although, once again, the judge in that case decided that the matter was not one on which he should give judgment. This case indicates, however, that the courts will treat very seriously any mistake on the face of the notice which might undermine the grounds upon which the right to suspend is claimed. It perhaps underlines that suspension is still viewed as a drastic remedy despite its validation by the 1996 Act, and that the notice needs to be unambiguous for the payer to avoid the consequences of any perceived inaction.

When to Suspend Performance?

15.25 Once the right to suspend has arisen and the notice period has expired, it appears that the payee may suspend performance immediately, or wait for a further period before suspending. The right continues to apply until payment of the full amount is made,[13] and therefore it is likely that the payee will retain the right to suspend even if it does not exercise it immediately.

What to Suspend?

The 1996 Act: Suspending Performance of its Obligations

15.26 Section 112(1) of the 1996 Act permits an unpaid party to 'suspend performance of his obligations under the contract'. However, it does not actually prescribe what this means in practice. Is it required to cease all work being performed, or can it keep certain services operating during the period of suspension? Can it operate a 'go slow' or perhaps cease cooperating in terms of attending meetings with the payer or ceasing to perform difficult items such as achieving planning permission? Is it possible for a payee to suspend its obligation to provide a parent company guarantee or collateral warranty?

15.27 In the absence of express wording in the contract, the payee may be free to do any of these things provided that they can properly be termed a 'suspension' of its obligations under the contract. Indeed, the payee may be within its rights to take the wording of s. 112(1) at face value and suspend all operations with immediate effect, including removing security guards and leaving the works in the condition they were in at the moment the notice period expired. The potential cost consequences of such a decision are discussed below.

15.28 Some adjudicators or courts may distinguish 'obligations under the contract' from obligations which arise under statute, such as Health and Safety laws or CDM Regulations. It is thought

10 See s. 116(3) and Ch. 17.
11 (2005) CILL 2321, [2005] EWHC 2943 (TCC). See Ch. 12 at 12.24.
12 [1999] BLR 426, 68 Con LR 52 (TCC).
13 See s. 112(3).

that a payee should exercise caution before suspending such statutorily imposed obligations, particularly without warning the payer of such an intention.

The 2009 Act: Suspending Performance of Any or All of its Obligations

15.29 The position may be different for contracts entered into after 1 October 2011 when the 2009 Act came into force. Amendments to s. 112(1) provide an express right for the contractor to suspend 'any or all' of its obligations under the contract which would appear to avoid the doubts set out above by expressly allowing a partial suspension.

15.30 Having suspended performance, the employer is liable to pay the contractor a reasonable amount in respect of costs and losses reasonably incurred as a result of the exercise of its right (s. 112(3A). It is important to note, however, that there is no express obligation of reasonableness inserted in the amendments to s. 112(1). Accordingly, it is suggested that there is no requirement that the act of suspension itself be subject to any test of reasonabless—it is an absolute right in the event of non-payment (subject to the necessary steps desicribed above being taken).

When Does the Right to Suspend Cease?

15.31 Section 112(3) states that the right to suspend performance 'ceases when the party in default makes payment in full' of the amount due. Therefore, if the payee fails to resume performance as soon as payment is made, it loses the protection of the Act (whether 1996 or 2009) and may be in breach of contract. The requirement for immediate re-mobilization remains untested however, and it may be the case that the courts, recognizing the difficulties involved in reassigning staff, subcontractors, and equipment quickly to the site, would allow some leeway to prevent the payer profiting from its own breach.

(4) The Consequences of Suspension

The Consequences of a Wrongful Suspension

15.32 Where any of the required elements discussed above are not in place, a suspension may be deemed invalid or unlawful. This may in turn put the payee purporting to exercise its statutory right to suspend into repudiatory breach of contract.[14] However, a bona fide but mistaken reliance on an express provision of a contract to suspend performance is not, alone, to be treated as a repudiatory breach—the suspension will be viewed in light of all the facts and circumstances of the case: *Mayhaven Healthcare Ltd v Bothma and Bothma* (2009) (Key Case).[15]

15.33 It is notable that the 2009 Act extends the right to claim costs and expenses only to the payee, and not the payer. This means that the payer has no statutory protection against the risk of wrongful suspension. However, the payer may revert to contractual or common law remedies to obtain compensation for the breach, including accepting what he considers to be a repudiatory breach of contract by termination.

The Consequences of a Valid Suspension

Entitlement to Loss and Expense

15.34 The 1996 Act does not expressly provide for compensation for the loss or expense incurred during the period of suspension. Such costs might include:

---

[14] *KNS Industrial Services (Birmingham) Ltd v Sindall Ltd* (2001) 75 Con LR 71, [2001] Const LJ 170 (TCC).

[15] [2010] BLR 154, [2009] EWHC 2634 (TCC).

1. the costs of standing or off-hiring equipment
2. costs of reassigning or laying off staff
3. the consequential costs of re-mobilization
4. compensation to subcontractors for periods of inactivity
5. the possible risk of dilapidations or defects resulting from works being left unfinished.

15.35 Whilst some standard-form contracts deal with this situation by providing that the payee is entitled to its additional costs *as a result of suspending works* under the Act,[16] many contracts do not.[17] As a result, whilst the payee might recover its payment due before the suspension, it might still face having to bring an action for breach of contract against the payer to recover its additional costs, with no certainty of recovery. Such costs should be recoverable as general damages for breach of the payment obligations in the contract, provided there are no provisions to the contrary such as an exclusive remedies clause. One specific cost is worth mentioning: if the payee suspends security to the site, and damage occurs to the works during the period of suspension, then under the 1996 Act, no claim would lie against the contractor for that damage as the contractor was under no obligation to the employer at the time the damage occurred. Although the contractor may be contractually obliged to rectify defects in the works, it should be able to claim the costs of doing so as damages for the payer's breach of contract.

15.36 The 2009 Act seeks to clarify s. 112, and ensures that the payee now has a statutory right to claim loss and expense which should apply even in cases where the contract contains an exclusive remedies clause. However, the obligation under the new s. 112(3A) extends only so far as 'a reasonable amount in respect of costs and expenses reasonably incurred as a result of the exercise of the right'. Whether the amendments have the effect of widening the recovery or whether they in fact limit the rights that the payee otherwise had at common law to claim loss and expense is yet to be seen.

15.37 In order to avoid a future charge that costs have been unreasonably incurred, it is suggested that payees should set out in their notice of intention to suspend the steps that they intend to take in furtherance of that right. If these include suspending the whole of their operations, then the notice should preferably state this. It would also be prudent to check the terms of the contract to see what provision is made for the care of the work and to keep these issues under review throughout the period of suspension. At least if the payer is placed on warning of the consequences of its breach, it may not later complain that it had no opportunity to arrange replacement security or take steps to prevent damage from occurring to the partially constructed works.

Entitlement to Extension of Time

15.38 Section 112(4) of the 1996 Act provides that for the purposes of calculating time for the completion of the works, '[a]ny period during which performance is suspended in pursuance of the right conferred by this section' is to be disregarded. This appears to mean that the payee is entitled to an extension of time equal to the period of suspension. It should be noted that the wording of this section does not allow the payee to claim the *actual* delay to the works, which may of course be much greater than the period of suspension itself. However, the contract administrator may be permitted by the terms of the contract to give an extension which takes into account the consequences of disruption if they exceed the period of the suspension itself. In addition, it is possible that a court might construe this period as including such matters as disruption to programme and de- and re-mobilization to prevent the payer profiting from its breach.

[16] See e.g. NEC 3, Option Y(UK)2; JCT Standard Building Contract 2005, Clause 4.24.3; JCT Design and Build Contact 2005, Clause 4.20.3; IChemE Red Book 4th edn (2001), Clause 39.8.

[17] See e.g. GC/Works/1 with Quantities (1998); ICE Conditions of Contract, 7th edn.

15.39 The amendments to s. 112 introduced by the 2009 Act now cover 'any period during which performance is suspended in pursuance of *or in consequence of* the exercise of the right conferred'. Whilst this amendment is clearly intended to permit the payee to claim periods of de- and re-mobilization as well as the actual period of suspension, some consider it doubtful that it has made the situation any clearer than before. In particular, it may be unclear during a period of re-mobilization at what stage the works might be said to have restarted for the purposes of assessing the period of suspension. It is perhaps unfortunate that the drafting of the 2009 Act did not take the opportunity to provide that it is 'the delay to the works consequent on the suspension' which determines the extension of time to which the payee is entitled. This would at least have avoided issues as to when the works have been fully recommenced and would have allowed the payee an extension of time in respect of any consequential disruption. However, as discussed above, it is possible that the payee may be able to claim for such consequences in any event.

15.40 Particular problems may arise where the suspension occurs after the contract completion date, and thus at a time when the payee is already in culpable delay. In such cases, depending on the terms of the contract, it seems that an extension of time may still be granted[18] and that the appropriate length of the extension remains the length of time that works are actually suspended whether before or after the contract completion date. In other words, the suspension should not deprive the payer from claiming liquidated damages in respect of the periods both before and after the payee's suspension.

15.41 Many standard forms contain particular clauses prescribing how events which are not the payee's responsibility will be assessed and it may be difficult in practice to reconcile the statutory entitlement with the contractual entitlement. The period following payment of the outstanding debt is likely to be fraught with difficulties in relation to agreeing the length of extension and loss and expense to which the payee is entitled. However, there is no statutory right for the payee to prolong the suspension whilst these matters are sorted out. The payee must decide either to return to work or maintain that the payer's continued intransigence in matters of payment amounts to a repudiation and terminate the contract accordingly.

Key Case: Suspension

Mayhaven Healthcare Ltd v David Bothma and Teresa Bothma (t/a DAB Builders) (2010) BLR 154, [2009] EWHC 2634 (TCC)

15.42 **Facts:** An adjudicator directed that certain sums should be paid to DAB under valuation number 9. A month later, DAB issued a notice to suspend work on the understanding that the amount had not been paid and would not be paid. In a letter from Mayhaven's solicitors it was stated that Mayhaven 'will not be making any payment to your client pursuant to the Decision'. In fact, by the time DAB suspended performance, the amount had been paid under subsequent valuation numbers 10 and 11. Mayhaven's solicitors wrote to DAB stating that the suspension was wrongful and constituted a repudiatory breach of contract which Mayhaven accepted by that letter. The matter went to arbitration where the arbitrator decided that DAB's actions did not amount to repudiatory breach. Mayhaven appealed on five points of law, including whether the wrongful suspension amounted to a repudiatory breach of contract.

[18] *Balfour Beatty Building Ltd v Chestermount Properties Ltd* [1993] 62 BLR 1, [1993] CILL 821 at paras. 28–34.

Held: The question was not capable of a simple answer as a matter of principle. The answer to the question would depend on the terms of the contract, the breach or breaches of contract, and all of the facts and circumstances of the case. In this case the arbitrator had been right to take into account the content of letters, including one which showed that DAB intended to return to site provided that they were paid the balance of valuation number 9: **15.43**

23. I do not accept Mr Pennicott's submission that a wrongful suspension of the work under Clause 4.4A of the Contract which gives rise to a failure to proceed regularly and diligently under Clause 2.1 amounts to a breach of a Condition or fundamental term so that every such breach amounts to a repudiation of the Contract. A wrongful suspension which gives rise to a failure to proceed regularly and diligently will vary in seriousness, depending on the circumstances. I do not accept that every wrongful suspension which leads to a breach of Clause 2.1 will automatically be a repudiatory breach. Rather, whether such a suspension and a consequent breach does amount to a repudiation depends on the breach and the facts and circumstances of the case.

24. Mr Pennicott refers to a passage in Keating on Construction Contracts (8th Edn.) at paragraph 6–070 where it states:

> '***Refusal or Abandonment.*** *An absolute refusal to carry out the work or an abandonment of the work before it is substantially completed, without any lawful excuse, is a repudiation.*'

25. As the cases cited in support of that proposition show, whether there had been a repudiatory breach will depend in each case on the breach and the facts and circumstances of the case. In this case there was no absolute refusal or abandonment of the type referred to in those cases.

26. In this case, can it be inferred from the way in which the Arbitrator applied the law to facts that he misunderstood the law? In my judgment it cannot. As Lord Wilberforce said in *Woodar v Wimpey* a party who bonafide relies on an express provision of the contract, in the present case to suspend performance, is not by that fact alone to be treated as having repudiated his contractual obligations if he turns out to be mistaken in his rights. Rather, that is one factor. The suspension must be viewed in the light of all the facts and circumstances of the case.

Table 15.1 Table of Cases: Suspension

Title	Citation	Issue
Palmers Ltd v ABB Power Construction Ltd	(1999) BLR 426, 68 Con LR 52 (TCC), Judge Thornton	**Validity of notice to suspend** Palmers' notice of intention to suspend named a final date for payment which it later appeared to agree (in evidence before the court) should have been a date some two weeks later. Despite the notice having been served after both possible dates had passed, ABB contended that the error made the notice invalid. The issue was not one for the courts to decide since the dispute resolution agreement between the parties required it to be referred back to the adjudicator.
KNS Industrial Services (Birmingham) v Sindall Ltd	(2001) 75 Con LR 71, [2001] Const LJ 170 (TCC), Judge LLoyd	**Premature suspension was repudiatory breach** The employer terminated the contract after the contractor suspended performance for alleged non-payment. An adjudicator decided that the final date for payment had not yet passed and that Sindall was entitled to set-off sums not included in its withholding notice.

(*Continued*)

Table 15.1 *Continued*

Title	Citation	Issue
		Consequently, KNS had suspended prematurely and Sindall was entitled to determine the contract. The point was not decided in court, but Judge LLoyd saw 'no significant error' in the adjudicator's reasoning and enforced the decision.
All in One Building & Refurbishments Ltd v Makers (UK) Ltd	(2005) CILL 2321, [2005] EWHC 2943 (TCC), Judge Wilcox	**Final date for payment** A dispute may arise between parties regarding payment even before the final date for payment has been passed if it is clear that the employer disputes that an entitlement to payment exists.
Leading Rule v Phoenix Interiors Ltd	[2007] EWHC 2293 (TCC), Akenhead J	**Payment not made in full, and period of notice of intention to suspend** Preliminary issues were referred to the court as to whether it was valid to stipulate a contractual period of 28 days for the notice of intention to suspend, and whether the employer's withholding of the VAT element of a payment amounted to a failure to make payment in full. Neither issue was decided since the judge considered that the ordering of preliminary issues in the case would not save time or money as required by the TCC Court Guide.
Mayhaven Healthcare Ltd v David Bothma and Teresa Bothma (t/a DAB Builders)	(2010) BLR 154, [2009] EWHC 2634 (TCC), Ramsey J	**Wrongful suspension not a repudiatory breach** Whether such a suspension and a consequent breach amounts to a repudiation depends on the breach and the facts and circumstances of the case. In this case the contractor had made a genuine mistake and had given notice of its intention to return to site once payments were made.

16

SECTIONS 113 AND 110(1): CONDITIONAL PAYMENT CLAUSES

(1) Overview

Introduction and Summary

16.01 This chapter examines the prohibition on certain types of conditional payment clauses, including pay-when-paid clauses, under the Housing Grants Construction and Regeneration Act 1996 ('the 1996 Act') and the Local Democracy, Economic Development and Construction Act 2009 ('the 2009 Act'). It identifies the types of provisions that are prohibited, and considers the interaction of ss. 113 and 110(1) of the Acts, both of which are relevant to assessing whether or not a payment provision may be deemed ineffective:

1. Section 113(1) prohibits payments being made conditional upon payment being received from a third party except in cases where the third party is insolvent as defined by s. 113(2)–(5).
2. Arrangements that make payment conditional upon some other event, such as a certificate being issued under the subcontract in question or under a superior contract may still be valid, although they may cease to be so for contracts entered into under the provisions of the 2009 Act.
3. The decision in *Midland Expressway* indicates that, even under the 1996 Act, the courts will not look favourably on such arrangements where they have *the effect* of a pay-when-paid clause.
4. Pay-when-certified arrangements may also be invalid where they provide an inadequate mechanism for determining the amounts or timing of payments or allow the paying party to profit from its own breach.
5. Pay-when-paid (and probably pay when certified) arrangements will be construed strictly according to their terms and may be subject to the reasonableness test where the Unfair Contract Terms Act 1977 applies.

6. If the contract contains an ineffective pay-when-paid clause then some or all of the payment provisions of the the Scheme for Construction Contracts ('the Scheme') will apply.

(2) What Types of Provision Are Prohibited?

Pay-when-Paid Clauses under the 1996 Act

16.02 Section 113(1) of the 1996 Act renders ineffective any contract term which makes payment conditional on the payer 'receiving payment from a third person'. This prohibition on pay-when-paid clauses applies in all situations except where the third person is insolvent, or where any other person is insolvent and payment by that other person is, under the contract (directly or indirectly), a condition of payment by that third person. Thus, in a chain of contracts the insolvency of a party up the line can mean that pay-when-paid provisions down the line are effective. The remainder of s. 113, in subclauses (2)–(5), describes what constitutes an insolvency situation for the purposes of the provision. Section 113(6) allows that, if a contract provision is rendered ineffective by sub-s. (1), the parties are free to agree other terms for payment. Failing that, the relevant terms of the Scheme apply to provide a mechanism for determining what payments are due and when (see Chapter 12).

16.03 Section 113 prohibits only those clauses which make payment conditional on the payer actually receiving payment from a third person. Clauses which made payment conditional on any other event, such as certificates being issued under contracts up the line, apparently remained valid under the 1996 Act, even though their effect might be the same in terms of delaying or frustrating payment. If a contractor was able to require that a subcontractor must wait until confirmation was received from the ultimate employer that the claim would be met, then s. 113(1) could be said to have very little protective effect.

Conditional Payment Clauses under the 1996 Act

16.04 There may have been a temptation following the enactment of the 1996 Act to view pay-when-paid clauses as a special type of clause singled out for statutory reprobation. Unsurprisingly, therefore, contract draftsmen appear to have been careful not to include clauses which made payment conditional on the receipt of payments in their contracts. However, as has been seen, pay-when-paid clauses are only one type of clause which may avoid or delay payment to a contractor or subcontractor.

16.05 One approach which has been taken by subcontractors faced with other types of conditional payment clause has been to demand that it should be treated as a restriction or exclusion clause and therefore construed strictly against the requirements of the 1996 and 2009 Acts to ensure not only that it does not offend against s. 113, but that it also provides an 'adequate mechanism' for payment as required by s. 110(1). In the cases of *Alstom Signalling Ltd v Jarvis Facilities Ltd* (2004)[1] and *Ringway Infrastructure Service Limited v Vauxhall Motors Limited* (2007),[2] the courts were asked to consider whether conditional payment clauses amounted to payment mechanisms which were 'inadequate' contrary to the requirements of s. 110(1), because payment could be frustrated by inaction of the contractor. These cases are considered in detail at 12.34–12.36 and 13.21–13.23 and relate to clauses which, in effect, made the timing of payments conditional upon certificates having been issued. In the *Alstom* case, payment dates were determined by the timing of certificates issued under a superior contract. In

---

[1] (2004) 95 Con LR, [2004] EWHC 1285 (TCC).

[2] [2007] All ER (D) 333, [2007] EWHC 2421 (TCC).

Ringway, the timing depended upon a payment notice having been issued under the contract in question. In both cases, however, the judges were willing to construe the contracts so that the payment mechanism would be adequate provided that the contractor did not breach its obligations and fail to issue or pursue certificates indefinitely.

16.06 A more robust view was taken in *Midland Expressway v Carillion Construction Ltd and ors (No. 2)* (2005) (Key Case).[3] The clause in question related to payment for varied work, and purported to limit the subcontractor's recovery to the amounts which the contractor was entitled to be paid by the Secretary of State under a corresponding change to the concession agreement. Jackson J considered the arrangement to be, in effect, the type of pay-when-paid arrangement against which Parliament had legislated. The use of the words 'entitled to be paid' rather than 'is paid' could not save it from being considered so and contracting parties could not escape the operation of s. 113 by the use of 'circumlocution'.[4] The judge gave six reasons for finding that the offending clause was caught by s. 113, one of which was the fact that the practical consequence of the clause was that the contractor would not be paid for variations unless the concessionaire had received a corresponding sum under the concession agreement. The clause was therefore deemed ineffective.

16.07 The approach taken by Jackson J was to construe the words of the contract strictly to determine whether they offended not only the provision but also the intent of the s. 113. Clauses which had the effect of preventing the subcontractor from being paid until the contractor was paid would be contrary to the Act.

16.08 A strict approach was also taken both at first instance and in the Court of Appeal in *William Hare Ltd v Shepherd Construction Ltd* (2009).[5] Here Coulson J identified that a pay-when-paid clause which operated in the event of the employer's insolvency was, in effect, a form of exclusion clause. Consequently, it was required to be construed strictly in accordance with its terms, and words should not be implied to give effect to other intentions, even where there appeared to have been a mistake. The Court of Appeal agreed:[6]

> 17. ... this clause was not truly 'sharing' the risk of insolvency, it was relieving Shepherd of a liability to pay which they otherwise had and it was for Shepherd to get a clause of this nature right if they wished to rely on it.

16.09 The implication of these latest decisions is that pay-when-paid clauses (and possibly other clauses which make payment conditional on matters outside of the subcontractor's control), will be construed strictly and treated as exclusion clauses. This would appear to allow scope for subcontractors employed both before and after 1 October 2011 to argue that conditional payment clauses do not comply strictly with the words or intent of the relevant Act, and/or should be subject to the provisions of the Unfair Contract Terms Act 1977.

Conditional Payment Clauses under UCTA

16.10 Where a subcontractor contracts on a contractor's standard terms, then the provisions of the Unfair Contract Terms Act 1977 may apply to require that clauses which exclude or restrict a party's rights are subjected to the test of reasonableness as set out in s. 3(2). As a result of the decision in *Hare v Shepherd* it would appear that pay-when-paid clauses, and possibly other types of conditional payment provisions, are now considered to be forms of exclusion clause which may be subject to UCTA.

[3] 106 Con LR 154, [2005] EWHC 2963 (TCC).
[4] At para. 71(5).
[5] [2009] BLR 447, (2009) EWHC 1603 (TCC).
[6] *William Hare Ltd v Shepherd Construction Ltd* [2010] All ER (D) 168; EWCA Civ 283 (CA).

Conditional Payment Clauses under the 2009 Act

16.11 The risk that the prohibition on pay-when-paid clauses could be rendered largely ineffective by the type of circumvention seen in *Midland Expressway* has been addressed under the 2009 Act. This introduces s. 110(1A) which prevents the amounts or timing of payments being made conditional on:

- the performance of obligations under another the contract or
- a decision by any person as to whether the obligations under another contract have been performed.

Section 110(1D) relates only to the timing, and not the amount, of payments, and renders ineffective any provision which requires the date on which the payment becomes due to be determined by the issue of a notice stating the amounts due under the contract.

16.12 In effect, s. 110(1A) and (1D) prohibits contractors from using the non-issuance of any certificate or notice under either the main contract with the employer, or under the subcontract itself, as a reason for avoiding or delaying payment to the subcontractor. Section 110(1A) also prevents matters being taken into account which affect only the performance of the contractor under the main contract as influencing the timing and amounts of payments due to the subcontractor. Such arrangements are now identified as examples which contravene the requirement of s. 110(1) of the original 1996 Act that every construction contract must contain an 'adequate mechanism' for assessing the amounts and timing of payments. This is not to say that a subcontractor could not argue that these types of arrangements provided an inadequate mechanism under contracts entered into prior to the 2009 Act. Such arguments have been made and, given the right circumstances, could be successful as discussed above.

16.13 Section 110(1B) and (1C) excludes from the ambit of s. 110(1A) pay-when-paid arrangements since these are already covered by s. 113, and certain types of management contracts because these are necessarily dependent upon obligations being performed under another contract, namely the works contract or subcontract which the contractor has been engaged to manage. Section 110(1D) continues to apply to management contracts so that the timing of the contractor's payment cannot be dependent on the issue of a notice under the management contract in question.

16.14 The 2009 Act applies to construction contracts entered into after 1 October 2011. However, s. 110(1A) does not apply to subcontracts entered into under PFI arrangements in England and Wales by virtue of the exclusion orders which came into force on the same date.[7] Thus, even though the amendments introduced by the 2009 Act answer directly the circumstances which gave rise to the decision in*Midland Expressway*, first-tier subcontracts under PFI arrangements (such as the one under discussion in that case) will not be caught by the new prohibition. That is not to say that *Midland Expressway* might have been decided differently under the 2009 Act as discussed immediately below.

Conditional Payment Clauses under PFI Contracts

16.15 The decision in *Midland Expressway* caused some alarm amongst those advising governments and PFI concessionaires. If variations were required to be assessed and paid without any equivalent funding relief from government and private funding partners, then the claims might not be met. Moreover this was a risk which PFI subcontractors understood and were generally in a better position than the concessionaire company to manage through their project-change controls.

[7] The Construction Contracts (England) Exclusion Order 201, SI 2332/2011; the Construction Contracts (Wales) Exclusion Order 2011, SI 1713/2011.

16.16 Parliament appears to have accepted these arguments by excluding first-tier PFI subcontracts from s. 110(1A) as imported by the 2009 Act.[8] This means that clauses which make payment conditional on the performance of obligations or certification under a superior contract will not automatically be deemed ineffective in first-tier PFI arrangements. It follows, however, that they may still be deemed ineffective in particular circumstances where a contractor can show that a clause offends either s. 113 or the wider provisions of s. 110(1), which are not limited by the examples given in s. 110(1A) and (1D). A clause which seeks to exclude or restrict a PFI subcontractor's entitlement to payment must still be construed strictly, and may be deemed ineffective if it does not provide an adequate mechanism for payment or contravenes the intent of s. 113. The decision in *Midland Expressway* has not been overturned, and it remains possible that equivalent funding-relief clauses of the type encountered in that case will not be enforced in future for the reasons given in the case.

The Insolvency Exemption under s. 113

16.17 The insolvency exemption casts a long shadow over the anti-pay-when-paid provisions of both the 1996 and 2009 Acts. It allows parties to agree clauses which limit the contractor's liability to pay subcontractors to amounts recovered from an insolvent third party.

16.18 The definition of insolvency contained in s. 113(2)–(5) is restricted and does not include voluntary arrangements with creditors or voluntary winding up (provided there is a declaration of solvency under s. 89 of the Insolvency Act 1986). However, pay-when-paid arrangements will remain effective in the case of a company entering into formal insolvency arrangements involving an administration or winding-up order, or the appointment of an administrative receiver.

16.19 Section 113(2)(a) has been amended by the Enterprise Act 2002 (Insolvency Order) 2003[9] to also include situations when a company enters administration within the meaning of Schedule B1 to the Insolvency Act 1986. Schedule B1 extends the circumstances in which companies can be placed into administration: as well as allowing for the court to make an administration order (as applied prior to 2002), an administrator can now be appointed by the holder of a floating charge, or by the company or its directors, merely by lodging the necessary forms. Whilst the stated intention of this is to streamline the process for administration, its effect is that an administrator can now be appointed through a process of self-certification without the need for a court hearing or for a court order to be made. The implications of this were considered in *William Hare Ltd v Shepherd Construction Ltd* (2009)[10] (Key Case) which is considered below.

16.20 In the case of partnerships and individuals, s. 113(3) and (4) provides that pay-when-pay clauses remain effective where a partnership is made subject to a winding-up order, or a bankruptcy order is made against an individual, or, in both cases, where their estates are subject to a sequestration order in Scotland or where they grant a trust deed for their creditors. Section 113(5) makes similar provision for companies, partnerships, and individuals in Northern Ireland and in any country outside of the United Kingdom on the occurrence of any event corresponding to those specified in sub-ss. (2), (3), or (4).

16.21 It is worth noting that the insolvency exemption applies only to pay-when-paid clauses and not to the clauses which are now prohibited by s. 110(1A) and (1D) of the 2009 Act. This appears to recognize that certification functions should continue regardless of an employer's

[8] The Construction Contracts (England) Exclusion Order 201, SI 2332/2011; the Construction Contracts (Wales) Exclusion Order 2011, SI 1713/2011.

[9] SI 2003/2096, Art. 4(30).

[10] [2010] All ER (D) 168, [2010] EWCA Civ 283 (CA).

insolvency, and it is only the risk of non-payment itself which may be passed on to subcontractors. If the contractor has already certified payment to a subcontractor when an employer becomes insolvent, then a clause permitting him to withhold payment may be effective without the need to serve a withholding or pay-less notice (*Melville Dundas Ltd v George Wimpey UK Ltd* [2007][11] as confirmed by s. 110(10) of the 2009 Act).

Key Cases: Conditional Payment Clauses

Midland Expressway v Carillion Construction Ltd and ors (No. 2) (2005) 106 Con LR 154, [2005] EWHC 2963 (TCC)

16.22 **Facts:** Midland Expressway Ltd ('MEL') was the concessionaire appointed by the Secretary of State for Transport to construct and operate the Birmingham Northern Relief Road. The defendant joint venture, known as CAMBBA, contracted with MEL (the 'D&C contract') for the design and construction of the road. CAMBBA commenced work in late 2000, and achieved completion in 2003. Numerous disputes broke out, relating, inter alia, to sums due under the contract mechanism for valuing additional works. Clause 39.6.2 of the construction contract related to payment for additional works ordered by the Department for Transport and provided as follows:

> Subject only to Clause Seven (Contractor's Rights) and notwithstanding any other provisions of this contract, the contractor's rights to any price adjustment under or in connection with Clause 39 (Changes) in respect of a department's change shall in no event exceed the amounts, if any, to which the employer is entitled to be paid by the Secretary of State in respect to a corresponding change pursuant to Clauses 8.1.3.1 and 8.1.3.3 of the Concession Agreement.

16.23 Section 7 of the construction contract was entitled 'Contractor's Rights', and provided further that if MEL was entitled to receive additional payments known as 'project relief' then CAMBBA would be entitled to a proportion of these funds subject to clause 7.1.3 which provided that MEL should either have received these funds from the Secretary of State or else certified that it had funds available to it for the purposes of payment.

16.24 **Held:** Clause 39.6.2 read on its own and in conjunction with clause 7.1.3 amounted to a pay-when-paid condition which was rendered ineffective by s. 113 of the Act. This provision could not be saved by the use of language which allowed payment subject to certification and/or entitlement, as opposed to actual payment, further up the contractual chain:

> 71. ... Mr. Streatfeild-James contends that CAMBBA has a properly arguable claim against MEL for interim payment ... If and insofar as clause 39.6.2 would debar that claim from being pursued in adjudication, clause 39.6.2 is ineffective by reason of s. 113 of the 1996 Act. In my view this submission is well founded for six reasons:
>
> ...
>
> (3) The effect of clause 39.6.2 is two-fold: (a) CAMBBA cannot be paid any money in respect of the department change until MEL has established its entitlement to be paid under clause 8 of the concession agreement. (b) If the original evaluation under clause 8 is in error, CAMBBA cannot be paid the correct sum due until the dispute resolution procedure under the concession agreement has been operated.
>
> (4) The practical consequence of clause 39.6.2 is that CAMBBA will not be paid for department's changes unless and until MEL has received a corresponding sum from the department. This is so even in cases where CAMBBA has established or could establish an

[11] [2007] 1 WLR 1136, [2007] 3 All ER 889, [2007] BLR 257, [2007] UKHL 18 (HL).

entitlement to payment or additional payment under the dispute resolution procedures of the D&C contract. This state of affairs is precisely what s. 113 of the 1996 Act is legislating against.

(5) Clause 39.6.2 uses the phrase 'the amounts ... to which the employer is entitled to be paid' rather than 'the amounts which the employer is paid.' In my view, this particular choice of language cannot save the clause. Contracting parties cannot escape the operation of s. 113 by the use of circumlocution.

(6) If I am wrong in the previous sub-paragraph, then I consider that clause 39.6.2 must be read in conjunction with clause 7.1.3. Save in those rare cases where the employer certifies that it has funds available, clause 7.1.3 in conjunction with clause 39.6.2 constitute express and ineluctable 'pay when paid' provisions.

William Hare Ltd v Shepherd Construction Ltd [2009] BLR 447, [2009] EWHC 1603 (TCC), and ***William Hare Ltd v Shepherd Construction Ltd*** [2010] All ER (D) 168; [2010] EWCA Civ 283 (CA)

16.25 **Facts:** The subcontract between the parties contained a pay-when-paid clause which operated in case of the ultimate employer's insolvency. Insolvency was defined by close reference to the terms of s. 113(2); however, despite the contract having been entered into several years after the 2003 amendments implementing the Enterprise Act 2002 came into force,[12] the clause was based on the unamended terms. The main contractor (Shepherd) argued that the reference to 'the making of an administration order under Part II of the Insolvency Act 1986' should be read to include the other circumstances in which a company could now enter into administration pursuant to the Schedule B1.

16.26 **Held:** Coulson J disagreed. The clause could still be read coherently since Schedule B1 retained the original 1986 provisions for an administrator to be appointed by court order. There was no reason to rewrite the clause in accordance with *Investors Compensation Scheme v West Bromwich Building Society* (1998)[13] to give effect to its intention, and the nature of the clause, as an exclusion clause, required that it was construed strictly.

16.27 Moreover, this route avoided the potential problems of the new self-certifying options in that it included safeguards such as the ability of a third party to be heard before making the decision as to whether or not to make an administration order. In the case of the self-certifying options, any hearing comes *after* the company has entered into administration, when very different priorities may be in play.

16.28 In the case of pay-when-paid clauses, Coulson J considered that a court should be particularly careful before making any changes to wording which might give such terms effect:

> 47. ... This is a pay when paid provision. It is endeavouring to identify the circumstances in which Hare can do a considerable amount of work for Shepherd under the sub-contract and then not be paid a penny for that work. It is attempting to pass on to Hare, who do not have a contract with Trinity or any obvious means of recovery against Trinity, the risk that Shepherd (who do have a contract with Trinity and have presumably done the necessary financial checks and ensured that proper warranties are in place) may not be paid under their main contract in respect of the sub-contract works. It is a form of exclusion clause.
>
> 48. In those circumstances the court is required to ensure that Shepherd are kept to the four corners of their bargain with Hare and that a clause of this nature is not rewritten to

[12] See 16.10 for details of the changes brought in by the Enterprise Act 2002.

[13] [1998] 1 WLR 886 (HL).

expand the circumstances in which Hare might find themselves (through no fault of their own) significantly out of pocket because of a financial failure up the contractual chain.

16.29 The Court of Appeal approved the judgment and Waller LJ added:

> 14. We pressed Mr Furst as to whether the principles on which he relied from the speeches of Lord Hoffmann[14] were applicable at all or in any event with the same force in the case of an exclusion clause inserted by one party entirely for his own benefit. His response was to argue that clause 32 was not an exclusion clause but a clause sharing the risk between sub-contractor and contractor of the employer becoming insolvent. In any event he argued that there was no reason for the principles not to apply if it could be demonstrated that a reasonable person would conclude that something had clearly gone wrong with the drafting.
>
> 15. I am very doubtful whether the principles relied on would apply at all in a case such as this. Pay when paid clauses were made ineffective unless the third party was insolvent and insolvency was defined by reference to the ways in which a company could become insolvent. If a main contractor wishes to have a pay when paid provision in a subcontract he would be bound, if it was to be effective, to identify a way in which the third party employer became insolvent as defined in the legislation. If he chose a way which was not in accordance with the legislation because he mis-drafted the provision, I can see no reason why, however obvious it was that he had mis-drafted the provision, the principles identified by Lord Hoffmann would come to his rescue.
>
> 16. If he has drafted his provision in a way which actually does work, even if a reasonable person would guess that it was not intended by the proferens to be so limited and that there has been an error, I see even less reason for the courts to come the rescue.
>
> 17. In this case there is in fact no evidence of any appreciation that an error had been made; the clause as worded does work although, as we were told, the number of court orders are miniscule as compared with self-certified administrations; this clause was not truly 'sharing' the risk of insolvency, it was relieving Shepherd of a liability to pay which they otherwise had and it was for Shepherd to get a clause of this nature right if they wished to rely on it.
>
> 18. The principle which the courts have always applied to clauses by which a party seeks to relieve itself from legal liability, i.e. that to do so they must use clear words, should, in my view, be the dominant principle. As Lord Bingham of Cornhill recently reiterated \ should be applied that, if a party otherwise liable is to exclude or limit his liability ... he must do so in clear words; unclear words do not suffice; any ambiguity or lack of clarity must be resolved against that party' (my underlining). It is not therefore in my view open to Shepherd to argue that there is a lack of clarity in a provision that they drafted so as to relieve themselves from liability, and that the court should use the principles identified by Lord Hoffmann as applicable in rare cases to rescue them.

What Types of Conditional Payment Provisions Are Likely to Be Effective and which Ineffective?

The 1996 Act

16.30 The limited guidance from the courts suggests that in cases where the amount or timing of a payment under a construction contract depends upon certification or the performance of obligations by others, whether under a superior contract or under the contract in question, it

[14] In *Investors Compensation Scheme v West Bromwich Building Society* [1998] 1WLR 886.

will be a matter for construction whether these will be effective under the 1996 Act. In view of the decisions considered above, it would appear that the following types of payment provision remain effective, for contracts entered into under the 1996 Act:[15]

1. provisions which require certificates under superior contracts to determine the *timing* of the payment, so long as the mechanism is not used to prevent a payment becoming due at all (for example if no certificate is issued): *Alstom Signalling Ltd v Jarvis Facilities Ltd* (2004)[16]
2. clauses which are dependent on the employer or contract administrator having issued a payment notice or certificate under the contract in question, provided that the clause is not used to allow the employer to avoid his responsibility to make payment altogether if the certificate is not issued: *Ringway Infrastructure Service Limited v Vauxhall Motors Limited* (2007)[17]
3. clauses setting payment intervals under the subcontract to coincide with the dates under the main contract when payment *should* have been received, provided they are not conditional upon payment *actually* having been received
4. clearly worded pay-when-paid clauses in cases of the employer's insolvency as covered by s. 113(2)–(5)
5. conditional payment provisions (including strict pay-when-paid clauses) under contracts not caught by the 1996 Act, for example agreements entered into by the government under PFI arrangements.

The 2009 Act

16.31 The following types of clause are, however, likely to be rendered ineffective under the 2009 provisions for contracts entered into after 1 October 2011:

1. clauses requiring the amount or timing of payments to be determined by reference to the performance of obligations under another contract, or a decision of another person as to whether obligations under another contract have been performed[18] (This is likely to be regardless of whether the subcontract contains other mechanisms which enable the amount or timing of a payment to be determined other than by reference to the other contract.)
2. clauses which make the timing of payments dependent upon the issue of a notice or certificate on or on behalf of the contractor.[19]

The 1996 and 2009 Acts

16.32 The following types of provision are likely to be ineffective under both the 1996 and 2009 Acts:

1. pay-when-paid clauses making payment conditional upon funds having been received from an employer in cases where the employer is not insolvent, or in cases of voluntary insolvency arrangements[20]
2. *Midland Expressway*-type clauses which have the effect of making payment conditional upon payment under a superior contract

[15] The position will be different for contracts entered into after the 2009 Act comes into force. See 16.31.
[16] 95 Con LR 55, [2004] EWHC 1285 (TCC).
[17] [2007] All ER (D) 333, EWHC 2421 (TCC).
[18] s. 110(1A) inserted by the 2009 Act.
[19] s. 110(1D) inserted by the 2009 Act.
[20] See 16.08–16.09.

3. any kind of conditional payment provision where the main contractor has, by his own default, caused the condition not to be met[21]
4. conditional payment provisions which are included in standard terms and fail to meet the requirements of reasonableness under the Unfair Contract Terms Act 1977.

(3) What Happens if a Conditional Payment Clause Is Rendered Ineffective?

16.33 If a contract contains a conditional payment provision which is rendered ineffective for any of the reasons set out above, ss. 113(6) and 110(3) apply to provide that the parties are free to agree other terms for payment, failing which the relevant provisions of the Scheme will apply. If the contract contains other provisions for payment, therefore, which are not conditional upon payment being made by a third party (such as, in the case of *Midland Expressway*, a mechanism for valuing and making payment for changes ordered under the construction contract), then these provisions should be effective to enable payment to proceed as normal.

16.34 However, if no such provisions exist, and if the parties fail to agree alternative arrangements, then paragraph 11 of the Scheme provides that the relevant provisions of paragraphs II.2, 4, 5, 7, 8, 9, and 10 shall apply in the case of construction contracts over 45 days' duration, whilst paragraphs II.6, 7, 8, 9, and 10 apply in the case of contacts of shorter duration. These provisions relate to the valuation and timing of payments and to the requirements for payment and withholding notices (see Chapter 12 above). In effect, they have the potential to substitute a whole new payment scheme into the contract, although it seems that this will only be the case where the contract includes none of the payment provisions required by the Act. If the contract already includes provisions for the timing of payments, but simply makes the valuation of payments subject to payment being received from a third party, then it is likely that only the valuation provisions will be substituted in place of the pay-when-paid arrangement.[22]

Table 16.1 Table of Cases: Conditional Payment Clauses under the 1996 Act

Title	Citation	Issue
Alstom Signalling Ltd v Jarvis Facilities Ltd	[2004] EWHC 1285 (TCC), Judge LLoyd	**Adequate mechanism: final date for payment determined by reference to superior contract** The final date for payment under a subcontract was linked to the dates upon which certificates were to be issued under the main contract. The mechanism was adequate since failure by the main client to make payment on the dates specified would not excuse payment under the subcontract.
Midland Expressway v Carillion Construction Ltd and ors (No. 2)	(2005) 106 Con LR 154, [2005] EWHC 2963 (TCC), Jackson J	**Pay-when-paid provision** The practical consequence of the clause was that the subcontractor would not be paid unless and until the main contractor had received a corresponding sum from the Department. The use of the phrase 'the amounts ... to which the employer is entitled to be

[21] See *Ringway Infrastructure Service Limited v Vauxhall Motors Limited* [2007] All ER (D) 333, [2007] EWHC 2421 (TCC).

[22] See 12.39–12.55 for a discussion of whether the payment provisions of the Scheme can apply in whole or in part.

Title	Citation	Issue
		paid' rather than 'the amounts which the employer is paid' made no difference. Contracting parties could not escape the operation of s. 113 by the use of 'circumlocution'.
Ringway Infrastructure Services Ltd v Vauxhall Motors Ltd	[2007] All ER (D) 333, [2007] EWHC 2421 (TCC).	**Adequate mechanism: payment mechanism adequate even if open to misapplication** A payment mechanism should not be deemed inadequate simply because it might be misapplied. The contract could be construed to avoid a precondition to payment which might otherwise enable the employer to profit from his breach.
William Hare Ltd v Shepherd Construction Ltd	[2009] EWHC 1603 (TCC), Coulson J	**Construction of pay-when-paid clauses** A pay-when-paid clause is a form of exclusion clause and should be construed strictly. In the circumstances, the clause made sense as drafted and there was no requirement to widen the application of the clause to include circumstances of insolvency which had not been mentioned in the contract, even though these were now part of s. 113(2) of the 1996 Act by virtue of a 2003 amendment.
William Hare Ltd v Shepherd Construction Ltd	[2010] BLR 358, [2010] All ER (D) 168, EWCA Civ 283	A pay-when-paid clause does not truly share the risk of insolvency; it is a limitation clause relieving the contractor of the liability to pay sums which would otherwise have been due. Accordingly clear words should be used and any ambiguity or lack of clarity must be construed against the contractor.
R&C Electrical Engineers Ltd v Shaylor Construction Ltd	[2012] BLR 373, [2012] EWHC 1254 (TCC), Edwards-Stuart J	The fact that payment under a subcontract was delayed pending issue of a certificate under the main contract did not mean that the payment machinery had broken down. The subcontractor was deemed to have knowledge of the main contract conditions and no evidence was produced to show that the certificate under the main contract would never be issued.

Part V

MISCELLANEOUS

17

NOTICES, RECKONING OF TIME, AND APPLICATION TO THE CROWN

(1) Summary: ss. 115–17 of the Act

17.01 This chapter deals with the 'supplementary' provisions at ss. 115–17 of the Act and sets out the relevant case law relating to service of notices, reckoning periods of time and the applicability of the provisions of the Act to the Crown.

17.02 The effect of ss. 115–17 of the Act is summarized below:

1. Construction contracts may contain an agreed manner of service (s. 115(1)), and if no such agreement is reached, the provisions in the Act will apply (s. 115(2)).
2. Where there is no agreed manner of service, any effective means may be used to ensure service (s. 115(3)) and, if delivered by post to the addressee's last known principal address, this shall be treated as effectively served (s. 115(4)).
3. Parties cannot deliberately send notices to incorrect addresses and expect the notice to be effectively served: *Rohde (t/a M Rohde Construction) v Markham-David* (2006).[1]
4. Parties may, in some circumstances, be required to take reasonable steps to ascertain the last known address of the other party: *Mersey Docks Property Holdings v Kilgour* (2004).[2]
5. The notice provisions do not apply to service of documents for legal proceedings (s. 115(5)).
6. The Civil Procedure Rules in relation to service do not apply to adjudication notices: *Cubitt Building & Interiors v Fleetglade Ltd* (2007)).[3]
7. Notices and documents referred to in Part II of the Act must be in writing (s.115(6)).

[1] [2006] BLR 291, [2006] EWHC 814 (TCC) (Key Case).
[2] [2004] BLR 412, [2004] EWHC 1638 (TCC).
[3] 110 Con LR 36, [2006] EWHC 3413 (TCC).

8. Periods of time are to be calculated from the day following service and exclude public holidays (s. 116).
9. Construction contracts entered by the Crown (except in private capacities) must also comply with the Act (s. 117).

(2) Section 115: Service of Notices etc.

17.03 The service of notices is a very important part of the contractual mechanism in construction contracts. The content of the notices (such as payment, withholding, or adjudication notices) is discussed in other chapters of this book. However, the manner in which such notices are served can determine whether those notices are effective. The legislative provisions relating to such service are discussed below. There are no amendments to these sections by the 2009 Act and no relevant provisions within the original or amended Scheme for Construction Contracts ('the Scheme').

Service of Notices by an Agreed Method of Service: s. 115(1)

17.04 Section 115(1) confirms the freedom of the parties to agree the manner of service of any notice (or other document) required or authorized to be served in pursuance of a construction contract. Such agreement may be in the form of a bespoke written agreement,[4] or by use of one of the many standard forms of construction contracts.[5] The advantage of agreeing the manner of service for notices is that it will reduce the likelihood of disputes as to whether notices have been validly served under the contract.

17.05 A contractual notice term may be in directory or mandatory terms. A mandatory provision must be fulfilled exactly according to the letter, whereas a directory provision is satisfied if it is met in substance according to the general intent.[6] Generally, the former do not prescribe a particular manner of service and may, for example, require merely 'actual delivery'.[7] Where a contract contains a mandatory term, it may specify that contractual notices are to be served in a particular way in order for service to be effective, for example by registered post, special delivery, 'delivered personally',[8] fax, or, increasingly, by email.

17.06 Generally, technical points as to service of documents do not find favour with the courts.[9] Where contractual requirements are mandatory and not permissive, the courts will insist upon strict compliance, even where there are significant adverse consequences for the non-complying party.[10]

---

[4] As to the requirement for writing, see s. 115(6).

[5] See examples of such clauses in Table 17.1.

[6] *Yates Building Company v R. J. Pulleyn & Co.* [1976] EG 123. This was a decision by Lord Denning concerning an option to purchase land by notice sent by registered or recorded delivery post to the registered office of the party or to the party's 'said solicitors'. Despite not being sent by the prescribed method, the notice was received by the right person by the right time. This case was quoted in *Anglian Water Services Ltd v Laing O'Rourke Utilities Ltd* (2010) 131 Con LR 94, [2010] EWHC 1529 (TCC).

[7] Such as JCT Building Contract 1998, which was found to permit service by fax which had actually been received: *Construction Partnership UK Ltd v Leek Developments Ltd* (2006) CILL 2357 (TCC), Judge Gilliland.

[8] *Primus Build Ltd v Pompey Centre Ltd* [2009] BLR 437, [2009] EWHC 1487 (TCC).

[9] *Nageh v Richard Giddings and anor* (2007) CILL 2420, [2006] EWHC 3240 (TCC); *Primus Build Ltd v (1) Pompey Centre Ltd and (2) Slidesilver Ltd* [2009] BLR 437, [2009] EWHC 1487 (TCC).

[10] *Von Essen Hotels 5 Ltd v Vaughan and anor* [2007] EWCA Civ 1349, referred to in *Primus Build Ltd v Pompey Centre Ltd* [2009] BLR 437, [2009] EWHC 1487 (TCC). See also *Anglian Water Services*

17.07 If the notice relates to adjudication, it is important to consider whether there are any additional rules relating to the service of notices under the applicable adjudication rules.[11]

Service of Notices in the Absence of an Agreed Method of Service

17.08 Where the parties have not agreed how notices will be served, s. 115(2) provides that the following default provisions shall apply.

Service by 'Any Effective Means': s. 115(3)

17.09 Section 115(3) permits a notice or other document to be served on a person by 'any effective means'. This provision is very broad and it is possible that a contractual term may validly be triggered and/or an adjudication commenced without the other party even being aware of it. This is markedly different from the strict notice regime for litigation which exists under the Civil Procedure Rules 1998.[12]

Deeming Provision: s. 115(4)

17.10 Without restricting the breadth of permissible service 'by any effective means' in s. 115(3), s. 115(4) contains a deeming provision which confirms that any notice which is addressed, pre-paid, and delivered by post to the last known principal address will be treated as effectively served. The relevant address must be:

1. The addressee's last known principal address;[13] or
2. If the addressee has been carrying on a trade, profession, or business, his last known principal business address;[14] or
3. If the addressee is a body corporate, the body corporate's registered or principal office.[15]

Delivered to Last Known Principal Address

17.11 Whether required by operation of the deeming provision or by the clause of the contract itself, many construction contract notices must be served to the last known principal address. Where there is an arguable case that adjudication notices have been deliberately sent to a previous residential address of the responding party, an adjudicator's decision may not be enforced.[16] This may be contrasted with cases where there is no suggestion that the claimant deliberately used the wrong address or that there was some other address known to the claimant where the documents could have been served on the defendants.[17]

17.12 Whilst the issue has not directly been considered in adjudication case law to date, it is possible that the courts will adopt the same approach as in *Mersey Docks Property Holdings v Kilgour* (2004),[18] which considered the CPR 6.5.4 requirement that service be made to the 'last known place of business' in litigation proceedings. The TCC confirmed that the proper construction is 'the last place of business known to the claimant' rather than 'the last known ascertainable place of business … known generally'. The court considered this requirement to be 'relatively onerous' on the claimant:

Ltd v Laing O'Rourke Utilities Ltd (2010) 131 Con LR 94, [2010] EWHC 1529 (TCC) (Key Case) in which the terms of clause 13.2 of NEC2 were considered mandatory.

[11] See Ch. 6 for consideration of the main contractual adjudication rules.
[12] See the discussion of s. 115(5) below at 17.15–17.16.
[13] s. 115(4)(a).
[14] s. 115(4)(a).
[15] s. 115(4)(b).
[16] *Rohde (t/a M Rohde Construction) v Markham-David* (2006) BLR 291 (TCC) (Key Case).
[17] *Nageh v Richard Giddings* (2007) CILL 2420, [2007] EWHC 3240 (TCC).
[18] [2004] BLR 412, [2004] EWHC 1638 (TCC).

> 63. … since in order to acquire the requisite knowledge a party must take reasonable steps to find out at the date of service what is the current place of business or the last place from which the party carried on its business. It will be a matter of evidence whether or not a party has discharged the obligation to have the requisite knowledge at the time of service. On balance, this seems to me to be a fairer and more workable test than one which refers to an objective standard of general knowledge or ascertainability.

Delivered to Last Notified Address

17.13 The question of whether a party has served an effective notice to the last notified address will be a question of fact in each case. Where a party to a construction contract is legally represented, service may be effected upon its legal representatives. However, parties should ensure that the notice is sent to a solicitor who has authority to receive the particular notice (for example a notice in relation to an adjudication).[19] There may be instances in which solicitors named in the contract are the solicitors last notified for effective service, notwithstanding the fact that other solicitors had been more recently appointed.[20]

Notices Must Be in Writing: s. 115(6)

17.14 Any notice or other document referenced in Part II of the Act must be in writing: s. 115(6). Thus, for example, a telephone call was found not to be an effective notice to set off certain sums against the balance owing.[21] It would be difficult to achieve certainty if such notices or other documents were permitted to be oral, or partly oral.

Section 115 Does Not Apply to Legal Proceedings: s. 115(5)

17.15 Section 115(5) confirms that the issuing of contractual notices is different from the service of documents in legal proceedings, for which there are rules of court.

17.16 There is no requirement for a notice served in compliance with s.115 to be served also in compliance with the Civil Procedure Rules: *Cubitt Building & Interiors v Fleetglade Ltd* (2007).[22] In that case it was argued that, because a notice of adjudication was served after 4.00 pm on a particular day, it was deemed to have been served on the following day due to the operation of CPR 6.7. The TCC recognized that the CPR 'is a set of commonsense, practical rules that govern the service of court documents, and there may be exceptional adjudications in which it might be appropriate to have regard to its terms'; however, it rejected the applicability of this argument to the enforcement proceedings in question, stating:

> 35. I am unattracted to the notion that the provisions of the CPR should be incorporated into the timetable and mechanisms of the adjudication process. There is no mention of such wholesale incorporation in the 1996 Act. Indeed, s.115, which contains a number of rules relating to the service of adjudication documents, makes no reference to the CPR save to say, at sub-section 115.5, that the rules of court do apply, following the production of an adjudicator's decision, to the service of enforcement proceedings and the like. This could therefore be said to be inconsistent with the suggestion that the CPR should be incorporated wholesale into the adjudication process: if that was the intention, s.115

---

[19] *Anglian Water Services Ltd v Laing O'Rourke Utilities Ltd* (2010) 131 Con LR 94, [2010] EWHC 1529 (TCC).

[20] *Von Essen Hotels 5 Ltd v Vaughan and anor* [2007] EWCA Civ 1349. Whilst this case does not deal with notice under a construction contract, it was referred to in *Primus Build Ltd v Pompey Centre Ltd* [2009] BLR 437, [2009] EWHC 1487 (TCC).

[21] *Strathmore Building Services Ltd v Colin Scott Greig (t/a Hestia Fireside Design)* (2001) 17 Const LJ 72, [2000] Scot CS 133.

[22] 110 Con LR 36, [2006] EWHC 3413 (TCC).

would have said so. In addition, I am aware of no authority in which the point has been successfully argued. I agree ... that complications could abound if the CPR was imported wholesale into the adjudication process. Take for example the present case, where the adjudicator's decision was e-mailed on 25th November. If the CPR provisions apply, then the relevant date for the service of that decision would be 27th November, which is not a result for which either party contends.

Table 17.1 Table of Construction Contract Clauses Relating to Service of Notices

Contract	Relevant Clause
JCT 1998 (with Contractor's Design)	**Clause 1.5:** where the contract does not specifically prescribe the manner of service, any notice given or served by any effective means to any agreed address will suffice. If none, effective service will be by pre-paid post to the addressee's last known principal business address or, where the addressee is a body corporate, to its registered or principal office.
JCT 2005 (Design & Build)	**Clause 1.7.1:** valid service may be made by any effective means. Where given by actual delivery or pre-paid post to the address in the contract particulars or such other address as agreed, it will be deemed to have been served. Where no address given, service is achieved by actual delivery or by pre-paid post to the party's last known principal business address, or, if a corporation, its registered or principal office (**clause 1.7.2**). The contract particulars make provision for addresses for service of notices etc. in para. 1.7 which provides for fax numbers to be entered. If no details are stated, the address in each case until agreed is that shown at the commencement of the agreement.
NEC2/NEC3	**Clause 13.1:** each communication that is required by the contract must be communicated in a form that can be read, copied and recorded in the language of the contract. **Clause 13.2:** a communication has effect when it is received at the last address notified by the recipient. However, if no address is notified, a communication will be effective if received at the address of the recipient stated in the contract data.
ICE 1999 (Measurement Version)	**Clause 1(6):** communications required to be 'in writing' under the contract may be hand-written, typewritten, or printed and sent by hand, post, telex, cable, fax, or other means resulting in a permanent record.
PPC 2000	**Clause 3.2:** except as otherwise agreed in writing, all communications between any partnering team members shall be in writing by receipted hand delivery, recorded delivery post, fax, or (if agreed) email, in each case delivered to the address of the relevant partnering team member set out in the project partnering agreement or any joining agreement or to such other address as a partnering team member shall notify.

17.17 Table 17.1 describes some of the relevant notice clauses from the most used standard forms of construction contract.

Key Cases: Service of Notices

Rohde (t/a M Rohde Construction) v Markham-David [2006] BLR 291, [2006] EWHC 814 (TCC)

17.18 **Facts:** The defendant claimed to be unaware of the existence of either the adjudication or the enforcement proceedings. The claimant argued that the notice of adjudication and other documents concerning the adjudication were sent to the defendant's last known principal address and that this constituted effective service under s. 115(4) of the Act.

However, the defendant could have easily been contacted through a quarry business he was operating (of which the claimant was aware because he had previously carried out work there); and his home telephone number was displayed on a notice board at the quarry. The defendant contended that the claimant deliberately avoided contacting him at the quarry and used a method of service that was unlikely to bring the documents to the defendant's attention. Accordingly, the defendant applied to set aside the default judgment entered against him.

17.19 **Held:** Jackson J accepted the defendant's arguments had a real prospect of success, especially in the absence of any evidence from the claimant. There was a serious issue to be tried as to whether there had been a breach of natural justice in relation to the service of the notice on the defendant. Jackson J set aside the default judgment entered against the defendant:

> 35. ... There is, however, a serious factual issue to be tried in this regard. I would formulate the issue in these terms: did the claimant have available during the adjudication a ready means of contacting the defendant, which the claimant chose neither to use nor to communicate to the adjudicator?
>
> 36. If it should turn out that the answer to this question is yes, then an issue of law will arise. That issue of law is whether section 108(3) of the 1996 Act and paragraph 23 of the Scheme require an adjudicator's award obtained in those circumstances to be enforced ...
>
> 38. If, after hearing evidence in the present case, it turns out that the claimant took a deliberate decision, which deprived the defendant of the opportunity to make representations in the adjudication, then I consider that this may be one of those rare and exceptional cases in which the court will decline to enforce an adjudicator's decision by reason of breach of natural justice.

Anglian Water Services Ltd v Laing O'Rourke Utilities Ltd 131 Con LR 94, [2010] EWHC 1529 (TCC)

17.20 **Facts:** An adjudication had taken place between the parties, both of whom were represented by solicitors. Where a party was dissatisfied with the adjudicator's decision, it could serve a notice of dissatisfaction within four weeks of the decision in accordance with clause 93.1 of the NEC 2. A notice of dissatisfaction was in fact served on Laing O'Rourke's (LOR's) solicitors who, during the course of the adjudication, had advised they would accept service of any 'documentation relevant to the adjudication'. LOR's solicitors immediately passed it on to the relevant personnel at LOR and confirmed safe receipt of the notice to Anglian Water's solicitors. A week later, they stated they needed to take instructions from their clients as to whether they were instructed to accept service, and sought confirmation that Anglian Water had served the relevant notice directly on their clients. Clause 13.2 made a communication effective 'when it is received at the last address notified by the recipient for receiving communications or, if none is notified, at the address of the recipient stated in the Contract Data'.

17.21 **Held:** Compliance with the mode of delivery specified in clause 13.2 is the only means of achieving or securing effective delivery of a communication under the contract, because the communication only takes effect when it is received at the prescribed address. The fact that the relevant personnel at LOR actually received the notice within the required four-week period was not determinative of compliance with clause 13.2.[23]

[23] However, it would be a relevant factor when considering an application under section 12 of the Arbitration Act 1996 to extend time.

17.22 The notification of a change of address under clause 13.2 is itself a communication that is required by the contract to be sent to the last address notified. There is no restriction on notifying addresses for different purposes under the contract. Email confirmation from LOR's solicitors that they could accept service was sufficient notification under clause 13.2 that its address was the notified address for documentation relating to the adjudication. Whilst it was not a document that formed part of the adjudication process itself, the notice of dissatisfaction was relevant to the adjudication because it prevented the adjudicator's decision from becoming final. Accordingly, it was a communication served in accordance with clause 13.2:

> 45. It seems to me that the probable commercial purpose of the clause is to enable each party to the contract to work on the basis that all communications in relation to the contract will be channelled through one particular office, with the obvious advantage of enabling every incoming document to be properly filed and its arrival properly recorded. I think that, as a specialist judge in this field, I can probably take judicial notice of the fact that for the best part of a decade or so electronic document management systems for providing a database of all communications generated in the course of the contract have been employed in many substantial civil engineering and construction projects. Service through one designated office may be more cumbersome but it does enable proper records to be kept. It is then for the project manager or co-ordinator, or a particular member or members of his staff, to ensure that incoming documents are then copied to all those individuals who have an interest in seeing them.
>
> 46. Turning to the wording of the clause itself, it seems to me that clause 13.2 is there to fix the moment in time when a communication takes effect. ... [I]n the case of certain types of communication the date of its receipt will trigger the start of the period in which a response or action is required. The answer to the rhetorical question ... – what is wrong with a document being handed over in a meeting? – is nothing, but I consider that the contract requires that a copy of the document should be sent also to that party's prescribed address.
>
> 47. It would be unsatisfactory, in my view, if in any case where there was a dispute about the time when a communication took effect, the parties had to investigate the circumstances in which the communication was made and received in order to determine whether the mode of delivery was actually as good as, or better than, the mode of delivery prescribed by clause 13.2. Apart from anything else, there might well be legitimate room for disagreement as to whether the actual mode of service was an improvement on the prescribed mode of service.

Table 17.2 Table of Cases: Service of Notices

Title	Citation	Issue
Strathmore Building Services Ltd v Colin Scott Greig(t/a Hestia Fireside Design)	[2000] Scot CS 133 (Outer House, Court of Session), Lord Hamilton	**Payment notice: phone call insufficient** A written withholding notice was not served. The payer sought to rely on a phone call resulting in a message left with a receptionist. This was not permitted, as written notice was required.
Mersey Docks Property Holdings v Kilgour	[2004] BLR 412, [2004] EWHC 1638 (TCC), Judge Toulmin	**Last known address: CPR 6.5.4** CPR 6.5.4 is a relatively onerous provision since, in order to acquire the requisite knowledge, a party must take reasonable steps to find out, at the date of service, what is the current place of business or the last place from which the party carried on its business.

Title	Citation	Issue
Rohde (t/a M Rohde Construction) v Markham-David	[2006] BLR 291, [2006] EWHC 814 (TCC), Jackson J	**Notices: deliberate non-service** It was arguable that adjudication notices were deliberately sent to a previous residential address of the responding party. The referring party could have easily made contact with the responding party. The adjudicator's decision was not enforced.
Construction Partnership UK Ltd v Leek Developments Ltd	(2006) CILL 2357 (TCC), Judge Gilliland	**Notices: 'actual delivery'** A notice of termination was served by fax on Friday 23 December at 8.46 am when the office closed at noon and was purportedly not seen by the relevant people. The JCT Building Contract 1998 required 'actual delivery' which means 'transmission by an appropriate means so that it is actually received'. There is nothing to suggest that a document sent by fax which is actually received, and is not disputed, cannot amount to actual delivery.
Collier v Williams and ors	(2006) 1 WLR 1945, [2006] EWCA Civ 20 (CA), Waller, Dyson, Neuberger LJJ	**Notices: CPR 6 – defendant had never resided at the address** Interpolating the words 'or reasonably believed' in the phrase 'the address known to be the last residence of the individual' does not add to certainty or reduce the risk of satellite litigation. Service was not effected on the defendant's last known residence for the simple reason that he had never resided there.
Nageh v Richard Giddings and anor	(2007) CILL 2420, [2007] EWHC 3240 (TCC), Judge Coulson	**Notices: no deliberate non-service** An adjudicator's decision had been enforced summarily. Many months after he found out about it, the defendant tried to have it set aside, claiming that he was unaware of the adjudication or court proceedings at the time either took place. There was no suggestion that the claimant deliberately used the wrong address, or that there was some other address known to the claimant where the documents could have been served on the defendant. In fact, the documents had been served at two different addresses. The defendant's application to set aside the summary judgment was dismissed.
Von Essen Hotels 5 Ltd v Vaughan and anor [24]	[2007] EWCA Civ 1349 (CA), Mummery, Hughes, David Richards LJJ	**Notices: sent to new solicitors** Notice was sent to the other side's new solicitors instead of the solicitors named in the contract. The notice requirements were mandatory not permissive and therefore the notice was not effective.
Primus Build Ltd v Pompey Centre Ltd	[2009] BLR 437, [2009] EWHC 1487 (TCC), Coulson J	**Notices: delivered personally** Technical points as to service do not generally find favour. The contract required notices to be 'delivered personally', which must be different from personal service. This meant actual delivery by an appropriate individual in one company to an appropriate individual in the other.

[24] Cited in *Primus Build Ltd v Pompey Centre Ltd* [2009] BLR 437, [2009] EWHC 1487 (TCC), Coulson J.

Title	Citation	Issue
Anglian Water Services Ltd v Laing O'Rourke Utilities Ltd	(2010) 131 Con LR 94, [2010] EWHC 1529 (TCC), Edwards-Stuart J	**Notice delivered to solicitors was effective** Clause 13.2 of NEC 2 is a clause that requires exact compliance. A notice of dissatisfaction served to solicitors who had advised they were instructed to accept service of documents relevant to the adjudication had been effectively served to the last notified address for such notices.

(3) Section 116: Reckoning Periods of Time

17.23 Section 116(2) of the Act provides that:

> Where an act is required to be done within a specified period after or from a specified date, the period begins immediately after that date.

17.24 Section 116(3) confirms that public holidays are excluded from the calculation. There is no case law considering this section of the Act, probably because its application is well understood within the industry. In practice, if a notice is served and received on 1 January, and the contract provides that, for example, the breach must be remedied within seven days of the notice, the first day for calculating the period will be 2 January. The seven-day period will therefore expire on 8 January—unless one of the days between 2–8 January is a public holiday, in which case the period will expire on 9 January.

17.25 Many construction contracts do not include any express provision about the reckoning periods as it is covered by the Act. However, the JCT suite of contracts expressly reflects the requirements of s. 117.[25]

(4) Section 117: Crown Application

17.26 The Act applies to construction contracts entered into by or on behalf of the Crown (otherwise than by or on behalf of Her Majesty in her private capacity),[26] or by or on behalf of the Duchy of Cornwall notwithstanding any Crown interest.[27] In adjudications or other proceedings arising out of construction contracts, Her Majesty is represented by the Chancellor of the Duchy or by such person as he may appoint[28] and the Duchy or Duke of Cornwall may appoint his own representatives.[29]

[25] See e.g. JCT 1998 (with Contractor Design), clause 1.6; JCT 1998 (Private with Quantities), clause 1.8; JCT 2005 Design & Build, clause 1.5; JCT 2005 (with Quantities), clause 1.5.

[26] s. 117(1) of the Act.

[27] s. 117(2).

[28] s. 117(3).

[29] s. 117(4).

Appendices: Materials

HOUSING GRANTS CONSTRUCTION AND REGENERATION ACT 1996, AS AMENDED BY THE LOCAL DEMOCRACY, ECONOMIC DEVELOPMENT AND CONSTRUCTION ACT 2009 (EXTRACTS)

Part II Construction Contracts

Introductory provisions

104 Construction contracts

(1) In this Part a "construction contract" means an agreement with a person for any of the following –
  (a) the carrying out of construction operations;
  (b) arranging for the carrying out of construction operations by others, whether under sub-contract to him or otherwise;
  (c) providing his own labour, or the labour of others, for the carrying out of construction operations.

(2) References in this Part to a construction contract include an agreement –
  (a) to do architectural, design, or surveying work, or
  (b) to provide advice on building, engineering, interior or exterior decoration or on the laying-out of landscape, in relation to construction operations.

(3) References in this Part to a construction contract do not include a contract of employment (within the meaning of the [1996 c. 18.] Employment Rights Act 1996).

(4) The Secretary of State may by order add to, amend or repeal any of the provisions of subsection (1), (2) or (3) as to the agreements which are construction contracts for the purposes of this Part or are to be taken or not to be taken as included in references to such contracts.
No such order shall be made unless a draft of it has been laid before and approved by a resolution of each of House of Parliament.

(5) Where an agreement relates to construction operations and other matters, this Part applies to it only so far as it relates to construction operations.
An agreement relates to construction operations so far as it makes provision of any kind within subsection (1) or (2).

(6) This Part applies only to construction contracts which –
  (a) are entered into after the commencement of this Part, and
  (b) relate to the carrying out of construction operations in England, Wales or Scotland.

(7) This Part applies whether or not the law of England and Wales or Scotland is otherwise the applicable law in relation to the contract.

105 Meaning of "construction operations"

(1) In this Part "construction operations" means, subject as follows, operations of any of the following descriptions –
  (a) construction, alteration, repair, maintenance, extension, demolition or dismantling of buildings, or structures forming, or to form, part of the land (whether permanent or not);
  (b) construction, alteration, repair, maintenance, extension, demolition or dismantling of any works forming, or to form, part of the land, including (without prejudice to the foregoing) walls, roadworks, power-lines, [electronic communications apparatus][1], aircraft runways, docks and harbours, railways, inland waterways, pipe-lines, reservoirs, water-mains, wells, sewers, industrial plant and installations for purposes of land drainage, coast protection or defence;
  (c) installation in any building or structure of fittings forming part of the land, including (without prejudice to the foregoing) systems of heating, lighting, air-conditioning, ventilation, power supply, drainage, sanitation, water supply or fire protection, or security or communications systems;

---

[1] Amended by the Communications Act 2003, s. 406(1), Sch. 17, para. 137.

- (d) external or internal cleaning of buildings and structures, so far as carried out in the course of their construction, alteration, repair, extension or restoration;
- (e) operations which form an integral part of, or are preparatory to, or are for rendering complete, such operations as are previously described in this subsection, including site clearance, earth-moving, excavation, tunnelling and boring, laying of foundations, erection, maintenance or dismantling of scaffolding, site restoration, landscaping and the provision of roadways and other access works;
- (f) painting or decorating the internal or external surfaces of any building or structure.

(2) The following operations are not construction operations within the meaning of this Part—

- (a) drilling for, or extraction of, oil or natural gas;
- (b) extraction (whether by underground or surface working) of minerals; tunnelling or boring, or construction of underground works, for this purpose;
- (c) assembly, installation or demolition of plant or machinery, or erection or demolition of steelwork for the purposes of supporting or providing access to plant or machinery, on a site where the primary activity is –
  - (i) nuclear processing, power generation, or water or effluent treatment, or
  - (ii) the production, transmission, processing or bulk storage (other than warehousing) of chemicals, pharmaceuticals, oil, gas, steel or food and drink;
- (d) manufacture or delivery to site of –
  - (i) building or engineering components or equipment,
  - (ii) materials, plant or machinery, or
  - (iii) components for systems of heating, lighting, air-conditioning, ventilation, power supply, drainage, sanitation, water supply or fire protection, or for security or communications systems, except under a contract which also provides for their installation;
- (e) the making, installation and repair of artistic works, being sculptures, murals and other works which are wholly artistic in nature.

(3) The Secretary of State may by order add to, amend or repeal any of the provisions of subsection (1) or (2) as to the operations and work to be treated as construction operations for the purposes of this Part.

(4) No such order shall be made unless a draft of it has been laid before and approved by a resolution of each House of Parliament.

106 Provisions not applicable to contract with residential occupier

(1) This Part does not apply –

- (a) to a construction contract with a residential occupier (see below)., [*or*
- *(b) to any other description of construction contract excluded from the operation of this Part by order of the Secretary of State.*][2]

(2) A construction contract with a residential occupier means a construction contract which principally relates to operations on a dwelling which one of the parties to the contract occupies, or intends to occupy, as his residence.

In this subsection "dwelling" means a dwelling-house or a flat; and for this purpose—

- "dwelling-house" does not include a building containing a flat; and
- "flat" means separate and self-contained premises constructed or adapted for use for residential purposes and forming part of a building from some other part of which the premises are divided horizontally.

(3) The Secretary of State may by order amend subsection (2).

(4) No order under this section shall be made unless a draft of it has been laid before and approved by a resolution of each House of Parliament.

[*106A Power to disapply provisions of this Part*

(1) The Secretary of State may by order provide that any or all of the provisions of this Part, so far as extending to England and Wales, shall not apply to any description of construction contract relating

[2] Repealed by the Local Democracy, Economic Development and Construction Act 2009, ss. 138(1), (2), 146(1), Sch. 7, Pt 5.

to the carrying out of construction operations (not being operations in Wales) which is specified in the order.

(2) The Welsh Minister may by order provide that any or all of the provisions of this Part, so far as extending to England and Wales, shall not apply to any description of construction contract relating to the carrying out of construction operations in Wales which is specified in the order.

(3) The Scottish Ministers may by order provide that any or all of the provisions of this Part, so far as extending to Scotland, shall not apply to any description of construction contract which is specified in the order.

(4) An order under this section shall not be made unless a draft of it has been laid before and approved by resolution of –
- (a) in the case of an order under subsection (1), each House of Parliament;
- (b) in the case of an order under subsection (2), the National Assembly for Wales;
- (c) in the case of an order under subsection (3), the Scottish Parliament.][3]

[*107 Provisions applicable only to agreements in writing*

(1) *The provisions of this Part apply only where the construction contract is in writing, and any other agreement between the parties as to any matter is effective for the purposes of this Part only if in writing.*
The expressions "agreement", "agree" and "agreed" shall be construed accordingly.

(2) *There is an agreement in writing –*
- (a) *if the agreement is made in writing (whether or not it is signed by the parties),*
- *(b) if the agreement is made by exchange of communications in writing, or*
- *(c) if the agreement is evidenced in writing.*

(3) *Where parties agree otherwise than in writing by reference to terms which are in writing, they make an agreement in writing.*

(4) *An agreement is evidenced in writing if an agreement made otherwise than in writing is recorded by one of the parties, or by a third party, with the authority of the parties to the agreement.*

(5) *An exchange of written submissions in adjudication proceedings, or in arbitral or legal proceedings in which the existence of an agreement otherwise than in writing is alleged by one party against another party and not denied by the other party in his response constitutes as between those parties an agreement in writing to the effect alleged.*

(6) References in this Part to anything being written or in writing include its being recorded by any means.][4]

Adjudication

108 Right to refer disputes to adjudication

(1) A party to a construction contract has the right to refer a dispute arising under the contract for adjudication under a procedure complying with this section.
For this purpose "dispute" includes any difference.

(2) The contract shall [include provision in writing so as to][5] –
- (a) enable a party to give notice at any time of his intention to refer a dispute to adjudication;
- (b) provide a timetable with the object of securing the appointment of the adjudicator and referral of the dispute to him within 7 days of such notice;
- (c) require the adjudicator to reach a decision within 28 days of referral or such longer period as is agreed by the parties after the dispute has been referred;
- (d) allow the adjudicator to extend the period of 28 days by up to 14 days, with the consent of the party by whom the dispute was referred;
- (e) impose a duty on the adjudicator to act impartially; and
- (f) enable the adjudicator to take the initiative in ascertaining the facts and the law.

---

[3] Inserted by the Local Democracy, Economic Development and Construction Act 2009, s. 138(1), (3).

[4] Repealed by the Local Democracy, Economic Development and Construction Act 2009, ss. 139(1), 146(1), Sch. 7, Pt 5.

[5] Amended by the Local Democracy, Economic Development and Construction Act 2009, s. 139(2)(a).

(3) The contract shall provide [in writing][6] that the decision of the adjudicator is binding until the dispute is finally determined by legal proceedings, by arbitration (if the contract provides for arbitration or the parties otherwise agree to arbitration) or by agreement.
The parties may agree to accept the decision of the adjudicator as finally determining the dispute.

[(3A) The contract shall include provision in writing permitting the adjudicator to correct his decision so as to remove a clerical or typographical error arising by act or omission.][7]

(4) The contract shall also provide [in writing][8] that the adjudicator is not liable for anything done or omitted in the discharge or purported discharge of his functions as adjudicator unless the act or omission is in bad faith, and that any employee or agent of the adjudicator is similarly protected from liability.

(5) If the contract does not comply with the requirements of subsections (1) to (4), the adjudication provisions of the Scheme for Construction Contracts apply.

(6) For England and Wales, the Scheme may apply the provisions of the Arbitration Act 1996 with such adaptations and modifications as appear to the Minister making the scheme to be appropriate.
For Scotland, the Scheme may include provision conferring powers on courts in relation to adjudication and provision relating to the enforcement of the adjudicator's decision.

[*108A Adjudication costs: effectiveness of provision*

(1) This section applies in relation to any contractual provision made between the parties to a construction contract which concerns the allocation as between those parties of costs relating to the adjudication of a dispute arising under the construction contract.
(2) The contractual provision referred to in subsection (1) is ineffective unless –
 (a) it is made in writing, is contained in the construction contract and confers power on the adjudicator to allocate his fees and expenses as between the parties, or
 (b) it is made in writing after the giving of notice of intention to refer the dispute to adjudication.][9]

Payment

109 Entitlement to stage payments

(1) A party to a construction contract is entitled to payment by instalments, stage payments or other periodic payments for any work under the contract unless –
 (a) it is specified in the contract that the duration of the work is to be less than 45 days, or
 (b) it is agreed between the parties that the duration of the work is estimated to be less than 45 days.
(2) The parties are free to agree the amounts of the payments and the intervals at which, or circumstances in which, they become due.
(3) In the absence of such agreement, the relevant provisions of the Scheme for Construction Contracts apply.
(4) References in the following sections to a payment [provided for by the contract][10] include a payment by virtue of this section.

110 Dates for payment

(1) Every construction contract shall –
 (a) provide an adequate mechanism for determining what payments become due under the contract, and when, and
 (b) provide for a final date for payment in relation to any sum which becomes due.
The parties are free to agree how long the period is to be between the date on which a sum becomes due and the final date for payment.

[6] Amended by the Local Democracy, Economic Development and Construction Act 2009, s. 139(2)(b).
[7] Inserted by the Local Democracy, Economic Development and Construction Act 2009, s. 140.
[8] Amended by the Local Democracy, Economic Development and Construction Act 2009, s. 139(2)(b).
[9] Inserted by the Local Democracy, Economic Development and Construction Act 2009, s. 141.
[10] Amended by the Local Democracy, Economic Development and Construction Act 2009, s. 143(1).

[(1A) The requirement under subsection (1)(a) to provide an adequate mechanism for determining what payments become due under the contract, or when, is not satisfied where a construction contract makes payment conditional on –

(a) the performance of obligations under another contract, or

(b) a decision by any person as to whether obligations under another contract have been performed.

(1B) In subsection (1A)(a) and (b) the references to obligations do not include obligations to make payments (but see section 113).

(1C) Subsection (1A) does not apply where –

(a) the construction contract is an agreement between the parties for the carrying out of construction operations by another person, whether under sub-contract or otherwise, and

(b) the obligations referred to in that subsection are obligations on that other person to carry out those operations.][11]

[(1D) The requirement in subsection (1)(a) to provide an adequate mechanism for determining when payments become due under the contract is not satisfied where a construction contract provides for the date on which a payment becomes due to be determined by reference to the giving to the person on whom the payment is due of a notice which relates to what payments are due under the contract.][12]

[*(2) Every construction contract shall provide for the giving of notice by a party not later than five days after the date on which a payment becomes due from him under the contract, or would have become due if –*

(a) the other party had carried out his obligations under the contract, and

(b) no set-off or abatement was permitted by reference to any sum claimed to be due under one or more other contracts, specifying the amount (if any) of the payment made or proposed to be made, and the basis on which that amount was calculated.][13]

(3) If or to the extent that a contract does not contain such provision as is mentioned in subsection (1) [*or (2)*][14], the relevant provisions of the Scheme for Construction Contracts apply.

[*110A Payment notices: contractual requirements*

(1) A construction contract shall, in relation to every payment provided for by the contract –

(a) require the payer or a specified person to give a notice complying with subsection (2) to the payee not later than five days after the payment due date, or

(b) require the payee to give a notice complying with subsection (3) to the payer or a specified person not later than five days after the payment due date.

(2) A notice complies with this subsection if it specifies –

(a) in a case where the notice is given by the payer –

(i) the sum that the payer considers to be or to have been due at the payment due date in respect of the payment, and

(ii) the basis on which that sum is calculated;

(b) in a case where the notice is given by a specified person –

(i) the sum that the payer or the specified person considers to be or to have been due at the payment due date in respect of the payment, and

(ii) the basis on which that sum is calculated.

(3) A notice complies with this subsection if it specifies –

(a) the sum that the payee considers to be or to have been due at the payment due date in respect of the payment, and

(b) the basis on which that sum is calculated.

(4) For the purpose of this section, it is immaterial that the sum referred to in subsection (2)(a) or (b) or (3)(a) may be zero.

---

[11] Inserted by the Local Democracy, Economic Development and Construction Act 2009, s. 142(1), (2).

[12] Inserted by the Local Democracy, Economic Development and Construction Act 2009, s. 142(1), (3).

[13] Repealed by the Local Democracy, Economic Development and Construction Act 2009, ss. 143(2)(a), 146(1), Sch. 7, Pt 5.

[14] Repealed by the Local Democracy, Economic Development and Construction Act 2009, ss. 143(2)(b), 146(1), Sch. 7, Pt 5.

(5) If or to the extent that a contract does not comply with subsection (1), the relevant provisions of the Scheme for Construction Contracts apply.

(6) In this and the following sections, in relation to any payment provided for by a construction contract –
"payee" means the person to whom the payment is due;
"payer" means the person from whom the payment is due;
"payment due date" means the date provided for by the contract as the date on which the payment is due;
"specified person" means a person specified in or determined in accordance with the provisions of the contract.][15]

[*110B Payment notices: payee's notice in default of payer's notice*

(1) This section applies in a case where, in relation to any payment provided for by a construction contract –
  (a) the contract requires the payer or a specified person to give the payee a notice complying with section 110A(2) not later than five days after the payment due date, but
  (b) notice is not given as so required.

(2) Subject to subsection (4), the payee may give to the payer a notice complying with section 110A(3) at any time after the date on which the notice referred to in subsection (1)(a) was required by the contract to be given.

(3) Where pursuant to subsection (2) the payee gives a notice complying with section 110A(3), the final date for payment of the sum specified in the notice shall for all purposes be regarded as postponed by the same number of days after the date referred to in subsection (2) that the notice was given.

(4) If –
  (a) the contract permits or requires the payee, before the date on which the notice referred to in subsection (1)(a) if required by the contract to be given, to notify the payer or a specified person of –
    (i) the sum that the payee considers will become due on the payment due date in respect of the payment, and
    (ii) the basis on which that sum is calculated, and
  (b) the payee gives such notification in accordance with the contract, that notification is to be regarded as a notice complying with section 110A(3) given pursuant to subsection (2) (and the payee may not give another notice pursuant to that subsection).][16]

[*111 Notice of intention to withhold payment*

(1) A party to a construction contract may not withhold payment after the final date for payment of a sum due under the contract unless he has given an effective notice of intention to withhold payment.
The notice mentioned in section 110(2) may suffice as a notice of intention to withhold payment if it complies with the requirements of this section.

(2) To be effective such a notice must specify–
  (a) the amount proposed to be withheld and the ground for withholding payment, or
  (b) if there is more than one ground, each ground and the amount attributable to it,
  and must be given not later than the prescribed period before the final date for payment.

(3) The parties are free to agree what that prescribed period is to be.
In the absence of such agreement, the period shall be that provided by the Scheme for Construction Contracts.

(4) Where an effective notice of intention to withhold payment is given, but on the matter being referred to adjudication it is decided that the whole or part of the amount should be paid, the decision shall be construed as requiring payment not later than–
  (a) seven days from the date of the decision, or
  (b) the date which apart from the notice would have been the final date for payment,
whichever is the later.][17]

[15] Inserted by the Local Democracy, Economic Development and Construction Act 2009, s. 143(3).
[16] Inserted by the Local Democracy, Economic Development and Construction Act 2009, s. 143(3).
[17] Repealed by the Local Democracy, Economic Development and Construction Act 2009, ss 144(1).

[*111 Requirement to pay notified sum*

(1) Subject as follows, where a payment is provided for by a construction contract, the payer must pay the notified sum (to the extent not already paid) on or before the final date for payment.
(2) For the purposes of this section, the "notified sum" in relation to any payment provided for by a construction contract means –
 (a) in a case where a notice complying with section 110A(2) has been given pursuant to and in accordance with a requirement of the contract, the amount specified in that notice;
 (b) in a case where a notice complying with section 110A(3) has been given pursuant to and in accordance with a requirement of the contract the amount specified in that notice;
 (c) in a case where a notice complying with section 110A(3) has been given pursuant to and in accordance with section 110B(2), the amount specified in that notice.
(3) The payer or a specified person may in accordance with this section give to the payee a notice of the payer's intention to pay less than the notified sum.
(4) A notice under subsection (3) must specify –
 (a) the sum that the payer considers to be due on the date the notice is served, and
 (b) the basis on which that sum is calculated.
 It is immaterial for the purposes of this subsection that the sum referred to in paragraph (a) or (b) may be zero.
(5) A notice under subsection (3) –
 (a) must be given not later than the prescribed period before the final date for payment, and
 (b) in a case referred to in subsection (2)(b) or (c), may not be given before the notice by reference to which the notified sum is determined.
(6) Where a notice is given under subsection (3), subsection (1) applies only in respect of the sum specified pursuant to subsection (4)(a).
(7) In subsection (5) "prescribed period" means –
 (a) such period as the parties may agree, or
 (b) in the absence of such agreement, the period provided by the Scheme for Construction Contracts.
(8) Subsection (9) applies where in respect of a payment –
 (a) a notice complying with section 110A(2) has been given pursuant to and in accordance with a requirement of the contract (and no notice under subsection (3) is given), or
 (b) a notice under subsection (3) is given in accordance with this section, but on the matter being referred to adjudication the adjudicator decides that more than the sum specified in the notice should be paid.
(9) In a case where this subsection applies, the decision of the adjudicator referred to in subsection (8) shall be construed as requiring payment of the additional amount not later than –
 (a) seven days from the date of the decision, or
 (b) the date which apart from the notice would have been the final date for payment, whichever is the later.
(10) Subsection (1) does not apply in relation to a payment provided for by a construction contract where –
 (a) the contract provides that, if the payee becomes insolvent the payer need not pay any sum due in respect of the payment, and
 (b) the payee has become insolvent after the prescribed period referred to in subsection (5)(a).
(11) Subsections (2) to (5) of section 113 apply for the purposes of subsection (10) of this section as they apply for the purposes of that section.][18]

112 Right to suspend performance for non-payment

(1) *[Where a sum due under a construction contract is not paid in full by the final date for payment and no effective notice to withhold payment has been given]*[19] [Where the requirement in section 111(1) applies in relation to any sum but is not complied with],[20] the person to whom the sum is due has the right (without prejudice to any other right or remedy) to suspend performance of [*any or all*

[18] Amended by the Local Democracy, Economic Development and Construction Act 2009, ss 144(2)(b).
[19] Amended by the Local Democracy, Economic Development and Construction Act 2009, ss 114(2)(b)
[20] Amended by the Local Democracy, Economic Development and Construction Act 2009, s. 144(2)(a).

of][21] his obligations under the contract to the party by whom payment ought to have been made ("the party in default").

(2) The right may not be exercised without first giving to the party in default at least seven days' notice of intention to suspend performance, stating the ground or grounds on which it is intended to suspend performance.

(3) The right to suspend performance ceases when the party in default makes payment in full of [*the amount due*][22] [the sum referred to in subsection (1).][23]

[(3A) Where the right conferred by this section is exercised, the party in default shall be liable to pay to the party exercising the right a reasonable amount in respect of costs and expenses reasonably incurred by that party as a result of the exercise of the right.][24]

(4) Any period during which performance is suspended in pursuance of[, or in consequence of the exercise of][25] the right conferred by this section shall be disregarded in computing for the purposes of any contractual time limit the time taken, by the party exercising the right or by a third party, to complete any work directly or indirectly affected by the exercise of the right.
Where the contractual time limit is set by reference to a date rather than a period, the date shall be adjusted accordingly.

113 Prohibition of conditional payment provisions

(1) A provision making payment under a construction contract conditional on the payer receiving payment from a third person is ineffective, unless that third person, or any other person payment by whom is under the contract (directly or indirectly) a condition of payment by that third person, is insolvent.

(2) For the purposes of this section a company becomes insolvent –
[(a) when it enters administration within the meaning of Schedule B1 to the Insolvency Act 1986,][26]
(b) on the appointment of an administrative receiver or a receiver or manager of its property under Chapter I of Part III of that Act, or the appointment of a receiver under Chapter II of that Part,
(c) on the passing of a resolution for voluntary winding-up without a declaration of solvency under section 89 of that Act, or
(d) on the making of a winding-up order under Part IV or V of that Act.

(3) For the purposes of this section a partnership becomes insolvent –
(a) on the making of a winding-up order against it under any provision of the Insolvency Act 1986 as applied by an order under section 420 of that Act, or
(b) when sequestration is awarded on the estate of the partnership under section 12 of the [1985 c. 66.] Bankruptcy (Scotland) Act 1985 or the partnership grants a trust deed for its creditors.

(4) For the purposes of this section an individual becomes insolvent –
(a) on the making of a bankruptcy order against him under Part IX of the [1986 c. 45.] Insolvency Act 1986, or
(b) on the sequestration of his estate under the Bankruptcy (Scotland) Act 1985 or when he grants a trust deed for his creditors.

(5) A company, partnership or individual shall also be treated as insolvent on the occurrence of any event corresponding to those specified in subsection (2), (3) or (4) under the law of Northern Ireland or of a country outside the United Kingdom.

(6) Where a provision is rendered ineffective by subsection (1), the parties are free to agree other terms for payment.
In the absence of such agreement, the relevant provisions of the Scheme for Construction Contracts apply.

---

[21] Amended by the Local Democracy, Economic Development and Construction Act 2009, s. 145(1), (2).
[22] Amended by the Local Democracy, Economic Development and Construction Act 2009, ss. 144(2)(b).
[23] Inserted by the Local Democracy, Economic Development and Construction Act 2009, s. 145(1), (3).
[24] Amended by the Local Democracy, Economic Development and Construction Act 2009, s. 145(1), (4).
[25] Amended by SI 2003/2096, Art. 4, Sch., Pt 1, para. 30.
[26] Amended by the Local Democracy, Economic Development and Construction Act 2009, ss 138(4)(a).

Supplementary provisions

114 The Scheme for Construction Contracts

(1) The Minister shall by regulations make a scheme ("the Scheme for Construction Contracts") containing provision about the matters referred to in the preceding provisions of this Part.
(2) Before making any regulations under this section the Minister shall consult such persons as he thinks fit.
(3) In this section "the Minister" means –
 (a) for England and Wales, the Secretary of State, and
 (b) for Scotland, the [Secretary of State].[27]
(4) Where any provisions of the Scheme for Construction Contracts apply by virtue of this Part in default of contractual provision agreed by the parties, they have effect as implied terms of the contract concerned.
(5) Regulations under this section shall not be made unless a draft of them has been approved by resolution of each House of Parliament.

115 Service of notices, &c

(1) The parties are free to agree on the manner of service of any notice or other document required or authorised to be served in pursuance of the construction contract or for any of the purposes of this Part.
(2) If or to the extent that there is no such agreement the following provisions apply.
(3) A notice or other document may be served on a person by any effective means.
(4) If a notice or other document is addressed, pre-paid and delivered by post—
 (a) to the addressee's last known principal residence or, if he is or has been carrying on a trade, profession or business, his last known principal business address, or
 (b) where the addressee is a body corporate, to the body's registered or principal office, it shall be treated as effectively served.
(5) This section does not apply to the service of documents for the purposes of legal proceedings, for which provision is made by rules of court.
(6) References in this Part to a notice or other document include any form of communication in writing and references to service shall be construed accordingly.

116 Reckoning periods of time

(1) For the purposes of this Part periods of time shall be reckoned as follows.
(2) Where an act is required to be done within a specified period after or from a specified date, the period begins immediately after that date.
(3) Where the period would include Christmas Day, Good Friday or a day which under the [1971 c. 80.] Banking and Financial Dealings Act 1971 is a bank holiday in England and Wales or, as the case may be, in Scotland, that day shall be excluded.

117 Crown application

(1) This Part applies to a construction contract entered into by or on behalf of the Crown otherwise than by or on behalf of Her Majesty in her private capacity.
(2) This Part applies to a construction contract entered into on behalf of the Duchy of Cornwall notwithstanding any Crown interest.
(3) Where a construction contract is entered into by or on behalf of Her Majesty in right of the Duchy of Lancaster, Her Majesty shall be represented, for the purposes of any adjudication or other proceedings arising out of the contract by virtue of this Part, by the Chancellor of the Duchy or such person as he may appoint.
(4) Where a construction contract is entered into on behalf of the Duchy of Cornwall, the Duke of Cornwall or the possessor for the time being of the Duchy shall be represented, for the purposes of any adjudication or other proceedings arising out of the contract by virtue of this Part, by such person as he may appoint.

---

[27] Amended by SI 1999/678, Art. 2(1).

Part V General Provisions

146 Orders, regulations and directions

(1) Orders, regulations and directions under this Act may make different provision for different cases or descriptions of case, including different provision for different areas.

(2) Orders and regulations under this Act may contain such incidental, supplementary or transitional provisions and savings as [*Secretary of State*][28] [the authority making them][29] considers appropriate.

(3) Orders and regulations under this Act shall be made by statutory instrument which, except for –

  (a) orders and regulations subject to affirmative resolution procedure (see sections 104(4), 105(4), 106(4)[, 106A][30] and 114(5)),

  (b) orders under section 150(3), or

  (c) regulations which only prescribe forms or particulars to be contained in forms, shall be subject to annulment in pursuance of a resolution of either House of Parliament.

[28] Amended by the Local Democracy, Economic Development and Construction Act 2009, ss. 138(4)(a).
[29] Amended by the Local Democracy, Economic Development and Construction Act 2009, s. 138(1), (4)(a).
[30] Amended by the Local Democracy, Economic Development and Construction Act 2009, s. 138(1), (4)(b).

STATUTORY INSTRUMENT 1998 NO. 648

Construction, England and Wales

The Construction Contracts (England and Wales) Exclusion Order 1998

Made 6th March 1998

Coming into force in accordance with article 1(1)

The Secretary of State, in exercise of the powers conferred on him by sections 106(1)(b) and 146(1) of the Housing Grants, Construction and Regeneration Act 1996[31] and of all other powers enabling him in that behalf, hereby makes the following Order, a draft of which has been laid before and approved by resolution of, each House of Parliament:

Citation, commencement and extent

1. (1) This Order may be cited as the Construction Contracts (England and Wales) Exclusion Order 1998 and shall come into force at the end of the period of 8 weeks beginning with the day on which it is made ("the commencement date").
   (2) This Order shall extend to England and Wales only.

Interpretation

2. In this Order, "Part II" means Part II of the Housing Grants, Construction and Regeneration Act 1996.

Agreements under statute

3. A construction contract is excluded from the operation of Part II if it is—
   (a) an agreement under section 38 (power of highway authorities to adopt by agreement) or section 278 (agreements as to execution of works) of the Highways Act 1980;[32]
   (b) an agreement under section 106 (planning obligations), 106A (modification or discharge of planning obligations) or 299A (Crown planning obligations) of the Town and Country Planning Act 1990;[33]
   (c) an agreement under section 104 of the Water Industry Act 1991[34] (agreements to adopt sewer, drain or sewage disposal works); or
   (d) an externally financed development agreement within the meaning of section 1 of the National Health Service (Private Finance) Act 1997[35] (powers of NHS Trusts to enter into agreements).

Other Agreements

3A. (1) A construction contract is excluded from the operation of Part II if it is an externally financed development agreement entered into by an NHS foundation trust.
   (2) In this article "externally financed development agreement" has the same meaning as in section 1 of the National Health Service (Private Finance Act) 1997 reading references in subsections (3) and (5) of that section to the trust as references to an NHS foundation trust.[36]

Private finance initiative

4. (1) A construction contract is excluded from the operation of Part II if it is a contract entered into under the private finance initiative, within the meaning given below.
   (2) A contract is entered into under the private finance initiative if all the following condition sare fulfilled—

---

[31] 1996 c.53.

[32] 1980 c.66: s. 38 was amended by and s. 278 substituted by the New Roads and Street Works Act 1991 (c.22) ss. 22 and 23.

[33] 1990 c.8: s. 106 was substituted and the other sections inserted by s. 12 of the Planning and Compensation Act 1991 (c.34).

[34] 1991 c.56.

[35] 1997 c.56.

[36] Added by Health and Social Care (Community Health and Standards) Act 2003 (Supplementary and Consequential Provision) (NHS Foundation Trusts) Order 2004/696 Sch.1 para.26 (April 1, 2004).

(a) it contains a statement that it is entered into under that initiative or, as the case may be, under a project applying similar principles;
(b) the consideration due under the contract is determined at least in part by reference to one or more of the following—
  (i) the standards attained in the performance of a service, the provision of which is the principal purpose or one of the principal purposes for which the building or structure is constructed;
  (ii) the extent, rate or intensity of use of all or any part of the building or structure in question; or
  (iii) the right to operate any facility in connection with the building or structure in question; and
(c) one of the parties to the contract is—
  (i) a Minister of the Crown;
  (ii) a department in respect of which appropriation accounts are required to be prepared under the Exchequer and Audit Departments Act 1866;[37]
  (iii) any other authority or body whose accounts are required to be examined and certified by or are open to the inspection of the Comptroller and Auditor General by virtue of an agreement entered into before the commencement date or by virtue of any enactment;
  (iv) any authority or body listed in Schedule 4 to the National Audit Act 1983[38] (nationalised industries and other public authorities);
  (v) a body whose accounts are subject to audit by auditors appointed by the Audit Commission;
  (vi) the governing body or trustees of a voluntary school within the meaning of section 31 of the Education Act 1996[39] (county schools and voluntary schools), or
  (vii) a company wholly owned by any of the bodies described in paragraphs (i) to (v).

Finance agreements

5. (1) A construction contract is excluded from the operation of Part II if it is a finance agreement, within the meaning given below.

(2) A contract is a finance agreement if it is any one of the following—
  (a) any contract of insurance;
  (b) any contract under which the principal obligations include the formation or dissolution of a company, unincorporated association or partnership;
  (c) any contract under which the principal obligations include the creation or transfer of securities or any right or interest in securities;
  (d) any contract under which the principal obligations include the lending of money;
  (e) any contract under which the principal obligations include an undertaking by a person to be responsible as surety for the debt or default of another person, including a fidelity bond, advance payment bond, retention bond or performance bond.

Development agreements

6. (1) A construction contract is excluded from the operation of Part II if it is a development agreement, within the meaning given below.

(2) A contract is a development agreement if it includes provision for the grant or disposal of a relevant interest in the land on which take place the principal construction operations to which the contract relates.

(3) In paragraph (2) above, a relevant interest in land means—
  (a) a freehold; or
  (b) a leasehold for a period which is to expire no earlier than 12 months after the completion of the construction operations under the contract.
  Signed by authority of the Secretary of State

Nick Raynsford

Parliamentary Under-Secretary of State, Department of the Environment, Transport and the Regions
6th March 1998

---

[37] 1866 c.39.

[38] 1983 c.44: amended by the Telecommunications Act 1984, (c.12) Sch. 7, Part III; the Oil and Pipelines Act (c.42) Sch. 20, para. 36, S.I. 1991/510, Art. 5(4); and the Coal Industry Act 1994, (c.21) Sch. 9, para. 29.

[39] 1996 c.56.

STATUTORY INSTRUMENT 2011 NO. 2332

CONSTRUCTION, ENGLAND

The Construction Contracts (England) Exclusion Order 2011

Made *19th September 2011*

Coming into force *1st October 2011*

The Secretary of State makes the following Order in exercise of the powers conferred by sections 106A(1) and 146 of the Housing Grants, Construction and Regeneration Act 1996 ("the Act"). In accordance with section 106A(4)(a) of the Act, a draft of this Order was laid before Parliament and approved by a resolution of each House of Parliament.

Citation, commencement and application

1.— (1) This Order may be cited as the Construction Contracts (England) Exclusion Order 2011 and comes into force on 1st October 2011.

(2) This Order does not apply to a construction contract to the extent that it relates to the carrying out of construction operations in Wales.

Interpretation

2. In this Order—

"the Act" means the Housing Grants, Construction and Regeneration Act 1996; and the reference to a "relevant contract" is to a contract excluded from the operation of Part 2 of the Act pursuant to Article 4 of the Construction Contracts (England and Wales) Exclusion Order 1998.

Private finance initiative sub-contracts

3. A construction contract is excluded from the operation of section 110(1A) of the Act if it is a contract pursuant to which a party to a relevant contract has sub-contracted to a third party some or all of its obligations under that contract to carry out, or arrange that others carry out, construction operations.

STATUTORY INSTRUMENT 1998 NO. 649, AS AMENDED BY STATUTORY INSTRUMENT 2011 NO. 1715 AND 2333—AMENDED SCHEME FOR CONSTRUCTION CONTRACTS (ENGLAND AND WALES) REGULATIONS 1998

Construction, England and Wales

The Scheme for Construction Contracts (England and Wales) Regulations 1998

Made—6th March 1998

Coming into force—1st May 1998

The Secretary of State, in exercise of the powers conferred on him by sections 108(6), 114 and 146(1) and (2) of the Housing Grants, Construction and Regeneration Act 1996,[40] and of all other powers enabling him in that behalf, having consulted such persons as he thinks fit, and draft Regulations having been approved by both Houses of Parliament, hereby makes the following Regulations:

Citation, commencement, extent and interpretation

1. (1) These Regulations may be cited as the Scheme for Construction Contracts (England and Wales) Regulations 1998 and shall come into force at the end of the period of 8 weeks beginning with the day on which they are made (the "commencement date").

   (2) These Regulations shall extend only to England and Wales.

   (3) In these Regulations, "the Act" means the Housing Grants, Construction and Regeneration Act 1996.

The Scheme for Construction Contracts

2. Where a construction contract does not comply with the requirements of section 108(1) to (4) of the Act, the adjudication provisions in Part I of the Schedule to these Regulations shall apply.
3. Where—
   (a) the parties to a construction contract are unable to reach agreement for the purposes mentioned respectively in sections 109, 111 and 113 of the Act, or
   (b) a construction contract does not make provision as required by section 110 [or by section 110A][41] of the Act, the relevant provisions in Part II of the Schedule to these Regulations shall apply.
4. The provisions in the Schedule to these Regulations shall be the Scheme for Construction Contracts for the purposes of section 114 of the Act.

Signed by authority of the Secretary of State

Nick Raynsford

Parliamentary Under-Secretary of State, Department of the Environment, Transport and the Regions

6th March 1998

Schedule

Regulations 2, 3 and 4

---

[40] 1996 c.53.

[41] Amended by Scheme for Construction Contracts (England and Wales) Regulations 1998. Regulations 2011/1715 and 2333 reg. 2 (1 Oct. 2011).

The Scheme for Construction Contracts

Part I—Adjudication

Notice of Intention to seek Adjudication

1. (1) Any party to a construction contract (the "referring party") may give written notice (the "notice of adjudication") [at any time][42]of his intention to refer any dispute arising under the contract, to adjudication.
   (2) The notice of adjudication shall be given to every other party to the contract.
   (3) The notice of adjudication shall set out briefly—
   (a) the nature and a brief description of the dispute and of the parties involved,
   (b) details of where and when the dispute has arisen,
   (c) the nature of the redress which is sought, and
   (d) the names and addresses of the parties to the contract (including, where appropriate, the addresses which the parties have specified for the giving of notices).
2. (1) Following the giving of a notice of adjudication and subject to any agreement between the parties to the dispute as to who shall act as adjudicator—
   (a) the referring party shall request the person (if any) specified in the contract to act as adjudicator, or
   (b) if no person is named in the contract or the person named has already indicated that he is unwilling or unable to act, and the contract provides for a specified nominating body to select a person, the referring party shall request the nominating body named in the contract to select a person to act as adjudicator, or
   (c) where neither paragraph (a) nor (b) above applies, or where the person referred to in (a) has already indicated that he is unwilling or unable to act and (b) does not apply, the referring party shall request an adjudicator nominating body to select a person to act as adjudicator.
   (2) A person requested to act as adjudicator in accordance with the provisions of paragraph (1) shall indicate whether or not he is willing to act within two days of receiving the request.
   (3) In this paragraph, and in paragraphs 5 and 6 below, an "adjudicator nominating body" shall mean a body (not being a natural person and not being a party to the dispute) which holds itself out publicly as a body which will select an adjudicator when requested to do so by a referring party.
3. The request referred to in paragraphs 2, 5 and 6 shall be accompanied by a copy of the notice of adjudication.
4. Any person requested or selected to act as adjudicator in accordance with paragraphs 2, 5 or 6 shall be a natural person acting in his personal capacity. A person requested or selected to act as an adjudicator shall not be an employee of any of the parties to the dispute and shall declare any interest, financial or otherwise, in any matter relating to the dispute.
5. (1) The nominating body referred to in paragraphs 2(1)(b) and 6(1)(b) or the adjudicator nominating body referred to in paragraphs 2(1)(c), 5(2)(b) and 6(1)(c) must communicate the selection of an adjudicator to the referring party within five days of receiving a request to do so.
   (2) Where the nominating body or the adjudicator nominating body fails to comply with paragraph (1), the referring party may—
   (a) agree with the other party to the dispute to request a specified person to act as adjudicator, or
   (b) request any other adjudicator nominating body to select a person to act as adjudicator.
   (3) The person requested to act as adjudicator in accordance with the provisions of paragraphs (1) or (2) shall indicate whether or not he is willing to act within two days of receiving the request.
6. (1) Where an adjudicator who is named in the contract indicates to the parties that he is unable or unwilling to act, or where he fails to respond in accordance with paragraph 2(2), the referring party may—

[42] Amended by Scheme for Construction Contracts (England and Wales) Regulations 1998. Regulations 2011/1715 and 2333 reg. 3(2) (1 Oct. 2011).

(a) request another person (if any) specified in the contract to act as adjudicator, or
(b) request the nominating body (if any) referred to in the contract to select a person to act as adjudicator, or
(c) request any other adjudicator nominating body to select a person to act as adjudicator.

(2) The person requested to act in accordance with the provisions of paragraph (1) shall indicate whether or not he is willing to act within two days of receiving the request.

7. (1) Where an adjudicator has been selected in accordance with paragraphs 2, 5 or 6, the referring party shall, not later than seven days from the date of the notice of adjudication, refer the dispute in writing (the "referral notice") to the adjudicator.

(2) A referral notice shall be accompanied by copies of, or relevant extracts from, the construction contract and such other documents as the referring party intends to rely upon.

(3) The referring party shall, at the same time as he sends to the adjudicator the documents referred to in paragraphs (1) and (2), send copies of those documents to every other party to the dispute. [Upon receipt of the referral notice, the adjudicator must inform every party to the dispute of the date that it was received.][43]

8. (1) The adjudicator may, with the consent of all the parties to those disputes, adjudicate at the same time on more than one dispute under the same contract.

(2) The adjudicator may, with the consent of all the parties to those disputes, adjudicate at the same time on related disputes under different contracts, whether or not one or more of those parties is a party to those disputes.

(3) All the parties in paragraphs (1) and (2) respectively may agree to extend the period within which the adjudicator may reach a decision in relation to all or any of these disputes.

(4) Where an adjudicator ceases to act because a dispute is to be adjudicated on by another person in terms of this paragraph, that adjudicator's fees and expenses shall be determined in accordance with paragraph 25.

9. (1) An adjudicator may resign at any time on giving notice in writing to the parties to the dispute.

(2) An adjudicator must resign where the dispute is the same or substantially the same as one which has previously been referred to adjudication, and a decision has been taken in that adjudication.

(3) Where an adjudicator ceases to act under paragraph 9(1)—

(a) the referring party may serve a fresh notice under paragraph 1 and shall request an adjudicator to act in accordance with paragraphs 2 to 7; and
(b) if requested by the new adjudicator and insofar as it is reasonably practicable, the parties shall supply him with copies of all documents which they had made available to the previous adjudicator.

(4) Where an adjudicator resigns in the circumstances referred to in paragraph (2), or where a dispute varies significantly from the dispute referred to him in the referral notice and for that reason he is not competent to decide it, the adjudicator shall be entitled to the payment of such reasonable amount as he may determine by way of fees and expenses reasonably incurred by him. The parties shall be jointly and severally liable for any sum which remains outstanding following the making of any determination on how the payment shall be apportioned.[44] *[The parties shall be jointly and severally liable for any sum which remains outstanding following the making of any determination on how the payment shall be apportioned.]*[45] [Subject to any contractual provision pursuant to section 108A(2) of the Act, the adjudicator may determine how the payment is to be apportioned and the parties are jointly and severally liable for any sum which remains outstanding following the making of any such determination].[46]

---

[43] Amended by Scheme for Construction Contracts (England and Wales) Regulations 1998. Regulations 2011/1715 and 2333 reg. 3(3) (1 Oct. 2011).

[44] Revoked by the Scheme for Construction Contracts (England and Wales) Regulations 1998. Regulations 2011/1715 and 2333 reg. 3(5) (1 Oct 2011).

[45] Revoked by the Scheme for Construction Contracts (England and Wales) Regulations 1998. Regulations 2011/1715 and 2333 reg. 3(5) (1 Oct 2011).

[46] Amended by the Scheme for Construction Contracts (England and Wales) Regulations 1998. Regulations 2011/1715 and 2333 reg. 3(4) (1 Oct. 2011).

10. Where any party to the dispute objects to the appointment of a particular person as adjudicator, that objection shall not invalidate the adjudicator's appointment nor any decision he may reach in accordance with paragraph 20.

11. (1) The parties to a dispute may at any time agree to revoke the appointment of the adjudicator. The adjudicator shall be entitled to the payment of such reasonable amount as he may determine by way of fees and expenses incurred by him. *[The parties shall be jointly and severally liable for any sum which remains outstanding following the making of any determination on how the payment shall be apportioned.]*[47] [Subject to any contractual provision pursuant to section 108A(2) of the Act, the adjudicator may determine how the payment is to be apportioned and the parties are jointly and severally liable for any sum which remains outstanding following the making of any such determination][48]
(2) Where the revocation of the appointment of the adjudicator is due to the default or misconduct of the adjudicator, the parties shall not be liable to pay the adjudicator's fees and expenses.

Powers of the adjudicator

12. The adjudicator shall—
(a) act impartially in carrying out his duties and shall do so in accordance with any relevant terms of the contract and shall reach his decision in accordance with the applicable law in relation to the contract; and
(b) avoid incurring unnecessary expense.

13. The adjudicator may take the initiative in ascertaining the facts and the law necessary to determine the dispute, and shall decide on the procedure to be followed in the adjudication. In particular he may—
(a) request any party to the contract to supply him with such documents as he may reasonably require including, if he so directs, any written statement from any party to the contract supporting or supplementing the referral notice and any other documents given under paragraph 7(2),
(b) decide the language or languages to be used in the adjudication and whether a translation of any document is to be provided and if so by whom,
(c) meet and question any of the parties to the contract and their representatives,
(d) subject to obtaining any necessary consent from a third party or parties, make such site visits and inspections as he considers appropriate, whether accompanied by the parties or not,
(e) subject to obtaining any necessary consent from a third party or parties, carry out any tests or experiments,
(f) obtain and consider such representations and submissions as he requires, and, provided he has notified the parties of his intention, appoint experts, assessors or legal advisers,
(g) give directions as to the timetable for the adjudication, any deadlines, or limits as to the length of written documents or oral representations to be complied with, and
(h) issue other directions relating to the conduct of the adjudication.

14. The parties shall comply with any request or direction of the adjudicator in relation to the adjudication.

15. If, without showing sufficient cause, a party fails to comply with any request, direction or timetable of the adjudicator made in accordance with his powers, fails to produce any document or written statement requested by the adjudicator, or in any other way fails to comply with a requirement under these provisions relating to the adjudication, the adjudicator may—
(a) continue the adjudication in the absence of that party or of the document or written statement requested,
(b) draw such inferences from that failure to comply as [the][49] circumstances may, in the adjudicator's opinion, [*justify*],[50] and

[47] Substituted for "be justified" by the Scheme for Construction Contracts (England and Wales) Regulations 1998. Regulations 2011/1715 and 2333 reg. 3(5) (1 Oct. 2011).

[48] Amended by the Scheme for Construction Contracts (England and Wales) Regulations 1998. Regulations 2011/1715 and 2333 reg. 3(6) (1 Oct. 2011).

[49] Substituted for "be justified" by Scheme for Construction Contracts (England and Wales) Regulations 1998. Regulations 2011/1715 and 2333 reg. 3(6) (1 Oct. 2011).

[50] Substituted for "the date" by the Scheme for Construction Contracts (England and Wales) Regulations 1998. Regulations 2011/1715 and 2333 reg. 3(6) (1 Oct. 2011).

(c) make a decision on the basis of the information before him attaching such weight as he thinks fit to any evidence submitted to him outside any period he may have requested or directed.

16. (1) Subject to any agreement between the parties to the contrary, and to the terms of paragraph (2) below, any party to the dispute may be assisted by, or represented by, such advisers or representatives (whether legally qualified or not) as he considers appropriate.

(2) Where the adjudicator is considering oral evidence or representations, a party to the dispute may not be represented by more than one person, unless the adjudicator gives directions to the contrary.

17. The adjudicator shall consider any relevant information submitted to him by any of the parties to the dispute and shall make available to them any information to be taken into account in reaching his decision.

18. The adjudicator and any party to the dispute shall not disclose to any other person any information or document provided to him in connection with the adjudication which the party supplying it has indicated is to be treated as confidential, except to the extent that it is necessary for the purposes of, or in connection with, the adjudication.

19. (1) The adjudicator shall reach his decision not later than—

(a) twenty eight days after [receipt][51] the date of the referral notice mentioned in paragraph 7(1), or

(b) forty two days after [receipt][52] of the referral notice if the referring party so consents, or

(c) such period exceeding twenty eight days after [receipt of][53] the referral notice as the parties to the dispute may, after the giving of that notice, agree.

(2) Where the adjudicator fails, for any reason, to reach his decision in accordance with paragraph (1)

(a) any of the parties to the dispute may serve a fresh notice under paragraph 1 and shall request an adjudicator to act in accordance with paragraphs 2 to 7; and

(b) if requested by the new adjudicator and insofar as it is reasonably practicable, the parties shall supply him with copies of all documents which they had made available to the previous adjudicator.

(3) As soon as possible after he has reached a decision, the adjudicator shall deliver a copy of that decision to each of the parties to the contract.

Adjudicator's decision

20. The adjudicator shall decide the matters in dispute. He may take into account any other matters which the parties to the dispute agree should be within the scope of the adjudication or which are matters under the contract which he considers are necessarily connected with the dispute. In particular, he may—

(a) open up, revise and review any decision taken or any certificate given by any person referred to in the contract unless the contract states that the decision or certificate is final and conclusive,

(b) decide that any of the parties to the dispute is liable to make a payment under the contract (whether in sterling or some other currency) and, subject to [section 111(9)][54] of the Act, when that payment is due and the final date for payment,

(c) having regard to any term of the contract relating to the payment of interest decide the circumstances in which, and the rates at which, and the periods for which simple or compound rates of interest shall be paid.

[51] Substituted for "the date" by the Scheme for Construction Contracts (England and Wales) Regulations 1998. Regulations 2011/1715 and 2333 reg. 3(6) (1 Oct. 2011).

[52] Amended by the Scheme for Construction Contracts (England and Wales) Regulations 1998. Regulations 2011/1715 and 2333 reg. 3(6) (1 Oct. 2011).

[53] Substituted for "section 111(4)" by the Scheme for Construction Contracts (England and Wales) Regulations 1998. Regulations 2011/1715 and 2333 reg. 3(8) (1 Oct. 2011).

[54] Revoked by the Scheme for Construction Contracts (England and Wales) Regulations 1998. Regulations 2011/1715 and 2333 reg. 3(9) (1 Oct. 2011).

21. In the absence of any directions by the adjudicator relating to the time for performance of his decision, the parties shall be required to comply with any decision of the adjudicator immediately on delivery of the decision to the parties [*in accordance with this paragraph*].[55]
22. If requested by one of the parties to the dispute, the adjudicator shall provide reasons for his decision.

[22A.—
   (1) The adjudicator may on his own initiative or on the application of a party correct his decision so as to remove a clerical or typographical error arising by accident or omission.
   (2) Any correction of a decision must be made within five days of the delivery of the decision to the parties.
   (3) As soon as possible after correcting a decision in accordance with this paragraph, the adjudicator must deliver a copy of the corrected decision to each of the parties to the contract.
   (4) Any correction of a decision forms part of the decision.][56]

Effects of the decision

[23. *(1) In his decision, the adjudicator may, if he thinks fit, order any of the parties to comply peremptorily with his decision or any part of it.*][57]
   (2) The decision of the adjudicator shall be binding on the parties, and they shall comply with it until the dispute is finally determined by legal proceedings, by arbitration (if the contract provides for arbitration or the parties otherwise agree to arbitration) or by agreement between the parties.

[24. *Section 42 of the Arbitration Act 1996 shall apply to this Scheme subject to the following modifications—*
   (a) *in subsection (2) for the word "tribunal" wherever it appears there shall be substituted the word "adjudicator",*
   (b) *in subparagraph (b) of subsection (2) for the words "arbitral proceedings" there shall be substituted the word "adjudication",*
   (c) *subparagraph (c) of subsection (2) shall be deleted, and*
   (d) *subsection (3) shall be deleted.*][58]

25. The adjudicator shall be entitled to the payment of such reasonable amount as he may determine by way of fees and expenses reasonably incurred by him. [*The parties shall be jointly and severally liable for any sum which remains outstanding following the making of any determination on how the payment shall be apportioned.*][59] [Subject to any contractual provision pursuant to section 108A(2) of the Act, the adjudicator may determine how the payment is to be apportioned and the parties are jointly and severally liable for any sum which remains outstanding following the making of any such determination][60]
26. The adjudicator shall not be liable for anything done or omitted in the discharge or purported discharge of his functions as adjudicator unless the act or omission is in bad faith, and any employee or agent of the adjudicator shall be similarly protected from liability.

Part II—Payment

Entitlement to and amount of stage payments

1. Where the parties to a relevant construction contract fail to agree—
   (a) the amount of any instalment or stage or periodic payment for any work under the contract, or

---

[55] Inserted by the Scheme for Construction Contracts (England and Wales) Regulations 1998. Regulations 2011/1715 and 2333 reg. 3(10) (1 Oct. 2011).

[56] Revoked by the Scheme for Construction Contracts (England and Wales) Regulations 1998. Regulations 2011/1715 and 2333 reg. 1(11) (1 Oct.2011).

[57] Revoked by the Scheme for Construction Contracts (England and Wales) Regulations 1998. Regulations 2011/1715 and 2333 reg. 3(12) (1 Oct. 2011).

[58] Revoked by the Scheme for Construction Contracts (England and Wales) Regulations 1998. Regulations 2011/1715 and 2333 reg. 3(13) (1 Oct 2011).

[59] Substituted by the Scheme for Construction Contracts (England and Wales) Regulations 1998. Regulations 2011/1715 and 2333 reg. 3(13) (1 Oct. 2011).

[60] Scheme for Construction Contracts (England and Wales) Regulations 1998. Regulations 2011/1715 and 2333 reg. 4(2) (1 Oct. 2011).

(b) the intervals at which, or circumstances in which, such payments become due under that contract, or

(c) both of the matters mentioned in sub-paragraphs (a) and (b) above, the relevant provisions of paragraphs 2 to 4 below shall apply.

2. (1) The amount of any payment by way of instalments or stage or periodic payments in respect of a relevant period shall be the difference between the amount determined in accordance with sub-paragraph (2) and the amount determined in accordance with sub-paragraph (3).

(2) The aggregate of the following amounts—

(a) an amount equal to the value of any work performed in accordance with the relevant construction contract during the period from the commencement of the contract to the end of the relevant period (excluding any amount calculated in accordance with sub-paragraph (b)),

(b) where the contract provides for payment for materials, an amount equal to the value of any materials manufactured on site or brought onto site for the purposes of the works during the period from the commencement of the contract to the end of the relevant period, and

(c) any other amount or sum which the contract specifies shall be payable during or in respect of the period from the commencement of the contract to the end of the relevant period.

(3) The aggregate of any sums which have been paid or are due for payment by way of instalments, stage or periodic payments during the period from the commencement of the contract to the end of the relevant period.

(4) An amount calculated in accordance with this paragraph shall not exceed the difference between—

(a) the contract price, and

(b) the aggregate of the instalments or stage or periodic payments which have become due.

Dates for payment

3. Where the parties to a construction contract fail to provide an adequate mechanism for determining either what payments become due under the contract, or when they become due for payment, or both, the relevant provisions of paragraphs 4 to 7 shall apply.

4. Any payment of a kind mentioned in paragraph 2 above shall become due on whichever of the following dates occurs later—

(a) the expiry of 7 days following the relevant period mentioned in paragraph 2(1) above, or

(b) the making of a claim by the payee.

5. The final payment payable under a relevant construction contract, namely the payment of an amount equal to the difference (if any) between—

(a) the contract price, and

(b) the aggregate of any instalment or stage or periodic payments which have become due under the contract, shall become due on [*the expiry of*]—

(a) [the expiry of] [61] 30 days following completion of the work, or

(b) the making of a claim by the payee, whichever is the later.

6. Payment of the contract price under a construction contract (not being a relevant construction contract) shall become due on

(a) the expiry of 30 days following the completion of the work, or

(b) the making of a claim by the payee, whichever is the later.

7. Any other payment under a construction contract shall become due

(a) on the expiry of 7 days following the completion of the work to which the payment relates, or

(b) the making of a claim by the payee, whichever is the later.

Final date for payment

8. (1) Where the parties to a construction contract fail to provide a final date for payment in relation to any sum which becomes due under a construction contract, the provisions of this paragraph shall apply.

[61] Revoked by the Scheme for Construction Contracts (England and Wales) Regulations 1998. Regulations 1022/1715 and 2333 reg 4(3) (1 Oct. 2011).

(2) The final date for the making of any payment of a kind mentioned in paragraphs 2, 5, 6 or 7, shall be 17 days from the date that payment becomes due.

*[9.— **Notice specifying amount of payment***
A party to a construction contract shall, not later than 5 days after the date on which any payment—
(a) becomes due from him, or
(b) would have become due, if—
(i) the other party had carried out his obligations under the contract, and
(ii) no set-off or abatement was permitted by reference to any sum claimed to be due under one or more other contracts, give notice to the other party to the contract specifying the amount (if any) of the payment he has made or proposes to make, specifying to what the payment relates and the basis on which that amount is calculated.][62]

[9. **Payment notice**
(1) Where the parties to a construction contract fail, in relation to a payment provided for by the contract, to provide for the issue of a payment notice pursuant to section 110A(1) of the Act, the provisions of this paragraph apply.
(2) The payer must, not later than five days after the payment due date, give a notice to the payee complying with sub-paragraph (3).
(3) A notice complies with this sub-paragraph if it specifies the sum that the payer considers to be due or to have been due at the payment due date and the basis on which that sum is calculated.
(4) For the purposes of this paragraph, it is immaterial that the sum referred to in sub-paragraph (3) may be zero.
(5) A payment provided for by the contract includes any payment of the kind mentioned in paragraph 2, 5, 6, or 7 above.][63]

[10. Notice of intention to withhold payment
Any notice of intention to withhold payment mentioned in section 111 of the Act shall be given not later than the prescribed period, which is to say not later than 7 days before the final date for payment determined either in accordance with the construction contract, or where no such provision is made in the contract, in accordance with paragraph 8 above.][64]

[10. **Notice of intention to pay less than the notified sum**
Where, in relation to a notice of intention to pay less than the notified sum mentioned in section 111(3) of the Act, the parties fail to agree the prescribed period mentioned in section 111(5), that notice must be given not later than seven days before the final date for payment determined either in accordance with the construction contract, or where no such provision is made in the contract, in accordance with paragraph 8 above.]

Prohibition of conditional payment provisions

11. Where a provision making payment under a construction contract conditional on the payer receiving payment from a third person is ineffective as mentioned in section 113 of the Act, and the parties have not agreed other terms for payment, the relevant provisions of—
(a) paragraphs 2, 4, 5, 7, 8, 9 and 10 shall apply in the case of a relevant construction contract, and
(b) paragraphs 6, 7, 8, 9 and 10 shall apply in the case of any other construction contract.

Interpretation

12. In this Part of the Scheme for Construction Contracts—

---

62 Substituted by the Scheme for Construction Contracts (England and Wales) Regulations 1998. Regulations 2011/1715 and 2333 reg. 4(3) (1 Oct. 2011).

63 Revoked by the Scheme for Construction Contracts (England and Wales) Regulations 1998. Regulations 1022/1715 and 2333 reg 4(4) (1 Oct. 2011).

64 Substituted by Scheme for Construction Contracts (England and Wales) Regulations 1998. Regulations 2011/1715 and 2333 reg.4(4) (1 Oct. 2011).

"claim by the payee" means a written notice given by the party carrying out work under a construction contract to the other party specifying the amount of any payment or payments which he considers to be due and the basis on which it is, or they are calculated;
"contract price" means the entire sum payable under the construction contract in respect of the work;
"relevant construction contract" means any construction contract other than one—
(a) which specifies that the duration of the work is to be less than 45 days, or
(b) in respect of which the parties agree that the duration of the work is estimated to be less than 45 days;
"relevant period" means a period which is specified in, or is calculated by reference to the construction contract or where no such period is so specified or is so calculable, a period of 28 days;
"value of work" means an amount determined in accordance with the construction contract under which the work is performed or where the contract contains no such provision, the cost of any work performed in accordance with that contract together with an amount equal to any overhead or profit included in the contract price;
"work" means any of the work or services mentioned in section 104 of the Act.

INDEX

Tables are referenced by page numbers in brackets